AF696285

Spark Plasma Sintering of Materials

Pasquale Cavaliere
Editor

Spark Plasma Sintering of Materials

Advances in Processing and Applications

Springer

Editor
Pasquale Cavaliere
Department of Innovation Engineering
University of Salento
Lecce, Italy

ISBN 978-3-030-05326-0 ISBN 978-3-030-05327-7 (eBook)
https://doi.org/10.1007/978-3-030-05327-7

Library of Congress Control Number: 2019932159

© Springer Nature Switzerland AG 2019
This work is subject to copyright. All rights are reserved by the Publisher, whether the whole or part of the material is concerned, specifically the rights of translation, reprinting, reuse of illustrations, recitation, broadcasting, reproduction on microfilms or in any other physical way, and transmission or information storage and retrieval, electronic adaptation, computer software, or by similar or dissimilar methodology now known or hereafter developed.
The use of general descriptive names, registered names, trademarks, service marks, etc. in this publication does not imply, even in the absence of a specific statement, that such names are exempt from the relevant protective laws and regulations and therefore free for general use.
The publisher, the authors, and the editors are safe to assume that the advice and information in this book are believed to be true and accurate at the date of publication. Neither the publisher nor the authors or the editors give a warranty, express or implied, with respect to the material contained herein or for any errors or omissions that may have been made. The publisher remains neutral with regard to jurisdictional claims in published maps and institutional affiliations.

This Springer imprint is published by the registered company Springer Nature Switzerland AG
The registered company address is: Gewerbestrasse 11, 6330 Cham, Switzerland

Preface

From the two last decades, elaboration techniques based on the use of high current intensities, among which the SPS technique is probably the most popular, have attracted considerable interest. Spark plasma sintering (SPS) is the process of consolidating powder (metallic or ceramic) constrained in a die (most often graphite) through the application of large amounts of pulsed DC electrical current.

Unlike conventional sintering, the SPS process affords rapid heating rates and reduced sintering temperatures that suppress microstructural coarsening. It has been demonstrated that the surface oxides at interparticle contact points can be mitigated by SPS. The SPS method has recently made significant advancements in materials science and materials processing thanks to its capability of fast and efficient densification of a great variety of materials and the suitability of the equipment for conducting solid-state syntheses and syntheses, in which liquid phases can form within limited volumes locally at the contacts between particles.

Spark plasma sintering (SPS) has garnered considerable attention for its ability to retain nanocrystalline microstructures and create materials with interesting functional properties. This is in part due to its ability to process materials at lower temperatures and in shorter amounts of time compared to conventional techniques, making SPS a true nonequilibrium processing technique. Decades of research have led to SPS finding success in nearly every category of material and property. SPS is uniquely positioned to take advantage of grain boundary strengthening due to its ability to process materials to high density with minimal grain growth.

Several scientists continue to develop the spark plasma sintering technology to fulfill the needs of high-performance applications. This book outlines current knowledge in SPS by reviewing the major fundamentals of the process and its features and applications, as well, provides an overview of recent and emerging developments in the process and the future potential of this technology. Despite SPS is still an emerging technology, investigations and applications have been widely reported in previous works, attracting great attentions from both scientific and industrial communities. Existing works mainly focused on applications, product properties, processing parameters optimization, and materials performances. So far, a systematic summarization and review on these topics is still lacking. In the present book, the different aspects of this emerging technology are addressed in the context of technology, materials, and applications. There are fundamental aspects of the SPS process that are not heavily researched but can be improved in order to enhance SPS transition into manufacturing. Many of these areas, including tooling design, tooling material, powder conditions, and current waveform, are evaluated in the present book to determine how changes could be used to improve the SPS process.

The book provides the SPS fundamentals and applications based on studies belonging to scientists from 12 countries and 40 universities and research centers. The book chapters focus on process fundamentals, nanocomposites, ceramics and hard metals, thermoelectric and magnetic materials, Ni- and Ti-based alloys, and high entropy alloys.

My special acknowledgments to the passion and cooperation of all the authors and reviewers who made possible the realization of the book and the reduction of the publication time with their hard work and prompt responses.

My special thanks to the professionalism of the editorial office assistants. Finally, I would like to dedicate the work to my brother, Andrea.

Lecce, Italy Pasquale Cavaliere

Contents

Part I
Process Fundamentals

Spark Plasma Sintering: Process Fundamentals

Pasquale Cavaliere, B. Sadeghi, and A. Shabani

1 Introduction

Sintering refers to the process of firing and consolidating powders at temperatures lower than their melting point, where diffusional mass transport leads to bonding between particles and the formation of a dense body. Bulk samples could be produced through sintering by using various powder materials. Generally, sintering can be done using various methods such as conventional sintering, microwave sintering, hot press (HP), hot isostatic press (HIP) and spark-plasma sintering (SPS). SPS is an emerging powder consolidating technique that provides significant advantages in the processing of high-temperature materials with poor deformability into configurations that were previously unattainable. It is known as a low-voltage, direct current, pulsed current–activated pressure technique. It can be applied to produce bulk samples that start from a broad range of powders (pure metals, alloys, ceramics and so on). During sintering, external pressure and an electric field are applied simultaneously to enhance the densification of the powder compacts. It is worth noting that SPS is similar to HP, but the main difference is in the procedure that produces heat and transmits it to the sintered material. In contrast to HP, which uses an external heating source, in SPS a pulsed direct current is allowed to pass through the conductive die and through the sample. So, the pulsed direct current produces an electric field during the sintering process. Therefore, the powders are heated from both outside and inside.

P. Cavaliere (✉)
Department of Innovation Engineering, University of Salento, Lecce, Italy
e-mail: pasquale.cavaliere@unisalento.it

B. Sadeghi · A. Shabani
Department of Material Engineering, Isfahan University of Technology, Isfahan, Iran
e-mail: b.sadeghi@ma.iut.ac.ir; ali.shabani@ma.iut.ac.ir

© Springer Nature Switzerland AG 2019

P. Cavaliere (ed.), *Spark Plasma Sintering of Materials*,
https://doi.org/10.1007/978-3-030-05327-7_1

In SPS, because no coarsening and grain growth is allowed, high relative densities are reached in a very short time, and nano-sized powders can be sintered without considerable grain growth, something that is not possible in a conventional sintering process.

Several theories have been proposed for physically modelling the SPS process; among them micro-spark/plasma theory is the most popular. In this theory, when spark discharge appears in the gap between the particles of a material, a local high-temperature state occurs as a result of the spark. Thus, during the SPS process, vaporization occurs and the powder particle surfaces melt; constricted shapes or "necks" form gradually around the contact area between the particles. Plastic transformation progresses during sintering, resulting in a sintered compact of over 99% density. Because only the surface temperature of the particles rises rapidly by self-heating, particle growth of the starting powder materials is controlled.

Spark-plasma sintering technique was first developed in 1930, but at that time the technology was not good enough to be commercialized. In 1980, the technique was developed further and various search laboratories—especially in Japan—began to use it. Sintering under electric current was first mention in a US patent published in 1913. The patent aimed to protect pressure- and current-assisted sintering system in a vacuum. More than 50 years—and about 20 additional patents—later, Kiyoshi Inoue invented "spark sintering", which is based on pulsed current; however, its commercialization was not a great success. In the late 1970s, even before Lockheed Missile and Space Co. bought parts of Inoue's work, the same technique had been used to produce near-net-shape products, very large components, functionally graded materials (FGMs) and non-equilibrium composites. But, there was little documentation of the process' successful implementation in industry. After Inoue's patents expired in the early 1990s, Japanese companies (in particular Sumitomo Coal Mining Co., Ltd.) began to produce SPS machines. Later, FCT Systeme GmbH, in Germany, and Thermal Technology LLC, Inc., in the USA, began producing similar equipment based on pulsed direct current. Nowadays, several companies in China, Korea and Japan offer FAST/SPS set-ups, and the number of FAST/SPS devices installed globally is estimated at 1750 (accounting for two-thirds of the industry).

2 SPS Components

Spark Plasma Sintering equipment is required to rapidly sinter conductive, non-conductive and/or composite powder materials to the required level of density, including full density. Rapid sintering/densification should take place by the generation of plasma due to sparking between the particles. Details of the SPS components follow.

2.1 Furnace Cum Pressing Chamber

Whole sintering operation will take place inside a furnace cum pressing chamber. The chamber should be capable of sintering the powder in various atmospheres such as vacuum, nitrogen or any inert atmosphere. It should withstand a vacuum of 1×10^{-4} mbar (at least in cold conditions) and a sintering temperature as high as 2200 °C. The size of the chamber should be sufficient to accommodate the largest mould set without obstruction. It should consist of:

- Stainless steel chamber
- Water-cooled, double-walled vacuum chamber
- Front-loading type
- Ports for viewing and placing pyrometer.
- Two additional blind flanges for installation of additional devices and gauges
- Inert gas inlet and air admittance
- An additional glass window for positioning lamp outside the chamber.

2.2 Power Supply for Furnace

The power supply to the furnace should be sufficient to reach a temperature of up to 2200 °C in vacuum/argon/nitrogen atmosphere at a heating rate variable of up to at least 800 °C/min. The input power supply should be 415 VAC ±10%, 50 Hz.

- Output DC pulse current (Fig. 1): Variable up to a minimum 8000 A (higher amperage up to 10,000 A is desirable), DC waveform (ON and OFF time) should be manually adjustable and programmable with a PC.
- Pulse duration: Programmable within a range from 5 to 255 ms. A higher range is acceptable. Figure 6 shows the ON-OFF pulsed DC used during the entire sintering cycle, with a standard on-off time relation of 12:2. Also, the ON pulses can be programmed from 1 to 99, whereas consecutive OFF pulses are limited to a maximum of 9. In most SPS machine catalogs, the manufacturer recommends a pulse sequence of 12:2 and that OFF pulses not exceed ON pulses. It is worth noting the height of the pulses varies, and, occasionally, the number of pulses can differ from the programmed sequence.
- Range of temperature: Selectable at any range from room temperature to 2200 °C; a higher range is acceptable. The temperature should be manually adjustable and programmable with a PC.
- Ammeter and voltmeter to read average output amps and volts.
- Heating rate variable up to 1000 °C/min. A higher rate is desirable. However, both the maximum temperature and heating rate are limited by the size of the die. However, a higher current is needed to heat a larger die to the same temperature level achieved by a smaller die. Additionally, the higher heating rates cause heterogeneous distribution of the temperature into the die and sample.

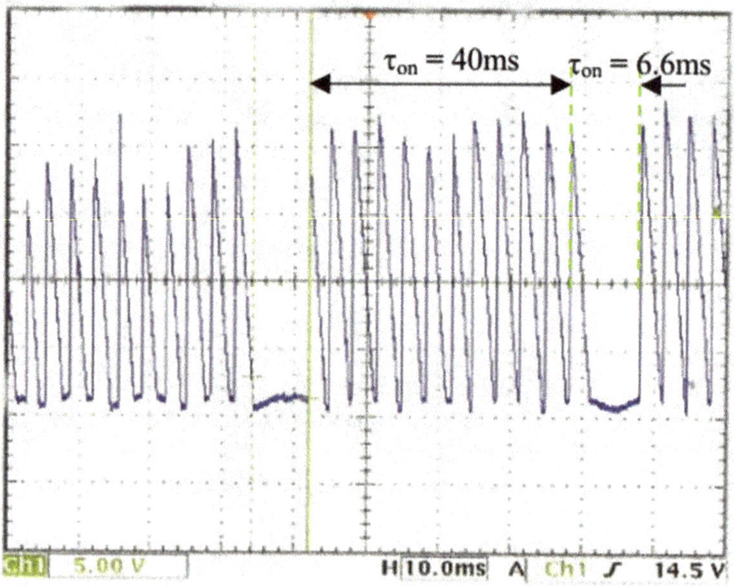

Fig. 1 DC pulsed current applied during sintering

- The power supply along with the mould design should be such that the temperature variation along a sample of 30 mm in diameter at a given set temperature should not exceed ±5 °C.

2.3 *Vacuum Pumping System*

A vacuum pumping system should be attached for evacuating the chamber and must be capable of attaining the following vacuum levels:

- Ultimate vacuum level: Minimum 5×10^{-2} mbar, higher vacuum as high as 10^{-4} mbar is preferable.
- Overall leak rate in the chamber: $<5 \times 10^{-3}$ mbar-l/s.
- Manually operated roughing valve, plumbing system to connect chamber to the pump, O rings, seals, connector, etc.
- Safety devices/interlocks: The unit should have safety devices/interlocks for the safety of the machine/operator like:
- Water pressure flow switch
- Automatic safety valve for releasing the pressure in case of emergency
- Vacuum-measuring gauge and gauge-heads: Digital display meter programmable with a PC.

2.4 *Pressing System*

A hydraulically (preferably) controlled pressing system should be provided to compact the powder at the desired pressure inside the furnace chamber. The pressure can be applied and released very quickly, and very high pressures compared to conventional HP can be used.

- Overall force: 25-ton-force or equivalent (~250 kN).
- The pressing frame should be mounted on four rigid posts to align the upper and lower ram.
- Both rams (anvil/piston) should be adequately water cooled to withstand the sintering temperature of 2200 °C in graphite die.
- The pressing cylinder should be driven servo-hydraulically and monitored manually as well as through programmable software using a PC for setting at a desired level of force value. Hydraulic power supply should include pump, oil reservoir, filters, etc.
- A pressure gauge should be provided for observing instantaneous compaction pressure.

2.5 *Water Circulation System*

Double-walled vacuum chamber, upper ram (anvil) and lower ram (piston), power feeder, electronics, etc. should have adequate water and air cooling for safe operation of equipment.

2.6 *Gas Inlet System*

A gas inlet system should be provided to purge the inert/nitrogen or other gas inside the chamber, and the gas inlet should be controlled using a solenoid valve and measured using a flow meter. Appropriate plumbing should be provided.

2.7 *Control and Instrumentation*

The input power (voltage and current), vacuum level, hydraulic press, closed cycle water circulation system and all safety interlocks should be controlled through the control console. The complete operation should be manually operated as well as automated through a PC using proper application software.

2.8 Temperature Measurement Tools

Temperature should be monitored at different places of the die using both thermocouple and IR pyrometer. For both cases, the observed temperature should be interfaced with a PC to record it. At least one of the temperature measurements should be very close (within 5 mm) to the centre axis of the sample. Appropriate thermocouples (K type) and an IR pyrometer for different ranges of temperatures should be provided. The accuracy of measurement for either case should be within ±0.2%. The IR pyrometer should measure and control the temperature within the range of 700–2200 °C.

Provision should be made to measure temperature at other places such as anvil and piston using appropriate thermocouples.

2.9 Mould Sets

One set of moulds consists of die and punch. Mould sets of high density graphite or tungsten carbide should be provided to obtain the different sintered specimen sizes.

3 SPS Mechanisms

Figure 2 shows a comparison between SPS and conventional sintering methods. HP and SPS have some similarities, i.e. loading of powder mixture into die, applying external pressure and holding it during sintering process. However, there is a remarkable difference in the nature and the 'how' of the generation of heat in both techniques. In SPS, unlike in HP that uses an external heating source, a pulsed direct current is passed through the graphite die and, in appropriate cases, also through the sample. Therefore, in SPS the sample is heated from both outside and inside (self-heating). It is of interest to note that the sintering temperature can be a few hundred degrees lower than what is typically used in a conventional sintering process such as HP.

Generally, spark plasma sintering is carried out in four main stages as shown in Fig. 2. Removing gases and creating vacuum is the first stage. Applying pressure during the second stage is followed by resistance heating in the third stage and, finally, cooling in the fourth stage (Saheb et al. 2012) (Fig. 3).

SPS is an under-pressure sintering technique based on electrical discharge of pulsed DC between the space of the powder particles and the generation of the spark. The propagation of the pulsed current occurs across particle surfaces as shown in Fig. 4.

When the spark generates into empty spaces or contact points of the powder particles, the local heat generated in the discharge column results in temporary

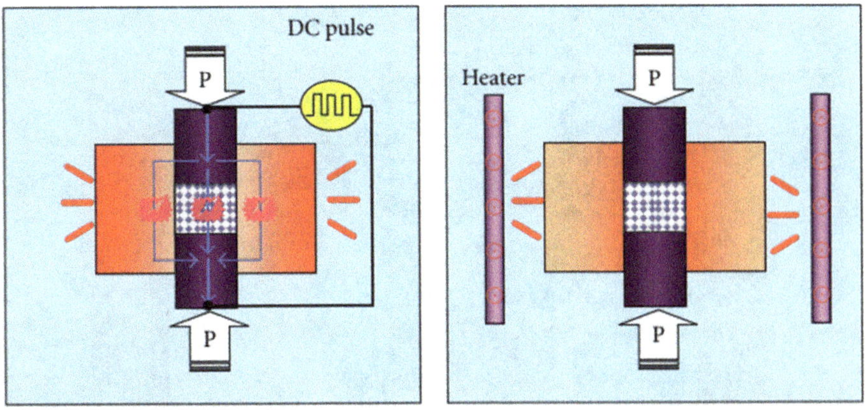

Fig. 2 Comparison between SPS and conventional sintering

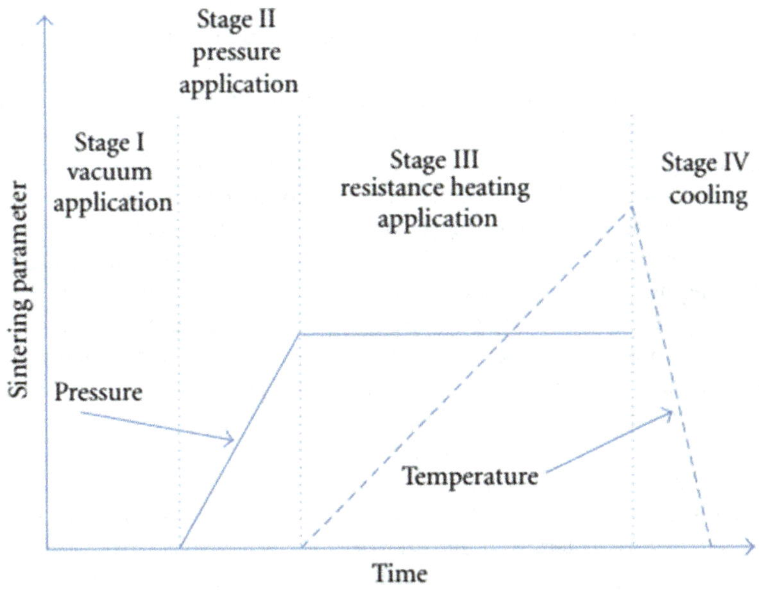

Fig. 3 SPS stages (Saheb et al. 2012)

enhancement of temperature to more than 1000 °C. This high temperature leads to evaporation of impurities and also evaporation and melting on the surface of powder particles. These melted regions are absorbed toward each other via electron current (ON time) and generation vacuum (OFF time), finally forming the necks between the powder particles.

It is known that metal powder particles show an oxide layer on the surface. The thickness of oxide layer and the applied pressure and particle size affect the

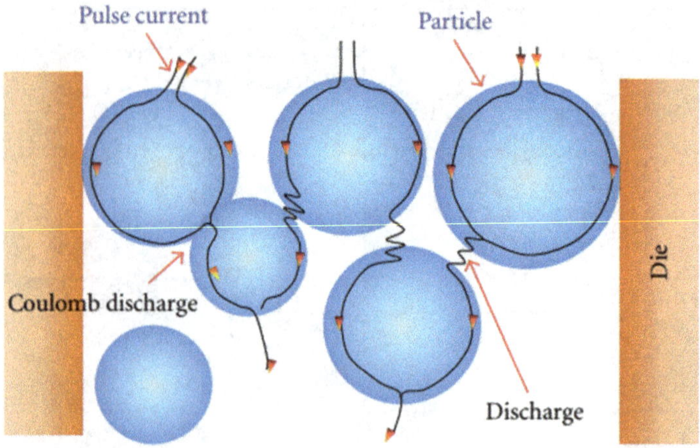

Fig. 4 Pulsed current that flows through powder particles

electrical conductivity of the powder volume and thus will be impressed at the initial stage of sintering. The compressive stress applied in SPS results in better contact between powder particles, and it changes the amount of those contacts and promotes the existing densification mechanisms or activates new mechanisms such as plastic deformation and grain boundary sliding (Guillon et al. 2014). Decreasing contact area at particle-to-particle contacts increases the local pressure there by a factor of 100, which results in theoretical local pressures of 1000–10,000 MPa. However, increasing the pressure in SPS powder agglomerates may break them, facilitating particle rearrangement at low temperatures and consequently increasing packing. Removing pores, especially large ones, allows for homogeneous grains and thus more densification and limited grain growth. In contrast, at high temperatures, although additional densification mechanisms such as plastic deformation or power law creep may be active, grain growth nevertheless increases. Consequently, sintering at low temperatures is more effective than at high temperatures due to significant limitation of grain growth.

During the SPS process, applied electrical potential causes a flow of electrons through conductive materials. In the case of non-conductive ceramics and conductive aluminum, a potential difference across aluminum particles separated by the ceramic phases may develop. The high pressure and electrical potential can easily break up the oxide layers and create direct metal-to-metal contact. This phenomenon can facilitate atom diffusion at the contacts. Mass transport mechanisms to the necks, or mechanisms involved in sintering for neck formation, are evaporation, surface diffusion, grain boundary and volume diffusion. During the initial stage, packing and necking of particles is 2–3% of total densification. During the intermediate stage, most of the porosity is still open, and the compact shrinks until 92% of the total densification is achieved. It should be noted that the active mechanisms during this stage are volume densification, grain boundary diffusion and probably dislocation climb. Isolation of pores from one another takes place during the last stage. In most cases, isolated pores are placed among three or more particles. Figure 4 shows the basic mechanism of neck formation by spark plasma. The formed pores achieve a spherical shape, and, as shown in Fig. 4 (V), the final densification takes place with the removal of the isolated pores by diffusion. The last densification stage duration is generally slow because it is governed by the diffusion phenomenon. However, by applying pressure, some additional mechanisms such as plastic deformation, dislocation and diffusional creep based on grain sliding accelerate the densification process thanks to the elimination of isolated pores. Figure 5 shows the different mechanisms involved during SPS.

On the other hand, when a spark discharge appears in a gap or at the contact point between the particles of a material, a local high-temperature state (discharge column) of several to tens of thousands of degrees centigrade is generated momentarily. This causes evaporation and melting on the surface of powder particles in the SPS process, and "necks" are formed around the area of contact between particles.

Figures 6 show a typical conventional material transfer path mechanism of evaporation, solidification, volume diffusion, surface diffusion and grain boundary diffusion on neck formation. It has been found that the ON-OFF DC pulse facilitates some phenomena such as formation of spark plasma, spark impact pressure, joule heating and electrical field diffusion effect (Suárez et al. 2013). More detail is shown in Fig. 7.

Moreover, with regard to the effect of cleaning the surface of powders during the SPS process, particle surfaces are more activated than in conventional electrical

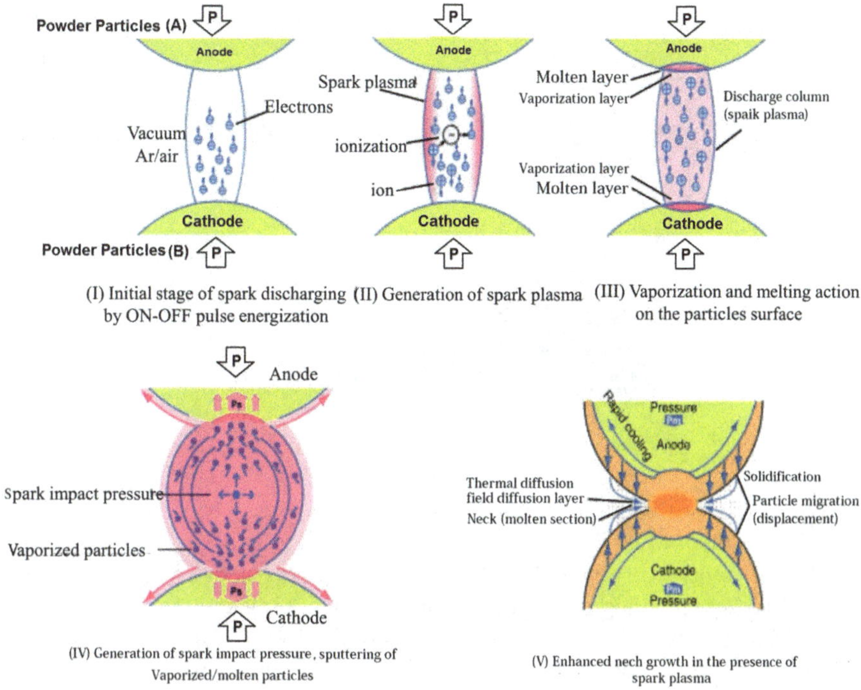

Fig. 5 Basic mechanism of neck formation by spark plasma

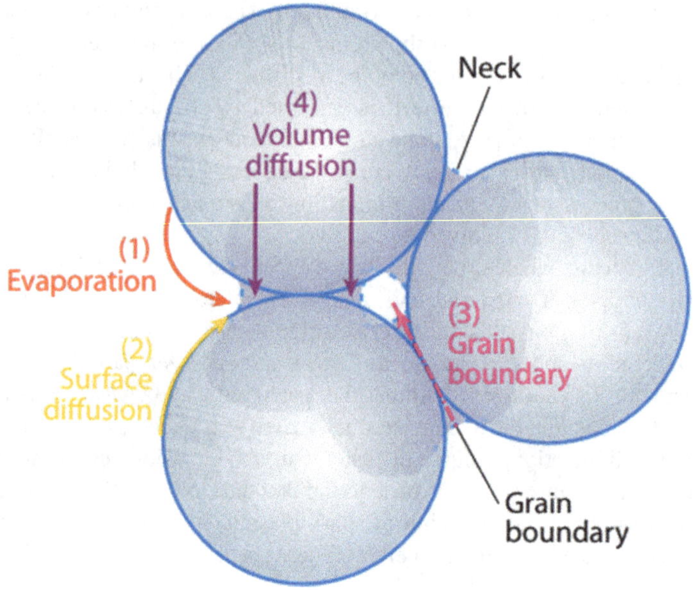

Fig. 6 Mass transport mechanisms that are involved in sintering

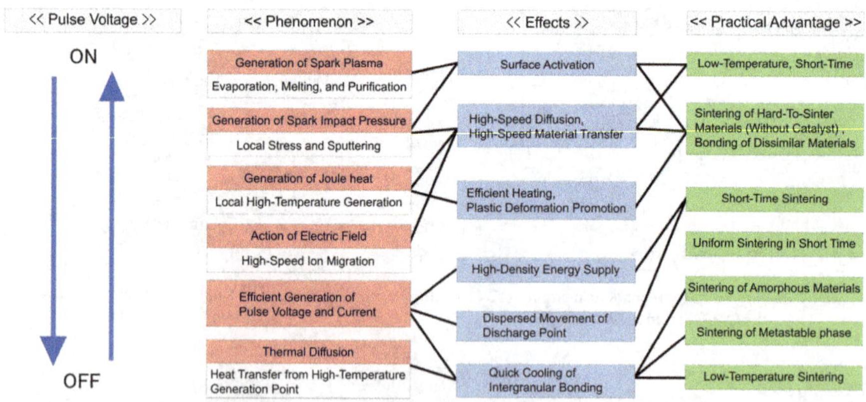

Fig. 7 Effects of ON-OFF DC

sintering processes, and, thus, diffusion is much more effective in comparison with the others (Tokita 1999; Tiwari et al. 2009).

The possibility of obtaining high heating rates during sintering is another advantage of SPS, and sintering of the powder mixture can occur within a short processing time. Heating rates during SPS are typically between 100 °C/min and 600 °C/min.

Figure 8 shows SEM micrographs belonging to SPS experiments performed at normal atmospheric pressure with graphite punches by consolidating spherical

Fig. 8 Particle deformation

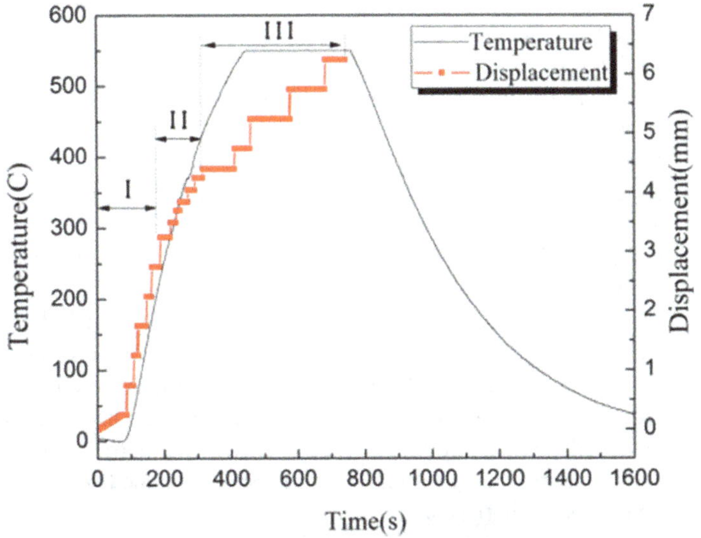

Fig. 9 Recorded shrinkage curve for Al

bronze alloy powder. The rearrangement of powder particles due to the initial press with localized deformation, bulk deformation and extensive sintering by mass transport phenomenon takes place at stages I, II and III, respectively (Fig. 9).

Rearrangement of aluminum particles is achieved by application of pressure through relative motion between powder particles. Then, with increasing temperature and time (region II), densification increases with densification rate compared with the previous stage. This is due to the localized deformation at particles contact points. At this stage, many necks between the surfaces of aluminum particles are formed by mass transport phenomena like grain boundary and volume diffusion. In fact, the localized deformation and consequent formation of necking results in slight densification. Neck formation, grain sliding and diffusion take place at the initial and intermediate stages.

Maximum displacement takes place at the end of region II, which could be due to the high displacement rate of region I. Finally, the main shrinkage (region III) starts. In this stage, the incipient propagation of deformation within the core of the aluminum particles begins. Bulk deformation and extensive sintering by the mass transport phenomenon cause more coalescence between aluminum particles, especially in necked areas. Diffusion is one of the main reasons for the lowest densification rate in region III. By holding the pressure at the maximum temperature, no distinct change in displacement and displacement rate occurs. This means in turn that the sintering process has been completed. The final stage occurs when isolated pores are removed by diffusion. It is worth noting that densification and formation of the necking depends on many factors, e.g. materials, shape and size of particles, particle size distribution, pressure, temperature and time. It should be noted that if particle size distribution is too narrow, packing becomes inefficient and thus densification rate decreases. In fact, proper particle size distribution enables more contact area between particles, which in turn increases neck formation and densification. It can be particularly important that small particles increase total contact area in the compact, and thus the driving force for densification is increased. It is worth noting that external pressure strongly affects diffusion processes in the sintering stages. As mentioned above, in the late stage of sintering, when shrinkage has practically come to an end, the remaining porosity consists of round, isolated pores. These pores are initially connected to grain boundaries. It is known that total elimination of the residual porosity is also assisted by external pressure.

Figure 7a shows the behaviour in the initial stage of neck formation due to sparks in the plasma. Heat is transferred immediately from the centre of the spark discharge column to the sphere surface and diffused so the intergranular bonding portion is quickly cooled. As seen in Fig. 7b, which shows several necks, the pulse energizing method causes spark discharges one after another between particles. Even with a single particle, the number of positions where necks are formed between adjacent particles increases as discharges are repeated.

Figure 7c shows the condition of an SPS-sintered grain boundary after sintering has progressed, making it plastic-deformed. This state is the result of the processing conditions in which applied pressure was 29 MPa, sintering temperature 500 °C, holding time 120 s, SPS current 850 A and voltage 3.9 V. Therefore, we note the applied pressure lends a positive effect to densification by increasing contact area and introducing additional densification mechanisms for SPS sintering.

Sintering dies and punches made of graphite are subject to joule heating according to the progress of the sintering of the internal powder material; the dies and punches function as heating elements to assume the role of maintaining the homogeneity of the sintering temperature for the densification.

4 Influence of Direct Current

The main difference between SPS and other sintering methods is that in SPS both die and powder are directly heated by the joule effect of DC. Such techniques make it possible to raise the temperature to 2000 °C at heating rates of up to 1000 °C/min or even higher. The role of the current and the sintering mechanism remain the subject of much debate between plasma formation (Zhang et al. 2014) and electro-migration supporters (Olevsky and Froyen 2009; Wang et al. 2011); proof of either phenomena is difficult to find. Olevsky and Froyen (2009) have studied possible mechanisms involved during SPS and categorized these effects into thermal and non-thermal phenomena. It is known that the ON and OFF DC pulse is energized to generate following effects: (1) spark plasma; (2) spark impact pressure; (3) joule heating; and (4) an electrical field, promoting material transfer and diffusion.

The plasma problem is complex. Makino has shown that only a very small part of the current (roughly 100 mA) crosses through the Al_2O_3 sample, while the machine injects some 1000 A (Tomino et al. 1997). In a similar study, Tomino et al. concluded that no current passes through the sample, and they eliminated the presence of a discharge in a dense insulator. According to Munir et al. (2011), the existence of such plasma or a discharge should be evaluated with different hypotheses, including applied pressure and development of sintering, since both parameters are linked with the formation of large contact surfaces between particles (Tokita 1999). In the case of conductive powders, there is a good probability discharges will occur between the particles at the beginning of the sintering; therefore, as sintering progresses, the probability of discharge decreases. In spite of the high current, 2000 A, the current density in a sample of conductive powder 20 mm in diameter should be around 400 A cm^{-2}, a level too low to explain the observed mass transport enhancement. It has, then, been proposed that due to the extremely small interfacial contact between particles, very high local current densities are present. It is worth noting that for each experiment all parameters—temperature, pressure, heating rate, holding time, etc.—will be extremely dependent on the nature of the sample and the size of its particles (Anselmi-Tamburini et al. 2005).

5 Advantages and Disadvantages of SPS

- Since the temperature used in SPS is usually 200–1000 °C lower than in conventional sintering techniques, it is categorized as a low temperature sintering technique. The process steps of SPS—applied pressure, increasing temperature and holding time—occur in a short time period (5–25 min) and thus provide a possibility of microstructure control as well grain growth, especially in nanostructured materials (Kubota 2007).
- As mentioned above, the existence of high-amperage current besides local pressures as high as 300 tons leads to formation of connections—and consequently necks—between powder particles. By increasing the pressure, plastic deformation of the particles leads to high densities. However, in low pressures and temperatures and also low holding time, materials can be fabricated which possess high porosities (Fig. 10).

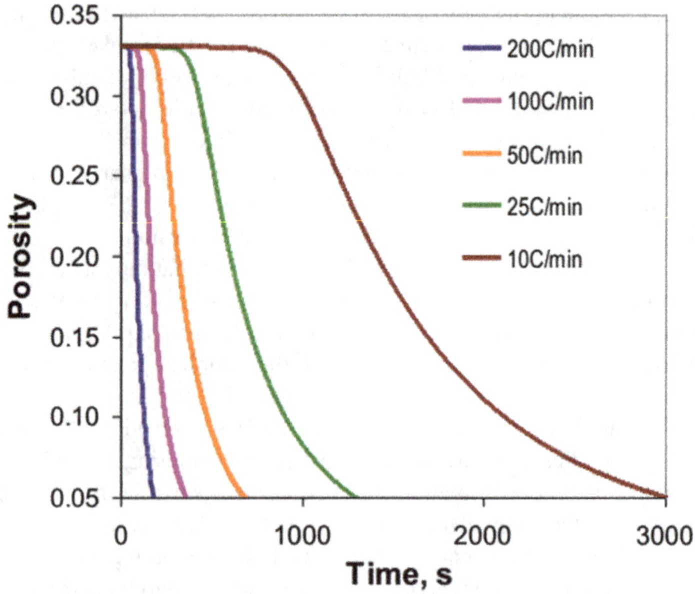

Fig. 10 Influence of high heating rates on porosity (Olevsky 1998)

- SPS is able to sinter the dissimilar materials without any melting. This technique is quite useful for the generation of composite materials with homogeneous microstructure, even though the density they experience will be low. Also, layered materials or FGMs can be produced with this technique.
- In fast sintering techniques such as microwave or SPS, enhancing of system temperature is caused by self-heating in the powder mixture.
- With regard to self-heating in SPS technique, the heating rate is high and thus the period of time for completing the SPS process is several minutes, whereas for conventional sintering techniques such as HP, the time extends from about several hours to some days. In addition, the dwelling time for SPS is typically about 5–15 min, while in conventional sintering techniques it is around several hours.
- Economically, the cost of SPS is around 50–80% lower than conventional sintering techniques. Such cost reductions have been attributed to high production rate, high heating rate and low sintering time.
- The concurrent use of temperature and pressure in SPS leads to high densification, and, consequently, the sintering temperature for attaining a high compact material is about 200–250 °C lower than that of conventional sintering techniques. In SPS, grain growth and coarsening is retarded due to the high heating rate, high relative densities are reached. Therefore, nanostructure ceramics and nanocomposites can be fabricated using SPS, materials which possess low porosity and high densities. These nanocomposites show excellent and unique mechanical properties such as strength and hardness.

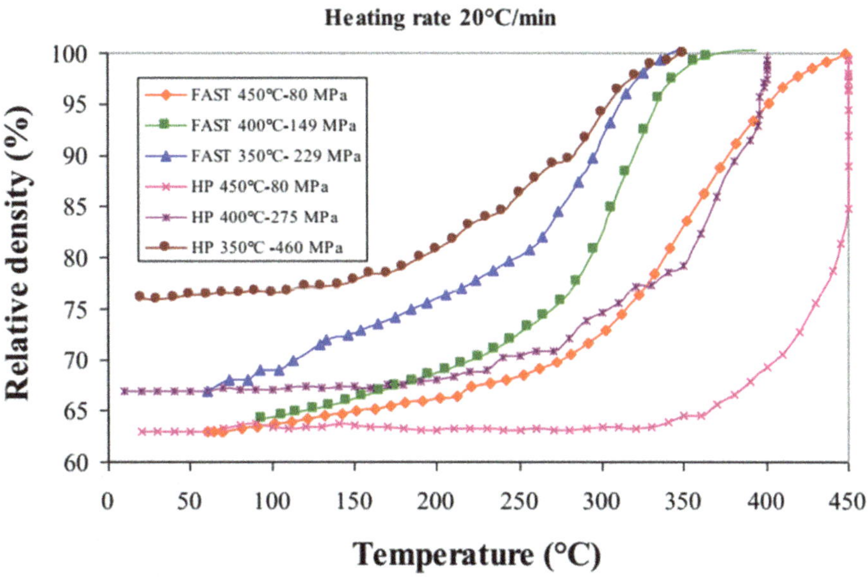

Fig. 11 Comparative study of relative density of SPS and HP (Olevsky 1998)

- Due to the high heating rate in SPS, unwanted reactions are restricted, and thus the formation of the undesirable phases and products is eliminated.
- One of the fundamental differences of SPS with other sintering techniques is apparent in the physical activation of the surface of powder particles. This surface activation, along with fast densification at lower temperatures, greatly restricts grain growth, compared to other techniques, and thus the microstructure possesses finer grains and improves mechanical and physical properties.
- Four main factors that aid increasing densification during SPS are: high heating and cooling rate; heat transfer from the die; mechanical pressure which causes the removal of porosities and pores and increases in the diffusion. Moreover, the use of DC current pulses results in increased diffusion rate.
- In SPS higher pressure can be used than in conventional sintering techniques, allowing them to reach high densities in a little time and at low temperatures (Fig. 11).
- Consolidation time of SPS is greatly decreased to a few minutes compared to the hours needed for conventional sintering, while the process' sintering temperature can be decreased by a few hundred degrees as shown in Fig. 12.
- SPS can be used to sinter a versatile and wide range of conductive and non-conductive materials.
- Fast densification by SPS reduces the time for the interaction between the sintering material and graphite die and consequently reduces the possibility of formation of undesired products. Figure 13 illustrates the impact of heating rate on shrinkage rate at the same holding time for all samples. It is clearly determined

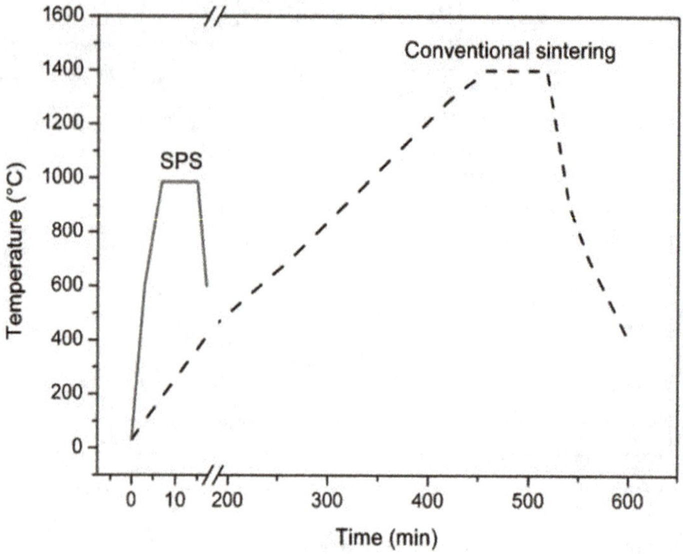

Fig. 12 Comparing time and temperature of SPS technique and conventional sintering

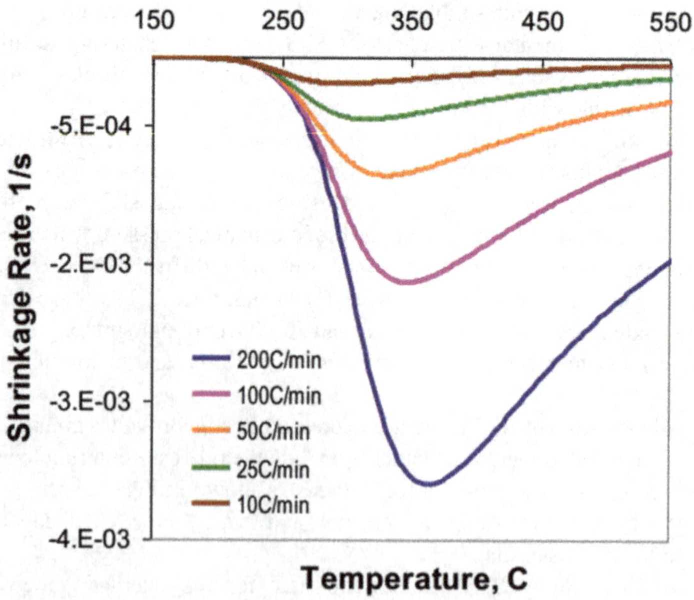

Fig. 13 Influence of heating rate versus shrinkage rate

that at high heating rates the duration of densification via surface diffusion becomes very low and subsequently intensifies the formation of necking and densification by grain boundary diffusion.

- Direction of current and pulse pattern have no effect on interaction between primary materials and growth rate, respectively.
- Other advantages of spark plasma sintering:

Reliable control of sintering process parameters and consequently control of its microstructure.
SPS can be considered a solid-phase sintering, hence, no melting occurs during the process.
Based on the die used, it is expected that temperature gradients will exist during sintering and subsequently fabrication of FGMs.
Possibility exists of combining sintering methods such as SPS-HP, SPS- microwave sintering, etc.
Low-temperature sintering via SPS due to use of temperatures below Curie temperature.
Fast sintering process.
Uniform sintering.
Low grain growth (nanograin materials may be prepared).
Compaction and sintering stages are combined in one operation.
Binders are not necessary.
Better purification and activation of powder particles surfaces.
Different materials (metals, ceramics, composites) may be processed.
High energy efficiency.
Easy operation.

- Disadvantages of spark plasma sintering:

Only simple symmetrical shapes may be prepared.
Expensive pulsed DC generator is required.
For materials powder with diameter smaller than 100 nm, it is possible to make a significant temperature gradient inside the sintered materials resulting in non-uniform densification.

6 Conclusions

SPS as a new technique for consolidation and sintering deserves serious attention from researchers. The SPS process' exclusive features have been recognized in many industrial sectors such as aerospace, automotive and military in comparison with other sintering techniques. Since duration of consolidation in SPS takes a maximum of 20 min, it brings time, cost and energy savings. Sintering of refractory materials, dense nanostructure materials and composite materials besides nanocomposites are a few examples of specimens fabricated by SPS. Specimens produced via SPS possess high strength, high wear resistance and strong magnetic properties. The main advantages of SPS in comparison with other sintering techniques can be summarized as: faster cooling rates, higher heating rates, generation of products with finer or nanostructure grains and higher densification efficiency.

References

Anselmi-Tamburini U, Gennari S, Garay JE, Munir ZA (2005) Fundamental investigations on the spark plasma sintering/synthesis process: II. Modeling of current and temperature distributions. Mater Sci Eng A394:139–148

Guillon O, Gonzalez-Julian J, Dargatz B, Kessel T, Schierning G, Räthel J, Herrmann M (2014) Field-assisted sintering technology/spark plasma sintering: mechanisms, materials, and technology developments. Adv Eng Mater 16:830–849

Kubota M (2007) Properties of nano-structured pure Al produced by mechanical grinding and spark plasma sintering. J Alloys Compd 434:294–297

Munir ZA, Quach DV, Ohyanagi M (2011) Electric current activation of sintering: a review of the pulsed electric current sintering process. J Am Ceram Soc 94:1–19

Olevsky EA (1998) Theory of sintering: from discrete to continuum. Mater Sci Eng R23:41–100

Olevsky EA, Froyen L (2009) Impact of thermal diffusion on densification during SPS. J Am Ceram Soc 92:S122–S132

Saheb N, Iqbal Z, Khalil A, Hakeem AS, Al Aqeeli N, Laoui T, Al-Qutub R, Kirchner R (2012) Spark plasma sintering of metals and metal matrix nanocomposites: a review. J Nanomater. https://doi.org/10.1155/2012/983470

Suárez M, Fernández A, Kessel H, Hennicke J, Menéndez J, Kirchner R, Torrecillas R, Kessel T (2013) Challenges and opportunities for spark plasma sintering: a key technology for a new generation of materials, INTECH Open Access Publisher. https://doi.org/10.5772/53706

Tiwari D, Basu B, Biswas K (2009) Simulation of thermal and electric field evolution during spark plasma sintering. Ceram Int 35:699–708

Tokita M (1999) Mechanism of spark plasma sintering. In: Proceeding of NEDO International Symposium on Functionally Graded Materials, Japan, p 22

Tomino H, Watanabe H, Kondo Y (1997) Electric current path and temperature distribution for spark sintering. J Jpn Soc Powder Powder Metall 44:974–979

Wang SW, Chen LD, Hirai T (2011) Densification of Al2O3 Powder Using Spark Plasma Sintering. J Mater Res 15:982–987

Zhang Z-H, Liu Z-F, Lu J-F, Shen X-B, Wang F-C, Wang Y-D (2014) The sintering mechanism in spark plasma sintering–proof of the occurrence of spark discharge. Scr Mater 81:56–59

Fundamentals of Spark Plasma Sintering (SPS): An Ideal Processing Technique for Fabrication of Metal Matrix Nanocomposites

N. Sharma, S. N. Alam, and B. C. Ray

1 Introduction

Spark plasma sintering (SPS), also known as pulsed electric current sintering (PECS), is a sintering technology developed in recent years. It is a novel processing technique for material development which utilizes a direct electrical current (DEC) to perform high-speed consolidation of powders in a very short period of time, using relatively high sintering pressures. During SPS the gap between the powder particles is charged with electrical energy, and a high-temperature spark plasma is applied momentarily which enables sintering at relatively low temperatures as compared to the conventional sintering and yet achieves almost full densification. Sintering assisted by plasma is expected to achieve very high levels of densification. The large current pulse energizing method generates spark plasma, spark impact pressure, Joule heating and an electrical field diffusion effect. The SPS process enjoys certain inherent advantages, such as (a) high thermo-efficiency, determined by the way and rate of heating; (b) spontaneous heating, mainly due to the high pulse power source; (c) better self-purification of the particle surface, as a strong enough electrical field is produced in the small gaps of the particles to make the electrons, cations and anions to strike the surface of the opposite particle and purify its surface; and lastly (d) fast sintering under low temperatures (Munir et al. 2006; Tokita 1993). The movement of these particles fulfils the material transportation, and a local high-temperature zone is created at the contacting areas of the particles, resulting in the quick formation of the neck, while the temperature still remains comparatively low. However, the temperature at the neck is much higher than inside the particles, as the electrical density of the current flowing through the neck is much higher than that of

N. Sharma (✉) · S. N. Alam · B. C. Ray
Metallurgical and Materials Engineering Department, National Institute of Technology Rourkela, Rourkela, India
e-mail: syedn@nitrkl.ac.in; bcray@nitrkl.ac.in

© Springer Nature Switzerland AG 2019
P. Cavaliere (ed.), *Spark Plasma Sintering of Materials*,
https://doi.org/10.1007/978-3-030-05327-7_2

the inside of the particles. The neck region grows quickly, and faster sintering is achieved. The reason why the SPS technique is so effective is because the heating power is not only distributed homogeneously over the entire volume of the sample in a macroscopic scale but at a microscopic scale, and also the heating power is dissipated exactly at the points of contact of the powder particles in the sample. The specimen is heated by the Joule heat due to the heat transferred from the pressing die. The SPS process is capable of introducing very high rates and high pressure due to the sintering at a relatively lower temperature and in much shorter time could be achieved (Mamedov 2002). SPS is an ideal technique for the development of nanocomposites reinforced with carbonaceous nanofillers like carbon fibres, graphite, diamond, carbon foams, carbon nanotubes (CNTs) and graphene as it is possible to achieve a high level of densification without damaging the structural integrity of the nanofillers. Achieving near full density, without damaging the nanofiller structure and morphology, is one of the major challenges during the processing of the metal matrix nanocomposites (MMNCs). This is why SPS is preferred as sintering can be done at lower temperatures and in a substantially short period of holding time. In these binary nanocomposites, CNTs/graphene are added to metal matrices, to noticeably modify and enhance their properties.

The idea to compact metallic materials by an electro-discharge process was originally proposed in the 1960s. Based on this concept, three sintering processes have been developed and commercialized during the recent years and named as "spark plasma sintering" (SPS), "plasma-activated sintering" (PAS) and "electroconsolidation" (Azarniya et al. 2017). These processes are similar to conventional hot pressing, in that the precursors are loaded in a die and a uniaxial pressure is applied during the sintering. However, instead of using an external heat source, a current (DC, pulsed DC or AC) is allowed to pass through the electrically conducting pressure die and, in appropriate cases, also through the sample. This implies that the die itself acts as a heat source and that the sample is heated from both outside and inside. The characteristics of the three processes are as follows:

(i) In the SPS process, a pulsed DC is applied from the beginning to the end of the sintering cycle.
(ii) In the PAS process, a pulsed DC is normally applied at room temperature for a short period of time and then a continuous DC during the remainder of the sintering process. This procedure is often referred to in the literature as a "single pulse cycle process. In a few cases the pulsed DC procedure is repeated during the sintering process, and is referred to as a "multiple pulse cycle process".
(iii) In the electroconsolidation process, various types of currents (DC, pulsed DC or AC) can be used, i.e. the system is designed to be flexible in terms of power supply.

These processes have now been developed beyond the production of small objects with simple shapes, as the continuous production of compacts of complex geometry and of pieces with diameters larger than 150 mm has been achieved.

Despite the fact that a uniaxial pressure is applied, green bodies of complex geometry can be exposed to a "pseudo-isostatic" pressure when embedded in free-flowing electrically conducting particulates that act as a pressure-transmitting medium inside the die (Lim et al. 2006).

SPS technique enables the manipulation of the kinetics of densification and grain growth, to yield various materials that hold significant fundamental and technological interest in composite systems. SPS provides a means by which metallic powders can be sintered very rapidly to full density. It is similar to hot pressing to the extent that graphite dies are used, but the heating is accomplished by spark discharges in voids between the particles, generated by an instantaneous pulsed direct current which is applied through electrodes at the top and bottom punches of the graphite die. Due to these discharges, the particle surface is activated and purified, and a self-heating phenomenon is generated between the particles, as a result of which heat transfer and mass transfer can be completed instantaneously (Goldberger and Merkle 2001). These advantages result in the superfast densification for metal powders through the SPS process. Four factors that contribute to the fast densification process in SPS are (i) rapid heat transfer, (ii) application of a mechanical pressure exceeding that used in normal hot pressing processes, (iii) the use of fast heating and cooling rates and (vi) the use of a pulsed direct current, implying that the samples are also exposed to an electric field. It is generally accepted that application of mechanical pressure is helpful in removing pores from compacts and enhancing diffusion. The heat transfer from the die to the compact is very efficient in these processes because the die itself (and the electrically conductive pressure-transmitting medium when used) acts as a heating element so that the thermal energy necessary for compaction is efficiently transmitted to the sample. However, it is frequently argued that the improved densification rates stem mostly from the use of DC pulses of high energy. Thus it was originally claimed by the process inventors that the pulses generate spark and even plasma discharges between the powder particles, which is why the processes were named spark plasma sintering and plasma-activated sintering. During the initial part of the sintering process, the spark and/or plasma discharges are said to clean the surfaces of the powders from adsorbed species such as CO_2 and H_2O. In subsequent stages, the presence of cleaned and activated surfaces is expected to enhance the grain boundary diffusion which, together with the proposed spark discharges and/or plasma processes, will promote the transfer of the material and thus facilitate densification and grain growth (Inoue 1966; Tamari et al. 1995).

Various materials have been compacted by spark plasma sintering. In this chapter, we will focus on how the kinetics of densification and grain growth can be manipulated by the use of SPS, to yield (i) dense compacts containing nanosized grains and (ii) fully compacted bodies.

2 An Overview of the SPS Technique

2.1 Principles and Mechanism of the SPS Process

The SPS process is a revolutionary high-speed powder consolidation process based on the electrical spark discharge phenomenon. It uses high-amperage pulsed DC for the consolidation of materials. A high-energy, low-voltage spark pulse current momentarily generates a spark plasma due to which the localized temperature between the particles reaches several thousands of degrees resulting in optimum thermal and electrolytic diffusion. SPS temperatures range from low to over 2000 °C which are ~200 °C to 500 °C lower than that for conventional sintering processes. Vaporization, melting and sintering are completed within approximately 5–30 min, including temperature rise and holding times. Unique to SPS process is the mechanism of internal heat generation due to DC as opposed to the other conventional methods of sintering, such as hot pressing, where external heat is supplied to the sample. Conventional sintering processes could require hours, whereas SPS would take only a few minutes to achieve the same or better consolidation. It combines the compaction and sintering stages into a single operation. The mechanical scheme of the process is very similar to the uniaxial die pressing. The load applied during SPS is transferred to the sample through the upper punch. The pulsed DC power supply is connected to the upper and lower punches/electrodes. The SPS process is a highly energy-efficient process and is easy to operate. It can be used to process conductive as well as non-conductive materials to attain very high levels of densification. SPS provides uniform sintering and also low grain growth which is why it is an ideal process for processing nanostructured materials (Yamazaki et al. 1996; Lukianova et al. 2017; Clemares et al. 2010).

2.1.1 Plasma Generation

The most widely accepted mechanism of the SPS process is based on the micro-spark discharge in the gap between the neighbouring powder particles. It was originally claimed by Inoue and several other inventors of the SPS process that the electrical pulses not only generated sparks but also generated plasma discharges between the particle contacts (Inoue 1965, 1966). They claimed that ionization at the particle contact due to the spark discharges created an impulsive pressure that facilitated diffusion of the atoms at the contacts. Groza et al. (1999) suggested that a pulsed current had a cleaning effect on the particle surfaces. This conclusion was based on the observation of a grain boundary without any oxidation between the particles. Whether plasma is generated or has not yet been confirmed directly by the experiments is still debated. However, it is widely accepted that occasional electric discharges may take place on a microscopic level. Plasma is a hot ionized gas consisting of an approximately equal number of positively charged ions and negatively charged electrons. Apart from the external forces that act on the plasma, it is

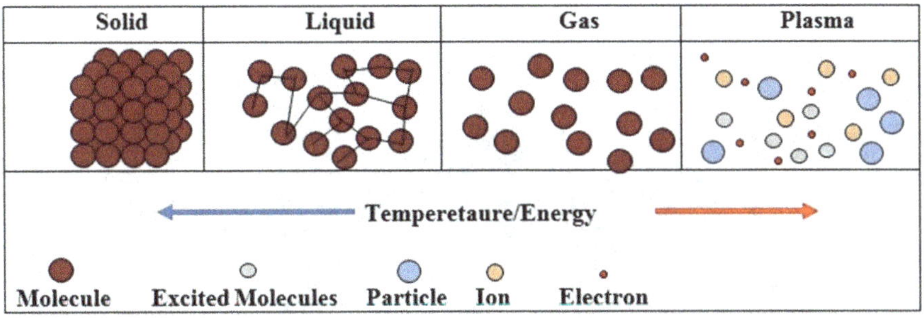

Fig. 1 States of matter

also acted upon by the electric and magnetic fields that are developed within it through localized charge concentrations and electric currents which result from the motion of the ions and electrons (Fig. 1). The forces exerted by these fields on the charged particles that constitute the plasma act over long distances and impart to these particles a coherent and collective property that is not seen in the case of neutral gases. The characteristics of plasma are significantly different from that of neutral gases, and it is considered as the fourth state of matter.

2.1.2 Joule Heating

Joule heating is the generation of heat by passing an electric current through a metal. It is also sometimes referred to as the resistance heating. Heating in electrical resistance heaters and stove-top burners are due to Joule heating. During the SPS process, the passage of electric current through the particles assists in the welding of the particles under mechanical pressure. In SPS the die and the sample are heated by Joule heating from a current passing through them. A volume heating resulting from the Joule effect in contrast to the conductive heat transport that takes place during conventional sintering enables a rapid rise in the temperature which in turn enhances the mass transport mechanisms responsible for the sintering phenomenon. The intense Joule heating effect at the particle surface can often be very high, and the boiling point of the particle could be attained and thereby lead to localized vaporization or cleaning of the powder surfaces. This, in turn, improves the consolidation rate. In processes like hot pressing, the die and the sample are heated by radiation from an enclosing furnace. The pulsed DC also creates plasma, which causes a cleaning effect on the surface of the particles leading to sintering enhancement. In order to achieve a homogeneous sintering behaviour, the temperature gradients inside the sample should be minimized. Adequate electrical conductivity of the powder samples and the achievement of homogeneous temperature distribution during sintering is a major concern of the SPS process. The important parameters which influence the temperature gradient within the sample are electrical conductivity of the materials constituting the sample, the pulsed current input and the wall

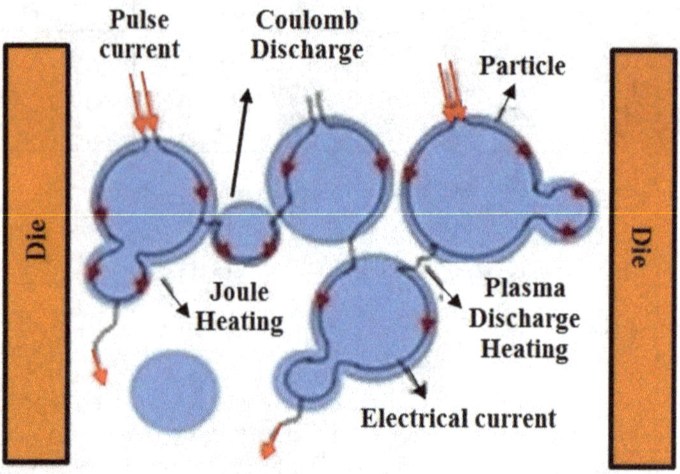

Fig. 2 Joule heating during SPS

thickness of the die. Graphite papers are used to avoid direct contact between the graphite parts and the sample and also to guarantee electrical contact between all parts (Conrad 2002; Tiwari et al. 2009). Figure 2 shows the mechanism of Joule heating during SPS.

2.1.3 Pulsed Current

The SPS process utilizes high-amperage pulsed DC to activate the consolidation during SPS. A pulsed DC or pulsating direct current is a periodic current which changes in value but never changes direction. A pulsed DC could be used to describe a signal consisting of one or more rectangular pulses rather than sinusoidal pulses. The on-off DC pulse energizing method generates spark plasma, spark impact pressure, Joule heating and an electrical field diffusion effect. During the SPS process, the powder particle surfaces are more easily purified and activated than in conventional electrical sintering processes, and material transfer at both the micro and macro levels is promoted, so a high-quality sintered compact is obtained at a lower temperature and in a much shorter time as compared to any other conventional processes. Figure 3 illustrates how pulse current flows through powder particles inside the SPS die. The SPS process is an electrical sintering technique which applies an on-off DC pulse voltage and current from a special pulse generator to the powder particles, and in addition to the factors promoting sintering, it also effectively generates electrical discharges between the particles which are essential for sintering during the initial stages. High temperature sputtering phenomenon generated by spark plasma and spark impact pressure eliminates adsorptive gas and impurities

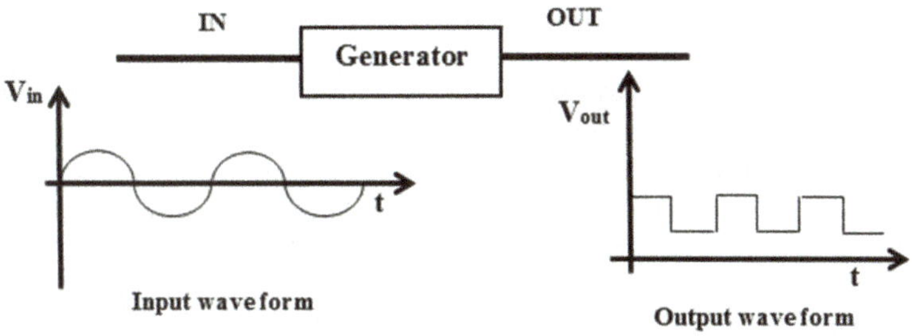

Fig. 3 Pulsed DC generator

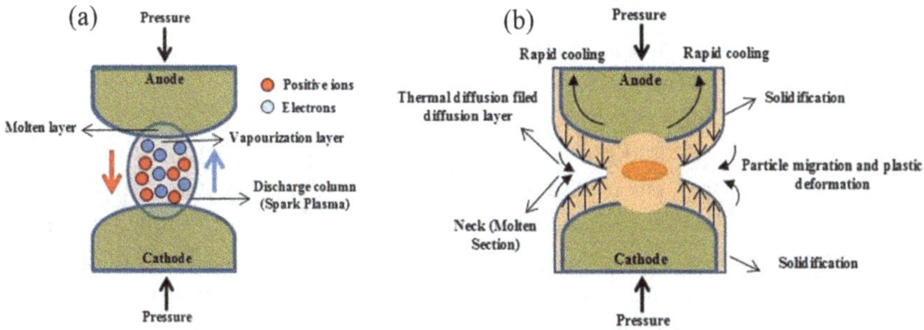

Fig. 4 (**a**) Vaporization and melting on particle surface, (**b**) neck growth in the presence of spark plasma

existing on the surface of the powder particles. The action of the electrical field causes high-speed diffusion due to the rapid migration of ions. The DC pulse time and pause time are few to tens of millions of milliseconds (ms). The voltage applied during the process is only a few volts, and the current that flows through the sintered powder material is typically a few thousands of amperes. The process is conducted in either vacuum, low pressure or an inert gas atmosphere.

When a spark discharge appears in a gap or at the contact point between the particles of a material, a local high-temperature state (discharge column) of several to ~10,000 °C is generated momentarily. This causes evaporation and melting on the surface of the powder particles, and necks are formed around the area of contact between the particles. Figure 4 shows the basic mechanism of neck formation during SPS. The heat is transferred immediately from the centre of the spark discharge column to the sphere surface and diffused so that the intergranular bonding portion is quickly cooled. The pulse energizing method causes spark discharges one after another between the particles. In a single particle itself, the number of positions where necks are formed between the adjacent particles increases as the discharges are repeated (Olevsky 2007).

2.1.4 Mechanical Pressure

Typically, the maximum loading pressure during SPS lies in the range of 80–140 MPa. Due to the simultaneous application of pressure and temperature during SPS, almost complete densification can be achieved at a sintering temperature which is lowered by 200–250 °C as compared to other conventional sintering processes. In the conventional sintering process, usually, a green compact is initially prepared externally using a suitable die and punch by the application of necessary pressure via hydraulic machine. On the other hand, along with the high DC pulse which is passed between the graphite electrodes an axial pressure is simultaneously applied right from the beginning of the sintering cycle in SPS. The actual contact area between the particles depends on the contact pressure and is usually only a small fraction of the apparent contact area. During the SPS process, imperfect contacts caused by surface roughness, impurities, insulating layers or oxides lead to complex and unpredictable thermal and electrical behaviour at the interfaces. The contact resistance is a function of the actual area of contact, electrical resistivity and interface temperature. A SPS-sintered grain boundary undergoes plastic deformation after the completion of sintering. The heated material becomes softer, and it undergoes plastic deformation under the uniaxial load. The SPS process is thus a combination of three stages. Plasma heating is the first stage of SPS. During this stage, the electrical discharge between the particles results in localized and momentary heating of the particles to several thousands of degrees. As the micro-plasma discharge forms throughout the volume of the sample, heat is generated uniformly. At high temperatures, the impurities concentrated on the particle surfaces are vaporized, and the surface of the particles is purified and activated. The purified surface layers of the particles then melt and fuse with each other forming necks between them. At this stage, the pulsed DC electrical current flows from the particles through the necks connecting them. The Joule heat generated by the electrical current increases the diffusion of atoms in the neck, enhancing their growth. The localized character of heating and its uniform distribution allows rapid temperature rise and drop which diminishes grain coarsening. Finally, the heated material becomes soft, and it undergoes plastic deformation under the uniaxial load which is applied during the sintering process. During SPS plastic deformation combined with diffusion results in densification of the powder compact to above 99% of its theoretical density (Olevsky and Froyen 2008).

2.1.5 The Spark Plasma Sintering (SPS) Machine

The basic SPS system consists of a vertical single-axis pressurization mechanism, single or dual vertical loadings, single-axis load frame, press with position measuring unit, load measuring unit, built-in water-cooled special energizing mechanism, specially designed punch electrodes incorporating water cooler, a water-cooled vacuum chamber, vacuum/air/argon gas atmosphere control unit, vacuum exhaust

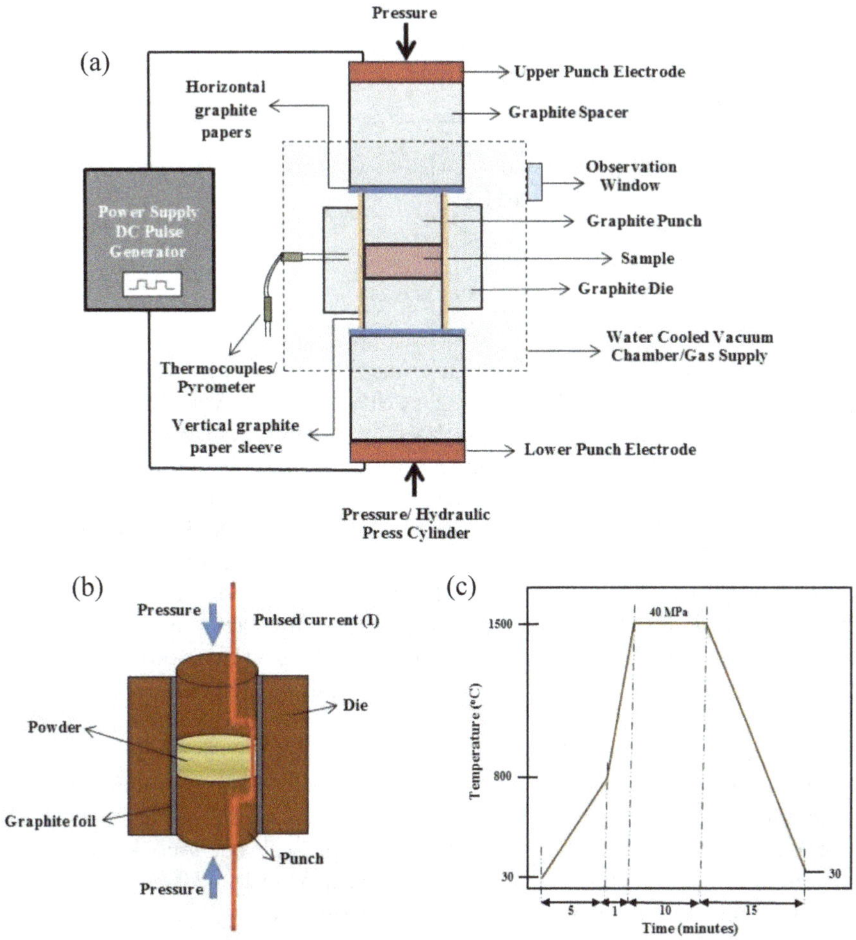

Fig. 5 (**a**) Configuration of the SPS machine; (**b**) die, punch and sample in the SPS machine; (**c**) a typical SPS profile

unit, a cooling water unit, temperature monitoring unit, DC pulse generator and a SPS controller. The basic configuration of a typical SPS system is shown in Fig. 5a, Table 1. The pulsed direct current (DC) power supply is connected to the upper and lower punches/electrodes. SPS utilizes a high-amperage pulsed DC which could vary in the range of 3000–60,000 amp. The powder sample is stacked between the die and punch on the sintering stage in the chamber and held between the electrodes. SPS is usually performed in a graphite die although depending on the pressure and temperature, and other dies can also be used. For instance, in case of very high loads along with an outer graphite die, an inner die made of silicon carbide (SiC) can also be used. The use of an outer and inner die both made of graphite is also possible. The punches can be made of other materials like carbon fibre composite or tungsten

Table 1 Configuration of a typical SPS system

Max. temperature attainable lies in the range of 1800–2200 °C
Hydraulic press capable of applying a force in the range of 10–400 tons
Heating rate of 5–1000 K/min
Holding time ~ 2–30 min
Pulsed DC power source providing a current in the range of 3000–40,000 A at 1–20 V
Max. power capacity rating of 37–1200 kW
Vacuum in cold oven 5 × 10^{-2} mbar
Process gases-Ar, N_2 (max. 5 bar)
Pulse duration of 1.255 ms and break duration of 0.255 ms

carbide (WC). The material and design of the die is chosen based on the level of pressure that one wants to achieve during the SPS process. A typical SPS sample can have a diameter of 10–15 mm and thickness of around 5 mm. Significantly, large samples having a diameter of 50 mm and pieces as large as 400 mm can also be sintered. In order to ensure good contact between the die/punches and the sample and for easy removal of the sample after sintering, interfacial materials like graphite foil, sprayed layer of graphite or boron nitride (BN) powders are generally used. The load applied ranges from 100 MPa to 500 MPa and is transferred to the powder through the upper punch. For the generation of the plasma, a high pressure must be applied, as a loose contact of the die alignment would lead to high resistance which would not allow the passage of electric current. Under pressure and pulse energized, the temperature can rise to 1000–2500 °C within a very short time resulting in a sintered compact having very high density. The entire process of vaporization, melting and sintering can be completed within a few minutes. The reason why the SPS technique is so effective is because the heat generated is not only distributed homogeneously over the entire volume of the sample at a macroscopic scale but is also dissipated at a microscopic scale exactly at the points of contact of the powder particles in the sample, where the energy is required for the sintering process (Giuntini et al. 2013). Figure 5b shows the die, punch assembly in the SPS machine with a typical SPS profile in Fig. 5c.

3 Sintering Techniques: Past, Present and Future of SPS

The application of external electric current to assist sintering was initiated by Taylor in 1933 (Taylor 1933), who incorporated the idea of resistance sintering during the hot pressing of cemented carbides. Later, Cremer (1944) patented a resistance sintering method to consolidate Cu and its alloys in 1944 in a spot welding machine. The concept of compacting metallic materials to a relatively high density (>90% of theoretical density) by an electric discharge process was originally proposed by Inoue in the 1960s (Inoue 1965). Inoue argued that a pulsed current was effective for densification at the initial sintering stages for low-melting-point metals (e.g. Bi, Cd, Pb, Sn) and at the later sintering stage for high-melting metals (e.g. Cr, Mo, W). In the United States, M. Suarez et al. (2013) used a spot welding machine for the sintering of metals in 1955. In addition to continuous pulse, some researchers also investigated a single discharge method, i.e. the powders were densified by a single discharge generated from a capacitor bank. In the late 1970s, Clyens et al. (1976), Raichenko et al. (1973) and Geguzin et al. (1975) studied the compaction of metal powders using electric discharge compaction (EDC) or electric discharge sintering (EDS). In all the methods mentioned, electrically conductive powders were heated by Joule heating, generated by an electric current. In 1990, Sumitomo Heavy Industries, Ltd. (Japan) developed the first commercially operated plasma-activated sintering (PAS) and spark plasma sintering (SPS) machines with punches and dies made from electrically conductive graphite (Yanagisawa et al. 1994). One of the salient features of these machines was that high density was achieved not only in electrically conductive powders but also in insulating materials.

3.1 Origin of Spark Plasma Sintering

As previously mentioned, the technology related to current-assisted sintering processes started in the late 1930s, when a sintering process using electrical heating was introduced in the United States. In Japan, a similar process based on the pulsed current applied sintering method was researched and patented in the 1960s and is known as spark sintering. However, due to the lack of application technology at that time, there were a limited number of fields where it could be applied. Unsolved problems associated with the industrial production, equipment cost and sintering efficiency were key points. There was little literature on research into this process until the latter half of the 1970s. The second generation was developed from the middle of the 1980s to the early 1990s. These units were small experimental systems with maximum sintering pressure of ~5 tons and pulse generators of up to 800 A and were used primarily for materials research. The emerging of SPS process took place along the 1990s when the third generation of the equipment was developed. These systems had large DC pulse generators of ~10 to 100 tons and ~2000 to 20,000 A and more. They gained a reputation as new industrial processes for synthetic processing of gradient and composite materials (Web of Knowledge 2012). The evolution of interest in SPS process can be followed by analysis of fair indicators such as patents or scientific publications. The explosion of scientific research on this technology started less than two decades ago, and it is continuing rapidly. The excellent properties which can be obtained when this technique is used for consolidating materials lead to an exponential growth of the research on this topic. It has passed from an exotic technique at the beginning of the 1990s as a fundamental tool for advanced material preparation, especially in the case of nanostructured materials. But the interest in this technology is not limited to the research area. Thus, 30% of patents related with SPS technology have priority dates of less than 10 years. Moreover, SPS has been progressively incorporated as one of the key steps in patents for obtaining advanced materials, especially for metallic and ceramic materials (Vanmeensel et al. 2005).

3.2 Industrial Production Requirements

As previously mentioned, an enormous amount of reported SPS applications are still in the area of material development, and drastic opportunities for an industrial implementation are now generated on a regular basis. Taking the next step forwards to an industrial production of novel materials by FAST (field-assisted sintering technique)/SPS is highly dependent on the availability of suitable equipment. The industrial applications of the SPS method for the rapid consolidation of novel materials require special features, which have to be fulfilled by the equipment and are different from the requirements of scientific work to some extent. The main constraints are electrical power output, flexible power supply, precise temperature measurement and control, optimized pressing tool systems, fast cooling system and automatic operation (Orru et al. 2009).

4 SPS of Different Types of Materials

The increasing importance of the electric current-assisted sintering method as a tool for consolidation of various metallic/non-metallic powders could be demonstrated by the large number of papers published in the recent years. The number of research articles in this area has shown an exponential growth since the late 1990s (Reis and Chaim 2008). Several metallic materials have been considered as matrix constituent for the development of MMNCs. In particular, the most interesting metals for industrial application are Al, Mg, Ti, Cu and their alloys. Pure and alloyed Al is the most investigated material with the largest number of published research articles. Al-based nanocomposites are potential candidates for various structural applications. Different species of nanosized oxides (Al_2O_3, Y_2O_3), nitrides (Si_3N_4, AlN), carbides (TiC, SiC), hydrates (TiH_2) and borides (TiB_2) have been employed as reinforcement agents. Apart from these, different allotropes of carbon (fullerenes, graphene and carbon nanotubes) have been investigated as nanofillers by several researchers. The most widely used carbonaceous nanofillers are CNTs and graphene as they impart excellent mechanical properties to the MMNCs. This chapter mainly focusses on the development of MMNCs by SPS.

4.1 SPS of Monolithic Metals

A literature survey of SPS of monolithic metals was done, and results reported by several researchers on SPS of Al, Cu, Ni and Mo and composites based on them have been given below. Kubota (2007) reported a significant improvement in the hardness and compressive proof stress of mechanically ground pure Al sintered by SPS. The hardness values of the SPS materials developed using 8-, 32- and 64-h mechanically grounded (MG) powders were found to be higher than that of the samples developed from powders without undergoing MG. They also reported that the compressive proof stress of the 8-h MG SPS sample (440 MPa) was ~2.5 times higher than that of the non-MG SPS material (173 MPa). The SPS behaviour of pure coarse Al powder was reported by Zadra et al. (2007). They reported that the presence of the current flow and applied sintering pressure with a certain delay in respect to the beginning of the cycle promoted better sintering and mechanical properties. With a heating rate of 100 K/min, a sintering temperature of 525 °C, dwell time of 1 min and pressure of 60 MPa, the mechanical properties and the fracture morphology were found to be very similar to those of pure annealed wrought Al. Diouf and Molinari (2012) investigated the densification mechanisms during SPS of commercial as-atomized Cu powder using three particle size ranges (<25, 25–45 and 45–90 m). They investigated and analysed the effect of particle size and temperature at which pressure is applied during the SPS. They reported that under low initial pressure, densification was due to particle rearrangement, localized deformation, bulk deformation and neck growth. However, with the increase in pressure, sintering temperature did not significantly influence the sintered density, although the tensile ductility of the samples improved with the applied pressure either during or after the bulk densification. They also reported that deformation and neck formation were enhanced by an increase in the particle size. The extremely high and localized temperature at the contact points between the particles leads to melting, giving rise to neck formation. Zhang et al. (2008) showed that high-quality bulk compact can be obtained through SPS of fine Cu powder in four stages, namely, activation and refining of the powder, formation and growth of the sintering neck, rapid densification and plastic deformation densification. The deformation behaviour of spark-plasma-sintered Ni powders with unimodal and bimodal grain size distributions was investigated by Holland et al. (2010). It was shown that it is possible to densify nanometric grain size Ni using fast heating rates and somewhat high uniaxial sintering pressures at low homologous temperatures (T/T_m in Kelvin scale) and retain the grain size with little to no grain growth. In another investigation, scanning tunnelling microscopy (STM) was used inside a transmission electron microscope (TEM) to apply an electric current directly to agglomerated nanometric Ni particles and a transformation from the initial soft agglomerate to a stage of early neck formation with relative particle rotation during the first sintering stage implying mass transfer was noticed. It was suggested that consolidation occurs in the absence of an external heat source. Neck formation between adjacent particles and increase in local Joule heat causes rapid densification. Kodash et al. (2004) investigated the influence of heating rate on densification and final grain structure of spark-plasma-sintered Ni nanopowders. It was found that a moderate heating rate was beneficial for the densification, whereas a high heating rate was conducive to a lower final density. Very high heating rates result in nonuniform densification of the samples and formation of cracks during sintering.

4.2 SPS of Intermetallics and Alloys

SPS is also a very efficient for the development of intermetallics and alloys. Skiba et al. (2010) studied the effect of varied sintering temperatures and heating rates during SPS of FeAl intermetallics. They concluded that sintering temperature of 1100 °C and heating rates of up to 400 °C/min were the best conditions. Matsugi et al. (1996) investigated the microstructural properties of titanium aluminide (Ti-53 mol% Al) developed by SPS at four temperatures, i.e. 1573, 1623, 1648 and 1673 K, after pulsed electrical discharge and found that the microstructure changed with sintering temperature and the grain growth was prevented. They reported that Vickers microhardness values for each phase in the sintered specimens were almost the same as those in specimens produced by other manufacturing methods. Feng et al. (2005) investigated the effects of sintering temperature on densification of TiB/Ti-4.0Fe-7.3Mo composites developed by mechanical alloying (MA) of Ti, Fe65Mo and B powders through SPS. They obtained a dense composite after sintering at 1000 °C for 5 min. Specimens with good shape memory effect and density were prepared from $Ti_{50}Ni_{50}$ nanopowder through SPS at a temperature of 800 °C. Below this sintering temperature, specimens had high porosity but with apparent shape memory effect. However, specimens sintered at higher temperatures had bulk density and experienced extensive oxidation which led to the loss of the shape memory effect. Couret et al. (2008) used SPS technique to densify pre-alloyed $Ti_{49}Al_{47}Cr_2Nb_2$ and $Ti_{51}Al_{44}Cr_2Nb_2B_1$ powders at temperatures ranging between 1100 °C and 1250 °C. They achieved full compaction in a short period of time which did not exceed 30 min and reported promising tensile properties at room temperature and a limited creep resistance at 700 °C. A fine-grained Ti-47%Al alloy was prepared by double mechanical milling (DMM) and SPS. The main phases observed were TiAl, Ti_3Al and Ti_2Al. Samples sintered at 1000 °C had a compressive strength of 2013 MPa, compression ratio of 4.6% and bending strength of 896 MPa. Samples sintered at 1100 °C had a compressive strength of 1990 MPa, compression ratio of 6% and bending strength of 705 MPa. Murakami et al. (1999) investigated the microstructure, mechanical properties and oxidation behaviour of Nb-Si-B system developed by SPS. It was found that the oxidation resistance of $Nb_5Si_3B_2$ compacts was better than that of Nb_5Si_3 compacts but poorer than that of $NbSi_2$ compacts. Also, it has been reported that compacts with compositions around the line of Nb_5Si_3-$Nb_5Si_3B_2$-NbB_2 exhibited high hardness at room temperature and high compressive strength at high temperatures in comparison to those with compositions away from this line. The strength of compacts containing $NbSi_2$ was found to decrease with the increase in the volume fraction of $NbSi_2$ phase. Akinrinlola et al. (2012) produced dense bimodal-grained Al-Mg alloys by SPS of cryomilled powders. It was found that the use of a two-stage sintering cycle did not induce grain growth or influence the hardness and flexural strength but doubled the fracture strength distribution statistic (Weibull modulus) from 13 to 25.

4.3 SPS of Metal Matrix Nanocomposites (MMNCs)

Metal matrix composites (MMCs) refer to materials in which rigid reinforcements are embedded in ductile metal or alloy matrix. MMCs combine metallic properties like ductility and toughness with reinforcement characteristics like high strength and modulus. Attractive physical and mechanical properties such as high specific modulus, strength-to-weight ratio, fatigue strength, temperature stability and wear resistance can be achieved with MMCs. Metal matrix nanocomposites (MMNCs) are the MMCs where the reinforcement is of nanometre dimensions, typically <100 nm in size. In MMNCs, the matrix can also be in nanometric dimension. Improvement in the mechanical properties of MMNCs can be attributed to the size and strength of the reinforcement as well as to the fine grain size of the matrix (Suryanarayana 2011). SPS has been extensively used over the past years to consolidate a wide range of materials including nanocomposites and is shown to be effective sintering method for obtaining fully dense materials with preserved nanostructure features. Recently, MMNCs received greater attraction due to their improved properties as compared to MMCs which can be attributed to the size and strength of the nanometric reinforcements. Achieving a uniform distribution of the nanosized reinforcement phase between the two components of the composites is not easy using the liquid-processing methods because of the difference in their densities and the non-wetting interface between the molten metal and the reinforcement, which makes mixing very difficult, leading to a heterogeneous structure that affects the overall properties of the nanocomposites. Also, grain growth during the conventional sintering remains a major problem during the development of MMNCs. The use of solid-state processing methods such as mechanical alloying (MA) enables the development of nanocomposites having a large nanosized reinforcement phase homogeneously dispersed in a nanostructured matrix. However, the use of conventional sintering methods such as hot pressing, high-temperature extrusion and hot isostatic pressing to consolidate these materials often results in grain growth which affects the properties of the final product. Preventing or at least minimizing grain growth to maintain the nanostructure of the matrix is possible through careful control of consolidation parameters, particularly heating rate, sintering temperature and time. As conventional processing techniques require a long holding time at high sintering temperatures, which deteriorates the nanofiller structure, processes like SPS are the ideal routes for the development of nanocomposites. Due to the pulsed electric current and the so-called spark plasma effect, sintering can be done at very high heating rates and short processing times. Therefore, grain growth can be inhibited and materials having submicron or nanosize microstructure or composites having unique compositions can be developed. In combination with the surface activation effect, sintered bodies having metastable microstructures or non-equilibrium compositions could also be obtained (Kim et al. 2003). Literature survey shows a large number of research papers on MMNCs developed by SPS. A few of the research papers have been reviewed, and the results reported by the authors have been summarized below.

4.3.1 Al Nanocomposites

Saheb et al. (2014) successfully produced carbon nanotube-reinforced Al_{6061} and Al_{2124} nanocomposites using ball milling and SPS. It was reported that CNTs were better dispersed, less agglomerated and had good adhesion to the matrix in composites containing 1 wt. % CNTs. The increase of CNT content to 2 wt. % led to the formation of CNT clusters which resulted in less uniform and homogeneous composite powders. Almost full densification of Al_{6061} reinforced with CNTs was achieved at 500 °C. A very high level of density was also attained for CNT reinforced Al_{2124} nanocomposites. Composites reinforced with 1 wt. % CNTs displayed better densification as compared to composites reinforced with 2 wt. % CNTs. The increase in the loading level of CNTs from 0.5 to 1 wt. % increased the hardness of the Al_{6061} and Al_{2124} alloys to maximum values. Further increase of CNT content to 2 wt. % decreased the hardness to values lower than that of the monolithic alloys. In another investigation, Al-Qutub et al. (2013) investigated the friction and wear behaviour of Al_{6061} monolithic alloy and 1 wt. % CNTs reinforced Al_{6061} nanocomposite developed through ball milling and SPS. It was reported that, under mild wear conditions, the composite displayed lower wear rate and friction coefficient compared to the monolithic alloy. However, for severe wear conditions, the composite displayed higher wear rate and friction coefficient compared to the monolithic alloy. Also, it was found that the friction and wear behaviour of Al-CNT composites is largely influenced by the applied load, and there exists a critical load beyond which CNTs could have a negative impact on the wear resistance of the Al alloy.

4.3.2 Cu Nanocomposites

Dash et al. (2012) reinforced Cu matrix with 5, 10 and 15vol. % Al_2O_3 particles having an average size of 5.71 microns to prepare microcomposites and conventionally sintered them under N_2, H_2 and Ar atmospheres. Also, they reinforced the same matrix with 1, 5 and 7vol. % Al_2O_3 particles having an average size <50 nm to develop nanocomposites and consolidated them through SPS. The authors reported a maximum Vickers hardness of 60, 75 and 80 for Cu-15vol.% Al_2O_3 conventionally sintered in N_2, Ar and H_2 atmosphere, respectively, and maximum hardness value of 125 for the Cu-5vol.% Al_2O_3 nanocomposite developed by SPS. Also, it was observed that Cu-Al_2O_3 composite showed poor mechanical properties when it was conventionally sintered in N_2 or Ar atmosphere compared to H_2 atmosphere. Kim et al. (2006) prepared Cu matrix nanocomposites by spark plasma sintering of high-energy ball-milled nanosized Cu powders and MWCNTs followed by cold rolling. The CNT/Cu nanocomposites showed a tensile strength of 281 MPa, which is approximately 1.6 times higher than that of monolithic Cu.

4.3.3 Fe Nanocomposites

Ravi et al. (2012) synthesized Fe-Al_2O_3 nanocomposite by reactive milling of Fe-Al_2O_3-Al-Fe powder mixture in toluene medium and consolidated by SPS. The nanocomposite showed heterogeneous grain structure of Fe consisting of nano-, submicron-, and micron-sized grains together with nanometre Fe-Al_2O_3 particles. The nanocomposites consolidated at 800 °C had a hardness of 795 MPa. The friction coefficient against an alumina ball was lower, because of the presence of the intergranular metal particles. However, $FeAl_2O_4$ grains formed during SPS were beneficial for a higher number of cycles. Libardi et al. (2007) produced iron-based nanocomposites reinforced with nanometric silica and used thermal analysis to calculate the activation energy of the grain growth process and extrapolate the temperature at which the powder could be consolidated by SPS without losing nanostructure. It was reported that without the addition of SiO_2, the hardness was constant up to 400 °C, whereas the dispersion of SiO_2 lead to the consistency of the hardness value up to 600 °C.

4.4 SPS of Other Nanocomposites

Udhayabanu et al. (2011) developed Ni-30 vol.% Al_2O_3 in situ nanocomposite through reaction milling of NiO-Al-Ni powder mixture followed by SPS. The nanocomposite had hardness and yield strength approximately two times higher than that of pure Ni of similar grain size. A nickel-matrix composite with MWCNT reinforcement was prepared by Yamanaka et al. (2007) through a slurry mixing process using ethanol as the solvent. The SPS-sintered relative densities of the composites containing up to 5vol% of MWCNT were above 99%. The thermal conductivity was found to increase by 10% for the composition with 3vol% MWCNT. Umeda et al. (2009) prepared magnesium reinforced with CNTs and Mg_2Si/MgO compounds. They used amorphous and porous silica particles originated from rice husks, which were coated with CNTs and contained nanotubes in the pores. The in situ synthesis of Mg_2Si and MgO via deoxidization and oxidation reaction occurred from the elemental mixture of pure magnesium and CNT-SiO_2 composite particles through SPS. A low and stable friction coefficient and a decreased wear rate with increased CNTs and Mg_2Si content were reported.

5 General Issues of Nanostructure Materials

Nanotechnology can be broadly be defined as the creation, processing, characterization and utilization of materials, devices and systems with dimensions in the order of 10–100 nm, exhibiting novel and significantly enhanced physical, chemical and biological properties, functions, phenomena and processes, due to their nanoscale size. A nanomaterial has a typical grain size <100 nm, whereas ultrafine grained materials are characterized by grain size 100–500 nm. The term nanocomposite comprises multiphase materials, where at least one constituent phase has a dimension less than 100 nm (Mayo 1996). In order to classify the nanocomposite structures, the different morphologies of the nanoscale reinforcements should be considered first. Nanofillers can be classified into three broad categories, namely, (a) 3D nanofillers such as nanoparticles and nanospheres, which are relatively equiaxed and have a diameter <100 nm; (b) 2D plate-like nanofillers, which can be layered materials with thickness of the order of 1 nm and aspect ratio in the other two directions being at least of 25; and (c) 1D fibres or tubes having diameter <100 nm and aspect ratio more than 100. It should be noted that a change in particle diameter or fibrous material diameter or the layer thickness from micrometre (μm) to nanometre (nm) range can change the surface area/volume ratio by one or two orders of magnitudes (Niihara 1991; Supova et al. 2011). This suggests that with the increase in interfacial area, the properties of the nanocomposites are more and more dominated by the properties of the interface between the matrix and the nanofiller. Most nanocomposites have a structure which consists of a micron-sized matrix in which nanofillers are embedded. In particular, the nanodispersions can be embedded within the matrix grains, located at the grain boundaries, or in other words, they can occupy both intergranular and intragranular positions. It is essential to understand how the nanoscale structure influences the bulk properties to enable the development of advanced and multifunctional composites. Figure 6 shows the SEM and HRTEM images of few nanofillers like multiwalled carbon nanotubes (MWCNTs), graphite nanoplatelets and graphene oxide (GO).

Agglomeration of nanofillers is the major issue while development of nanocomposites. Nanofillers tend to agglomerate when added at a higher loading level in the matrix. Figure 7 shows the optical micrographs of alumina-based composites using exfoliated graphite nanoplatelets (xGnP) as the nanofiller. Large agglomerates of xGnP could be clearly seen in the optical micrograph of Al_2O_3-5 vol. % xGnP composite, whereas the optical micrographs of Al_2O_3-0.2 vol. % xGnP show a more uniform distribution of xGnP in the Al_2O_3 matrix (Alam et al. 2017).

6 Sintering and Grain Growth of Nanostructure Powders

Sintering is the process of transforming a powder into a solid body using heat. Sintering is done to join particles together without melting them. It is a heat treatment technique applied to a powder compact in order to impart strength and

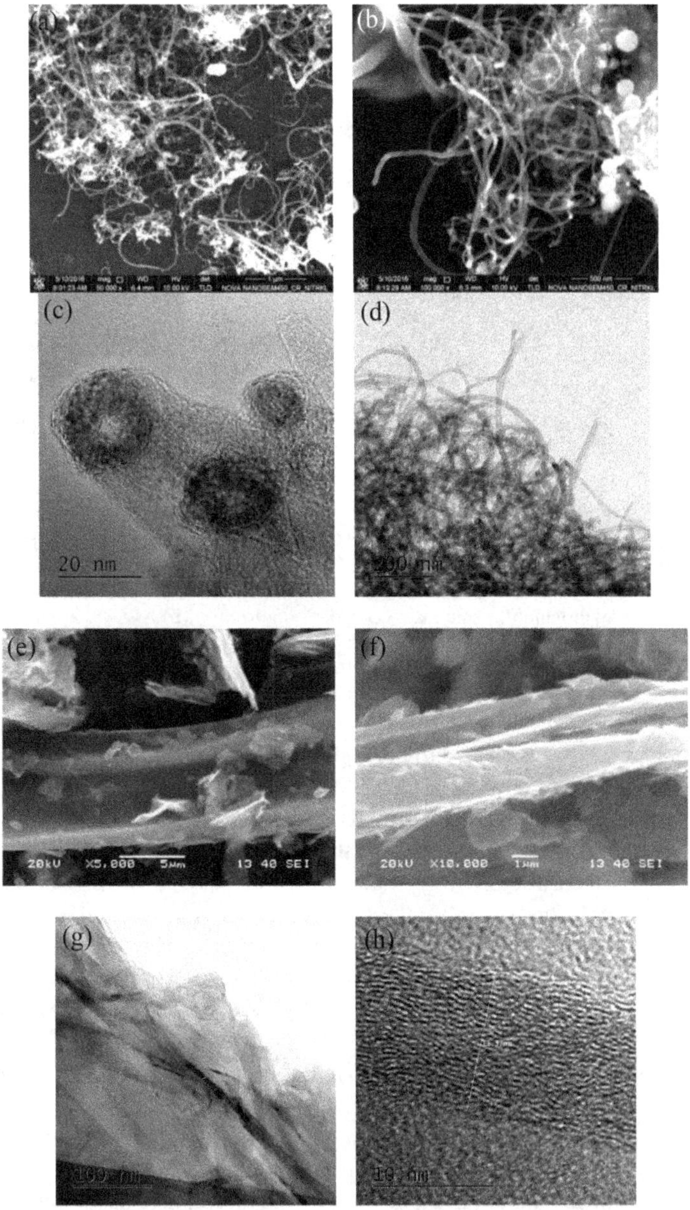

Fig. 6 (**a**, **b**) SEM and (**c**, **d**) HRTEM images of MWCNT, (**e**, **f**) SEM images of graphite nanoplatelets and (**g**, **h**) HRTEM images of rGO (reduced graphite oxide)

integrity and occurs during firing of a piece that has been pressed. The temperature used for sintering is below the melting point of the major constituents of the powder compact. After compaction, neighbouring powder particles are held together by cold

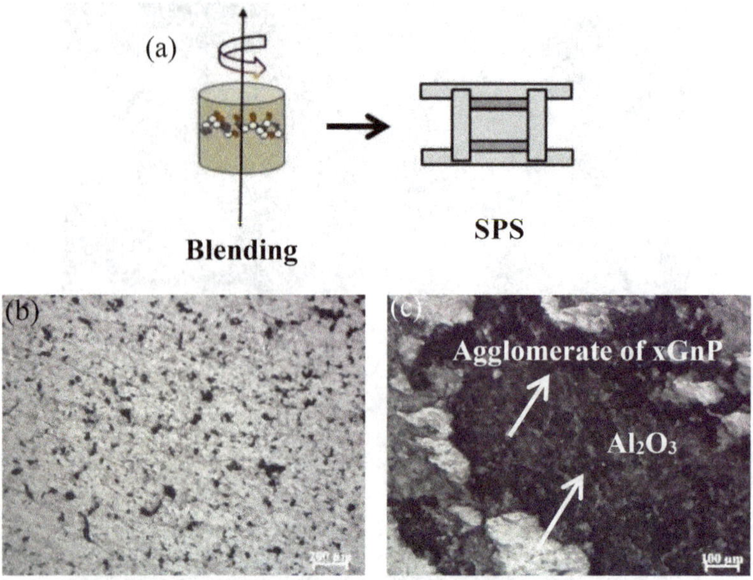

Fig. 7 (**a**) Blending of nanofiller and matrix powder prior to SPS, (**b**) Al_2O_3-0.2 vol. % xGnP and (**c**) Al_2O_3-5 vol. % xGnP composites developed by SPS carried out at 1450 °C for 5 min with a pulse on-off ratio of 12:2. The heating rate was 100 °C/min and a pressure of 50 MPa was applied

weld, which gives the compact sufficient green strength to be handled. When heated to the sintering temperature, diffusion processes cause the neck to form and grow at these contact points. Although sintering is still one of the most important processes in developing bulk metallic structures, the phenomenon is not similar as to the ceramics. One of the reasons for processing metals by sintering is to control the grain size. In solid-state sintering, the green compact is sintered at around 0.5–0.9 T_m. Although sintering explicitly involves solid powder grains, a liquid may form if a component has a melting temperature lower than the sintering temperature. This is referred to as liquid phase sintering (LPS). Figure 8a shows the various stages of sintering. There is a direct particle to particle contact during solid-state sintering. On the other hand, during liquid phase sintering, the presence of a liquid film ensures adherence of the solid particles together. Grain growth takes place during and after sintering. It could be called as a process by which a powder compact is transformed into a strongly bonded monolithic material by the removal of pores between particles. The schematics diagram in Fig. 8b shows the mechanism of formation of pores during sintering. The solid body formed by sintering is most likely to have a density less than 100% of its theoretical density and contain pores. Pores are a structural component of the solid sintered body. Sintering is a complex process in which several processes occur simultaneously (Thostenson et al. 2005).

Refining the material grain size in metals can induce significant changes on both mechanical and physical parameters. To take advantage of the unique properties of

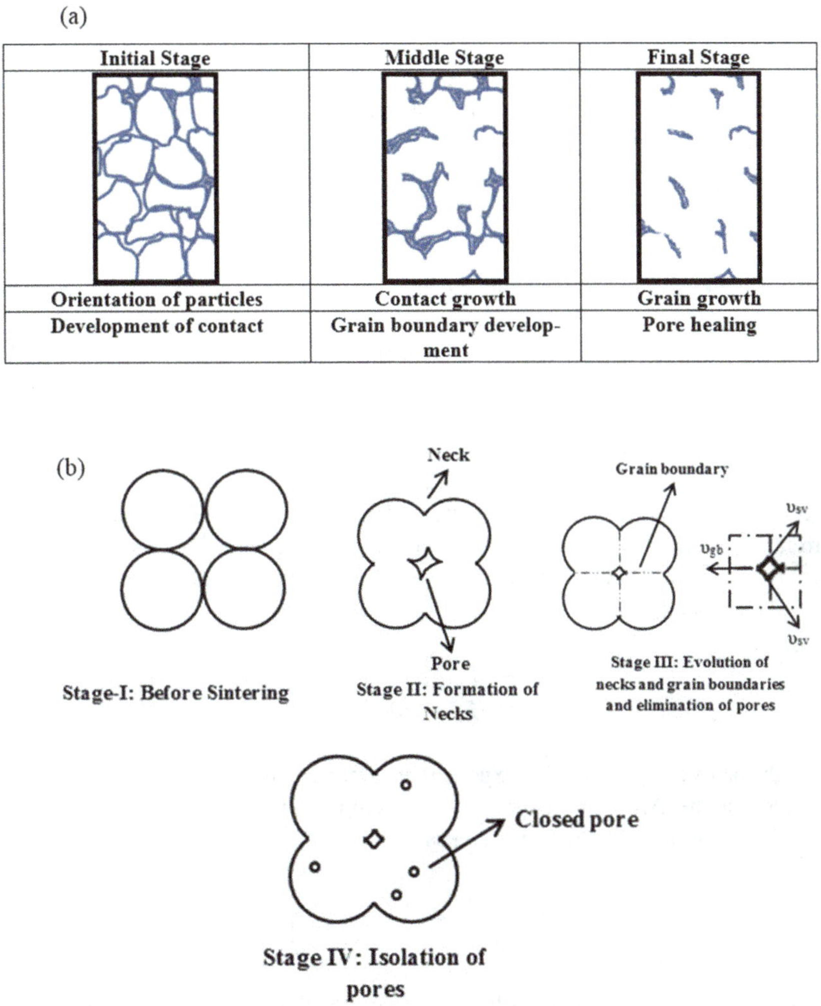

Fig. 8 (**a**) Three stages of solid-state sintering, (**b**) formation of pores during sintering

bulk nanocrystalline materials, the nanometre range powders have to be densified into parts of certain properties, geometry and size. The key to the nanopowder consolidation process is to achieve densification with minimal microstructural coarsening and/or undesirable microstructural transformations. While the densification process for conventional powders is well known, the densification of nanopowders poses significant additional challenges, by both theoretical and practical point of views. Firstly, due to the very large surface area of nanoparticles, agglomeration is an ubiquitous issue, affecting the densification process in a negative way. In addition to this, the other question which arises is whether the sintering mechanism remains the same at the nanoscale or not. As reviewed by Groza (1999), nanopowders are thermodynamically highly unstable. The tendency to reduce the excessively large surface area per unit volume of the nanocrystalline powders is the mechanism that drives the sintering process. The surface energy (γ) of a particle having a radius of curvature (R) can be expressed in the terms of stress (σ) by the Laplace equation given below.

$$\sigma = \frac{\gamma}{R} \tag{1}$$

From the kinetics point of view, we expect enhanced densification for processes that have a direct dependence on grain size. Sintering of nanopowders is expected to be accomplished at significantly lower temperatures and shorter time as compared to conventional powders. The other major problem associated with nanocrystalline powders is their tendency to cluster and form aggregates. Therefore, in order to maximize the benefits from the unique properties of the nanostructured materials, the agglomeration of the nanoparticles should be avoided, and they should be loosely dispersed in the matrix. In 1950, Herring (1950), developed the classical scaling rule, correlating the effect of the particle size on the sintering time. The time, t, to achieve the same sintering condition is correlated with the powder particle size, d, given by the equation below.

$$\frac{T_1}{\mathrm{T}_2} = \left(\frac{d_1}{d_2}\right)^n \tag{2}$$

In the above equation, the exponent n depends on the sintering mechanism. Considering the Arrhenius expression for temperature, the sintering temperature dependence on the particle size becomes

$$\ln\left(\frac{d_1}{d_2}\right) = \frac{Q}{R}\left[\frac{1}{T_1} - \frac{1}{T_2}\right] \tag{3}$$

where Q is the activation energy for the predominant sintering mechanism, R is the gas constant, d_1 and d_2 are the different powder particle sizes for the matrix and reinforcement and T_1 and T_2 are their respective sintering temperatures. Figure 9a shows the material transfer paths during diffusion. Figure 9b, c shows the SEM images of the sintering of Al_2O_3 particles in Al_2O_3-3 vol. % xGnP (exfoliated graphite nanoplatelets) at 1450 °C for 3 h.

7 Strengthening Mechanisms in Nanocomposites

The superior mechanical properties of MMNCs is the result of several strengthening mechanism contributions, namely, Hall-Petch strengthening, Orowan strengthening, coefficient of thermal expansion (CTE) and elastic modulus (EM) mismatch

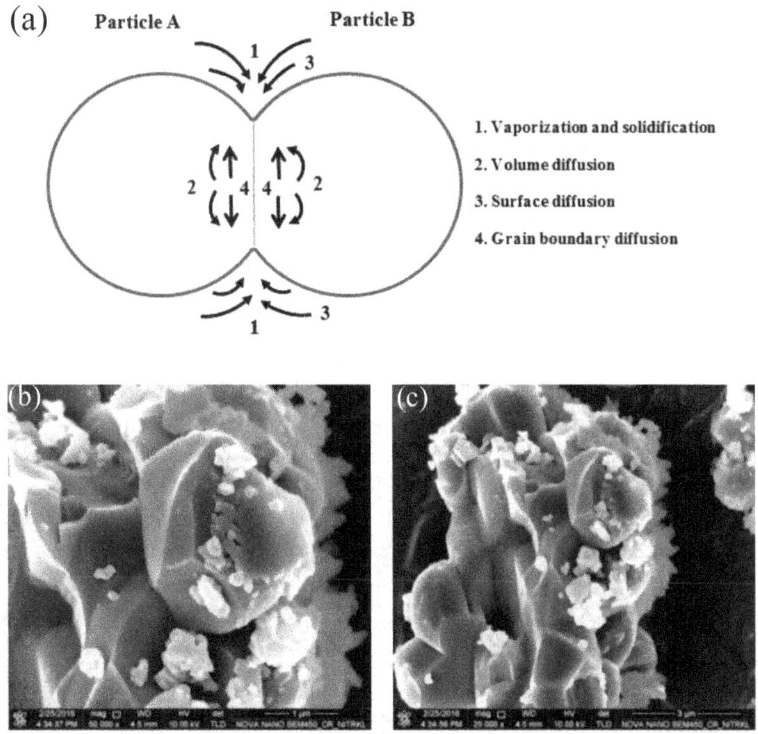

Fig. 9 (**a**) Material transfer path during diffusion and (**b, c**) sintering of alumina particles in Al_2O_3-3 vol. % xGnP (exfoliated graphite nanoplatelets) at 1450 °C for 3 h

(Hughes et al. 1986; Queyreau et al. 2010; Hiremath and Hemanth 2017; Zadeh 2012). The addition of nanocrystalline second phases to ordinary grain-sized material is very effective in reducing the particles' grain growth during sintering. In fact, dopants or second-phase particles, which are insoluble in the matrix phase, can decrease the coarsening kinetics of the matrix by reducing the grain boundary mobility through a particle pinning effect. This behaviour was described in Zener's theory in 1948 (Zener 1948), showing that the inclusion exerts a drag force on the grain boundary which can be expressed as,

$$F = 2\pi r\gamma \sin\theta \cos\theta \tag{4}$$

where F is the pinning force, r is the particle radius and γ is the average boundary energy of the system. If particles are assumed to be spherical and γ to be isotropic, the maximum value of the force, F_{max}, is obtained at $\theta = \pi/4$.

$$F_{\max} = \pi r \gamma \tag{5}$$

The number of particles per area of boundary, N_s, can be approximated by considering the particles to be distributed randomly on the boundary. Therefore

$$N_s = \frac{3f}{4\pi^2} \tag{6}$$

where f is the volume fraction of particles. Combining the above equations, the pinning stress (σ_z) can be expressed by the equation given below.

$$\sigma_z = \frac{3f\gamma}{4r} \tag{7}$$

In this theory, we should consider that numerous assumptions have been made, for instance, it applies to spherical and randomly distributed particles. It is assumed that each particle in contact with the boundary exerts the maximum force, whereas the pinning force exerted by each particle depends on the position of the boundary (Anderson et al. 1989).

8 Sintering, Consolidation, Reaction and Crystal Growth by SPS

The term spark plasma system includes spark plasma sintering (SPS), spark plasma consolidation (SPC), spark plasma joining (SPJ), spark plasma growth (SPG) and spark plasma reaction (SPR) (Guillon et al. 2014). Therefore, SPS is an appropriate description of the process as it enables sintering, joining, crystal growth and chemical reactions. Compared to traditional sintering processes, SPS has several features which could offer a wide range of possibilities for the development of new materials. Few of the features of SPS are listed below:

1. Rapid sintering
2. Reliable control of sintering process parameters which enables control of material microstructure
3. Generation of temperature gradient during sintering which allows sintering of functionally graded materials
4. Solid state sintering
5. Low-temperature sintering
6. Possibility of affecting sintered material with electromagnetic field energy
7. Possibility of combining sintering methods like spark plasma sintering and hot pressing (SPS-HP), SPS in a magnetic field and SPS-SHF (super high microwave) sintering

SPS has been mainly characterized by spark plasma created by a pulse DC during heat treatment of powders in a graphite die. Gaps are assumed between the two electrodes in the electric discharge machine, and high-energy plasma is generated in the gaps. However, the die alignment of SPS does not allow any gaps, and high-energy plasma cannot be generated without such gaps. Therefore, it is reasonable that the plasma of SPS has not been identified directly. The electrical noise that has been observed is thought to correspond to the plasma generation (Furuya et al. 1998). The generation of spark plasma during SPS is considered as a means for the fabrication of new materials. The effect of electric field and electric current in the conductor or skin current in the semiconductor and insulator had not been considered to be significant for sintering and reactions. An impact is felt within the sample due to the mechanical pressure caused by the spark plasma. There are no insulators and heating elements having very large heat capacity. These conditions result in rapid heating and cooling rates which are one of the reasons behind the high efficiency of the SPS process (Boesel et al. 1969). The effect of the plasma is very significant at the initial stage of the sintering as there are many gaps in the powder. The plasma sites are gradually reduced with decreasing porosity during the middle stage of sintering and are not generated when the gaps disappear. The heat treatment with SPS is controlled as observing the shrinkage of the samples. Samples shrink until the densification is achieved. After the shrinkage stops at a certain temperature, the sample starts to melt simultaneously in the small area between the sample and the graphite punch. Gaps appear in the melt, and the spark plasma is regenerated there. As the site of plasma generation is limited, the plasma is concentrated only in the melted area. The plasma increases the melt and leaves some patterns on the surface. Ionized ions and radicals are created in the plasma. High heating rates, using pressure and electrical current, differentiate the SPS technique from the other sintering methods. In addition to the reduction in grain growth, the high heating rate also increases condensation by the elimination of surface diffusion mechanism and creates an extra driving force by the high-temperature gradient. Pressure applied during the heating can increase the driving force of process and facilitate the sintering process. Electrical current can condense the powder in the mould by creating many sparks between particles and creation of plasma environment (Omori 2000). Unlike conventional methods of sintering where large grain growth is observed, sintering through SPS shows very little grain growth, so the crystals remain in nanoscale. Ultrafine grained materials have gained considerable scientific attention and technological interest due to potential benefits associated with fine grain size. Grain size influences several properties of the materials like the thermal conductivity and the mechanical behaviour. The changes in the thickness of the sample with respect to time containing different amounts of reinforcement include three zones. The first zone can be considered as the zone in which the sample is heated and little expansion takes place. This is followed by a zone where there is an increase in the temperature when the sintering process occurs and the samples contracts quickly. The sintering process is complete by the end of the second zone, and the samples are densified. This is followed by a period when the change of displacement is constant. Due to the lower sintering temperature, short temperature zone and holding times, it is possible to develop nanocomposites having near theoretical density and with little crystal growth (McTaggart 1967; Raju and Basu 2014). To attain full densification by using conventional sintering techniques, a higher sintering temperature and soaking time is required. SPS can produce concentrated heating, which allows energy saving and time efficient densification of metals compared to other conventional sintering techniques. Apart from being a tool to densify materials, SPS can also be exploited to engineer microstructures. The rapidity of the SPS process can also allow new temperature profiles which can suppress atomic diffusion, resulting in thermodynamically metastable materials and microstructures. The directionality of electric field (DC or pulsed DC), which is a common feature in most of the electric current-assisted sintering (ECAS) techniques, can also be reflected in the microstructure of the sintered materials. When a DC field is used, strong directional polarity-dependent effects can be produced due to the thermal gradients generated within the sample. These directional effects might be engineered to produce textured or functionally graded microstructures (Razavi et al. 2008; Grasso et al. 2016).

9 Densification Mechanisms During SPS of Nanocomposites

Coarsening of powder particles and their subsequent grain growth lead to a reduction of the total surface energy within the compact which in turn reduces the capillary driving forces for the consolidation of the compact. The smaller the powder particle size, the more active and significant these processes become (Wang et al. 2017). Therefore it is essential to take coarsening and grain growth processes into consideration, for the calculation of SPS densification maps when using nanocrystalline powders. Coarsening and grain growth can be expressed by the following equation.

$$D_t^n - D_0^n = \mathrm{K}t \tag{8}$$

where D_0 and D_t are the particle/grain diameters at time $t = 0$ and $t = t$, respectively, n is the coarsening/growth exponent and K is a temperature-dependent coarsening/growth constant given by

$$\mathrm{K} = \mathrm{K}_0 \exp\left(-\frac{Q}{RT}\right) \tag{9}$$

where K_0 is the pre-exponent constant and Q is the activation energy.

The original HIP model (Olevsky and Froyen 2006) neglected the effects of particle coarsening and grain growth since their effects on the model geometry were unknown. However, we assumed that in the nanocrystalline powder compact subjected to densification, particle size and the resulting grain size are equivalent, as the particles undergo grain growth during densification and become grains. Consequently, the effect of particle coarsening and grain growth on densification kinetics is considered by letting the diameter D increase with time during densification. This is done by calculating the new grain size according to the growth equation, at the beginning of every step of the numerical integration. The high pressure applied during SPS also has a significant negative effect on the grain growth, which is closely related with the particular sintering conditions. During the SPS process, the ultra-high heating rate plays an important role in reducing the grain growth thereby preserving the large specific surface area of the nanoparticles until the peak temperature is reached. Due to the small area of coalescence, the effective stress of the nanocomposites would be very high and is likely to cause plastic deformation of the coalescence so as to suppress the grain growth based on surface diffusion by disrupting the channel of mass transfer. The high pressure applied during sintering also inhibits grain growth by reducing their mutual driving force of sintering. Furthermore, as the densification process based on plastic flow stagnates, the retained nanostructure enables further densification by a diffusion mechanism. This suppresses the final-stage grain growth by decreasing the sintering temperature and reducing the effective soaking time required for sintering of nanocrystalline materials (Casellas et al. 2003). Under the application of a low initial pressure, densification is caused due to particle rearrangement, localized deformation, bulk deformation and neck growth. Densification by particle rearrangement and localized deformation are enhanced by an increase in particle size. Bulk deformation is not affected by a decrease in particle size, and on the other hand, neck growth is instead enhanced by a decrease in particle size. If the pressure is increased during the SPS cycle, the temperature at which it is applied does not influence sintered density significantly, but tensile ductility increases if pressure is applied either during or after the bulk densification (Randall 2010).

The generation of plasma during the SPS process is still not confirmed. The formation of plasma, proposed by Tokita (1999), is still an open point although in a recent paper, Hulbert et al. (2009) gave an experimental demonstration of the absence of plasma during SPS. A very high temperature above the melting temperature of the material may be attained at the contact area which can lead to localized melting and enhance interparticle bonding. Overheating is localized at the surface as the temperature gradient along the particle radius is very steep and the bulk of the particle remains practically unaffected by the surface phenomena. The sintering mechanism can be considered as a sequence of stages. The first stage involves activation and refining of the powder which is followed by the formation and growth of the sintering neck. The final stage involves plastic deformation and densification. In micrometric powders, the mass transport mechanism is a power-law creep. The first two stages are promoted by the spark discharge between particles, which removes the surface oxide because of the voltage breakdown effect and heats up the powder surface, leading to the formation of the neck by evaporation, condensation and diffusion. The third and fourth steps are promoted by the current flow through the necks which heats up the material by Joule effect. At this point, fast densification is promoted and enhanced by the application of the pressure.

Deformation and densification depend on the pressure and on the temperature at which it is applied since they are enhanced by the thermal softening of the powder. Similar to the traditional hot pressing (HP) and hot isostatic pressing (HIP) processes, application of high pressure during SPS leads to high initial relative density and promotes diffusional processes which thereby decrease the densification duration. However, in contrast to HP and HIP, the rapid heating rates during SPS limit the dwell time during the heating where coarsening usually takes place. Consequently, the sintering duration at the SPS temperature, which is largely particle and grain size-dependent, is also reduced. The use of nanocrystalline powders leads to substantially shorter densification durations and lower sintering temperatures as compared to that required for conventional size particles.

Recently, MMNCs using graphene/MWCNTs as nanofillers have emerged as the material of interest due to their excellent structural and functional properties. The high thermal conductivity and very low coefficient of thermal expansion of these nanocomposites have made them very attractive in a wide range of applications. However, achieving full densification of these materials is still a challenge due to the oxidation of the matrix and difficulty in dispersing the nanofillers homogeneously in the matrix. Addition of an optimum amount of the nanofiller can lead to near full densification as well as in obtaining excellent properties compared to the monolithic materials.

10 SPS Techniques: Prospects and Limitations

The consolidation of nanocomposites presents a challenge using conventional sintering techniques. Although success has been achieved at laboratory scale by SPS for research purposes, there are still a few challenges to commercialize the technology for the production of large components having complex geometry. Using conventional sintering processes, densification is always accompanied by grain growth; however SPS separates grain growth from densification. As discussed above, the occurrence of a plasma discharge during SPS is still debated, but it seems to be widely accepted that an electric discharge process takes place on a microscopic level. Since no current or only a very weak current passes through non-conducting samples, the discharge ought to originate from the electric field set up by the pulsed direct current used. The intensity of this discharge is presumably not only dependent on the intensity of the applied pulses but also on factors such as particle size, pore size and relative density of the compact. Anyhow, it seems plausible that it is only during the initial part of a sintering process, i.e. up to the point where the system reaches a closed porosity state, such discharge processes can make any major contribution to the densification. From this stage onwards, grain boundary diffusion and grain boundary migration are the rate-determining processes. An electric discharge process would certainly clean the surfaces of the powders from adsorbed species and create various types of surface defects that enhance the grain boundary diffusion. A few limitations of the SPS process have been listed below.

10.1 Sample Size

As sample sizes are increased to industry relevant diameters, the current SPS practices face real limitations. These limitations are primarily a function of the power requirements necessary to achieve high temperatures and heating rates of the dies, plungers and parts themselves. As the sample diameter grows beyond a few centimetres, the magnitude of the current required to maintain the typical heating rates used in SPS processing (>100 °C/min) becomes impractical. As a result, the heating rates for large samples in SPS processing become comparable with the one used in hot pressing. Further, any contribution related to mechanisms involving fields or current densities tend to become less evident, as it is readily apparent that very large die assemblies result in low current densities and low voltages across the samples. The major concern as the diameters of the samples are increased is to maintain a reasonably even temperature distribution across the sample. It has been shown by numerical modelling that large samples might present significant macroscopic temperature gradients, depending on the geometry of the die assembly and on the electrical properties of the sample. In the case of non-conducting samples, complex temperature distributions are expected, and a careful control of the die geometry is required to avoid significant sample in homogeneities (Shi et al. 2010).

10.2 Thermal and Athermal Aspects

The role of electric fields on powder densification has been and to a certain degree remains unclear. However, some recent works demonstrate how an applied electric field may have significant influences during SPS. As Olevsky et al. (2007; Olevsky and Froyen 2006, 2008) have pointed out the basic mechanisms active in SPS processing are either thermal or athermal. Thermal effects stem from heating rates or thermal gradients present within the samples. Athermal effects stem from the electric fields present during SPS processing. Additionally, it is important to understand that these effects play different roles during each stage of sintering and cannot be expected to contribute ubiquitously throughout the sintering process. The thermal effects active during SPS have been known to contribute most strongly to the excellent consolidation behaviours observed. However, in the case of nanometric powders, the contributions of thermal effects may be overestimated. For large powder grain sizes (>1 μm), the benefits of fast heating rates are very clear and well founded. It was also observed that when more time is spent during heating, when evaporation, condensation or surface diffusion mechanisms dominate, there are increases in the pore spheroidization. This results in stiff, low-energy neck forming between powders particles, which decrease the overall densification rates and final densities to a great extent. However, it was also found that the effects are inverse functions of grain size and contribute less as the powder sizes are reduced. Generally, as the powder diameters are decreased to the nanometre size range, spheroidization of the neck region has less influence due to the high curvature. Further, the assumption that surface diffusion is nondensifying may, in fact, be inaccurate for nanometric powders. Another thermal mechanism expected to contribute to SPS processing is the Ludwig-Soret effect (Jung et al. 2011). This effect occurs in the presence of a thermal gradient, in which the differently mobile species within a material result in the establishment of concentration gradients. There are two types of thermal gradients which occur during SPS, namely, local- and sample-wide.

10.3 The Role of Electric Field

The role of the electric field in SPS processing has largely been discounted due to the low applied field strength generated across the specimen. Several possible mechanisms have been proposed to be active in SPS plasma formation between particles, such as electromigration effects, dielectric breakdown, increased kinetics from the applied field and thermodynamic changes due to the applied field. Early explanations of the effectiveness of SPS were based upon the presence of plasma generated between the loosely packed powder particles. Very little experimental, or theoretical, evidence has been presented to support this theory. Additionally, all credible attempts to locate the presence of any plasma have failed. The first step in evaluating the potential role of a field during SPS is to understand the local field strengths involved which refer to the field strengths at interfaces and grain boundaries between individual powder particles. Using numerical models it has been established that the local field strength in dielectric materials at point contacts and neck region is much above the applied external field strengths for SPS. These field strengths are exponentially dependent on the relative permittivity of the materials. An analysis of the mechanisms active during SPS suggests that thermal processes in the absence of an applied field do not contribute to the densification of nanometric powders, whereas electric fields, in both contacting and non-contacting modes, are found to contribute significantly to the densification (Voisin et al. 2013).

11 Effect of Various Nanofillers on Microstructural Behaviour and Strengthening Mechanism of Al-Based Nanocomposites

Aluminium (Al) is the second most widely used metal in the world today after iron (Fe). It has a low density (2.7 g/cc), superior malleability, excellent corrosion resistance, good thermal conductivity (237 W/m K), very low electrical resistivity (2.65×10^{-8} Ω m) and good formability. Its Young's modulus is 70 GPa, and its Vickers hardness is 160–350 MPa. Al has a melting point of 660.32 °C. However, at high temperatures, the strength of Al decreases. The demand for Al and its alloys having a much higher strength is increasing. Therefore, to overcome these shortcomings and to meet the ever increasing demand of modern day technology, strengthening of Al by introducing nanofillers is an excellent option to overcome the limitations of monolithic Al. This could enable us to achieve a good combination of strength, stiffness, toughness and density (Al-Qutub et al. 2013; Dash et al. 2012; Kumar and Xavior 2014; Chunfeng et al. 2007; Saboori et al. 2017). A few research papers on Al-based nanocomposites developed by SPS have been reviewed, and the results reported by the authors have been reported below.

Sadeghi et al. (2018) reported the effect of bimodal micro- and nanosized Al_2O_3 particles on the properties of Al matrix composites developed by SPS. The percentages of micro- and nanosized Al_2O_3 particles added to the Al matrix are given in Table 2.

Samples were sintered by SPS technique using model 10 ton equipment. A 40-mm-diameter graphite die was used for sintering. A thermocouple hole was drilled in the punch. For easy removal of the sintered pellet, a graphite paper sheet was placed between the punch and the powder as well as between the die and the powder in order to prevent sticking of the sintered pellet with the graphite die/punch due to the possible reaction between them. Sintering was conducted under a very low pressure of ~ 0.0075 MPa, and the heating and cooling rates were 75 and 30 °C/min, respectively. Pulsed DC, with a 36 ms on and 8 ms off pulsed profile, was applied. After reaching the chamber internal pressure, the furnace cavity was filled with Ar gas. The gas removal takes place during the initial stages of heating. The powder sample and the mould were heated to the sintering temperature of 550 °C at a heating rate of 75 °C/min and then held at that temperature for 5 min. A uniaxial pressure of 50 MPa was applied throughout the entire heating and cooling cycle. The total time for sintering including heating and cooling time was less than 25 min. Figure 10 shows the temperature and pressure profile applied during the SPS process.

By increasing the ratio of the nano- to micron-sized particles, the agglomeration trend of Al_2O_3 particles in Al matrix increases. The specimen with ratio nano to micron 2:8 shows the minimum agglomeration among all the specimens. As the weight percent of nano to micron particles increases, grain refinement effect decreases. The matrix porosity increases by increasing the content of nanoparticles. Agglomerates were found to produce an increase in porosity. The density was found to increase with an increase in nanoparticles percentage of up to 4 wt. %, then the density decreases with further increasing. Figure 11 shows neck formation in the

Table 2 Different composites produced for the present study

Sample	Al	Wt.% nanopowder	Wt.% micron powder
Al-10mAl_2O_3	90	0	10
Al–2n8mAl_2O_3	90	2	8
Al–4n6mAl_2O_3	90	4	6
Al–5n5mAl_2O_3	90	5	5
Al–6n4mAl_2O_3	90	6	4

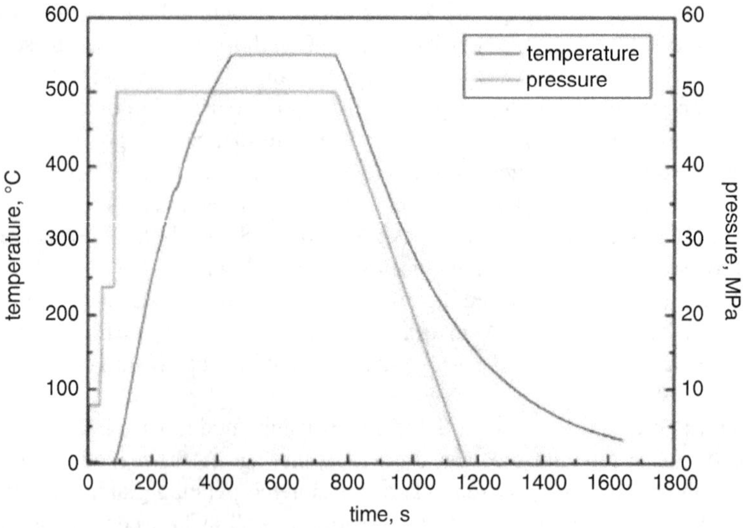

Fig. 10 Temperature and applied pressure employed during SPS (Sadeghi et al. 2018)

various Al-Al_2O_3 nanocomposites developed by SPS. At the particle-particle contact zones, the local pressure can increase upto 1000 times. The high localized pressure causes neck formation around the contact area between the particles.

From Fig. 12 it is evident that the yield strength increases up to 4 wt. % of the nanoparticles, and then it decreases with further increment of the bimodal Al_2O_3 particles. The enhanced strength of the Al-2n4mAl_2O_3 and Al-4n6mAl_2O_3 nanocomposites can be attributed to the stronger diffusional bonds and homogeneous distribution of nano- and micron-sized Al_2O_3 particles in Al matrix during SPS. Sadeghi et al. (2017) also studied the influence of 2 wt. % of Al_2O_3 nanoparticles on microstructure and strengthening mechanism of Al-based nanocomposites developed by SPS. The material density and the mechanical properties of the SPSed materials were found to strongly improve with the addition of a low percentage of nanosized Al_2O_3 particles. The hardness and elastic modulus (Fig. 13a) of the composite showed a noticeable increase as compared to the pure SPSed Al. The tensile strength (Fig. 13b) was also found to increase although no difference in the material ductility was observed. The composite was mainly found to deform in a transgranular way, while pure Al showed higher local ductility.

In another paper, Cavaliere et al. (2017) reported the effect of 0.5 and 1 wt. % of CNTs addition on the mechanical and microstructural behaviour of Al-based MMCs developed by SPS. The results show that the CNTs addition increases the porosity and tensile strength with respect to the pure SPSed Al. The powder mixtures were sintered at 550 °C at a heating rate of 75 °C/min for the holding time of 5 min. The microstructural behaviour and the final mechanical properties of the Al matrix composites with 0.5 and 1 wt. % of CNTs developed by SPS have been found to be strongly dependent on the nanosized reinforcements blended with pure Al before

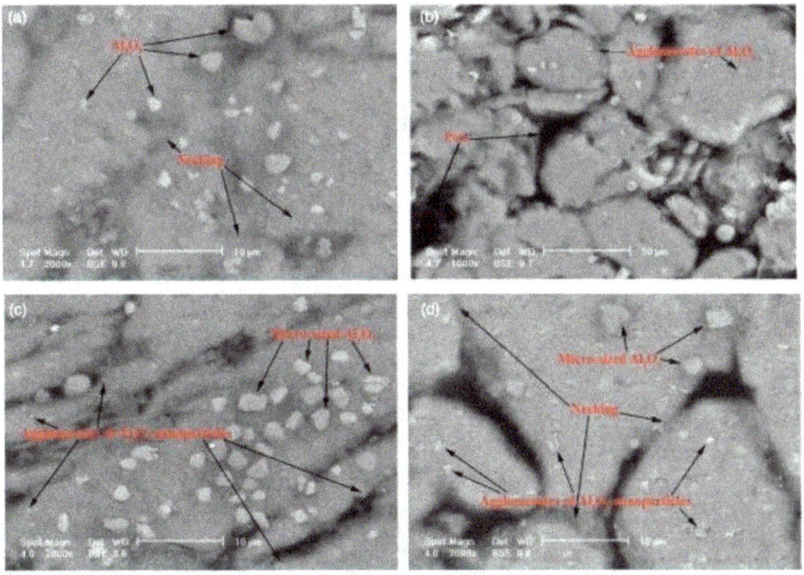

Fig. 11 Neck formation in Al_2O_3 in Al-2n8mAl_2O_3 (**a**) Al-4n6mAl_2O_3 (**b**) Al-5n5mAl_2O_3 (**c**) Al-6n4mAl_2O_3 (**d**) SPSed nanocomposites (Sadeghi et al. 2018)

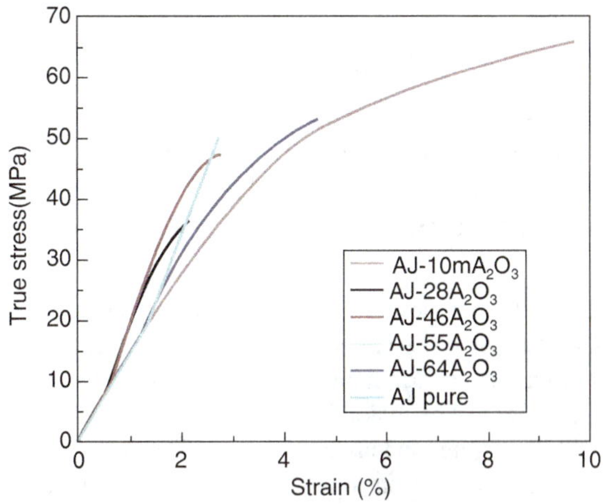

Fig. 12 Tensile curves of SPSed Al-Al_2O_3 having different ratios of the nano- to micron-sized Al_2O_3 particles (Sadeghi et al. 2018)

processing. The density of the Al matrix composites was found to strongly increase with the addition of a low percentage of nanosized CNTs. The reduction of porosity shows a noticeable increase in hardness and elastic modulus as compared to the pure SPSed Al. The tensile strength was also found to increase although a reduction in ductility was observed. The fracture surfaces show that composites mainly deform in a transgranular mode, while pure Al exhibits higher local ductility (Fig. 14).

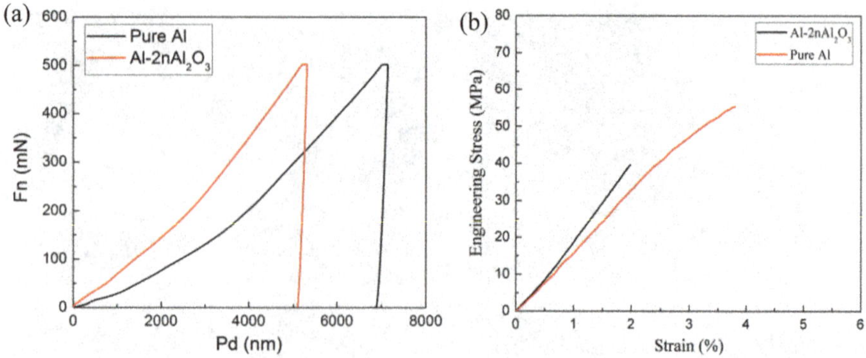

Fig. 13 (**a**) Comparison of nanoindentation results and (**b**) tensile curves for pure Al and Al-2wt. % Al_2O_3 nanocomposite after SPS (Sadeghi et al. 2017)

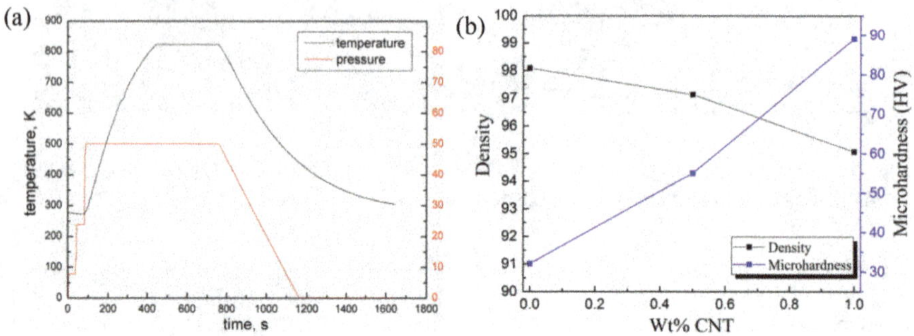

Fig. 14 (**a**) Temperature and applied pressure employed during SPS. (**b**) Variation of relative density and hardness for pure Al and for Al-CNTs composites (Cavaliere et al. 2017)

12 The Future of SPS

The SPS technique in its modern form was developed by Sumitomo Coal Mining Co., Ltd. in 1989, and it has altogether with the concept of simultaneous application of high pulse current and high mechanical pressure become a new phase in powder technology. The major feature of the SPS method is sintering in a conducting mould with the simultaneous application of mechanical pressure and DC pulses, which makes densification and sintering run simultaneously. To date, SPS has enjoyed much success in materials research and is expanding widely in industrial applications. Recently, the method has transformed into a new technology for the production of powder materials with various advantages as high-energy efficiency, little effect on the environment and high production capacity. The SPS method showed high efficiency in the consolidation of ceramic and metal nanomaterials, nanocomposites, functionally graded materials, hard materials, electronic materials and biomaterials (Grasso et al. 2009). SPS makes it possible to create a gradient inside the sintered material, providing the possibility for sintering various multilayer samples. Though SPS-related research at laboratory scale has been quite prolific, yet industrialization has been limited. One of the largest issues with moving the SPS technique from the laboratory scale to large-scale production is the problem with scalability (Olevsky et al. 2012). The rapid heating and short sintering times that are used during SPS are the major obstacles in the path of scaling the SPS process. These problems are not as important for samples with diameters in the range of 15–50 mm, for which it is easy to achieve uniform heating of the sample. However, if the sample diameter is in the range of 100–300 mm, a significant temperature gradient inside the sample may occur which results in nonuniform densification. Due to the nonuniform heat distribution between the sintered material and moulds, the transfer of powder particles during sintering, the nonuniform distribution of stresses and the homogeneous heating become complicated. In SPS scalability is a major problem from both technical and practical point of view. The maintenance of uniform temperature with minimal sample temperature gradient and nonuniform pressure distribution are a few problems associated with SPS of large-sized samples. The necessity of very high current to achieve very high temperatures and high heating rates has significant implications towards the efficiency of SPS. The various mechanisms contribute differently at different stages of sintering. Identification of the contribution of each mechanism will enable us to design more practicable and efficient processing techniques in the future(Feest 1986).

13 Conclusions

A key target in MMNC research has been the identification and assessment of ways to improve the mechanical properties, especially flexural strength, which still hinder the wider application of metallic composites. There is a great potential for tailoring the microstructures and improvement in mechanical properties of MMNCs by utilizing SPS technology. The critical requirements for achieving this goal include control of grain size, particle morphology, particle size distribution, and the volume fraction of the second-phase particles. Densification by conventional sintering processes is always accompanied by grain growth. However, the SPS process is unique and can separate grain growth from densification. The outstanding feature of the SPS process is the short sintering time and the low sintering temperature needed to obtain fully dense compacts. In most of the reported work, the microstructures of the SPS-sintered nanocomposites are excellent, but the mechanical properties are still unsatisfactory and inconsistent. It is commonly accepted that nanocomposite structural materials have great potential for wider applications from the scientific point of view, but their development at the industrial level is held back due to a number of reasons like the high cost of producing bulk nanoscale powders and the lack of a suitable processing technology for the homogeneous mixing of the nanopowders. The SPS technology has shown great potential for mitigating some of these problems. However, much research is still needed in order to obtain a clear understanding of the fundamental mechanisms of sintering and the effect of the electrical current on mass transport, reactivity, microstructure evolution and final properties, as well as to scale-up the process for developing bulk samples with more flexible geometries.

Acknowledgement We gratefully acknowledge the support provided by the XRD, SEM laboratories of Metallurgical and Materials Engineering Department and FESEM and Thermal analysis laboratory, Ceramic Engineering Department, NIT Rourkela. We also thank the Physics Department and the Central Research Facility, IIT Kharagpur, for their sincere support. The use of SPS facility at IIT Kanpur, procured with partial funding from Department of Science and Technology, Government of India as well as CARE funding from IIT Kanpur, is gratefully acknowledged. We would also like to acknowledge the Advanced Centre for Materials Science (ACMS), IIT Kanpur.

References

Akinrinlola B, Gauvin R, Brochu M (2012) Improving the mechanical reliability of cryomilled Al-Mg alloy using a two-stage spark plasma sintering cycle. Scr Mater 66(7):455–458

Alam SN, Sharma N, Ray BC, Yadav S, Biswas K (2017) Effect of graphite nanoplatelets on the mechanical properties of alumina-based composites. Ceram Int 43(14):11376–11389

Al-Qutub AM, Khalil A, Saheb N, Hakeem AS (2013) Wear and friction behavior of Al_{6061} alloy reinforced with carbon nanotubes. Wear 297(1–2):752–761

Anderson MP, Grest GS, Doherty RD, Li K, Srolovitz DJ (1989) Inhibition of grain growth by second phase particles: three dimensional Monte Carlo computer simulation. Scr Metall 23:753–758

Azarniya A, Sovizi S, Boyuk MRRT et al (2017) Physicomechanical properties of spark plasma sintered carbon nanotube-containing ceramic matrix nanocomposites. Nanoscale 9 (35):12779–12820

Boesel RW, Jacobson MI, Yoshioka IS (1969) Spark sintering tames exotic P/M materials. Mater Eng 70:32–35

Casellas D, Nagl MM, Llanes L, Anglada M (2003) Fracture toughness of alumina and ZTA ceramics: microstructural coarsening effects. J Mater Process Technol 143–144:148–152

Cavaliere P, Sadeghi B, Shabani A (2017) Carbon nanotube reinforced aluminum matrix composites produced by spark plasma sintering. J Mater Sci 52:8618–8629

Chunfeng D, Xuexi Z, Yanxia M, Dezun W (2007) Fabrication of aluminum matrix composite reinforced with carbon nanotubes. Rare Metals 26(5):450–455

Clemares IA, Osoro GM, Fernandez A et al (2010) Transparent alumina/ceria nanocomposites by spark plasma sintering. Adv Eng Mater 12(11):1154–1160

Clyens S, Al-Hassani STS, Johnson W (1976) The compaction of powder metallurgy bars using high voltage electrical discharges. Int J Mech Sci 18(1):37–40

Conrad H (2002) Thermally activated plastic flow of metals and ceramics with an electric field or current. Mater Sci Eng A 322(1–2):100–107

Couret A, Molenat G, Galy J, Thomas M (2008) Microstructures and mechanical properties of TiAl alloys consolidated by spark plasma sintering. Intermetallics 16(9):1134–1141

Cremer GD (1944) Sintering together powders metals such as bronze, brass or aluminum. US Patent No. 2,355,954. https://patents.google.com/patent/US2355954

Dash K, Ray BC, Chaira D (2012) Synthesis and characterization of copper-alumina metal matrix composite by conventional and spark plasma sintering. J Alloys Compd 516:78–84

Diouf S, Molinari A (2012) Densification mechanisms in spark plasma sintering: effect of particle size and pressure. Powder Technol 221:220–227

Feest E (1986) A metal matrix composites for industrial application. Mater Des 7(2):58–64

Feng H, Zhou Y, Jia D, Meng Q (2005) Rapid synthesis of Ti alloy with B addition by spark plasma sintering. Mater Sci Eng A 390(1–2):344–349

Furuya Y, Watanabe R, Yaguchi H, Saito M, Kubo T, Abe T, Omori M (1998) Preparation of smart composites by SPS. In: Proceedings of the Third Symposium on Spark Plasma Sintering, Kure, pp 42–45

Geguzin YAE, Kaganovsky YUS, Onoprienko AA, Zung FN (1975) Behaviour of metallic granules at the surface of an ionic crystal in an external electric field. Sov Phys Solis State 17 (3):457–458

Giuntini D, Olevsky EA, Cardona CG et al (2013) Localized overheating phenomena and optimization of spark-plasma sintering tooling design. Materials 6:2612–2632

Goldberger WM, Merkle BD (2001) Electroconsolidation offers fast, low-cost densification. Met Powder Rep 56(2):30–33

Grasso S, Sakka Y, Maizza G (2009) Electric current activated/assisted sintering (ECAS): a review of patents. Sci Technol Adv Mater 10:053001

Grasso S, Kim EY, Saunders T, Yu M et al (2016) Ultra-rapid crystal growth of textured SiC using flash spark plasma sintering route. Cryst Growth Des 16:2317–2321

Groza JR (1999) Nanosintering. Nanostruct Mater 12:987–992

Groza JR, Anderson KR, Fendorf M, Echer CJ (1999) Surface oxide debonding in field assisted powder sintering. Mater Sci Eng A 270(2):278–282

Guillon O, Julian JG, Dargatz B, Kessel T, Schierning G et al (2014) Field-assisted sintering technology/spark plasma sintering: mechanisms, materials, and technology developments. Adv Eng Mater 16(7):830–849

Herring C (1950) Effect of change of scale on sintering phenomena. J Appl Phys 21:301–303

Hiremath A, Hemanth J (2017) Experimental evaluation of the coefficient of thermal expansion of chilled aluminum alloy-borosilicate glass (P) composite. J Mater Environ Sci 8(12):4246–4252

Holland TB, Ovidko IA, Wang H, Mukherjee AK (2010) Elevated temperature deformation behavior of spark plasma sintered nanometric nickel with varied grain size distributions. Mater Sci Eng A 528(2):663–671

Hughes GD, Smith SD, Pandey CS et al (1986) Hall-petch strengthening for the microhardness of twelve nanometer grain diameter electrodeposited nickel. Scr Metall 20(1):93–97

Hulbert DM, Anders A, Andersson J, Lavernia EJ, Mukherjee AK (2009) A discussion on the absence of plasma in spark plasma sintering. Scr Matell 60:835–838

Inoue K (1965) Electric discharge heat treatment of metals in electrolytes. US Patent No. 3,188,245. https://patents.google.com/patent/US3198675

Inoue K (1966) Method of and apparatus for controlling the porosity of electrically sintered bodies. US Patent No. 3241-956. https://patents.google.com/patent/US3317705

Jung H, Gusev VE, Baek H, Wang Y, Diebold GJ (2011) Ludwig-Soret effect in a linear temperature field: theory and experiments for steady state distributions. Phys Lett A 375 (19):1917–1920

Kim KH, Chae JH, Park JS, Ahn JP, Shim KB (2003) Sintering behavior and mechanical properties of B4C ceramics fabricated by spark plasma sintering. Mater Charact 50:31–37

Kim KT, Cha SI, Hong SH (2006) Microstructures and tensile behavior of carbon nanotube reinforced Cu matrix nanocomposites. Mater Sci Eng A 430(1–2):27–33

Kodash VY, Groza JR, Cho KC et al (2004) Field-assisted sintering of Ni nanopowders. Mater Sci Eng A 385(1–2):367–371

Kubota M (2007) Properties of nano-structured pure Al produced by mechanical grinding and spark plasma sintering. J Alloys Compd 434-435:294–297

Kumar HGP, Xavior MA (2014) Graphene reinforced metal matrix composite (GRMMC): a review. Proc Eng 97:1033–1040

Libardi S, Leoni M, Facchini L, Incau MD et al (2007) Effect of the dispersion of nanometric silica particles on the thermal stability of a nanostructured iron based powder. Mater Sci Eng A 445–446:244–250

Lim BS, Kim CJ, Kim BJ (2006) The effects of interfacial bonding on mechanical properties of single-walled carbon nanotube reinforced copper matrix nanocomposites. Nanotechnology 17:5759–5764

Lukianova OA, Novikov VY, Parkhomenko AA, Sirota VV, Krasilnikov VV (2017) Microstructure of spark plasma-sintered silicon nitride ceramics. Nanoscale Res Lett 12:293–299

Mamedov V (2002) Spark plasma sintering as advanced PM sintering method. Powder Metall 45:322–328

Matsugi K, Ishibashi N, Hatayama T, Yanagisawa O (1996) Microstructure of spark sintered titanium-aluminide compacts. Intermetallics 4(6):457–467

Mayo MJ (1996) Processing of nanocrystalline ceramics from ultrafine particles. Int Mater Rev 41:85–115

McTaggart FK (1967) Plasma chemistry in electrical discharges. Elsevier Scientific Publishing Company, Amsterdam

Munir ZA, Tamburini UA, Ohyanagi M (2006) The effect of electric field and pressure on the synthesis and consolidation of materials: a review of the spark plasma method. J Mater Sci 41:763–777

Murakami T, Xu CN, Kitahara A (1999) Microstructure, mechanical properties and oxidation behavior of powder compacts of the Nb-Si-B system prepared by spark plasma sintering. Intermetallics 7(9):1043–1048

Niihara K (1991) New design concept of structural ceramics-ceramic nanocomposites. J Ceram Soc Jpn 99:974–982

Olevsky E, Froyen L (2006) Constitutive modelling of spark plasma sintering of conductive materials. Scr Mater 55:1175–1178

Olevsky EA, Froyen L (2008) Impact of thermal diffusion on densification during SPS. J Am Ceram Soc 92:122–132

Olevsky EA, Kandukuri S, Froyen L (2007) Consolidation enhancement in spark plasma sintering: impact of high heating rates. J Appl Phys 102(11):114913

Olevsky EA, Bradbury WL, Haines CD, Martin DG, Kapoor D (2012) Fundamental aspects of spark plasma sintering: experimental analysis of scalability. J Am Ceram Soc 95:2406–2413

Omori M (2000) Sintering, consolidation, reaction and crystal growth by the spark plasma system (SPS). Mater Sci Eng A 287:183–188

Orru R, Richeri R, Locci AM, Cincotti A, Cao G (2009) Consolidation/synthesis of materials by electric current activated/assisted sintering. Mater Sci Eng R 63:127–287

Queyreau S, Monnet G, Devincre B (2010) Orowan strengthening and forest hardening superposition examined by dislocation dynamics simulations. Acta Mater 58(17):5586–5595

Raichenko AI, Burenkov GL, Leshchinsky VI (1973) Theoretical analysis of the elmentary act of electric discharge sintering. Phys Sinter 5(2):215–225

Raju G, Basu B (2014) Spark plasma sintering of nanoceramic composites. Compr Hard Mater 2:177–205

Randall GM (2010) Coarsening in sintering: grain shape distribution, grain size distribution, and grain growth kinetics in solid-pore systems. Crit Rev Solid State Mater Sci 35:263–305

Ravi KR, Murugesan A, Udhayabanu V, Subramanian R, Murty BS (2012) Microstructure and mechanical property of Fe-Al_2O_3 nanocomposites synthesized by reactive milling followed by spark plasma sintering. Mater Sci Forum 710:291–296

Razavi M, Rahimipour MR, Mansoori R (2008) Synthesis of TiC–Al_2O_3 nanocomposite powder from impure Ti chips, Al and carbon black by mechanical alloying. J Alloys Compd 450:463–467

Reis J, Chaim R (2008) Densification maps for spark plasma sintering of nanocrystalline MgO ceramics: particle coarsening and grain growth effects. Mater Sci Eng A 491(1–2):356–363

Saboori A, Pavese M, Badini C, Fino P (2017) Development of Al- and Cu-based nanocomposites reinforced by graphene nanoplatelets: fabrication and characterization. Front Mater Sci 11 (2):171–181

Sadeghi B, Shamanian M, Ashrafizadeh F, Cavaliere P, Rizzo A (2017) Influence of Al_2O_3 nanoparticles on microstructure and strengthening mechanism of Al-based nanocomposites produced via spark plasma sintering. J Mater Eng Perform 26(6):2928–2936

Sadeghi B, Shamanian M, Ashrafizadeh F et al (2018) Microstructural behaviour of spark plasma sintered composites containing bimodal micro- and nano-sized Al_2O_3 particles. Pow Met 61 (1):50–63

Saheb N, Khalil A, Hakeem AS et al (2014) Spark plasma sintering of CNT reinforced Al_{6061} and Al_{2124} nanocomposites. Adv Mater Sci Eng 2014(535640):1–9

Shi X, Xu F, Zhang Z, Dong Y, Tan Y, Wang L, Yang J (2010) Mechanical properties of hot-pressed Al_2O_3/SiC composites. Mater Sci Eng A 527:4646–4649

Skiba T, Hausild P, Karlík M et al (2010) Mechanical properties of spark plasma sintered FeAl intermetallics. Intermetallics 18(7):1410–1414

Suarez M, Fernandez A, Menendez JL et al (2013) Challenges and opportunities for spark plasma sintering: a key technology for a new generation of materials. In: Ertug B (ed) Sintering applications. InTech, Rijeka, pp 319–342

Supova M, Martynkova GS, Barabaszova K (2011) Effect of nanofillers dispersion in polymer matrices: a review. Sci Adv Mater 3:1–25

Suryanarayana C (2011) Synthesis of nanocomposites by mechanical alloying. J Alloys Compd 509 (1):229–234

Tamari N, Tanaka T, Tanaka K (1995) Effect of spark plasma sintering on densification and mechanical properties of silicon carbide. J Ceram Soc Jpn 103(7):740–742

Taylor GF (1933) Apparatus for making hard metal compositions. US Patent No. 1,896,854. https://patents.google.com/patent/US1896854

Thostenson ET, Li C, Chou TW (2005) Nanocomposites in context. Compos Sci Technol 65:491–516

Tiwari D, Basu B, Biswas K (2009) Simulation of thermal and electrical field evolution during spark plasma sintering. Ceram Int 35(2):699–708

Tokita M (1993) Trends in advanced SPS spark plasma sintering systems and technology. J Soc Powder Technol Jpn 30(11):790–804

Tokita M (1999) Mechanism of spark plasma sintering. In: Miyake S, Samandi M (eds) Proceedings of the International Symposium on Microwave, Plasma and Thermochemcial Processing of Advanced Materials, JWRI. Osaka Universities, Osaka, pp 69–76

Udhayabanu V, Ravi KR, Murugan K (2011) Development of Ni-Al_2O_3 in-situ nanocomposite by reactive milling and spark plasma sintering. Metall Mater Trans A 42(7):2085–2093

Umeda J, Kondoh K, Imai H (2009) Friction and wear behavior of sintered magnesium composite reinforced with CNT-Mg_2Si/MgO. Mater Sci Eng A 504(1–2):157–162

Vanmeensel K, Laptev A, Hennicke J et al (2005) Modelling of the temperature distribution during field assisted sintering. Acta Mater 53(16):4379–4388

Voisin T, Durand L, Karnatak N, Gallet SL et al (2013) Temperature control during spark plasma sintering and application to up-scaling and complex shaping. J Mater Process Technol 213:269–278

Wang D, Yuan H, Qiang J (2017) The microstructure evolution, mechanical properties and densification mechanism of TiAl-based alloys prepared by spark plasma sintering. Metals 7:201

WOK(Web of Knowledge) citation report (2012) http://wok.mimas.ac.uk/news/2012/20120924

Yamanaka S, Gonda R, Kawasaki A et al (2007) Fabrication and Thermal Properties of Carbon Nanotube/Nickel Composite by Spark Plasma Sintering Method. Mat. Trans. 48(9):2506–2512

Yamazaki K, Risbud SH, Aoyama H, Shoda K (1996) PAS (Plasma activated sintering): transient sintering process control for rapid consolidation of powders. J Mater Process Technol 56 (1–4):955–965

Yanagisawa O, Hatayama T, Matsugi K (1994) Recent research on spark sintering. Materia Jpn 33 (12):1489–1496

Zadeh AS (2012) Comparison between current models for the strength of particulate-reinforced metal matrix nanocomposites with emphasis on consideration of Hall–Petch effect. Mater Sci Eng A 531:112–118

Zadra M, Casari F, Girardini L, Molinari A (2007) Spark plasma sintering of pure aluminium powder: mechanical properties and fracture analysis. Powder Metall 50(1):40–45

Zener C (1948) Theory of Strain Interaction of Solute Atoms. Phys. Rev. 639:74

Zhang ZH, Wang FC, Wang L et al (2008) Microstructural characteristics of large-scale ultrafine grained copper. Mater Charact 59(3):329–333

The Effect of Current Pathways on Spark Plasma Sintering

F. Kellogg and C. Hofmeister Mock

1 Introduction

At its core, spark plasma sintering (SPS) is the process of consolidating powder (metallic or ceramic) constrained in a die (most often graphite) through the application of large amounts of pulsed DC electrical current (Omori 2000) as shown schematically in Fig. 1. Research interest in SPS started building in the late 1990s/ early 2000s (Orrù et al. 2009), primarily as a tool for consolidation of materials that required excessive times or temperatures to sinter with normal hot-pressing techniques or nanostructured powders in which the high heating and cooling rates were used to maintain nanostructure from powder to bulk. An empirical overview of the volume of SPS literature currently available to researchers is primarily focused on the materials being processed and/or modeling of temperature development and flow during processing. However, there are fundamental aspects of the SPS process that are not heavily researched but can be improved in order to enhance SPS transition into manufacturing. Four of these areas, including tooling design, tooling material, powder conditions, and current waveform, will be evaluated further to determine how changes could be used to improve the SPS process.

Tooling design and its influence over SPS processing have not been well examined. The vast majority of the literature is based on simple geometries with punches and dies made of graphite, with foils and spacers for diffusion barriers, lubricants, or temperature control. Little research has been published that examines how tooling design can be used to improve the SPS process. SPS tooling requires redesign if SPS

F. Kellogg (✉)
SURVICE Engineering Company, Belcamp, MD, USA
e-mail: Franklin.R.Kellogg.ctr@mail.mil

C. Hofmeister Mock
ORISE, Belcamp, MD, USA
e-mail: Clara.M.Mock.ctr@mail.mil

© Springer Nature Switzerland AG 2019

P. Cavaliere (ed.), *Spark Plasma Sintering of Materials*,
https://doi.org/10.1007/978-3-030-05327-7_3

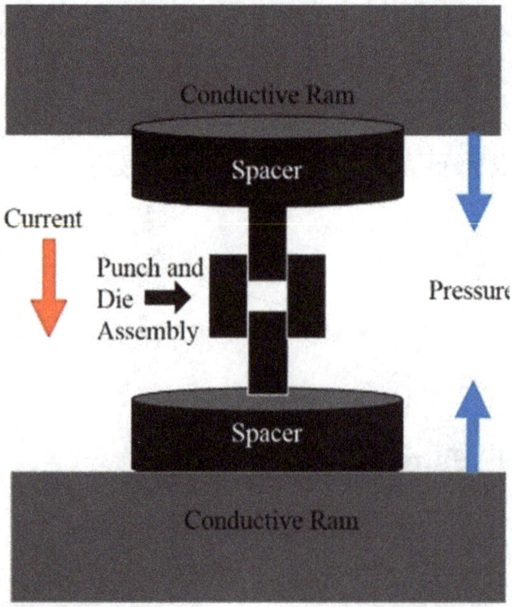

Fig. 1 Schematic of generic SPS setup

Fig. 2 Schematic of possible die configurations for either single or multiple samples

is to be used for the manufacturing of complex parts or production of multiple parts (Fig. 2). The models used to predict the temperature profiles during SPS also predict current flow and current density within an SPS part. Redesigned tooling has been shown (Giuntini et al. 2015, 2016) as an effective means to alter the current pathways during SPS and produce parts and samples under more uniform temperatures, leading to more uniform microstructures and densities.

In conjunction with redesigning SPS tooling, it is critical to examine the materials used to make the tooling and determine whether or not they are ideal for the process. Graphite has been the material of choice for SPS tooling chiefly due to its high conductivity and operating temperatures. However, as understanding of the SPS process increases, these more desirable properties of graphite have become less important and some of the drawbacks (weak flexural strength, low wear resistance, relatively high chemical reactivity) more pronounced. The limited research focused on the influence of tooling materials over the SPS process indicates that significant improvements can be made by replacing with other materials. The conductivity of the tooling material can influence redirection of current pathways through the tooling or the sample and can be used to increase the already high heating rates of SPS. In addition to modifying tooling to alter current pathways, the powder being processed can also be manipulated to influence the SPS process.

While the majority of SPS literature emphasizes powder consolidation, limited research is focused on the manipulation of powder and its effect on the processing behavior. For example, Cho et al. (2004) looked at how different grades and particle sizes of tungsten powder sinter, while Kellogg et al. (2017) examined the processing differences between aluminum alloys in the as-atomized state as opposed to after cryomilling. Although sintering aids are commonly used in more traditional hot-pressing operations, they are not prevalent for use in SPS processing. While SPS is often chosen as a means of avoiding traditional sintering aids, specialized sintering aids could potentially be developed to alter the conductivity and current pathways of the sample powder. In addition to altering current pathways to improve upon the SPS process, the current waveform itself could also be manipulated.

The review paper by Orrù et al. in 2009 showed that more than half of the literature dedicated to resistance sintering techniques mentioned pulsed DC current as the current source, while approximately one-fifth of the papers did not specify the current waveform. Over 75% of the published research either defaulted to the established normal pulsed DC current or did not report the current waveform

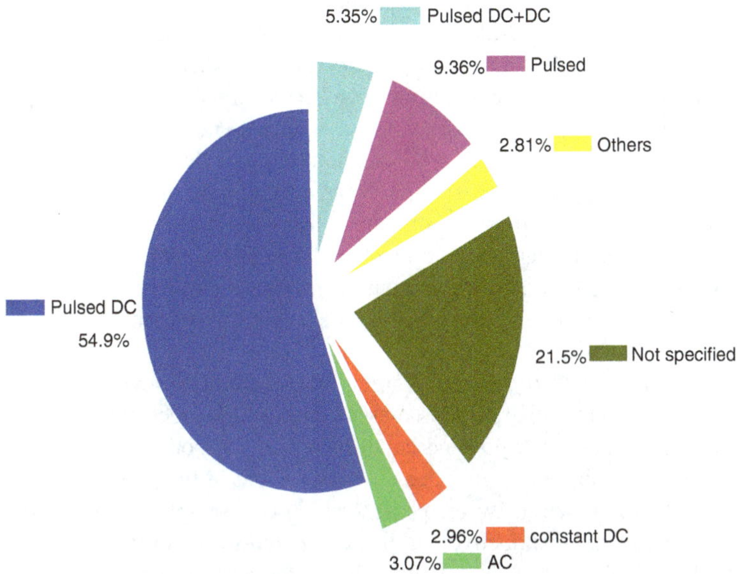

Fig. 3 Breakdown of current waveforms used in SPS literature. (Reprinted by permission from Elsevier, Mater Sci Eng. R, Orrù et al. 2009)

(Fig. 3). While this focus likely contributed to an understanding of the basic concepts of SPS, it ignored other possible avenues for improving the process. For example, an AC current with comparable power levels to a DC current produces an electric field $\sqrt{2}$ times stronger (Yang and Conrad 2010). In addition, it is possible that the polarity reversals of AC currents can hinder electromigration, thus retaining grain size better during sintering than comparable DC currents (Narayan 2013; Yang and Conrad 2010). An AC current waveform could provide beneficial changes to the SPS process but has yet to be a focus of research efforts.

This chapter focuses on how changes to the four basic components of any SPS process, including tooling design, tooling material, powder, and current waveform, have been implemented to expand the use of SPS beyond that of a simple research tool.

2 Tooling Designs

The die and punch tooling required to enclose the powder sample makes it very difficult to acquire an accurate temperature measurement during SPS without specialized equipment. Temperature measurements made on die walls have been shown to vary greatly compared to actual sample temperatures (Wang et al. 2002). Attempts to make measurements by embedded thermocouples in either the die wall or sample (Wang et al. 2002; Zavaliangos et al. 2004), or by drilling holes through the tooling for pyrometer measurements (Vanmeensel et al. 2005), are not universal methods for making direct sample measurements. Thermocouples can react with the sample or tooling material, and holes drilled into the die for taking measurements can threaten mechanical integrity, limiting the amount of pressure that can be applied to the sample. Furthermore, thermocouples and pyrometers are simple point measurement techniques, and may not provide representative measurements of large or complex-shaped parts. Using dies with thin walls can help mitigate the difference between die wall and sample temperature measurements but also limit the amount of pressure that can be applied. Further complications arise from the sample material itself and whether or not it is insulating or conductive (Wang et al. 2002). Because of this difficulty in obtaining direct, representative sample measurements, much of the understanding on sample temperatures depends on computer simulations and modeling of the SPS process.

The model developed by Anselmi-Tamburini et al. (2005b) sought to simulate the current and temperature distributions present during SPS processing by comparing 3 mm disks of insulating Al_2O_3 and conductive Cu. For both materials, the model calculated that the highest current density was present in the punches and that the current density differences between the punches and the samples were much lower in both directions for the conductive Cu disk compared to the insulating Al_2O_3 disk. Their model calculated that the conductivity of the sample did not influence heat generation and that for both the insulating and conducting materials, heat was generated only within the graphite tooling. In contrast, sample conductivity did have an influence on radial temperature gradients between the sample and the die wall, as the gradient was slightly lower for the conductive Cu sample compared to the insulating Al_2O_3 sample. The radial gradients in the Al_2O_3 sample were verified experimentally by drilling a hole through the bottom punch and inserting a thermocouple, placing it in contact with the sample. The experimental results corresponded well with the calculations in the model (Fig. 4).

The presence of radial temperature gradients was attributed to current distribution and radiative heat loss. The current density was highest in the punches, while the temperature and heat generation were highest at the punch-die interface. Radiative heat kept the die wall at lower temperatures, as the heat gradient decreased to zero when the current was turned off. The current density distribution was also the source of axial thermal gradients. It was observed that if the sample was not exactly centered (e.g., the two punches were not evenly inserted into the die during loading or uneven punch movement occurred during sintering), the top and bottom surfaces of the sample experienced different temperatures.

Similar results were presented while modeling sintering behavior of nanocrystalline tungsten via SPS (McWilliams et al. 2006). It was found that the highest temperatures were generated in the punches and that shortening the punches (from 1 inch long to ½ inch long) did reduce the difference between punch and sample temperatures, though it was unable to eliminate the temperature gradient (Fig. 5). The model also predicted that punch length did not influence the temperature gradient between the sample and the die wall, which was calculated to be approximately 600 °C.

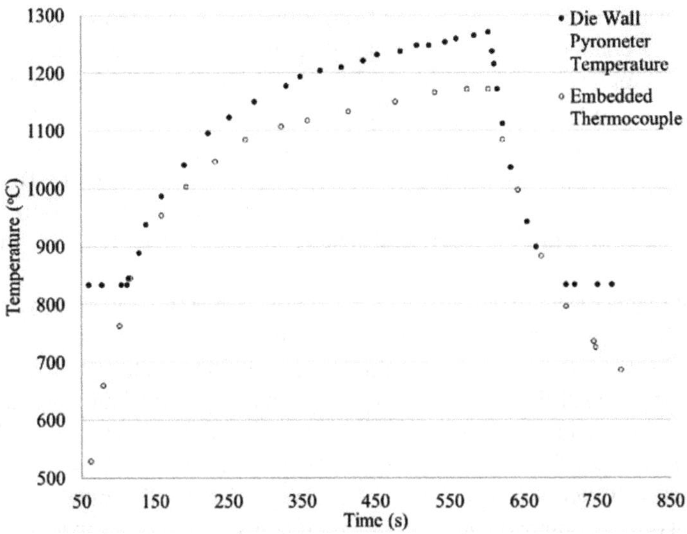

Fig. 4 Die wall temperature measured via pyrometer compared to temperature measured via a thermocouple embedded within a sample. (Graph generated from data published in Anselmi-Tamburini et al. 2005b)

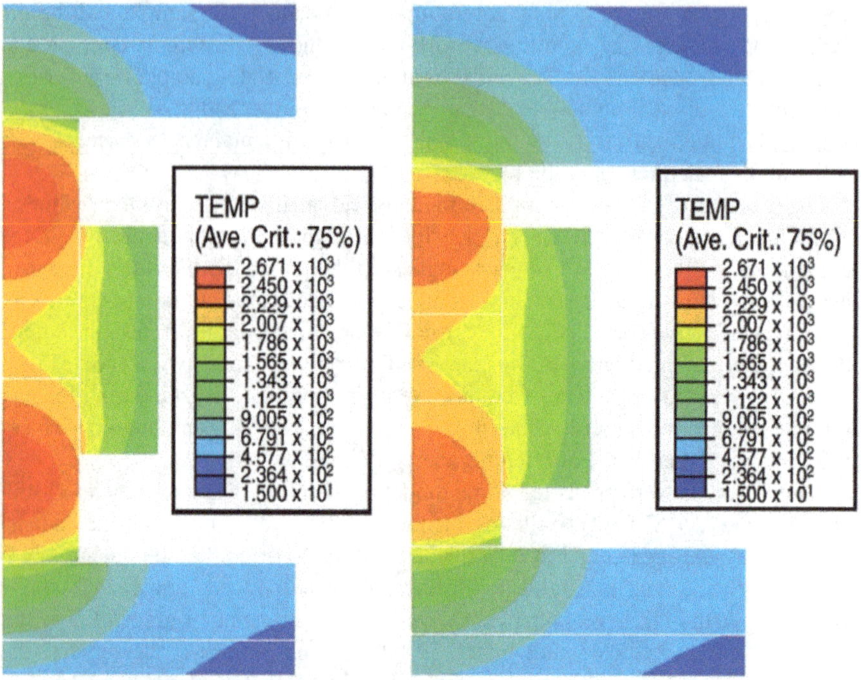

Fig. 5 Temperature gradients in SPS punch and die setup. (Reprinted by permission from Springer Nature, JOM, McWilliams et al. 2006)

While Anselmi-Tamburini et al. (2005a) showed that the thermal gradients were reduced in a conductive sample compared to an insulating material, Vanmeensel et al. (2005) found that the radial gradients were larger for conductive TiN than for insulating 3Y-ZrO_2. Experimental differences between the two setups may have resulted in the different conclusions, as Vanmeensel's group used a pulsed current and had graphite foil between all contacting surfaces, while Anselimi-Tamburini's group did not use graphite foil and used a constant current. They both, however, calculated that the highest current densities and temperatures were within the punches.

The presence of thermal gradients led to inhomogeneous sintering (Vanmeensel et al. 2005), which resulted in parts with inhomogeneous microstructures (Anselmi-Tamburini et al. 2005a) and lower mechanical properties, highlighting the need for proper process controls to ensure sample repeatability (Zavaliangos et al. 2004). While the temperature gradients could be minimized to some extent (Anselmi-Tamburini et al. 2005a; McWilliams et al. 2006; Wang et al. 2002) by using slower heating rates, it would also offset the advantages of the SPS process over more traditional powder compaction methods. Another way to minimize thermal gradients would be by using thin-walled dies. Direct sample temperature measurements could be used to reduce gradients (Vanmeensel et al. 2005), but drilling holes in the tooling could also weaken the tooling. A hybrid system could introduce secondary heating via separate resistive heaters that could heat the sample while negating the radiative heat loss in a normal SPS system. Tungsten heavy alloy samples processed with a hybrid SPS had a more uniform grain structure and improved mechanical properties when compared to the same powder processed with a normal SPS under the same temperature and current profiles (Shongwe et al. 2016a). The downside was that a hybrid system could lead to increased cost of an already expensive SPS system, and could not be retrofitted to an existing system without significant challenges.

Another method for removing the thermal gradients is to alter the current distribution through modification of the punch design. As shown above (Fig. 5), changing the punch length reduces the axial thermal gradients, but did not have an effect on the radial gradients (McWilliams et al. 2006). Voisin et al. experimented with different die and punch lengths, along with different spacer configurations and sizes to control thermal gradients (Voisin et al. 2013). Their model calculated that longer dies and punches, along with smaller diameter spacers, would minimize thermal gradients. This was experimentally proven by producing TiAl samples with improved microstructural homogeneity from the center to the edge (Fig. 6).

Giuntini et al. (2013) examined how changing spacer sizes and configurations, adding smaller graphite spacers between the sample and the punches, could also be used to minimize thermal gradients. Alumina spacers placed between the sample

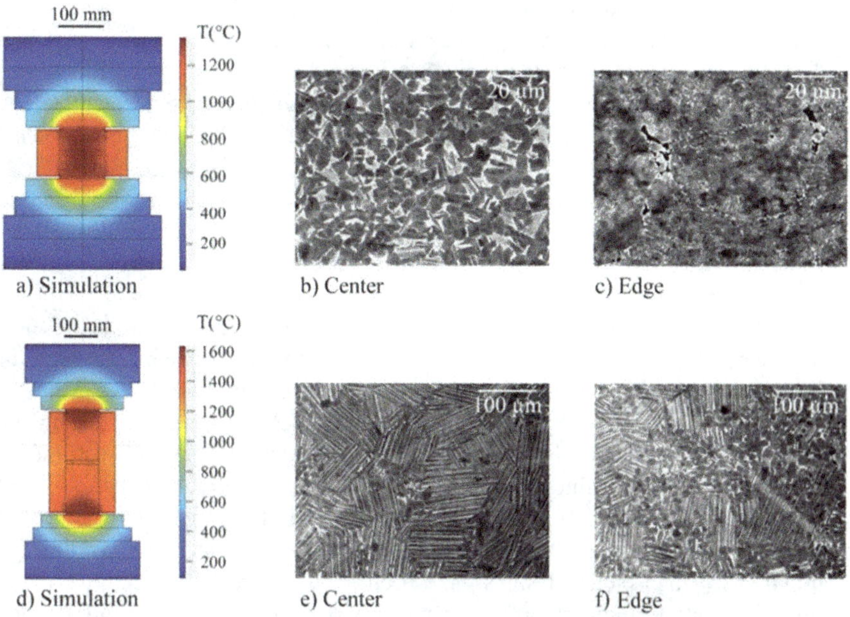

Fig. 6 Model predictions for short die and punches (**a**) and resulting micrographs (**b** and **c**) compared to model predictions for long die and punches (**d**) and resulting micrographs (**e** and **f**). (Reprinted by permission from Elsevier, Journals of Materials Processing Technology, Voisin et al. 2013)

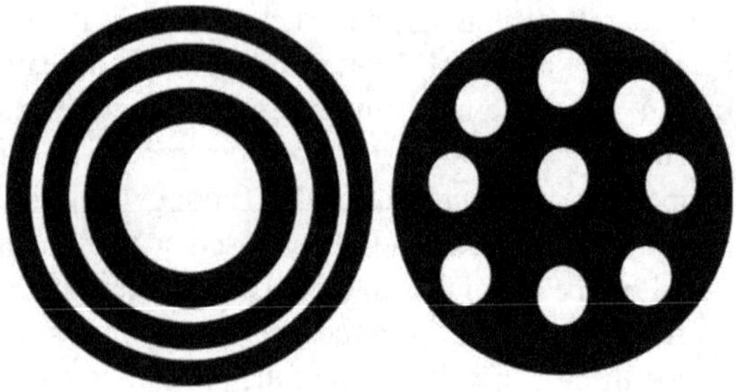

Fig. 7 Schematics for alternative punch designs inspired by Giuntini et al.'s (2016) ring design (left) and holes (right). Not drawn to scale

powder and the punches have been used for gradient minimization (Voisin et al. 2013; Zadra et al. 2007). It was found that the alumina spacers were detrimental to the tensile properties of the consolidated aluminum samples (Zadra et al. 2007), and Balomenos et al. (2011) concluded that the spacers increased the risk of carbothermal reduction at elevated temperatures.

In several follow-ups to their 2013 paper, Olevsky's research team conducted modeling and experiments with novel punch designs in which portions of the graphite were machined out of the punch surface, leaving holes or rings (Fig. 7) (Giuntini et al. 2015, 2016).

Since the rings and holes did not go all the way through the punch, much of the mechanical integrity could be maintained. By removing material and leaving voids behind, the current pathways were directed away from the center of the punch and could be more evenly distributed through the sample. The ring configuration was found to be more effective at minimizing radial and axial thermal gradients than the hole configuration, and the radial gradients were calculated to be much smaller when compared to normal, unaltered punches (Giuntini et al. 2016). Boron nitride (BN) coating (Giuntini et al. 2015) on the punch edge (outermost ring) was also found to be beneficial for reducing radial thermal gradients. Since BN is electrically insulating, the current pathways were steered away from the edges of the punch without changing the thermal profiles.

It is difficult to produce uniform parts during the SPS process due to the presence of large thermal gradients, which can be magnified due to inaccurate measurement of die wall temperatures and an inability to obtain actual sample temperatures. Multiple modeling calculations have predicted that the punches are the regions with the highest current densities during SPS processing. This leads to uneven distribution of the current density, which is the root cause of the thermal gradients. Axial gradients can be controlled by proper placement of the punches, while radial gradients are more difficult to mitigate. Alterations to current pathways through modified spacer and punch designs have been shown to reduce or remove radial thermal gradients by redistributing current flow through the tooling and sample, resulting in higher current densities. Current flow could also be redistributed by using materials other than graphite in the dies.

3 Die Materials

Graphite is the most common material used to produce dies and punches for SPS processing. Graphite is electrically conductive at room temperature and can be used in sintering operations at temperatures up to 2400 °C in vacuum or inert atmosphere (graphite reacts with oxygen at a temperature of approximately 600 °C) while maintaining mechanical integrity (Guillon et al. 2014). However, graphite tooling is not without its drawbacks. Graphite has poor flexural and compressive strength (Aalund 2010) that can be reduced further if thin-walled dies are used to reduce thermal gradients. Some of the previously discussed machining operations for punch modification may also compromise mechanical strength and/or load distribution. The relative softness of graphite also makes it susceptible to wear (Aalund 2010), limiting the number of samples that can be produced per die. While this may not be a significant issue in a research community where mass production is not typically required, it does work against the potential scale-up of SPS to eventually manufacture large quantities of parts.

Another downside of using graphite is that it is chemically reactive with a wide range of materials. Graphite readily reacts with oxygen, which could potentially create a reducing atmosphere between the die material and the SPS atmosphere or the die material and the powder sample (Guillon et al. 2014). Graphite is an oxygen getter and can reduce the powders being sintered. During SPS processing of tungsten powder, the graphite die was able to reduce tungsten oxide present in the powder creating an area around the circumference of the sample that had a significant reduction in tungsten oxide compared to the interior of the sample (Fig. 8) (Klotz et al. 2009). While the inhomogeneous powder reduction was detrimental in this case, the reducing atmosphere produced by processing in graphite could also be beneficial for protecting samples from oxidation (Aalund 2010).

Carbothermal reactions (Balomenos et al. 2011) have prevented SPS from consolidating alumina powders without carbon contamination. SPS has also been credited with preventing chemical reactions, as in this case of B_6O, where rapid consolidation was achieved while avoiding the formation of boron carbide (Guillon

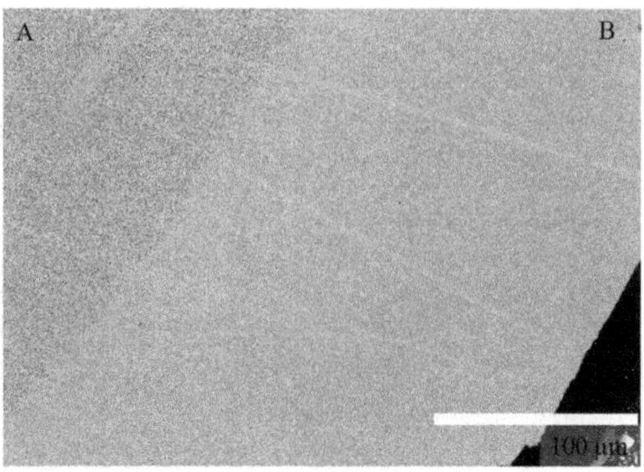

Fig. 8 Tungsten powder consolidated via SPS in a graphite die. Region A is the interior of the sample while region B was against the die wall. The dark gray in A that is absent in B was identified as WO2 (Previously unpublished)

et al. 2014). Carburization has been identified as another possible reaction during SPS processing of metallic powders. Graphite die-powder reactions resulted in the formation of a molybdenum carbide (Mo_2C) layer when SPS processing pure Mo powder (Fig. 9) (Ohser-Wiedemann et al. 2010). While in this circumstance, the formation of the Mo_2C layer was not intended, strategic design of in situ carburization could be beneficial. For example, contact with the graphite punches during SPS could be used to recover the WC phase that transformed to W_2C during plasma spraying of WC-Co coatings (Yu et al. 2004). SPS processing has been used to design a titanium carbide Ti6Al4V composite for in situ carburization by adding carbon fiber to the system, as opposed to contact with the tooling (Hao et al. 2015), and to bond yttria-doped aluminum nitride with graphite disks (Violeta et al. 2015).

Some of the negative aspects of graphite tooling could be mitigated through changes in design. Mechanical properties could be improved by using thicker-walled dies, or a hybrid system could be used to control thermal gradients. The reactivity of graphite could be controlled by manipulating the temperature range or processing time, using a sacrificial layer of graphite foil between the sample powder and the punches or spray-on coatings (e.g., boron nitride) to form diffusion barriers. Since SPS processing is still in its relative infancy, the majority of the research has been based on work done with graphite tooling rather than experimentation with different tooling materials. For example, to increase applied loads during SPS, albeit at relatively low temperatures of 1000 °C or less, alternative die materials such as tungsten carbide or TZM (titanium-zirconium-molybdenum alloys) have been suggested (Guillon et al. 2014). At higher temperatures, ceramic materials (i.e., boron nitride, boron carbide, silicon carbide, or silicon nitride) would be possible alternatives to graphite, although they may be inferior to graphite in handling sintering pressures.

Liu et al. used tungsten carbide tooling (5 weight% Co binder) to consolidate nanostructured aluminum via SPS (Liu et al. 2012). The impetus of the study was that WC tooling could enable larger pressures for application during processing, but the effect of tooling change on current pathways was not addressed. By implementing identical thermal and loading conditions (with identical tooling dimensions), it was found that while both samples produced similar densities, the

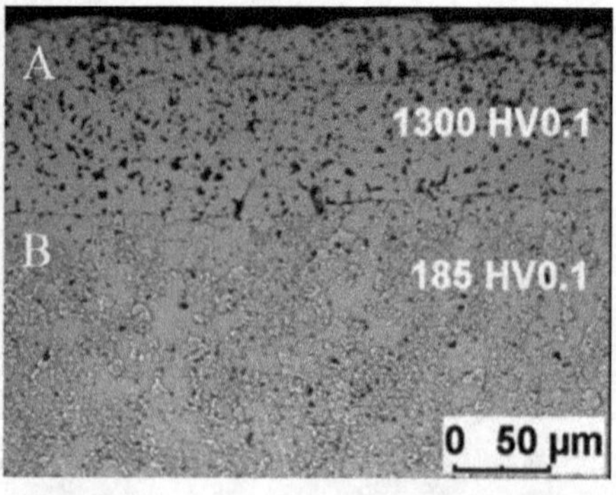

Fig. 9 Mo powder was consolidated via SPS producing a very hard carbide layer (**a**) where the bulk of the material (**b**) contacted the graphite tooling. (Reprinted by permission from Elsevier, Int. J. Refrac. Met. Hard Mater., Ohser-Wiedemann et al. 2010)

sample consolidated using WC-5Co tooling had a smaller average grain size compared to the sample consolidated using graphite tooling. The explanation of these results was that the powder sample in the graphite tooling experienced a higher heating rate relative to the WC-5Co tooling due to the higher electrical resistivity of graphite compared to WC-5Co. In other words, more current passed through the aluminum powder loaded into the graphite die, leading to higher (relative) temperatures that, when combined with a higher relative current density, led to increased grain growth when using the graphite tooling. Arnaud's team conducted a similar study, comparing copper dog-bone specimens consolidated via SPS with either graphite or WC-Co tooling (Arnaud et al. 2015). It was found that under the same conditions, samples produced with WC-based tooling had a lower hardness and tensile strength compared to samples produced in the graphite tooling, primarily due to grain size differences. The study found that the WC tooling enabled the copper sample to reach a higher temperature, leading to larger grains. Simulations showed that in the graphite tooling, the current density was maximum within the bulk of the copper sample, while in the WC tooling, the current density reached its maximum along the edges of the sample (Fig. 10). The distribution of current density was comparable to the study by Liu et al., though with different powder samples and results. Both Liu et al. and Arnaud et al. showed that WC-based tooling was viable

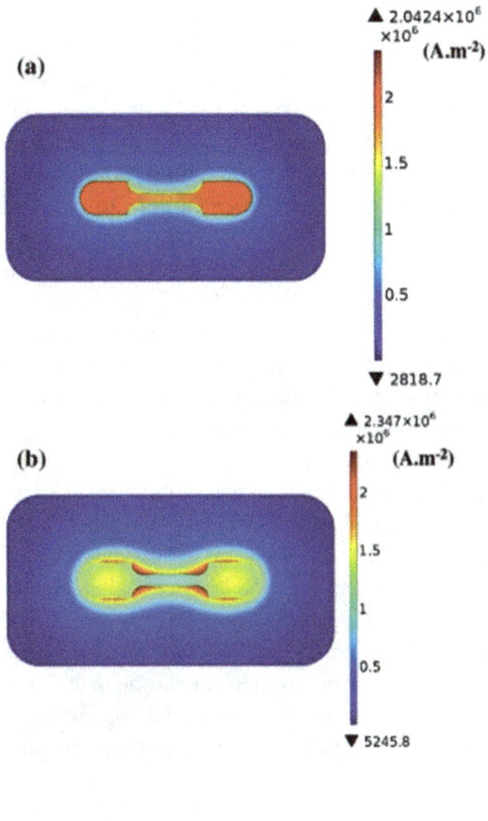

Fig. 10 Simulation of current density for a copper dog-bone produced in (**a**) graphite tooling and (**b**) WC-Co tooling. (Reprinted by permission from Springer Nature, J. Mater. Sci., Arnaud et al. 2015)

for SPS processing. In addition to allowing larger applied loads and higher wear resistance, the switch to WC tooling resulted in alteration of the current pathways during SPS processing. This was evidenced by the current density reaching a maximum for the samples processed using graphite tooling as opposed to WC tooling (which would be expected for any tooling material with electrical properties different from graphite).

TZM was used as SPS tooling material for consolidation of tantalum powder, but the work did not focus on the effect of TZM on powder consolidation (Zhang et al. 2011). Steel dies with TZM spacers were used in SPS experiments on aluminum scrap and powder, but again without investigating how the tooling may have impacted processing (Paraskevas et al. 2015).

Replacing graphite with alternative tooling materials to accommodate higher pressures during SPS has also been shown to change the sintering behavior and the resultant material properties. When a graphite environment was desired, graphite inserts were placed inside higher strength dies with graphite punches to enable sintering at higher loads while maintaining an all-graphite SPS environment for the powder (Aalund 2010). Another possibility was to select the tooling material to deliberately influence the current pathways. For example, both aluminum and copper powders consolidated in WC reached higher temperatures than in graphite tooling (Liu et al. 2012; Arnaud et al. 2015). This method could potentially be used to produce samples under lower currents with WC (as opposed to graphite) tooling, leading to cost (i.e., more durable parts) and energy (i.e., lower current required) savings. Modeling of the SPS process indicated that the current travels through the tooling and the powder samples (Anselmi-Tamburini et al. 2005a; McWilliams et al. 2006; Wang et al. 2002; Zavaliangos et al. 2004). A more conductive die, such as WC, would result in more current flow through the die, as the resistivity of WC is approximately $5*10^{-5}$ ohm-cm at room temperature, compared to $6*10^{-3}$ ohm-cm at room temperature for graphite. On the other hand, switching to an insulating ceramic die would force all of the current to go through the powder.

Ceramic dies have traditionally been more wear-resistant and less reactive than graphite (Aalund 2010) and could generally function and maintain strength at much higher temperatures than their metallic alternatives. However, most non-carbide ceramic materials have resistivity values that are several orders of magnitude higher than graphite and room temperature strengths that are significantly low. To work around the conductivity issue, doping of ceramic materials with conductive phases and the use of graphite inserts have been suggested (Aalund 2010; Guillon et al. 2014). Another work-around could be to pair ceramic dies with fully conductive powder samples. Cryomilled aluminum alloy 5083 (AA5083) powder was processed in both graphite and ceramic (BN) dies (Kellogg et al. 2017). At the same sample size, applied load, and constant current, the sample processed in the BN die exhibited higher heating rates than the sample processed in the graphite die. One downside of ceramic tooling was that the punches required a conductive material to transfer the current from the electrode to the sample. Additionally, powders that were considered conductive may not be conductive under the processing conditions used during SPS. When experimenting with boron nitride (BN) dies, it was found that as-atomized AA5083 was not conductive at room temperature under the processing pressure (18 MPa applied to the sample) used (Kellogg et al. 2016). However, the insulating nature of the ceramic dies and sample powders could be used to turn the SPS process into a flash sintering-like process.

Flash sintering occurs when an electrical field triggers a thermal runaway event, resulting in near instantaneous sintering at temperatures much lower than required for traditional hot pressing (Cologna et al. 2010). The flash refers to a light emission that is given off at the moment the sample becomes conductive and densifies. While the majority of materials processed under flash sintering have consisted of ceramic oxides (Yu et al. 2016), at least some metals (i.e., aluminum alloys) have been successfully produced as well (McWilliams et al. 2016). Intrinsically, the methodologies of flash sintering and SPS can be considered as opposites, as flash sintering depends on high voltages and low currents, while SPS uses high currents with low voltages. However, it has been shown (Olevsky et al. 2016) that the high voltages used in flash sintering are not always necessary and that the SPS process and equipment are capable of being used for flash sintering. Additionally, the SPS equipment can provide operating variables, such as applied loads, that are not found in typical flash sintering setups.

The flash sintering setup used in Olevsky et al. (Fig. 11) was designed using graphite tooling and a sacrificial copper insert for flash SPS of a pre-sintered silicon carbide (SiC) sample (Olevsky et al. 2016). The models indicated that SiC undergoes thermal runaway around 1085 °C. Before the flash event, the graphite punch was supported by a copper sleeve. All of the current passed through the sleeve and punches until the melting point of copper was reached at 1085 °C. When the copper insert melted, the upper punch came into contact with the sample, effectively applying current to the SiC at the flash sintering temperature. Upon application of the current, the sample underwent a nearly instantaneous increase in density. This setup enabled heating of SiC to its flash sintering temperature without being subjected to any current. However, it was also limited to any materials that flashed at the melting point of copper (such as SiC in this case). The matching of

Fig. 11 The flash SPS tooling setup, with copper insert, before and after processing. (Originally published in Olevsky et al. 2016)

flash sintering temperatures and insert melting points severely limits the number of applicable materials for flash SPS.

A way to circumvent this limitation is to use a hybrid SPS system (Kellogg et al. 2016). Experiments determined that at low pressures (18 MPa), an atomized aluminum alloy 5083 (AA5083) powder packed into a BN die could not be processed via typical SPS, as the current could not flow through either the powder or the die (Kellogg et al. 2016). By heating the aluminum powder while applying current (via the hybrid system), flash sintering occurred, consolidating the powder almost instantaneously, as evidenced by the significant change in displacement at approximately 310 °C (Fig. 12). This flash temperature corresponded well with the flash event temperature reported in McWilliams et al. (2016) (306 °C under a DC field of 56 V/cm) for flash sintering of AA5083 (McWilliams et al. 2016). While Olevsky et al. initially used the copper insert to separate the current from the powder, in this case, the insulating die was directly used to keep the powder isolated. While Olevsky et al. reached the flash temperature by passing current through the die to simulate oven-like conditions, in this case, the hybrid system was directly used as an oven to heat the powder without the need for sacrificial copper inserts (Olevsky et al. 2016).

In a follow-up paper, Olevsky's group redesigned their tooling and removed the need for either sacrificial copper inserts or a hybrid SPS system for flash SPS (Maniere et al. 2017). In those flash SPS experiments, the setup presented in Olevsky et al. (2016) was used, but the copper inserts were replaced with a BN-coated graphite foil (Fig. 13). The BN-coated foil functioned as an electric insulator, preventing the sample from heating due to the current. The graphite die was heated by the current and behaved like an oven, assuming the role that the hybrid system had served in Kellogg et al. (2016), therefore producing a non-hybrid flash SPS system.

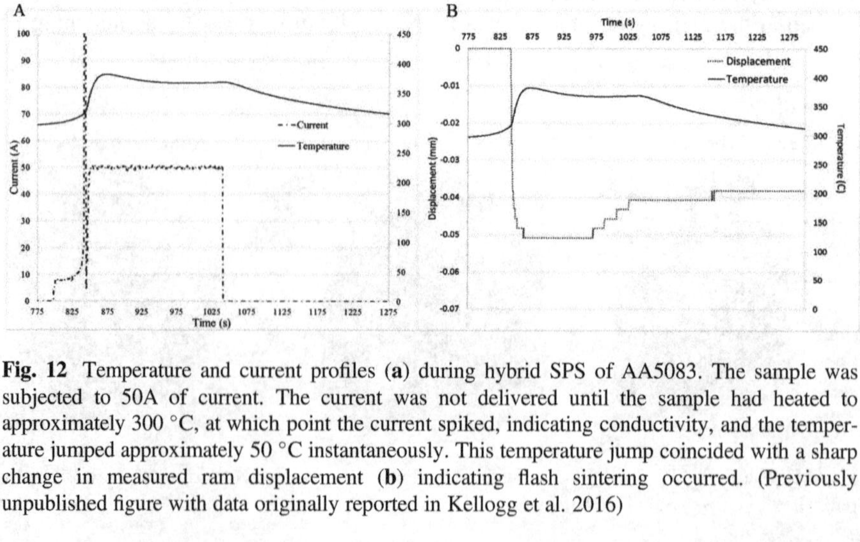

Fig. 12 Temperature and current profiles (**a**) during hybrid SPS of AA5083. The sample was subjected to 50A of current. The current was not delivered until the sample had heated to approximately 300 °C, at which point the current spiked, indicating conductivity, and the temperature jumped approximately 50 °C instantaneously. This temperature jump coincided with a sharp change in measured ram displacement (**b**) indicating flash sintering occurred. (Previously unpublished figure with data originally reported in Kellogg et al. 2016)

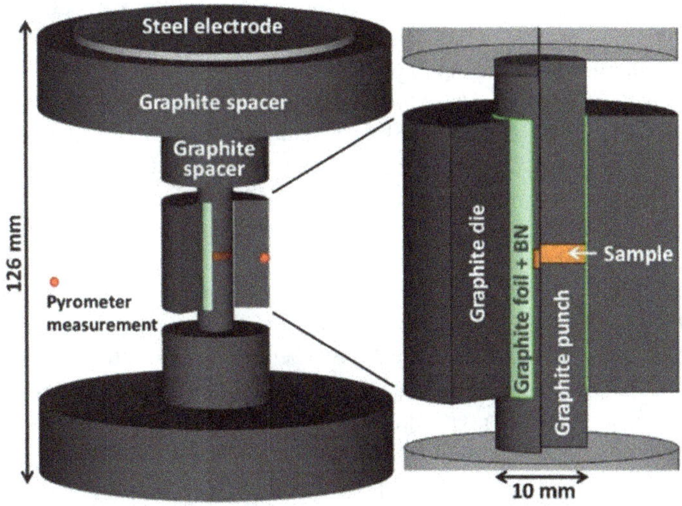

Fig. 13 Flash SPS setup using graphite foil coated with BN. (From Manière et al. 2017)

Just as the SPS originally provided improved processing over traditional hot pressing, the adaptation of flash sintering could lead to faster sintering and lower temperatures for improving the capabilities of SPS. Flash sintering via SPS consists of current pathway diversion away from the sample until the thermal runaway temperature required for flash is reached. It is also dependent on the adaptation of nonconductive ceramic tooling to accommodate the SPS procedure. While the development of flash SPS could be used to make ceramic dies functional for nonconductive samples, it could also be utilized for rapid sintering of conductive materials by forcing all of the current into the sample. Using tooling designs and materials to alter current pathways has been identified as a possible method for improving SPS processing and transitioning it beyond academic research.

4 Powder Manipulation

Another option for altering current pathways is changing the type of powder that is used. In addition to selecting the material and tooling design, powder properties will also affect sinterability. In this section, choice of powder material, particle size, cleanliness (in terms of presence or absence of an oxide layer), and addition of a ball milling step will be briefly discussed.

There have been extensive studies in the literature on the effect of particle size on sinterability of specimens, but the results have been conflicting. For the sake of this work, particle size will refer to the individual powder particle or agglomerate size, and grain size will refer to the average size of the crystallites within the material. The majority of literature has been focused on the densification process during SPS as a function of particle size, using the density and/or particle size to account for differences in mechanical properties. The effect of particle size on sinterability seems to change depending on which densification mechanisms are active at that stage of the consolidation process. Before exploring this further, the typical densification mechanisms during SPS will be reviewed. A characteristic of well-densified material is that the displacement rate versus temperature curve should contain three peaks (Fig. 14). These peaks represent the densification mechanisms, including rearrangement of particles, localized deformation, and core deformation (Diouf and Molinari 2012). The rearrangement of particles occurs due to the application of pressure during the initial stages of SPS. During the second stage of densification, particles are slightly deformed at locations where particles contact one another and localized necking occurs. During core or bulk deformation, the voids shrink between particles as they become increasingly packed. While additional inter-particle necking occurs at this stage, the part still lacks good particle cohesion. After the majority of displacement occurs, sintering completes the SPS process, and the result is a well-densified product with excellent inter-particle bonding.

The particle size of the material can affect the displacement rate during each stage of densification. Initially, powder is added to the die. Just as fine sand poured into a container is less porous (contains fewer air pockets) than the same container filled with rocks, it is expected that a powder with finer particle size would be less porous than one with a larger particle size. Upon application of pressure during the initial stage of particle rearrangement, the particles slide and rotate around one another to become more densely packed. This effect is more pronounced in samples with larger particles that have more porosity (room to move around). In a study by Diouf and Molinari, a commercially available copper powder was sieved into three different

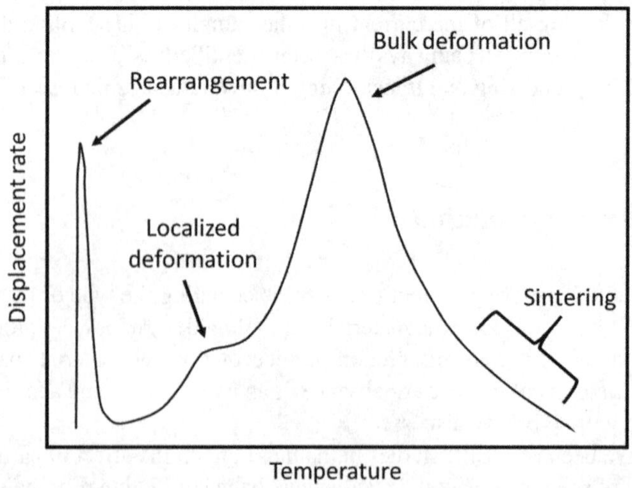

Fig. 14 Schematic of a typical displacement rate versus temperature curve illustrating the mechanisms of densification during SPS processing. (Based on Fig. 1 from Diouf and Molinari 2012)

sizes (<25 μm, 25–45 μm, and 45–90 μm) (Diouf and Molinari 2012). It was found that the height of the initial particle rearrangement peak increased with particle size. The second stage of densification was evidenced by localized deformation and necking. The size and frequency of localized necking between adjoining particles were also related to the particle size. In order to explore this further, it was assumed that the powder particles could be modeled as smooth spheres experiencing elastic contact. The maximum pressure the particles could endure was directly related to the point at which they were in contact along the pressure direction. The maximum, or Hertz, pressure can be calculated by (Johnson 1987):

$$P_{\max} = \frac{3L}{2\pi a^2} \tag{1}$$

where L is the applied load and a is the radius of the contact area given by:

$$a = \left(\frac{3LR}{4E}\right)^{1/3} \tag{2}$$

where R and E are the equivalent curvature and elastic modulus of the two contacting spheres, respectively. The Hertz pressure was originally derived for the case of a sphere contacting a plane of material. Therefore, the calculation of a radius of equivalent curvature for the two contacting spheres was required. The equivalent curvature, R, for two contacting spheres can be determined by:

$$\frac{1}{R} = \frac{1}{r_1} + \frac{1}{r_2} \tag{3}$$

where r_1 and r_2 are the radii of the respective top and bottom contacting particles. Similarly, the equivalent elastic modulus is given by:

$$\frac{1}{E} = \frac{1 - v_1^2}{E_1} + \frac{1 - v_2^2}{E_2} \tag{4}$$

where E_1 and E_2 are the elastic moduli of the top and bottom contacting particles, respectively. Based on these equations, a few observations can be made: (1) The contact area increases with applied load; (2) in the case of a single-phase material with a uniform particle size, it can be observed from Eq. (2) that the contact radius scales with the particle size (i.e., larger particles will have a larger contact radius than a smaller particle under the same conditions). In general, the localized pressure, as calculated by Eq. (1), is not significantly affected by particle size (Diouf et al. 2012; Diouf and Molinari 2012). However, for a conductive material, the contact area between particles provides the conductive path through the bulk of the powder. The formation of necks is instigated by localized overheating at the particle contacts. Song et al. derived a model to calculate the temperature distribution associated with particle contacts during SPS (Song et al. 2006). The particle contact temperature exceeded the boiling point of the material and rapidly decreased toward the center of the particle. The formation of necks in this material was suggested to be a result of boiling or melting the particles followed by rapid solidification to bridge the two particles. The rapid solidification process created a refined grain structure within and surrounding the necks of the bridged particles. The larger contact radius afforded by larger particles allowed for a higher current density, creating higher localized temperatures and faster densification compared to smaller particles during the second stage of densification. Diouf and Molinari observed an increase in the peak height and width of the secondary peak associated with deformation due to localized pressure and larger particle sizes (Diouf and Molinari 2012). The third displacement rate peak, resulting from bulk deformation, was unaffected by particle size (Diouf and Molinari 2012). During the final stage of densification, power-law creep was responsible for powder sintering. The influence of the power-law creep portion of densification is more pronounced at smaller grain sizes (Olevsky and Froyen 2006). In general, this resulted in higher densities for samples with finer initial particle sizes (Diouf and Molinari 2012; Shongwe et al. 2016b). When comparing four different tungsten powders and their SPS sintering behavior, it was found that under identical processing parameters, powders with smaller particle sizes produced the highest density compacts (Cho et al. 2004).

Olevsky and Froyen developed densification maps for conductive powders during SPS processing (Olevsky and Froyen 2006). An example of a typical densification map is provided in Fig. 15. Initially, when porosity was higher than a few tenths of a percent ($\geq \sim 0.3$), the external load served as the driving force for densification. At lower porosity values, densification was driven by either surface tension at very small particle sizes (below ~1 μm) or by diffusional or

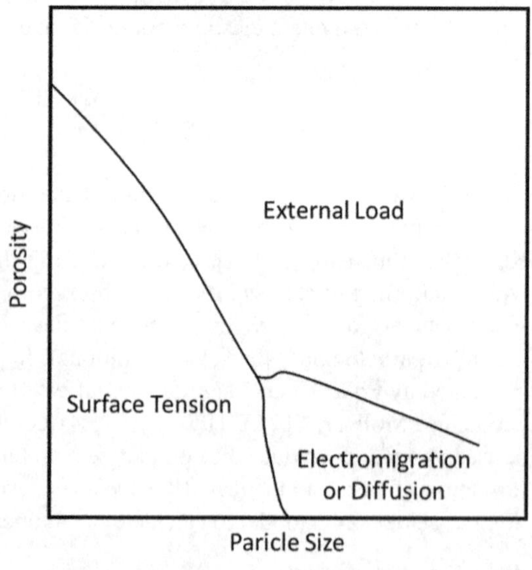

Fig. 15 Schematic of densification for a conductive powder under a constant temperature and load. (Based on Fig. 2 of Olevsky and Froyen 2006)

electromigration processes at larger particle sizes. Recently, there has been some discussion in the literature as to whether electromigration, Joule heating, or a combination of both acts as the densification mechanism. The active mechanism or mechanisms are highly likely to be dependent on the material system being studied. In any case, these mechanisms should become dominant at particle sizes above ~1 μm. A thorough discussion regarding the exact roles of thermal (Joule heating) and current (electromigration) effects for the aforementioned densification mechanisms is beyond the scope of this text.

Nonconductive powders exhibit different sintering mechanisms than conducting powders. Anselmi-Tamburini et al. modeled the current distribution for an aluminum oxide powder (Anselmi-Tamburini et al. 2005b). While none of the current passed through the nonconductive powder, it was instead distributed inhomogeneously throughout the surrounding die. This resulted in thermal gradients and nonuniform sintering of the part (Zhao et al. 2008). The presence of current effects in nonconductive materials is still being investigated in the literature. However, several authors have successfully modeled the SPS process using pressure-assisted analytical sintering models, such as those used for hot isostatic pressing applications (Chaim and Margulis 2005; Dobedoe et al. 2013). In the case of nonconductive ceramics, it was generally accepted that samples with smaller initial particle sizes experienced higher densification rates at lower temperatures compared to samples with larger particle sizes (Chaim and Margulis 2005; Reis and Chaim 2008).

During the SPS process, surface oxides are broken up to form stronger interparticle bonds. Le et al. observed that the particle surface oxides (varying from 10 to 32 nm in thickness) changed from solid layer to nanoscale spherical oxides as a result of the SPS process (Le et al. 2013). These oxides helped pin the grain boundaries and preserve the ultrafine grain size in the material (Fig. 16). Thicker

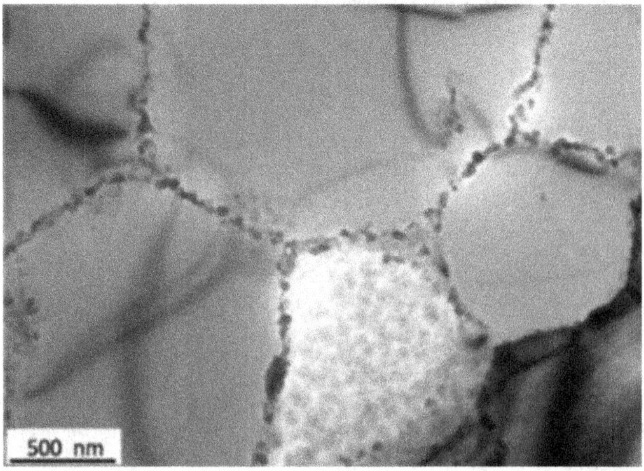

Fig. 16 Breakup of the native oxide scale into spherical oxide particles at the former particle boundaries in an SPS-processed Al. (Originally published in Le et al. used with permission Le et al. 2013)

oxides have been studied in copper by Romaric et al. (Romaric et al. 2017), who started with plasma-treated Cu powder that contained only 0.04 wt.% oxygen. The powder was oxidized at 150 °C and 170 °C, creating 75-nm- and 160-nm-thick oxide layers on the surfaces of the powders, respectively. During SPS consolidation, the oxides were first repelled (accredited to the application of pressure and temperature) to create clean Cu surfaces. As the Cu surfaces started to consolidate, the oxides remained as collars around the necks forming between the Cu particles, resulting in metallic grain boundary formation between the particles. When the oxide layers were not present, consolidation of the necks occurred at lower temperatures.

Preprocessing of powders by mechanical milling is becoming an increasingly popular option for influencing mechanical, optical, and dielectric properties and, in some case, the chemistry of the final product. Ball milling of powders prior to SPS has been used for mixing, mechanical alloying, nanostructuring, and mechanical activation. Ball milling is commonly used in ceramic or ceramic and binder systems to thoroughly mix the powders. A uniform distribution of ceramic powders is especially important during reactive sintering. TiC-TiB_2 composites have been formed through the reactive sintering of Ti, B_4C, and C starting materials via high-energy ball milling (Locci et al. 2006). Ball milling is commonly used for dispersing binders in a ceramic matrix. For example, cobalt binder was uniformly distributed in a WC matrix via ball milling (Liu et al. 2008, 2010; Zhao et al. 2008, 2009). When mixed with WC, the ball-milled Co formed a thin film around the WC powder, leading to improved conductivity throughout the compact and producing a more uniform microstructure (Zhao et al. 2008).

Mechanical alloying of various materials has been extensively studied (Suyanarayana 2001), with SPS being utilized as an attractive route for consolidation. Ball milling for mechanical alloying has been conducted on metals, semiconductor materials, and ceramics. Sb-doped Bi_2Te_3-type alloys are important materials to the thermoelectric community but are difficult to produce. It was possible to create bulk $(Bi,Sb)_2Te_3$ doped with varying amounts of Sb through mechanical alloying of pure Bi, Te, and Sb powder reinforced with SiC nanoparticles and processing by SPS (Chen et al. 2010). β-sialon nanoceramics were produced through mechanical alloying of β-Si_3N_4 with AlN and Al_2O_3 via high-energy ball milling and SPS (Xu et al. 2005). Mechanical alloying of the powders produced a homogenous amorphous phase which crystallized into nanocrystalline β-sialon during SPS sintering.

Nanostructuring often occurs during the milling process. One of the most common reasons for nanostructuring materials is to improve their strength via the Hall-Petch effect. SPS processing is commonly used to produce fully dense nanocrystalline materials, as the high heating rates and short sintering times have been shown to effectively minimize grain growth in the material (Keller et al. 2016). Additionally, nanopores or nano-sized second-phase particles could help pin the grain boundaries and retain the grain size (Chaim et al. 2013).

It has been noted in the literature that ball milling of powders results in mechanical activation of the powders, leading to better sinterability at lower temperatures than conventionally produced powders. For example, Guyon et al. found that milled

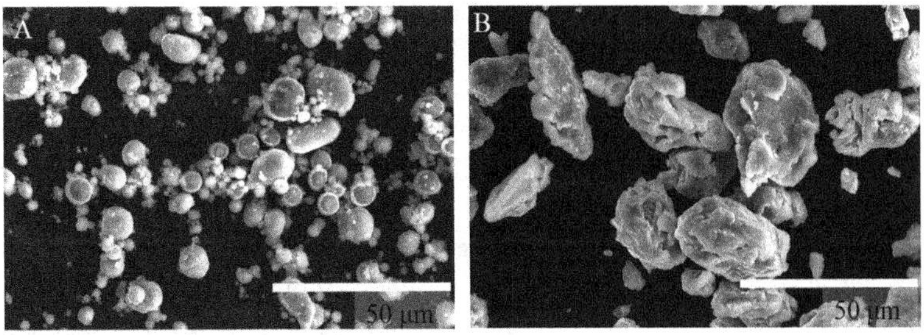

Fig. 17 SEM micrographs of as-atomized AA 5083 (**a**) and cryomilled AA 5083 (**b**). (Previously unpublished)

powders of TiAl densified at much lower temperatures than gas atomized powders (Guyon et al. 2013). It was postulated that different densification mechanisms were active in coarse-grained (gliding and twinning) versus fine-grained materials (grain boundary sliding and surface/grain boundary diffusion).

Enhanced densification of mechanically milled powders has also been observed in aluminum alloys. As-atomized AA5083 and cryomilled AA5083 powders (Fig. 17) have been processed via SPS using both graphite and BN dies under the same loads and constant current profiles (Kellogg et al. 2017). Kellogg et al. determined that under an applied pressure of 18 MPa, the as-atomized powder was not conductive at room temperature, but the cryomilled powder was conductive. While the authors could only speculate on the reason for the change in conductivity, three hypotheses were provided. Cryomilling may have led to a change in powder packing due to morphological modifications, the introduction and dispersion of a small amount of graphite (0.15 weight % as a processing aid), or possible changes to the surface chemistry. Regardless of the reason, the cryomilled powders exhibited higher conductivity than the as-atomized materials and enabled the use of BN dies without a need for an external heat source. This led to higher heating rates and faster consolidation than possible with the as-atomized material. Although there could be many reasons for the enhanced sinterability of ball-milled powder over conventionally processed powder, it is currently not well-understood and likely to be material specific. Additional studies will need to be conducted to clarify the sintering mechanisms of milled powder.

5 Current Waveforms

It is easy to dismiss the role of the electric current as just a heat source and to attribute everything in SPS to Joule heating, as evidenced by the number of papers that do not specify current used (Orrù et al. 2009). However, using Joule heating as a singular explanation for the enhanced consolidation during SPS processing is a simplification that ignores the different current and field effects present from atomistic to bulk length scales (Anselmi-Tamburini and Groza 2017). The field and current effects that occur during resistive heating have been well discussed and mathematically explained in Anselmi-Tamburini et al. (2013) and will only be summarized here. The field effects during SPS were generally small due to the low voltages used (Guillon et al. 2014) but have been shown to influence diffusion (particularly grain boundary and defect migration) (Antolovich and Conrad 2004; Raj et al. 2011). The field effects are material-dependent and may be much more significant for materials with high-dielectric permittivity values (Holland et al. 2012).

The current effects can be broken down to four general components: percolation effects, Peltier effects, electrochemical reactions, and electromigration (Anselmi-Tamburini et al. 2013; Guillon et al. 2014). Due to the nature of the powders used during SPS, the initial powder bed is not uniformly dense. As pressure and heat are applied to the powder, the particles in contact with each other begin to form necks and bond. This necking provides the percolation pathways that enable the current to travel through the powder. The flow of current along these percolation paths causes Joule heating and is credited with generating the fast heating rates and reduced sintering times that make SPS attractive. Peltier heating occurs as electrical current passes through an interface between two different materials (i.e., between the punch and sample during SPS) (Guillon et al. 2014). These effects are generally negligible when processing metallic powders but can become quite significant when processing semiconductor materials and can serve as a contributing source to the large thermal gradients observed at the interface between the punch and sample. This interface is also the source of potential electrochemical reactions between the punch and powder, as discussed in section three on the reactivity of graphite. The fourth current effect, electromigration, has been difficult to resolve during SPS experimentation (Guillon et al. 2014; Trzaska and Monchoux 2015), often due to the high current densities or long experimental times required to measure this effect (Frei et al. 2007; Toyofuku et al. 2011). However, electromigration effects have been credited with providing enhanced growth for intermetallic formation in Ni-Ti systems (Garay et al. 2003) and Al-Au diffusion couples (Bertolino et al. 2001). It has also been stated that electromigration can have an influence on grain growth (Yang and Conrad 2010). The percolation and electrochemical effects occur whenever electrical current is used during sintering. Peltier effects, electromigration, and electric field effects can be altered by changing current waveforms. Under a constant DC current, these effects are likely at a maximum, as the current is driving in one direction. While a pulsed DC current should reduce these effects (due to the on/off nature of pulsing), switching to an AC current waveform could potentially minimize these effects. Figure 18, reproduced from Orrù et al. (2009), provides a graphical breakdown of the different current waveforms. Since an AC current reverses flow direction periodically, it is hypothesized that electromigration effects do not occur with an AC current (Narayan 2013; Yang and Conrad 2010). These polarity shifts could also mitigate any Peltier effects observed at the interface between the punch and powder (Becker et al. 2012; Guillon et al. 2014). An AC current at the same RMS strength as a comparable DC current has a stronger electrical field, by a factor of $\sqrt{2}$, (Yang and Conrad 2010) that could provide enhanced field effects during SPS for improved sintering (Antolovich and Conrad 2004; Raj et al. 2011).

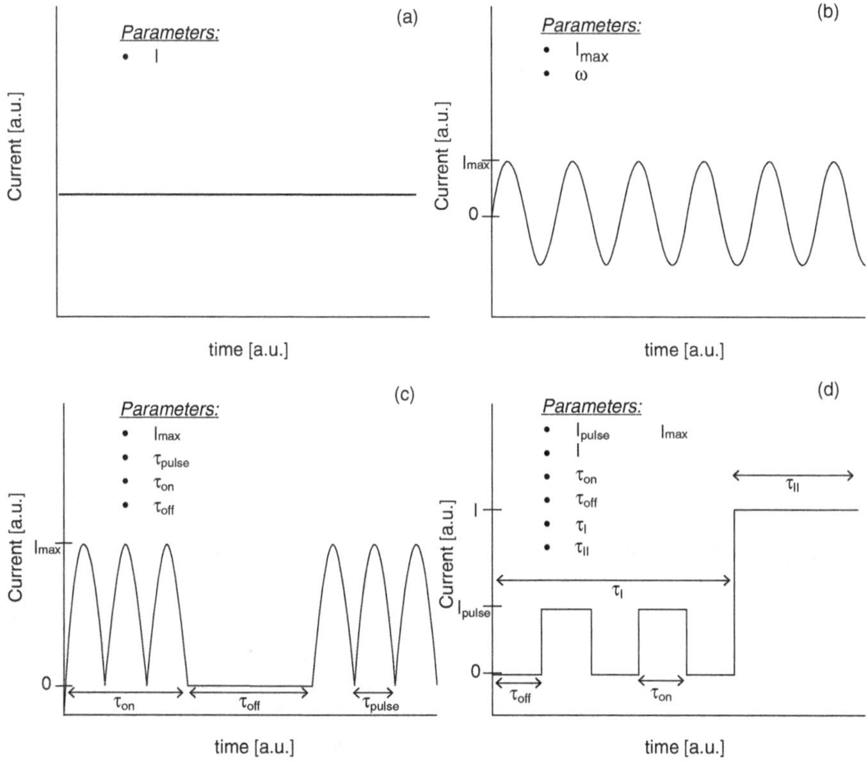

Fig. 18 Graphical illustrations of typical waveforms including constant current DC (**a**), constant current AC (**b**), pulsed DC (**c**), and pulsed and constant current DC (**d**). (Reprinted by permission from Elsevier, Mater. Sci. Eng., R, Orrù et al. 2009)

Despite the different characteristics of each current waveform, it can be hard to attribute the waveform effects to the microstructural and mechanical properties of the resulting parts. Anselmi-Tamburini and Groza noted that AC currents produce results that are often indistinguishable from pulsed DC currents, although they did not elaborate on whether the results were based on comparisons of consolidated samples or processing variables (currents, times, temperatures) collected during the consolidation process (Anselmi-Tamburini and Groza 2017). Part of the reason that Joule heating has been considered as a "catch-all" for field-assisted sintering techniques is that it is very difficult to design experiments that can isolate current effects from thermal effects while demonstrating how they contribute to SPS sintering.

In a pair of related papers, the effects of current on the reactivity of Mo foils and Si wafers during SPS were studied (Anselmi-Tamburini et al. 2005a; Chen et al. 2005). Included in the findings was that the growth of Mo_xSi_y layers was more rapid in the presence of a current than without. The layer growth, however, was not affected by pulsing direction or pulse patterns (on/off times and cycles). Despite showing improved layer growth during the application of a current, the activation energies required for growth of Mo_xSi_y layers were calculated to be the same with or without current (within statistical noise) (Anselmi-Tamburini et al. 2005a). It was concluded that this means the current does not alter the reaction mechanisms and that the relative enhanced growth is due to increased defect concentration or mobility, i.e., electromigration. The same results were found for Nb-C diffusion couples (Kondo et al. 2008). Under constant DC currents of varying strengths, field-enhanced diffusional layer growth between Ni-Ti diffusion couples was observed (Fig. 19) and attributed to increased defect concentration and mobility (Garay et al. 2003). Unlike the previous work, Garay et al. did find a relationship between the current strength and activation energies. As the current density increased, the activation energy for each Ni_xTi_y intermetallic layer decreased.

As opposed to diffusion couples and foils, Xie et al. sintered aluminum powder at different pulse frequencies and found that it had no effect on the final density, electrical resistivity, or tensile properties of the sintered samples (Xie et al. 2003). While examining the roles of oxide layers on the sintering of copper powders, no differences in sample density were found when comparing a pulsed DC current to a constant DC current (Romaric et al. 2017). The purpose of the research was to maximize current densities and to create plasmas for removing the oxide layers during processing. By using a hot-pressed sample as a control, it was found that the oxide layer dissolution was controlled by pressure and temperature, and not any current effects. In addition no evidence of plasma or spark generation was found.

There is a prevalent belief that a pulsed current in SPS generates a spark and a plasma inside the powder bed that can be used to clean contaminants such as oxide layers on powder surfaces (Guillon et al. 2014; Hulbert et al. 2009; Munir et al. 2006; Orrù et al. 2009). However, up to this point, the presence of sparking or plasma generation has yet to be demonstrated, leading to the publication of SPS research

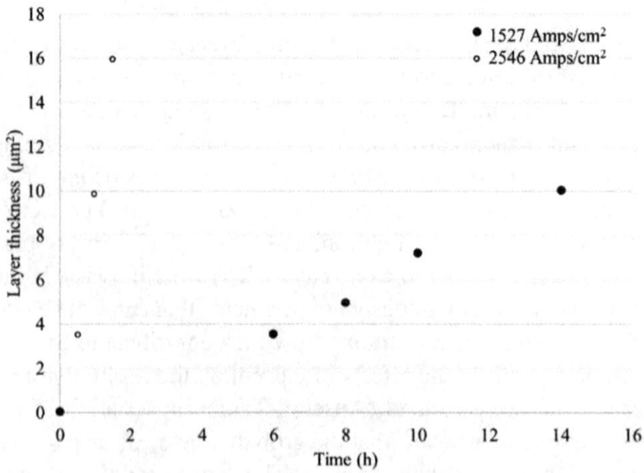

Fig. 19 Intermetallic layer thickness in Ni-Ti diffusion couples under different constant DC current densities. (Previously unpublished figure with data originally reported in Garay et al. 2003)

under many different pseudonyms. Research that has claimed to have found evidence of a plasma has either used experimental conditions or processing parameters that are not consistent with SPS. In addition, no evidence has been shown to indicate that pulsed currents produce different samples than constant currents.

Few papers have investigated the use of AC current for SPS, and even fewer fundamental investigations have compared AC to the other conventional current types. Heian et al. created a custom system to consolidate B_4C using an AC current, but the justification for utilizing AC current was not specified (Heian et al. 2004). Similarly, Zhou and Kwon used an AC current to consolidate W-Cu powder composites, but did not specify the current parameters or indicate why an AC current source was chosen (Zhou and Kwon 2005). Carrilo-Heian et al. produced Nb_5Si_3-Nb composites using either pulsed DC current or AC current sources (Carrillo-Heian et al. 2001). The processes were designated as SPS and a field-activated pressure-assisted synthesis apparatus (FAPASA), respectively. Smaller (19 mm in diameter and 5 mm thick) and larger samples (45 mm diameter by 5 mm thick) were consolidated via FAPASA and SPS. Processing parameters (pressures, heating rates, and hold times) differed between the two consolidation methods. While the smaller samples produced via AC current were slightly denser than the larger samples produced with a DC current, the processing parameter differences made it difficult to draw any conclusions in terms of current. The authors attributed the density differences to slower heating rates, allowing more controlled chemical reactions during consolidation. An AC current was used to consolidate SiC-Cu composites through a technique referred to as electric conduction sintering, claiming that it could consolidate samples at lower sintering temperatures, pressures, and times compared to SPS or hot pressing. However, no data was provided to back up this claim (Wang et al. 2007).

Evidence exists in other processing systems that an AC field provides benefits not observed with a DC field. In particular, the research of Conrad on the 3Y-TZP system (Conrad and Wang 2014; Conrad and Yang 2013; Yang and Conrad 2010) (Fig. 20) demonstrated that an AC field, in contrast to a DC field, prevented grain growth during sintering and allowed for sintering at lower temperatures (up to a critical field strength).

There is a fair amount of research that can be used to theorize that an AC current could lead to improvements in resulting SPS parts. However, research showing a direct one-to-one correlation between AC and DC currents is virtually nonexistent. Closer examinations of changes to the current waveforms could pay large dividends for SPS processing.

6 Conclusions

The goal of this chapter was to examine how changes to some of the most basic fundamental principles of the SPS process, including tooling design, die material, powder form, and current waveform, could be used to improve part production. Each

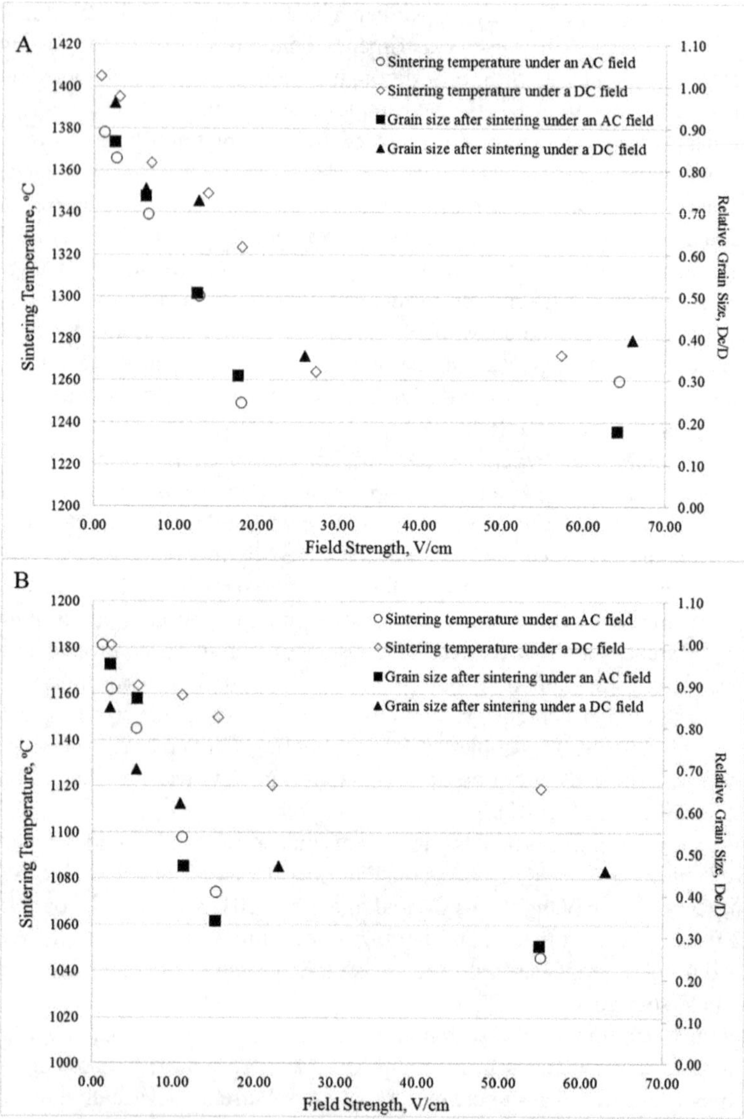

Fig. 20 The effects of electric fields on sintering temperatures and grainsizes. The samples in (**a**) were sintered to 0.95 relative density, while the samples in (**b**) were sintered to 0.60 relative density. Relative grain size was calculated via an equation in the original source. In both cases, the samples sintered under an AC field reached the targeted densities at lower sintering temperatures and had smaller final grain sizes than when sintered under a DC field. (Data extracted from Conrad and Yang 2013)

of these factors can be modified to alter the pathways that the electrical current uses to pass through the tooling, the powder sample, or both. These current pathways represent the most fundamental aspect of SPS. Manipulation of these pathways could be used in numerous ways to enhance the SPS process by increasing the number of parts being processed in a single run, improving the grain structure uniformity of parts and samples, introducing better tooling options, and strategically speeding up or slowing down the sintering process. Alterations to these pathways will be the key to improving SPS processing and expanding it beyond the laboratory for more widespread industrial production of larger and more complex parts.

Acknowledgments The research reported in this document was performed in connection with contract/instrument W911QX-16-D-0014 with the US Army Research Laboratory. The views and conclusions contained in this document are those of SURVICE Engineering and the US Army Research Laboratory. Citation of manufacturer's or trade names does not constitute an official endorsement or approval of the use thereof. The US government is authorized to reproduce and distribute reprints for government purposes notwithstanding any copyright notation hereon.

This research was supported in part by an appointment to the Postgraduate Research Participation Program at the US Army Research Laboratory administered by the Oak Ridge Institute for Science and Education through an interagency agreement between the US Department of Energy and USARL.

Figure 2 was reproduced from an open-access article "under the terms of the Creative Commons Attribution Non-Commercial License, which permits use, distribution and reproduction in any medium, provided the original work is properly cited and is not used for commercial purposes." The caption has been altered from the original.

Figures 11 and 13 were reproduced from open-access papers published under a Creative Commons Attribution 4.0 International License. In these cases, the figures were unaltered and credited to the authors in new captions. For more information, and for a copy of the license, please visit http://creativecommons.org/licenses/by/4.0/.

References

Aalund R (2010) Unveiling spark plasma sintering high-throughput processing. In: Bansal N, Sing J, Lamon J, Choi SR, Mahmouad M (eds) Processing and properties of advanced ceramics and composites II, Ceramic transactions, vol 220. Wiley, Hoboken

Anselmi-Tamburini U, Groza JR (2017) Critical assessment: electrical field/current application – a revolution in materials processing/sintering? Mater Sci Technol 33(16):1855–1862. https://doi.org/10.1080/02670836.2017.1341692

Anselmi-Tamburini U, Garay JE, Munir ZA (2005a) Fundamental investigations on the spark plasma sintering/synthesis process III. Current effect on reactivity. Mater Sci Eng A 407 (1–2):24–30. https://doi.org/10.1016/j.msea.2005.06.066

Anselmi-Tamburini U, Gennari S, Garay JE, Munir ZA (2005b) Fundamental investigations on the spark plasma sintering/synthesis process II modeling of current and temperature distributions. Mater Sci Eng A 394(1–2):139–148. https://doi.org/10.1016/j.msea.2004.11.019

Anselmi-Tamburini U, Spinolo G, Maglia F, Tredici I, Holland T, Mukherjee A (2013) Field assisted sintering mechanisms. In: Castro R, van Benthem K (eds) Sintering mechanisms of convention nanodensification and field assisted processes, Engineering materials, vol 35. Springer, New York

Antolovich SD, Conrad H (2004) The effects of electric currents and fields on deformation in metals, ceramics, and ionic materials: an interpretive survey. Mater Manuf Process 19 (4):587–610. https://doi.org/10.1081/amp-200028070

Arnaud C, Manière C, Chevallier G, Estournès C, Mainguy R, Lecouturier F, Mesguich D, Weibel A, Durand L, Laurent C (2015) Dog-bone copper specimens prepared by one-step spark plasma sintering. J Mater Sci 50:7364–7373. https://doi.org/10.1007/s10853-015-9293-5

Balomenos E, Paniasa D, Paspaliaris I (2011) Energy and exergy analysis of the primary aluminum production processes: a review of current and future sustainability. Miner Process Extr Metall Rev 32(2):69–89. https://doi.org/10.1080/08827508.2010.530721

Becker A, Angst S, Schmitz A, Engenhorst M, Stoetzel J, Gautam D, Wiggers H, Wolf DE, Schierning G, Schmechel R (2012) The effect of Peltier heat during current activated densification. Appl Phys Lett 101(1):013113. https://doi.org/10.1063/1.4731272

Bertolino N, Garay J, Anselmi-Tamburini U, Munir ZA (2001) Electromigration effects in Al-Au multilayers. Scr Mater 44:737–742. https://doi.org/10.1016/S1359-6462(00)00669-2

Carrillo-Heian E, Unuvar C, Gibeling J, Paulino G, Munir Z (2001) Simultaneous synthesis and densification of niobium silicide/niobium composites. Scr Mater 45:405–412. https://doi.org/10.1016/S1359-6462(01)01023-5

Chaim R, Margulis M (2005) Densification maps for spark plasma sintering of nanocrystalline MgO ceramics. Mater Sci Eng A 407(1–2):180–187. https://doi.org/10.1016/j.msea.2005.07.024

Chaim R, Marder R, Estournés C, Shen Z (2013) Densification and preservation of ceramic nanocrystalline character by spark plasma sintering. Adv Appl Ceram 111(5–6):280–285. https://doi.org/10.1179/1743676111y.0000000074

Chen W, Anselmi-Tamburini U, Garay JE, Groza JR, Munir ZA (2005) Fundamental investigations on the spark plasma sintering/synthesis process I effect of dc pulsing on reactivity. Mater Sci Eng A 394(1–2):132–138. https://doi.org/10.1016/j.msea.2004.11.020

Chen C, Liu D-W, Zhang B-P, Li J-F (2010) Enhanced thermoelectric properties obtained by compositional optimization in p-Type Bi x Sb2−x Te3 fabricated by mechanical alloying and spark plasma sintering. J Electron Mater 40(5):942–947. https://doi.org/10.1007/s11664-010-1463-2

Cho K, Woodman R, Klotz B, Dowding R (2004) Plasma pressure compaction of tungsten powders. Mater Manuf Process 19(4):619–630. https://doi.org/10.1081/LMMP-200028077

Cologna M, Rashkova B, Raj R (2010) Flash sintering of nanograin zirconia in <5s at 850°C. J Am Ceram Soc 93(11):3556–3559. https://doi.org/10.1111/j.1551-2916.2010.04089.x

Conrad H, Wang J (2014) Equivalence of AC and DC electric field on retarding grain growth in yttria-stabilized zirconia. Scr Mater 72-73:33–34. https://doi.org/10.1016/j.scriptamat.2013.10.010

Conrad H, Yang D (2013) Effect of the strength of an AC electric field compared to DC on the sintering rate and related grain size of zirconia (3Y-TZP). Mater Sci Eng A 559:591–594. https://doi.org/10.1016/j.msea.2012.08.146

Diouf S, Molinari A (2012) Densification mechanisms in spark plasma sintering: effect of particle size and pressure. Powder Technol 221:220–227. https://doi.org/10.1016/j.powtec.2012.01.005

Diouf S, Menapace C, Molinari A (2012) Study of effect of particle size on densification of copper during spark plasma sintering. Powder Metall 55(3):228–234. https://doi.org/10.1179/1743290111y.0000000019

Dobedoe RS, West GD, Lewis MH (2013) Spark plasma sintering of ceramics: understanding temperature distribution enables more realistic comparison with conventional processing. Adv Appl Ceram 104(3):110–116. https://doi.org/10.1179/174367605X16662

Frei JM, Anselmi-Tamburini U, Munir ZA (2007) Current effects on neck growth in the sintering of copper spheres to copper plates by the pulsed electric current method. J Appl Phys 101(11):114914. https://doi.org/10.1063/1.2743885

Garay JE, Anselmi-Tamburini U, Munir ZA (2003) Enhanced growth of intermetallic phases in the Ni–Ti system by current effects. Acta Mater 51(15):4487–4495. https://doi.org/10.1016/s1359-6454(03)00284-2

Giuntini D, Olevsky EA, Garcia-Cardona C, Maximenko AL, Yurlova MS, Haines CD, Martin DG, Kapoor D (2013) Localized overheating phenomena and optimization of spark-plasma sintering tooling design. Materials (Basel) 6(7):2612–2632. https://doi.org/10.3390/ma6072612

Giuntini D, Raethel J, Herrmann M, Michaelis A, Olevsky EA, Bouvard D (2015) Advancement of tooling for spark plasma sintering. J Am Ceram Soc 98(11):3529–3537. https://doi.org/10.1111/jace.13528

Giuntini D, Raethel J, Herrmann M, Michaelis A, Haines CD, Olevsky EA (2016) Spark plasma sintering novel tooling design: temperature uniformization during consolidation of silicon nitride powder. J Ceram Soc Jpn 124(4):403–414. https://doi.org/10.2109/jcersj2.124.P4-1

Guillon O, Gonzalez-Julian J, Dargatz B, Kessel T, Schierning G, Räthel J, Herrmann M (2014) Field-assisted sintering technology/spark plasma sintering: mechanisms, materials, and technology developments. Adv Eng Mater 16(7):830–849. https://doi.org/10.1002/adem.201300409

Guyon J, Hazotte A, Monchoux JP, Bouzy E (2013) Effect of powder state on spark plasma sintering of TiAl alloys. Intermetallics 34:94–100. https://doi.org/10.1016/j.intermet.2012.11.005

Hao Y, Liu J, Li J, Li S, Zou Q, Chen X (2015) Rapid preparation of TiC reinforced Ti6Al4V based composites by carburizing method through spark plasma sintering technique. Mater Des 65:94–97. https://doi.org/10.1016/j.matdes.2014.09.008

Heian EM, Khalsa SK, Lee JW, Munir ZA (2004) Synthesis of dense, high-defect-concentration B4C through mechanical activation and field-assisted combustion. J Am Ceram Soc 87 (5):779–783. https://doi.org/10.1111/j.1551-2916.2004.00779.x

Holland TB, Anselmi-Tamburini U, Quach DV, Tran TB, Mukherjee AK (2012) Local field strengths during early stage field assisted sintering (FAST) of dielectric materials. J Eur Ceram Soc 32(14):3659–3666. https://doi.org/10.1016/j.jeurceramsoc.2012.03.012

Hulbert DM, Anders A, Andersson J, Lavernia EJ, Mukherjee AK (2009) A discussion on the absence of plasma in spark plasma sintering. Scr Mater 60(10):835–838. https://doi.org/10.1016/j.scriptamat.2008.12.059

Johnson KL (1987) Normal contact of elastic solids: Hertz theory. In: Contact mechanics. Cambridge University Press, Cambridge

Keller C, Tabalaiev K, Marnier G, Noudem J, Sauvage X, Hug E (2016) Influence of spark plasma sintering conditions on the sintering and functional properties of an ultra-fine grained 316L stainless steel obtained from ball-milled powder. Mater Sci Eng A 665:125–134. https://doi.org/10.1016/j.msea.2016.04.039

Kellogg F, McWilliams B, Cho K (2016) Effect of current pathways during spark plasma sintering of an aluminum alloy powder. Metall Mater Trans A 47(12):6353–6367. https://doi.org/10.1007/s11661-016-3803-1

Kellogg F, McWilliams B, Sietins J, Giri A, Cho K (2017) Comparison of SPS processing behavior between as atomized and cryomilled aluminum alloy 5083 powder. Metall Mater Trans A 48 (11):5492–5499. https://doi.org/10.1007/s11661-017-4286-4

Klotz B, Kellogg F, Klier E., Dowding R, Cho K (2009) Characterization, processing, and consolidation of tungsten powder. Army Research Laboratory technical report 5045, publicly accessible at http://www.dtic.mil/docs/citations/ADA538261

Kondo T, Yasuhara M, Kuramoto T, Kodera Y, Ohyanagi M, Munir ZA (2008) Effect of pulsed DC current on atomic diffusion of Nb–C diffusion couple. J Mater Sci 43(19):6400–6405. https://doi.org/10.1007/s10853-008-2758-z

Le GM, Godfrey A, Hansen N (2013) Structure and strength of aluminum with sub-micrometer/micrometer grain size prepared by spark plasma sintering. Mater Des 49:360–367. https://doi.org/10.1016/j.matdes.2013.01.018

Liu X, Song X, Zhang J, Zhao S (2008) Temperature distribution and neck formation of WC–Co combined particles during spark plasma sintering. Mater Sci Eng A 488(1–2):1–7. https://doi.org/10.1016/j.msea.2008.01.048

Liu X, Song X, Zhao S, Zhang J (2010) Spark plasma sintering densification mechanism for cemented carbides with different WC particle sizes. J Am Ceram Soc 93(10):3153–3158. https://doi.org/10.1111/j.1551-2916.2010.03862.x

Liu D, Xiong Y, Li Y, Topping TD, Zhou Y, Haines C, Paras J, Martin D, Kapoor D, Schoenung JM, Lavernia EJ (2012) Spark plasma sintering of nanostructured aluminum: influence of tooling material on microstructure. Metall Mater Trans A 44(4):1908–1916. https://doi.org/10.1007/s11661-012-1533-6

Locci AM, Orru R, Cao G, Munir ZA (2006) Simultaneous spark plasma synthesis and densification of TiC-TiB2 composites. J Am Ceram Soc 89(3):848–855. https://doi.org/10.1111/j.1551-2916.2005.00832.x

Maniere C, Lee G, Olevsky EA (2017) All-materials-inclusive flash spark plasma sintering. Sci Rep 7(1):15071. https://doi.org/10.1038/s41598-017-15365-x

McWilliams B, Zavaliangos A, Cho KC, Dowding RJ (2006) Modeling of electric-current-assisted sintering to produce bulk nanocrystalline tungsten. JOM 58(4):67–71. https://doi.org/10.1007/s11837-006-0218-2

McWilliams B, Yu J, Kellogg F, Kilczewski S (2016) Enhanced sintering kinetics in aluminum alloy powder consolidated using DC electric fields. Metall Mater Trans A 48(2):919–929. https://doi.org/10.1007/s11661-016-3861-4

Munir ZA, Anselmi-Tamburini U, Ohyanagi M (2006) The effect of electric field and pressure on the synthesis and consolidation of materials: a review of the spark plasma sintering method. J Mater Sci 41(3):763–777. https://doi.org/10.1007/s10853-006-6555-2

Narayan J (2013) A new mechanism for field-assisted processing and flash sintering of materials. Scr Mater 69(2):107–111. https://doi.org/10.1016/j.scriptamat.2013.02.020

Ohser-Wiedemann R, Martin U, Seifert HJ, Müller A (2010) Densification behaviour of pure molybdenum powder by spark plasma sintering. Int J. Refract Met Hard Mater 28 (4):550–557. https://doi.org/10.1016/j.ijrmhm.2010.03.003

Olevsky E, Froyen L (2006) Constitutive modeling of spark-plasma sintering of conductive materials. Scr Mater 55(12):1175–1178. https://doi.org/10.1016/j.scriptamat.2006.07.009

Olevsky EA, Rolfing SM, Maximenko AL (2016) Flash (ultra-rapid) spark-plasma sintering of silicon carbide. Sci Rep 6:33408. https://doi.org/10.1038/srep33408

Omori M (2000) Sintering, consolidation, reaction and crystal growth by the spark plasma system (SPS). Mater Sci Eng A 287:183–188

Orrù R, Licheri R, Locci AM, Cincotti A, Cao G (2009) Consolidation/synthesis of materials by electric current activated/assisted sintering. Mater Sci Eng R 63(4–6):127–287. https://doi.org/10.1016/j.mser.2008.09.003

Paraskevas D, Vanmeensel K, Vleugels J, Dewulf W, Duflou JR (2015) The use of spark plasma sintering to fabricate a two-phase material from blended aluminium alloy scrap and gas atomized powder. Procedia CIRP 26:455–460. https://doi.org/10.1016/j.procir.2014.07.074

Raj R, Cologna M, Francis JSC, Green DJ (2011) Influence of externally imposed and internally generated electrical fields on grain growth, diffusional creep, sintering and related phenomena in ceramics. J Am Ceram Soc 94(7):1941–1965. https://doi.org/10.1111/j.1551-2916.2011.04652.x

Reis J, Chaim R (2008) Densification maps for spark plasma sintering of nanocrystalline MgO ceramics: particle coarsening and grain growth effects. Mater Sci Eng A 491(1–2):356–363. https://doi.org/10.1016/j.msea.2008.02.011

Romaric C, Sophie LG, Foad N, Frederic C, Guillaume B, Gilbert F, Jean-Marc C, Frederic B (2017) Effect of current on the sintering of pre-oxidized copper powders by SPS. J Alloys Compd 692:478–484. https://doi.org/10.1016/j.jallcom.2016.08.191

Shongwe MB, Diouf S, Durowoju MO, Olubambi PA, Ramakokovhu MM, Obadele BA (2016a) A comparative study of spark plasma sintering and hybrid spark plasma sintering of 93W–4.9Ni–2.1Fe heavy alloy. Int J Refract Met Hard Mater 55:16–23. https://doi.org/10.1016/j.ijrmhm.2015.11.001

Shongwe MB, Ramakokovhu MM, Diouf S, Durowoju MO, Obadele BA, Sule R, Lethabane ML, Olubambi PA (2016b) Effect of starting powder particle size and heating rate on spark plasma sintering of Fe Ni alloys. J Alloys Compd 678:241–248. https://doi.org/10.1016/j.jallcom.2016.03.270

Song X, Liu X, Zhang J (2006) Neck formation and self-adjusting mechanism of neck growth of conducting powders in spark plasma sintering. J Am Ceram Soc 89(2):494–500. https://doi.org/10.1111/j.1551-2916.2005.00777.x

Suryanarayana C (2001) Mechanical alloying and milling. Prog Mater Sci 46(1–2):1–184. https://doi.org/10.1016/S0079-6425(99)00010-9

Toyofuku N, Kuramoto T, Imai T, Ohyanagi M, Munir ZA (2011) Effect of pulsed DC current on neck growth between tungsten wires and tungsten plates during the initial stage of sintering by the spark plasma sintering method. J Mater Sci 47(5):2201–2205. https://doi.org/10.1007/s10853-011-6026-2

Trzaska Z, Monchoux J-P (2015) Electromigration experiments by spark plasma sintering in the silver–zinc system. J Alloys Compd 635:142–149. https://doi.org/10.1016/j.jallcom.2015.02.122

Vanmeensel K, Laptev A, Hennicke J, Vleugels J, Vanderbiest O (2005) Modelling of the temperature distribution during field assisted sintering. Acta Mater 53(16):4379–4388. https://doi.org/10.1016/j.actamat.2005.05.042

Violeta T, Wihelm K, Georgeta V, Cristian S, Dorinel T, Florentina C, Carmen H, Virgil M (2015) Joining of Y2O3-Doped aluminum nitride with high density graphite by spark plasma sintering. Adv Mater Res 1111:260–265. https://doi.org/10.4028/www.scientific.net/AMR.1111.260

Voisin T, Durand L, Karnatak N, Le Gallet S, Thomas M, Le Berre Y, Castagné J-F, Couret A (2013) Temperature control during spark plasma sintering and application to up-scaling and complex shaping. J Mater Process Technol 213(2):269–278. https://doi.org/10.1016/j.jmatprotec.2012.09.023

Wang YC, Fu ZY, Zhang QJ (2002) SPS temperature distribution of different conductivity materials. Key Eng Mater 224-226:717–720. https://doi.org/10.4028/www.scientific.net/KEM.224-226.717

Wang C, Guan S, Li S, Zhang R (2007) Mechanical properties of SiCp/Cu composites prepared by electric conduction sintering. Key Eng Mater 353-358:1402–1405. https://doi.org/10.4028/www.scientific.net/KEM.353-358.1402

Xie G, Ohashi O, Chiba K, Yamaguchi N, Song M, Furuya K, Noda T (2003) Frequency effect on pulse electric current sintering process of pure aluminum powder. Mater Sci Eng A 359(1–2):384–390. https://doi.org/10.1016/s0921-5093(03)00393-9

Xu X, Nishimura T, Hirosaki N, Xie R-J, Yamamoto Y, Tanaka H (2005) Fabrication of β-sialon nanoceramics by high-energy mechanical milling and spark plasma sintering. Nanotechnology 16(9):1569–1573. https://doi.org/10.1088/0957-4484/16/9/027

Yang D, Conrad H (2010) Enhanced sintering rate of zirconia (3Y-TZP) by application of a small AC electric field. Scr Mater 63(3):328–331. https://doi.org/10.1016/j.scriptamat.2010.04.030

Yu LG, Khor KA, Li H, Pay KC, Yip TH, Cheang P (2004) Restoring WC in plasma sprayed WC–Co coatings through spark plasma sintering (SPS). Surf Coat Technol 182(2-3):308–317. https://doi.org/10.1016/j.surfcoat.2003.08.065

Yu M, Grasso S, McKinnon R, Saunders T, Reece MJ (2016) Review of flash sintering: materials, mechanisms and modelling. Adv Appl Ceram 116(1):24–60. https://doi.org/10.1080/17436753.2016.1251051

Zadra M, Casari F, Girardini L, Molinari A (2007) Spark plasma sintering of pure aluminium powder: mechanical properties and fracture analysis. Powder Met 50(1):40–45. https://doi.org/10.1179/174329007x186417

Zavaliangos A, Zhang J, Krammer M, Groza JR (2004) Temperature evolution during field activated sintering. Mater Sci Eng A 379(1–2):218–228. https://doi.org/10.1016/j.msea.2004.01.052

Zhang YS, Zhang XM, Wang G, Bai XF, Tan P, Li ZK, Yu ZT (2011) High strength bulk tantalum with novel gradient structure within a particle fabricated by spark plasma sintering. Mater Sci Eng A 528(29-30):8332–8336. https://doi.org/10.1016/j.msea.2011.08.026

Zhao S, Song X, Zhang J, Liu X (2008) Effects of scale combination and contact condition of raw powders on SPS sintered near-nanocrystalline WC-Co alloy. Mater Sci Eng A 473:323–329. https://doi.org/10.1016/j.msea.2007.04.094

Zhao S, Song X, Wei C, Zhang L, Liu X, Zhang J (2009) Effects of WC particle size on densification and properties of spark plasma sintered WC–Co cermet. Int. J. Refract. Met. Hard Mater. 27(6):1014–1018. https://doi.org/10.1016/j.ijrmhm.2009.07.017

Zhou ZJ, Kwon YS (2005) Fabrication of W–Cu composite by resistance sintering under ultra-high pressure. J Mater Process Technol 168(1):107–111. https://doi.org/10.1016/j.jmatprotec.2004.11.008

Sintering Mechanisms of Metals Under Electric Currents

Jean-Philippe Monchoux

1 Introduction

From the two last decades, elaboration techniques based on use of high current intensities, among which the SPS technique is probably the most popular, have attracted considerable interest. In the case of the SPS, the reason of the fast development of this technique throughout the word is its unprecedented rapidity, which allows obtaining materials in typically ~30 min. For metals, this feature is particularly interesting, because short processing durations allow keeping fine microstructures and consequently elevated strengths. This remarkable processing rapidity was however poorly understood, and many studies have focused on investigating its physical bases. In particular, because the metallic systems exhibit, among the other classes of materials, the highest electrical conductivity, the influence of electric current on the microscopic mechanisms and on the densification kinetics has been actively investigated. The hypothesis is that the SPS current (~100 A/cm^2) could accelerate the elementary metallurgical mechanisms involved in densification and thus account for the rapidity of this technique.

Hence, a rapid review of the effects of electric currents in metallic systems will first be carried out. Then, relevant theoretical relations of densification under uniaxial pressure will be recalled, and typical SPS densification cycles of metallic systems will be given. Finally, a last section will be devoted to studies on the densification mechanisms at the microscopic scale, both by experiments and simulations.

J.-P. Monchoux (✉)
CEMES CNRS UPR 8011, Université de Toulouse, Toulouse, France
e-mail: monchoux@cemes.fr

© Springer Nature Switzerland AG 2019

P. Cavaliere (ed.), *Spark Plasma Sintering of Materials*,
https://doi.org/10.1007/978-3-030-05327-7_4

2 Effects of Electric Currents in Metallic Systems

The effects of electric current in metallurgical processes have been studied since at least the 1930s. The influence of current on plasticity, diffusion, and phase transformations has then been investigated, the most known effect being probably electromigration because of its role in microelectronics. The development of the SPS in the 1990s–2000s induced studies focusing on the influence of electric currents on diffusion, because this elementary mechanism is involved in the sintering processes. As for studies on plasticity, also involved in the densification mechanisms, they are somewhat scarcer.

2.1 *Electric Effects on Mass Transport and Reactivity*

The motivation to study the effect of electric current on mass transport was primarily fundamental, because these investigations allowed to precise the interactions between the electrons and the point defects. The first studies, reviewed in (Adda and Philibert 1966), treated the problem in the framework of the phenomenological theory of diffusion and then on the basis of atomistic calculations of momentum transfer by the electrons. These works considered that the diffusing species exhibit an "effective charge" z^*, which allows to take into account that the actual force applied on the diffusing species is composed of the regular electrostatic force, plus a contribution coming from the momentum transfer of the electrons. Values of z^* are often elevated and negative (typically between −5 and −20, e.g., Ni, −5; Au, −8.8; Ag, −24; etc.), leading to migration toward the anode, but can also be positive (e.g., Co, +1.6; Cu, +12; Fe, +43; etc.) for migration toward the cathode. This effect leads to the following differential equation for the diffusion profiles (Adda and Philibert 1966; Shewmon 1989):

$$\frac{\partial c}{\partial t} = D\frac{\partial^2 c}{\partial x^2} - \frac{D}{RT}Fz^*\rho \cdot J \cdot \frac{\partial c}{\partial x} \tag{1}$$

where D is the diffusion coefficient (supposed independent on composition); J, current density; c, concentration; F, Faraday's constant; R, the gas constant; T, temperature; x, spatial coordinate; and t, time. The resolution of this equation, which is a modification of the second Fick's law, leads to a temporal drift of the diffusion profiles for current densities ~10^3–10^4 A/cm^2 and durations ~100–500 h (Fig. 1).

But the most popular effect of an electric current on materials is indisputably the failure of electronic interconnects by electromigration. Due to the reduction of the interconnect sections to the micrometer scale, which allows fast dissipation of the heat generated by Joule effect, very strong current densities (~10^6–10^7 A/cm^2) can be achieved. These considerable current densities generate mass transfer (Fig. 2) which can provoke rupture of connections after typical durations of ~10^2–10^3 h (Frankovic 1996).

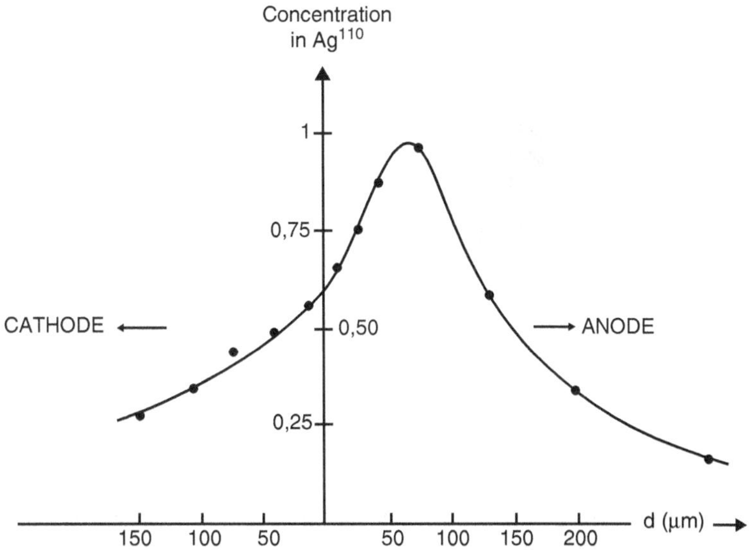

Fig. 1 Migration in Ag of Ag^{110} markers initially positioned at $x = 0$ in the presence of a current. The influence of the current is seen by the displacement of the maximum of the profile toward the anode. Conditions: J, 6800 A/cm^2; T, 820 °C; and t, 160 h (Adda and Philibert 1966)

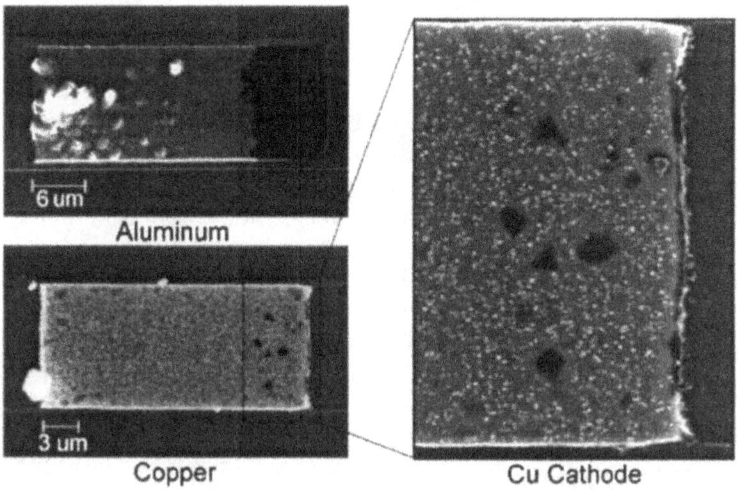

Fig. 2 Al and Cu conductor segments submitted to a current density of 6×10^5 A/cm^2 in the following conditions: 250 °C, 336 h for Cu, and 200 °C, 500 h for Al (electron sense, from right to left). Note the depletion of matter at the cathode and its accumulation at the anode (Frankovic 1996)

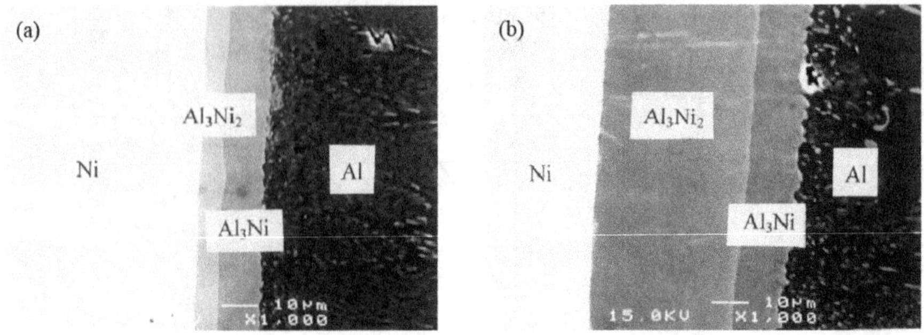

Fig. 3 Micrographs of a Al-Ni diffusion couple annealed at 400 °C for 336 h (**a**) without current and (**b**) with a current of 1000 A/cm^2 passing through the sample (Liu e al. 1998)

Finally, the influence of an electric current of ~10^2–10^3 A/cm^2 in experiments lasting t ~ 10–500 h on reactivity of many metallic systems has been studied around the year 2000 in the context of the development of sintering and synthesis techniques under electrical solicitations. Thus, experiments with electric currents passing through diffusion couples have been performed in the following systems: Al-Ni (Liu et al. 1998), Sn-Cu and Sn-Ni (Chen et al. 1998; Wang et al. 2011), Sn-Ag (Chen and Chen 1999), Bi-Ni (Chen and Chen 2000), Al-Au (Bertolino et al. 2001), Ni-Ti (Garay et al. 2003; Zhou et al. 2011), Cu-Ni (Zhao et al. 2007), and Ag-Zn (Friedman et al. 2004). In all these studies, electric currents were reported to significantly modify the growth kinetics of intermetallic layers and to modify diffusion profiles. For example, Fig. 3 shows the example of a Al-Ni diffusion couple annealed at 400 °C for 336 h and submitted to a current of 1000 A/cm^2. It can be seen that the layer is wider with current passing through the sample than without. The effect can be either dependent ("vectorial effect") or independent ("scalar effect") of the current direction. The interpretation of these phenomena remains largely unclear. There could be an influence of electromigration mechanisms (momentum transfer of the electrons), but other interpretations are based on influence of the current on equilibrium point defect concentration (Conrad 2000; Munir et al. 2006) or on mobility activation energy (Munir et al. 2006). However, in electromigration experiments in which the temperature profiles were carefully controlled, no effect of high currents (~10^3 A/cm^2) was detected in the Ag-Zn system (Trzaska and Monchoux 2015), contradicting a previous study (Friedman et al. 2004).

2.2 *Electric Effects on Plasticity and on Other Metallurgical Mechanisms*

Some studies have focused on the influence of current on metallurgical phenomena such as plasticity, recrystallization, phase transformations, etc. These studies are interesting in the framework of the understanding of the SPS processes, to account for the impact of the current on the final microstructures obtained by this technique. Moreover, plasticity can also be involved in sintering, as described later, and the influence of a current on this mechanism could allow to deeper understand the SPS kinetics. Unfortunately, studies on electric effects on these metallurgical phenomena are rather scarce and most of the time controversial. Here is a short report about the main results.

The influence of an electric current on plasticity has been investigated in the former Soviet Union in the years 1960–1970 by Conrad et al. (1990) and by Sprecher et al. (1986). For example, Fig. 4 shows the effect of single high-intensity

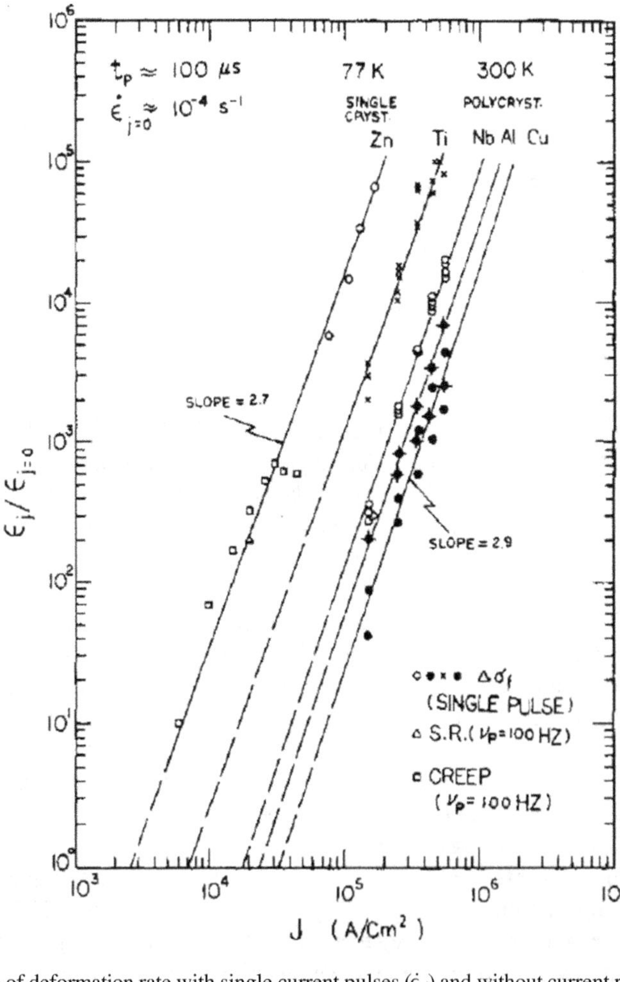

Fig. 4 Ratio of deformation rate with single current pulses ($\dot{\epsilon}_j$) and without current pulses ($\dot{\epsilon}_{j=0}$) of single and polycrystals of various pure metals, in function of the intensity of the current pulses (Conrad et al. 1990)

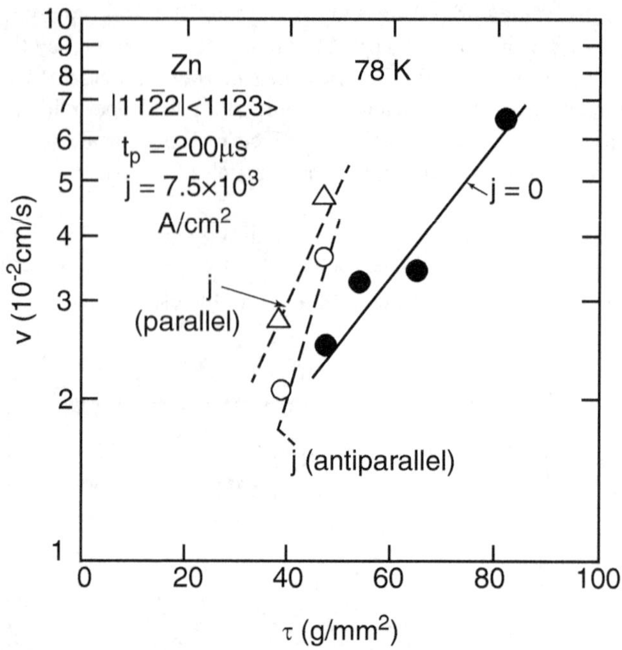

Fig. 5 Dislocation velocity in Zn at 78 K measured by etch-pits, under the influence of short current pulses ($J = 7.3 \times 10–3$ A/cm2, pulse duration = 200 μs), in function of an applied stress (τ) (Conrad et al. 1990)

(~10^5–10^6 A/cm^2) current pulses of short duration (~100 μs) on the strain rate of pure metals. It was verified that the current pulses did not increase the temperature by Joule effect by more than ~0.5 °C. It can be seen that these high-intensity current pulses can increase the deformation rate by orders of magnitude. Figure 5 shows an interesting result of dislocation velocity (v) measurements by the etch-pits technique in Zn at 78 K under the influence of short current pulses ($J = 7.3 \times 10^3$ A/cm^2, pulse duration = 200 μs), in function of an applied stress (τ). The current produced an increase in dislocation velocity for both the parallel and antiparallel directions (scalar effect), but larger for the first case, suggesting a vectorial contribution (electron wind effect). Again, the Joule heating generated by the current pulses was negligible ($\Delta T < 0.5$ °C). A more recent study showed that the elongation of a Zn-Al alloy increased by a factor of ~4 under current pulses, this being in relation with TEM observations of acceleration of dislocation motion (Zhu et al. 2009).

Because of these effects, processing of metals by electroplastic manufacturing has been developed, as reviewed by Guan et al. (2010). This paper reports on many studies showing that under high-intensity electric current pulses (10^3–10^5 A/cm^2, 100 μs), deformation resistance reduces markedly and, at the same time, plasticity increases significantly. For example, the elongation at room temperature of a Zn-Al alloy is reported to increase by a factor of more than 4 under electropulsing (1.6×10^3 A/cm^2, 2300 μs, 100 Hz).

However, the increase in dislocation velocity by electroplastic effect seems to be rather moderate (factor of 2 or so, Fig. 5), and this hardly accounts for the strong increase in deformation rates (Fig. 4). Thus, it appears that these electroplastic effects are still far from being well understood.

An influence of electric currents on other metallurgical mechanisms (precipitation, recrystallization, etc.) has also been reported (Conrad 2000). For example, current pulses (8×10^4 A/cm^2, 90 μs, 2 Hz) decreased the time to reach 50% of recrystallization by a factor of ~2. This was interpreted by the fact that current pulses of the order of 10^5 A/cm^2 cause about an order of magnitude increase in plastic strain rate and presumably in dislocation velocity (Conrad et al. 1988).

2.3 Conclusions

In summary, strong currents are reported to exhibit significant effects on diffusion and plasticity kinetics, leading to higher deformability of metals. Therefore, the current pulses of the SPS technique can potentially increase the kinetics of sintering, through their impact on the elementary diffusion and plasticity mechanisms involved in densification. However, to observe significant current effects on diffusion and on plasticity, currents of at least 10^2–10^3 A/cm^2 in the first case, and of 10^3–10^6 A/cm^2 in the second, seem necessary. These values are at the upper limit or well above the average current passing through powders in typical SPS experiment (~10^2 A/cm^2). Thus, in that case, the current effects in the SPS process would be very limited. However, the local current densities at the contact areas between powder particles could be much higher than this average value, due to narrowing of the conducting section close the interparticle contacts (see Sect. 5.2). Consequently, current effects on diffusion and plasticity can potentially occur at the first sintering stage, when the contact areas between the powder particles are very small. We will discuss these questions in the next sections.

3 Micromechanical Models of Densification Under Pressure

For powders under pressure, densification can occur either by mass transport from the volume to the surface of the powder particles or by deformation of the particles. In the second case, the particles can be deformed by dislocation mechanisms (glide, climb, twinning) or by mass transport *within* the volume of polycrystalline particles by volume (Nabarro-Herring creep) or grain boundary (Coble creep) diffusion. Roughly speaking, deformation is the dominant mechanism for “big” powder particles, that is, for particles larger than a few micrometers. For “small” particles of diameters below a few micrometers, the capillary forces become strong enough to exceed the mechanical pressure, and, in that case, densification occurs by mass transport. Micromechanical models have been derived to theoretically calculate the densification rates resulting from these mechanisms, and several reviews give the relevant expressions, for example, Arzt et al. (1983), Helle et al. (1985), Bernache-Assollant (1993), Davies and Jones (1997), Duszczyk et al. (1998), Atkinson and Davies (2000), and Rahaman (2003). However, in most practical cases, metallic powders processed by SPS are of sizes larger than about 10 μm, and the densification mechanism is deformation of the particles by the mechanical pressure and the deformation occurring either by dislocation mechanisms or by Nabarro-Herring or Coble creep. Tables 1 and 2 give the theoretical expression for densification in the two classical sintering stages: discrete particles touching at necks (stage 1, $D < 0.9$) and closure of porosity (stage 2, $D > 0.9$), D being the relative density (density ratio of porous and fully dense states. Table 3 gives the notations employed in Tables 1 and 2.

Table 1 Theoretical expressions for densification in stage 1 ($D < 0.9$), from Davies and Jones (1997)

Mechanism	Equation	
Plastic yield	$D = \left(\frac{(1-D_0)P}{1.3\sigma_y}\right)^{1/3}$	(2)
Power law creep	$\dot{D} = 5.3\left(D^2 D_0\right)^{1/3} \frac{1}{\sqrt{3}} \left(\frac{D-D_0}{1-D_0}\right)^{1/2} A \left(\frac{P_{1\mathrm{eff}}}{3}\right)^n$	(3)
Nabarro-Herring and Coble creep	$\dot{D} = \frac{14.3}{D} \left(\frac{1-D_0}{D-D_0}\right)^{1/2} \left(\frac{D_v}{G^2} + \frac{\pi \delta D_b}{G^3}\right) \frac{\Omega}{kT} P_{1\mathrm{eff}}$	(4)
Diffusion at the interparticle boundaries	$\dot{D} = 43\left(\frac{1-D_0}{D-D_0}\right) \frac{\delta D_b}{R^3} \frac{\Omega}{kT} P_{1\mathrm{eff}}$	(5)
Diffusion in the volume of the particles	$\dot{D} = 32(1-D_0) \frac{D_v}{R^2} \frac{\Omega}{kT} P_{1\mathrm{eff}}$	(6)
Effective pressure	$P_{1\mathrm{eff}} = \frac{P(1-D_0)}{D^2(D-D_0)} + \frac{3\gamma_{sv}}{R} D^2 \left(\frac{2D-D_0}{1-D_0}\right)$	(7)

Table 2 Theoretical expressions for densification in stage 2 ($D > 0.9$), from Davies and Jones (1997)

Mechanism	Equation	
Plastic yield	$D = 1 - \exp\left[\frac{-3\left(P - P_0 \frac{(1-D_c)D}{(1-D)D_c}\right)}{2\sigma_y}\right]$	(8)
Power law creep	$\dot{D} = \frac{3}{2} \cdot \frac{D(1-D)}{\left(1-(1-D)^{1/n}\right)^n} A \left(\frac{3P_{2\mathrm{eff}}}{2n}\right)^n$	(9)
Nabarro-Herring and Coble creep	$\dot{D} = 32(1-D)\left(\frac{D_v}{G^2} + \frac{\pi \delta D_b}{G^3}\right) \frac{\Omega}{kT} P_{2\mathrm{eff}}$	(10)
Diffusion at the interparticle boundaries	$\dot{D} = 4 \frac{\delta D_b}{R^3} \frac{\Omega}{kT} P_{2\mathrm{eff}}$	(11)
Diffusion in the volume of the particles	$\dot{D} = 3\left(\frac{1-D}{6D}\right)^{1/2} \frac{D_v}{R^2} \frac{\Omega}{kT} P_{2\mathrm{eff}}$	(12)
Effective pressure	$P_{2\mathrm{eff}} = P + \frac{2\gamma_{sv}}{R} \left(\frac{6D}{1-D}\right)^{1/3} - P_0 \frac{(1-D_c)D}{(1-D)D_c}$	(13)

Table 3 Symbols used in Tables 1 and 2

Symbol	Description
A	Creep constant in Norton law: $\dot{\varepsilon} = A\sigma^n$, $A = A_0 \exp\left(-\frac{Q}{RT}\right)$
A_0	Constant
D	Relative density
$\dot{D}$	Densification rate (=dD/dt)
D_0	Initial relative density
D_c	Relative density at which pores close
D_v	Volume diffusion coefficient
δD_v	Grain boundary thickness × grain boundary diffusion coefficient
$\dot{\varepsilon}$	Deformation rate
G	Grain size
k	Boltzmann's constant
n	Stress exponent
P	External pressure
P_0	Outgassing pressure
P_{1eff}, P_{2eff}	Effective pressure on necks during stage 1 or 2
Q	Activation energy
R	Powder particle radius
T	Temperature
Ω	Molecular volume
γ_{sv}	Specific surface energy
σ	Stress
σ_y	Yield stress

To conveniently analyze the densification curves, a rearrangement of Eqs. (2–7) for stage 1 has been proposed (Bernard-Granger and Guizard 2007):

$$\frac{\dot{D}}{D} = \mu_{eff} \cdot K \frac{\exp\left(\frac{-Q}{RT}\right)}{T} \left(\frac{b}{G}\right)^p \left(\frac{P_{1eff}}{\mu_{eff}}\right)^n \tag{14}$$

with

$$\mu_{eff} = \frac{E}{2(1+\nu)} \cdot \frac{D - D_0}{1 - D_0} \tag{15}$$

Q being the apparent activation energy of the mechanism controlling densification, p the grain size exponent, b the Burgers' vector, ν the Poisson's coefficient, E the elastic modulus, and R the gas constant. Determination of stress exponent n and of activation energy Q is then possible by transformation of Eq. (14) as follows:

$$\ln\left[\frac{1}{\mu_{eff}} \frac{\dot{D}}{D}\right] = n \ln\left[\frac{P_{1eff}}{\mu_{eff}}\right] + K_1 \tag{16}$$

$$\ln\left[\frac{T}{\mu_{\text{eff}}}\left(\frac{\mu_{\text{eff}}}{P_{\text{1eff}}}\right)^{n}\frac{\dot{D}}{D}\right] = -\frac{Q}{RT} + K_2 \quad (17)$$

The slope of the curves in suitable coordinates gives thus n and Q.

However, for implementing the densification behavior of powders for elaborating complex shapes by SPS, the trend seems now to use theories based on continuum mechanics of compressible bodies, as reviewed by Olevsky (1998). As this subject is beyond the scope of the present review, the readers are invited to refer to the above paper.

4 Densification by SPS: Cycles and Kinetics

We indicate in this section typical SPS conditions employed for metallic systems and the resulting densification cycles. Then, we report on studies analyzing the densification kinetics by the mechanical behavior of the powder at high temperature.

4.1 Typical SPS Conditions

Typical SPS densification cycles of metals are as follows. The powder is enclosed into molds constituted of an electrically conducting material, in general graphite. The diameter of the molds is generally of a few centimeters in diameter, but dedicated SPS machines allow densifying samples of diameter up to 50 cm. A graphite foil is introduced between the powder and the mold walls to insure lubrication and sealing. At the beginning of the SPS cycle, the chamber is evacuated to a vacuum of $\approx$10 Pa. The densification cycle begins ($t = 0$) when the evacuation procedure is completed. A uniaxial pressure of 10–100 MPa is applied rapidly (1 min), unless outgassing procedure under low applied loads is necessary. Simultaneously, the samples are heated up by increasing the intensity of the pulsed current applied to the graphite molds. If a pyrometer is employed for measuring the temperature, a special procedure is used for heating the samples below the minimum temperature that the pyrometer is able to detect. Then, temperature is increased at a rate of typically 100 °C/min, sometimes more slowly for large specimen, rarely more rapidly. When temperature approaches the desired final value, the heating rate is frequently reduced, to avoid overtaking of this value (overshoot). Finally, the temperature is kept constant for a few minutes, sometimes for longer durations (up to 1 h) if metallurgical transformations are necessary. At the end of the cycle, the heating current and the uniaxial pressure are released, and the samples cool down in the vacuum of the SPS chamber, at 100–500 °C/min depending on temperature and on mold size. Most of the time, the total duration of the SPS cycle is ~30 min. Figure 6 shows an example of SPS cycle for TiAl powder (Trzaska et al. 2016).

The heating current is measured by a sensor which gives the total intensity (in amperes). Because metallic powders exhibit most of the time a very high electric conductivity, a fraction of this current goes through the sample, the other fraction passing through the surrounding graphite mold. Experimental measurements of these fractions have been performed by Misawa et al. (2009). They show that, with Cu powder, about 30% of the total current goes through the samples and then 70% through the molds (with 20 mm molds). Using this estimate, Fig. 6 shows that typical order of magnitude of currents passing through samples for electrically conducting metallic powders (here, TiAl) is ~100 A/cm^2. Concerning the electric tension, its value is low, ~2–5 V.

Accurate determination of sample temperature is an important issue in the SPS experiments, because of the inhomogeneity of the current density in the samples and in the molds, which gives rise to strong temperature gradients (Munir et al. 2006). Temperature is measured by optical pyrometers or by thermocouple. The inconvenient of pyrometers is that they work only from a minimum temperature, which can be elevated (e.g., ~600 °C). This can be not suitable for some metals, and, in that case, thermocouple are employed. However, the latter have the drawback to be very sensitive to thermal contact resistances, which can lead to underestimation of temperature and scattering of the measurements of ~50 °C. Moreover, due to the temperature gradients in the molds and in the samples, and because the temperature is measured at the surface of the graphite molds, or inside them, there is a difference between the measured value and the sample temperature. There can also be significant temperature gradients within the samples (Voisin et al. 2013). To minimize these effects, modern SPS systems allow temperature to be measured very close to

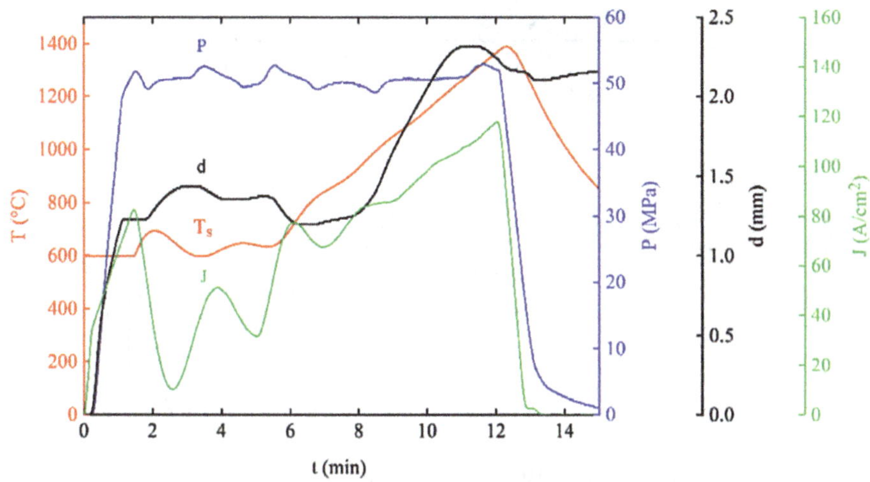

Fig. 6 Evolution of pressure (*P*), sample temperature (T_s), current density (*J*, supposing that 30% of the current of the SPS go through the sample (Misawa et al. 2009)), and displacement of the lower punch (*d*, positive when the punch goes up), during densification of a TiAl powder in a 20 mm graphite matrix (Trzaska et al. 2016)

the sample by axial pyrometers. Nevertheless, temperature differences still remain, and calibration procedure based on melting of metals (e.g., Al, Cu (Trzaska et al. 2017)) or on solid-state phase transitions (e.g., TiAl a transition (Voisin et al. 2013)) can be necessary.

4.2 Densification Curves of Ductile and Brittle Metallic Systems

Figure 7 gives the evolution of the relative density of powders of metallic systems exhibiting emblematic mechanical behavior: brittle (TiAl) and ductile (Ni) (Trzaska et al. 2018). The striking feature is that in the two cases, densification occurs within minutes, that is, quite rapidly. In comparison, typical hot isostatic pressing (HIP) cycles for TiAl last several hours (Choi et al. 1990). However, the curves are also quite different. For TiAl, densification begins at ~900 °C and finishes at ~1250 °C; the entire process thus takes place in a temperature range of ~350 °C. Conversely, the densification of Ni extends over a much wider temperature range, beginning near room temperature and ending above 1300 °C (full density could not be reached before melting of the sample). This difference results mainly on the temperature dependence of the deformation rate of the powder particles, that is, on the activation energy. Thus, the lower activation energy of Ni powder compared to TiAl (164 ± 30 kJ/mol (Trzaska et al. 2018) and

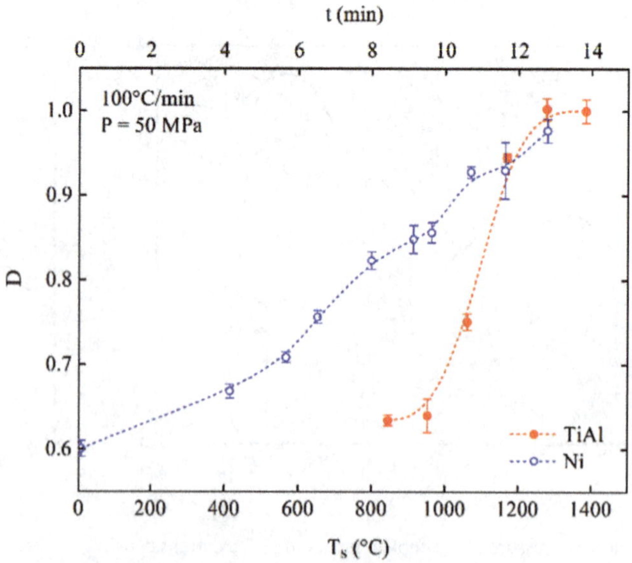

Fig. 7 Relative density as a function of time (t) and of sample temperature (T_S) of TiAl and Ni processed by SPS using the conditions shown in Fig. 6 (Trzaska et al. 2018)

308 ± 20 kJ/mol (Trzaska et al. 2017), respectively) accounts for the wider temperature range over which densification of Ni occurs as compared to TiAl.

4.3 Determination of Densification Curve from Bulk Constitutive Relations

The SPS densification curves have been theoretically determined using the constitutive relation of the bulk material:

$$\dot{\varepsilon} = A\sigma^n \tag{18}$$

with

$$A = A_0 \exp\left(-\frac{Q}{RT}\right) \tag{19}$$

These relations have been used in micromechanical models (Eqs. (3) and (9)) to calculate the densification curve of TiAl (Trzaska et al. 2017) or in continuum mechanics models to calculate the densification curves of ZrC (Wei et al. 2015) and of TiAl (Martins et al. 2017). Figure 8 shows that the densification curve of TiAl powder, for example, can be quite accurately reproduced using the bulk parameters. The difficulty is to select the suitable A_0, n, and Q values, which can vary a lot for a given material, even if the microstructures are similar. For near γ TiAl alloy of similar microstructures (grain size, fraction of residual α_2 phase, etc.), the deformation rate $\dot{\varepsilon}$ can vary over one order of magnitude at least for given conditions (temperature, applied stress) depending on the authors (Trzaska et al. 2017). This impacts severely the calculated densification curves. Finally, the order of magnitude

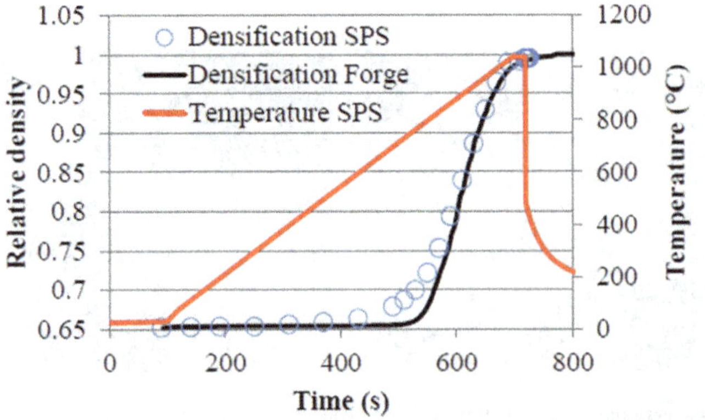

Fig. 8 Experimental SPS densification curve of a TiAl alloy, compared with calculated densification kinetics (using Forge software) using constitutive relation of bulk TiAl (parameters: $Q = 414$ kJ/mol; $n = 2.81$; $A_0 = 3.14 \times 10^{-8}$ Pa^{-n}s^{-1}) (Martins et al. 2017)

of typical deformation rate ($\dot{\varepsilon}$) of the powder particles in SPS experiments is 10^{-3} s^{-1} (Trzaska et al. 2016), which gives bulk deformation (ε) of the order of 30% at full densification.

4.4 Effect of Current on Densification Rate

Using dedicated sinter-forging devices, allowing densification in absence or in presence of SPS current (0 or ~1000 A/cm^2) passing directly through highly conducting (Cu) powders, it was reported that the final density was significantly higher with current (Fig. 9). This was attributed to electroplasticity effects (Aleksandrova et al. 2015). On the contrary, with Cu powder processed by SPS (~200 A/cm^2 passing through the powder by use of alumina dies) and by HP (0 A/cm^2), it was found that the densification cycle was essentially unaffected by the current (Collet et al. 2017). The same observation was made in SPS (~100 A/cm^2) and HP (0 A/cm^2) experiments with a TiAl powder (Trzaska et al. 2017). The effect of the current on densification kinetics remains then controversial. However, it seems that increase in densification kinetics are observed only for high current densities (~1000 A/cm^2 or so), that is, close to those for which electroplastic effects are observed (see Sect. 2.2).

4.5 Conclusions

These results show that, under high currents (~1000 A/cm^2), current-induced acceleration of densification is observed. With typical SPS currents (~100 A/cm^2), the densification rate is identical than without current and can be deduced from the high-temperature mechanical behavior of the bulk material, meaning that densification is unaffected by the current.

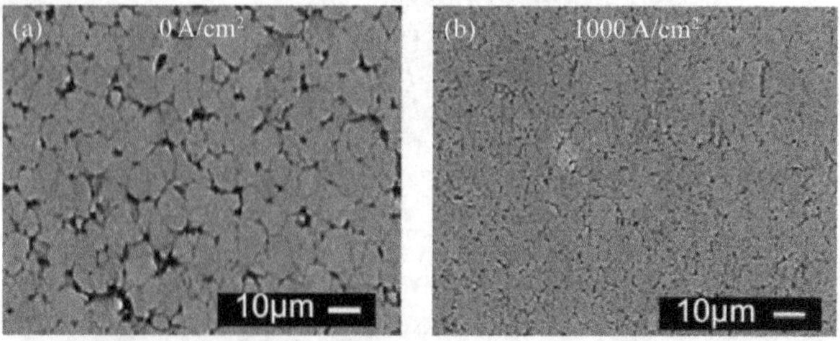

Fig. 9 Cross-sectional microstructure of Cu powder of 10 μm in size densified without current (**a**) and with a current of ~1000 A/cm^2 (**b**). Conditions: max. temperature, 575 °C; heating rate, 50 °C/min; pressure, 10 MPa

5 SPS Microscopic Mechanisms

Because metallic powder particles are crossed by electric currents during densification by SPS, many modeling and experimental works have been carried out to investigate possible original electric phenomena, which can be classified as follows: (i) dielectric breakdown of the thin oxide layer at the surface of the powder particles due to electric fields exceeding the dielectric strength of the oxide, (ii) arcs and plasma due to ionization of the residual gas in the gaps between the powder particles, and (iii) increase in current density due to striction of the current lines at the small contact areas between the powder particles. Hypotheses (i) to (iii) were believed to generate local overheating in the vicinity of the necks between the powder particles, leading to local softening of the material, to easier welding, and to faster densification. Finally, a last class of phenomena was (iv) activation of electromigration and/or electroplastic effects due to the strong current densities at the contact areas between the powder particles. These effects were supposed to modify the elementary plasticity and diffusion phenomena or their kinetics and consequently to a modification of the densification rate. Simulation works mainly focused on point (iii), whereas most of the other phenomena have been investigated experimentally.

5.1 *Local Overheating Generated by Arcs, Plasma, and Dielectric Breakdown*

Concerning the local generation of heat by arcs, plasma, and dielectric breakdown, a complex scenario involving these phenomena was first proposed in (Tokita 1999). Then, experimental works on electric breakdown of oxide layers at the surface of the powder particles were reported by several authors (Omori 2000; Okazaki 2000; Groza and Zavaliangos 2000). These phenomena were believed to induce fast grain growth (Ye et al. 2006). Yanagisawa et al. (2003) experimentally observed electric arcs between Cu powder particles 550 μm in diameter subjected to macroscopic high-intensity current pulses (1700–17,000 A/cm^2) of 500 ms duration, in low-load (10 MPa) dedicated experiments (Fig. 10). These arcs provoked local melting at the necks between the powder particles. Saunders et al. (Saunders et al. 2015) have shown plasma formation in W particles 0.5 μm in size for low applied tension (<50 V) and low applied pressure in atomic emission spectroscopy (AES) experiments. However, an opposite conclusion has been obtained by Hulbert et al. (2009), using also AES for investigating plasma formation with metallic and ceramic powders. Other studies on SEM observations of neck regions between Cu powder particles showed fine grains in these regions, which were interpreted as local melting followed by rapid solidification (Song et al. 2006). However, fine grains in the neck region can also result from recrystallization following intense plastic deformation by mutual indentation of the particles (Trzaska et al. 2016).

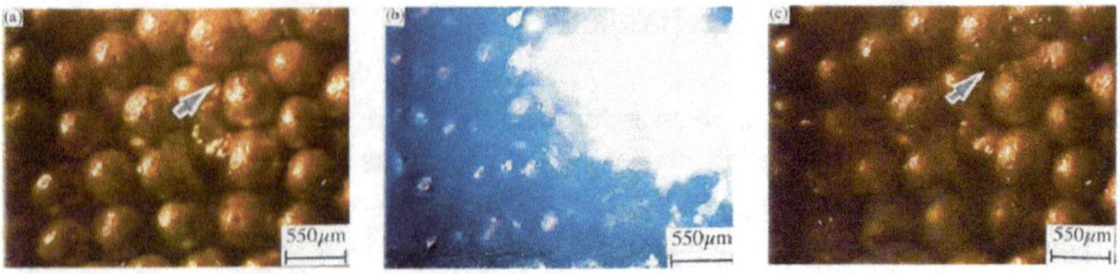

Fig. 10 Image sequence (time interval between each image, 0.02 s) of an electric arc between Cu particles 550 μm in diameter submitted to a single current pulse of 6700 A/cm^2 and of 100 ms. Note the welding of the particles (image (**c**), arrow) resulting from local melting (Yanagisawa et al. 2003)

5.2 Local Overheating Generated by Strong Current Densities at the Necks

This point is treated independently of the preceding, because many modeling studies have been performed around these ideas, which were first proposed in 1944 by Cramer (Orru et al. 2009). Due to the striction of the current lines at the small contact areas between the powder particles, very elevated heat generation by Joule effect (reaching 10,000 °C) has been analytically calculated by several authors (Yanagisawa et al. 2003; Song et al. 2006; Diouf et al. 2013; Xiong et al. 2012). However, early works (Yanagisawa et al. 2003; Song et al. 2006) did not take into account heat conduction within the particles. More recent works have been conducted taking into account heat conduction, for millimetric or centimetric particles, and also calculated elevated temperatures at the particle contacts (Diouf et al. 2013; Xiong et al. 2012). Trapp and Kieback (2015) evaluated by experiments and by simulation a Joule overheating of 0.05–5 °C at the contacts between Cu and steel spheres of 1 mm in diameter in capacitor discharge experiments and concluded that, in normal operating conditions, essentially no temperature difference between contact and center of the particles should occur during SPS experiments. As can be seen, these studies predict temperature elevations (large or small) at the necks, but they concern particle sizes of millimeters to centimeters, i.e., much larger than the typical particle sizes used in SPS experiments (around 10–100 μm). However, a recent modeling study (Collard et al. 2017) showed that, despite current densities in the neck regions ~100 times higher than the average current, very low-temperature elevations would occur (<1 °C). This was explained by the very fast heat diffusion at the scale of metallic powder particles 100 μm in size, which lead to temperature homogenization by heat diffusion in only milliseconds (Fig. 11).

To experimentally investigate the hypothesis of local overheating, microstructure characterizations have been performed. For example, local melting in FeAl powder (Fig. 12) (Ji et al. 2007) and local devitrification in $Zr_{57}Cu_{20}Al_{10}Ni_8Ti_5$ metallic glass powder (Nowak et al. 2011) have been reported.

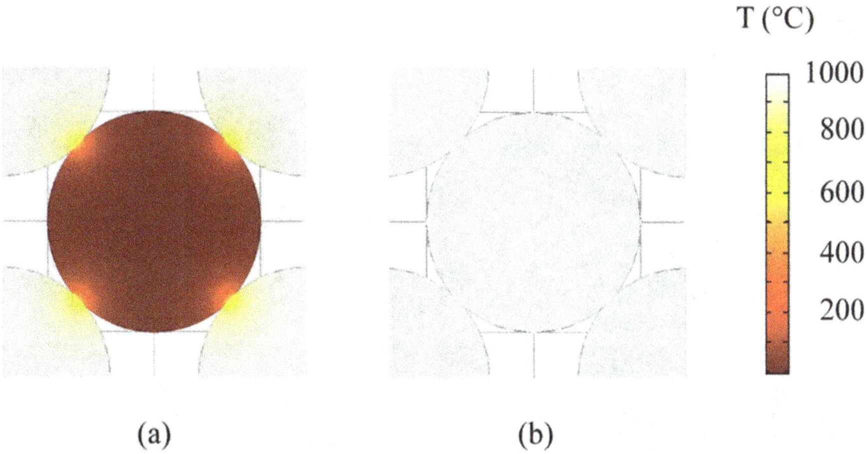

Fig. 11 FEM temperature maps in metallic powder particles 100 μm in diameter heated by hot spots ($T \approx 1000$ °C) at their periphery, showing temperature homogenization in the particles within milliseconds. (**a**) Temperature map at $t = 0$. (**b**) Temperature map at $t = 4 \times 10^{-3}$ s (Collard et al. 2017)

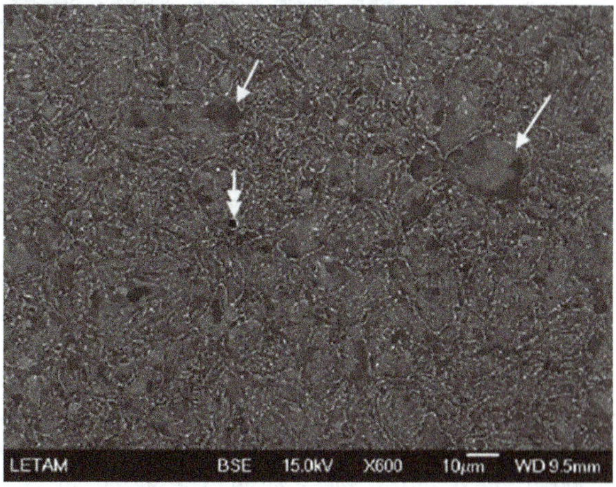

Fig. 12 SEM image of fairly large FeAl grains (single arrows) which formation is believed to result from melting due to local overheating (Ji et al. 2007)

5.3 *Activation of Electroplasticity and Electromigration*

Another consequence of strong current densities at the necks between the powder particles is the potential activation of electroplastic and electromigration phenomena, which would accelerate densification. Values of 30–50 × 10^4 A/cm^2 can be reached at the necks between particles at early sintering stages according to modeling (Frei et al. 2007; Collard et al. 2017), an order of magnitude which is close to that

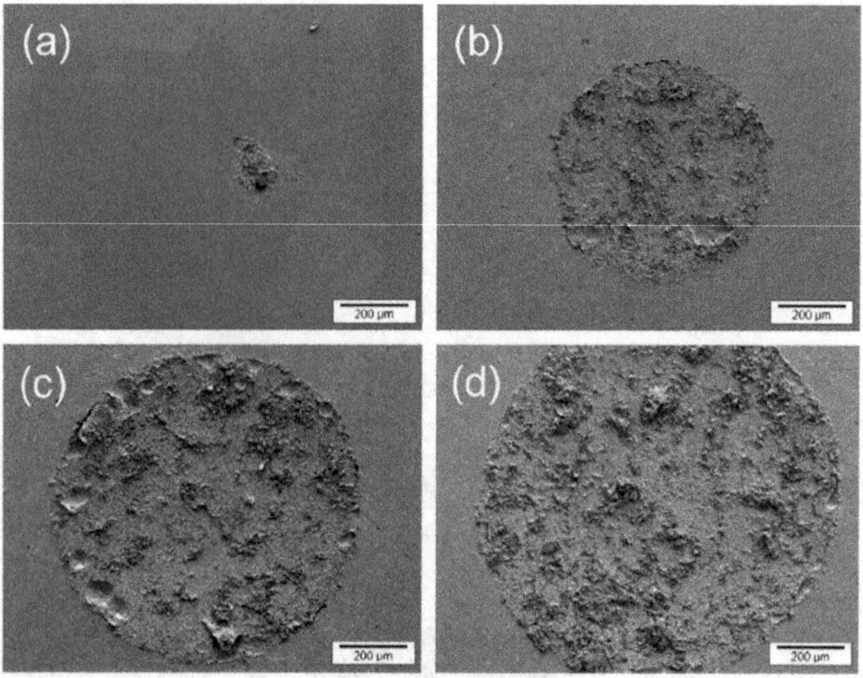

Fig. 13 Imprints of necks between Cu spheres and plates annealed 900 °C for 60 min, with average current density of (**a**) 0 A/cm^2, (**b**) 350 A/cm^2, (**c**) 420 A/cm^2, and (**d**) 520 A/cm^2

necessary for electroplasticity (10^3–10^6 A/cm^2) and even above that for which electromigration is reported (10^2–10^3 A/cm^2) (see Sect. 2).

Thus, Frei et al. (2007) report model experiments by SPS of Cu spheres 3 mm in diameter in contact with Cu plates, to simulate the contacts between powder particles. Observation of enhanced neck growth was attributed to electromigration and evaporation-condensation phenomena activated by average currents of 350–520 A/cm^2 passing through the Cu spheres but which could be much higher at the contact zones due to striction of the current lines (Fig. 13). Moreover, to investigate the mechanisms occurring under strong current densities at the necks of metallic particles, microstructure characterizations at fine scale (by EBSD and TEM) have been performed, involving notably careful procedure of TEM thin foil lift-out at the necks by FIB (Fig. 14; Trzaska et al. 2016, 2018). Thus, in Ni, the neck regions exhibited intense plastic deformation by mutual indentation of the powder particles, leading to formation of cell boundaries (Fig. 14a and c). Analysis of elementary microscopic mechanisms in TiAl showed screw dislocations, loops, and probably helices, indicating glide and climb mechanisms (Fig. 14d). In both cases, the proposed elementary mechanisms controlling the dislocation processes were diffusion in volume for TiAl and in dislocations for Ni. This plastic deformation was followed by recovery

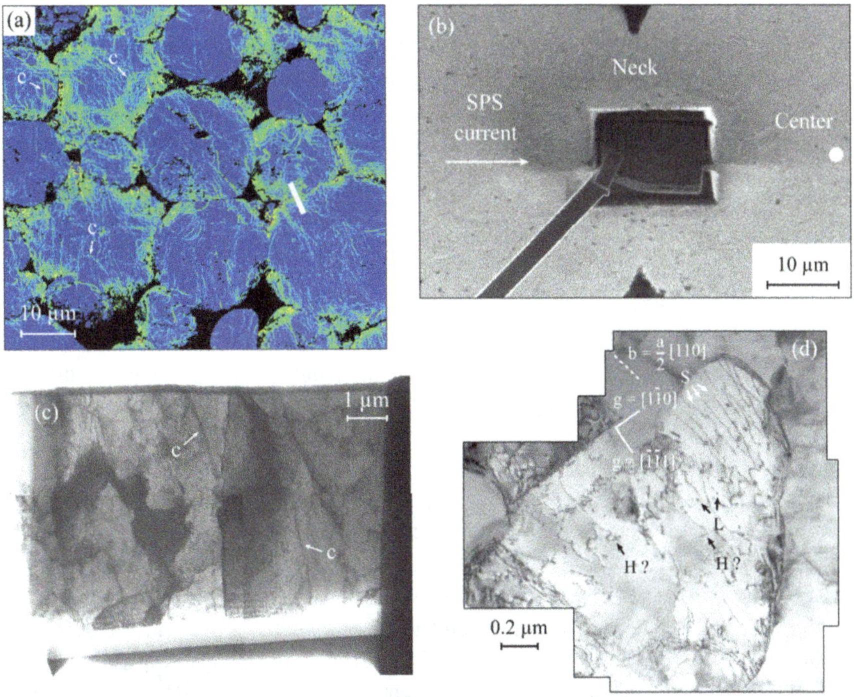

Fig. 14 (**a**) EBSD local misorientation map of Ni powder during densification by SPS (512 °C, D = 0.7). Typical location of FIB extraction is indicated by a white segment. Deformation cells are labeled as c. (**b**) Extraction of a TEM thin foil by FIB in a neck region. (**c**) TEM image of an entire FIB thin foil extracted at the junction between two Ni particles in a sample densified at 727 °C (D ≈ 0.75). Dislocation walls are labeled as c. (**c**) TEM bright-field image of a TiAl γ grain in a neck region (densification interrupted at 1040 °C, D = 0.7). Screw dislocations, helices, and loops are indicated by letters S, H, and L, respectively (Trzaska et al. 2016, 2018)

and recrystallization in TiAl. These mechanisms appeared classical, without any obvious modification induced by the strong current at the necks.

5.4 Conclusions

In summary, the existence of significant heat generation by spark and plasma, as well as by striction of the current lines at the necks between powder particles, is still controversial. However, the recent theoretical works tend to demonstrate that overheating by striction of the current lines is probably low. Arcing phenomena by ionization of the residual gas between the powder particles, or by dielectric breakdown of the superficial oxide layers, seem to occur in particular conditions (low load ~10 MPa, current pulses ~10,000 A/cm^2), not really representative of typical SPS conditions. However, increase in neck growth rate by electrically enhanced mass transport was observed for lower average current densities (350–520 A/cm^2), close to typical currents of SPS experiments. At the microscopic scale, the elementary mechanisms which were observed in the neck regions were dislocation (glide, climb) and recovery-recrystallization mechanisms, that is, classical high-temperature phenomena. Thus, the SPS currents seem not to activate new mechanism but could modify their kinetics.

6 Conclusions

Many of the microscopic phenomena described in literature to account for the high SPS kinetics remain controversial, but some conclusions can be drawn. It seems then established that the densification rate of metallic powders can be deduced directly from the high-temperature mechanical behavior (power-law creep) of the bulk material. The rapidity of the SPS process is then a consequence of the high deformation rate of the powder particles ($\dot{\varepsilon}^{\sim}10^{-3}$ s^{-1}). Concerning the role of the current, typical SPS average currents are ~100 A/cm^2. At the early sintering stages, the contact areas between powder particles are very small. This could give rise to strong striction of the current lines, and local current densities could reach much higher values (10^4–10^5 A/cm^2). Thus, activation of arcing, of electroplasticity, or of electromigration phenomena (triggered by current densities of ~10^4 A/cm^2, of 10^3–10^6 A/cm^2, and of 10^2–10^3 A/cm^2, respectively) is in principle possible. But, because most of metallic systems exhibit significant plasticity at high temperature, the contact zones between particles are quickly deformed under the action of the applied pressure. This drastically reduces the phenomenon of striction of the current lines, and the local current density comes back close to the average value (~100 A/cm^2). Thus, electric phenomena can potentially play a short transitory role in the very early stages of densification, but then most of the densification probably occurs without influence of the electric current and involves classical microscopic mechanisms.

Acknowledgments This work is partially supported by the MF2-ANR-2011-PBS09-020 project of the French "Agence Nationale de la Recherche." The authors also gratefully acknowledge the helpful comments and suggestions of the reviewers, which have improved the presentation.

References

Adda Y, Philibert J (1966) La diffusion dans les solides. Institut national des sciences et techniques nucléaires, Saclay

Aleksandrova EV, Ilyina AM, Grigoryev EG, Olevsky EA (2015) Contribution of electric current into densification kinetics during spark plasma sintering of conductive powder. J Am Ceram Soc 98:3509–3517

Arzt E, Ashby MF, Easterling KE (1983) Practical applications of hot-isostatic pressing diagrams: four case studies. Metall Trans A 14:211–221

Atkinson HV, Davies S (2000) Fundamental aspects of hot-isostatic pressing: an overview. Metall Mater Trans A 31:2981–3000

Bernache-Assollant D (1993) Chimie-physique du frittage. Hermès, Paris

Bernard-Granger G, Guizard C (2007) Spark plasma sintering of a commercially available granulated zirconia powder: I. Sintering path and hypotheses about the mechanism(s) controlling densification. Acta Mater 55:3493–3504

Bertolino N, Garay J, Anselmi-Tamburini U, Munir ZA (2001) Electromigration effects in Al-Au multisayers. Scr Mater 44:737–742

Chen CM, Chen SW (1999) Electric current effects on Sn/Ag interfacial reactions. J Electron Mater 28:902–906

Chen CM, Chen SW (2000) Electromigration effect upon the Zn/Ni and Bi/Ni interfacial reactions. J Electron Mater 29:1222–1228

Chen SW, Chen CM, Liu WC (1998) Electric current effects upon the Sn/Cu and Sn/Ni interfacial reactions. J Electron Mater 27:1193–1199

Choi BW, Deng YG, McCullough C, Paden B, Mehrabian R (1990) Densification of rapidly solidified titanium aluminied powders – I. Comparison of experiments to hiping models. Acta Metall Mater 38:2225–2243

Collard C, Trzaska Z, Durand L, Chaix JM, Monchoux JP (2017) Theoretical and experimental investigations of local overheating at particle contacts in spark plasma sintering. Powder Technol 321:458–470

Collet R, Le Gallet S, Naimi F, Charlot F, Lay S, Bonnefont G, Fantozzi G, Chaix JM, Bernard F (2017) Current effect on the sintering by SPS of a pre-oxidized copper powder. J Alloys Compd 692:478–484

Conrad H (2000) Effects of electric current on solid state phase transformations in metals. Mater Sci Eng A 287:227–237

Conrad H, Karam N, Mannan S, Sprecher AF (1988) Effect of electric current pulses on the recrystallization kinetics of copper. Scr Metall 22:235–238

Conrad H, Sprecher AF, Cao WD, Lu XP (1990) Electroplasticity – the effect of electricity on the mechanical properties of metals. JOM 42:28–33

Davies GC, Jones DRH (1997) Creep of metal-type organic compounds – IV. Application to hot isostatic pressing. Acta Mater 45:775–789

Diouf JS, Fedrizzi A, Molinari A (2013) A fractographic and microstructural analysis of the neck regions of coarse copper particles consolidated by SPS. Powder Technol 221:220–227

Duszczyk J, Zhuang LZ, Buekenhout L (1998) Densification of a rapidly solidified nickel aluminide powder – I. Application of hot-isostatic pressing diagrams. J Mater Sci 33:2735–2743

Frankovic R (1996) Electromigration drift and threshold in Cu thin-film interconnect. IEEE Trans Electron Devices 43:2233–2239

Frei JM, Anselmi-Tamburini U, Munir ZA (2007) Current effects on neck growth in the sintering of copper spheres to copper plates by the pulsed electric current method. J Appl Phys 101:114914

Friedman JR, Garay JE, Anselmi-Tamburini U, Munir ZA (2004) Modified interfacial reactions in the Ag-Zn multilayers under the influence of high DC currents. Intermetallics 12:589–597

Garay JE, Anselmi-Tamburini U, Munir ZA (2003) Enhanced growth of intermetallic phases in the Ni-Ti system by current effects. Acta Mater 51:4487–4495

Groza JR, Zavaliangos A (2000) Sintering activation by external electrical field. Mater Sci Eng A 287:171–177

Guan L, Tang G, Chu PK (2010) Recent advances and challenges in electroplastic manufacturing processing of metals. J Mater Res 25:1215–1224

Helle AS, Easterling KE, Ashby MF (1985) Hot-isostatic pressing diagrams: new developments. Acta Metall 33:2163–2174

Hulbert DM, Anders A, Andersson J, Lavernia EJ, Mukherjee AK (2009) A discussion on the absence of plasma in spark plasma sintering. Scr Mater 60:835–838

Ji G, Grosdidier T, Bozzolo N, Launois S (2007) The mechanisms of microstructure formation in a nanostructured oxide dispersion strengthened FeAl alloy obtained by spark plasma sintering. Intermetallics 15:108–118

Liu WC, Chen SW, Chen CM (1998) Finite-element analysis of current-induced thermal stress in a conducting sphere. J Electron Mater 27:L5

Martins D, Grumbach F, Manière C, Sallot P, Mocellin K, Bellet M, Estournès C (2017) In-situ creep law determination for modeling spark plasma sintering of TiAl 48-2-2 powder. Intermetallics 86:147–155

Misawa T, Shikatani N, Kawakami Y, Enjoji T, Ohtsu Y, Fujita H (2009) Observation of internal pulsed current flow through the ZnO specimen in the spark plasma sintering method. J Mater Sci 44:1641–1651

Munir ZA, Anselmi-Tamburini U, Ohyanagi M (2006) The effect of electric field and pressure on the synthesis and consolidation of materials: a review of the spark plasma sintering method. J Mater Sci 41:763–777

Nowak S, Perrière L, Dembinski L, Tusseau-Nenez S, Champion Y (2011) Approach of the spark plasma sintering mechanism in Zr57Cu20Al10Ni8Ti5 metallic glass. J Alloys Compd 509:1011–1019

Okazaki K (2000) Electro-discharge consolidation applied to nanocrystalline and RSP/MA powders. Mater Sci Eng A 287:189–197

Olevsky EA (1998) Theory of sintering: from discrete to continuum. Mater Sci Eng R 23:41–100

Omori M (2000) Sintering, consolidation, reaction and crystal growth by the spark plasma system (SPS). Mater Sci Eng A 287:183–188

Orru R, Licheri R, Locci AM, Cincotti A, Cao G (2009) Consolidation/synthesis of materials by electric current activated/assisted sintering. Mater Sci Eng R 63:127–287

Rahaman MN (2003) Ceramic processing and sintering. Taylor and Francis, Boca Raton

Saunders T, Grasso S, Reece MJ (2015) Plasma formation during electric discharge (50 V) through conductive powder compacts. J Eur Ceram Soc 35:871–877

Shewmon P (1989) Diffusion in solids. The Minerals, Metals and Materials Society, Warrendale

Song X, Liu X, Zhang J (2006) Neck formation and self-adjusting mechanism of neck growth of conducting powders in spark plasma sintering. J Am Ceram Soc 89:494–500

Sprecher AF, Mannan SL, Conrad H (1986) On the mechanisms for the electroplastic effect in metals. Acta Metall 34:1145–1162

Tokita M (1999) Development of large-size ceramic/metal bulk FGM fabricated by spark plasma sintering. Mater Sci Forum 308–311:83–88

Trapp J, Kieback B (2015) Temperature distribution in metallic powder particles during initial stage of field-activated sintering. J Am Ceram Soc 98:3547–3552

Trzaska Z, Monchoux JP (2015) Electromigration experiments by spark plasma sintering in the silver-zinc system. J Alloys Compd 635:142–149

Trzaska Z, Couret A, Monchoux JP (2016) Spark plasma sintering mechanisms at the necks between TiAl powder particles. Acta Mater 118:100–108

Trzaska Z, Bonnefont G, Fantozzi G, Monchoux JP (2017) Comparison of densification kinetics of a TiAl powder by spark plasma sintering and hot pressing. Acta Mater 135:1–13

Trzaska Z, Cours R, Monchoux JP (2018) Densification of Ni and TiAl by SPS: kinetics and microscopic mechanisms. Metall Mater Trans A. https://doi.org/10.1007/s11661-018-4775-0. in press

Voisin T, Durand L, Karnatak N, Le Gallet S, Thomas M, Le Berre Y, Castagné JF, Couret A (2013) Temperature control during Spark Plasma Sintering and application to up-scaling and complex shaping. J Mater Process Technol 213:269–278

Wang CH, Kuo CY, Chen HH, Chen SW (2011) Effect of current density and temperature on Sn/Ni interfacial reactions under current stressing. Intermetallics 19:75–80

Wei X, Back C, Izhvanov O, Khasanov OL, Haines CD, Olevsky EA (2015) Spark plasma sintering of commercial zirconium carbide powders: densification behavior and mechanical properties. Materials 8:6043–6061

Xiong Y, Liu D, Li Y, Zheng B, Haines C, Paras J, Martin D, Kapoor D, Lavernia EJ, Schoenung JM (2012) Spark plasma sintering of cryomilled nanocrystalline Al alloy – Part I: microstructure evolution. Metall Mater Trans A 43:327–339

Yanagisawa O, Kuramoto H, Matsugi K, Komatsu M (2003) Observation of particle behavior in copper powder compact during pulsed electric discharge. Mater Sci Eng A 350:184–189

Ye J, Ajdelsztajn L, Schoenung J (2006) Bulk nanocrystalline aluminum 5083 alloy fabricated by a novel technique: cryomilling and spark plasma sintering. Metall Mater Trans A 37:2569–2579

Zhao J, Garay JE, Anselmi-Tamburini U, Munir ZA (2007) Directional electromigration-enhanced interdiffusion in the Cu-Ni system. J Appl Phys 102(2007):114902

Zhou Y, Wang Q, Sun DL, Han XL (2011) Co-effect of heat and direct current on growth of intermetallic layers at the interface of Ti-Ni diffusion couples. J Alloys Compd 509:1201–1205

Zhu YH, To S, Lee WB, Liu XM, Jiang YB, Tang GY (2009) Effects of dynamic electropulsing on microstructure and elongation of a Zn–Al alloy. Mater Sci Eng A 501:125–132

Part II
SPS of Nanocomposites

Surface-Modified-CNTs/Al Matrix Nanocomposites Produced via Spark Plasma Sintering: Microstructures, Properties, and Formation Mechanism

Farhad Saba, Faming Zhang, Seyed Abdolkarim Sajjadi, and Mohsen Haddad-Sabzevar

1 Introduction

Carbon is the most important element in the periodic table. It can form one-, two-, and three-dimensionally bond structured materials providing a wide range of applications, due to various bond structures such as sp3, sp2, sp hybrids, and multiple pπ–pπ bonds (Morisada and Miyamoto 2006). Carbon materials such as graphite, diamond, activated carbons, carbon fibers, and C–C composites have been extensively investigated and used for many years. Discovery of carbon nanotubes (CNTs) has been widely attributed to Iijima in 1991 (Iijima 1991). However, this finding has been debated because several other researchers had reported carbon structures similar to those reported by Iijima. Most significant researches, which were done before Iijima, are the filamentous tubes synthesized by Radushkevich et al. (Radushkevich and Lukyanovich 1952), Bacon (1960), and Oberlin et al. (1976). Discovery of CNTs has excited interest in producing advanced nanomaterials such as nanotubes, nanorod, nanowhiskers, and so on for many applications, because of their remarkable properties (Taguchi et al. 2005; Son et al. 2017; Li et al. 2016; Deng et al. 2007; Singu et al. 2017; Liu et al. 2014; Phuong et al. 2014; Annu et al. 2017). Because of the extraordinary mechanical, thermal, and electrical properties, CNTs have been used to improve the properties of many different materials such as polymers (Lahiri et al. 2012; Tjong 2006), metals (Nam et al. 2012; Morsi et al. 2010; Yoo et al. 2013a, b), and ceramics (Wang et al. 2004; Mei et al. 2015). Some difficulties arise from distinct differences

F. Saba · S. A. Sajjadi · M. Haddad-Sabzevar
Department of Materials Science and Engineering, Engineering Faculty, Ferdowsi University of Mashhad, Mashhad, Iran

F. Zhang (✉)
Jiangsu Key Lab for Advanced Metallic Materials, School of Materials Science and Engineering, Southeast University, Nanjing, China
e-mail: fmzhang@seu.edu.cn

© Springer Nature Switzerland AG 2019

P. Cavaliere (ed.), *Spark Plasma Sintering of Materials*,
https://doi.org/10.1007/978-3-030-05327-7_5

between physical properties of metals and CNTs (Landry et al. 1998). Non-wetting nature of CNTs to most molten metals results in their clustering. For instance, molten aluminum (with the surface tension of 955 mN/m (Molina et al. 2007)) can't sufficiently wet the CNT surface (with the surface tension of 45.3 mN/m (Nuriel et al. 2005)). Another major challenge is to ensure the structural and chemical stability of nanotubes in the metal matrix. Carbon nanotubes might react with metals to form carbides and hence be destroyed. For example, strong chemical reaction between CNTs and aluminum matrix destroys the structure of CNTs, forming an unstable Al_4C_3 phase, which deteriorates the mechanical properties (Agarwal et al. 2011). These challenges could be solved by decorating or coating the surfaces of CNTs with stable ceramics (So et al. 2013; Kim et al. 2010).

Carbide ceramics, especially TiC, have high melting points, good resistance to corrosion and oxidation in harsh environments, and significant mechanical properties such as ultrahigh hardness and high elastic modulus (Kosolapova 2012; Yamanoglu et al. 2016). TiC has better thermal stability and wettability to the metal matrices when compared with pure CNTs (Saba et al. 2016b). Therefore, this type of ceramic is a promising material for applying in novel CNT-reinforced hybrid composites. Aluminum and its alloys, being the most abundantly used nonferrous structural materials, are the favorite choice for reinforcement with carbon nanotubes (Agarwal et al. 2011; Kuzumaki et al. 1998). The contact angle of aluminum on a TiC substrate was found to be significantly lower (about 65° (Selcuk and Kennedy 2006)) than that on a carbon substrate (165° before aluminum and carbon reaction (Landry et al. 1998)). Because of good wetting between Al and TiC (Landry et al. 1998; Rhee 1970), TiC decorating of CNTs could decrease the contact angle and improve the wettability to Al matrix as well as provide a good interfacial bonding between CNTs and metal matrix. Furthermore, the stable TiC particles can limit structural degradation of CNTs from Al-C reaction. Matrix needs to strongly adhere to the reinforcements so that the load can be transferred from matrix to CNTs, preventing the interface slipping (Lao et al. 2002). Moreover, the exceptional strength of inner graphene walls in multiwalled carbon nanotubes (MWCNTs) cannot be exploited owing to extremely weak inner wall shear resistance (Estili and Kawasaki 2010). This important issue causes that just the outermost wall of MWCNTs carry the applied load. A specific method causing load transfer to the inner walls of MWCNTs has not yet been experimentally proposed by carbide decorating nanotubes. Some authors (Peng et al. 2008) reported that the inter-wall load transfer can be improved with increasing the defect density. However, the net defective surface of MWCNTs may deteriorate the other properties of resultant composites.

In addition, many studies have been carried out to prepare new one-dimensional (1-D) nanostructured advanced materials such as nanotubes, nanowires, nanorods, and nanowhiskers for potential applications. In particular, ceramic nanotubes/nanorods represent an interesting and important class of nanomaterials with wide applications such as nano-electronics or quantum devices and drug delivery due to their singular shape and properties. To the best of our knowledge, there are just few researches focused on coating CNTs with nano-ceramic particles. Morisada et al.

(Morisada and Miyamoto 2004) coated CNTs with SiC using SiO vapor via gaseous method. Tungsten carbide was coated on carbon nanotubes through microwave-assisted method (Liang et al. 2009). In other studies (So et al. 2013), SiC layers were produced by high temperature annealing at 1100 °C for several hours. Recently, core-shell structured MWCNT-Ti (TiC) was produced using a one-pot reaction from a mixture of $TiCl_3$, TiH_2, and MWCNTs (Zang et al. 2010). Lao (Lao et al. 2002) reported that boron carbide nanolumps were formed on the surface of MWCNTs by a solid-state reaction between boron and carbon nanotubes through a thermal decomposition of MgB_2 powders at temperatures of 1100–1150 °C for 2 h. However, almost all applied methods for coating need complicated procedure and the particle size of the ceramic coatings is very large in micrometer size. In addition, the coating processes need almost more than1–2 h(s) with some long and frustrating procedures. Taguchi et al. synthesized TiC nanotubes by heating CNTs and Ti powders at 1300 °C in a vacuum for 30 h (Taguchi et al. 2007). In addition, they also produced SiC-CNTs hybrid structures in which carbon nanotubes were sheathed with a SiC layer by heating the materials in a crucible at 1200 °C for 100 h (Taguchi et al. 2005). Moreover, MWCNTs were coated with SiC by heating nanotubes with Si powder at 1200 °C for 5 h in a vacuum (~10^{-3} Pa) (Taguchi et al. 2013). In order to overcome the problems of the mentioned techniques with long durations and frustrating stages, spark plasma sintering (SPS) was used in this study. SPS process is a pulsed current field-assisted sintering technique based on electrical spark discharge phenomenon and can be used as a remarkable method for synthesizing and consolidating a large variety of materials (Saba et al. 2016a, b). The main advantage of this method is preventing grain growth in nanostructured materials (Zhang et al. 2008). The presence of the electric field accelerates diffusion due to the enhanced ion migration speed. Modification or synthesizing process can be done rapidly through SPS, and, therefore, it can be employed to control grain growth by using sintering times shorter than those typical of conventional processes (Yamanoglu et al. 2016). There are few studies regarding pressureless spark plasma sintering (SPS) just for porous materials (Yamanoglu et al. 2016; Bokhonov and Dudina 2016). To address the abovementioned challenges, the objective of this study is to use a novel and simple method for coating CNTs with nanocrystalline TiC particles by using spark plasma sintering (SPS) technique at a pressureless condition. The SPS process is a pulsed current field-assisted sintering technique based on electrical spark discharge phenomenon and is a remarkable technique to synthesize and consolidate a large variety of materials (Zhang et al. 2013; Ibrahim et al. 2011; Bradbury and Olevsk 2010). The pressureless-SPS has been firstly proposed by Zhang et al. for producing of porous nano-ceramic scaffolds in 2008 (Zhang et al. 2008). It was also applied for fabrication and heat treatment of porous titanium foams (Ibrahim et al. 2011; Quan et al. 2013) and production of SiC-C composites (Bradbury and Olevsk 2010).

A simple, effective, and novel three-stage method for modifying the surface of carbon nanotubes by nanocrystalline TiC was introduced and primarily investigated in our previous study (Saba et al. 2016a, b). Formation mechanism of the nano-TiC on the surface of CNTs in this novel approach was investigated (Saba et al. 2017). In addition, knowledge about the preferential nucleation sites of TiC nanoparticles can elucidate formation mechanism of the carbide layers and also can be useful for controlling the heterostructure of TiC/CNT with the well-defined crystalline interface. Understanding the formation mechanism and preferential nucleation sites of nanocrystalline TiC on the surface of MWCNTs was studied, as well as its effect on Al-matrix composites (Saba et al. 2016a, b, 2017, 2018). It is of great importance for a deep understanding of the formation mechanism of TiC on the surface of CNTs, as well as the effect on Al-matrix composites with improved load-bearing contribution of the inner walls. This chapter summarized the processing method, microstructures, properties and their mechanisms of the nano-TiC-modified-CNTs reinforced Al matrix nanocomposites.

2 Experimental

Diameter and length of the MWCNTs (Shenzhen Nanotech Port Co. Ltd., China) were in the range of 10–20 nm and more than 5 micrometers, respectively. Titanium (99.9% purity, 300 mesh) and gas-atomized Al powders (99.9% purity, 300 mesh) were manufactured by Xingye Metallic Materials Co. Ltd. (China). At the first stage, a 1:1 atomic ratio powder mixture of Ti to CNTs was prepared by using ultrasonication and wet milling (QM-3SP2). At stage 2, additional raw-CNTs (2 g) were added to the mixture prepared at the previous stage. Different mixing ratios of the crushed mixture to the raw-CNTs (TiC/CNT) were investigated (1:2, 2:2, 3:2, 4:2, and 5:2). In order to simplify the parameters involved in the present study, pressureless-SPS technique (SPS, FCT-HP-D5) was performed at the temperature and holding time of 1050 °C and 5 min, respectively. It should be noted that these values are the lowest possible ones for temperature and time required to form TiC through the pressureless-SPS (Saba et al. 2016a, b). After surface modifying process, TiC-modified-CNTs were mixed with Al powders by using ultrasonication method and wet milling for a short duration of time (1 h). In order to find the optimal mixing ratio, the amount of TiC-modified-CNTs in aluminum matrix was kept constant (1.0 wt.%). After finding the optimal mixing ratio, the weight fraction of modified-CNTs in Al matrices were changed within the range of 0.5–5 wt.%. The obtained composite powders were sintered at 520 °C, with a holding time of 10 min, a heating rate of 50 °C/min, and a pressure of 50 MPa. As a comparison, a pure Al and Al-reinforced with 1.0 and 1.5 wt.% unmodified CNT bulk specimens were also prepared by the same way.

The density of the specimens was measured at room temperature according to the Archimedes principle. The morphology of the coated layer was investigated with a SEM using field-emission gun (SEM, Sirion, FEI) and a high-resolution transmission electron microscope (HRTEM, Tecnai, FEI) at an accelerating voltage of 200 keV. The phase structure was identified by X-ray diffraction (XRD) using a Bruker D8 diffractometer with Cu kα radiation. Thermogravimetric analysis (TGA, Netzsch) was performed in air condition on the coated and uncoated carbon nanotubes. Raman spectroscopy (Thermo Scientific DXR) was performed by using a 532 nm argon laser as the excitation source. Nanoindentation (Micro Materials-NanoTest) was conducted using indenter equipped with a standard Berkovich tip indenter in order to deduce the nanohardness (H_n) and Young's modulus (E) of the nanocomposite samples. A constant loading rate, maximum load, and holding time on maximum load were 0.25 mN/s, 5 mN, and 5 s, respectively. A total of 20 indents per samples were made at a separation of 20 μm to prevent the hardening effects. The average values of H_n and Young's modulus (E) were calculated. In order to evaluate the tribological behavior of the composites, wear tests were performed by a pin-on-disk-type tribometer at room temperature, using a sliding speed of 0.5 m/s and a normal load of 20 N. An AISI 52100 steel disk (12.6 mm in diameter) with a hardness of 63 RC (or ~ 800 Hv) was used as a counterpart. Prior to each test, surface of the samples (cylindrical pins, 10 mm in diameter, 10 mm in length) was carefully polished and then cleaned ultrasonically. The sliding distance was kept constant (500 m).

3 Results and Discussions

3.1 CNT Surface Modification Method

TiC-modified-CNTs were prepared through a three-stage process: (1) mechanical crushing of Ti and MWCNTs powders via ultrasonication methods and wet milling, (2) adding additional pristine MWCNTs into the crushed mixture of Ti/MWCNT via ultrasonication and wet-milling methods, and (3) applying pressureless-SPS in order to coat or decorate the pristine MWCNTs with TiC. Schematic illustration of these stages is shown in Fig. 1.

At stage 1, four different milling durations were used to crush the powders in order to get the finest particles. In this regard, a portion of the mechanically crushed powders was extracted at the specified milling times (5, 10, 15, and 20 h) for analyzing by SEM. Figure 2 shows SEM micrographs of the Ti/CNT crushed

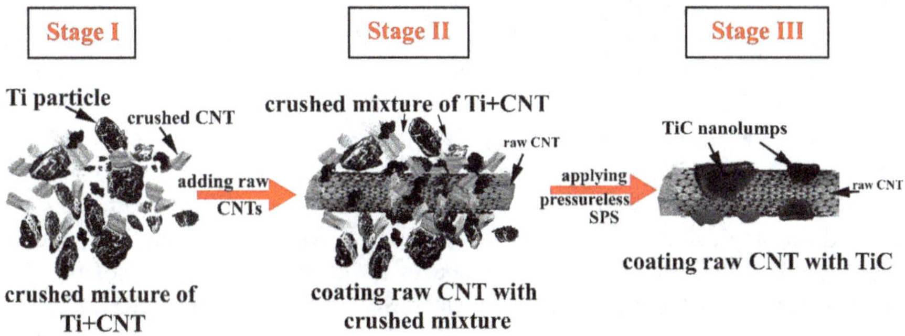

Fig. 1 Schematic of TiC formation on the CNT surface through the mentioned three stages

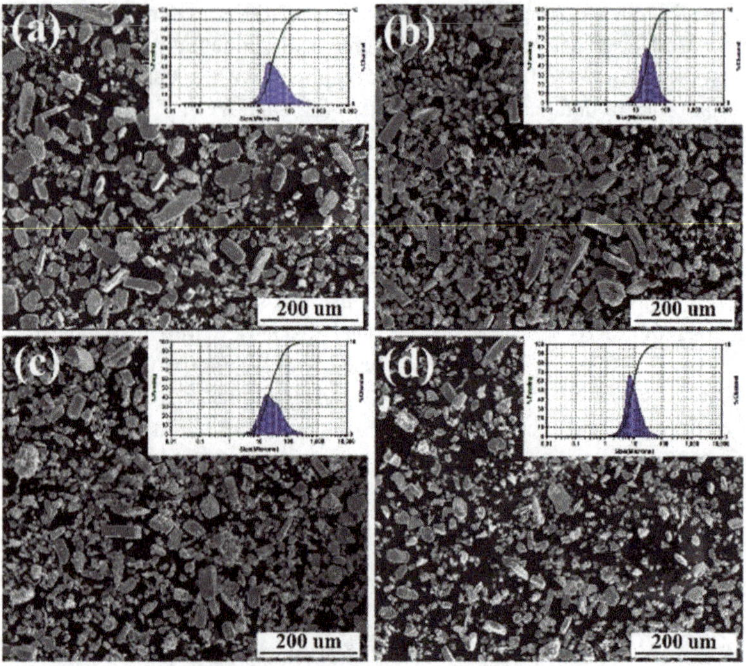

Fig. 2 SEM micrographs of the Ti/CNT crushed mixture prepared after 5 (**a**), 10 (**b**), 15 (**c**), and 20 (**d**) hours of wet-milling process. Insets reveal particle size distribution results

mixture prepared after 5 (Fig. 2a), 10 (Fig. 2b), 15 (Fig. 2c), and 20 (Fig. 2d) hours of milling process. As can be seen, particle size of the crushed mixture decreases with increasing milling time from 5 to 20 h. In addition, evolution of the particle size distribution is also shown in the insets of Fig. 2. Based on the particle size distribution results, the average particle size of the mixture is 29.6, 24.9, 24, and 8 um for 5, 10, 15, and 20 h of milling durations, respectively. The competition between cold welding and fracturing always takes place during milling process (Saba et al. 2016a, b). As can be concluded, cold welding may be a leading phenomenon at the durations less than 5 h, and after that fracturing is dominant over the rest durations from 5 to 20 h. In addition, CNTs which were used as carbon sources for TiC formation can enhance crushing of Ti to fine particles (So et al. 2013; Kwon et al. 2011).

During stage 2, the crushed mixture of Ti and MWCNTs was coated on the surface of pristine CNTs. One hour milling duration was selected for this stage in order to prevent damage to the pristine MWCNTs. At stage 3, pressureless-SPS method was applied. In order to achieve pressureless conditions during SPS process (used here), some modifications are made to the conventional die assembly. To better understanding the differences between SPS and pressureless-SPS methods, schematic of these processes are provided in Fig. 3. Based on Fig. 3a, T-shaped punches were fabricated from the same graphite material used for the die. There is a

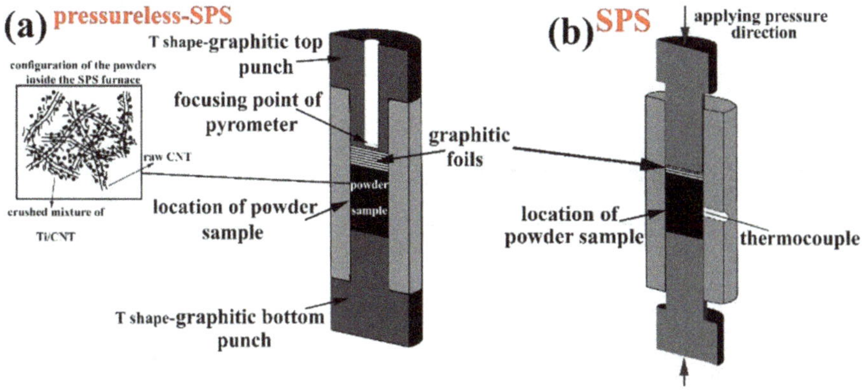

Fig. 3 Configuration of die assembly for (**a**) pressureless-SPS and (**b**) SPS (Saba et al. 2018)

working space above the sample which can produce no external load condition during processing. The working space was filled by several graphite foils between the sample and upper punch of the graphite mold to ensure the pulsed current can pass through the powders. A minimum pressure was applied in order to provide sufficient contact resistance through the punch exterior and die wall interior. In the case of pressureless-SPS, the temperature was measured by a pyrometer focusing on the bottom of the central borehole in the upper graphite punch, as can be seen in Fig. 3a. For a comparison, Fig. 3b shows a configuration of the die assembly for the conventional SPS (i.e., SPS with pressure). For measuring the temperature in SPS, two types of instruments are used; thermocouple (Fig. 3b) for sintering temperatures below 1000 °C and pyrometer (Fig. 3a) for sintering temperatures in excess of 1000 °C (Saba et al. 2018). Pressureless-SPS was used to synthesize modified nanotubes, while the conventional SPS with the pressure was used to produce bulk samples. Simply, it can be said that the reaction between the powders could be thermally assisted by direct heating of the die and direct Joule heating of the samples. The newly modified pressureless-SPS can provide an appropriate chamber for the solid-state chemical reaction (Saba et al. 2018).

3.2 X-Ray Diffraction (XRD) Analysis of the Modified-CNTs at Different Mixing Ratios

Formation of nanocrystalline TiC at the temperature of 1050 °C and holding time of 5 min was verified by XRD for different mixing ratio of 1:2 to 5:2, as shown in Fig. 4. According to the previous investigation (Saba et al. 2016a, b), Ti can fully reacts with carbon nanotubes to form TiC at the minimal temperature of 1050 °C for just 5 min. Therefore, pressureless-SPS was applied at 1050 °C for a very short time (5 min) in order to form TiC on the raw-CNTs (TiC/CNT). Comparing XRD results

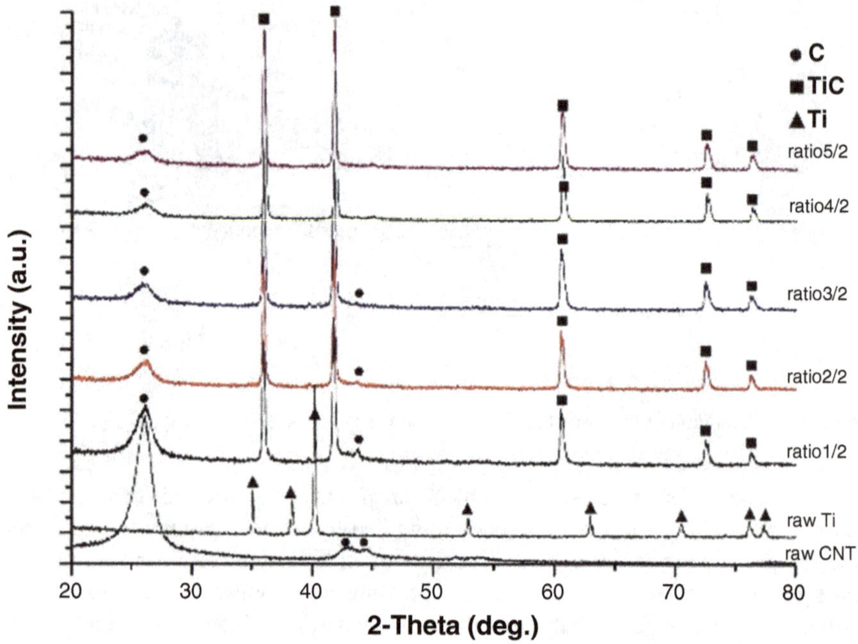

Fig. 4 XRD analysis of TiC formation at different mixing ratios

of the raw-CNTs and Ti with the modified-CNTs, it can be concluded that at the mentioned conditions all peaks correspond to TiC for different mixing ratios. Formation of crystalline TiC is confirmed by the peaks at 35.96°, 41.76°, 60.51°, 72.43°, and 76.21° corresponding to [111], [200], [220], [311], and [222] planes of face-centered-cubic TiC (black squares), respectively.

3.3 *TEM and HRTEM Observations of the Modified-CNTs at Different Mixing Ratios*

Structure and morphology of the modified-CNTs were also investigated by transmission electron microscopy (TEM) in detail. Figure 5 shows high-resolution TEM (HRTEM) and TEM images of TiC-modified-CNTs prepared at the different mixing ratios of TiC/CNT. Morphology of the raw-CNTs is also shown in Fig. 5a. Moreover, the sidewalls of CNTs are shown in the HRTEM inset image of Fig. 5a. As can be seen, raw-CNTs exhibit a bamboo-like structure. Figure 5a also shows opened tip of a single CNT with some structural defects (white arrows). As is obvious, some parts of the CNT surfaces are covered with amorphous impurities. Direct evidences for the formation of TiC nanoparticles on the surface of the raw-CNTs are provided in Fig. 5b–i. The average size of the TiC particles attached on the surface of CNTs is

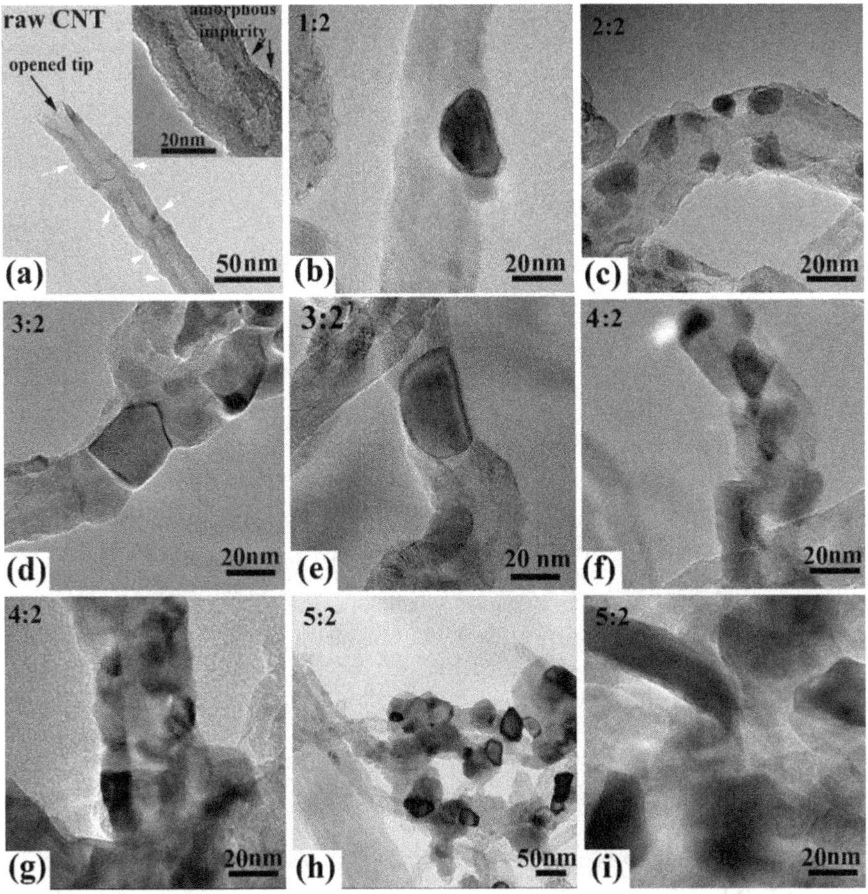

Fig. 5 Typical TEM and HRTEM images of the raw (**a**) and modified-CNTs at different mixing ratios (**b–i**)

in the range of ~40–50 nm at all mixing ratios. It should be noted that the reaction between Ti and carbon nanotube is highly confined and the main structure of CNTs remains almost unchanged. It is clear that the amount of attached TiC nanoparticles on the surface of CNTs for the mixing ratios of 1:2 and 2:2 (Fig. 5b, c) is not as much as those in the other mixing ratios. In other words, as the mixing ratio of TiC/CNT is increased from 1:2 to 4:2, the occupied surface areas of CNTs increase. However, the mixing ratio increment from 4:2 to 5:2 provides more free TiC particles separating from the raw-CNTs. These free particles could not contribute to the load-bearing and transferring phenomenon, compared to the attached carbides, and may deteriorate the properties of the produced composite. At the ratio of 5:2 (Fig. 5h, i), the covered areas of nanotubes decrease. From the TEM images of TiC-modified-CNTs prepared at the mixing ratio of 4:2 (Fig. 5f, g), it can clearly be seen that almost all nanotubes are coated with TiC nanoparticles, indicating a good and effective coating conditions in this case. As a result, the amount of nanoparticles attached on the surface of CNTs can be controlled by adjusting the process parameters, such as the mixing ratio of the crushed mixture to the raw-CNTs.

3.4 Formation Mechanisms of Nanocrystalline TiC

Information about the nature of the nanocrystalline TiC and carbon sources in the modification method can elucidate the formation mechanism of the decorated regions. Figure 6a schematically shows the different carbon sources involved in the modification method along with the corresponding TEM observations of TiC formed on the surface of CNTs (Fig. 6b–e).

In the applied process, there are two sources of carbon for producing a carbide layer: (1) carbon materials mostly provided from the sidewall of the raw CNTs (the area marked with 1 in Fig. 6a) and (2) carbon mainly obtained from crushed CNTs produced at the first stage (the area marked with 2 in Fig. 6a). Based on the nature of carbon sources, two different morphologies of nanostructured TiC can be produced by this method. When Ti reacts mostly with the raw CNTs, block-like TiC is produced (region 1 in Fig. 6b–e). It should be noted that during the second stage Ti particles/atoms can be embedded into the raw CNTs through the structural defects. These Ti particles/atoms as well as Ti particles attached on the surface of the raw CNTs may act as Ti source to provide a block-like TiC. On the other hand, simultaneous reaction mainly between crushed CNTs and Ti along with the sidewall of the raw CNTs can form TiC particles or TiC nanolumps on the surface of nanotubes (region 2 in Fig. 6c–e).

As can be seen from Fig. 6b–e, TiC particles and also TiC blocks are formed at both TiC/CNT ratios of 3:2 and 4:2. Dark field TEM inset image of Fig. 6c illustrates TiC nanoparticles attached on the surface of CNT. In addition, magnified image of the marked area in Fig. 6d, presented as a dark field TEM image in Fig. 6e, clearly shows the different morphologies of nanocrystalline TiC deposited on the surface of nanotubes. Finally, it can be concluded that the final morphology of TiC is highly related to the nature of carbon sources. Therefore, for producing block-like and nanolump carbides, the main carbon sources are raw- and crushed-CNTs, respectively.

Figure 7 shows TiC nanoparticles/nanolumps (a–d) and nanoblocks (e) attached to the surface of the raw CNTs. It is evident from Fig. 7b, c that the nanolumps with crossed {111} and {200} planes of TiC nucleate on the surface of CNTs, making a good interfacial bonding. In addition, corresponding selected area electron diffraction (SAED) patterns of Fig. 7b, c further confirm the presence of (111) and (200) planes of TiC. Interestingly, the inner walls of the raw CNTs retain the structure of graphene. Based on the HRTEM image, presented in Fig. 7b, the measured angle between these two crossed-lattice planes is around 57°, which is consistent with the result provided by the following equation; $\cos\theta = \frac{h_1.h_2+k_1.k_2+l_1.l_2}{\sqrt{h_1^2+k_1^2+l_1^2}\sqrt{h_2^2+k_2^2+l_2^2}} = \frac{\sqrt{3}}{3} \rightarrow \theta = 54.73^{\circ}$.

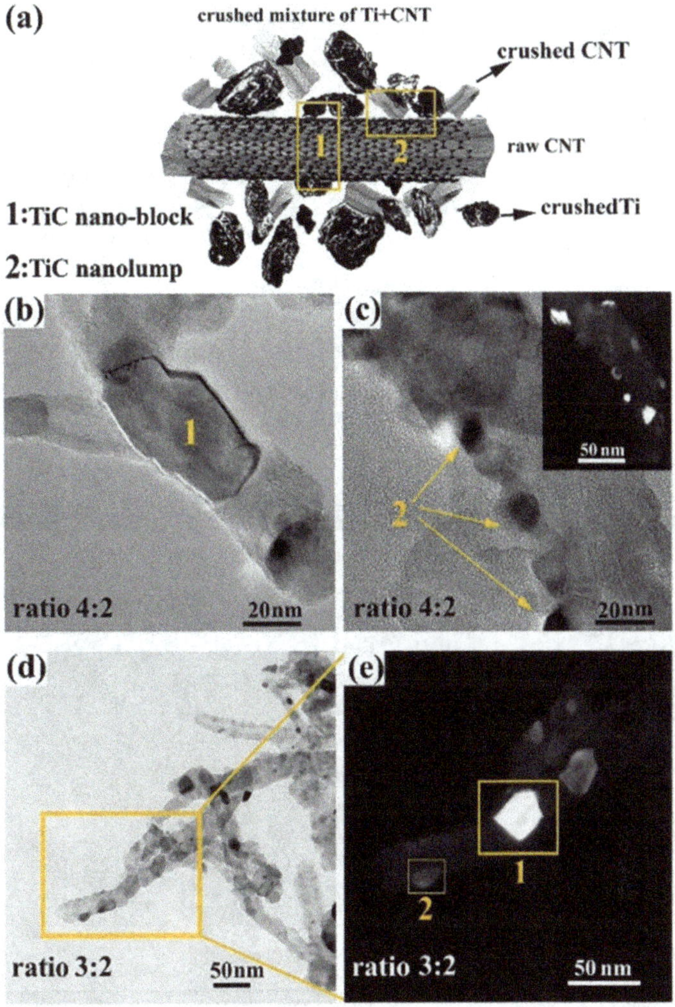

Fig. 6 (**a**) Schematic illustration of a raw CNT coated with crushed mixture powder showing different carbon sources. (**b–e**) TEM observations of TiC nanoblocks (regions marked by 1) and nanolumps (regions marked by 2) produced mainly from the sidewall of the raw CNTs and crushed CNTs, respectively (Saba et al. 2017)

Moreover, as can be seen in Fig. 7b, an area of disordered zone, marked by d.z., is formed at the TiC/CNT interface, indicating the atomic diffusion between CNTs and TiC at the interface during the TiC formation process.

The measured interspacing between the CNT walls is 0.34 nm, corresponding to the (002) plane of CNTs, which is similar to the initial raw CNTs (Fig. 7c). Furthermore, a 0.25 nm d-spacing of TiC (111) and 0.21 nm d-spacing of TiC (200) planes can be clearly observed in the HRTEM images of Fig. 7c, respectively.

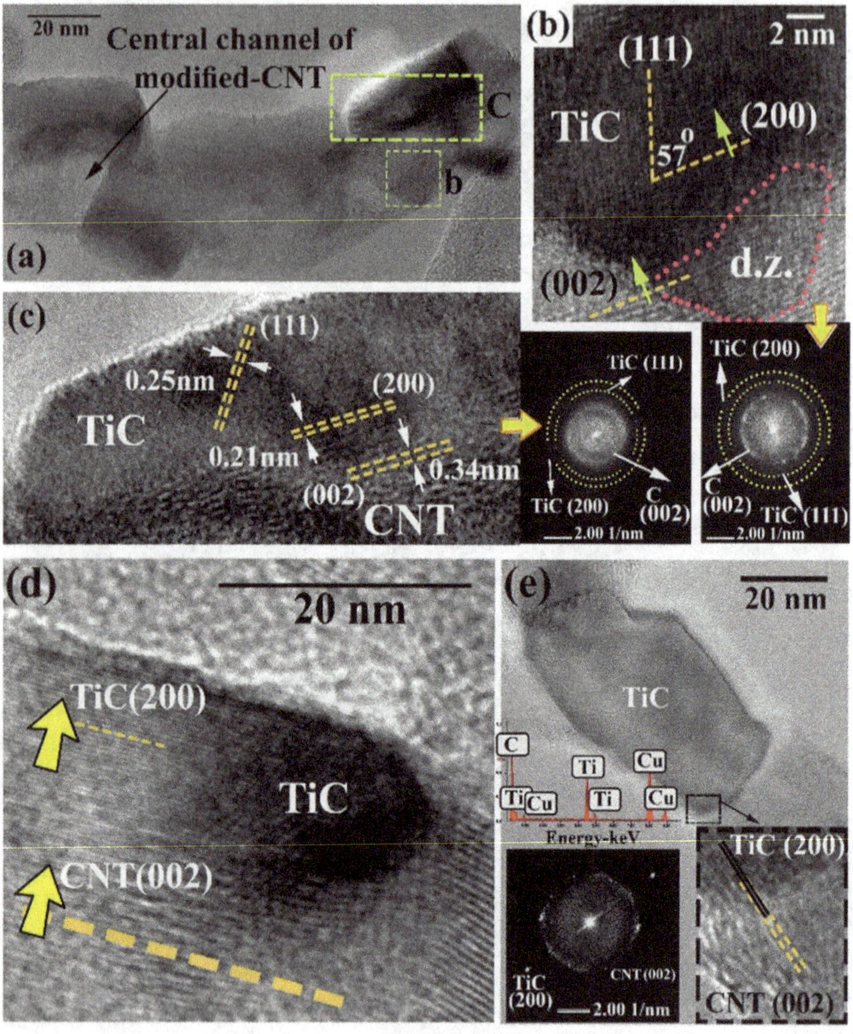

Fig. 7 HRTEM images along with electron diffraction patterns and lattice fringes of (**a–d**) TiC nanoparticles/nanolumps and (**e**) TiC nanoblocks; EDS analysis result is shown as an inset. The Cu peaks originate from the copper support grid (Saba et al. 2017)

Combining the SAED analysis results and measured d-spacing values of 0.25 nm and 0.21 nm, it can be confirmed that (111) and (200) planes of TiC closely contact with the surface and sidewalls of the nanotubes. Based on HRTEM observations provided in Fig. 7b–d, it is found that the (200) planes of TiC are parallel to the sidewall of the raw CNTs. Therefore, the typical orientation relationship between the raw CNT and TiC at their interface is CNT (002)/TiC (200), indicating an epitaxial growth of TiC on the surface of the raw nanotubes. Furthermore, such relationship between (200) planes of TiC and (002) planes of CNT can also be observed in

Fig. 7e. This nanoblock-like morphology of TiC, which is extended along nanotubes, reveals that Ti atoms may directly react with the raw CNT (as a main source) to form TiC without any substantial changes to the original shape of CNT during the process. It should be noted that TiC nanoblocks have no central hollow channel. The result of EDS analysis (inset of Fig. 7e) further confirms the TiC formation over CNT surface.

Figure 8a shows schematically the arrangement of (111) and (200) planes of titanium carbide. As is known, TiC has a face-centered cubic (FCC) crystal structure, and their {100} (or {200}) and {111} surfaces are the most stable and close-packed crystalline planes (Johansson 1995). {100} planes are composed of Ti and carbon atoms together, while {111} planes are made of a series of alternating atomic layers containing only Ti and carbon atoms. It is generally said that {100} planes of TiC are more stable than {111} planes (Johansson 1995; Dudiy and Lundqvist 2004). The notably higher energy activity of {111} planes is due to the polar character caused by alternating arrangement of Ti and carbon layers along <111> direction. Therefore, because of the energy difference between {111} and {200} planes, TiC particles are supposed to have an equilibrium cubic shape growth in theory (Zhou et al. 2013). Based on the above explanation, it can be concluded that the formation of (200) planes of TiC parallel to the CNT axis (Fig. 8b) may produce a stable interface with the lowest energy. It should be some defects on the surface of the raw CNTs through which Ti and crushed CNTs atoms diffuse into the raw CNTs, causing formation of TiC nanoblocks and nanolumps. In conclusion, carbide formation can occur via two possible reactions in the applied technique. One reaction occurs between raw CNTs (as a main carbon source) and deep-diffused/attached titanium atoms, forming TiC nanoblocks. The other reaction happens between the crushed CNTs (as a main carbon source) and shallow-diffused/attached Ti on the surface of the raw CNTs

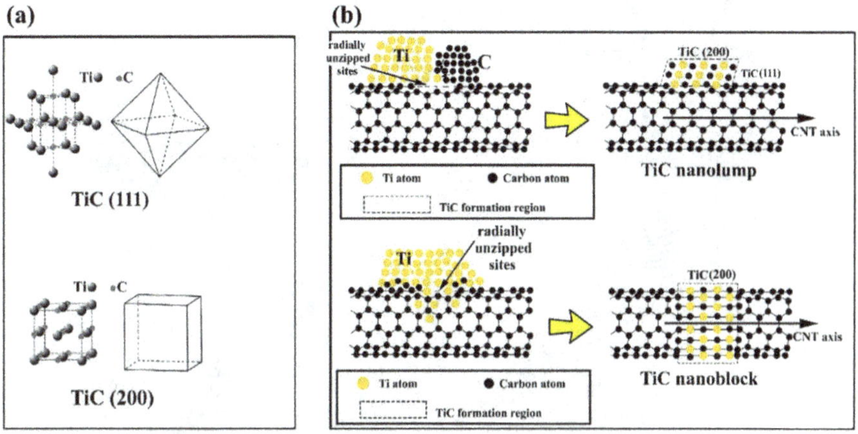

Fig. 8 Schematic representation of (**a**) (111) and (200) planes of TiC and (**b**) TiC nanolumps and nanoblocks formed at the surface of CNTs with the orientation relationship of TiC (200) // CNT axis (i.e., CNT (002)) (Saba et al. 2017)

producing TiC nanolumps. It is worth noting that dangling carbon atoms at the defect sites, such as radially unzipped defects (dashed lines on the surface of raw CNTs in Fig. 8b), have higher energy and more reactivity with Ti atoms to form a carbide layer.

The as-received raw CNTs, used in this study, have some types of defects and disorders, such as amorphous layer, radially unzipped sites, and "bamboo" defects. In addition, after applying mechanical milling, more defects are made such as ruptured regions on the CNT surface, additional radially unzipped defect sites and so on. Figure 9 shows HRTEM images of bamboo-shaped MWCNTs where the unique structural defects of nanotubes can be observed clearly. It should be noted that bamboo-shaped nanotubes are periodically closed along their lengths by bamboo knots, dividing the tubes into compartments (Banks et al. 2006). Bamboo-shaped carbon nanotubes exhibit a higher reactivity owing to the high density of edge-plane-like defects along the wall (Pujado 2012). In addition, as can be seen in Fig. 9b (magnified image of the marked area in Fig. 9a) and schematic illustration of Fig. 9c, weak-bonded areas near the bamboo defects provide a preferential path for diffusing Ti atoms into the CNT walls to make carbide nanoparticles or nanoblocks (Fig. 9d).

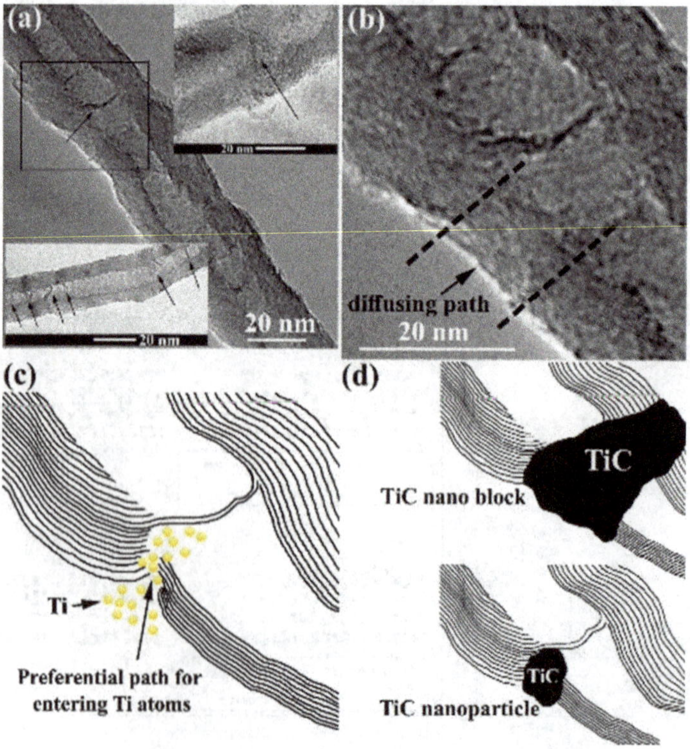

Fig. 9 (**a**, **b**) HRTEM observations and (**c**, **d**) schematic illustrations of bamboo-like defects. Preferential diffusing path is also shown in this figure (Saba et al. 2017)

Figure 10 shows HRTEM micrographs of TiC nanoblocks (Fig. 10a, b) and nanolumps (Fig. 10c, d) nucleated at some preferential bamboo sites (marked regions). The images confirm that the Ti-CNT interaction is enhanced by the presence of bamboo knots. As is obvious, TiC nanolumps or nanoblocks prefer to deposit at bamboo sites where defects are located. As mentioned before, bamboo-like MWCNTs provide a large number of edge-plane-like defects because the graphite planes are formed at an angle to the axis of the tube (Banks and Compton 2006). A plane composed of sp^2-bonded hexagonal carbon atoms is known as the basal plane, while the irregular surface perpendicular to the basal plane is known as the edge plane. The edge planes are more reactive than the basal planes (Chen et al. 2013). Figure 10e schematically shows edge and basal plane of MWCNT.

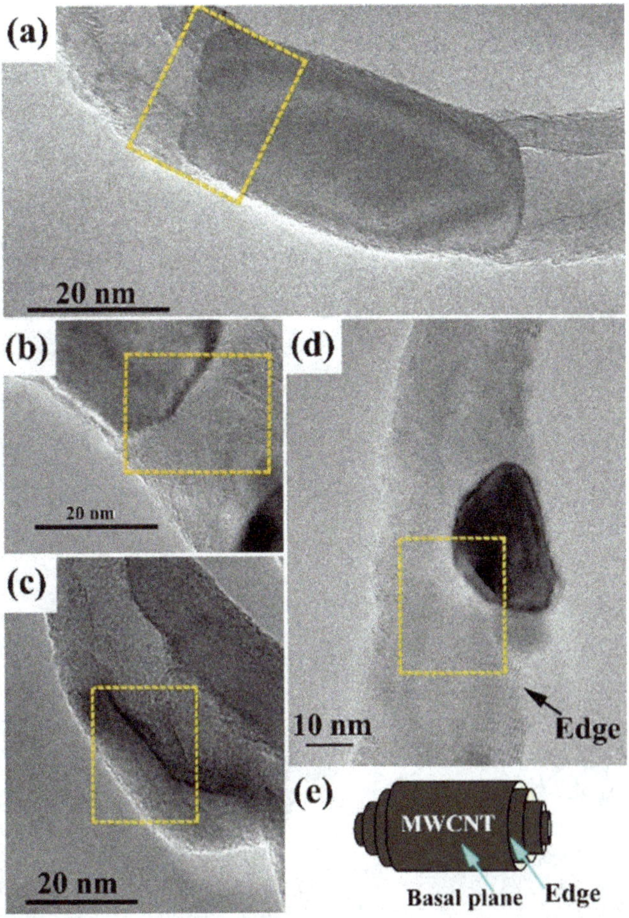

Fig. 10 HRTEM images of (**a**, **b**) TiC nanoblocks and (**c**, **d**) TiC nanolumps nucleated at the preferential bamboo sites (marked regions). (**e**) Schematic illustration of edge and basal plane of MWCNT (Saba et al. 2017)

Figure 11 shows other defects of CNTs which are preferential sites for TiC nucleation. During the mechanical milling process, impact energy of the balls influences on MWCNTs structure, and it can induce the infiltration of Ti atoms through the atomic defects on the surface of the carbon nanotubes. Such atomic defects, namely, ruptured regions, as a result of ball impact energy during the milling, are shown in Fig. 11a–c. During the application of SPS at a relatively high temperature, Ti atoms can easily diffuse into the sidewall of the CNTs through these ruptured defective regions. Figure 11b shows a schematic illustration of the ruptured defect through which Ti atoms can diffuse into the CNT. However, it is worth noting that the atomic radius of Ti is around 1.47 A, while the diameter length of the hexagonal graphene is 2.84 A, indicating that Ti atoms can also diffuse directly into the CNTs through the graphene hexagonal lattice. HRTEM image of

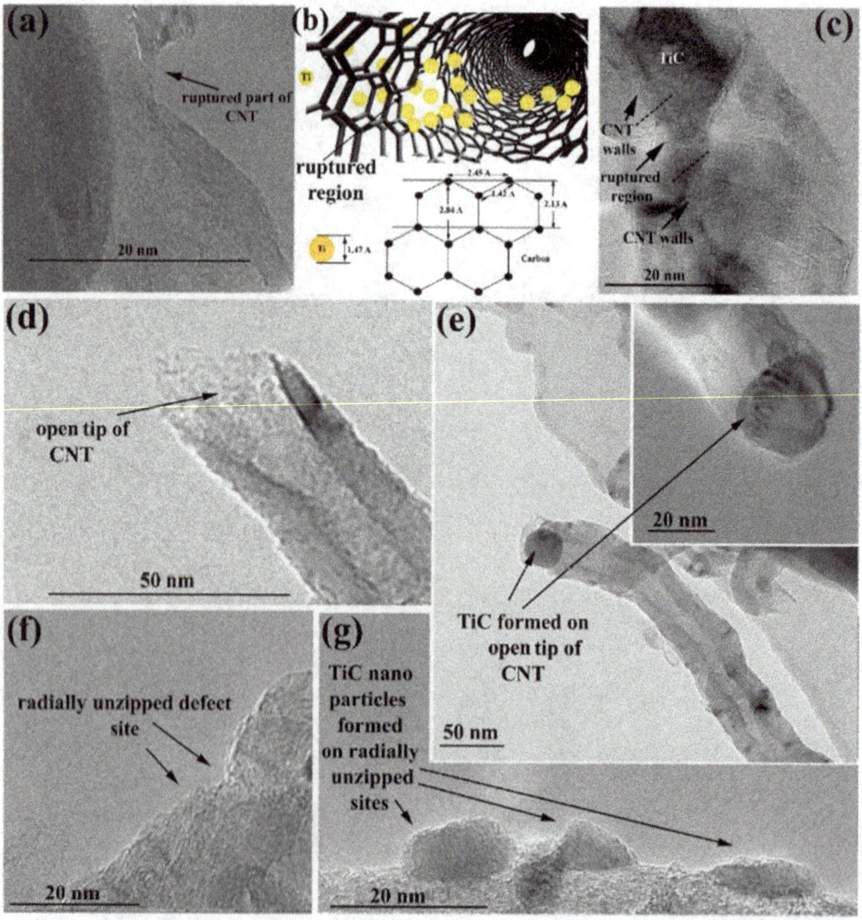

Fig. 11 Preferential sites for TiC nucleation; (**a–c**) ruptured regions, (**d, e**) open tip of CNTs and (**f, g**) radially unzipped defects (Saba et al. 2017)

a ruptured region on the surface of a modified-CNT is shown in Fig. 11c. As is obvious, formation of a cubic TiC nanolump at the central hollow part of the CNT is very likely due to the diffusion of Ti atoms through the ruptured region. Another preferential site existed in CNT is an open tip, in which many dangled carbon atoms can be found (Fig. 11d). TiC nanoparticles formed on the open tip of CNTs are shown in Fig. 11e. Moreover, radially unzipped sites located on the surface of the CNTs are also suitable places for growing carbide nanoparticles. Dangling carbon atoms at the unzipped defect sites (Fig. 11f) have a sufficient energy to react with Ti atoms forming carbide nanoparticles/nanolumps (Fig. 11g).

3.5 *Thermal Stability of the Modified-CNTs*

After applying the modification process, thermal analysis of the modified-CNTs is the most important and should be taken into consideration. In this regard, thermal properties of TiC-modified MWCNTs were analyzed by thermogravimetric analysis (TGA), as shown in Fig. 12. Figure 12a shows TGA data taken from the raw and modified-CNTs at different mixing ratios of TiC/CNT. As can be seen, raw-MWCNTs start to burn, and their overall weight decreases from around 490 °C with a sharp slope. On the other hand, modified-CNTs produced at different mixing ratios start burning at temperatures higher than around 570 °C, due to the formation of protective carbide layer on the surface of MWCNTs. It is interesting to note that the nanocrystalline TiC particles and TiC blocks formed on the surface of MWCNT can act as thermal insulator coating and also protect them from oxygen improving thermal properties of CNTs. In addition, as can be seen from Fig. 12a, there is an obvious weight gain during the test for all modified-CNTs at around 450 °C, indicating the oxidation of TiC coated layer (Saba et al. 2016a, b). Figure 12b also shows TGA curves of raw CNTs, crushed mixture of TiC+CNT (first stage), crushed mixture of the powders together with raw CNTs (TiC/CNT = 4:2, second stage), and TiC-modified-CNTs produced at the mixing ratio of 4:2 (third stage).

According to the results of TGA curves, the amount of burned CNTs in the crushed mixture is 13.36 wt.%. The amount of CNTs burned during TG analysis increases to 38.52 wt.% after mixing with additional raw CNTs (stage 2). This amount is smaller than the calculated input amount of CNTs (46.66 wt.%). This small difference may be due to the partial disintegration of carbon materials during the wet milling process. As indicated by TGA measurements, after applying the pressureless-SPS process (stage 3), the amount of burned CNTs is decreased to 20.52 wt.%. In addition, the onset of the CNT burning temperature increases up to around 610 °C, compared to 490 °C for the raw CNTs. Figure 12c presents the weight loss of the unmodified and TiC-modified-CNTs as a function of the mixing ratios of TiC/CNT. As can clearly be seen, the oxidation weight loss decreases from 96.64 wt.%, for the raw CNTs, to 20.46 wt.%, for the mixing ratio of 5:2, revealing that the TiC coating process improves the oxidation resistance of CNTs efficiently. This can be explained by the fact that the TiC coating can protect the CNTs from

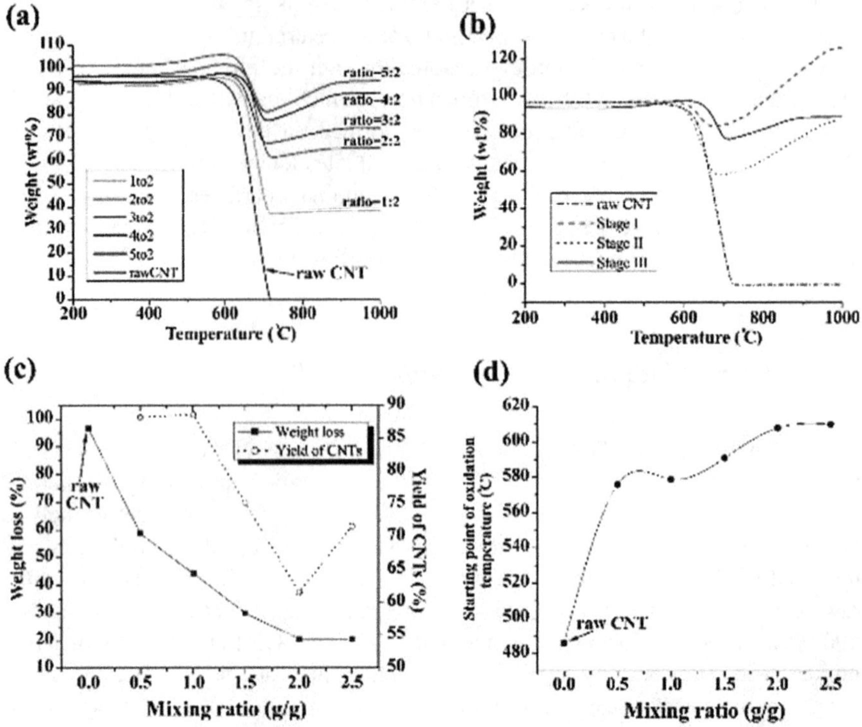

Fig. 12 Thermal analysis of TiC formation by using TGA. (**a**) TGA data from modified-CNTs at different mixing ratios, (**b**) TGA data taken from raw CNTs and powders prepared at stages 1, 2, and 3 for mixing ratio of 4:2, (**c**) yield of CNTs and weight loss percentage amounts of CNTs for different mixing ratios, and (**d**) variation of oxidation temperature onsets as a function of mixing ratios (Saba et al. 2017)

direct exposure to air. It should be noted that the weight loss differences are more distinct at lower mixing ratios. However, the oxidation weight loss at the mixing ratios of 4:2 and 5:2 are 20.52 wt.% and 20.46 wt.%, respectively, indicating a negligible difference between these ratios of TiC/CNT. In the present study, the yield of CNTs was calculated from the burned amount of CNTs per input non-crushed CNTs. Figure 12c shows the yield of CNTs versus the input TiC/CNT mixing ratio. According to the results, when the mixing ratio of TiC/CNT increases to 4:2, the yield of CNTs decreases rapidly and thereafter increases again as the ratio increases to 5:2. This indicates that increasing the mixing ratio from 4:2 to 5:2 is no longer useful for the applied modification process. Figure 12d shows the onset of oxidation as a function of the mixing ratio. Compared to the raw CNTs, which are oxidized at around 490 °C, the oxidation temperature of the TiC-modified-CNTs starts at 576 °C, 579 °C, 591 °C, 608 °C, and 610 °C for the mixing ratios of 1:2, 2:2, 3:2, 4:2, and 5:2, respectively. Therefore, the results show that the present method is so effective for protecting CNTs.

3.6 *Effect of TiC/CNT Mixing Ratio on the Mechanical Properties of the Composites*

In order to evaluate the effect of mixing ratio on the mechanical properties of the samples, the mixed Al and 1 wt.% modified-CNT powders, prepared at different TiC/CNT mixing ratios of 1:2, 2:2, 3:2, 4:2, and 5:2, were consolidated by SPS. In order to get the highest load-bearing effect of the reinforcements, the matrix and the modified-CNTs must be in direct contact and the sintered composite should be fully densified. Figure 13a shows the relative density of pure Al along with 1 wt.% unmodified and modified-CNT/Al composites as a function of mixing ratio. As can be seen, almost all the modified samples as well as pure Al are fully dense, with a relative density of more than 99%. Density is vigorously affected by CNT distribution in matrix materials (Saba et al. 2016a, b). The density of 1 wt.% CNT/Al is less than 99% (98.89%) which can be attributed to the poor distribution of CNTs

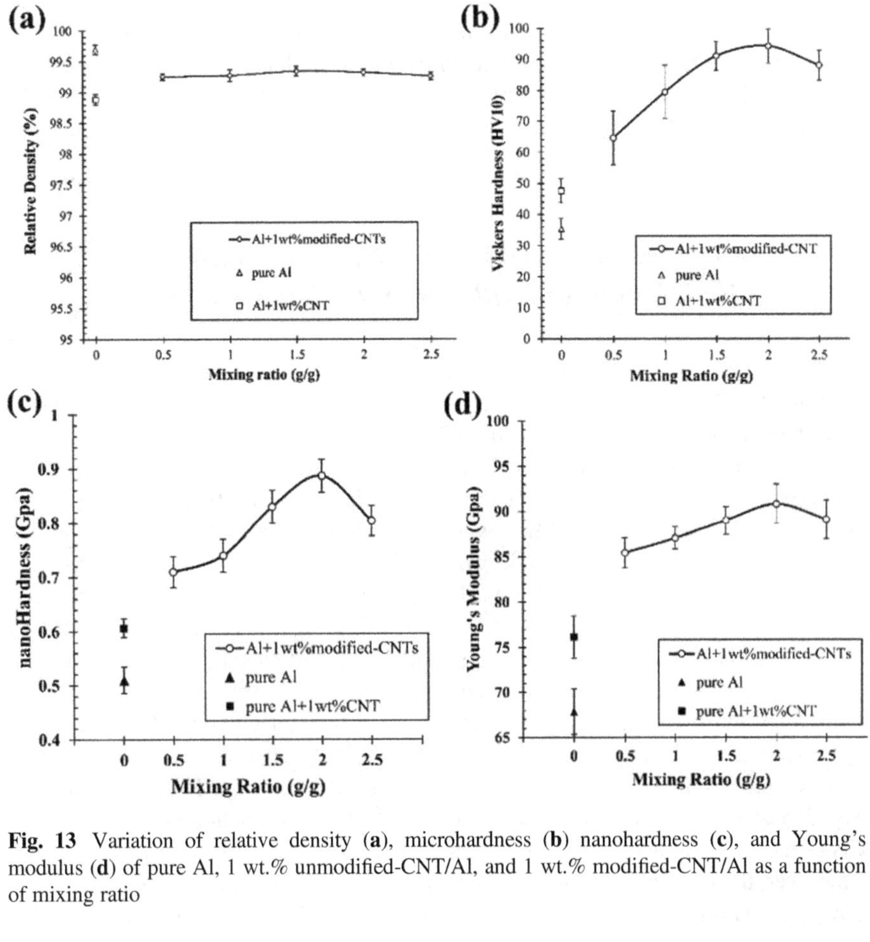

Fig. 13 Variation of relative density (**a**), microhardness (**b**) nanohardness (**c**), and Young's modulus (**d**) of pure Al, 1 wt.% unmodified-CNT/Al, and 1 wt.% modified-CNT/Al as a function of mixing ratio

in this case. It can be said that the modified-CNTs have better dispersion in Al matrix. Microhardness, nanohardness, and Young's modulus variation of pure Al and nanocomposites reinforced with 1.0 wt.% modified- and unmodified-CNT as a function of the mixing ratio are also presented in Fig. 13b–d, respectively. As is obvious, with increasing the mixing ratio from 1:2 to 4:2, all the mechanical properties increase and thereafter decrease from 4:2 to 5:2. This reveals that with increasing the ratio up to 4:2 the amount of TiC nanoparticles attached on the surface of CNTs increases, improving the mechanical properties of the composites. As mentioned before, at the ratio of 5:2, there are many free agglomerated TiC particles separating from the surface of carbon nanotubes. Therefore, the reduction of mechanical properties at the ratio of 5:2 can be attributed to the free agglomerated carbides which could not contribute to the load-bearing phenomenon. However, at the other ratios, attached nanoparticles or nano-layers can link the inner walls of CNTs and involve them to the load-bearing phenomena, improving the mechanical properties (Lao et al. 2002). As can be seen in Fig. 13b–d, there is an optimum point at the mixing ratio of 4:2 in which the microhardness (94.17 HV), nanohardness (0.88 GPa), and Young's modulus (90.82 GPa) are all at their highest values. Based on the abovementioned results, ratio of 4:2 was selected as the optimal TiC/CNT mixing ratio. Therefore, after finding the optimal value, the weight fraction of modified-CNTs in Al matrices were changed within the range of 0.5–5 wt.% at the constant mixing ratio of 4:2.

3.7 SEM Observations of Al Nanocomposite Powder and Bulk Reinforced with Different CNT Contents

SEM images of Al nanocomposite powders containing 0.5, 1.5, 2, and 5 wt.% modified-CNT are shown, respectively, in Fig. 14a–d. In addition, Fig. 14e represents TEM image of 1.5 wt.% modified-CNT/Al mixture powders. As can be seen in Fig. 14a, there are a few number of CNTs attached to the surface of Al particles in the sample reinforced with 0.5 wt.% modified-CNT. It may be attributed to the very low amount of CNT in this case. Moreover, micrograph of Fig. 14b shows embedded CNTs into Al particle with the pulled out tips. More importantly, the figure illustrates the uniform dispersion of CNTs on Al surface in the case of 1.5 wt.% reinforced composite. Yellow dashed arrows, provided in Fig. 14b, clearly show the locations of individually dispersed CNTs on the surface of Al particles. It is important to note that no CNT clusters were observed in the microstructure of the composites reinforced with 1.5 wt.% modified-CNT. The elemental composition of a single TiC-modified-CNT, which was determined by EDS line scan, shows the spectral counts corresponding to Ti and C (top-right inset in Fig. 14b), indicating a TiC-deposited layer on the surface of CNT. On the other hand, as the amount of CNTs increases up to 2 wt.%, CNT clusters appear on the surface of Al (Fig. 14c). Moreover, the CNT clustering become more aggravated in the 5 wt.% CNT mixture powder, as shown in Fig. 14d. A large CNT cluster with a size of around 12 um can

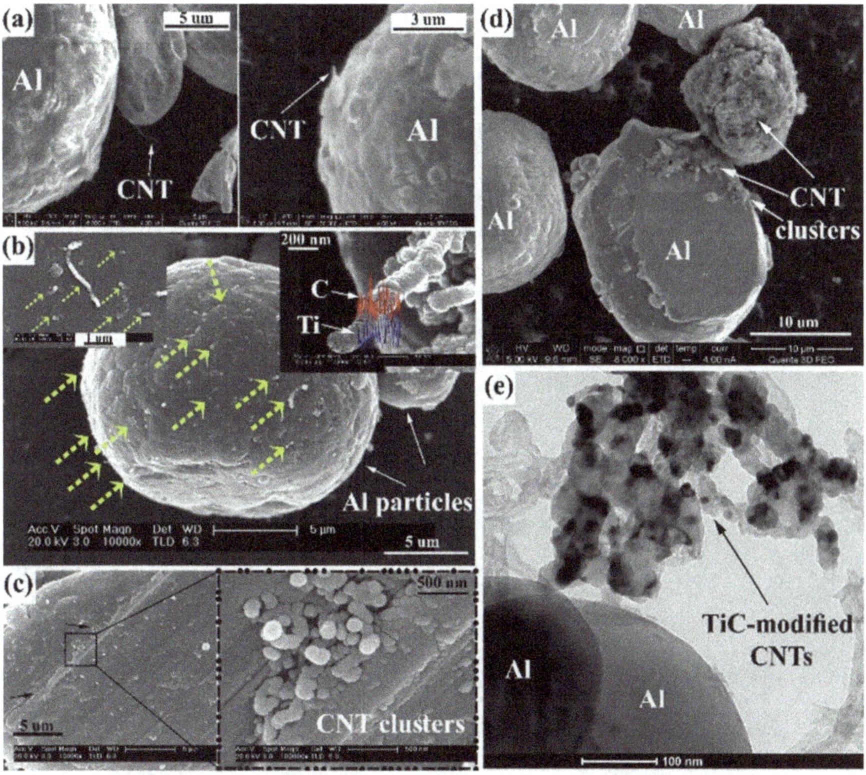

Fig. 14 SEM images of Al nanocomposite powders containing (**a**) 0.5, (**b**) 1.5, (**c**) 2, and (**d**) 5 wt.% modified-CNT. (**e**) TEM image of 1.5 wt.% modified-CNT/Al mixture powders

be seen in Fig. 14d. TEM image of Fig. 14e shows TiC-modified-CNTs-dispersed region on the surface of Al particles in the 1.5 wt.% modified-CNT/Al composite powder.

Microstructure of the sintered Al nanocomposites reinforced with 0.5, 1.5, 2, and 5 wt.% TiC-modified-CNT contents can be observed, respectively, in the SEM images of Fig. 15a–d. Despite initial dispersion of CNTs, a significant proportion of nanotubes is still concentrated in the form of agglomerates (dark phases in Fig. 15), which may originate from the clusters in the powder mixture. Very small quantity of CNT clusters is found in the sintered 0.5 wt.% modified-CNT/Al composite (Fig. 15a). According to the figure, as the amount of CNT increases from 0.5 to 5 wt.%, the dark phases increase clearly. Larger amounts of CNTs are strongly inclined to tangle together in the grain boundary of Al matrix, impeding the densification of the specimens and thus leading to the decrease of the relative density and mechanical properties of the composites. It can be concluded that nanotube content considerably affects the size of CNT clusters and dispersion quality. A higher magnification SEM image of a cluster can be seen in the inset of Fig. 16 along with EDS analysis of the marked area, which further confirms the presence of TiC-modified-CNTs in the dark phases.

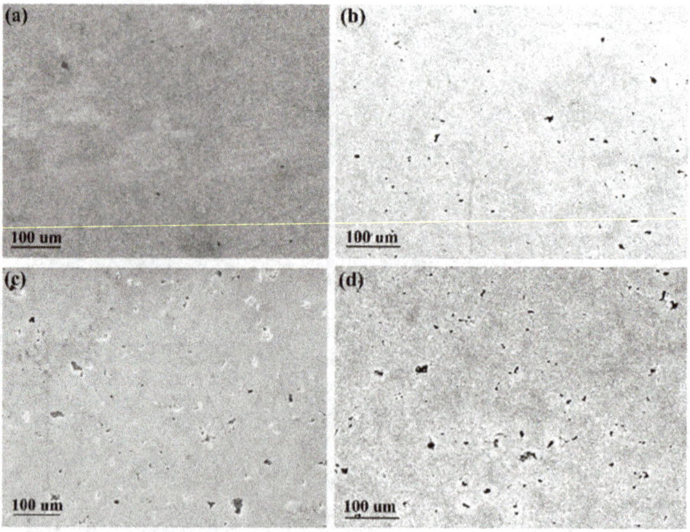

Fig. 15 Typical SEM images of the sintered Al nanocomposites containing (**a**) 0.5, (**b**) 1.5, (**c**) 2, and (**d**) 5 wt.% TiC-modified-CNT

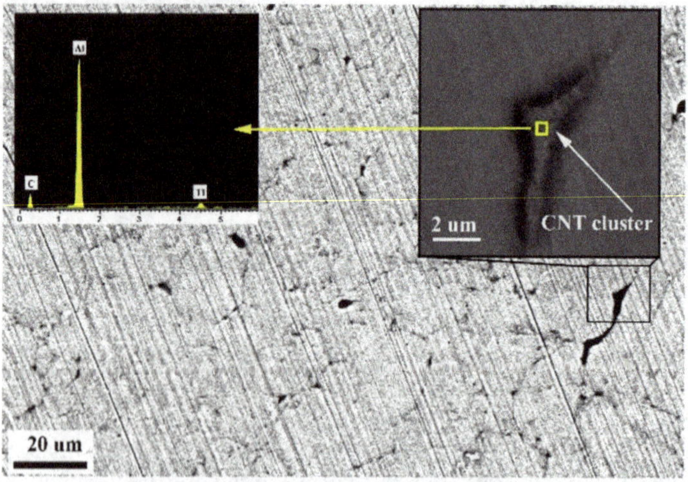

Fig. 16 SEM image of a sintered sample showing a CNT cluster located at the grain boundary of Al as well as EDS analysis taken from the marked region

3.8 *Raman Spectroscopy Analysis*

Typical Raman spectra of the composites containing different amounts of the modified-CNT along with the spectrum of the raw-CNTs are provided in Fig. 17a. The bands around 1580 cm^{-1} are called G bands, corresponding to carbon-carbon stretching modes (Stein et al. 2012; Dresselhaus et al. 2005; Liu et al. 2016). The

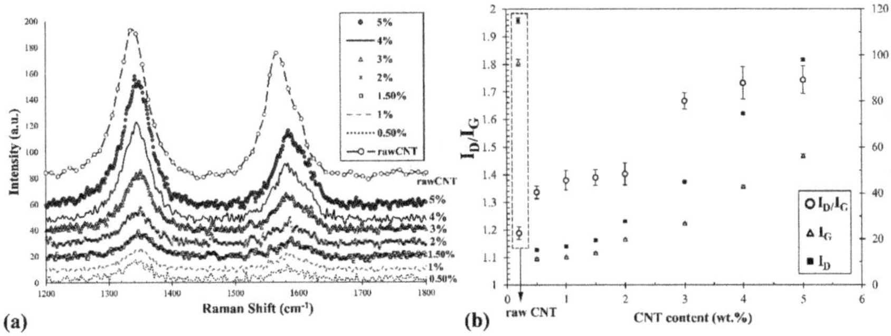

Fig. 17 (**a**) Raman spectra of the composites reinforced with different amounts of modified-CNTs. A Raman spectrum of the raw-CNTs is given as a comparison. (**b**) Variation of I_D/I_G, I_D, and I_G values as a function of CNT content

so-called D bands observed at around 1350 cm^{-1} correspond to zone edge phonons and are activated through double resonance mechanism involving elastic scattering of the electrons by structural defects (Dresselhaus et al. 2005). Accordingly, I_D/I_G intensity ratios are commonly used to evaluate the defect density in carbonaceous structures (Stein et al. 2012). The high I_D/I_G ratio of the raw-CNTs (1.18) confirms that they possess a relatively defective structure. Variations of I_D/I_G ratio, I_D, and I_G as a function of CNT content are shown in Fig. 17b. The values of I_D/I_G, I_D, and I_G for the raw-CNTs have been added to the figure as a comparison (the area marked by dashed rectangle). As can be seen, the intensity of G and D bands increases with increasing CNT content. The trend of G-band intensity can be attributed to the fact that the amount of graphitic carbon, which is presented after the milling process, is intensified with increasing CNT content. On the other hand, as carbon nanotubes suffer damage during milling and SPS process, they possess defective structures. Therefore, the intensity of D band should increase with increasing CNT content. It should be noted that the intensity of G and D bands in the case of raw-CNTs is higher due to the larger exposure area during measurement, in comparison with the bulk composite samples. The ratio of D and G band (I_D/I_G) increases from 1.33 for 0.5 wt.% to 1.74 for 5 wt.% modified-CNT/Al composite. Comprising the I_D/I_G value of the raw-CNTs (I_D/I_G) with those of modified-composites, it can be concluded that the defective structure of CNTs become more significant after milling and compacting of SPS process.

3.9 Effect of CNT Content on the Mechanical Properties of the Composites

In order to evaluate the effect of CNT content on the mechanical properties of the composites, Al with different amounts of modified-CNT powders was consolidated by SPS. Figure 18 illustrates the variation of relative density (Fig. 18a), Vickers

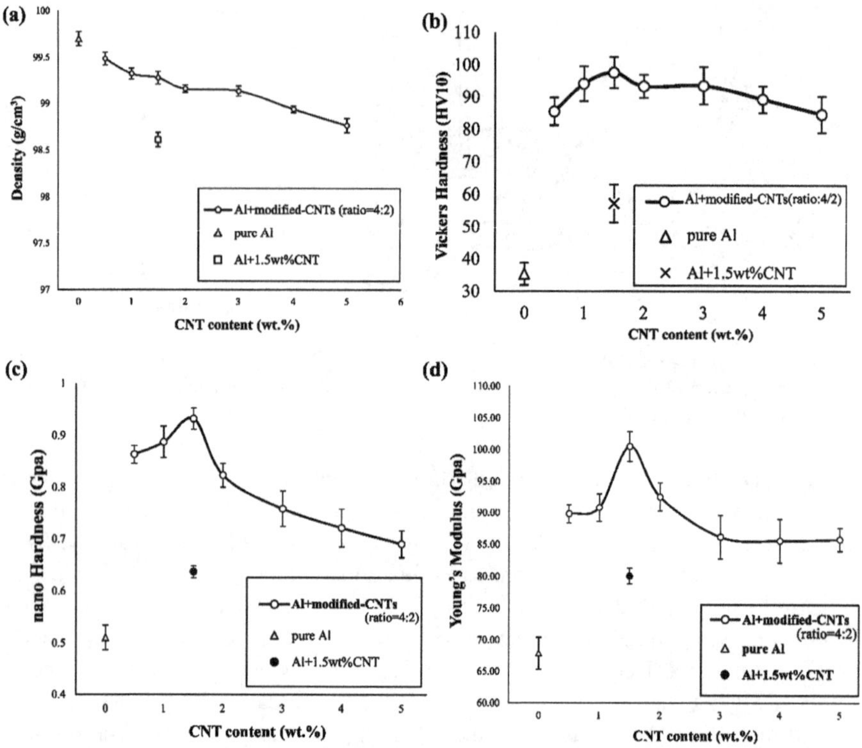

Fig. 18 Variation of (**a**) relative density, (**b**) microhardness, (**c**) nanohardness and (**d**) Young's modulus of pure Al, 1.5 wt.% unmodified-CNT/Al and 1.5 wt.% modified-CNT/Al as a function of CNT content

hardness (Fig. 18b), nanohardness (Fig. 18c), and Young's modulus (Fig. 18d) of the composite samples as a function of modified-CNT content (0.5, 1, 1.5, 2, 3, 4, and 5 wt.%). For a comparison, the values of relative density and mechanical properties of pure Al and 1.5 wt.% unmodified CNT/Al composites are also added to the curves in Fig. 18. As can be observed in Fig. 18a, almost all the samples are highly consolidated with the relative density of more than 98.5%. However, as mentioned before, increasing the amount of CNTs leads to an increase in CNT clustering. Therefore, as CNT content increases, the density is kept decreasing mainly due to the increase in clustering which impedes the composite samples from being fully densified. In other word, lower densification of the composite at higher CNT contents can be attributed to the CNT clustering in the resultant samples, as is shown in Fig. 18. Moreover, with increasing the modified-CNT content from 0.5 wt.% to 1.5 wt.%, all the mechanical properties increase and thereafter decrease. The reduction in the mechanical properties for the samples containing more than 1.5 wt.% modified-CNTs can be attributed to the presence of CNT clusters which represent sources of weakness in the samples and may act as a porosity. As can be observed in Fig. 18b–d, there are optimum points for the mechanical properties corresponding to the CNT content of 1.5 wt.% in which the microhardness, nanohardness, and Young's modulus have the highest values of 97.61 HV, 0.93 GPa, and 100.48 GPa, respectively. Therefore, it can be concluded that the composite sample containing 1.5 wt.% modified-CNTs possesses the best mechanical properties. In this regard, 1.5 wt.% modified-CNT/Al composite shows 176% and 71% increase in Vickers hardness, compared to pure Al and 1.5 wt.% unmodified-CNT/Al composite sample, respectively. In addition, 82% and 45% increase in nanohardness as well as 48% and 25.5% increase in Young's modulus was achieved for 1.5 wt.% modified sample compared to pure Al and 1.5 wt.% unmodified sample, respectively.

3.10 Compressive Behavior and Strengthening Mechanism of Al-Matrix composites Reinforced with TiC-Modified CNTs

Figure 19 shows the effect of addition of modified-CNTs (0.5, 1, 1.5, 2, and 5 wt.%) on the compressive behavior of Al-matrix composites. As can be seen from Fig. 19a, incorporation of modified-CNTs into Al-matrix composites leads to an increase in the compressive strength of the nanocomposites, compared with pure Al sample. In addition, the compressive yield strength of the modified nanocomposites is improved with increasing the weight amount of TiC-modified-CNTs and reaches a maximum value of 84.36 MPa for 1.5 wt.% modified-CNTs. However, the yield strength decreases beyond 1.5 wt.% of the modified-CNTs and reaches to 69.79 and 58.14 MPa for 2 and 5 wt.% of the modified-CNTs, respectively. The decrease in compressive strength can be explained by the presence of the modified-CNTs agglomerates in the nanocomposites. It has been reported that the presence of carbon nanotube agglomerates in the resultant composites has mostly resulted in the reduction of strength and stiffness (Esawi et al. 2010).

In Fig. 19b, Al + 1.5 wt.% modified-CNTs sample with the highest value of compressive yield strength is compared with three reference samples, namely, pure Al, Al + 1.5 wt.% unmodified-CNTs, and Al + 1.5 wt.% (pure TiC+CNTs). Compressive yield strength values of these samples are listed in the table presented in Fig. 19b. As is obvious, the sample containing 1.5 wt.% modified-CNTs shows a drastic difference in yield strength (YS), compared with other samples. It can be attributed to the effective load-bearing contribution of the linked inner walls of the modified-CNTs (Saba et al. 2017).

In general, MWCNTs can bear applied loads just by their outermost walls because the weak Van der Waals inter-wall interactions make the inner walls totally useless (Kurita et al. 2015). If the inner walls of MWCNTs are bonded together and to the matrix, load transfer effect would be increased dramatically (Lao et al. 2002; Cho et al. 2013). TiC nanolumps/nanoparticles or nanoblocks, in the present study, can link the walls, improving the interaction between nanotubes and matrix. It is worth noting that isolated carbides on the surface of carbon nanotubes are more

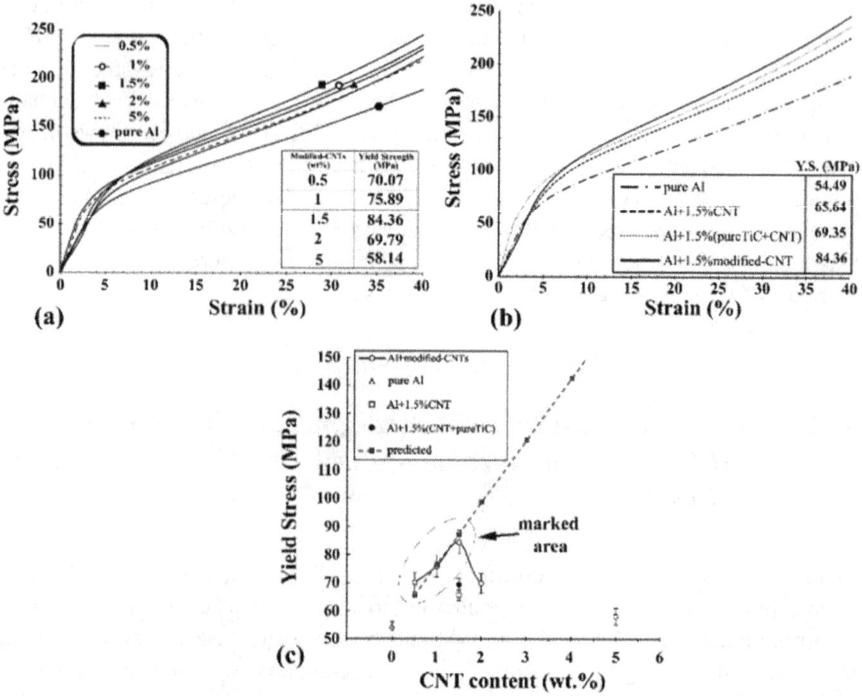

Fig. 19 Engineering stress-strain curves obtained from compression tests for Al composites reinforced with (**a**) different amounts of CNTs and (**b**) 1.5 wt.% CNTs. (**c**) Comparison between experimental and predicted values of yield stress as a function of CNT weight fraction. Marked area shows the well-matched region (Saba et al. 2017)

favorable than uniform coating layer covering whole the tubes (Lao et al. 2002). Schematic image in Fig. 20 illustrates linking of CNT inner walls by a TiC nanolump. In addition, HRTEM images in Fig. 20 show the inner walls of CNTs which are linked through TiC formation. As can be seen, inner walls of MWCNTs, at the reaction area, are bonded to titanium carbides. As the bonding between Ti and carbon atoms in the carbide is a covalent type, the bonding between TiC and carbon nanotubes is most likely covalent. Strong interlocked bonding can prevent interface fracturing during the load transfer. Indeed, these are the first evidences for linking the inner walls by titanium carbide formation, in which inner walls of CNTs, in addition to the outermost wall, can contribute to load-bearing phenomenon. Some researchers reported such inner-walls contribution, in an uncontrollable manner, induced by MWCNTs defects (Chen et al. 2015a, b; Estili and Sakka 2014; Estili et al. 2013). However, the net defective surface of CNTs can deteriorate the properties of resultant samples. Based on the above explanations, the modification technique, proposed here, suggests that MWCNTs with carbide-linked inner walls provide ideal reinforcements with high load-bearing capacity. These ideal reinforcements open up a window to the new area in which many types of matrices, such as metals, polymers, and so on, can be strengthened.

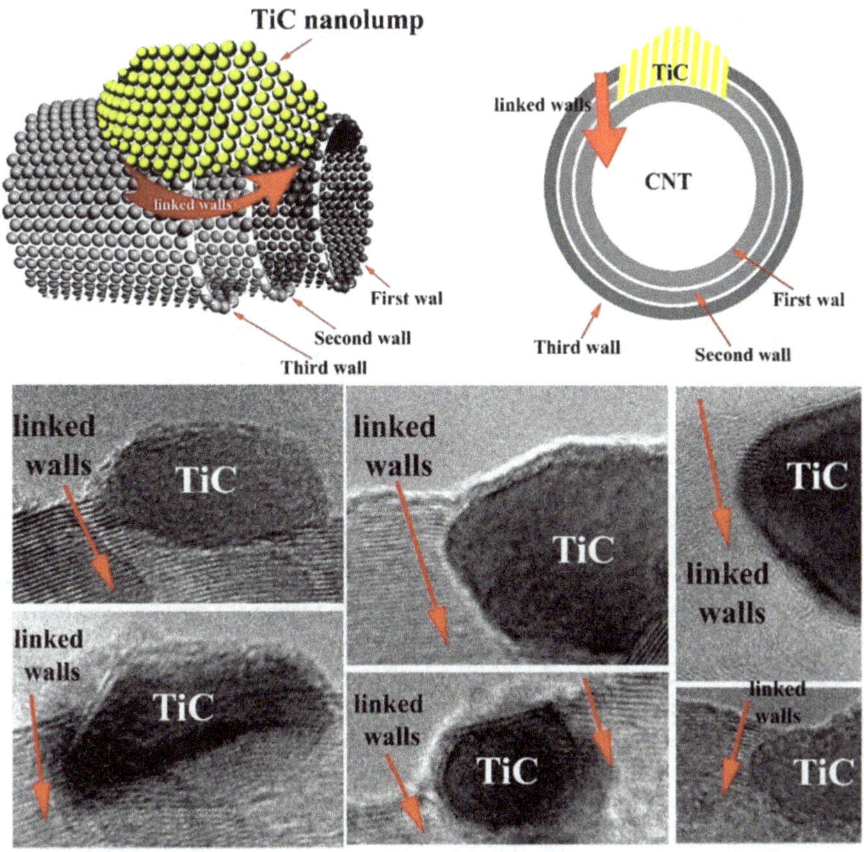

Fig. 20 Schematic illustration as well as HRTEM observations of the CNT walls which are bonded together through a TiC formation (Saba et al. 2017)

There are several possible strengthening mechanisms in metal-matrix composites reinforced by CNTs including load transfer from matrix to CNTs (George et al. 2005), grain refining and texture strengthening by pinning effect of CNTs (Wei et al. 2014), dispersion strengthening of CNTs (Yoo et al. 2013a, b), solution strengthening of carbon atoms (Li et al. 2013), and thermal mismatch between CNTs and matrix (George et al. 2005). Among the mechanisms, load transfer is the most desirable because in this mechanism high loads can be tolerated by CNTs during composite failure, employing the excellent mechanical properties of CNT in the composite (Chen et al. 2015a, b). For the composites reinforced with discontinuous fibers, an applied force can be transferred from the matrix to the fibers through the shear stresses which are developed along the fiber-matrix interface. Therefore, variation of fiber strength along the fiber length (l) is generated by the shear stresses (Choi et al. 2008). As a result, strength of the composites is highly related to the fiber length (l) and can be estimated by Kelly-Tyson formula based on the shear-lag model

(Park et al. 2015). According to the mentioned formula, the strength of the composites is given by the following equations (Bakshi and Agarwal 2011):

$$\sigma_c = \sigma_{re} V_{re} \left(\frac{l_{CNT}}{l_c} \right) + \sigma_m (1 - V_{re}) \quad \text{for} \quad l \leq l_c \tag{1}$$

$$\sigma_c = \sigma_{re} V_{re} \left(1 - \frac{l_c}{2 l_{CNT}} \right) + \sigma_m (1 - V_{re}) \quad \text{for} \quad l > l_c \tag{2}$$

where σ_{re} (=3.6 GPa (Xie et al. 2000)) and σ_m (=54.49 MPa, calculated here) are the strength of the fiber reinforcements and matrix, respectively. V_{re} is the volume fraction of reinforcements, and l_c is the critical length, which is the minimum length of the reinforcement carrying a maximum stress at its midpoint. The critical length can be defined as:

$$l_c = \frac{d \sigma_{re}}{2 k \tau_m} \tag{3}$$

where d and τ_m are the average diameter of the fibers and shear strength of the matrix (=$0.5\sigma_m$), respectively. In addition, k indicates the efficiency of the load transfer and is equal to 1, assuming that the matrix completely transfers the load to the fiber reinforcements. The calculated l_c value is around 0.98 um which is lower than the length of carbon nanotube used in this study (which is more than 5 um). Thus, the yield strength of the composites was estimated in the condition of $l > l_c$ (Eq. 1). Figure 19c shows the experimental and theoretical variation of yield strength as a function of CNTs weight fraction (wt.%). As is obvious, the discrepancies of YS between the two curves increase with an increase of the modified-CNT amounts, especially more than 1.5 wt.%. This drastic difference may be attributed to the aggravation of TiC-modified-CNTs clustering in the samples with increasing the amount of CNTs. However, the experimentally obtained values reasonably agree with the predicted strength for composites reinforced with 0.5, 1, and 1.5 wt.% modified-CNTs (marked area in Fig. 19c). As mentioned before, it is assumed that the applied load can be transferred entirely from the matrix to the fiber reinforcement. Generally, the load transfer efficiency is below one due to the weak interfacial bonding between matrix and fibers. However, the reasonable match at the marked region of Fig. 19c indicates that the load transfer efficiency from the aluminum matrix to the modified-CNTs is almost one. Therefore, the ideal reinforcements, produced here, have an appropriate interfacial bonding with the matrix through the formation of carbide zones, leading to the significant strengthening in the modified-composites.

During applied compression test, the composite samples were not broken because of high ductility of the aluminum matrix. However, there were some severe and deep cracks as well as small fractured parts, especially, on the surface of the samples containing higher amounts of modified-CNTs. There were also some microcracks on the surface of the specimens containing low amount of CNTs. Figure 21a shows the

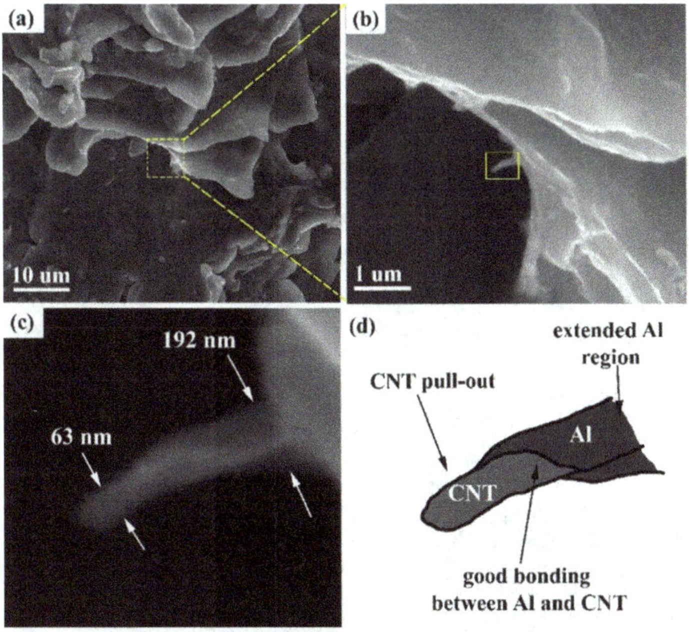

Fig. 21 (**a–c**) SEM micrographs at different magnifications showing a pulled-out CNT. (**d**) Schematic representation of one carbon nanotube covered by Al sheath as a result of strong CNT/Al interface (Saba et al. 2017)

SEM image taken from a microcrack located on the surface of the sample reinforced with 1.5 wt.% modified-CNTs. Figure 21b illustrates a high magnification SEM image of the marked region in Fig. 21a. In addition, high magnification image of the pulled-out CNT, taken from the marked area in Fig. 21b, is shown in Fig. 21c. As is obvious, the pulled-out carbon nanotube is covered by aluminum sheath because this part of the matrix was extended by the tube during deformation. This is an evidence for the strong bonding between the reinforcement and Al matrix, indicating a load transfer process. The schematic image, provided in Fig. 21d, also illustrates the extended aluminum region as a result of a good bonding at the CNT-matrix interface.

Figure 22a, b shows nano-cracks located at the surface of the samples after applying compression test. As can be seen, nanotubes can act as a bridge preventing propagation of the cracks. This phenomenon is called bridging which can be accounted as a main strengthening mechanism. Figure 22c and d are some examples of CNT pulling out. Moreover, as is shown in Fig. 22d by the arrow, the inner tube of the CNT is pulled out due to the interlayer sliding. This behavior can be referred to the telescopic fracture at the CNT tip. It can be said that the presence of nanocarbides on the surface of nanotubes links the walls causing the mentioned phenomenon (Saba et al. 2017). In addition, accumulation of dislocations behind carbon nanotubes is shown in Fig. 23. It should be noted that the dark regions in the figure demonstrate highly deformed parts of the sample due to the presence of high dislocations density (Fig. 23a–d).

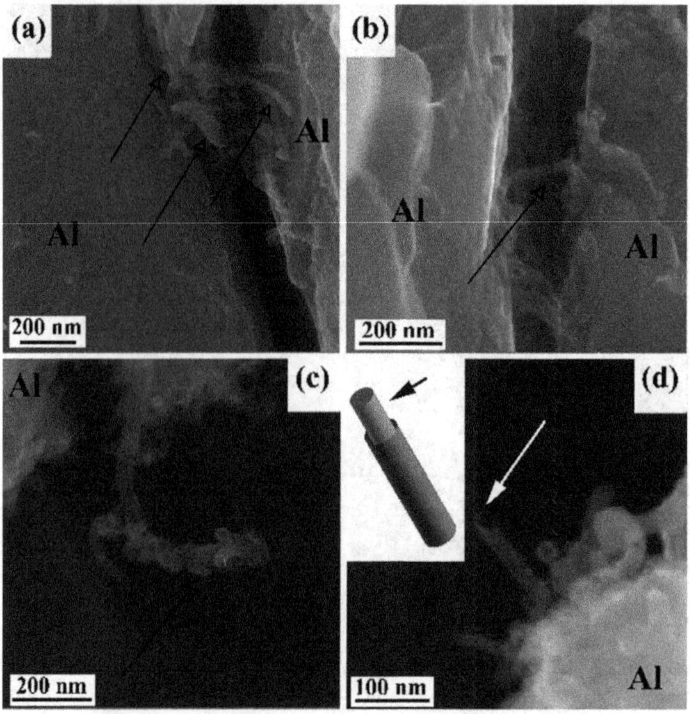

Fig. 22 CNT bridging (**a**, **b**) and pulling out (**c**, **d**)

Figure 24 shows TEM results of the individually dispersed modified-CNTs (red dashed lines) in the bulk composite containing 1.5 wt.% CNTs. It should be noted that the regular CNT walls were not observed. The reason is that in the composite, the overlapping of carbon nanotube and matrix atoms can induce difficulties in observing CNT structure during TEM study. On the other hand, infiltration of Al atoms into the CNT structure increases the irregularity and diameter of nanotubes (Chen et al. 2016; Choi et al. 2009). Choi (Choi et al. 2009) reported that during milling course, carbon nanotubes can be filled with Al atoms, getting enlarged in diameter up to around $3.5 \times d_o$. The similar phenomenon might be the reason for CNT diameter increase observed in Fig. 21c. The Raman G-band peak shift may be originated from the infiltration of Al atoms in CNTs, causing slight distortion of sp^2 bonding structure of CNTs (Chen et al. 2016). Nano-sized TiC can be observed clearly on the surface of the tubes. The carbide morphologies are in the form of nanolumps (Fig. 24a, b, d) and nanoblocks (Fig. 24c) with the size less than around 100 nanometers. Electron diffraction pattern of Fig. 24a also confirm the TiC formation on the surface of CNTs, indicating an epitaxial growth of the carbides. In addition, HRTEM image along with EDS analysis provided as insets in Fig. 24c can be regarded as another proof-validating TiC formation on the surface of CNT. As mentioned before, it is expected to have preferred nucleation sites for titanium

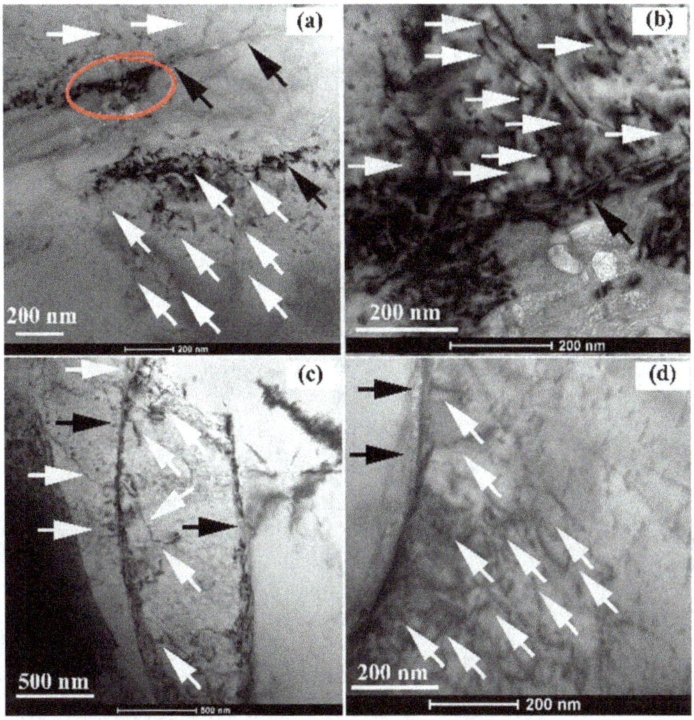

Fig. 23 Accumulation of dislocations behind CNTs

carbides at the defects sites of the MWCNTs. It can be assumed that the carbides are preferentially generated at the defect sites of the tubes by chemical reaction between titanium and dangled carbon atoms and then grow up. These attached particles can link the inner walls of the MWCNTs and by that means contribute to the load-bearing and load transfer phenomena (see inset in Fig. 24d). Combining the TEM observations with the compressive strength results, it can be confirmed that the formation of titanium carbide on the surface of the modified-CNTs can drastically enhance the load-bearing capacity, improving the mechanical properties of the resultant composites.

3.11 Effect of CNT Content on the Tribological Behavior of the Composites

Figure 25a, b, respectively, shows the mean friction coefficient and wear weight loss of the modified-composites as a function of CNT content. As can be seen, in the case of 1.5 wt.% modified-CNT, the mean friction coefficient and wear weight loss are at

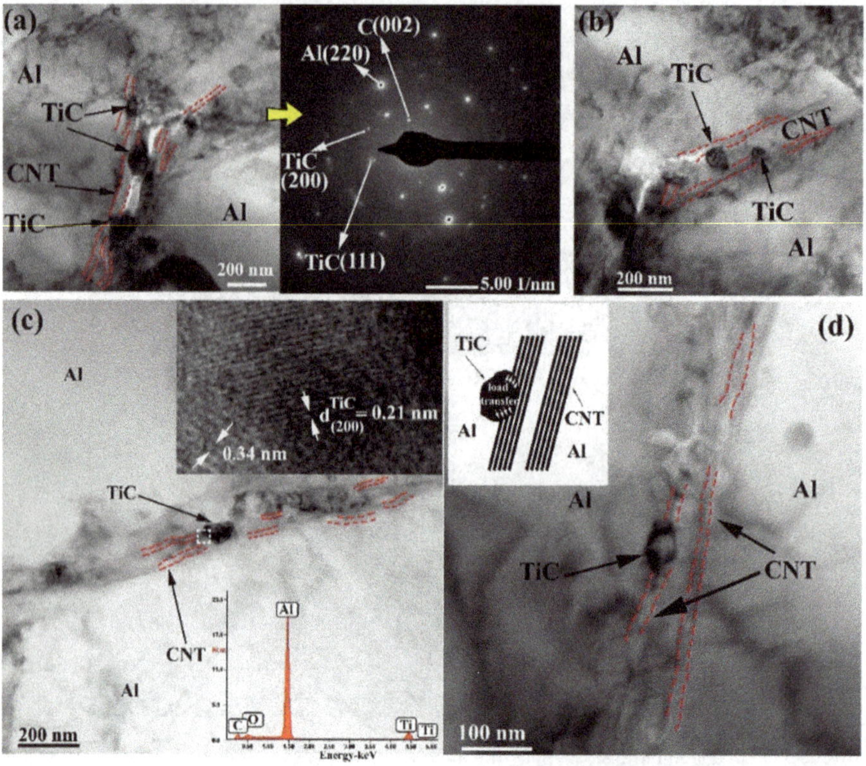

Fig. 24 (**a–d**) Typical TEM images of the bulk sample containing 1.5 wt.% modified-CNTs. Schematic illustration of multilayer load distribution by nanostructured TiC at the radially unzipped defect sites of CNTs (inset in Section **d**). Dashed lines show the nanotubes (Saba et al. 2017)

their lowest values. As compared with pure Al and 1.5 wt.% unmodified-CNT/Al samples, tribological properties are highly improved by using TiC-modified CNTs. A reduction of around 68% and 61% in mean friction coefficient as well as a reduction of about 88% and 84.5% in wear weight loss was achieved for 1.5 wt.% modified-CNT/Al composite compared to pure Al and 1.5 wt.% unmodified sample, respectively. In addition to the before-mentioned explanation of the mechanical properties, it can also be said that when a little CNT content is added, mechanical properties of the composite are improved because nanotubes can fill microvoids of the metal particles. However, if a large amount of CNT is added, the excess CNT which remains after filling the microvoids may form agglomerations and clusters, interrupting the sintering process and producing defects in the composite samples (Kim et al. 2009).

SEM micrographs of representative worn surfaces of the composites reinforced with different amounts of modified-CNTs along with pure aluminum sample and 1.5 wt.% unmodified-CNT/Al composite are presented in Fig. 26. As can be clearly

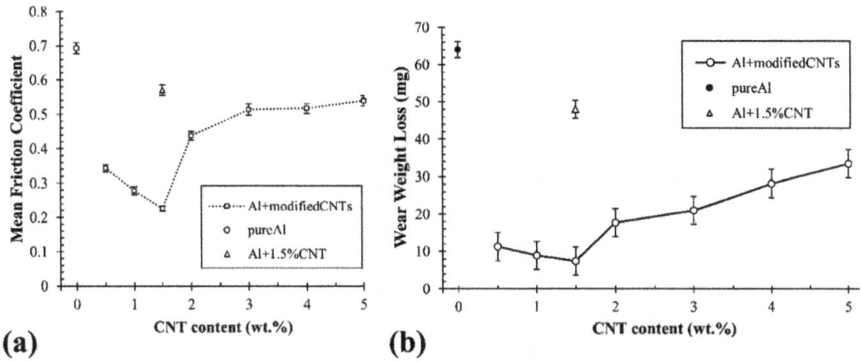

Fig. 25 Variation of mean friction coefficient (**a**) and wear weight loss (**b**) as a function of CNT content

seen, the surface roughness of the composites containing 1 wt.% (Fig. 26a, b) and 1.5 wt.% (Fig. 26c, d) modified-CNTs, compared to the other samples, is the lowest with minimum plastic deformations, having no evidence of delamination, exfoliation, and microploughing that can be observed in the Al composites reinforced with 2 wt.% (Fig. 26e, f), 3 wt.% (Fig. 26g, h), and 5 wt.% (Fig. 26i, j) modified-CNTs as well as pure Al (Fig. 26k) and 1.5 wt.% unmodified-CNT/Al (Fig. 26l) samples. Evaluating the worn surface of the 1.5 wt.% modified-CNTs/Al composite reveals less severe scars and material delamination compared to the others, which is consistent with the results provided by the tribological experiments (Fig. 25). In addition, an obvious transition in the wear mechanism can be observed with increasing the CNT content. The dominant wear mechanism changes from adhesion (for 2, 3, and 5 wt.% modified-CNTs) to abrasion (for 1 and 1.5 wt.% modified-CNTs), generating smooth surfaces without any exfoliation and flake-like wear scars on the worn surfaces at the lower CNT contents which are typical characteristics of abrasive wear. As mentioned before, at higher amounts of CNTs ($\geq$ 2 wt.%), nanotube clusters serve as a source of crack nucleation and cause rigorous micro-fragmentation resulting in poor wear resistance of the composite. The presence of a void as well as a microcrack, which were produced during the wear experiment of 5 wt. % modified-CNTs/Al composite, is illustrated in Fig. 26i, j, respectively. Moreover, a long crack can also be observed in the SEM image of the worn surface of the pure Al (Fig. 26k). It seems that presence of the brittle oxide inclusions, which may be initiated from the grinding steel disk, could facilitate the propagation of the microcracks in the subsurface regions of the pure Al and causes the apparent delamination of the worn surface (Zhou et al. 2007). It was reported that wear resistance of a sample produced by powder metallurgy route is very sensitive to porosity. As porosity increases, the wear rate increases rapidly because they serve as a source of crack nucleation (Al-Qutub et al. 2013). Therefore, when the amount of CNTs is more than a critical extent (1.5 wt.%, here), clustering would increase porosity in the samples, resulting in the reduction of mechanical and tribological

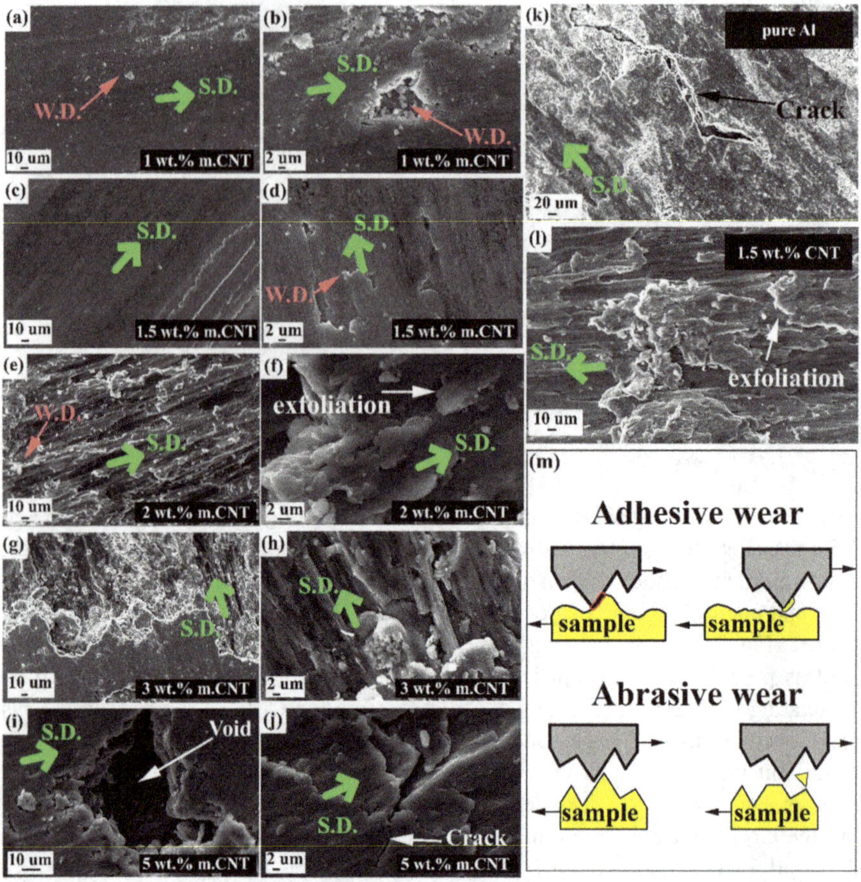

Fig. 26 SEM micrographs of the worn surfaces with varying modified-CNT content; (**a–b**) 1 wt.%, (**c–d**) 1.5 wt.%, (**e–f**) 2 wt.%, (**g–h**) 3 wt.%, (**i–j**) 5 wt.%, (**k**) pure Al, (**l**) 1.5 wt.% unmodified-CNT. Abbreviations of m.CNT, S.D., and W.D. refer to the modified-CNT, sliding direction, and wear debris, respectively. (**m**) Schematic representation of adhesive and abrasive wear mechanisms

properties. Comparing the worn surfaces of Fig. 26c, d with that of Fig. 26l, it was revealed that the distinct wear characteristics can be made by adding modified-CNTs into the composite. In the case of 1.5 wt.% unmodified-CNTs/Al (Fig. 26l), the dominant wear mechanism is the adhesion with a significant exfoliation. However, after adding 1.5 wt.% modified-CNTs to the aluminum matrix, the dominant wear mechanism changes to the abrasion wear (Fig. 26c). Higher magnification SEM image of Fig. 26d also reveals that the wear tracks are featureless and smooth along the sliding direction (S.D.). It can be concluded that the modified-CNT addition extremely reduced the severity of the adhesion mechanism compared to the unmodified samples. Figure 26m also shows schematic representation of the adhesive and abrasive wear mechanisms.

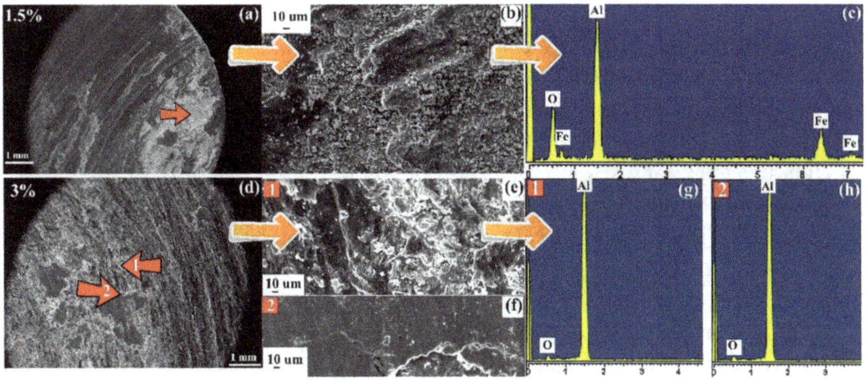

Fig. 27 SEM images along with corresponding EDX analyses of the worn surfaces of (**a–c**) 1.5 wt.% and (**d–h**) 3 wt.% modified-CNT/Al composites

Figure 27 presents the SEM micrographs of typical worn surfaces of the 1.5 wt.% (Fig. 27a, b) and 3 wt.% (Fig. 27d–f) modified-CNT/Al nanocomposites with corresponding EDS analysis (Fig. 27c, g, h). According to Fig. 27a, compared with Fig. 27d, an area of continuous whitish phase can be observed on the worn surface of the sample containing 1.5 wt.% modified-CNTs (marked by a red arrow). EDS elemental mapping analysis was conducted to identify the chemical composition of the whitish phase, and the result is shown in Fig. 27c. The EDS result of the higher magnification SEM image of Fig. 27b, which was taken from the red arrow marked area of very low magnification image of Fig. 27a, shows that the whitish phase is mainly composed of Fe and O. The iron base oxide, which is observed as a whitish phase, may be deduced to be hematite (Fe_2O_3), generating in an air-rich environment (Lyu et al. 2016). As the samples and steel abrasive discs were carefully cleaned prior to the tests, the iron base oxide is thought to be self-generated during the wear experiment. It was reported that the worn surfaces containing large amount of self-generated oxides indicate mild wear, decreasing wear between the composite samples and contacting steel discs (Lyu et al. 2016). The lubrication effect of the oxide layers can be ascribed to their physical properties. Iron base oxides, with the hardness in the range of around 800–1000 HV (Stevenson et al. 2002), are much harder than the composites, enhancing the tribological properties of the samples. The mentioned characteristic enables iron base oxide to hinder the adhesion mechanism between two contacting surfaces, and thus abrasion mechanism is also reduced. Based on the SEM image of 3 wt.% modified-CNTs/Al composite (Fig. 27d–f) as well as corresponding EDX analysis results provided in Fig. 27g, h, it can be deduced that the abrasive wear was very severe so as the oxide film did not form in the composites containing more than 1.5 wt.% CNTs. Owing to the absence of iron in the both regions 1 (Fig. 27e) and 2 (Fig. 27f) with the evidences provided by EDS results (Fig. 27g, h), it appears that the oxide transition did not occur in this sample. Comparing the higher magnification SEM image of Fig. 27b with that of Fig. 27e, f, it can be observed that oxide debris appear just on the wear surface of the sample reinforced with 1.5 wt.% modified-CNTs. As mentioned earlier, these debris would be transferred from the steel disk. The iron base oxide tribo-layer can protect the worn surface of the sample, contributing to an increase the wear resistance. The higher strength of 1.5 wt.% modified-CNTs/Al composite sample can make the substrate more stable to support the oxides formed on the wear surface and therefore can improve the wear resistance of the sample (Zafari et al. 2014). On the other hand, owing to the higher plastic deformation at the contact zone of the composite containing 3 wt.% modified-CNTs, no oxide phases can be supported on the wear surface, leading to the higher wear rate at the higher amounts of nanotubes. The observations suggest that, in addition to the abrasive wear mechanism, oxidation wear also contributes to the mechanisms that take place in the samples reinforced with lower CNT contents. Therefore, abrasive and oxidation are identified as the main wear mechanisms of the composites containing 1.5 wt.% modified-CNTs (mild wear regime), while adhesion and severe plastic deformation were operative wear mechanisms above the mentioned content (severe wear regime).

4 Conclusions

MWCNTs were modified by nanocrystalline TiC particles using pressureless-SPS method for a short time. Nano-TiC layer was synthesized through a three-step process: (1) mechanical crushing of Ti and CNT powder via wet-milling and ultrasonication methods, (2) adding a pristine CNT into the crushed mixture of Ti/CNT via wet-milling and ultrasonication methods, and (3) applying pressureless-SPS in order to decorating the pristine CNTs with TiC. XRD results showed that the formed layer is TiC at all ratios. Formation mechanism of nano-TiC on the surface of MWCNTs and its effect on the properties of Al composites was studied. It was shown that the morphology of the deposited carbides was highly related to the carbon sources. Two types of TiC were mostly nanolumps and nanoblocks with the interfacial orientation relationship of TiC (200) // MWCNT (002). Different preferential nucleation sites for the formation of TiC, such as bamboo-like defects, radially unzipped sites, and so on, were evaluated. It was observed that the inner walls of the modified-CNTs were linked together by TiC, leading to the effective load transfer between the walls which was in agreement with the shear-lag model. TG analysis showed that a significant improvement in the thermal stability was achieved for modified-CNTs, compared with unmodified-CNTs. It was found that the optimal TiC/CNT mixing ratio of 4:2 can highly improve mechanical properties of the nanocomposites. The results revealed that the composite sample containing 1.5 wt.% modified-CNT with the mixing ratio of 4:2 shows the greatest mechanical and tribological properties. The optimal sample exhibited 176% and 71% increase in Vickers hardness, compared to pure Al and 1.5 wt.% unmodified-CNT/Al composite sample, respectively. 82% and 45% increase in nanohardness as well as 48% and 25.5% increase in Young's modulus were achieved for the modified sample compared to pure Al and 1.5 wt.%

unmodified sample, respectively. According to the compression test results, 55%, 28.5%, and 21.6% increment in yield strength was obtained for Al + 1.5 wt.% modified-CNTs (84.36 MPa) with the TiC/CNT mixing ratio of 4:2 compared with pure Al (54.49 MP), Al + 1.5 wt.%CNT (65.64 MPa), and Al + 1.5 wt.% [pure TiC +CNT] (69.35 MPa). The experimental yield strength values of the modified-composites were consistent with the theoretical values estimated by the shear-lag model. In the case of tribological characteristics, reduction of around 68% and 61% in mean friction coefficient and reduction of about 88% and 84.5% in wear weight loss were obtained for 1.5 wt.% modified-CNT/Al composite, compared to the pure Al and 1.5 wt.% unmodified sample, respectively. It was found that the combination of abrasive and oxidation wear contribute to the mechanisms of the samples with the lower amount of CNTs (less than 1.5 wt.%), while adhesive wear mechanism dominates in the composites with the CNT content of more than 2 wt.%.

Acknowledgments The authors appreciate the financial supports of Ferdowsi University of Mashhad regarding the PhD program with the number of 3/31889 (adopted on 29 October 2014). In addition, the authors gratefully acknowledge the financial supports from the Natural Science Foundation of Jiangsu Province (No. BK20161419), Jiangsu Key Laboratory for Advanced Metallic Materials (No. BM2007204) at Southeast University, and Scientific Research Foundation for the Returned Overseas Chinese Scholars at the State Education Ministry (No. 2015–1098).

References

Journal Article

Al-Qutub AM, Khalil A, Saheb N et al (2013) Wear and friction behavior of Al6061 alloy reinforced with carbon nanotubes. Wear 297:752–761. https://doi.org/10.1016/j.wear.2012.10.006

Annu A, Bhattacharya B, Singh PK et al (2017) Carbon nanotube using spray pyrolysis: recent scenario. J Alloys Compd 691:970–982. https://doi.org/10.1016/j.jallcom.2016.08.246

Bacon R (1960) Growth, structure, and properties of graphite whiskers. J App Phys 31:283–290. https://doi.org/10.1063/1.1735559

Bakshi SR, Agarwal A (2011) An analysis of the factors affecting strengthening in carbon nanotube reinforced aluminum composites. Carbon 49:533–544. https://doi.org/10.1016/j.carbon.2010.09.054

Banks CE, Compton RG (2006) New electrodes for old: from carbon nanotubes to edge plane pyrolytic graphite. Analyst 131:15–21. https://doi.org/10.1039/B512688F

Banks CE, Ji X, Crossley A et al (2006) Understanding the electrochemical reactivity of bamboo multiwalled carbon nanotubes: the presence of oxygenated species at tube ends may not increase electron transfer kinetics. Electroanalysis 18:2137–2140. https://doi.org/10.1002/elan.200603633

Bokhonov BB, Dudina DV (2016) Preparation of porous materials by spark plasma sintering: peculiarities of alloy formation during consolidation of Fe@ Pt core-shell and hollow Pt (Fe) particles. J Alloys Compd 707:233–237. https://doi.org/10.1016/j.jallcom.2016.09.146

Bradbury WL, Olevsk EA (2010) Production of SiC-C composites by free pressureless spark plasma sintering (FPSPS). Scr Mater 63:77–80. https://doi.org/10.1016/j.scriptamat.2010.03.009

Chen R, Li Y, Huo K et al (2013) Microelectrode arrays based on carbon nanomaterials: emerging electrochemical sensors for biological and environmental applications. RSC Adv 3:18698–18715. https://doi.org/10.1039/C3RA43033B

Chen B, Li S, Imai H et al (2015a) Inter-wall bridging induced peeling of multi-walled carbon nanotubes during tensile failure in aluminum matrix composites. Micron 69:1–5. https://doi.org/10.1016/j.micron.2014.10.005

Chen B, Li S, Imai H et al (2015b) Load transfer strengthening in carbon nanotubes reinforced metal matrix composites via in-situ tensile tests. Compos Sci Technol 113:1–8. https://doi.org/10.1016/j.compscitech.2015.03.009

Chen B, Kondoh K, Imai H et al (2016) Simultaneously enhancing strength and ductility of carbon nanotube/aluminum composites by improving bonding conditions. Scr Mater 113:158–162. https://doi.org/10.1016/j.scriptamat.2015.11.011

Cho S, Kikuchi K, Miyazaki T et al (2013) Epitaxial growth of chromium carbide nanostructures on multiwalled carbon nanotubes (MWCNTs) in MWCNT-copper composites. Acta Mater 61:708–716. https://doi.org/10.1016/j.actamat.2012.10.022

Choi HJ, Kwon GB, Lee GY et al (2008) Reinforcement with carbon nanotubes in aluminum matrix composites. Scr Mater 59:360–363. https://doi.org/10.1016/j.scriptamat.2008.04.006

Choi H, Shin J, Min B et al (2009) Reinforcing effects of carbon nanotubes in structural aluminum matrix nanocomposites. J Mater Res 24:2610–2616. https://doi.org/10.1557/jmr.2009.0318

Deng C, Zhang X, Wang D et al (2007) Preparation and characterization of carbon nanotubes/aluminum matrix composites. Mater Lett 61:1725–1728. https://doi.org/10.1016/j.matlet.2006.07.119

Dresselhaus MS, Dresselhaus G, Saito R et al (2005) Raman spectroscopy of carbon nanotubes. Phys Rep 409:47–99. https://doi.org/10.1016/j.physrep.2004.10.006

Dudiy SV, Lundqvist BI (2004) Wetting of TiC and TiN by metals. Phys Rev B 69:125421–125516. https://doi.org/10.1103/PhysRevB.69.125421

Esawi AMK, Morsi K, Sayed A et al (2010) Effect of carbon nanotube (CNT) content on the mechanical properties of CNT-reinforced aluminum composites. Compos Sci Technol 70:2237–2241. https://doi.org/10.1016/j.compscitech.2010.05.004

Estili M, Kawasaki A (2010) Engineering strong intergraphene shear resistance in multi-walled carbon nanotubes and dramatic tensile improvements. Adv Mater 22:607–610. https://doi.org/10.1002/adma.200902140

Estili M, Sakka Y (2014) Recent advances in understanding the reinforcing ability and mechanism of carbon nanotubes in ceramic matrix composites. Sci Technol Adv Mater 15:64902–64925

Estili M, Sakka Y, Kawasaki A (2013) Unprecedented simultaneous enhancement in strain tolerance, toughness and strength of Al_2O_3 ceramic by multiwall-type failure of a high loading of carbon nanotubes. Nanotechnology 24:155702–155709

George R, Kashyap KT, Rahul R et al (2005) Strengthening in carbon nanotube/aluminium (CNT/Al) composites. Scr Mater 53:1159–1163. https://doi.org/10.1016/j.scriptamat.2005.07.022

Ibrahim A, Zhang F, Otterstein E et al (2011) Processing of porous Ti and Ti_5Mn foams by spark plasma sintering. Mater Des 32:146–153. https://doi.org/10.1016/j.matdes.2010.06.019

Iijima S (1991) Helical microtubules of graphitic carbon. Nature 354:56. https://doi.org/10.1038/354056a0

Johansson LI (1995) Electronic and structural properties of transition-metal carbide and nitride surfaces. Surf Sci Rep 21:177–250. https://doi.org/10.1016/0167-5729(94)00005-0

Kim I, Lee JH, Lee GS et al (2009) Friction and wear characteristics of the carbon nanotube–aluminum composites with different manufacturing conditions. Wear 267:593–598. https://doi.org/10.1016/j.wear.2008.12.096

Kim IH, Lee W, Lee CW et al (2010) Effect of SiC coating on interfacial reaction between carbon short fiber and Al melt. Surf Interface Anal 42:743–747. https://doi.org/10.1002/sia.3378

Kurita H, Estili M, Kwon H et al (2015) Load-bearing contribution of multi-walled carbon nanotubes on tensile response of aluminum. Compos Part A 68:133–139. https://doi.org/10.1016/j.compositesa.2014.09.014

Kuzumaki T, Miyazawa K, Ichinose H et al (1998) Processing of carbon nanotube reinforced aluminum composite. J Mater Res 13:2445–2449. https://doi.org/10.1557/JMR.1998.0340
Kwon H, Bradbury CR, Leparoux M (2011) Fabrication of functionally graded carbon nanotube-reinforced aluminum matrix composite. Adv Eng Mater 13:325–329. https://doi.org/10.1002/adem.201000251
Lahiri D, Ghosh S, Agarwal A (2012) Carbon nanotube reinforced hydroxyapatite composite for orthopedic application: a review. Mater Sci Eng C 32:1727–1758. https://doi.org/10.1016/j.msec.2012.05.010
Landry K, Kalogeropoulou S, Eustathopoulos N (1998) Wettability of carbon by aluminum and aluminum alloys. Mater Sci Eng A 254:99–111. https://doi.org/10.1016/S0921-5093(98)00759-X
Lao JY, Li WZ, Wen JG et al (2002) Boron carbide nanolumps on carbon nanotubes. Appl Phys Lett 80:500–502. https://doi.org/10.1063/1.1435062
Li S, Sun B, Imai H et al (2013) Powder metallurgy Ti-TiC metal matrix composites prepared by in situ reactive processing of Ti-VGCFs system. Carbon 61:216–228. https://doi.org/10.1016/j.carbon.2013.04.088
Li S, Su Y, Ouyang Q et al (2016) In-situ carbon nanotube-covered silicon carbide particle reinforced aluminum matrix composites fabricated by powder metallurgy. Mater Lett 167:118–121. https://doi.org/10.1016/j.matlet.2015.12.155
Liang C, Ding L, Wang A et al (2009) Microwave assisted preparation and hydrazine decomposition properties of nanostructured tungsten carbides on carbon nanotubes. Ind Eng Chem Res 48:3244–3248. https://doi.org/10.1021/ie801591x
Liu W, Chai S, Mohamed AR et al (2014) Synthesis and characterization of graphene and carbon nanotubes: a review on the past and recent developments. J Ind Eng Chem 20:1171–1185. https://doi.org/10.1016/j.jiec.2013.08.028
Liu ZY, Zhao K, Xiao BL et al (2016) Fabrication of CNT/Al composites with low damage to CNTs by a novel solution-assisted wet mixing combined with powder metallurgy processing. Mater Des 97:424–430. https://doi.org/10.1016/j.matdes.2016.02.121
Lyu Y, Bergseth E, Olofsson U (2016) Open system tribology and influence of weather condition. Sci Rep 6:32455–32466. https://doi.org/10.1038/srep32455
Mei H, Bai Q, Dassios KG et al (2015) Oxidation resistance of aligned carbon nanotube reinforced silicon carbide composites. Ceram Int 41:12495–12498. https://doi.org/10.1016/j.ceramint.2015.06.002
Molina JM, Voytovych R, Louisa E et al (2007) The surface tension of liquid aluminum in high vacuum: the role of surface condition. Int J Adhes Adhes 27:394–401. https://doi.org/10.1016/j.ijadhadh.2006.09.006
Morisada Y, Miyamoto Y (2004) SiC-coated carbon nanotubes and their application as reinforcements for cemented carbides. Mater Sci Eng A 381:57–61. https://doi.org/10.1016/j.msea.2004.03.055
Morsi K, Esawi AMK, Borah P et al (2010) Properties of single and dual matrix aluminum-carbon nanotube composites processed via spark plasma extrusion (SPE). Mater Sci Eng A 527:5686–5690. https://doi.org/10.1016/j.msea.2010.05.081
Nam DH, Cha SI, Lim BK et al (2012) Synergistic strengthening by load transfer mechanism and grain refinement of CNT/Al-Cu composites. Carbon 50:2417–2423. https://doi.org/10.1016/j.carbon.2012.01.058
Nuriel S, Liu L, Barber AH et al (2005) Direct measurement of multiwall nanotube surface tension. Chem Phys Lett 404:263–266. https://doi.org/10.1016/j.cplett.2005.01.072
Oberlin A, Endo M, Koyama T (1976) Filamentous growth of carbon through benzene decomposition. J Cryst Growth 32:335–349. https://doi.org/10.1016/0022-0248(76)90115-9
Park JG, Keum DH, Lee YH (2015) Strengthening mechanisms in carbon nanotube-reinforced aluminum composites. Carbon 95:690–698. https://doi.org/10.1016/j.carbon.2015.08.112
Peng B, Locascio M, Zapol P et al (2008) Measurements of near-ultimate strength for multiwalled carbon nanotubes and irradiation induced crosslinking improvements. Nat Nanotechnol 3:626–631. https://doi.org/10.1038/nnano.2008.211

Phuong DD, Van Trinh P, Van An N et al (2014) Effects of carbon nanotube content and annealing temperature on the hardness of CNT reinforced aluminum nanocomposites processed by the high pressure torsion technique. J Alloys Compd 613:68–73. https://doi.org/10.1016/j.jallcom.2014.05.219

Quan Y, Zhang F, Rebl H et al (2013) Ti_6Al_4V foams fabricated by spark plasma sintering with post-heat treatment. Mater Sci Eng A 565:118–125. https://doi.org/10.1016/j.msea.2012.12.026

Radushkevich L, Lukyanovich V (1952) About the structure of carbon formed by thermal decomposition of carbon monoxide on iron substrate. J Phys Chem 26:88–95

Rhee SK (1970) Wetting of ceramics by liquid aluminum. J Am Ceram Soc 53:386–389. https://doi.org/10.1111/j.1151-2916.1970.tb12138.x

Saba F, Kabiri E, Vahdati-Khaki J et al (2016a) Fabrication of nanocrystalline TiC coating on AISI D2 steel substrate via high-energy mechanical alloying of Ti and C. Powder Technol 288:76–86. https://doi.org/10.1016/j.powtec.2015.10.030

Saba F, Zhang F, Sajjadi SA et al (2016b) Pulsed current field assisted surface modification of carbon nanotubes with nanocrystalline titanium carbide. Carbon 101:261–271. https://doi.org/10.1016/j.carbon.2016.02.012

Saba F, Sajjadi SA, Haddad-Sabzevar M et al (2017) Formation mechanism of nano titanium carbide on multi-walled carbon nanotube and influence of the nanocarbides on the load-bearing contribution of the nanotubes inner-walls in aluminum-matrix composites. Carbon 115:720–729. https://doi.org/10.1016/j.carbon.2017.01.062

Saba F, Sajjadi SA, Haddad-Sabzevar M et al (2018) TiC-modified carbon nanotubes, TiC nanotubes and TiC nanorods: synthesis and characterization. Ceram Int. https://doi.org/10.1016/j.ceramint.2018.01.233

Selcuk C, Kennedy AR (2006) Al-TiC composite made by the addition of master alloys pellets synthesized from reacted elemental powders. Mater Lett 60:3364–3366. https://doi.org/10.1016/j.matlet.2006.03.021

Singu BS, Male U, Srinivasan P et al (2017) Preparation and performance of polyaniline–multiwall carbon nanotubes–titanium dioxide ternary composite electrode material for supercapacitors. J Ind Eng Chem 49:82–87. https://doi.org/10.1016/j.jiec.2017.01.010

So KP, Jeong JC, Park JG et al (2013) SiC formation on carbon nanotube surface for improving wettability with aluminum. Compos Sci Technol 74:6–13. https://doi.org/10.1016/j.compscitech.2012.09.014

Son SY, Um SH, Jang H-S et al (2017) Effective disentangling method of bundled multi-walled carbon nanotubes into individual multi-walled carbon nanotubes by magnetic-field induction. J Ind Eng Chem 46:28–34. https://doi.org/10.1016/j.jiec.2016.09.030

Stein J, Lenczowski B, Frety N et al (2012) Mechanical reinforcement of a high-performance Al alloy AA5083 with homogeneously dispersed multi-walled carbon nanotubes. Carbon 50:2264–2272. https://doi.org/10.1016/j.carbon.2012.01.044

Stevenson ME, Kaji M, Bradt RC (2002) Microhardness anisotropy and the indentation size effect on the basal plane of single crystal hematite. J Eur Ceram Soc 22:1137–1148. https://doi.org/10.1016/S0955-2219(01)00415-0

Taguchi T, Igawa N, Yamamoto H et al (2005) Synthesis of silicon carbide nanotubes. J Am Ceram Soc 88:459–461. https://doi.org/10.1111/j.1551-2916.2005.00066.x

Taguchi T, Yamamoto H, Shamoto S (2007) Synthesis and characterization of single-phase TiC nanotubes, TiC nanowires, and carbon nanotubes equipped with TiC nanoparticles. J Phys Chem C 111:18888–18891. https://doi.org/10.1021/jp0756909

Taguchi T, Kibria AF, Shamoto S (2013) Morphology change of multi-walled carbon nanotubes with SiC coating by electron irradiation. J Phys Conf Ser, IOP Publishing, 417:012037–012043.

Tjong SC (2006) Structural and mechanical properties of polymer nanocomposites. Mater Sci Eng R 53:73–197. https://doi.org/10.1016/j.mser.2006.06.001

Wang X, Padture NP, Tanaka H (2004) Contact-damage-resistant ceramic/single-wall carbon nanotubes and ceramic/graphite composites. Nat Mater 3:539–544. https://doi.org/10.1038/nmat1161

Wei H, Li Z, Xiong DB et al (2014) Towards strong and stiff carbon nanotube-reinforced high-strength aluminum alloy composites through a microlaminated architecture design. Scr Mater 75:30–33. https://doi.org/10.1016/j.scriptamat.2013.11.014

Xie S, Li W, Pan Z et al (2000) Mechanical and physical properties on carbon nanotube. J Phys Chem Solids 61:1153–1158. https://doi.org/10.1016/S0022-3697(99)00376-5

Yamanoglu R, Gulsoy N, Olevsky E et al (2016) Production of porous $Ti_5Al_{2.5}Fe$ alloy via pressureless spark plasma sintering. J Alloys Compd 680:654–658. https://doi.org/10.1016/j.jallcom.2016.04.176

Yoo SJ, Han SH, Kim WJ (2013a) A combination of ball milling and high-ratio differential speed rolling for synthesizing carbon nanotube/copper composites. Carbon 61:487–500. https://doi.org/10.1016/j.carbon.2013.04.105

Yoo SJ, Han SH, Kim WJ (2013b) Strength and strain hardening of aluminum matrix composites with randomly dispersed nanometer-length fragmented carbon nanotubes. Scr Mater 68:711–714. https://doi.org/10.1016/j.scriptamat.2013.01.013

Zafari A, Ghasemi HM, Mahmudi R (2014) An investigation on the tribological behavior of AZ91 and AZ91+3wt% RE magnesium alloys at elevated temperatures. Mater Des 54:544–552. https://doi.org/10.1016/j.matdes.2013.08.073

Zang JB, Lu J, Wang YH et al (2010) Fabrication of core-shell structured MWCNT-Ti(TiC) using a one-pot reaction from a mixture of $TiCl_3$,TiH_2, and MWCNTs. Carbon 48:3802–3806. https://doi.org/10.1016/j.carbon.2010.06.043

Zhang F, Lin K, Chang J et al (2008) Spark plasma sintering of macroporous calcium phosphate scaffolds from nanocrystalline powders. J Eur Ceram Soc 28:539–545. https://doi.org/10.1016/j.jeurceramsoc.2007.07.012

Zhang F, Reich M, Kessler O et al (2013) Potential of rapid cooling spark plasma sintering for metallic materials. Mater Today 16:192–195. https://doi.org/10.1016/j.mattod.2013.05.005

Zhou S, Zhang X, Ding Z et al (2007) Fabrication and tribological properties of carbon nanotubes reinforced Al composites prepared by pressureless infiltration technique. Compos A Appl Sci Manuf 38:301–306. https://doi.org/10.1016/j.compositesa.2006.04.004

Zhou D, Shenbao J, Yanjun L et al (2013) Effect of stoichiometry on the surface energies of {100} and {111} and the crystal shape of TiC_x and TiN_x. Cryst Eng Comm 15:643–649. https://doi.org/10.1039/C2CE26419F

Book and Chapter

Agarwal A, Bakshi SR, Lahiri D (2011) Carbon nanotubes reinforced metal matrix composites, 2nd edn. Taylor & Francis, New York

Kosolapova TY (2012) Carbides: properties, production, and applications. Springer Science & Business Media, Springer US, New York

Morisada Y, Miyamoto Y (2006) Nanostructured coatings on advanced carbon materials. In: Ceramic-matrix composites: microstructure, properties and applications, Woodhead Publishing, Cambridge CB1 6AH, England

Pujado MP (2012) Carbon nanotubes as platforms for biosensors with electrochemical and electronic transduction, 1st edn. Springer

Al-Based Nanocomposites Produced via Spark Plasma Sintering: Effect of Processing Route and Reinforcing Phases

Pasquale Cavaliere, B. Sadeghi, M. Shamanian, and F. Ashrafizadeh

1 Introduction

Particles-reinforced aluminum-based composites can be produced by employing numerous processing routes based on the general processing techniques of casting or powder metallurgy. In general, powder metallurgy techniques allow to obtain components in net shape through pre-alloyed particles applying pressure and heat in controlled atmosphere. In general, they allow the production of components and materials which are difficult to be realized with conventional routes. Among the powder metallurgy technologies, spark plasma sintering represents the scientific and technological novelty able to lead to the obtaining of sound components with reduced porosity and enhanced mechanical properties. An interesting example is shown in Ghasali et al. (2015) where the same material (Al-10wt%ZrO_2-1wt%Co) was produced with two different processing routes, microwave and spark plasma sintering. The SPSed material showed higher density and mechanical strength if compared with those of the microwave sintered one. The authors performed the same type of comparison for Al-1.5 BC-1.5Co nanocomposite (Ghasali et al. 2016a), showing the best properties for the SPSed composite. Here the improved microstructural and mechanical behavior was thought to be due to the lack of formation of Al_3BC phase due to the interfacial reaction between aluminum and boron carbide particles with the decrease in sintering temperature in SPS process. Some experimental evidences show the potential of solid-state reactions during sintering leading to strong strength enhancement (Ghasali et al. 2016a; Kubota et al. 2008). Here Al-15 at.% MgB_2 composite was SPSed and then heated at 600 °C for 24 h. The

P. Cavaliere (✉)
Department of Innovation Engineering, University of Salento, Lecce, Italy
e-mail: pasquale.cavaliere@unisalento.it

B. Sadeghi · M. Shamanian · F. Ashrafizadeh
Department of Material Engineering, Isfahan University of Technology, Isfahan, Iran
e-mail: shamanian@cc.iut.ac.ir; ashrafif@cc.iut.ac.ir

© Springer Nature Switzerland AG 2019
P. Cavaliere (ed.), *Spark Plasma Sintering of Materials*,
https://doi.org/10.1007/978-3-030-05327-7_6

strengthening phases were recognized in the precipitation of AlB_2, Al_3BC, and spinel $MgAl_2O_4$. The authors also underline that for a given composition, solid-state reaction occurs at relatively lower heating temperatures when the milling time is increased. The detailed description of the solid-state reactions leading to reinforcements precipitation is described in details in Kubota (2010). In Ghasali et al. (2016b) the authors prepared Al-10wt%VC metal matrix composite by spark plasma, microwave, and conventional sintering processes. Compared to microwave and conventional sintering methods, spark plasma sintering had significant enhancement on the microstructure and mechanical properties of aluminum-VC composite. The nearly full dense sample with highest bending strength (295 ± 15 MPa) and microhardness (232 ± 16 Vickers) was obtained by SPS. Microwave sintering led to the decomposition of VC and formation of Al_3V compounds. However, porosities in the microstructure of composite prepared by microwave heating resulted in weak mechanical properties. In Ghasali et al. (2016c) the authors demonstrate the improved properties of SPSed Al-15wt%Zr with the same material produced with traditional powder metallurgy techniques, also in the case of low-temperature SPS (450 °C). SPS is conducted by applying pulsed high current and pressure in order to compact the pre-mixed powders. The process is characterized by high heating rates (in the order of 1000 °C/min) and consequently reduced sintering times (in the order of few minutes). The heating power is not only distributed over the volume of the powder compact homogeneously in a macroscopic scale, but moreover the heating power is dissipated exactly at the locations in the microscopic scale, where energy is required for the sintering process, namely, at the contact points of the powder particles. In Olevsky et al. (2007), it is reported that sufficiently high heating rates would essentially skip the low-temperature surface diffusion mechanism that is readily apparent in conventional sintering. Surface diffusion primarily allows the development of a neck between particles and is classified as a non-densifying mechanism of sintering. Avoiding this neck development maintains the small radius of curvature of pore tips at higher temperatures. When processing temperatures reach that sufficient for grain boundary diffusion, this smaller pore-tip radius drives higher diffusion rates and ultimately more rapid densification kinetics. The effect of thermal diffusion can be significant, especially for small particle sizes. Its contribution is considerable not only in the case of free sintering and low-pressure-assisted sintering, but it competes with the contribution of external pressure-driven power-law creep within certain temperature ranges (Olevsky and Froyen 2009). The very short sintering time in SPS makes this technique very attractive because of the very reduced crystallite growth during heating. In Saheb et al. (2014) the authors demonstrate how the crystallite growth of Al-SiC composites is very low if compared with the one recorded for similar materials sintered with traditional techniques. In Kwon et al. (2010) the authors hypothesized that a pinning effect from the pre-existing aluminum oxide surface (not destroyed during sintering) shell inhibited any grain boundary movement. Anyway the starting powders microstructure results fundamental in retaining a very fine grain size. Through cryomilling and SPS, it has been observed that powders with a grain size on the order of ~25 nm prior to SPS were able to maintain a final grain size of some 50 nm (Kubota and Wynne 2007).

Using this milling and sintering technique, the tensile strength of Al5083 has been shown to exceed 700 MPa; significantly higher than that achievable through any conventional means. For the Al-5356/B_4C composite produced with the same processing route, a nanostructured matrix and high material strength are achieved (Vintila et al. 2011). The same evidences are reported for Al 5083/SiCp (Bathul et al. 2012). The enhanced consolidation of the cryomilled powders may be attributed to the following two factors. First, the nanostructure may facilitate the densification process. Studies have shown that nanoparticles have higher surface energy. Therefore, they exhibit increased activity such as faster diffusion and lower onset sintering temperature in comparison with coarse grains. Accordingly, dense samples can be prepared at relatively low temperatures, which is beneficial to grain growth control. In addition, the nanostructured grains in the cryomilled material are likely to promote increased diffusive processes during SPS. The consolidation process is closely related to the grain boundary diffusion or lattice diffusion at the intermediate and final stages of sintering.

2 Effect of Processing Parameters

The pulsed voltage and currents provide the joule heating of the particles contact points, thanks to the spark discharge between the powders and the die (Anselmi-Tamburini et al. 2005a). As demonstrated in Olevsky and Froyen (2009), the thermal effects can be divided into four categories, as follows: high heating rates, high local temperature gradients, highly nonuniform local temperature distributions, and highly nonuniform macroscopic temperature distributions. As previously discussed, high heating rates allow for increased sinterability. The other effects can be summarized as temperature differentials within a compact. Together, they promote thermal diffusion while simultaneously producing internal stress fields that cause dislocation creep. In instances of very small contact area between adjacent particles, localized melting has been shown, which instantaneously welds particles.

The compaction is due to the high pressure applied during heating (Chen et al. 2005). The pressure has a mechanical role as well as an intrinsic role. Mechanically the pressure has a direct effect on particle rearrangement and the destruction of agglomerates, particularly in the case of nanometric powders. When the particle size is small, the relative contribution of the pressure is small but becomes significant as the particle size increases (Munir et al. 2006). An increased driving force for sintering is the result of the application of a pressure. This leads to a decrease in the sintering temperature and a limitation of grain growth. However, densification through these early-stage mechanisms is limited, as the pressures permissible/applied using a graphite die are typically an order of magnitude smaller than that those routinely attained with conventional tool steel dies and well below the yield strength of the material being consolidated. However, with increasing process temperature, the yield strength of most metallic powders drops significantly. In this weakened state, the applied SPS pressure is then sufficient to invoke significant plastic deformation and, in turn, densification (Garay 2010).

Obviously, the materials properties are dependent on the tunable processing parameters such as heating rate, sintering temperature, pressure rate, maximum pressure, and cooling rate. Heating rate has a direct consequence on mechanisms which are fundamental in the evolution of microstructure; such mechanisms are surface diffusion, grain boundary diffusion, and power-law creep influencing the final sintered materials grain size (Olevsky and Froyen 2006). Rearrangement and localized deformation are enhanced by an increase in the particle size, due to its effect on the local overheating (Diouf and Molinari 2012). For this reason, low temperature and high pressure are indicated as optimal combinations in SPS of pure aluminum in order to retain nanostructure in the sintered material (Liu et al. 2014). Grain size decreases with increasing the heating rate (Fig. 1). Figure 1 refers to particle starting dimensions of 30 μm. Obviously, a reduction in the starting particle dimensions enhances the heating rate effect on the grain size reduction. Maximum temperature has a direct influence on the sintered materials density. Basing on the studies of Garay (2010), a linear relationship is established between the sintering temperature and the final materials density (Eq. 1).

$$\rho = s\frac{T}{T_{\mathrm{m}}} + b \tag{1}$$

where ρ is the sinter density, T the sintering temperature, T_{m} the material melting point, s the material temperature sensitivity, and b is the intercept on density axis.

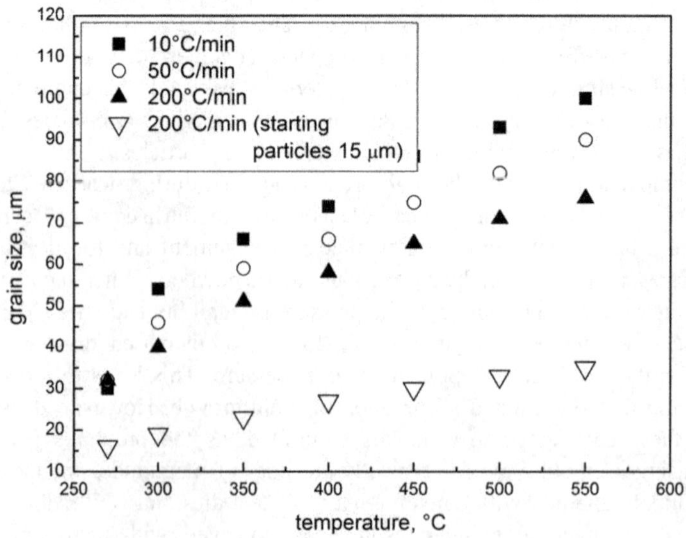

Fig. 1 Aluminum grain size as a function of heating rate of spark plasma sintering

The sensitivity is high for carbon-based materials and ceramics (1.5–3), while it is lower for metals (0.5–1). Increasing the sintering temperature is helpful to improve the final density of the consolidated compact. Relative density increases significantly with sintering temperature until a nearly fully dense condition can be realized. In Saheb et al. (2015) this aspect is demonstrated for Al2124 + 1wt%CNT nanocomposites. Here the materials density and hardness linearly increase as both temperature and holding time increase. Normally, with the same temperature and pressure, an increase in density and hardness is recorded for SPSed Al-based materials (Saheb et al. 2015; Milligan et al. 2013). When a high pressure is applied at a lower temperature, the stability of the hard reinforcing particles prevents diffusion and plastic flow, thus leaving more closed pores. The application of a high pressure will reduce the distances required for mass transport, enhancing diffusion processes (Liu et al. 2012). Experiments performed on Al-1wt%SiC (with ceramic particles in the range 38–53 μm) report that, for increasing density and hardness, the influence of applied pressure was more pronounced than heating rate while the influence of sintering temperature was more significant than sintering time (Aliyu et al. 2015). In Lalet et al. (2014) the authors show how the pulse condition of spark plasma sintering affected the aluminum particle boundaries and aluminum/carbon fiber interfaces. Our studies performed on different types and percentages of reinforcing phases in SPSed Al-based nanocomposites show that the type of reinforcement has a direct influence on materials hardness and density (Fig. 2).

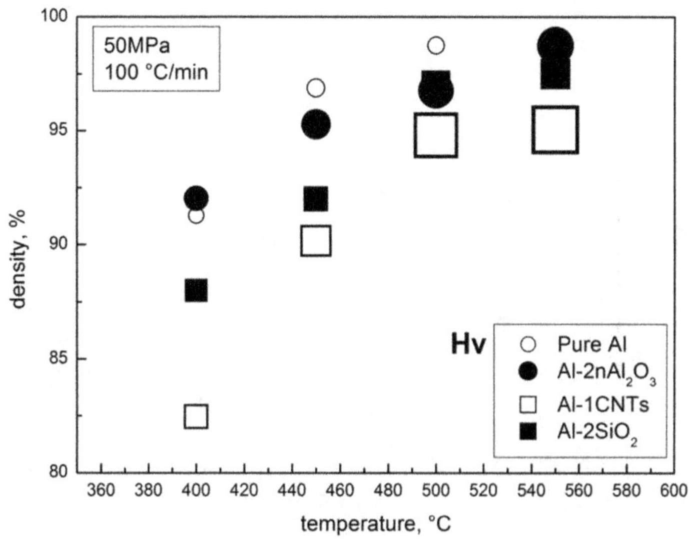

Fig. 2 Density and hardness vs. sintering temperature for different reinforced nanocomposites SPSed at a pressure of 50 MPa employing a heating rate of 100 °C/min

Nanosized alumina particles result very effective in increasing the materials density; CNTs addition seems to be the most effective in increasing materials hardness.

By observing the effect of monomodal- and bimodal-sized particles addition on the materials properties (Fig. 3), it is clear how the addition of 10 wt% of microsized alumina particles is very effective in increasing the materials density at the highest sintering temperature; on the contrary the density results very low at the lowest sintering temperatures. Among the bimodal reinforced composites, the Al-4n6mAl_2O_3 one shows the best properties in terms of density and hardness. This aspect will be clarified in the following discussion on the effect of particles size and percentages on the ceramic phase agglomeration during sintering.

By observing the effect of the heating rate for different types of reinforcements (Fig. 4), it is clear how carbon nanotubes (CNTs) are very effective in increasing the material hardness for a broad range of heating rates. For all the composites, the maximum hardness is achieved for a heating rate of 100 °C/min. Among all monomodal- or bimodal-reinforced composites, the Al-4n6mAl_2O_3 composite shows the highest hardness values for all the employed heating rates (Fig. 5).

For all the materials, the highest hardness is experienced when a heating rate of 100 °C/min is employed during sintering.

The pressure effect is well known in its mechanical and intrinsic behavior. SPS involves the simultaneous application of a uniaxial pressure and a high-intensity electric discharge. Surface energy provides an inherent sintering stress that is assisted by the external pressure but degraded by entrapped gas (Olevsky and Froyen

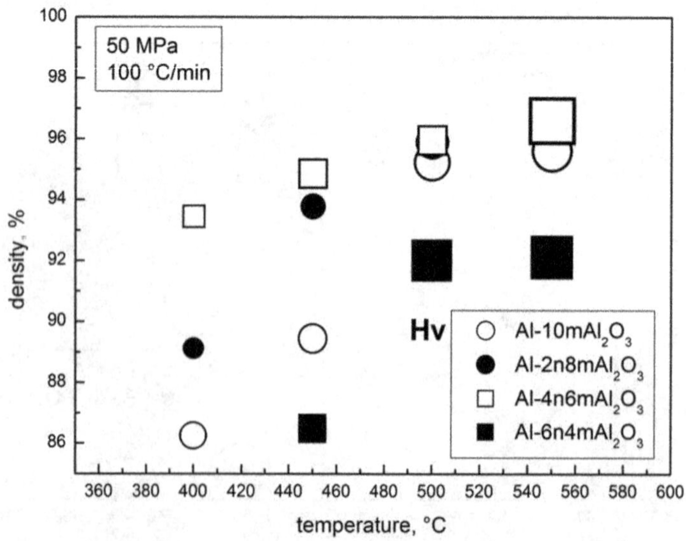

Fig. 3 Density and hardness vs. sintering temperature for monomodal- and bimodal-sized alumina reinforcing aluminum-based nanocomposites SPSed at a pressure of 50 MPa employing a heating rate of 100 °C/min

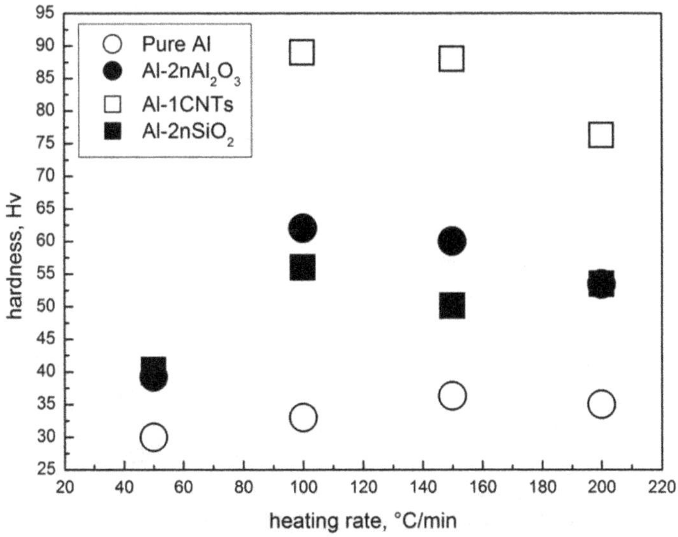

Fig. 4 Hardness as a function of heating rate for different reinforced nanocomposites SPSed at a sintering temperature of 550 °C

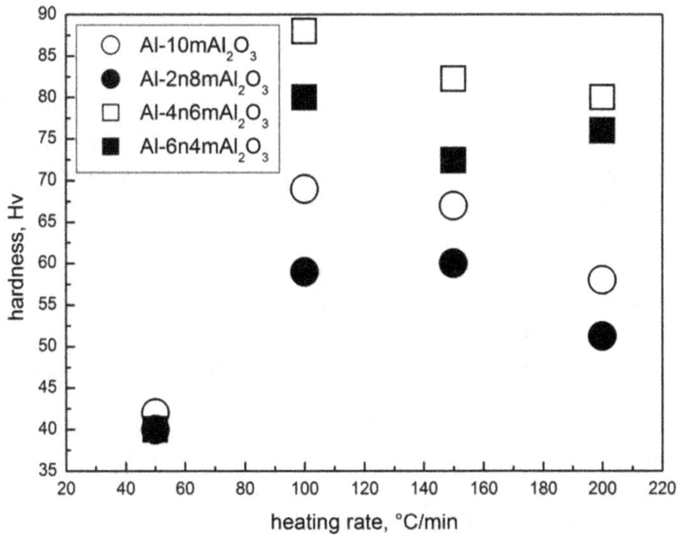

Fig. 5 Hardness as a function of heating rate for monomodal- and bimodal-sized alumina reinforcing aluminum-based nanocomposites SPSed at a sintering temperature of 550 °C

2009). The external pressure not only causes the rearrangement of the particles but also can be amplified in the microstructure to directly supplement the inherent sintering stress. Powder compacts can be densified by plastic flow. When a pressure is applied to loosely compacted powders, the powders will deform as the initial contact areas between them are small, causing an increase in the contact area and a decrease in effective pressure. However, the strength of the powders decreases rapidly with increasing temperature.

Consequently, pressure in the range of MPa accelerates sintering by plastic flow at an increased temperature. Liu et al. (2016) demonstrated that near full dense SPSed aluminum can be obtained at lower temperatures (450 °C) by employing high pressures in the order of 500 MPa. The SPS-consolidated material exhibits a microstructure consisting of nanometric grains embedded inside a matrix of ultrafine grains. The average grain size is about 125 nm, and no coarse grains were observed. It is suggested that localized heating and electromigration play important roles in the formation of larger ultrafine grains (>200 nm) during SPS. As a matter of fact, traditional processing conditions (600 °C and 5 MPa) lead to mean grain size around 7 μm (Le et al. 2013). Normally a linear increase in density is recorded as the pressure increase. In SPS the pressure is limited by the resistance of graphite die.

The current pulsing generating the spark during sintering represents the uniqueness of SPS technique. In general, aluminum densification does not vary in a broad range of frequency (0–300 kHz) (Xie et al. 2003; Anselmi-Tamburini et al. 2005b).

Aluminum matrix composites make a distinct category of advanced engineering materials that provide unique properties and advantages over conventional Al alloys. In general, these materials exhibit higher hardness, stiffness, and strength as well as good wear and corrosion resistance. Obviously their properties depend on the processing route, reinforcing type and percentage, and microstructure based on the matrix-reinforcement interaction (Cavaliere 2004). As mentioned, among all the available processing routes, SPS represents optimal potential for providing net shape MMCs components through the elimination of the necessity of machining.

Various types of reinforcements are available for the improvement of mechanical and microstructural properties of SPSed composites.

Carbon nanotubes (CNTs) are very interesting for the large aspect ratio reinforcement/filler materials for improving strength, stiffness, toughness, and high temperature stability of structural materials. Sintering processing of CNT-reinforced aluminum matrix composites has attracted significant interests due to the potential of achieving high density, uniform reinforcement dispersion, and net shaping. It can be demonstrated that improvement in mechanical properties of Al-CNT composites can be obtained with relatively lower CNT content (<2 wt%). Morsi et al. (2010a) reported about SPS processing of Al-CNT composites (with 2.5 and 5 wt% CNTs) using ball-milled powder mixture. About 17% increase in hardness (from 91 to 107 HV) with increasing CNT content from 2.5 to 5 wt% was reported. The spark plasma sintered composite samples also exhibited minor residual porosities. Jin-Zhi et al. (2010) also reported fabrication of Al-CNT composites (with up to 2 wt% CNTs) using SPS of blended powder mixture followed by hot extrusion. The improvement in the hardness with CNT reinforcement was not significant (~10%) due to pronounced agglomeration of CNTs in the composites (Esawi et al. 2010). Morsi et al. (2010b) also investigated the effect of spark plasma extrusion (modified SPS process to allow extrusion during SPS) on the mechanical properties of the

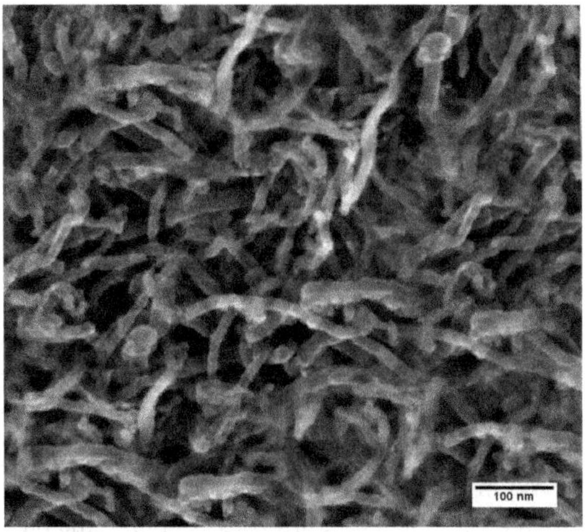

Fig. 6 Multiwalled carbon nanotubes employed in the present study

Al-2.5 wt% CNT composites. They reported higher hardness (~33%) for the Al-CNT composites. It is clear that the dispersion of CNTs in the starting powder mixture is important to gain the advantages of SPS processing. Furthermore, most of these investigations reported effect of CNT reinforcement on microhardness of the composites. Due to multi-scale microstructures of these Al-CNT composites, it is important to understand the length-scale dependence of hardness. It is also important understand other properties, especially wear properties, of these spark plasma sintered Al-CNT composite before their potential utilization in structural applications. The CNTs employed in the present study are shown in Fig. 6.

3 Effect of Reinforcing Type

The materials final properties are strongly related to the reinforcement types as well as to the processing parameters employed during synthesis. In this part of the paper, we analyze the effect of the reinforcement type by fixing the reinforcement percentage. 2 wt% of Al_2O_3, SiO_2, and carbon nanotubes (CNTs) were mixed with the aluminum powder, and samples were prepared via SPS. Alumina and silica particles measured 25 and 10 nm, respectively. The beneficial effect of minor addition of ceramic particles to Al_5Mg alloys was demonstrated by Babu et al. (2016). They show that alumina particles addition is very effective in reducing the matrix grain size after SPS. The high hardness and strength were attributed to the dislocation, Orowan strengthening due to the presence of Al_2O_3 reinforcement particles in the matrix, and Hall-Petch effect as a result of fine grain structure. This was also showed by Li et al. (2016). Alumina nanoparticles are active for strengthening also in the case of SPSed high-entropy alloys (Yang et al. 2016). The theoretical calculations revealed that the Orowan, dislocation and grain boundary, and solid solution strengthening contributed more to increase mechanical properties. Different simulations were performed by Siddiqui et al. (2017) finding the main contribution to the strengthening of SPS nanocomposites is due to the Hall-Petch effect. In our studies, the mean grain size of Al-Al_2O_3 composite is very similar to the Al-CNTs ones, here a very effective grain refinement is observed with respect to the SPSed pure aluminum (Fig. 7). It seems that the SiO_2 addition does not have the same effect resulting in a coarser structure with respect to the other two composites. It seems that below a certain particle dimension, there is no effect on grain refinement due to ceramic addition. In particular, the mean grain size of the SPSed pure aluminum measured 14 μm, the mean grain size of Al-Al_2O_3 and Al-CNTs was around 10 μm, and the mean grain size of Al-SiO_2 composite measured 34 μm. In addition, Al-Al_2O_3 and Al-CNT composites reveal a more equiaxed microstructure with respect to the Al-SiO_2 material where a well-clear bimodal structure with fine and coarse grains can be seen.

The materials densities are shown in Fig. 8.

Here, composites reinforced with alumina particles and carbon nanotubes show a density higher with respect to the pure Al one; in this case the reinforcement addition leads to a beneficial effect on the materials properties. By adding SiO_2 nanoparticles, the materials density decreases with respect to the SPSed pure aluminum one. The relative density of different samples is simply attributed to the wettability and chemical reactivity between the reinforcement and the matrix (Olevsky and Froyen 2009). Generally larger particles lead to higher densification in the composites; the difference between the Al_2O_3-reinforced composite and the SiO_2-reinforced one is due to the difference in particle dimensions. Al_2O_3 particles (larger than SiO_2 ones)

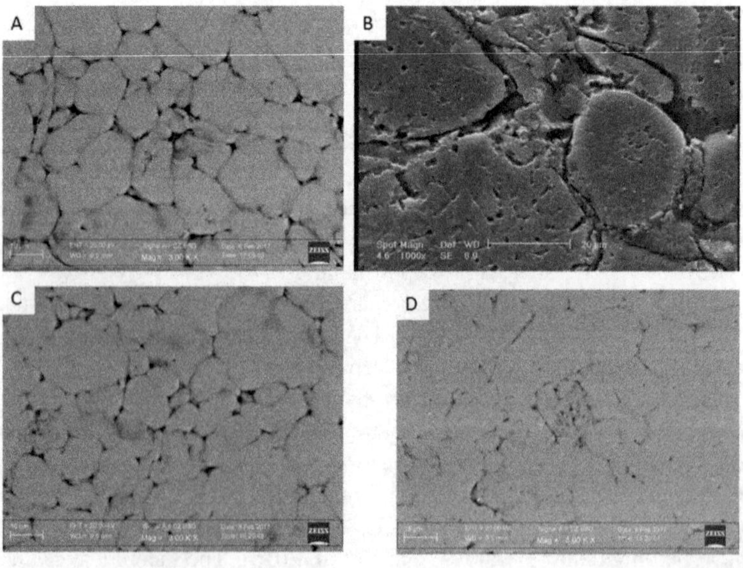

Fig. 7 SEM microstructure of aluminum matrix nanocomposites. (**a**) Al-Al_2O_3, (**b**) Al-SiO_2, (**c**) Al-CNTs, and (**d**) pure Al

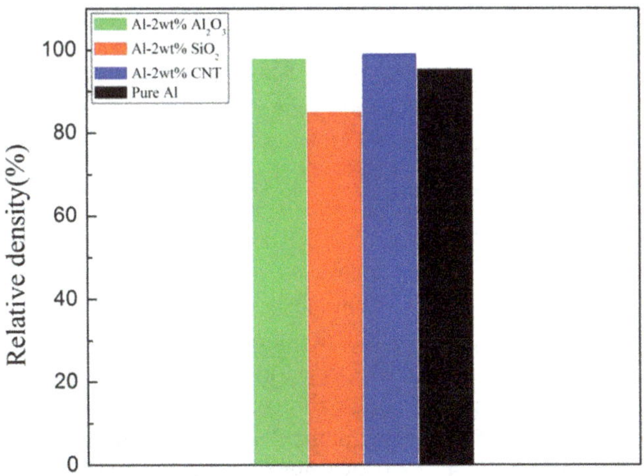

Fig. 8 Density of the different studied materials

are more effective in the reduction of porosity. This behavior can be attributed to two factors: first, a smaller particle can accelerate diffusion of powder scales with powder size, and second, the densification rates involving lattice and GB diffusion are inversely proportional to the particle size. In addition, the percentage of finer grains in smaller particles may be higher than that in larger powder, which facilitates the densification (Liu et al. 2012).

The results are consistent with those belonging to the studies of Sweet et al. (2015). Here the authors performed SPS of AlN, Si_3N_4, or BN particles leading to the conclusion that the average particle size and particle size distribution of the ceramic particulate influenced densification. Sources that invoked the formation of clusters of small ceramic particles were the most difficult to densify, while coarser particles with a narrow size distribution favored full densification. The average size, narrow particle size distribution, high hardness, and angular morphology of SiC (D10 = 12.8 μm; D50 = 21.3 μm; D90 = 35.4 μm) are all believed to have contributed to the highest mechanical properties.

The addition of nanoparticles to aluminum matrix causes an enhancement in the density of all nanocomposites in comparison with the pure aluminum (Sadeghi et al. 2017; Cavaliere et al. 2017). It is ascribed to the nanoparticles filling of large pores between large particles of the matrix. But in some cases, based on the content of agglomeration of nanoparticles, density falls down; this is due to the nanoparticles dimension decrease. The type of nanoparticles affects the density of nanocomposites. The amount of porosity changes as a function of the type of reinforcements which can cause large modifications in mechanical properties such as hardness and tensile strength. Regarding to the high densities for Al-Al_2O_3 and Al-CNTs, it can be concluded that SPS process employed in this study is a suitable technique for densifying these kinds of nanocomposites. In the case of Al-SiO_2 composite, the presence of large number of SiO_2 agglomerates (due to the very fine size of the particles) causes that the interparticle distances become larger than the theoretical one. The density of Al-CNTs nanocomposites is highly dependent on the dispersion of CNTs (Ostovan et al. 2015). The studied materials hardness is shown in Fig. 9. The composites hardness follows the same trend observed in the density values; obviously pure aluminum shows the lowest value. The hardness of the SPSed Al-CNT nanocomposite is close to that obtained by Ostovan et al. (2015), and the hardness of the Al-Al_2O_3 nanocomposite is higher than that of Al-5wt%Al_2O_3 nanocomposite produced by Garbiec et al. (2015). The application of high sintering temperature and short dwelling time allows to achieve higher hardness and density with respect to those belonging to the composites fabricated by Dash et al. (2013). In fact, a homogeneous and uniform distribution of nanoparticles or CNTs in the aluminum matrix leads reinforcements to act as a barrier to the dislocations movement strengthening the matrix through the creation of high dislocations density regions during cooling at room temperature. The agglomerates formation increases the interparticles distance, and then Orowan mechanism cannot be dominant. This behavior is clearly seen in the Al-2wt%SiO_2 nanocomposite.

The highest hardness is measured for the Al-CNT nanocomposite due to high uniform dispersion of CNTs and to the intrinsic hardening effect. In fact, it causes tighter bonding between the CNTs and aluminum matrix particles and so resulting in an improvement in hardness and tensile strength. Also in the case of CNTs reinforced SPSed nanocomposites, agglomeration and drop of mechanical properties are observed if the CNTs percentage exceeds 1.5% (Wang et al. 2017). In Dong Hoon et al. (2016), the authors demonstrate that also the thermal conductivity properties of Al-Cu-CNTs composites, produced via SPS, show a pronounced decrease as increasing the content of CNTs because of the interface thermal resistance between CNTs and Al–Cu matrix. In Wu et al. (2012) the authors find the maximum thermal conductivity in Al-0.5wt%MWNTs. The enhanced thermal

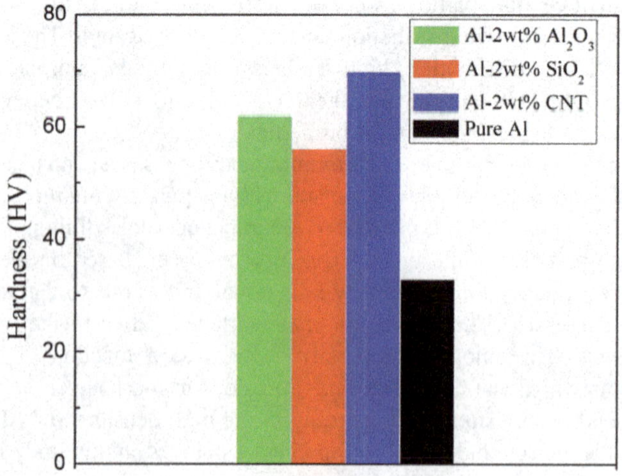

Fig. 9 Nanoindentation hardness of the studied materials

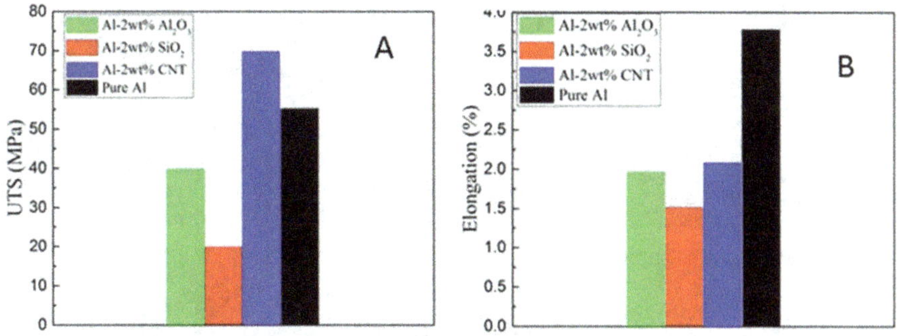

Fig. 10 Tensile strength and ductility for the studied materials

conductivity and tensile strength were ascribed to the bridging of CNTs in the composites. The thermal conductivity and tensile strength were decreased to lower than that of pure Al when >1.0 wt% MWNTs are added. Also the addition of graphite leads to the enhancement of thermal conductivity in such kinds of composites (Kwon et al. 2016). Generally, nanosized alumina particles are located at grain boundaries and also distributed in the grain interiors. It causes the lock of grain boundaries and the hindering of dislocations movements. The highest tensile strength is observed for the Al-CNTs nanocomposite, while the lowest is for the Al-SiO_2 nanocomposite. A significant role of CNTs in dislocations piling, intensified by the uniform and homogeneous distribution, explains such behavior. On the contrary, a large number of agglomerate clusters of SiO_2 in aluminum matrix produce the lowest necking between the aluminum particles. Then, the required energy for dislocation movements decreases, and consequently tensile strength drops (Fig. 10a).

The highest ductility is obviously shown for the pure SPSed aluminum, and a small reduction is observed for the Al-Al_2O_3 and Al-CNTs composites. A strong decrease elongation to failure is recorded for the Al-SiO_2 nanocomposite (Fig. 10b).

4 Effect of Reinforcement Percentage

Figure 11 shows the effect of different percentages of CNTs on the microstructure of SPSed Al matrix composites.

Figure 12 Shows the TEM microstructure of the two composites. In this case the addition of minor percentages of CNTs leads to a decrease of the materials density and to an increase in the hardness (Fig. 13)

Another important effect of carbon nanotubes is grain refinement. In the case of 0.5% of carbon nanotubes in the aluminum matrix, the grain size measured 28.5 μm ± 6. By increasing the carbon nanotubes percentage to 1, the grain size measured 17 μm ± 4. Guo et al. (2017) demonstrated that Al-CNTs SPSed

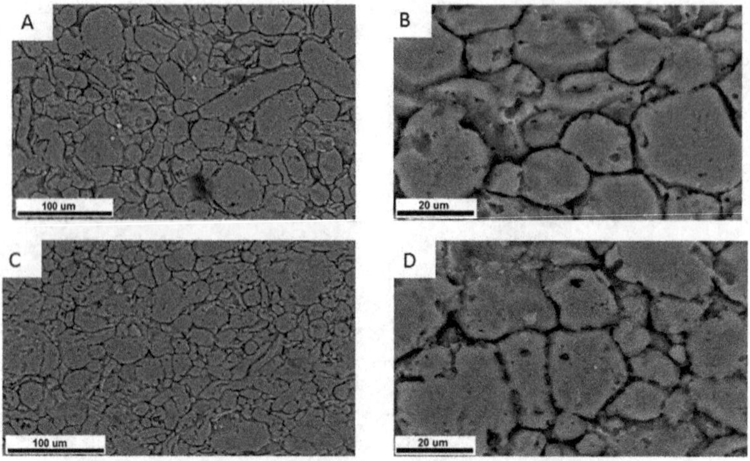

Fig. 11 Microstructure of the SPSed Al-0.5% CNTs composite (**a**) and (**b**); Al-1%CNTs composite (**c**) and (**d**)

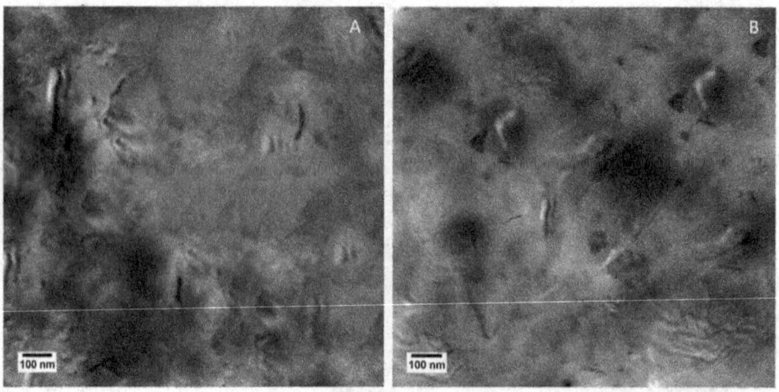

Fig. 12 TEM microstructure of the SPSed Al-0.5%CNTs composite (**a**); Al-1%CNTs composite (**b**)

composites strength can be increased by refining the aluminum microstructure and the formation of a non-equilibrium interface between the metal and the nanotubes. Such properties were achieved through the much higher cooling rate and much shorter holding time. The approach is interesting even if a too much high porosity can be induced in the material after the processing. Chen et al. (2015) demonstrate that SPSed Al-0.75wt%CNTs show increased strength, thanks to the combination of refined grains and oxygen-rich interfaces with respect to SPSed pure aluminum. The formation of Al_4C_3 at the aluminum-CNTs interfaces was recognized as responsible for the strength increase in Kwon et al. (2014). On the other hand, in Chen et al. (2017) the authors find that with the increase of sintering temperature and holding time, the strength decreased. Conversely, the ductility and toughness noticeably increased. In Singh et al. (2016), the authors show how pristine and physically

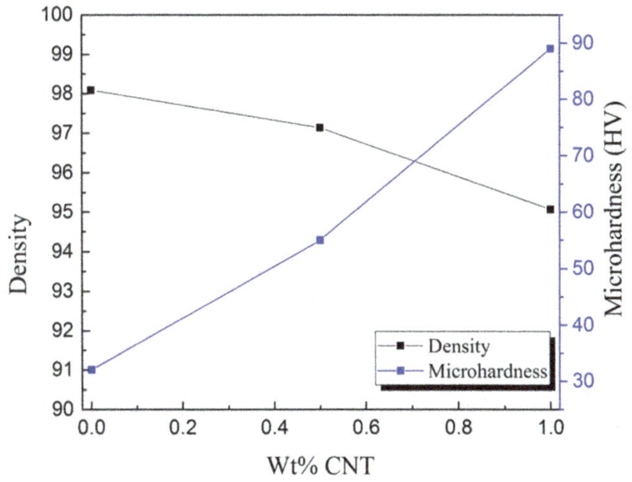

Fig. 13 Comparison of the relative density and hardness for pure Al and for A-CNTs composites

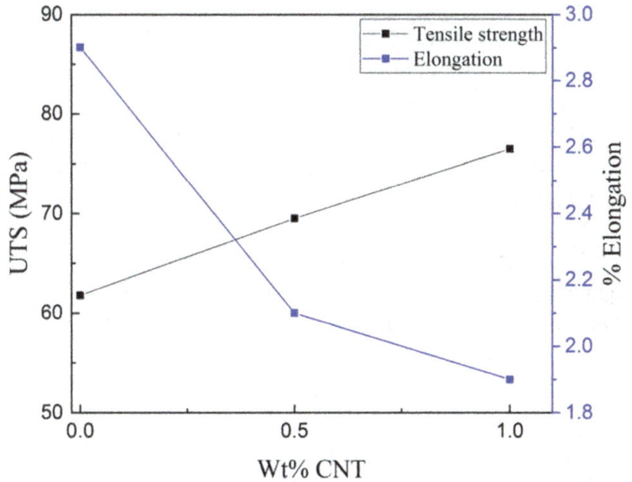

Fig. 14 Tensile properties of the CNTs-reinforced materials

functionalized multiwalled carbon nanotubes lead to the improvement of density and hardness of Al-based nanocomposites.

From previous results, it is clear that nanosized reinforcements have significant effect on tensile strength of nanocomposites. The tensile strength of the composite was 68 MPa for the addition of 0.5% of CNTs; the strength increases to a value of 76 for a percentage of carbon nanotubes of 1, while the pure Al one was 60 MPa. The addition of nanosized reinforcements influences also the material ductility as shown in Fig. 14.

The results are consistent with those reported by Boesl et al. (2014); the in situ tensile tests performed on pure Al and 1wt% CNTs reinforcing SPSed Al result in an enhancement of mechanical strength due to the presence of carbon nanotubes with a reduction in ductility. Anyway they report that the materials density is the same (99.9% of the theoretical one) without apparent effect of the reinforcing phase of such a parameter. The enhanced strength of the nanocomposite is attributed to the stronger diffusional bonds and homogeneous distribution of nanosized carbon tubes in the aluminum matrix during spark plasma sintering. A homogeneous and uniform distribution of CNTs in the aluminum matrix leads reinforcements to act as a barrier to the dislocations movement strengthening the matrix through the creation of high dislocations density regions during cooling at room temperature. The pure aluminum strength could be as a result of the coarse grain structure due to the absence of reinforcement for impeding grain boundaries movement during grain growth. The agglomerates formation increases the interparticles distance, and then Orowan mechanism cannot be dominant. The highest hardness is measured for the Al-CNT nanocomposite due to high uniform dispersion of CNTs. In fact, it causes tighter bonding between the CNTs and aluminum matrix particles and so resulting in an improvement in hardness and tensile strength. Also Liao et al. (2010) reported a similar behavior affirming that it was due to the agglomeration of the multiwalled carbon nanotubes and to the consequent insufficient interfacial bonding. The same limit to the percentage of reinforcement addition is reported in Tian et al. (2016) for graphene reinforcing 7075 aluminum alloy.

Generally, CNTs are located at grain boundaries and also distributed in the grain interiors. It causes the lock of grain boundaries and the hindering of dislocations movements. This aspect was also underlined by Grácio et al. (2013); here the final composites had grains of approximately 100 nm dimensions, with SiC particles located mostly at grain boundaries. The largest increase in strength is attributed to grain size reduction. Adding 0.5 vol pct SiC increases the yield stress by an additional 47 pct, while the addition of 1 vol pct SiC leads to 50 pct increase relative to the nanocrystalline Al without SiC. Moreover, homogeneous distribution of reinforcements in aluminum matrix causes a decrease in CNTs distances, and so higher energy is required for the movement of dislocations for deformation. So, an increase in yield and tensile strength is experienced by the composites. The presence of CNTs networks can impede the deformability and necking between aluminum particles. On the other hand, load partitioning effect of the reinforcements can decrease the densification and consequently increase the mass centers of powder particles. Accordingly, it can thus be referred that diffusion distances for volume and grain boundary diffusion, which cause necking formation, increase and consequently do not produce appropriate necking between particles. The more the CNTs are placed at grain boundaries and in the interior of aluminum particles, the more they can contribute to strength enhancement of composites by Orowan and grain boundary strengthening mechanisms. Orowan strengthening depicts the effect of dimensions and inter-particulate spacing of secondary-phase dispersoids. This mechanism, caused by the resistance of closely spaced CNTs to the passing of dislocations, is important for aluminum matrix nanocomposite but is not a significant strengthening factor for traditional metal matrix composites at room temperature. The Orowan mechanism was also recognized for graphene nanoplatelets reinforcing aluminum in different percentages from 0 to 5 wt% (Bisht et al. 2017). The strain hardening behavior of the pure aluminum and the composites is completely different. The pure Al had a brittle fracture and almost no hardening occurs in this sample. However, strain hardening behavior of the composites has two steps. First is the elastic-plastic deformation of the composites. In this step composites have higher strain hardening due to resistance of the composites to deform especially near the CNT particles. Second is the plastic deformation of the composites. Due to enhancement in the force, composite deforms and consequently reveals less strain hardening amount compared to the first step. On the other hand, enhancement in the CNT amount leads to increase in agglomeration and structure defects and also reduction in density. These phenomena are the main reasons for the less amount of strain hardening in the composite with 1 wt% of CNT comparing to the other one.

A further deep analysis has been conducted by analyzing the effect of different sizes and percentages of Al_2O_3 particles mixed to the aluminum ones. The bimodal size addition of reinforcements has been underlined for diamond reinforcing Al composites by Mizuuchi et al. (2014a, b). The authors employ very high diamond percentages, but they find a beneficial effect on the density and thermal conductivity by adding bimodal-sized particles with respect to the case on monomodal diamond particles. In Mizuuchi et al. (2012), they demonstrate the improvement of thermal conductivity through SPS of Al/AlN composites. Zhang et al. (2014) studied similar materials demonstrating that the addition of Si to the Al improves the diamond wettability with the consequent improvement of density and thermal properties. The same improvement was demonstrated in Mizuuchi et al. (2014a, b) for high cBN percentages reinforcing SPSed aluminum. In the case of bimodal-reinforced aluminum (Mizuuchi et al. 2015), an increase in density and strength is recorded also for very high cBN contents.

The materials property improvement with the bimodal size addition of reinforcing particles is reported in literature also for Al-WC by Pakdel et al. (2017). Here it is found that too high nanoparticles percentages lead to agglomeration and drop in mechanical properties with respect to the same percentage of microsized particles. Once bimodal-sized reinforcing WC are added to the Al, both density and hardness show the highest levels for the same volume fraction percentage.

The bimodal size ceramic effect on the microstructural and mechanical behavior of the composites was studied by producing six different types of materials with different combinations of micro- and nanoparticles. The percentages of the nano- and microparticles in the aluminum matrix are summarized in Table 1.

Figure 15 shows the morphology of the starting alumina powders. The preliminary SEM observations on the as-received powders depict dispersed regions of Al_2O_3 nanoparticle agglomeration. Figure 15b shows that the shape of aluminum particles is circular. By focusing on Fig. 15a, b, it can be affirmed that both aluminum and ceramic particle dimensions are in narrow size ranges. In Fig. 16a–d, SEM micrographs of the powder mixtures with the ratio of the nano- to micron-sized reinforcement 2:8, 4:6, 5:5, and 6:4 are presented. The milling leads to a

Table 1 Different composites produced for the present study

Sample	Al	wt% nano-powder	wt% micron-powder
Al-10mAl_2O_3	90	0	10
Al-2n8mAl_2O_3	90	2	8
Al-4n6mAl_2O_3	90	4	6
Al-5n5mAl_2O_3	90	5	5
Al-6n4mAl_2O_3	90	6	4

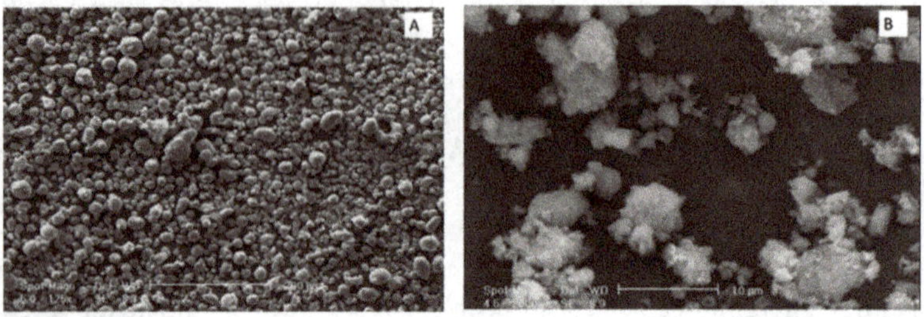

Fig. 15 SEM morphology of (**a**) nanoparticles of Al_2O_3 in the as-received condition; (**b**) as-received Al powder

Fig. 16 SEM micrographs of (**a**)Al-2n8mAl_2O_3, (**b**) Al-4n6mAl_2O_3, (**c**) Al-5n5mAl_2O_3, (**d**) Al-6n4mAl_2O_3 powder mixtures; note that the black and red arrows indicate the agglomeration of the Al_2O_3 nanoparticles and aluminum particles, respectively

decrease particle dimension. During milling, the Al_2O_3 microparticles and nanoparticles are homogeneously distributed between and on the surface of aluminum particles.

Samples were sintered by SPS technique using a Model 10 Ton equipment manufactured by KHPF Inc. The desired amount of milled powder blend (approximately 15 g) was loaded into a graphite die of 40 mm in diameter. The graphite die was ISG-M303 with a thermocouple hole drilled into the punch. For sintered pellet removal, a graphite paper sheet was placed between the punch and the powder as well as between the die and the powder to prevent sticking due to the possible reaction between the powder and graphite die/punch. Sintering was conducted under very low pressure (approximately 75 mbar), and the heating and cooling rates were 75 and 30 °C/min, respectively. Pulsed DC current, with a 36 ms on and 8 ms off pulsed profile, was applied. The gas removal takes place during the initial stages of heating in SPS. The mixture powders and mold were heated from ambient temperature to the sintering temperature of 550 °C, with mentioned heating rate and then holding at that temperature for 5 min. A uniaxial pressure of 50 MPa was applied throughout the entire heating and cooling cycle. The whole sintering period (including heating and cooling time) was less than 25 min. Samples were then furnace cooled to room temperature in vacuum atmosphere.

Figure 17a–d shows that by increasing the ratio of the nano- to micron-sized particles, agglomeration trend of alumina particles in aluminum matrix increases. Figure 16a shows that the shape and size of aluminum particles are mainly equiaxed. In addition, as the weight percent of nano- to micron particles increases, the size of aluminum particles increases. SEM micrographs of the SPSed specimens with the different ratio of the nano- to micron-sized Al_2O_3 particles are depicted in Fig. 18. The lighter zones in black circles represent agglomerates of Al_2O_3 with high content of nanoparticles. The Al_2O_3 nanoparticles are also located at aluminum grain

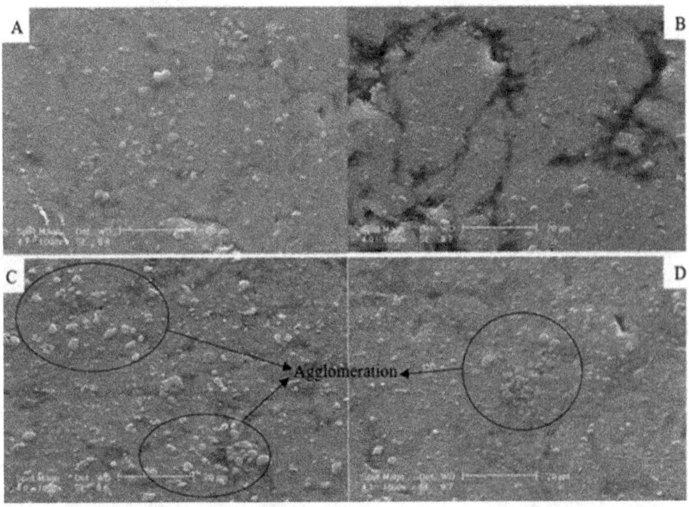

Fig. 17 SEM micrographs of (**a**) Al-2n8mAl2O3, (**b**) Al-4n6mAl2O3, (**c**) Al-5n5mAl2O3, and (**d**) Al-6n4mAl2O3 nanocomposites sintered using SPS

boundaries. Figure 17a, b shows a more homogeneous and uniform distribution of reinforcement particles in the aluminum matrix.

It can be affirmed that the specimen with ratio nano to micron 2:8 shows the minimum agglomeration among all the specimens. However, the increased particle agglomeration with increased nanosized Al_2O_3 particle percentage leads to the decrease in many properties such as density, hardness, and tensile strength as well as the inability to homogeneously pass the electrical current flow through material. This aspect is clarified by the TEM microstructure of the studied composites compared with the material reinforced with 10% of microsized particles (Fig. 18). Garbiec et al. (2015) produced SPSed Al-Al_2O_3 composites with reinforcing percentages of 5, 10, 15, and 20 wt%. The mean alumina particles dimension was 300 nm. Here the authors fin an increase in hardness with the increase of reinforcing phase percentage even if a linear increase in porosity (up to 5%) is recorded as the alumina content increases.

As the composite powder starts to sinter, current flows through the graphite die. As soon as the first electrical paths are established, current starts to flow within the powder before reaching full density; the resistance might drop to very low values in the case of particular combinations of ceramic particle dimensions (Guillon et al. 2014).

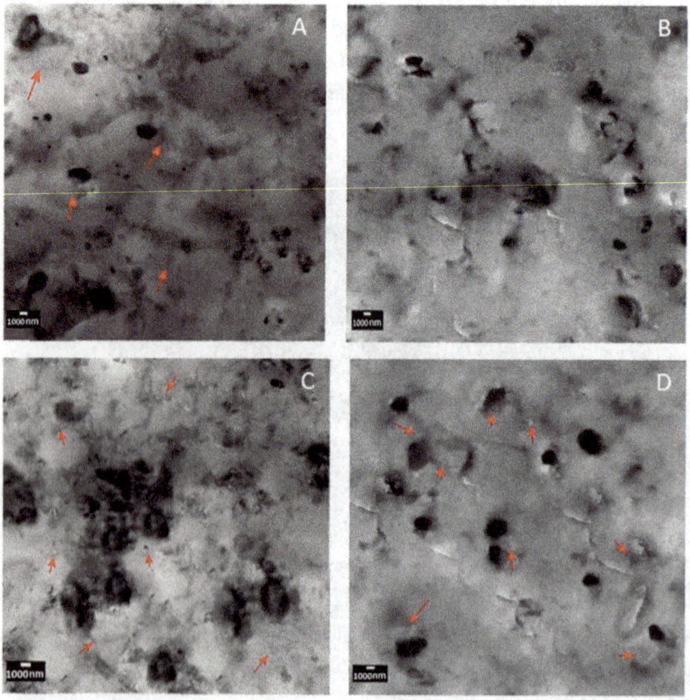

Fig. 18 TEM microstructure of (**a**) Al-2n8mAl2O3, (**b**) Al-4n6mAl2O3, (**c**) Al-6n4mAl2O3, and (**d**) Al-10mAl2O3 nanocomposites sintered using SPS

Concerning the particle-particle contact zones, here the local pressure can increase up to 1000 times. The application of pressure and current in addition to the high localized temperature and pressure causes neck formation around the contact area between particles (Fig. 19a–d). This aspect is amplified as the ratio of the nano- to micron-sized Al_2O_3 particles increases. The heat is transferred immediately from the center of the spark discharge column to the sphere surface and diffused so that intergranular bonding portion is quickly cooled. The pulse energizing method causes spark discharges one after another between particles. Even with a single particle, the number of positions where necks are formed between adjacent particles increases as the discharges are repeated. On the other hand, the thickness of oxide layers, compacting pressure, and particle size affect the electrical current in the initial stage of sintering. During the SPS process, the applied electrical potential causes a flow of electrons through conductive materials. In the case of nonconductive ceramics and conductive aluminum, a potential difference across aluminum particles separated by the ceramic phases may develop. These high pressures and electrical potential can easily break up the oxide layers and create direct metal-to-metal contacts. This phenomenon can facilitate atom diffusion at the contacts. The mass transport mechanisms, involved in sintering for neck formation, are evaporation, surface diffusion, grain boundary, and volume diffusion. Clustering and agglomeration increase as the ratio of the nano- to micron-sized particles increases (Fig. 19). As the content of nanoparticles increases and the content of microparticles decreases, for the same specimen, both the neck size and number of necks decrease.

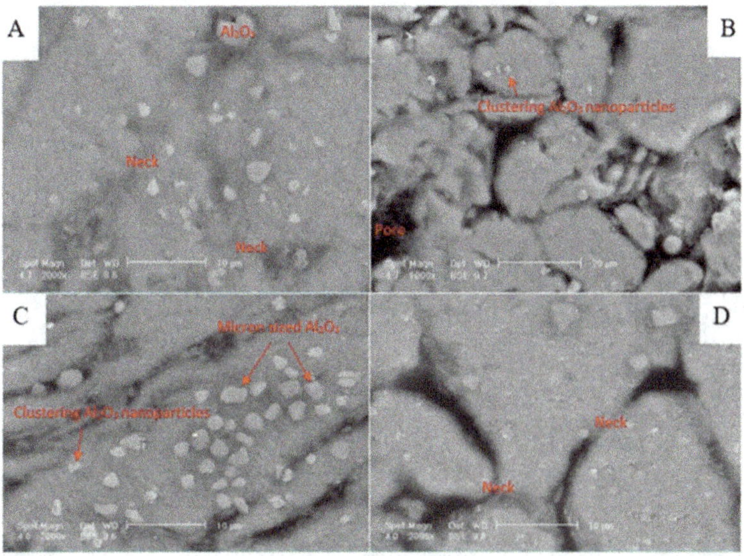

Fig. 19 Neck formation in (**a**) Al-2n8mAl2O3, (**b**) Al-4n6mAl2O3, (**c**) Al-5n5mAl2O3, and (**d**) Al-6n4mAl2O3 nanocomposites sintered using SPS

Figure 19a confirmed suitable necking and intensified connections between aluminum particles, as well as the homogeneous and bimodal distribution of nano- and micron-sized Al_2O_3 particles throughout the microstructure within pores and along aluminum particle boundaries. The size and number of pores between aluminum particles increase by increasing the content of nanoparticle (Fig. 19b–d); this can be attributed to the clusters and agglomeration of nanoparticles in aluminum matrix. It has been found that when the amount of nanoparticles increases, the uniform dispersion of the particulate becomes difficult (Fig. 19c, d). These observations demonstrated that agglomerates produce an increase in porosity in terms of inter- and intra-particle porosity. In addition, Al_2O_3 nanoparticles can saturate the aluminum grain boundaries. As a consequence, Al_2O_3 brittle nanoparticles phase, along the aluminum grain boundaries, causes a decrease in the potential grain boundary pinning and the weakening of the final nanocomposites.

Nevertheless, the crystallite size of aluminum for Al-2n8mAl_2O_3 and Al-4n6mAl_2O_3 nanocomposites decreases by increasing content of nano- to micron-sized alumina particles (Fig. 20).

A further content of alumina nanoparticles had no significant effect on crystallite size. The formation of hard agglomerates, by increasing the content of nanoparticles, was unavoidable and consequently hinders the local plastic deformation of aluminum particles. It may cause smear on the aluminum particle surfaces and form a network of alumina nanoparticles on matrix particles. By observing the material grain size as a function of the nano- to micron-sized alumina particles, it was found that the grain size of pure aluminum measured 14 μm (Fig. 22a), then it increases to 22.5 μm for Al-2n8mAl_2O_3 (Fig. 21b) decreasing to 11 μm for Al4n6mAl_2O_3 (Fig. 21c), and then grain sizes increase as the nanoparticles reinforcement

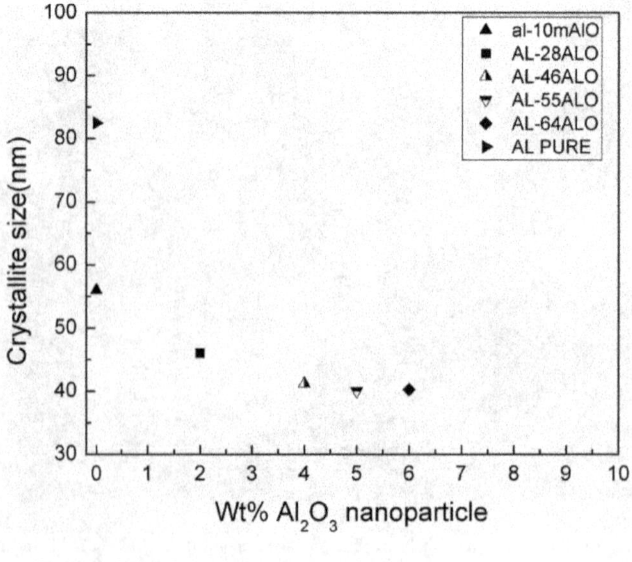

Fig. 20 Crystallite size of aluminum in nanocomposites

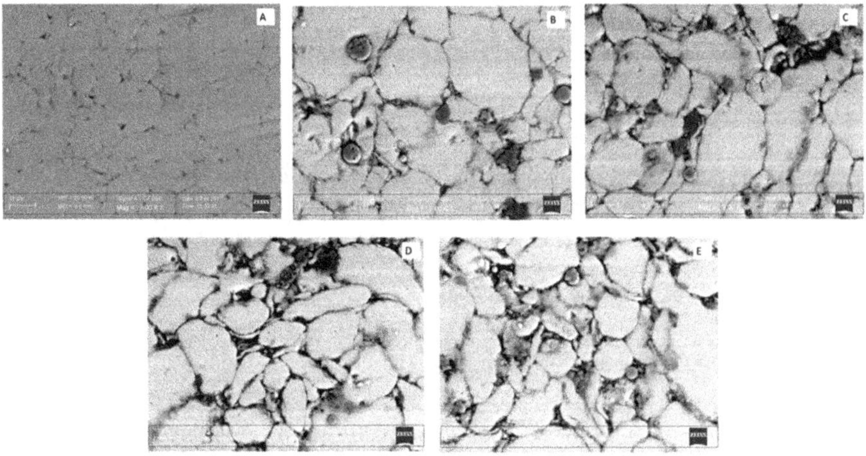

Fig. 21 SEM pictures showing the grain size variation between the different studied composites, (**a**) pure Al, (**b**) 2:8, (**c**) 4:6; (**d**) 5;5, (**e**) 6:4

percentages increased to 17 μm for the Al-5n5mAl_2O_3 (Fig. 21d) and to 15 mm for the Al-6n4mAl_2O_3 (Fig. 21e).

The density increases with the increasing ratio of nano- to micron-sized alumina particles up to around 96.6% in the case of Al-4n6mAl_2O_3. The density decreases with further increasing ratio. Such behavior can be attributed to SPS parameters and microstructural characteristics. The heating rate, pulsed current, and applied pressure are the most important parameters which influence the sample density. The high sensitivity of alumina and aluminum to temperature and applied pressure during SPS, respectively, can be the cause of significant changes in density of SPSed samples.

The applied pressure leads to an increase in the formation of necking between aluminum particle surfaces and consequently to full densification at lower temperatures. In addition, the use of a high heating rate and efficient heat transfer in the SPS process resulted in highly dense materials in short times. In the latter case, the explanation is that the metals like aluminum are more sensitive to mechanical deformation mechanisms such as plastic deformation than the ceramics like Al_2O_3.

As mentioned, among the mechanisms involved in mass transfer for necking, only grain boundary and volume diffusion can cause densification of powder mixtures. In general, taking the powder mixtures to high temperature (especially above 0.5 Tm) quickly minimizes the time that a material spends at a temperature at which coarsening mechanisms like surface diffusion are active, favoring densification. Thus, it is expected that spark plasma, generated during the SPS process, can activate the surfaces of aluminum powder particles, increasing the formation of necking between aluminum particles which consequently results in a density increase. On the other hand, with a further alumina nanoparticles content increase, the density falls. This behavior can be attributed to both agglomeration and compressibility of hard nanoparticles in a ductile aluminum matrix. In general, the compressibility of hard and non-deformable particles in a ductile matrix decreases with the increasing content of the hard particles (Leon et al. 2009). This can be due to the fact that the specific surface of nanosized particles is higher than that of the microsized ones, so with the increasing of the nanoparticle content, the total specific surface of the alumina nanoparticles in the aluminum matrix increases, and consequently there is more interaction and attraction forces between reinforcement nanoparticles. This results in the increase of the actual interparticle distance with respect to the theoretical one. It can cause more agglomeration as well as the tendency to form interconnected networks of nanoparticles around aluminum particles. These two phenomena in nanocomposites lead to lesser densification, whereas this does not impair the densification of Al-10mAl_2O_3 microcomposites to a larger extent. In Ghasali et al. (2017), the authors studied the microstructural and mechanical behavior of SPSed aluminum reinforced with 5, 10, and 15% of TaC (mean particle dimensions of 200 nm). The maximum bending and relative density of 438 ± 17 MPa and 99.13 ± 0.5%, respectively, were measured for Al-5 wt% TaC composite. The maximum microhardness of 295 ± 11 (HV) was obtained for Al-10 wt% TaC composite. Anyway, the SEM investigations showed the agglomeration of TaC reinforcement particles in Al-10 and Al-15 wt% TaC composites, while a uniform distribution of reinforcement particles in Al-5 wt% TaC composite was found. Based on the hardness results (Fig. 22), it was shown that as the ratio of nano- to micron-sized alumina particle increases, both nano- and microhardness values first strongly increase and then slightly decrease. Many mechanisms are involved in such behavior. First of all, the level of alumina distribution and

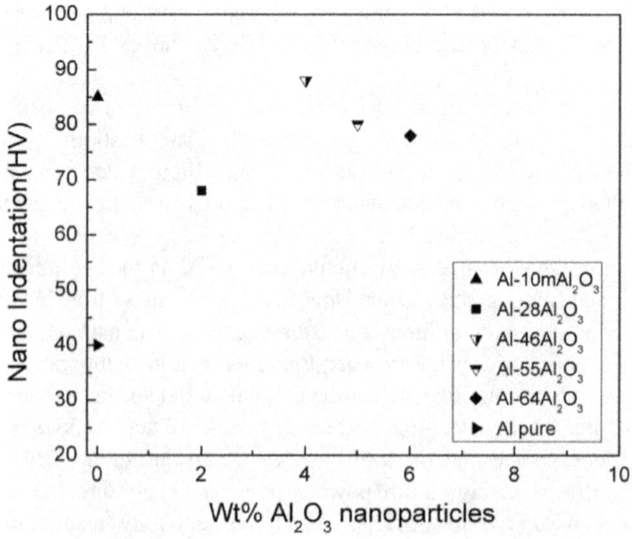

Fig. 22 Nanohardness of nanocomposites with different ratio of nano- to micron-sized alumina particles

agglomeration affects hardness and tensile strength. Then, the nanoparticles influence the strength mechanisms like Orowan strengthening, Hall-Petch effect, and dislocation strengthening. For content of alumina nanoparticles lower than 4 wt%, the agglomeration is reduced.

This resulted in a noteworthy distribution of alumina in aluminum matrix. Accordingly, the interparticle distances of alumina in aluminum matrix is lower than that of agglomerated clusters, inducing the increase of particle-dislocation interaction. Dash et al. (2013) have concluded that the enhanced microhardness and strength of the Al-Al_2O_3 nanocomposites can be attributed to the stronger diffusional bonds and structural integrity achieved after SPS process. By increasing the ratio of nano- to micron-sized alumina particles, generally more than 4 wt% of nanoparticles, agglomeration phenomena of nanoparticles can cause an inhomogeneous distribution of alumina nanoparticles in aluminum matrix. Consequently pores and interparticle distance increases are found. Such behavior, in high ratios of the nano- to micron-sized alumina particles, can cause the weakening of mechanical properties. It was shown that the strengths of the nanocomposites first increase with the nanometric particulates contents; however, when the content of the nanoparticles exceeds 4 wt%, the strengths decrease. The enhanced strength of the Al-2n4mAl_2O_3 and Al-4n6mAl_2O_3 nanocomposites is attributed to the stronger diffusional bonds and homogeneous distribution of nano- and micron-sized alumina particles in aluminum matrix during spark plasma sintering. Moreover, homogeneous distribution of reinforcements in aluminum matrix cause a decrease in alumina particles distances, and so higher energy is required for the movement of dislocations for deformation. First, as mentioned above and shown in SEM micrographs, by increasing the nano- to micron-sized ratio, an increase in alumina nanoparticles agglomeration is observed. The effect is amplified with percentages higher than 4, and consequently this results in the presence of agglomerates and network of the alumina particles within and intergranular space of the aluminum matrix. Second, the bimodal particle-sized systems have a high grain growth rate because of the different curvature of small and large particles. The presence of a network of alumina can cause increases in alumina-alumina contact which impedes the deformability and necking between aluminum particles. On the other hand, a load partitioning effect of the alumina nanoparticles can decrease the densification and consequently increase the mass centers of powder particles. Accordingly, it can thus be referred that diffusion distances for volume and grain boundary diffusion, which cause necking formation, increase and consequently do not produce appropriate necking between particles. The same behavior was observed by Firestein et al. (2015, 2017); they analyzed the room- and high-temperature tensile properties of SPSed Al-based composites reinforced with BN nanoparticles added to the al matrix with different percentages from 0.5 to 7.5%. The highest tensile properties were recorded for the composite reinforced with 4.5 wt% on nanosized BN particles. Also in this case particles agglomeration leads to the drop of mechanical properties.

Therefore, in our case, the densification and mechanical property trends show a drop with the increase in alumina nanoparticles of more than 4 wt%. The more the micro- and nanosized alumina placed grain boundaries and grain interior of aluminum particles, the more they can contribute to strength enhancement of micro- and nanocomposites by Orowan and grain boundary strengthening mechanisms. When the particle size becomes finer than 1 μm, the interaction between nanoparticle and dislocations increases so that dislocation motion becomes difficult and therefore causes an increase in the nanocomposite strength. In fact, Orowan strengthening depicts the effect of dimensions and inter-particulate spacing of secondary-phase dispersoids. This mechanism, caused by the resistance of closely spaced alumina particles to the passing of dislocations, is important for aluminum matrix nanocomposites but is not a significant strengthening factor for traditional metal matrix composites.

5 Conclusions

The most recent results belonging to the studies performed on aluminum matrix composites produced via spark plasma sintering have been analyzed and described in the present paper. It was demonstrated how the precise tuning of process parameters during sinter can lead to the optimal control of the matrix grain size and of the composites density. Obviously, the materials mechanical and microstructural behavior is influenced by the reinforcing phase addition and depends on their size, dimension, and distribution. The nanosized reinforcement addition must be precisely controlled in order to avoid agglomeration leading to porosity and brittleness in the SPSed composites. Carbon nanotubes addition seems to be an excellent choice for improving mechanical properties and thermal conductivity; anyway, its potential percentage addition is limited. By adding ceramic reinforcements, the potential percentage can be increased, but the particles size must be precisely controlled in order to avoid undesirable microstructure and mechanical properties. The possibility of mixing bimodal-sized (nano and micro) ceramic particles results very promising to improve mechanical properties of SPSed aluminum matrix composites.

References

Aliyu IK, Saheb N, Fida Hassan S, Al-Aqeeli N (2015) Microstructure and properties of spark plasma sintered aluminum containing 1 wt.% SiC nanoparticles. Materials 5:70–83

Anselmi-Tamburini U, Garay JE, Munir ZA (2005a) Fundamental investigations on the spark plasma sintering/synthesis process: III. Current effect on reactivity. Mater Sci Eng A 407:24–30

Anselmi-Tamburini U, Gennari S, Garay JE, Munir ZA (2005b) Fundamental investigations on the spark plasma sintering/synthesis process II. Modeling of current and temperature distributions. Mater Sci Eng A 394:139–148

Babu NK, Kallip K, Leparoux M, AlOgab KA, Maeder X, Rojas Dasilva YA (2016) Influence of microstructure and strengthening mechanism of AlMg5–Al2O3 nanocomposites prepared via spark plasma sintering. Mater Des 95:534–544

Bathul S, Anandani RC, Dhar A, Srivastava AK (2012) Microstructural features and mechanical properties of Al 5083/SiCp metal matrix nanocomposites produced by high energy ball milling and spark plasma sintering. Mater Sci Eng A 545:97–102

Bisht A, Srivastav M, Manoj Kumar R, Lahiri I, Lahiri D (2017) Strengthening mechanism in graphene nanoplatelets reinforced aluminum composite fabricated through spark plasma sintering. Mater Sci Eng A 695:20–28

Boesl B, Lahiri D, Behdad S, Agarwal A (2014) Direct observation of carbon nanotube induced strengthening in aluminum composite via in situ tensile tests. Carbon 69:79–85

Cavaliere P (2004) Isothermal forging of AA2618 reinforced with 20% of alumina particles. Composites Part A 35:619–629

Cavaliere P, Sadeghi B, Shabani A (2017) Carbon nanotube reinforced aluminum matrix composites produced by spark plasma sintering. J Mater Sci 52:8618–8629

Chen W, Anselmi-Tamburini U, Garay JE, Groza JR, Munir ZA (2005) Fundamental investigations on the spark plasma sintering/synthesis process I. Effect of dc pulsing on reactivity. Mater Sci Eng A 394:132–138

Chen B, Li S, Imai H, Jia L, Umeda J, Takahashi M, Kondoh K (2015) Carbon nanotube induced microstructural characteristics in powder metallurgy Al matrix composites and their effects on mechanical and conductive properties. J Alloys Compd 651:608–615

Chen B, Imai H, Umeda J, Takahashi M, Kondoh K (2017) Effect of spark plasma sintering conditions on tensile properties of aluminum matrix composites reinforced with multiwalled carbon nanotubes (MWCNTs). JOM 69(4):669–675

Dash K, Chaira D, Ray BC (2013) Synthesis and characterization of aluminium–alumina micro-and nano-composites by spark plasma sintering. Mater Res Bull 48(7):2535–2542

Diouf S, Molinari A (2012) Densification mechanisms in spark plasma sintering: effect of particle size and pressure. Powder Technol 221:220–227

Dong Hoon N, Seung Il C, Kyung Moon L, Jun Ho J, Hoon Mo P, Jong Kook L, Soon Hyung H (2016) Thermal properties of carbon nanotubes reinforced aluminum-copper matrix nanocomposites. J Nanosci Nanotechnol 16(11):12013–12016

Esawi AMK, Morsi K, Sayed A, Taher M, Lanka S (2010) Effect of carbon nanotube (CNT) content on the mechanical properties of CNT-reinforced aluminium composites. Compos Sci Technol 70:2237–2241

Firestein KL, Steinman AE, Golovin IS, Cifre J, Obraztsova EA, Matveev AT, Kovalskii AM, Lebedev OI, Shtansky DV, Golberg D (2015) Fabrication, characterization, and mechanical properties of spark plasma sintered Al–BN nanoparticle composites. Mater Sci Eng A 642:104–112

Firestein KL, Corthay S, Steinman AE, Matveev AT, Kovalskii AM, Sukhorukova IV, Golberg D, Shtansky DV (2017) High-strength aluminum-based composites reinforced with BN, AlB2 and AlN particles fabricated via reactive spark plasma sintering of Al-BN powder mixtures. Mater Sci Eng A 681:1–9

Garay J (2010) Current-activated, pressure-assisted densification of materials. Annu Rev Mater Res 40:445–468. https://doi.org/10.1146/annurev-matsci-070909-104433

Garbiec D, Jurczyk M, Levintant-Zayonts N, Mościcki T (2015) Properties of Al–Al 2 O 3 composites synthesized by spark plasma sintering method. Archiv Mater Manuf Eng 15(4):933–939

Ghasali E, Pakseresht A, Safari-kooshali F, Agheli M, Ebadzadeh T (2015) Investigation on microstructure and mechanical behavior of Al–ZrB2 composite prepared by microwave and spark plasma sintering. Mater Sci Eng A 627:27–30

Ghasali E, Alizade M, Ebadzadeh T (2016a) Mechanical and microstructure comparison between microwave and spark plasma sintering of Al-B4C composite. J Alloys Compd 655:93–98

Ghasali E, Nouraniana H, Rahbari A, Majidian H, Alizadeh M, Ebadzadeh T (2016b) Low temperature sintering of aluminum-zircon metal matrix composite prepared by spark plasma sintering. Mater Res 19(5):1189–1192

Ghasali E, Pakseresht AH, Alizadeh M, Shirvanimoghaddam K, Ebadzadeh T (2016c) Vanadium carbide reinforced aluminum matrix composite prepared by conventional, microwave and spark plasma sintering. J Alloys Compd 688:527–533

Ghasali E, Shirvanimoghaddam K, Pakseresht AH, Alizadeh M, Ebadzadeh T (2017) Evaluation of microstructure and mechanical properties of Al-TaC composites prepared by spark plasma sintering process. J Alloys Compd 705:283–289

Grácio JJ, Picu CR, Vincze G, Mathew N, Schubert T, Lopes A, Buchheim C (2013) Mechanical behavior of AlSiC nanocomposites produced by ball milling and spark plasma sintering. Met Trans A 44(11):5259–5269

Guillon O, Gonzalez-Julian J, Dargatz B, Kessel T, Schierning G, Räthel J, Herrmann M (2014) Field-assisted sintering technology/spark plasma sintering: mechanisms, materials, and technology developments. Adv Eng Mater 16(7):830–849

Guo B, Song M, Yi J, Ni S, Shen T, Du Y (2017) Improving the mechanical properties of carbon nanotubes reinforced pure aluminum matrix composites by achieving non-equilibrium interface. Mater Des 120:56–65

Jin-Zhi L, Ming-Jen T, Idapalapati S (2010) Spark plasma sintered multi-wall carbon nanotube reinforced aluminum matrix composites. Mater Des 31:S96–S100

Kubota M (2010) Solid-state reaction in mechanically milled and spark plasma sintered Al–B4C composite materials. J Alloys Compd 504S:S319–S322

Kubota M, Wynne PB (2007) Electron backscattering diffraction analysis of mechanically milled and spark plasma sintered pure aluminum. Scripta Mater 57:719–722

Kubota M, Kaneko J, Sugamata M (2008) Properties of mechanically milled and spark plasma sintered Al–AlB2 and Al–MgB2 nano-composite materials. Mater Sci Eng A 475:96–100

Kwon H, Park DH, Park Y, Silvain JF, Kawasaki A, Park Y (2010) Spark plasma sintering behavior of pure aluminum depending on various sintering temperatures. Met Mater Int 16(1):71–75

Kwon H, Leparoux M, Kawasaki A (2014) Functionally graded dual-nanoparticulate-reinforced aluminium matrix bulk materials fabricated by spark plasma sintering. J Mater Sci Technol 30 (8):736–742

Kwon H, Park J, Joo S, Hong S, Mun J (2016) Spark plasma sintering behavior and heat dissipation characteristics of the aluminum matrix composite materials with the contents of graphite. J Korean Powder Metall Inst 23(3):195–201

Lalet G, Kurita H, Miyazaki T, Kawasaki A, Silvain JF (2014) Microstructure of a carbon fiber reinforced aluminum matrix composite fabricated by spark plasma sintering in various pulse conditions. J Mater Sci 49(8):3268–3275

Le GM, Godfrey A, Hansen N, Liu W, Winther G, Huang X (2013) Influence of grain size in the near-micrometre regime on the deformation microstructure in aluminium. Acta Mater 61:7072–7086

Leon C, Rodriguez-Ortiz G, Aguilar-Reyes E (2009) Cold compaction of metal–ceramic powders in the preparation of copper base hybrid materials. Mater Sci Eng A 526(1):106–112

Li M, Mab K, Jiang L, Yang H, Lavernia EJ, Zhang L, Schoenung JM (2016) Synthesis and mechanical behavior of nano structured Al5083/n-TiB2 metal matrix composites. Mater Sci Eng A 656:241–248

Liao J-Z, Tan M-J, Sridhar I (2010) Spark plasma sintered multi-wall carbon nanotube reinforced aluminum matrix composites. Mater Des 31:S96–S100

Liu D, Xiong Y, Topping TD, Zhou Y, Haines C, Paras J, Martin D, Kapoor D, Schoenung JM, Lavernia EJ (2012) Spark plasma sintering of cryomilled nanocrystalline Al alloy – part II: influence of processing conditions on densification and properties. Metall Mater Trans A 43:340–350

Liu Z-F, Zhang Z-H, Lu J-F, Korznikov AV, Korznikova E, Wang F-C (2014) Effect of sintering temperature on microstructures and mechanical properties of spark plasma sintered nanocrystalline aluminum. Mater Des 64:625–630

Liu D, Xiong Y, Li P, Lin Y, Chen F, Zhang L, Schoenung JM, Lavernia EJ (2016) Microstructure and mechanical behavior of NS/UFG aluminum prepared by cryomilling and spark plasma sintering. J Alloys Compd 679:426–435

Milligan J, Gauvin R, Brochu M (2013) Consolidation of cryomilled Al–Si using spark plasma sintering. Philos Mag 93(19):2445–2464

Mizuuchi K, Inoue K, Agari Y, Nagaoka T, Sugioka M, Tanaka M, Takeuchi T, Tani J-I, Kawahara M, Makino Y, Ito M (2012) Processing and thermal properties of Al/AlN composites in continuous solid–liquid co-existent state by spark plasma sintering. Composites Part B 43:1557–1563

Mizuuchi K, Inoue K, Agari Y, Sugioka M, Tanaka M, Takeuchi T, Tani J, Kawahara M, Makino Y, Ito M (2014a) Bimodal and monomodal diamond particle effect on the thermal properties of diamond-particle-dispersed Al–matrix composite fabricated by SPS. Microelectron Reliab 54:2463–2470

Mizuuchi K, Inoue K, Agari Y, Sugioka M, Tanaka M, Takeuchi T, Tani J, Kawahara M, Makino Y, Ito M (2014b) Thermal properties of cBN particle dispersed Al matrix composites fabricated by SPS. J Jpn Soc Powder Metall 61(12):549–555

Mizuuchi K, Inoue K, Agari Y, Sugioka M, Tanaka M, Takeuchi T, Tani J, Kawahara M, Makino Y, Ito M (2015) Effect of bimodal cBN particle size distribution on thermal conductivity of Al/cBN composite fabricated by SPS. J Jpn Soc Powder Metall 62(5):263–270

Morsi K, Esawi AMK, Borah P, Lanka S, Sayed A (2010a) Characterization and spark plasma sintering of mechanically milled aluminum-carbon nanotube (CNT) composite powders. J Compos Mater 44:1991–2003

Morsi K, Esawi AMK, Lanka S, Sayed A, Taher M (2010b) Spark plasma extrusion (SPE) of ball-milled aluminum and carbon nanotube reinforced aluminum composite powders. Composites Part A 41:322–326

Munir ZA, Anselmi-Tamburini U, Ohtanagi M (2006) The effect of electric field and pressure on the synthesis and consolidation of materials: a review of the spark plasma sintering method. J Mater Sci 41:763–777

Olevsky E, Froyen L (2006) Constitutive modeling of spark-plasma sintering of conductive materials. Scripta Mater 55:1175–1178

Olevsky EA, Froyen L (2009) Impact of thermal diffusion on densification during SPS. J Am Ceram Soc 92:S122–S132

Olevsky EA, Kandurkuri S, Froyen L (2007) Consolidation enhancement in spark-plasma sintering: impact of high heating rates. J Appl Phys 102(11):114913

Ostovan F, Matori KA, Toozandehjani M, Oskoueian A, Yusoff HM, Yunus R, Ariff AHM, Quah HJ, Lim WF (2015) Effects of CNTs content and milling time on mechanical behavior of MWCNT-reinforced aluminum nanocomposites. Mater Chem Phys 166:160–166

Pakdel A, Witecka A, Rydzek G, Awang Shri DN (2017) A comprehensive microstructural analysis of Al–WC micro- and nano-composites prepared by spark plasma sintering. Mater Des 119:225–234

Sadeghi B, Shamanian M, Ashrafizadeh F, Cavaliere P, Rizzo A (2017) Influence of Al2O3 nanoparticles on microstructure and strengthening mechanism of Al-based nanocomposites produced via spark plasma sintering. J Mater Eng Perform 26:2928–2936

Saheb N, Aliyu IK, Hassan SF, Al-Aqeeli N (2014) Matrix structure evolution and nanoreinforcement distribution in mechanically milled and spark plasma sintered Al-SiC nanocomposites. Materials 7:6748–6767

Saheb N, Khan MS, Hakeem AS (2015) Effect of processing on mechanically alloyed and spark plasma sintered Al-Al2O3 nanocomposites. J Nanomater 2015:609824

Siddiqui MU, Arif AFM, Nouari S, Shahzeb Khan M (2017) Constitutive modeling of elastoplasticity in spark-plasma sintered metal matrix nanocomposites. Mater Sci Eng A 689:176–188

Singh LK, Maiti A, Maurya RS, Laha T (2016) Al alloy nanocomposite reinforced with physically functionalized carbon nanotubes synthesized via spark plasma sintering. Mater Manuf Process 31(6):733–738

Sweet GA, Brochu M, RL HJ, Donaldson IW, Bishop DP (2015) Consolidation of aluminum-based metal matrix composites via spark plasma sintering. Mater Sci Eng A 648:123–133

Tian W, Li S, Wang B, Chen X, Liu J, Yu M (2016) Graphene reinforced aluminum matrix composites prepared by spark plasma sintering. Int J Min Met Mater 23(6):723–729

Vintila R, Charest A, Drew RAL, Brochu M (2011) Synthesis and consolidation via spark plasma sintering of nanostructured Al-5356/B4C composite. Mater Sci Eng A 528:4395–4407

Wang L, Liu Y, Wu J, Zhang X (2017) Mechanical properties and friction behaviors of CNT/AlSi Mg composites produced by spark plasma sintering. Int J Min Met Mater 24(5):584–593

Wu J, Zhang H, Zhang Y, Wang X (2012) Mechanical and thermal properties of carbon nanotube/aluminum composites consolidated by spark plasma sintering. Mater Des 41:344–348

Xie G, Ohashi O, Chiba K, Yamaguchi N, Song M, Furuya K, Noda T (2003) Frequency effect on pulse electric current sintering process of pure aluminum powder. Mater Sci Eng A 359:384–390

Yang S, Yan X, Yang K, Fu Z (2016) Effect of the addition of nano-Al2O3 on the microstructure and mechanical properties of twinned Al0.4FeCrCoNi1.2Ti0.3 alloys. Vacuum 131:69–72

Zhang Y, Li J, Zhao L, Zhang H, Wang X (2014) Effect of metalloid silicon addition on densification, microstructure and thermal–physical properties of Al/diamond composites consolidated by spark plasma sintering. Mater Des 63:838–847

Spark Plasma Sintering of Aluminium-Based Materials

M. A. Awotunde, A. O. Adegbenjo, M. B. Shongwe, and P. A. Olubambi

1 Introduction

The growing technological demand for lightweight and high-strength multifunctional materials in service has fuelled the intensive research on improving material properties. Automotive and aerospace industries are not left out of this quest as environmental sustainability concerns have propelled them to fabricate lighter, environmental friendly vehicles and airplanes with reduced fuel consumption (Evans 2001). Metal matrix composites (MMC) have increasingly attracted interest in both scientific and industrial communities. The incorporation of a strong reinforcing phase in a soft and ductile matrix yields various combinations of properties not feasible in traditional alloys (Steinman et al. 2018). One of the most commonly utilized metals for MMC is aluminium (Casati and Vedani 2014), primarily owing to its light weight. Furthermore, aluminium-based matrix composites (AMCs) have attracted this much considerable attention because of their high stiffness, excellent resistance to corrosion and good thermal and wear resistance (Liu et al. 2013). They have found applications in the fields of aerospace, aeronautics, automobiles and sporting equipment (Ram et al. 2017). However, despite all these advantages, the low strength of aluminium restricts their use as structural materials. For structural

M. A. Awotunde (✉) · P. A. Olubambi
Department of Metallurgy, Center for Nanoengineering and Tribocorrosion
University of Johannesburg, Johannesburg, South Africa
e-mail: mary.awotunde@uniben.edu

A. O. Adegbenjo
Department of Metallurgy, Center for Nanoengineering and Tribocorrosion
University of Johannesburg, Johannesburg, South Africa

Mechanical Engineering Department, The Ibarapa Polytechnic, Eruwa, Nigeria

M. B. Shongwe
Department of Chemical, Metallurgical and Materials Engineering, Institute for NanoEngineering Research, Tshwane University of Technology, Pretoria, South Africa

© Springer Nature Switzerland AG 2019

P. Cavaliere (ed.), *Spark Plasma Sintering of Materials*,
https://doi.org/10.1007/978-3-030-05327-7_7

aluminium alloys, strength and ductility are two of the most important mechanical properties. Generally speaking, high-strength aluminium alloy entails strength above 500 MPa at room temperature, while a tensile elongation higher than 5% is required for structural use (Valiev 2002). To make up for the strength deficit, various elements have been successfully incorporated in pure aluminium and aluminium matrices to improve the mechanical properties of AMMCs, including Al–SiC (Chu et al. 2012; Tan et al. 2013; Zhang et al. 2016), Al–carbon nanotube (Mokdad et al. 2016), Al–Al_2O_3 (Schultz et al. 2011; Wu et al. 2012), Al–diamond (Chu et al. 2009; Dash et al. 2013), Al–ZrB_2 (Zhang et al. 2014), Al–metallic glass (Ghasali et al. 2015; Marko et al. 2014; Khoramkhorshid et al. 2016), Al–B_4C (Perriere and Champion 2012), Al–SiO_2 (Vintila et al. 2011), BN (Steinman et al. 2018) and Al-intermetallic composites (Zhou et al. 2014). These various elements are added to impart diverse properties on aluminium matrices, but with the primary emphasis on strength enhancement.

Powder metallurgy (PM) route has been developed over the years as a viable method for producing dense powder compacts with superior mechanical properties (Sarmasti et al. 2011). It primarily involves mixing and (or) distribution of reinforcing phases in powder matrices followed by the sintering process. The core strength of powder metallurgy lies in its capacity to fabricate near net shape components without material wastage. Powder metallurgy also has the added advantage of facilitating the alloying of two or more immiscible elements in powder form consolidating a compound that would have otherwise been impossible via other manufacturing methods. Different PM methods have been investigated over the years, namely, conventional sintering, microwave sintering, hot isostatic pressing, hot pressing and spark plasma sintering (Bonache et al. 2011).

Among these PM methods, spark plasma sintering (SPS) has attracted significant attention over the past few years. Its uniqueness is in its faster heating rates, lower sintering temperatures, shorter sintering times and better microstructural control in comparison to orthodox PM techniques. The SPS technique works largely on the mechanism of electric spark discharge, where a high energy pulsated current briefly generates and discharges spark plasma between the powder particles resulting in high localized temperatures (Kwon et al. 2009; Xu et al. 2017). The heat generated at the point contacts of the powder particles presents a surface treatment effect on the particle surfaces, removing contaminants and breaking the characteristic oxide layer on the aluminium powders prior to local welding or neck formation (Yang et al. 2006; Kwon et al. 2010). This results in cleaner particle surfaces, increase in densification of the powders and obvious comparative improvement in the properties of the resulting composites (Munir et al. 2006; Liu et al. 2014; Park et al. 2017a, b; Lorrette et al. 2013). Metallurgical bonding between Al powders is thus facilitated by this disruption of oxide films which enables the molten Al around the neighbouring Al powders to fuse with each other (Xu et al. 2017).

SPS has been the choice route for consolidation and sintering of Al powders for obtaining near theoretical densification while retaining a fine-grained structure which positively influences the mechanical properties (Orru et al. 2009). It can maintain the nano- and submicron-structures of powder-based materials after consolidation while creating a strong adhesion between the reinforcement and the matrix, thereby yielding fully dense materials (Sajjadi et al. 2011). The rapid heating rates and shorter sintering cycles during SPS with simultaneous application of pressure have the peculiar attribute of minimizing grain growth while maximizing densification. Moreover, high heating rates up to 1000 °C/min can be achieved in heating rate intervals due to the high current utilized (Sun et al. 2015) which significantly aids densification kinetics.

2 Types of Reinforcements

Though the major reason for reinforcement of aluminium is for strength purposes, other properties such as hardness, wear resistance, thermal resistance and so on can be imparted on aluminium matrix composites (AMCs). These reinforcements can come in different morphologies. The mechanical properties of the reinforced aluminium alloys are primarily dependent not only on the intrinsic nature and properties of the reinforcement, but also on the processing route chosen. These two factors largely determine the major reason for reinforcement of aluminium is for strength purposes, and other properties such as hardness, wear resistance, thermal resistance and so on, can be imparted on aluminium matrix composites (AMCs). These reinforcements can come in different morphologies. The mechanical properties of the reinforced aluminium alloys are primarily dependent not only on the intrinsic nature and properties of the reinforcement the quality of the interface formed between the matrix and the reinforcement, which in turn affects the resulting mechanical properties of the fabricated composites (Jafari et al. 2018). SPS has been the researchers' favourite for the unique advantage of producing clean and strong interfaces between reinforcement and matrix hence yielding better mechanical properties of composites in comparison with other sintering methods (Tian et al. 2016a, b).

2.1 *Particulate Reinforcement*

Particle-reinforced AMCs have huge potentials in aerospace and automotive industry because they not only have good mechanical and wear properties but are also economically viable (Yan et al. 2008). The interest in particle-reinforced composites (PRCs) has increasingly grown over the last few decades owing to their high elastic modulus, increased strength and improved ductility (Tan et al. 2016). It is perceived that adding these second-phase particles into the Al-based alloys would impede the propagation of cracks and increase the resistance to fracture of the reinforced alloys. Garbiec et al. (2015) reinforced aluminium with varying contents of Al_2O_3 between 5 and 20 wt% and investigated the resulting mechanical properties. It was observed that the microhardness value of the sintered composites increased with increase in reinforcement content with the 20 wt% reinforcement achieving about 240% increase over the unreinforced aluminium composite at a microhardness value greater than 1355 MPa. However, Li et al. (2016) observed that the ductility of the as-rolled AA6061–SiC composites decreased with increase in SiC addition.

Ghasali et al. (2017) successfully incorporated 5, 10 and 15 wt% TaC into aluminium matrix using spark plasma sintering method. The choice for tantalum carbide as the reinforcement was due to its high melting point and chemical stability. Results show that the maximum bending and relative density of 438 ± 17 MPa and 99.13 ± 0.5%, respectively, were measured for Al-5 wt% TaC composite. Though maximum microhardness of 295 ± 11 HV was obtained for Al-10 wt% TaC composite, investigations showed agglomeration of TaC reinforcement particles in Al-10 and 15 wt% TaC composites, while the Al-5 wt% TaC composite exhibited a uniform distribution of reinforcement particles. The 10 wt% TaC composite exhibited reduced toughness in comparison with the 5 wt% TaC composite. Authors attribute this to the presence of agglomerates of TaC which led to increase in porosity levels and consequently lower bending strength. In the same manner, the low relative density and larger amount of agglomerates in the 15 wt% TaC composite accelerated the propagation of cracks and subsequently fracture, yielding the lowest toughness value for this composite.

During spark plasma sintering, some authors have observed the formation of intermetallic phases which affect the resulting mechanical properties of the AMCs. Due to the effect of the high temperatures generated during SPS, powder particles locally react with one another; hence, it can be anticipated that occasionally a compound will form as a result of the reaction between the matrix and the reinforcement. Gu et al. (2012) incorporated iron into aluminium matrix in 2.5, 3.5, 5.0 and 10.0 at% using mechanical alloying and SPS. Hardness and yield strength were observed to increase with increasing volume fraction of Fe with Al-2.5%Fe and Al-5%Fe sintered samples exhibiting 1.44 and 1.61 GPa, respectively, and the Al-10%Fe exhibiting the highest hardness and yield strength of 1.97 GPa and 1130.9 MPa, respectively. Plastic strain however was observed to decrease with the increase of Fe content with the least value recorded for the Al-5%Fe of 14.7% which is still quite high. This remarkable strength was attributed to the fine dispersion of the needle-like in situ intermetallic phase – Al13Fe4 due to dispersion strengthening mechanism. This intermetallic phase was observed to increase with increase in the Fe content, hence the observed pattern of the hardness of the composites.

Similarly Park et al. (2017a, b) developed AMCs having a fine distribution of $Al_{13}Fe_4$ and $AlFe_3$ intermetallic phases during the SPS consolidation of stainless steel SUS316L being incorporated into aluminium matrix powders using 20, 50 and 80 vol% compositions. Intermetallic compounds were formed from the reaction between SUS316L and aluminium by means of micro-plasma during the SPS process. The Al-50 vol% SUS316L composite showed the highest Vickers hardness of 630 HV which was more than 7 times higher than pure SUS316L bulk and 22 times higher than pure aluminium bulk prepared in this study as well. The strengthening mechanism responsible for this increased hardness was precipitation strengthening owing to the precipitation of the intermetallic phases. The Al-50 vol% SUS316L composite contained more intermetallic phases which were well dispersed in the aluminium matrix and as such exhibited the highest hardness.

Jafari et al. (2018) investigated the effect of SiCp reinforcement content on the wear rate of aluminium composites from 0 to 15%. Results show that hardness increased with increase in reinforcement content, with the highest hardness obtained for the 15% reinforced composite. Wear rate was also observed to decrease with increase in reinforcement content. This is due to the fact that with increase in reinforcement content, the interparticle distance between reinforcement particles decreased which facilitated more reinforcement particles to be in contact with the wear surface, leading to a reduction in the wear rate. Additionally, a direct relationship exists between wear rate and hardness; hence with increase in hardness, wear rate was progressively reducing. Comparing the results of this study with literature, authors observed a higher wear resistance of the reinforced samples than those documented in literature. This was attributed to the uniqueness of the SPS technique used in this study, which obtained near full densification and also improved the quality of the interface formed between the matrix and reinforcement as against the sintering technique used in Vencl et al. (2010).

2.2 *Nano-reinforcement*

Mechanical properties of reinforced AMCs are majorly dependent on the uniform dispersion of the reinforcements in the aluminium matrix. Good dispersion of reinforcement and good interfacial bonding between the matrix and reinforcement are essential conditions for improved mechanical properties of composites (Bisht et al. 2017). Dispersion of nano-dimensioned reinforcements has proven to be more challenging than dispersion of their particulate counterparts. Apparently, dispersion gets more difficult with reduction in the reinforcement size (Bartolucci et al. 2011). Hence, agglomeration effects are more obvious in nano-reinforcement dispersion than in particulates exclusively because of their smaller diameters (Wang et al. 2017).

In the work of Ogawa et al. (2018), a significantly high strength and unprecedented ductility of 37.2% were observed which the authors attributed to the success of their meticulous choice of parameters with the sole aim of preserving the CNT structural integrity during the milling and sintering processes. They documented that the dominant strengthening mechanisms were grain size refinement and dispersion strengthening due to the shortened carbon nanotubes during the milling process. A small amount of Al4C3 was observed which the authors observed was beneficial in enhancing the strength of the composites as it helped to improve the bonding conditions between aluminium matrix and the CNTs promoting good interfacial bonding and strong interaction between them. It is documented that the formation of Al4C3 increases the interfacial strength by improving bonding conditions, but that the formation of nanorods of Al4C3 decreases strengthening efficiency.

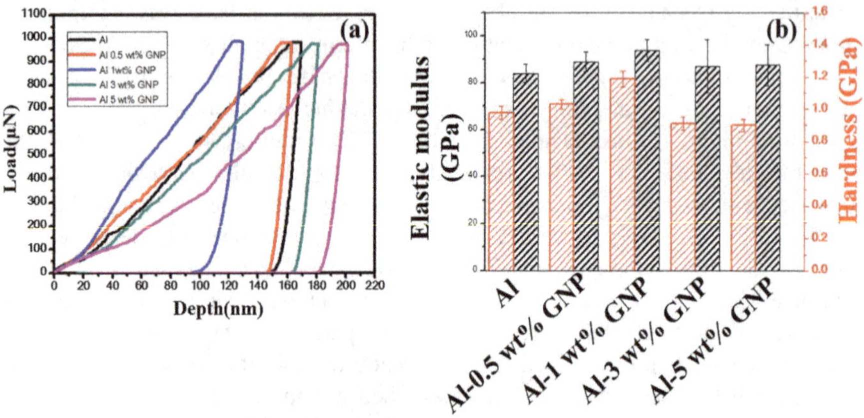

Fig. 1 (**a**) The representation of load displacement curve and (**b**) variation inelastic modulus and hardness of different composites based on different reinforcement contents (Bisht et al. 2017)

Bisht et al. (2017) incorporated graphene nanoplatelets (GNPs) into aluminium matrix in concentrations up to 5 wt%. The mechanical properties of the 1 wt% reinforcement composites surpassed the others with an 84.5%, 54.8% and 21.4% increase in yield strength, tensile strength and hardness, respectively. These properties however fell off at higher concentrations of GNPs owing to the increased agglomeration of the reinforcing phase at such concentrations. Good dispersion and adequate interfacial bonding were observed in the fracture surfaces of the 0.5 and 1.0 wt% composites which showed no free or weakly bonded GNPs on the fracture surfaces of these composites. At higher volume fractions greater than 1 wt% however, agglomerations were observed along grain boundaries of the aluminium matrix, leaving these two crucial conditions unmet; hence deterioration sets in. The re-agglomeration of reinforcement particles during sintering favours irregularities like porosities leading to propagation of cracks and subsequently, failure of the composites (Khosroshahi et al. 2015). Indentation depth was seen to increase with increase in GNP content above 1 wt% as seen in Fig. 1. Ductility on the other hand was observed to reduce with increase of the GNPs even in the 0.5 and 1 wt % composites. The dominant strengthening mechanism was observed to be Orowan strengthening mechanism. This was achieved by the uniform dispersion of smaller particles of GNP reinforcement obstructing the free movement of dislocations. The structure of the GNPs was preserved even after sintering, and authors attribute this to the unique mechanisms of the SPS technique employed.

Boron nitride (BN) nanostructures have also drawn researchers' attention due to their ability to significantly improve mechanical properties of Al-MMCs. They have been incorporated in composites leading to notable strength improvements up to three times for microribbons with ultimate tensile strength (UTS) of 420 MPa (Xue et al. 2015). Firestein et al. (2015) fabricated aluminium composites with 0.5, 1.5, 3, 4.5, 6 and 7.5 wt% boron nanoparticles (BNNPs) – a cheaper alternative to their boron nanotube counterparts. The composite with the 4.5 wt% exhibited a 50 and 75% increase in the tensile strength at room temperature and elevated temperature

(300 °C), respectively, over the unreinforced aluminium. The yield stress was about 115 MPa at 300 °C which was a 190% increase over the yield stress of the pristine aluminium fabricated under the same conditions. Results showed that the BNNPs helped to restrict dislocation movement which conferred strength on the Al-BNNP composites; hence strengthening was due to Orowan's mechanism via homogenous dispersion of BNNPs in the Al matrix and the presence of a large number of strong Al-BNNP interfaces with high critical shear stress inhibiting the beginning of plastic deformation and fracture. Additionally, chemical reaction between the matrix and the reinforcement also led to a strong interface which yielded an effective load transfer mechanism from matrix to the reinforcement.

Steinman et al. (2018) in his work developed boron nitride nanophases by in situ formation of this reinforcement via reactive ball milling and reactive spark plasma sintering. The best microhardness value was obtained for the 7 wt% boron nitride reinforced aluminium composites. An increase in the room temperature tensile strength of 135% was also observed in the Al-5 wt% BN which exhibited the highest UTS of 380 MPa compared to that of pure Al (162 MPa). Interestingly the 5 wt% reinforced composite which showed the highest UTS also displayed the highest plasticity of approximately 12%. The high strength observed in the Al/BN composites was attributed to the uniform distribution of in situ formed BN, AlN and AlB2 phases during the reactive milling and reactive sintering. Due to the SPS sintering, the porosity levels were reported to be between 1 and 2%.

Lahiri et al. (2013) incorporated boron nitride nanotubes into aluminium matrix in varying proportions of 0, 2 and 5 wt% volume fractions with a view to evaluating BNNTs as potential reinforcements for aluminium composites. Compression testing of the composites showed that the composite with Al-5 vol% BNNT exhibited yield strength and compressive strength of 88 MPa and 216 MPa, respectively, which was 54% higher than that of the unreinforced aluminium. Fractographic examination of the fracture surfaces revealed that the reinforced composites still showed ductile failure mechanism just like the unreinforced composites. The significance of this was that the BNNT addition did not impair the ductility of the aluminium. Crack bridging was observed which indicated the strengthening of the aluminium matrix by the load bearing capacity of the BNNTs exhibiting an effective stress transfer mechanism between the matrix and the BNNTs. The BNNT reinforced composite also showed high deformability evidenced by the successful 75% reduction by cold rolling of the sintered pellets without fracturing. SPS was identified as an effective sintering route for fabricating Al-BNNT composites with enhanced mechanical properties.

2.3 *Hybrid Reinforcement*

The interest in dual and hybrid reinforcements is growing, owing to the fact that some individual reinforcements are still incapable of imparting the total desired properties in the composites. For example, the inverse relationship between strength and ductility has put researchers in a dilemma as the two properties appear to be in conflict with each other (Valiev 2002). In view of this, some researchers have incorporated an additional element to solve this problem, so that the additional reinforcement would provide what was lacking in the main reinforcement in a complementary manner. Zhang et al. (2018) reinforced pure aluminium with particulate SiC in combination with CNTs and investigated the effect of these on the mechanical properties. Reinforcements were dispersed in aluminium matrix via ultrasonication and low energy planetary ball milling. Admixed powders were consolidated by SPS followed by hot extrusion. Results showed that composites with dual combination of 0.5SiC-0.5CNTs showed the best mechanical properties with YS, UTS and elongation of 185 MPa, 247 MPa and 20.7% respectively. Compared to the reference aluminium composites, this composition showed a 94% improvement in UTS with just 9.1% reduction in elongation due to the presence of the silicon carbide particulate reinforcements. It was observed that though the SiC was not the main load bearing phase and hence its presence did not significantly improve the strength of the AMCs, it strongly enhanced the plasticity of the resulting AMCs. A trend was observed in the grain sizes owing to the pinning effect of the CNTs; hence there was a marked decrease in grain sizes with increase in CNT content. A thin layer of Al4C3 was observed between the aluminium and the CNTs, the presence of which improved the interfacial bonding and made the load transfer mechanism effective. The strengthening effect was documented to be the synergistic effect of both the dispersion hardening of the SiC particulates and the grain refinement of the CNTs.

Durowoju et al. (2015) observed that the addition of SiC and Si particles into an aluminium graphite matrix led to a reduction in the hardness values obtained. This they attributed to the higher level of porosities obtained in the reinforced Al-Gr composites during the sintering process. Hence the unreinforced Gr-Al composite showed the best properties in terms of the relative density, microhardness and microstructure. It has been observed that increasing the quantity of ceramic particles in a composite leads to an increase in the porosity due to difficulty of compressing the hard ceramic particles in a ductile matrix. The reason for the hybrid reinforcement in this study was to improve the wettability and the infiltration of the metal into the graphite matrix with the additions of SiC and Si particles. Results showed a relative density of 97.1% achieved for Al-Gr sample, while 96.1 and 95.4% were obtained for Al-Gr 10 wt% Si and Al-Gr 10 wt% SiC, respectively. The microhardness values obtained were 26 $\pm$ 5, 26 $\pm$ 4 and 20 $\pm$ 3 HV for Al-Gr, Al-Gr 10 wt% Si and Al-Gr 10 wt% SiC, respectively.

Rana et al. (2009) reinforced AA6061 aluminium alloy with both magnesium and silicon ball milling and SPS. Nanocrystalline powders of AA6061 alloy with resultant grain size of $\sim$30 nm was successfully obtained after 20 h of ball milling. SPS was carried out which yielded over 99.5% densification at 773 K with grain size still less than 100 nm even after SPS sintering. Compression tests revealed that the consolidated samples exhibited exceptionally high compressive strength of 800 MPa, due to the fine grain size with corresponding strain to fracture of 12%. However, annealing these samples at 350 °C led to an increase in grain size from 85 to 100 nm; it also led to an increase in compression strength to 827 MPa and a corresponding increase in strain to fracture of 15%. This simultaneous increase in strength and ductility was attributed to the precipitation of fine needle-like particles following the annealing procedure; therefore the loss of strength due to grain growth was compensated for by the strengthening contribution of the precipitated phases.

Al_2O_3 is one of the most commonly used reinforcements in aluminium due to its chemical inertness and thermal stability. Magnesium is also used specifically for its high specific strength. Babu et al. (2016) employed dual reinforcement of magnesium and Al_2O_3 to benefit from the strengths of the two reinforcements via ball milling and SPS. For comparison purposes, singly reinforced milled and unmilled Al-5wt%Mg was also consolidated using the same process parameters. Though the dual reinforced composite showed the least relative density among all three samples, it exhibited the highest hardness of 144 ± 6 HV showing that the hardness was not majorly dependent on the densification but on the reinforcement.

High entropy alloys (HEAs) are a relatively new group of advanced metals that are solid solution alloys containing more than five major elements in approximately equal atomic percent (Zhang et al. 2011; Gorr et al. 2015). Tan et al. (2016) incorporated aluminium alloy $Al_{65}Cu_{16.5}Ti_{18.5}$ with high entropy alloy $Al_{0.6}CoCrFeNi$ with a combined face-centred cubic and body-centred cubic dual phase for a synergistic reinforcement effect of both good strength and ductility. Results showed that the reinforced composite exhibited a strength of 3120 ± 80 MPa over the 1700 MPa of the unreinforced alloy. The characteristic localized high temperature generated during SPS stimulated the rapid diffusion of elements between the amorphous alloy powders and the HEA reinforcing particles, hence leading to the formation of an interdiffusion layer around the HEA particles. During deformation, the interface between the interdiffusion layer and the HEA particle is a highly stressed region due to the wide difference in the hardness between them. It was observed however, that the HEA particle could carry more load through plastic deformation without fracture. It was therefore assumed that the fracture mainly occurred within the matrix and the interdiffusion layer. Consequently, the combined action of the stable propagation of cracks and the observed low content of sintering defects in the matrix were key contributors to the remarkably high strength of the composite.

A common denominator in the incorporation of these various reinforcements into metal matrices is the uniform distribution of these reinforcements, as the resulting mechanical properties of the composites are largely dependent not only on the properties of the matrix and reinforcement, but also on the distribution of the latter. Agglomerations of reinforcement phases have led to declining mechanical properties (Firestein et al. 2015) sometimes even less than the unreinforced composites (Backshi and Agarwal 2011; Tian et al. 2016a, b). Different types of dispersion techniques have been employed over the years to effectively solve the agglomeration problem. Ball milling has been the most popularly used method of dispersion owing to simplicity of operation (Esawi et al. 2010). Different variants of ball milling techniques have evolved over the years, namely, dry milling (Mokdad et al. 2016), wet milling (Rikhtegar et al. 2017) and cryomilling (milling under cryogenic conditions) (He et al. 2017). Researchers have also explored a combination of methods to achieve homogeneity of the reinforcements, like ultrasonication in combination with ball milling (Guo et al. 2017), while other more sophisticated routes like nanoscale dispersion (NSD) have been developed especially for the distribution of nano-reinforcements like CNTs (Noguchi et al. 2004). Possible ways of circumventing the agglomeration problem is by developing the reinforcing phases in situ within the matrix. Mechanical alloying via reactive ball milling of the constituent elements or reactive sintering for a deliberate in situ formation of the reinforcing phases and agents has shown good mechanical properties due to a good distribution of the in situ formed phases within the matrix (Steinman et al. 2018). He et al. (2007) also used an in situ grown method of dispersing CNTs by growing the CNTs on aluminium powders making them substrates for CNT growth using nickel as a catalyst. Al-5wt% CNT was prepared via this method which exhibited 200% and 180% increase in hardness and tensile strength over the composites obtained by conventional blending method. Cooke et al. (2016) also strengthened aluminium by alloying with scandium for the sole purpose of precipitating Al_3Sc phase evenly in the matrix, thus enhancing the strength of this alloy via the precipitation hardening strengthening mechanism. Similarly, Gu et al. (2012) enhanced the strength of aluminium by achieving a fine distribution of $Al_{13}Fe_4$ precipitates within the aluminium matrix. Additionally, in situ formed phases can also solve the strength-ductility trade-off in composites by the precipitation of small second-phase materials in the matrix (Shanmugasundaram et al. 2006). This was observed in Rana et al. (2009) who documented a simultaneous increase in strength and ductility as a result of phase precipitation.

Park et al. (2018) investigated the role of the intermetallics formed during sintering in the strengthening of Al-Mg composites fabricated by SPS. It was observed that the intermetallic compounds – Al_3Mg_2 and $Al_{12}Mg_{17}$ – act as obstacles to dislocation movements within the matrix, hence improving their strength. Moreover, it was observed that the distribution and amount of intermetallic compounds in the Al-Mg composite materials may also be an essential factor. For example, in the Al-50 vol% Mg composite material, the amount of intermetallic compounds was larger than in the other composite materials and uniformly dispersed, resulting in the highest Vickers hardness for this composition at 189 HV compared to the Al-20vol%Mg and Al-80vol%Mg with 78 and 102 HV, respectively. The study confirmed that the intermetallics formed was responsible for the improved strength observed as the hardness values for the composites were significantly higher than that of pure aluminium (29 HV) and magnesium (37 HV).

3 Reinforcement Type and Content

As previously established, different reinforcements impart different properties and varying levels of strength on aluminium matrices. Also, a limiting or critical volume fraction of reinforcement has been documented by several researchers (Liu et al. 2010), above which the mechanical properties of the AMCs start to deteriorate. To investigate the effect of different reinforcement types and contents, Sweet et al. (2015) incorporated different ceramic particulate reinforcements in aluminium matrix, namely, SiC, AlN, Si_3N_4 and BN in volume fractions 2, 5 and 10 vol%. Full densities were obtained at about 550 °C for all the differently reinforced composites at low concentrations though with increasing reinforcements, composites were unable to attain near theoretical densities. Si_3N_4 showed inferior properties

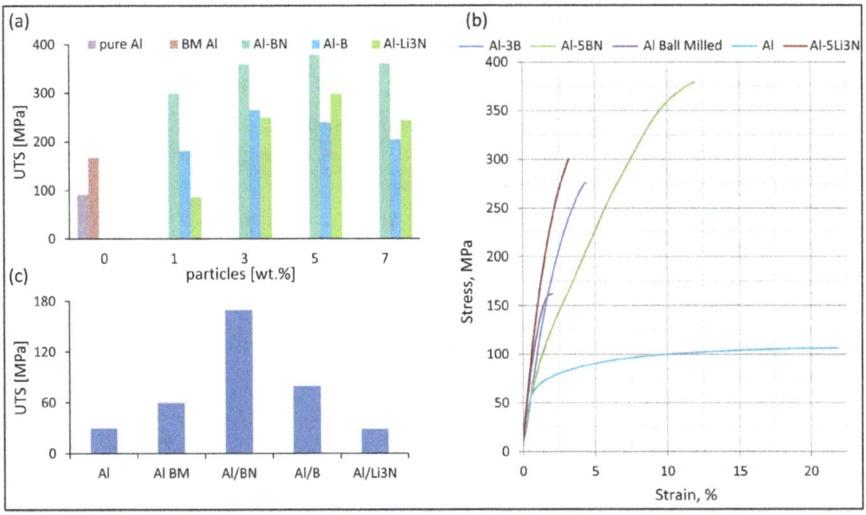

Fig. 2 (**a**) UTS of all samples in dependence of a reinforcing phase content, (**b**) stress-strain curves for the samples revealing best mechanical properties – the results for pure Al are shown for comparison – and (**c**) UTS of the representative samples tested at 500 °C (Steinman et al. 2018)

to SiC and AlN with only the 2 vol% achieving near theoretical densification. Increase in reinforcement content further led to higher densification losses. BN exhibited a weak dependence of densification on temperature even at low concentrations of 2 vol% which became progressively weaker with increase in volume fraction. Baskut et al. (2017) observed a limiting volume fraction of 1 wt% graphene platelet (GPL) reinforcement in aluminium nitride matrix beyond which the mechanical properties began to diminish due to the agglomeration effects.

Steinman et al. (2018) observed the effect of different reinforcement contents on the mechanical properties of AMCs as seen in Fig. 2. This shows the limiting reinforcement content above which the mechanical properties begin to fall off. Al-BN shows a limiting reinforcement content at 5 wt%, Al-B at 3 wt% and Al-Li3N at 5 wt%.

Tian et al. (2016a, b) reinforced aluminium 7055 with graphene in varying contents and observed a limiting volume fraction at 1 wt% after which the mechanical properties deteriorated due to the agglomeration of graphene in the matrix at higher concentrations. Vickers hardness of the unreinforced aluminium 7055 alloy was observed to increase from 131.5 to 151.2 HV with 1 wt% graphene plates. However, further increment in graphene content reduced the Vickers hardness to 128.3 and 98.6 HV with 3 and 5 wt% graphene, respectively. These values were significantly lower than that of the unreinforced aluminium 7055, substantiating the disadvantages of agglomerations in reinforced composites.

Though Tian et al. (2016a, b) documented a limiting content of 1 wt%, Pillari et al. (2018) documented a limiting content of 0.5 wt% for graphene in aluminium AA2219. Studies show diverse limiting reinforcement content even for the same reinforcement type. This is possibly due to the dispersion method of these reinforcements, thus with higher dispersion success, the higher the limiting content observed.

4 Strengthening Mechanisms

The strength of the composite is anticipated to be the addition of the operative strengthening mechanisms in combination with the intrinsic strength of the matrix (Park et al. 2015). Studies have shown a number of strengthening mechanisms in AMCs responsible for the improved strength.

Firestein et al. (2015) attributed the improved strength observed in their study to the strong interfacial bonding formed between Al and boron nanoparticles which they observed led to an effective load transfer system between the matrix and reinforcement. Similarly, Ogawa et al. (2018) also observed a small amount of Al_4C_3 which accomplished the same feat during the course of their mechanical testing. The load transfer system works in this manner; when composites are subjected to external stress, the load is borne primarily by the matrix, while the reinforcements help in holding the matrix together. After a while, the load borne by the matrix reduces as the matrix begins to give way, and the reinforcements kick in and bear the major loads. Good interfacial bonding is required at this phase for the effective transfer of this load or stress from matrix to the reinforcement, thus improving the strength of the composite (Poirier et al. 2009).

Improvement in strength due to thermal mismatch is basically the generation of dislocations in the matrix due to the disparity between the coefficients of thermal expansion of the matrix and reinforcement. This gives rise to an increase in dislocation density within the matrix, leading to strengthening (Park et al. 2015) due to work hardening effect (George et al. 2005).

In Hall-Petch strengthening (or grain boundary strengthening), the free movement of dislocations during deformation is hindered by grain boundaries due to the misorientation between neighbouring grains. In this form of strengthening, the grain boundaries are the obstacles to dislocation movement. Hence, the larger the grain size, the fewer the grain boundaries and vice versa. With smaller grain size, more grain boundaries are present, leading to a more difficult dislocation movement and consequently a higher flow stress (Mokdad et al. 2016). This is due to the fact that grain boundaries act as "pinning points" prohibiting the dislocation movement leading to a "pile up" of dislocation clusters being unable to move past a grain boundary (Itoh et al. 2008).

Bisht et al. (2017) attributed the strength enhancement he observed in his study to the uniform dispersion of graphene nanoplatelets in the aluminium matrix which impeded the movement of dislocations thereby strengthening the composite. Here the obstacles are the reinforcements in whatever form whether precipitates, dispersoids, particulates or nanoreinforcements. This is known as Orowan strengthening mechanism, which facilitates the bending of dislocations between reinforcements, generating a back stress which will then further inhibit dislocation movement (Mirza and Chen 2015). Steinman et al. (2018) observed this strengthening due to the uniform distribution of in situ formed BN, AlN and AlB_2 precipitates within the matrix.

5 Processing Parameters

5.1 *Milling Parameters*

During milling, the milling balls impact the powder particles causing them to plastically deform and subsequently work harden, which is then followed by fracture and rewelding of the particles. Though a lot of the resulting mechanical properties are dependent on the intrinsic properties of the reinforcement, a lot also depends on the processing routes and parameters that led to the fabrication of the AMCs. Novak et al. (2014) investigated the effect of milling parameters like ball-to-powder ratio on the phases formed during milling by milling elemental powders of aluminium, copper and iron together at 400 rpm. Al-Cu-Fe quasicrystalline phase was successfully formed after 2 h of ball milling due to the utilized high ball to powder ratio of 70:1 and high rotational velocity. This significantly reduced the milling time by more than 90% of the milling time documented using lower ball to powder ratio. Aside from the increase in kinetic energy of the milling balls, the increase in the ball to powder ratio and rotational velocity led to increased frequency between the powder particles, balls and milling vessel. This correspondingly led to increased frictional forces between the balls and milling vessel which significantly raised the localized temperature. This was followed by rapid cooling when the heat is transferred to the milling vessel leading to the formation of quasicrystalline $Al_{60}Cu_{30}Fe_{10}$ phase.

5.2 *Grain Size Reduction*

In attempting to further enhance the strength of aluminium alloys and composites, grain size refinement has been explored intensively over the past few decades. The Hall-Petch relationship between grain size and strength gives researchers the viewpoint of achieving better mechanical properties by reducing grain size of powders to nanoscale levels (Shaeri et al. 2014). The uniqueness of grain size strengthening by mechanical alloying lies in the achievement of higher strength values without the use of costly alloying elements (Le et al. 2013). Eldesouky et al. (2014) investigated the influence of grain size of starting powders on the mechanical properties of the resulting composites after consolidation using micro- and nanocrystalline powders of AA2124 aluminium alloy. It was observed that microhardness and compressive strength values of the nano-powder composites were significantly higher than micro-powder strength values. Nano-powders showed microhardness and compressive strength values of 250 HV and 703 MPa as compared to their micro-powder counterparts with 155 HV and 300 MPa, respectively. This marked variance was attributed to the intrinsic properties of nano-powders according to Hall-Petch's inverse relationship between grain size and strength and to the presence of surface oxides on the surface of the nano-powders due to high-surface-area-to-volume ratio. This increase in hardness value was apparent, despite the lower final density of the nano-powders as compared to the micro-powders. Thirty-six hours of milling time led to the refinement of the starting micro-powders (45 μm average particle size and 87 nm average grain size) down to ultrafine nanoscale powder (<100 nm in size) which was responsible for the remarkable strength observed in the bulk nanocomposite.

Sasaski et al. (2009) fabricated Al-0.5at%Fe composites by mechanical alloying and SPS. Mechanical alloying was used to successfully refine the grain sizes of the elemental powders of iron and aluminium. Milling time varied between 60 and 150 h. The room temperature strengths of the composites were found to increase with increase in milling time due to a progressive reduction of the grain size as milling progressed. After 100 h of milling, the composites showed high compression strength of >1000 MPa at ambient temperatures and 500 MPa at 350 °C. However, this high temperature strength dropped to 400 MPa with increase in milling time to 150 h with the room temperature strength largely unchanged. The increase in strength was attributed to grain size refinement as a result of the mechanical alloying and the uniform dispersion of iron aluminide intermetallic (Al6Fe) particles in the matrix.

Eldesouky et al. (2014) observed in his study between as-received micro-powders (AMP) and ball milled nano-powders (BNP) that the consolidated BNP specimens showed significant enhancement in hardness and compressive strength over the AMP due to the presence of oxide dispersoids. The inherent thin oxide layers or films on aluminium powder particles have been established in literature (Salem and Sadek 2010), and these oxide layers have a role to play in the strengthening of sintered aluminium composites. Though it is said that the presence of these oxide films is known to hinder the sintering process, the unique surface cleaning feature of SPS can prevent it from hindering the sintering process during SPS processing (Le et al. 2013; Park et al. 2017a, b). Le et al. (2013) observed an increase in strength with decreasing grain size and attributed the enhanced strength to the presence of Al oxide particles such that the strength contribution from oxide dispersion strengthening increased with a decrease in grain size.

Bathula et al. (2012) incorporated 10 vol% SiC particulates into Al5083 aluminium alloy using ball milling and SPS. During milling, reduction of crystallite size to 25 nm led to an improvement of 824 MPa and 280 HV over the 305 MPa and 98 HV of the unreinforced in terms of compression strength and hardness, respectively.

Though a major challenge in the processing of composites through nanocrystalline powder consolidation has been grain growth of these powders during sintering, a bimodal grain size distribution has been recommended by researchers as a good way of achieving combined strength and ductility. SPS seems to offer this huge benefit as the localized high temperatures generated during the sintering process sometimes lead to random coarse grains in the composite. Bimodal grain size distribution (Sadeghi et al. 2017) is fast becoming a tool for a simultaneous enhancement of ductility and strength especially in single-phase nanocrystalline materials due to the strength ductility trade-off dilemma that persists in these composites (Lee et al. 2005).

Ye et al. (2006) milled Al 5083 under cryogenic conditions to obtain nanocrystalline powders that was consolidated by SPS. Results showed a bimodal grain size distribution, namely, 90–10% of fine grains to coarse grains with the average grain size in the fine grained region recorded to be around 47 nm. This bimodal grain size distribution was actually desirable as authors concluded that it enhanced the mechanical properties of the bulk material, as the fine grains contributed to the strength and the coarse grains enhanced the ductility. The average grain growth during the sintering in the fine grained region was observed to be 25 nm, that is, a 25 nm increase in the grain size of the as-milled powders during sintering. Nanoindentation studies revealed there was minimal variation but not a gradient in microstructure across the samples from centre to edge and from top to bottom as hardness varied slightly from point to point. The hardness of the SPSed samples was found to be 165 $\pm$ 6 HV, which is almost twice the value for conventional Al 5083. This improvement was attributed to its refined grain structure as a result of the combination of cryomilling and SPS which prevented excessive grain growth during sintering. Though 10% of the grains are coarse grains, the overall average grain size in the SPS-consolidated cryomilled 5083 Al was still small, finer than most of that which is documented in literature via PM routes. In comparison, Park et al. (2004) observed an average grain size of 116 nm in a hot isostatic pressed Al-Mg alloy, while the grain size in Al 5083 consolidated via cold isostatic pressing and extrusion was 207 nm (Han et al. 2005). The smaller grain size observed in Ye et al. (2006) was undeniably due to the short sintering time, 5mins, required for SPS consolidation which does not allow sufficient time for grain growth.

Eldesouky et al. (2014) observed in one of their studies that increasing the milling time to 6 h resulted in a substantial refinement of the crystallite size from an average of 78 nm to 32 nm which was more than 50% reduction. Further milling did not lead to such rapid reductions though as it took an additional 30 h of milling to refine the crystallite size to 15.6 nm. Increasing the milling time above 36 h however did not result in additional refinement. This can be termed as the limiting milling time beyond which no apparent refinement is observed. Sometimes, beyond this point some grain coarsening may actually occur which is due to the heat generated during long milling hours.

Zheng et al. (2017) after 20 h of milling and SPS sintering of AA2024 aluminium powders obtained composites with YS, UTS and tensile elongation of 375 MPa, 456 MPa and 5%, respectively. It was observed that with increase in milling time came increase in strength but decrease in ductility; hence the composites exhibited a strength-ductility trade-off tendency that is common in composites. The increasing strength with milling time was attributed to the continuous reduction in grain size with progressive milling, while the reduction in ductility was attributed to micropores and the impurities introduced during the milling process which led to inadequate bonding.

6 Sintering Parameters

6.1 Sintering Temperature

Porosity levels are seen to reduce with increase in temperature. Novak et al. (2014) observed a drastic reduction in porosity level to less than 1 vol% at sintering temperature of 800 °C from a 62 vol% porosity level achieved with sintering temperature of 450 °C.

With the exception of holding time, the sintering temperature and pressure are the two crucial parameters on which the densification of the bulk composite largely depends on. Varying sintering parameters like pressure and temperature have varying effects on the mechanical properties of the consolidated materials. Vintila et al. (2010) in their study observed during spark plasma sintering of cryomilled Al2024 powders that nearly the same relative density was achieved when sintering temperature and pressure were set at 400 °C and 200 MPa in one test and 500 °C with an applied pressure of 50 MPa in another test. Hence with higher sintering pressure, sintering can be carried out at lower temperatures with a view to restricting grain growth. Additionally, micro-pores may be effectively eliminated with the application of higher sintering pressures which would enhance the mechanical properties of the resulting composites as a lot of properties depend on the extent of densification in sintered composites.

Chen et al. (2016) fabricated Al-CNT composites using ball milling and SPS. Their study demonstrated the simultaneous enhancement of strength and ductility in Al-CNT composites with increasing sintering temperature. This was attributed to the improved bonding conditions between the aluminium and CNT, favouring good interfacial contact between the matrix and reinforcement. The influence of the sintering temperature on microstructures and mechanical properties of the compact aluminium bulks was studied by Liu et al. (2014) using SPS sintered nanocrystalline aluminium powders at 673, 723, 773 and 823 K, respectively. Figure 3 portrays the effect of sintering temperature on the mechanical properties of the composites, while Fig. 4 shows the relative densities of the consolidated composites increasing with increasing temperature up until 723 K beyond which the increase in density slowed down drastically. Microhardness was observed to increase initially, but beyond 823 K it began to decline. The probable explanation for this was the rapid diffusion at such elevated temperatures which favours grain growth. The microhardness of sintered nanocrystalline aluminium is always higher than that of coarser grained powders. Microhardness is largely dependent on three features – the relative density, grain size (Najimi and Shahverdi 2017) and reinforcement content (Ghasali et al. 2017). Samples sintered at sintering temperatures less than 773 K exhibited lower microhardness values than those sintered at 773 K due to reduced density and consequently higher porosities which is detrimental to the mechanical properties of composites. Thus, the nanocrystalline Al sintered at 773 K showed the highest microhardness of 3.06 GPa, the best compressive strength of 665 MPa and the highest tensile strength of 282 MPa. The grain growth phenomenon led to a reduction in microhardness and strength of the nanocrystalline Al sintered at 823 K.

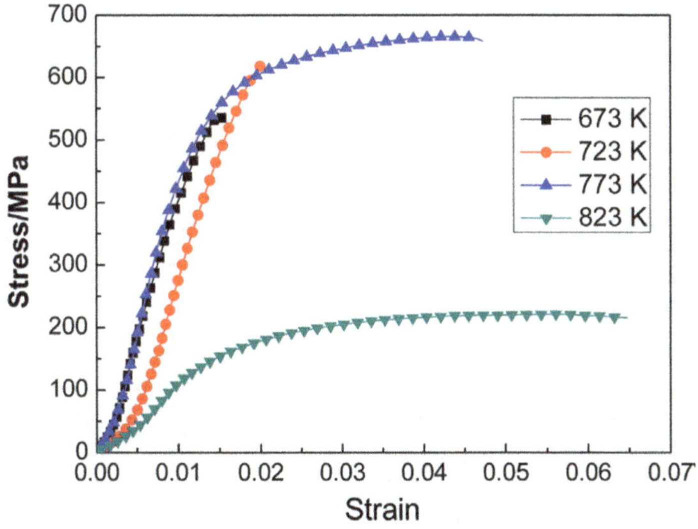

Fig. 3 Stress-strain curves of the nanocrystalline Al showing the effect of sintering temperature on the properties of the composites (Liu et al. 2014)

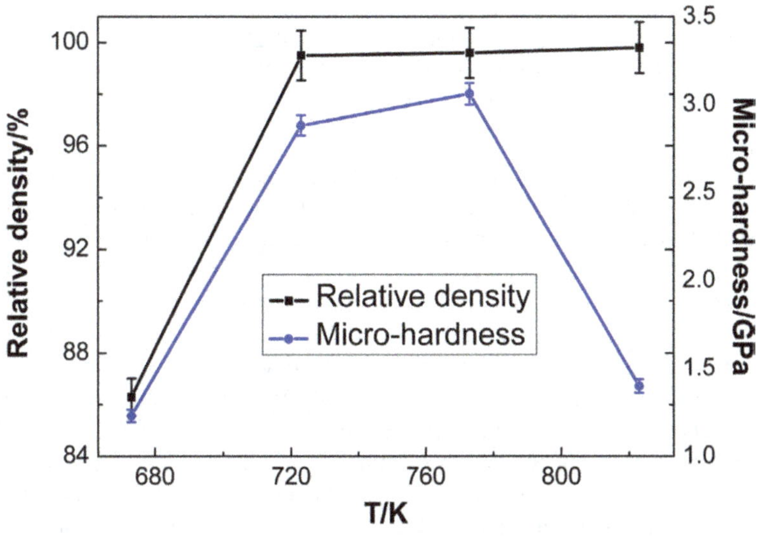

Fig. 4 Effect of sintering temperature on relative density and microhardness of the nanocrystalline Al prepared by SPS (Liu et al. 2014)

Interesting results were obtained by Najimi and Shahverdi (2017) who investigated the same sintering temperature phenomenon using AA6061 alloy and CNTs. Though significant grain growth was observed for the consolidated composite of unreinforced AA6061, the grain growth of the AA6061-CNT was greatly reduced as seen in Fig. 5. This is due to the grain pinning properties of CNTs

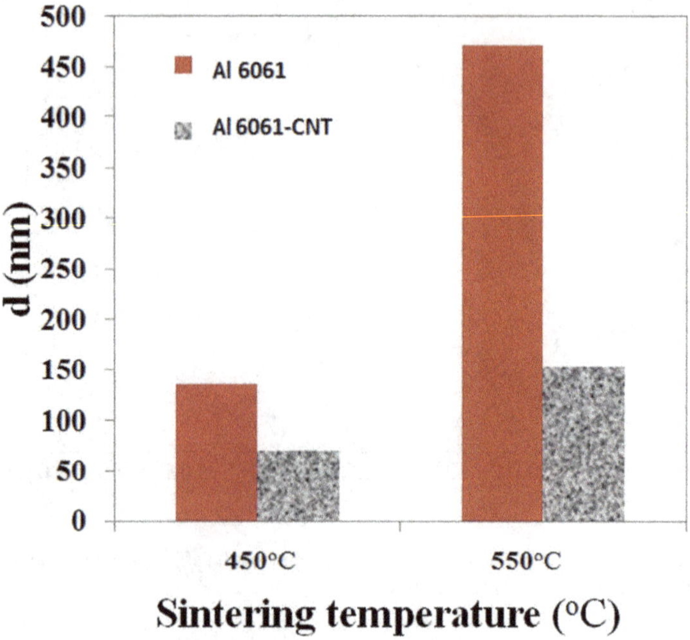

Fig. 5 Plot of the mean equivalent diameter of SPSed Al6061 alloy and Al6061-CNT composite at different sintering temperatures. The grain growth during sintering of CNT reinforced Al6061 alloy is very much reduced (Najimi and Shahverdi 2017)

(Mokdad et al. 2016). Usually, CNTs are located at the grain boundaries; hence during sintering due to their high thermal conductivities, they serve as heat sinks (Munir et al. 2015) and thus dissipate heat quite readily, preventing grain growth in the reinforced composites. Thus an increase in the sintering temperature from 450 to 550 °C led to a reduction in the yield strength and UTS but an increase in ductility in the unreinforced alloy, while for the CNT reinforced AA6061, the same increase in sintering temperature led to a simultaneous increase in all three properties from 310 MPa, 375 MPa and 4.1% to 360 MPa, 430 MPa and 4.9% for yield strength, UTS and ductility, respectively. Moreover, the presence of aluminium carbide was also observed at the grain boundaries seemingly to restrict grain growth as well. Also the Al4C3 was seen to improve the bonding conditions between the aluminium and the CNTs, hence enhancing the load transfer mechanism from the matrix to the reinforcement.

Cooke et al. (2016) compared properties between SPS sintering temperatures of 400 and 550 °C. As seen in Fig. 6, the samples sintered at 550 °C showed far less porosity than those sintered at 400 °C of the same composition. It can also be observed that the level of porosity increased with increase in scandium reinforcement content. Three reinforcement contents were varied – 0.1, 0.2 and 0.4 wt% scandium (Sc). It was observed that the sample with the least reinforcement content showed the least porosity irrespective of the sintering temperature. This was attributed to the creep power law that is known to control densification mechanism during

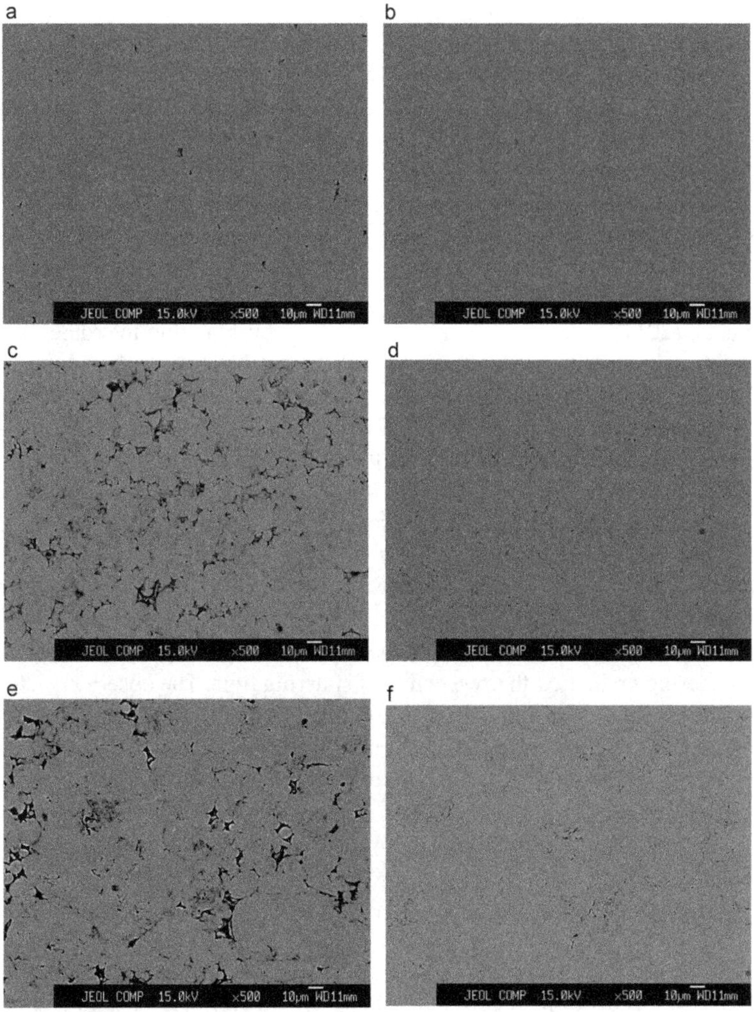

Fig. 6 Evolution in the residual porosity within (**a–b**) Al-0.1Sc, (**c–d**) Al-0.2Sc and (**e–f**) Al-0.4Sc materials as a result of SPS processing temperatures. Peak temperatures were 400 °C (**a**, **c**, **e**) and 550 °C (**b**, **d**, **f**) (Cooke et al. 2016)

SPS sintering. Hence, the Al–0.1%Sc, being the softest sample showed highest densification while those with increased scandium content became more irrepressible to the creep-induced deformation. Interestingly, though the hardness values increased with increase in Sc volume fraction, they decreased with increase in sintering temperature in spite of the improved density observed. This inverse relation between densification and hardness was proof that densification was not the major factor on which the hardness of the Al-Sc samples depended on. Authors explain this as a direct result of the coarsening of the Al_3Sc strengthening precipitates at higher temperatures which led to the observed loss of hardness with increased temperature.

Xu et al. (2017) sintered micrometric grains of aluminium using six different temperatures to evaluate the effect of sintering temperature on the densification during sintering. Densification was investigated at 350 °C, 450 °C, 500 °C, 550 °C, 580 °C and 600 °C. It was observed that at sintering temperatures between 350 °C and 500 °C, the average relative density was marginally improved from 77.4% to 79.2%. However, as the sintering temperature increases from 500 °C to 550 °C, the average relative density remarkably increases from less than 80% to ~99.1%. Further increase in the sintering temperature from 550 °C to 600 °C, however, barely led to an improvement, as the relative density obtained at this temperature was just about ~99.4%.

Xu et al. (2017) opted for SPS in their study of consolidating micrometric grains of aluminium in order to prevent grain growth and to disintegrate the inherent oxide films on the aluminium powder particle surfaces. Furthermore, low ductility has been associated with nanocrystalline (NC) or ultra-fine grained (UFG) composites, hence the use of micrometric grains in this work to enhance the ductility of the developed composite. Results showed that there was no grain growth during sintering due to the presence of nanoscale oxide particles at grain boundaries retarding the grain growth and the short sintering time used in the present study which was 1 min. Further analysis showed that the high volume fraction of the nanoscale oxide particles led to negative driving forces for grain boundary migration and grain rotation (which are the two mechanisms of grain growth), hence effectively hindering grain growth irrespective of sintering time. The observed increase in densification with increase in sintering temperature is justifiable with the fact that creep flow is a diffusion-driven process and therefore depends majorly on temperature; hence an increase in sintering temperature directly results in higher densification (Liu et al. 2011).

6.2 *Pressure*

Usually, there seems to be a correlation between sintering parameters and densification. Sintering at higher temperatures or higher pressures tends to give better densification during the sintering process. To validate this point and evaluate the effect of pressure on densification mechanics, Liu et al. (2016) varied 100 and 500 MPa of sintering pressure during SPS. It was observed that the relative density of the sample sintered at 450 °C and 500 MPa for 3 min was 98.1%, whereas the sample sintered at 500 °C and 100 MPa for 10 min showed a relative density of 96.5% as shown in Table 1. This indicated further that with higher pressures, higher densification can be achieved in less time.

In conventional sintering, the densification is dependent on the applied pressure which also increases the rate at which the grains grow. Apparently, the enhanced diffusion under the applied pressure has a substantial effect on grain growth. In Ye et al. (2006), it was observed that the aluminium grains in the core of the consolidated material remained very fine (47 nm), even with the application of the current

Table 1 SPS processing parameters and relative densities of the consolidated samples

Sample ID	Sintering temperature (°C)	Dwell time	Pressure (MPa)	Relative density (%)
CM 500	450	3	500	98.1
CM 100	500	10	100	96.5
CM 100b	400	3	100	95.8
CM 100c	380	3	100	94.5

Liu et al. (2016)

and pressure indicating that these two parameters had minimum impact on the grain growth in the core. The significance of this is that the effect of the applied pressure may actually progressively reduce from the surface or edges to the core. Furthermore, since the temperature and pressure on the surface of the material are very high, it may lead to softening or melting to liquid phase thereby damping the actual strain in the core and thus limiting the effect of the load on grain growth in the core region.

7 Comparison Between SPS and Other Sintering Methods

Ghasali et al. (2015) compared SPS with microwave sintering and observed that SPS produced aluminium-ZrB2 composites with higher relative density even at lower sintering temperatures. Eldesouky et al. (2014) observed the same phenomenon. Comparison between SPS and HP sintered composites revealed significantly higher microhardness values in the SPSed composites due to faster heating rate and much shorter holding time during SPS sintering (holding time of 10 min in SPS compared to 60 min in HP) which prevents grain coarsening. Additionally, the spark plasma sintered composites exhibited higher relative density (98 ± 0.3% of theoretical density) as against the microwave sintered composites (97 ± 0.3% of theoretical density) even at a lower sintering temperature. Pulse currents among powder particles in contact seemed to give better sintering results than volumetric heating in microwave sintering, leading to reduced grain growth and higher densification. Some researchers also observed new nanostructured phase formations developed during SPS which could also lead to the enhancement in the hardness of the composites (Rana et al. 2009). Additionally, the consolidated specimens fabricated via SPS showed significant increase in both compressive yield strength and ductility by 1.2 and 2.5 times that of the hot pressed counterparts. Le et al. (2013) also observed the same phenomenon in SPSed samples as compared to samples of similar grain sizes prepared by thermomechanical processing.

Though development of composites through nanocrystalline powders has become a popular route for researchers, the control of grain growth during its consolidation remains a crucial factor in the process selection. SPS has proved its grit by yielding less grain growth in comparison to other traditional sintering techniques. Rana et al. (2009) in their study comparing vacuum hot pressing and SPS observed this

Table 2 Relative density, microhardness, bending and compressive strength of samples

Sample sintered by	Relative density (%)	Microhardness (Vickers)	Bending strength (MPa)	Compressive strength (MPa)
Microwave	97 ± 0.3	65 ± 15	179 ± 10	262 ± 15
SPS	98 ± 0.3	110 ± 12	275 ± 10	381 ± 16

Ghasali et al. (2015)

phenomenon, and it is usually attributed to the enhanced sintering kinetics due to the electric field. Moreover, full densification can be achieved at much lower temperatures and in much shorter times (Chaim 2007).

Ghasali et al. (2015) compared microwave sintering of Al-ZrB2 with SPS of same. Results show higher density of SPS compared to microwave sintered samples related to the nature of the former method in improvement of the sintering process. SPS simultaneously applying pulsed electrical current and pressure directly on the sample led to densification at relatively lower temperatures (Figiel et al. 2011). Table 2 presents microhardness, bending and compressive strength values of SPS sintered samples which are higher than those of microwave sintered samples due to a higher relative density, good interfacial bonding between Al matrix and ZrB2 particles and a microstructure free of microcrack and porosity in SPS sintered samples. In the SPS process, the heating power is not only distributed over the volume of the powder compact homogeneously in a macroscopic scale, but also more heating power is dissipated exactly at the locations in the microscopic scale, where energy is required for the sintering process, namely, at the contact points of the powder particles. This fact results in a favourable sintering behaviour with less grain growth, more densification and suppressed powder decomposition (Suarez et al. 2013). The formation of plasma during SPS has a direct effect on enhancement of the diffusion rate by the vaporization and melting of powder particle surfaces during the SPS process which result in a sintered compact to a high density (Saheb 2013). Mechanical properties of specimens are extremely dependent on the relative density and microstructure. Table 2 depicts the higher relative density and more uniform microstructure of SPS samples compared to microwave samples resulted in higher mechanical properties.

Likewise, Saheb (2013) also compared SPS to microwave sintering of pre-alloyed powders of Al6061 and Al2124 using the same sintering temperatures of 400, 450 and 500 °C with dwell time of 20 min for effective comparison between the two techniques. SPS of the Al6061 yielded a relative density of 94.07% at 400 °C, while 100% was obtained at 450 °C. Microwave sintering (MS) on the other hand yielded a relative density of 91.07% at 400 °C with only a marginal rise to 91.08% at 450 °C.

SPS of Al2124 alloy yielded a relative density of 93.82% at 400 °C, and at 450 °C, densification of 99.53% was obtained. Full densification was achieved at 500 °C. Microwave sintering however yielded a relative density of 91.98% at 400 °C, 92.02% at 450 °C and 93.52% at 500 °C. These results show reduced densification in the microwave sintered composites as compared to that observed in

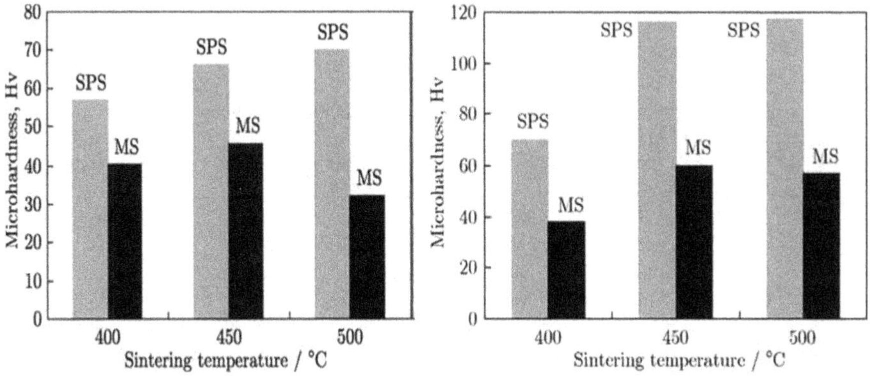

Fig. 7 L–R: Microhardness of spark plasma and microwave sintered Al6061 alloy; microhardness of spark plasma and microwave sintered Al2124 alloy (Saheb 2013)

SPSed samples. This disparity in the achieved relative densities led to the variation in microhardness values obtained from both techniques as can be seen in Fig. 7.

8 Conclusion

1. The intrinsic nature of the reinforcements is not sufficient in assessing the mechanical properties of the composites, as the processing routes and parameters also have a crucial role to play in the resulting properties of the composites.
2. Regardless of the morphology of the reinforcements, uniform dispersion is key to obtaining good mechanical properties as agglomeration of these reinforcements is counterproductive, thereby leading to loss of beneficial properties of the composites.
3. In situ precipitation of the strengthening phase within the matrix of the AMCs may possibly preclude both the strength-ductility dilemma and the dispersion problem as a fine distribution of a secondary phase eliminates the agglomeration observed while dispersing a reinforcing phase.
4. A bimodal grain size distribution within the matrix can help solve the lack of ductility observed in nanocrystalline composites as the fine grains would yield high strength while the coarse grains would help enhance ductility of the AMCs.
5. SPS is a novel sintering method that yields superior mechanical properties compared to other sintering methods owing to its unique sintering mechanism which improves densification and interfacial bonding which are key determinants of the composite properties.

Acknowledgement Authors are grateful to National Research Foundation (NRF) and Global Excellence and Stature (GES) for financing the research that produced this review.

References

Babu NK, Kallip K, Leparoux M, AlOgab KA, Maeder X, Dasilva YA (2016) Influence of microstructure and strengthening mechanism of AlMg5–Al2O3 nanocomposites prepared via spark plasma sintering. Mater Des 95:534–544

Bakshi SR, Agarwal A (2011) An analysis of the factors affecting strengthening in carbon nanotube reinforced aluminium composites. Carbon 49(2):533–544

Bartolucci SF, Paras J, Rafiee MA, Rafiee J, Lee S, Kapoor D, Koratkar N (2011) Graphene–aluminium nanocomposites. Mater Sci Eng A 528(27):7933–7937

Baskut S, Cinar A, Turan S (2017) Directional properties and microstructures of spark plasma sintered aluminium nitride containing graphene platelets. J Eur Ceram Soc 37(12):3759–3772

Bathula S, Anandani RC, Dhar A, Srivastava AK (2012) Microstructural features and mechanical properties of Al 5083/SiCp metal matrix nanocomposites produced by high energy ball milling and spark plasma sintering. Mater Sci Eng A 545:97–102

Bisht A, Srivastava M, Kumar RM, Lahiri I, Lahiri D (2017) Strengthening mechanism in graphene nanoplatelets reinforced aluminium composite fabricated through spark plasma sintering. Mater Sci Eng A 695:20–28

Bonache V, Salvador MD, Fernández A, Borrell A (2011) Fabrication of full density near-nanostructured cemented carbides by combination of VC/Cr3C2 addition and consolidation by SPS and HIP technologies. Int J Refract Met Hard Mater 29(2):202–208

Casati R, Vedani M (2014) Metal matrix composites reinforced by nano-particles—a review. Metals 4(1):65–83

Chaim R (2007) Densification mechanisms in spark plasma sintering of nanocrystalline ceramics. Mater Sci Eng A 443(1–2):25–32

Chen B, Kondoh K, Imai H, Umeda J, Takahashi M (2016) Simultaneously enhancing strength and ductility of carbon nanotube/aluminium composites by improving bonding conditions. Scr Mater 113:158–162

Chu K, Jia C, Liang X, Chen H, Gao W, Guo H (2009) Modelling the thermal conductivity of diamond reinforced aluminium matrix composites with inhomogeneous interfacial conductance. Mater Des 30(10):4311–4316

Chu K, Jia CC, Li WS (2012) On CTE of SPS consolidated SiCp/Al composites with various particle size distributions. Mater Sci Technol 28(12):1397–1401

Cooke RW, Kraus NP, Bishop DP (2016) Spark plasma sintering of aluminium powders pre-alloyed with scandium additions. Mater Sci Eng A 657:71–81

Dash K, Chaira D, Ray BC (2013) Synthesis and characterization of aluminium–alumina micro-and nano-composites by spark plasma sintering. Mater Res Bull 48(7):2535–2542

Durowoju MO, Sadiku ER, Diouf S, Shongwe MB, Olubambi PA (2015) Spark plasma sintering of graphite–aluminium powder reinforced with SiC/Si particles. Powder Technol 284:504–513

Eldesouky A, Johnsson M, Svengren H, Attallah MM, Salem HG (2014) Effect of grain size reduction of AA2124 aluminium alloy powder compacted by spark plasma sintering. J Alloys Compd 609:215–221

Esawi AM, Morsi K, Sayed A, Taher M, Lanka S (2010) Effect of carbon nanotube (CNT) content on the mechanical properties of CNT-reinforced aluminium composites. Compos Sci Technol 70(16):2237–2241

Evans AG (2001) Lightweight materials and structures. MRS Bull 26(10):790–797

Figiel P, Rozmus M, Smuk B (2011) Properties of alumina ceramics obtained by conventional and non-conventional methods for sintering ceramics. J Achiev Mater Manuf Eng 48(1):29–34

Firestein KL, Steinman AE, Golovin IS, Cifre J, Obraztsova EA, Matveev AT, Kovalskii AM, Lebedev OI, Shtansky DV, Golberg D (2015) Fabrication, characterization, and mechanical properties of spark plasma sintered Al–BN nanoparticle composites. Mater Sci Eng A 642:104–112

Garbiec D, Jurczyk M, Levintant-Zayonts N, Mościcki T (2015) Properties of Al–Al2O3 composites synthesized by spark plasma sintering method. Arch Civ Mech Eng 15(4):933–939

George R, Kashyap KT, Rahul R, Yamdagni S (2005) Strengthening in carbon nanotube/aluminium (CNT/Al) composites. Scr Mater 53(10):1159–1163

Ghasali E, Pakseresht A, Safari-Kooshali F, Agheli M, Ebadzadeh T (2015) Investigation on microstructure and mechanical behaviour of Al–ZrB2 composite prepared by microwave and spark plasma sintering. Mater Sci Eng A 627:27–30

Ghasali E, Shirvanimoghaddam K, Pakseresht AH, Alizadeh M, Ebadzadeh T (2017) Evaluation of microstructure and mechanical properties of Al-TaC composites prepared by spark plasma sintering process. J Alloys Compd 705:283–289

Gorr B, Azim M, Christ HJ, Mueller T, Schliephake D, Heilmaier M (2015) Phase equilibria, microstructure, and high temperature oxidation resistance of novel refractory high-entropy alloys. J Alloys Compd 624:270–278

Gu J, Gu S, Xue L, Wu S, Yan Y (2012) Microstructure and mechanical properties of in-situ Al13Fe4/Al composites prepared by mechanical alloying and spark plasma sintering. Mater Sci Eng A 558:684–691

Guo B, Ni S, Yi J, Shen R, Tang Z, Du Y, Song M (2017) Microstructures and mechanical properties of carbon nanotubes reinforced pure aluminium composites synthesized by spark plasma sintering and hot rolling. Mater Sci Eng A 698:282–288

Han K, Van Tyne CJ, Levy BS (2005) Effect of strain and strain rate on the Bauschinger effect response of three different steels. Metall Mater Trans A 36(9):2379–2384

He C, Zhao N, Shi C, Du X, Li J, Li H, Cui Q (2007) An approach to obtaining homogeneously dispersed carbon nanotubes in Al powders for preparing reinforced Al-matrix composites. Adv Mater 19(8):1128–1132

He T, He X, Tang P, Chu D, Wang X, Li P (2017) The use of cryogenic milling to prepare high performance Al2009 matrix composites with dispersive carbon nanotubes. Mater Des 114:373–382

Itoh D, Terada Y, Sato T (2008) Effect of calcium additions on creep properties of a die-cast AM50 magnesium alloy. Mater Trans 49(9):1957–1962

Jafari F, Sharifi H, Saeri MR, Tayebi M (2018) Effect of reinforcement volume fraction on the wear behaviour of Al-SiCp composites prepared by spark plasma sintering. Silicon 10:1–9

Khoramkhorshid S, Alizadeh M, Taghvaei AH, Scudino S (2016) Microstructure and mechanical properties of Al-based metal matrix composites reinforced with Al84Gd6Ni7Co3 glassy particles produced by accumulative roll bonding. Mater Des 15(90):137–144

Khosroshahi NB, Mousavian RT, Khosroshahi RA, Brabazon D (2015) Mechanical properties of rolled A356 based composites reinforced by Cu-coated bimodal ceramic particles. Mater Des 83:678–688

Kwon H, Estili M, Takagi K, Miyazaki T, Kawasaki A (2009) Combination of hot extrusion and spark plasma sintering for producing carbon nanotube reinforced aluminium matrix composites. Carbon 47(3):570–577

Kwon H, Park DH, Silvain JF, Kawasaki A (2010) Investigation of carbon nanotube reinforced aluminium matrix composite materials. Compos Sci Technol 70(3):546–550

Lahiri D, Hadjikhani A, Zhang C, Xing T, Li LH, Chen Y, Agarwal A (2013) Boron nitride nanotubes reinforced aluminium composites prepared by spark plasma sintering: microstructure, mechanical properties and deformation behaviour. Mater Sci Eng A 574:149–156

Le GM, Godfrey A, Hansen N (2013) Structure and strength of aluminium with sub-micrometer/micrometer grain size prepared by spark plasma sintering. Mater Des 49:360–367

Lee Z, Witkin DB, Radmilovic V, Lavernia EJ, Nutt SR (2005) Bimodal microstructure and deformation of cryomilled bulk nanocrystalline Al–7.5 Mg alloy. Mater Sci Eng A 410:462–467

Li XP, Liu CY, Ma MZ, Liu RP (2016) Microstructures and mechanical properties of AA6061–SiC composites prepared through spark plasma sintering and hot rolling. Mater Sci Eng A 650:139–144

Liu ZY, Wang QZ, Xiao BL, Ma ZY, Liu Y (2010) Experimental and modelling investigation on SiCp distribution in powder metallurgy processed SiCp/2024 Al composites. Mater Sci Eng A 527(21–22):5582–5591

Liu P, Mao SC, Wang LH, Han XD, Zhang Z (2011) Direct dynamic atomic mechanisms of strain-induced grain rotation in nanocrystalline, textured, columnar-structured thin gold films. Scr Mater 64(4):343–346

Liu CY, Wang Q, Jia YZ, Zhang B, Jing R, Ma MZ, Jing Q, Liu RP (2013) Evaluation of mechanical properties of 1060-Al reinforced with WC particles via warm accumulative roll bonding process. Mater Des 43:367–372

Liu ZF, Zhang ZH, Lu JF, Korznikov AV, Korznikova E, Wang FC (2014) Effect of sintering temperature on microstructures and mechanical properties of spark plasma sintered nanocrystalline aluminium. Mater Des 64:625–630

Liu D, Xiong Y, Li P, Lin Y, Chen F, Zhang L, Schoenung JM, Lavernia EJ (2016) Microstructure and mechanical behavior of NS/UFG aluminium prepared by cryomilling and spark plasma sintering. J Alloys Compd 679:426–435

Lorrette C, Réau A, Briottet L (2013) Mechanical properties of nanostructured silicon carbide consolidated by spark plasma sintering. J Eur Ceram Soc 33(1):147–156

Markó D, Prashanth KG, Scudino S, Wang Z, Ellendt N, Uhlenwinkel V, Eckert J (2014) Al-based metal matrix composites reinforced with Fe49.9Co35.1Nb7.7B4.5Si2. 8 glassy powder: mechanical behaviour under tensile loading. J Alloys Compd 615:S382–S385

Mirza FA, Chen DL (2015) A unified model for the prediction of yield strength in particulate-reinforced metal matrix nanocomposites. Materials 8(8):5138–5153

Mokdad F, Chen DL, Liu ZY, Xiao BL, Ni DR, Ma ZY (2016) Deformation and strengthening mechanisms of a carbon nanotube reinforced aluminium composite. Carbon 104:64–77

Munir ZA, Anselmi-Tamburini U, Ohyanagi M (2006) The effect of electric field and pressure on the synthesis and consolidation of materials: a review of the spark plasma sintering method. J Mater Sci 41(3):763–777

Munir KS, Qian M, Li Y, Oldfield DT, Kingshott P, Zhu DM, Wen C (2015) Quantitative analyses of MWCNT-Ti powder mixtures using Raman spectroscopy: the influence of milling parameters on nanostructural evolution. Adv Eng Mater 17(11):1660–1669

Najimi AA, Shahverdi HR (2017) Microstructure and mechanical characterization of Al6061-CNT nanocomposites fabricated by spark plasma sintering. Mater Charact 133:44–53

Noguchi T, Magario A, Fukazawa S, Shimizu S, Beppu J, Seki M (2004) Carbon nanotube/aluminium composites with uniform dispersion. Mater Trans 45(2):602–604

Novák P, Kubatík T, Vystrčil J, Hendrych R, Kříž J, Mlynár J, Vojtěch D (2014) Powder metallurgy preparation of Al–Cu–Fe quasicrystals using mechanical alloying and spark plasma sintering. Intermetallics 52:131–137

Ogawa F, Yamamoto S, Masuda C (2018) Strong, ductile, and thermally conductive carbon nanotube-reinforced aluminium matrix composites fabricated by ball-milling and hot extrusion of powders encapsulated in aluminium containers. Mater Sci Eng A 711:460–469

Orru R, Licheri R, Locci AM, Cincotti A, Cao G (2009) Consolidation/synthesis of materials by electric current activated/assisted sintering. Mat Sci Eng R Rep 63(4–6):127–287

Park YS, Chung KH, Kim NJ, Lavernia EJ (2004) Microstructural investigation of nanocrystalline bulk Al–Mg alloy fabricated by cryomilling and extrusion. Mater Sci Eng A 374(1–2):211–216

Park JG, Keum DH, Lee YH (2015) Strengthening mechanisms in carbon nanotube-reinforced aluminium composites. Carbon 95:690–698

Park K, Park J, Kwon H (2017a) Effect of intermetallic compound on the Al-Mg composite materials fabricated by mechanical ball milling and spark plasma sintering. J Alloys Compd 739:311–318

Park K, Park J, Kwon H (2017b) Fabrication and characterization of Al-SUS316L composite materials manufactured by the spark plasma sintering process. Mater Sci Eng A 691:8–15

Park K, Park J, Kwon H (2018) Effect of intermetallic compound on the Al-Mg composite materials fabricated by mechanical ball milling and spark plasma sintering. Journal of Alloys and Compounds, 739:311–318

Perrière L, Champion Y (2012) Phases distribution dependent strength in metallic glass–aluminium composites prepared by spark plasma sintering. Mater Sci Eng A 548:112–117

Pillari LK, Shukla AK, Murty SN, Umasankar V (2018) On the comparison of graphene and multi-wall carbon nanotubes as reinforcements in aluminium alloy AA2219 processed by ball milling and spark plasma sintering. Trans Indian Inst Metals 71(5):1099–1112

Poirier D, Gauvin R, Drew RA (2009) Structural characterization of a mechanically milled carbon nanotube/aluminium mixture. Compos A: Appl Sci Manuf 40(9):1482–1489

Ram SC, Chattopadhyay K, Chakrabarty I (2017) High temperature tensile properties of centrifugally cast in-situ Al-Mg2Si functionally graded composites for automotive cylinder block liners. J Alloys Compd 724:84–97

Rana JK, Sivaprahasam D, Raju KS, Sarma VS (2009) Microstructure and mechanical properties of nanocrystalline high strength Al–Mg–Si (AA6061) alloy by high energy ball milling and spark plasma sintering. Mater Sci Eng A 527(1–2):292–296

Rikhtegar F, Shabestari SG, Saghafian H (2017) Microstructural evaluation and mechanical properties of Al-CNT nanocomposites produced by different processing methods. J Alloys Compd 723:633–641

Sadeghi B, Shamanian M, Cavaliere P, Ashrafizadeh F, Sanayei M, Szpunar JA (2017) Microstructural and mechanical behaviour of bimodal reinforced Al-based composites produced by spark plasma sintering and FSP. Int J Adv Manuf Technol 94:1–14

Saheb N (2013) Spark plasma and microwave sintering of Al6061 and Al2124 alloys. Int J Miner Metall Mater 20(2):152–159

Sajjadi SA, Ezatpour HR, Beygi H (2011) Microstructure and mechanical properties of Al–Al2O3 micro and nano composites fabricated by stir casting. Mater Sci Eng A 528(29–30):8765–8771

Salem HG, Sadek AA (2010) Fabrication of high performance PM nanocrystalline bulk AA2124. J Mater Eng Perform 19(3):356–367

Sarmasti AB, Yazdanirad M, Nouri Khezrabad M, Karbalaie M (2011) Effect of alumina particle size and thermal condition of casting on microstructure and mechanical properties of stir cast Al–Al_2O_3 composites. Mater Sci Technol 27(11):1653–1656

Sasaki TT, Ohkubo T, Hono K (2009) Microstructure and mechanical properties of bulk nanocrystalline Al–Fe alloy processed by mechanical alloying and spark plasma sintering. Acta Mater 57 (12):3529–3538

Schultz BF, Ferguson JB, Rohatgi PK (2011) Microstructure and hardness of Al_2O_3 nanoparticle reinforced Al–Mg composites fabricated by reactive wetting and stir mixing. Mater Sci Eng A 530:87–97

Shaeri MH, Salehi MT, Seyyedein SH, Abutalebi MR, Park JK (2014) Microstructure and mechanical properties of Al-7075 alloy processed by equal channel angular pressing combined with aging treatment. Mater Des 57:250–257

Shanmugasundaram T, Murty BS, Sarma VS (2006) Development of ultrafine grained high strength Al–Cu alloy by cryorolling. Scr Mater 54(12):2013–2017

Steinman AE, Corthay S, Firestein KL, Kvashnin DG, Kovalskii AM, Matveev AT, Sorokin PB, Golberg DV, Shtansky DV (2018) Al-based composites reinforced with AlB2, AlN and BN phases: experimental and theoretical studies. Mater Des 141:88–98

Suárez M, Fernández A, Menéndez JL, Torrecillas R, Kessel HU, Hennicke J, Kirchner R, Kessel T (2013) Challenges and opportunities for spark plasma sintering: a key technology for a new generation of materials. Sinter Appl: 320–338. https://doi.org/10.5772/53706

Sun Y, Haley J, Kulkarni K, Aindow M, Lavernia EJ (2015) Influence of electric current on microstructure evolution in Ti/Al and Ti/TiAl3 during spark plasma sintering. J Alloys Compd 648:1097–1103

Sweet GA, Brochu M, Hexemer Jr RL, Donaldson IW, Bishop DP (2015). Consolidation of aluminum-based metal matrix composites via spark plasma sintering. Materials Science and Engineering: A. 648:123-33

Tan Z, Li Z, Fan G, Kai X, Ji G, Zhang L, Zhang D (2013) Fabrication of diamond/aluminium composites by vacuum hot pressing: process optimization and thermal properties. Compos Part B 47:173–180

Tan Z, Wang L, Xue Y, Zhang P, Cao T, Cheng X (2016) High-entropy alloy particle reinforced Al-based amorphous alloy composite with ultrahigh strength prepared by spark plasma sintering. Mater Des 109:219–226

Tian W, Li S, Chen X, Liu J, Yu M (2016a) Intergranular corrosion of spark plasma sintering assembled bimodal grain sized AA7075 aluminium alloys. Corros Sci 107:211–224

Tian WM, Li SM, Wang B, Chen X, Liu JH, Yu M (2016b) Graphene-reinforced aluminium matrix composites prepared by spark plasma sintering. Int J Miner Metall Mater 23(6):723–729

Valiev R (2002) Materials science: nanomaterial advantage. Nature 419(6910):887

Vencl A, Bobic I, Arostegui S, Bobic B, Marinković A, Babić M (2010) Structural, mechanical and tribological properties of A356 aluminium alloy reinforced with Al2O3, SiC and SiC+ graphite particles. Journal of Alloys and Compounds, 506(2):631–639

Vintila RR, Drew RR, Brochu M (2010) Nanostructured NS Al-2024 alloy fabricated by cryogenic milling and consolidated via spark plasma sintering. In Proceedings of the 2010 international conference on powder metallurgy & particulate material, pp 948–960

Vintila R, Charest A, Drew RA, Brochu M (2011) Synthesis and consolidation via spark plasma sintering of nanostructured Al-5356/B4C composite. Mater Sci Eng A 528(13–14):4395–4407

Wang LZ, Liu Y, Wu JJ, Zhang X (2017) Mechanical properties and friction behaviours of CNT/AlSi 10 Mg composites produced by spark plasma sintering. Int J Miner Metall Mater 24(5):584–593

Wu J, Zhang H, Zhang Y, Wang X (2012) Mechanical and thermal properties of carbon nanotube/aluminium composites consolidated by spark plasma sintering. Mater Des 41:344–348

Xu P, Feng Y, Xu B, Chen F, Lin Y (2017) Disruption of oxide films during spark plasma sintering of micrometric-sized Al powders and its effect on microstructure and tensile properties of the consolidated Al. J Alloys Compd 12:822–832

Xue Y, Jiang B, Bourgeois L, Dai P, Mitome M, Zhang C, Yamaguchi M, Matveev A, Tang C, Bando Y, Tsuchiya K (2015) Aluminium matrix composites reinforced with multi-walled boron nitride nanotubes fabricated by a high-pressure torsion technique. Mater Des 88:451–460

Yan C, Lifeng W, Jianyue R (2008) Multi-functional SiC/Al composites for aerospace applications. Chin J Aeronaut 21(6):578–584

Yang M, Zhang D, Gu X, Zhang L (2006) Effects of SiC particle size on CTEs of SiCp/Al composites by pulsed electric current sintering. Mater Chem Phys 99(1):170–173

Ye J, Ajdelsztajn L, Schoenung JM (2006) Bulk nanocrystalline aluminium 5083 alloy fabricated by a novel technique: cryomilling and spark plasma sintering. Metall Mater Trans A 37 (8):2569–2579

Zhang H, Pan Y, He Y, Jiao H (2011) Microstructure and properties of 6FeNiCoSiCrAlTi high-entropy alloy coating prepared by laser cladding. Appl Surf Sci 257(6):2259–2263

Zhang Y, Li J, Zhao L, Zhang H, Wang X (2014) Effect of metalloid silicon addition on densification, microstructure and thermal–physical properties of Al/diamond composites consolidated by spark plasma sintering. Mater Des 63:838–847

Zhang L, Xu H, Wang Z, Li Q, Wu J (2016) Mechanical properties and corrosion behaviour of Al/SiC composites. J Alloys Compd 678:23–30

Zhang X, Li S, Pan D, Pan B, Kondoh K (2018) Microstructure and synergistic-strengthening efficiency of CNTs-SiCp dual-nano reinforcements in aluminium matrix composites. Compos A: Appl Sci Manuf 105:87–96

Zheng R, Ma F, Xiao W, Ameyama K, Ma C (2017) Achieving enhanced strength in ultrafine lamellar structured Al2024 alloy via mechanical milling and spark plasma sintering. Mater Sci Eng A 687:155–163

Zhou W, Sasaki S, Kawasaki A (2014) Effective control of nanodefects in multi-walled carbon nanotubes by acid treatment. Carbon 78:121–129

Processing of a Ternary Al-Sc-Zr Powder Metallurgy Alloy via Spark Plasma Sintering and Hot Asymmetric Rolling

M. Y. Amegadzie, I. W. Donaldson, and D. P. Bishop

1 Introduction

Aluminum powder metallurgy (PM) is a net-shape metal-forming technology in which metallic powders are converted into engineered products with tight dimensional tolerances. This manufacturing process offers advantages such as the capacity for high-volume production, ability to fabricate lightweight products, favorable process economics, as well as the efficient use of energy and raw materials. The automotive industry has been particularly receptive to this process as it has been implemented in the fabrication of a growing list of structural parts for this sector dating back to 1992 (Capus 2005). Exemplary parts fabricated through aluminum PM have included shock absorber parts, piston rods, gears, oil pump rotors, camshaft bearing caps, and, more recently, transmission components such as an award-winning planetary reaction carrier.

The core consolidation phases of the process are die compaction followed by liquid phase sintering. Combined, these steps aim to convert the raw powder into a coherent product that offers a high level of metallurgical integrity and a shape approaching that of the intended component design. However, the surface film present on aluminum-based powder particles poses certain technological challenges in this context. Studied by many authors, this feature has been frequently characterized as a hydrated film of Al_2O_3 that contains adsorbed molecules of H_2O (Yamasaki and Kawamura 2006) and CO_2. The alumina component exhibits a high thermodynamic stability that cannot be eliminated by conventional tactics such as gaseous reduction with hydrogen or carbon monoxide gas. Consequently, attempts to

M. Y. Amegadzie · D. P. Bishop (✉)
Department of Mechanical Engineering, Dalhousie University, Halifax, NS, Canada
e-mail: Mark.Amegadzie@dal.ca; Paul.Bishop@dal.ca

I. W. Donaldson
Advanced Engineering, GKN Sinter Metals, Auburn Hills, MI, USA
e-mail: Ian.Donaldson@gkn.com

© Springer Nature Switzerland AG 2019
P. Cavaliere (ed.), *Spark Plasma Sintering of Materials*,
https://doi.org/10.1007/978-3-030-05327-7_8

remediate this challenge have emphasized alternate practices including chemical modification with magnesium (Kondoh et al. 2000), degassing studies (Kirchoff et al. 1986), as well as hot pressing in a vacuum atmosphere (Quist and Lewis 1986).

More recently, the technology of spark plasma sintering (SPS) has emerged as a new means to overcome the surface film constraint (Xie et al. 2001; Sweet et al. 2014). This process typically applies a high amperage DC current and uniaxial pressure simultaneously to compact and sinter the powder in a single operation (Song et al. 2013). Unlike conventional sintering, the SPS process affords rapid heating rates and reduced sintering temperatures that suppress microstructural coarsening. It has been demonstrated that the surface oxides at interparticle contact points can be mitigated by SPS (Xie et al. 2001). However, a semicontinuous network of residual oxides still prevails in the sintered product.

A residual oxide network is undesirable as it can be detrimental to mechanical properties, in particular, tensile ductility and dynamic properties such as fatigue. As such, it was hypothesized that if severe shear deformation was applied to a SPS product, this could physically disrupt the remnants of this feature and, in turn, improve mechanical properties. One means of deformation in this manner is asymmetric rolling (ASR). In this practice, the thickness of a bar is reduced by passing it through a pair of rollers with each run at a different rotational speed (Simões et al. 2009; Zuo et al. 2008). This is known to induce exceptionally high levels of shear stress within the material which could present an effective means to fracture and distribute residual oxides within the final product. As such, ASR was applied to SPS-consolidated preforms made from a prealloyed aluminum powder (Al-0.3Sc-0.2Zr in wt %) in this study in an effort to gather some preliminary data on this concept.

2 Materials

The raw powder utilized in this study was manufactured via nitrogen gas atomization at Kymera International in Germany. The morphology of the powder was mainly spherical and had a D_{50} of 75 μm as measured through laser scattering. Inductively coupled plasma spectroscopy was used to determine the chemical composition of the powder as shown in Table 1. This confirmed that the powder was in reasonable

Table 1 Measured assay (weight %) of the starting prealloyed powder

Composition (weight %)		
Element	Nominal target	Measured
Sc	0.30	0.33
Zr	0.20	0.17
Fe	–	0.08
Si	–	0.02
Mg	–	0.01
Al	Bal	Bal

agreement with the nominal target of Al-0.3Sc-0.2Zr (weight %), a chemistry of Al-0.2Sc-0.05Zr in atomic %. Trace concentrations of iron, silicon, and magnesium were also detected.

3 Experimental Procedures

The as-received powder was spark plasma sintered in a SPS Model 10-3 unit manufactured by Thermal Technologies Inc. operated with pulsed DC current. Specimens were processed in a graphite tool set; at the start of each run, the die was lined with a new sheet of graphite foil. The thermal cycle involved heating the powder to the set temperature (500 °C, 550 °C, or 600 °C) at 50 K/min under a mechanical vacuum atmosphere and an applied pressure of 5 MPa. Samples were then isothermally held at temperature for 120 s before pressure was ramped to 50 MPa. Once the maximum pressure was reached, each sample was cooled to room temperature, depressurized, and removed from the furnace. The resultant product was a cylindrical puck 40 mm in diameter and 7 mm in height. All sintered specimens were ground slightly to remove any remnants of the graphite foil. Sintered densities were measured using Archimedes principle coupled with vacuum-assisted infiltration in accordance with MPIF standard 42. Rectangular bars with dimensions 31.7 mm × 12.7 mm × 6.4 mm were machined from sintered pucks, preheated for 20 min, and then asymmetrically rolled in a Reusch rolling mill manufactured by International Rolling Mills, USA. Preheat temperatures of 200 °C, 300 °C, 400 °C, and 500 °C were considered. Each specimen was subjected to two rolling passes and was rotated by 180° about the rolling direction between passes. The net thickness reduction achieved after rolling was approximately 80%.

Tensile specimens were machined from the rolled bars and tested with an Instron Satec model 5594 200HVL hydraulic load frame equipped with a 50 kN load cell, wedge grips, and an Epsilon model 3542 extensometer which remained affixed to the specimen through to fracture. Hardness measurements were made using a LECO Vickers hardness testing machine operated with an applied load of 10 kg. Average hardness readings were determined from a minimum of five measurements. Microstructural characterization included optical microscopy (OM), electron backscatter diffraction (EBSD), and transmission electron microscopy (TEM). For OM/EBSD work, cross sections of the material in question were mounted using a Struers CitoPress-5 automatic mounting press. Mounts were then loaded in a Struers Tegramin autopolisher, planed with a 1200 grit silicon carbide disk, and subsequently polished in stages using cloth pads and diamond suspensions from 15, 3, and 1 μm, before finishing with 0.05 μm colloidal silica suspension. EBSD studies employed a Hitachi S-4700 cold-field emission scanning electron microscope (SEM) equipped with a HKL system and Oxford Instruments Nordlys detector. Sections of samples for TEM analysis were first thinned to ~100 microns using a 400 grit sandpaper. This was followed by using a puncher to make 3 mm disks prior to dimpling. Finally, the foils were electropolished using a Tenupol 3 twin-jet polisher in a solution comprising of 90% methyl alcohol and 10% perchloric acid at a temperature of −30 °C. TEM studies were completed with a FEI Talos F200X system operated at 200 KV.

4 Results and Discussions

4.1 Metallurgical Attributes of SPS-Processed Specimens

The as-received powder was first processed via SPS at temperatures of 500 °C, 550 °C, and 600 °C. The specimen sintered at the highest temperature considered appeared to be fully dense, and the prior particle boundaries were less visible thereby implying a more acute level of interparticle bonding. The visual trend in densification agreed with bulk density measurements (Fig. 1). In this sense, the densities of all specimens were relatively high, while those sintered at 550 °C and 600 °C were effectively fully dense. Representative micrographs of the as-sintered microstructures are shown in Fig. 2. At the lowest temperature, the general quality of sinter was reasonable as interparticle bonding had transpired and the level of residual porosity was moderate. Both traits improved with further increases in SPS temperature.

EBSD was employed to better understand the evolution of the microstructure. Figures 3 and 4 show EBSD maps of as-sintered specimens processed at the lowest and highest temperatures investigated, respectively. The black and white lines represent grain boundaries with misorientation angles >15° and < 15°, respectively. The former are referred to as high-angle grain boundaries (HAGBs) while the latter as low-angle grain boundaries (LAGBs) or subgrains. EBSD confirmed that both materials were largely comprised of equiaxed grains that spanned a comparable

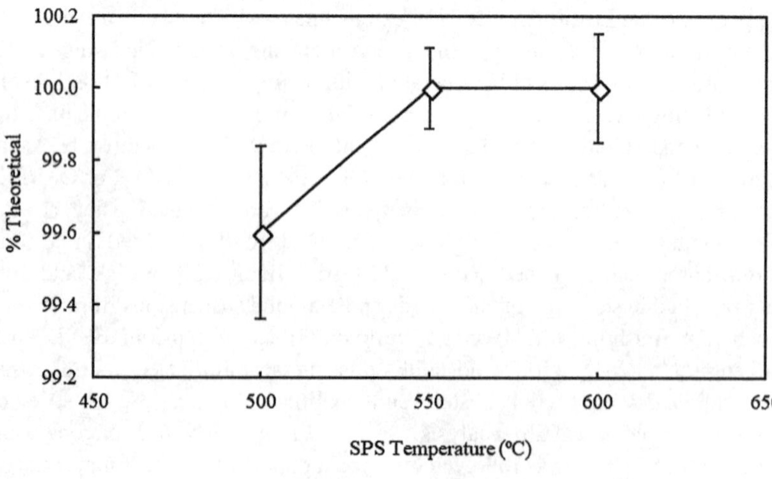

Fig. 1 Final product density as a function of sintering temperature

range of grain sizes. The average grain size in each case was found to be ~11 μm. Interestingly, the use of progressively higher SPS temperatures did not appear to invoke any obvious coarsening of the grains. This was potentially attributable to the presence of the $Al_3(Sc,Zr)$ phase which effectively pins grain boundaries (Jia et al. 2012). This was consistent with optical images as no obvious differences in the extent of grain coarsening were observed. Quantitative analyses revealed that in both specimens an equivalent ratio of HAGB-LABG prevailed (Fig. 5). Hence, the general nature of grain boundaries in the specimens was also unaffected by differences in SPS temperature.

Next, tensile properties of the as-sintered products were assessed (Fig. 6). Here it was noted that the yield strength decreased with increasing sintering temperature. This attribute declined by 16% as SPS temperature was raised from 500 °C to 550 °C. The loss then increased to 35% when a SPS processing temperature of 600 °C was employed. The ultimate tensile strengths declined in a comparable fashion although the extent of the transition was less acute. In contrast, the ductility improved with increasing SPS temperature from a starting value of ~1% to nearly 16% when sintered at 600 °C. Akin to yield strength and UTS, the hardness data presented in Fig. 7 also demonstrated a downward trend with increasing SPS temperature.

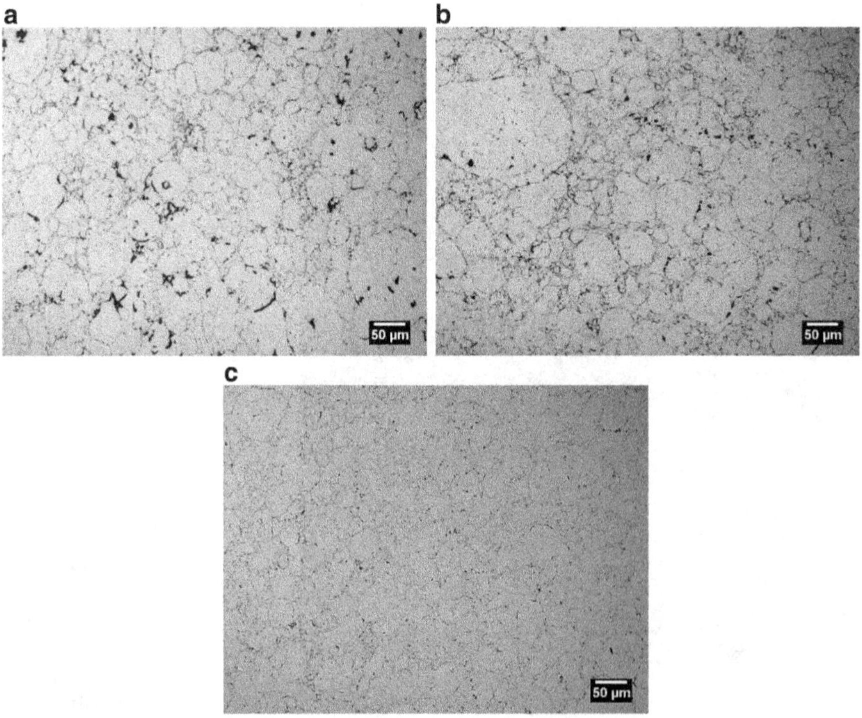

Fig. 2 Optical images of specimens SPS processed at (**a**) 500 °C, (**b**) 550 °C, and (**c**) 600 °C

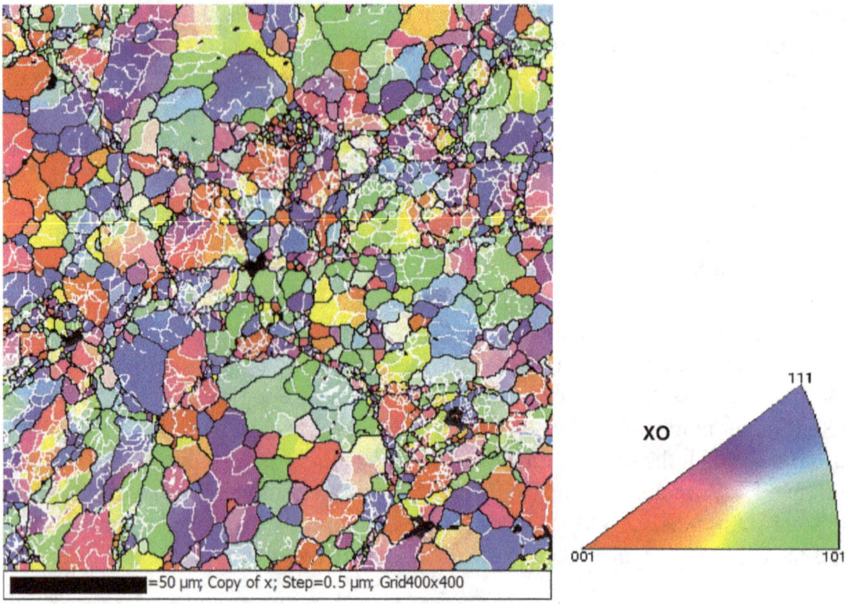

Fig. 3 EBSD results for a specimen SPS processed at 500 °C. Black lines indicate HAGBs; white lines indicate LAGBs

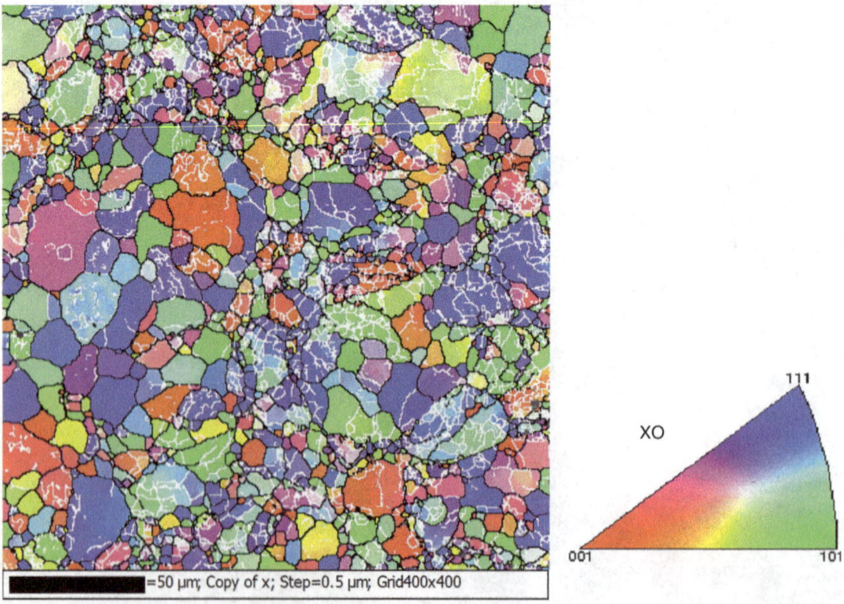

Fig. 4 EBSD results for a specimen SPS processed at 600 °C. Black lines indicate HAGBs; white lines indicate LAGBs

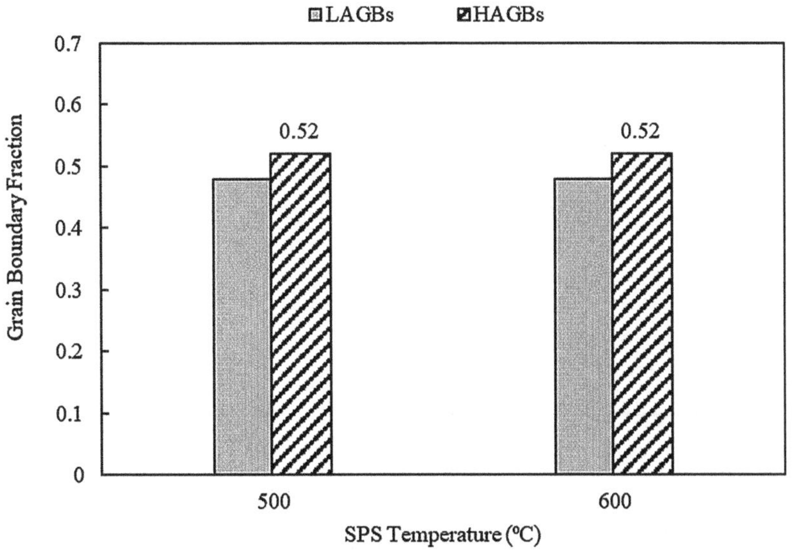

Fig. 5 Fractions of LAGBs and HAGBs in samples SPS processed at 500 °C and 600 °C

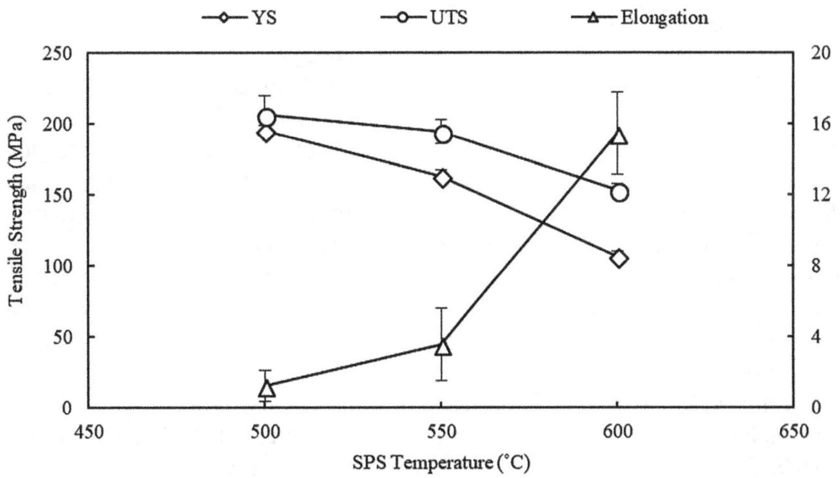

Fig. 6 Tensile properties of SPS-processed specimens

Data in Figs. 3, 4, and 5 confirmed that fundamental changes to the underlying grain structure which could potentially explain the noted transitions in hardness and tensile properties had not occurred. Conversely, gradual improvements in sinter quality (Fig. 2) and density (Fig. 1) had transpired that should have been influential. Gains in these traits would certainly be expected to improve tensile ductility but not diminish yield strength. Hence, TEM studies were completed to assess the presence and potential evolution of secondary phases that existed within the materials.

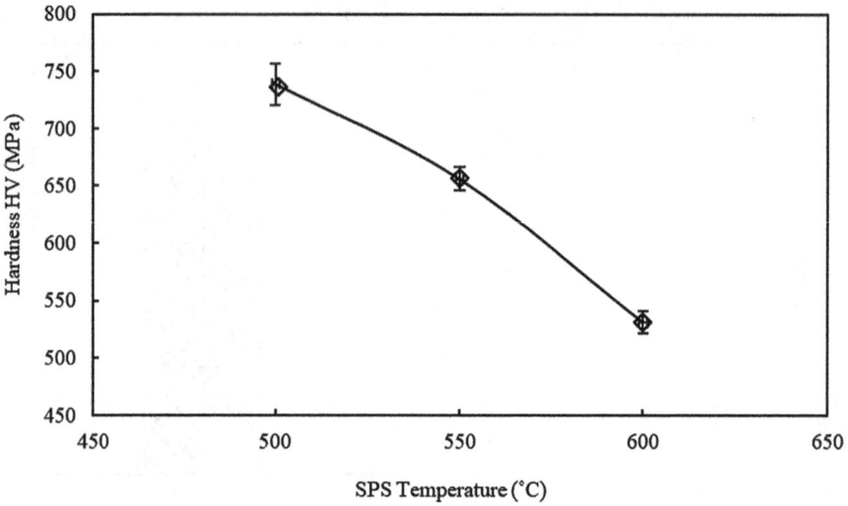

Fig. 7 Hardness of products sintered at different temperatures

Bright-field images (Fig. 8a, b) confirmed that a rather abundant secondary phase with a spherical morphology was homogeneously distributed within samples sintered at the lowest and highest temperatures investigated. The phase appeared to remain coherent with the α-aluminum matrix in both instances as evidenced by coffee bean contrast (Fuller et al. 2005). Dark-field images (Fig. 8c, d) confirmed that particles of the secondary phase had a diameter of ~4–10 nm in the specimen sintered at 500 °C but had grown to a size of ~20 to 35 nm when SPS was completed at 600 °C. Selected area diffraction patterns recorded near the <100> zone axis are shown in Figs. 8e, f. Both patterns were comprised of bright reflection spots from the FCC α-aluminum matrix coupled with an identical array of superlattice reflections. The only noted difference was that the superlattice reflections were somewhat diffuse in the 500 °C sample yet sharp and focused in the product consolidated at 600 °C.

Collectively, TEM findings indicated that the secondary phase was most likely $L1_2$ precipitates of $Al_3(Sc,Zr)$ in both samples (Vlach et al. 2013). It is known that the $Al_3(Sc,Zr)$ phase can impart significant strengthening in Al-Sc-Zr alloy systems (Vlach et al. 2013; Rostova et al. 2000). Furthermore, in precipitation-hardened alloys, the size and distribution of the precipitates present typically underpin mechanical properties. In this sense, fine, coherent precipitates are generally synonymous with relatively high yield strength yet moderate tensile ductility. This situation then begins to reverse as the precipitates coarsen and concurrently, interparticle spacing increases. TEM revealed that $Al_3(Sc,Zr)$ precipitates had clearly coarsened as the SPS temperature was increased from 500 to 600 °C. Hence, the evolution of this phase was thereby viewed as the dominant contributor toward the noted transitions in tensile properties. It is also noteworthy that in a recent study coherent $Al_3(Sc,Zr)$ precipitates were found to hinder recrystallization (Jia et al. 2012).

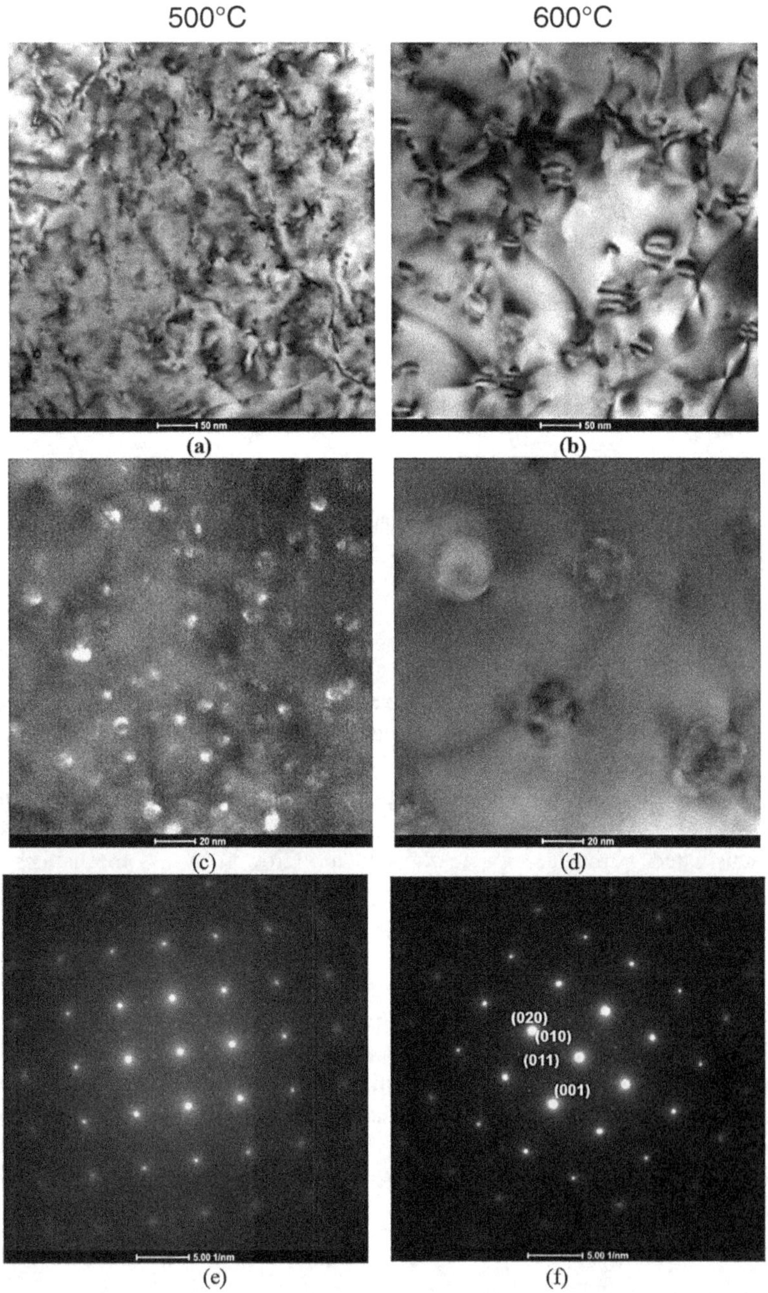

Fig. 8 TEM results acquired from SPS specimens consolidated at 500 °C and 600 °C. Bright-field images (**a** and **b**), dark-field images (**c** and **d**), and selected area diffraction patterns recorded with the beam oriented closely parallel to the <100> zones axis (**e** and **f**)

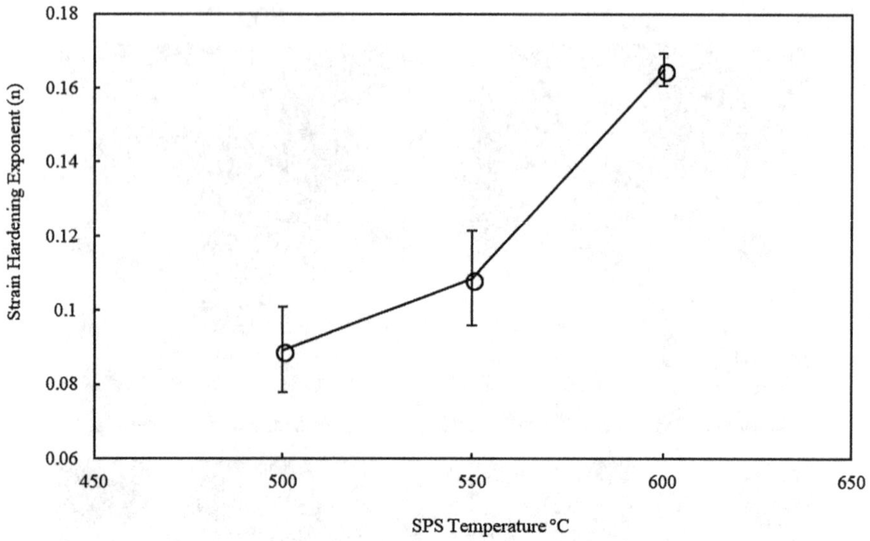

Fig. 9 Plot of the strain-hardening exponent (n) measured for Al-0.3Sc-0.2Zr specimens consolidated at 500 °C, 550 °C, and 600 °C

It appeared that this beneficial effect had also transpired in the PM systems studied given that EBSD revealed a highly comparable grain structure devoid of excessive coarsening regardless of the sintering temperature employed.

In light of the measurable change in $Al_3(Sc,Zr)$ precipitate size and coherency, it would be expected that the yield strength of the material would decline as noted earlier. However, a transition in how such precipitates interact with dislocations was also anticipated – from precipitate shear to the Orowan bypass mechanism. The former is typically associated with a heightened yield strength and represents a scenario wherein dislocations slice through the small precipitates. Precipitate shearing is the dominant strengthening mechanism when precipitates are about 5 nm or less (Jia et al. 2012). Conversely, the latter is synonymous with an inferior yield strength but an enhanced capacity for strain hardening as dislocations cannot cut through the large precipitates but are forced to loop around them instead. Evidence of this phenomenon can be gleaned from stress-strain data as an increase in the strain-hardening exponent (n). Data on this attribute are presented in Fig. 9. This confirmed that a progressive and statistically relevant increase in the strain-hardening exponent occurred with increasing SPS temperature.

4.2 *Effects of Hot Asymmetric Rolling (ASR)*

After SPS, the sintered specimens were preheated to 200 °C, 300 °C, 400 °C, and 500 °C in a furnace before ASR. The deformed microstructures are presented in the optical micrographs shown in Fig. 10. In all instances, the starting equiaxed grain

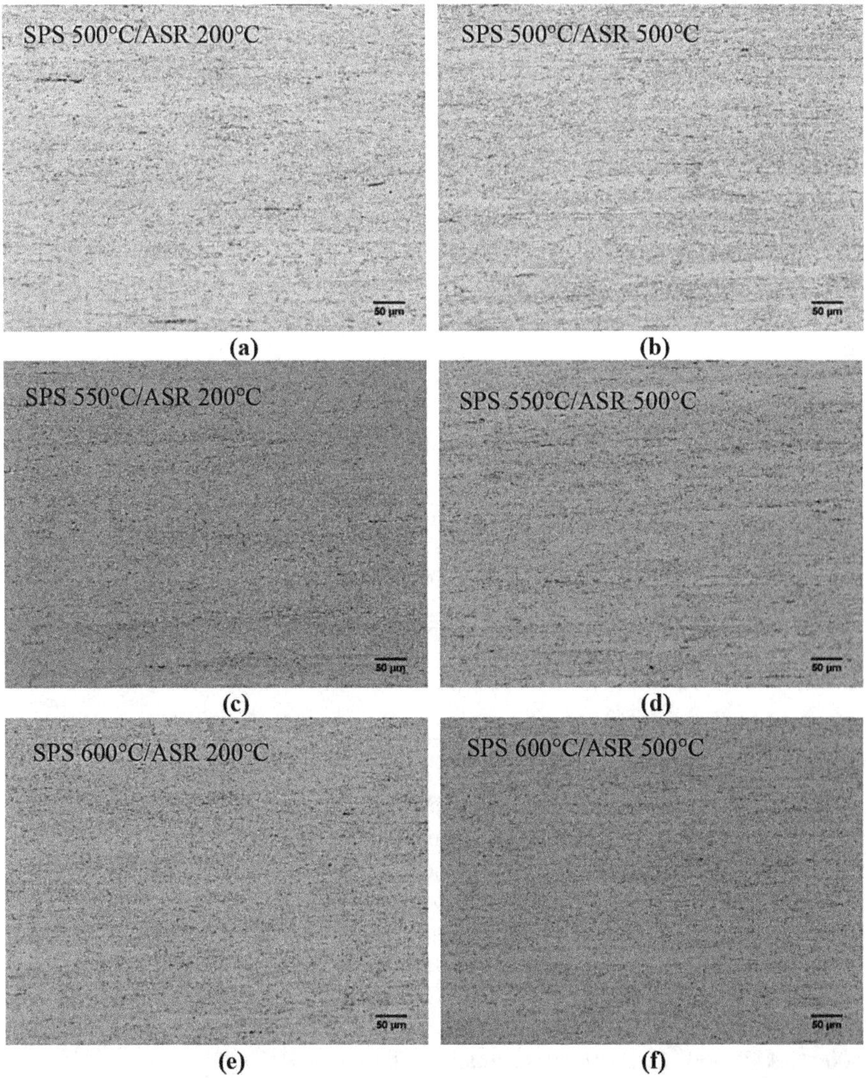

Fig. 10 Optical micrographs of specimens processed through various combinations of SPS and ASR

structure (Figs. 3 and 4) was heavily deformed and now elongated in the rolling direction. All rolled products were seemingly at full density, including the preform initially sintered at 500 °C which contained a measurable fraction of residual porosity (Fig. 1). This observation indicated that the ASR process was effective at eliminating this microstructural defect from the SPS products.

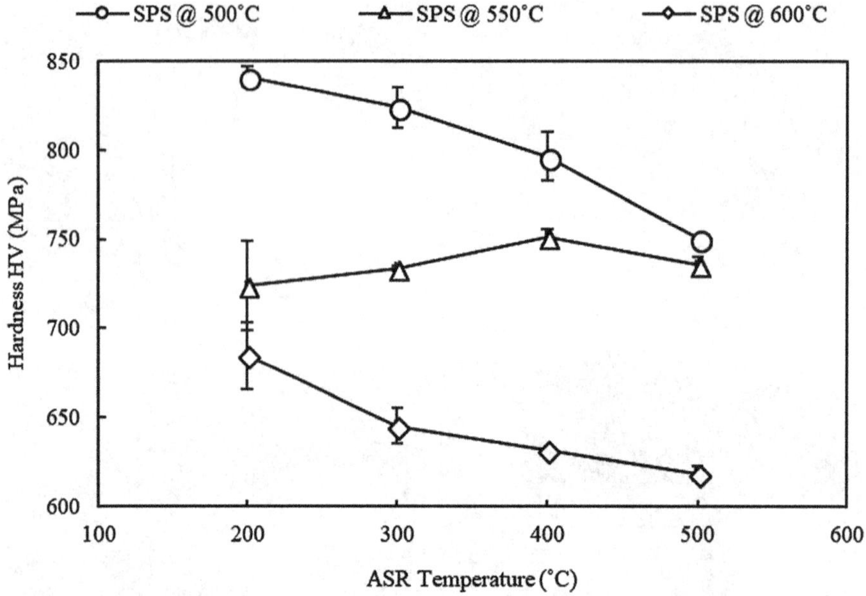

Fig. 11 Comparison of the hardness of sintered preforms asymmetrically rolled at temperatures of 200 °C, 300 °C, 400 °C, and 500 °C

Data on the hardness of the rolled products are shown in Fig. 11. Generally, hardness decreased with increasing ASR temperature except for the specimen sintered at 550 °C wherein hardness largely remained static. Relative to their as-sintered counterparts (Fig. 7), specimens subjected to ASR at 200 °C were found to incur the greatest gains in hardness. For example, in the material sintered at 600 °C, hardness increased from 531 to 684 HV because of ASR. Net improvements of 102 HV and 68 HV were noted for samples sintered at 500 and 550 °C, respectively. When higher ASR temperatures were utilized, the gains became less pronounced. Relative to their as-sintered counterparts, the hardness of the SPS 600 °C/ASR 500 °C sample was now only 88 HV higher, while that of the SPS 500 °C/ASR 500 °C sample was actually softer by 11 HV.

As shown in Figs. 12 and 13, the yield strength/UTS decreased subtly as the ASR temperature increased. Meanwhile, the percent elongation to failure increased considerably with increasing roll temperature except for the preform sintered at 600 °C which recorded ~4% ductility at 200 °C and then remained stagnant at 6% from 300 °C to 500 °C (Fig. 14). The preform sintered/rolled at 500 °C/200 °C had outperformed the other preforms sintered at 550 and 600 °C and then rolled at 200 °C by exhibiting a nominal yield strength of 241 MP compared to values of 216 MPa and 178 MPa, respectively. These results implied that the preform sintered at 500 °C was more responsive to hot ASR than those sintered at 550 °C and 600 °C.

In general, the as-sintered products exhibited lower tensile strengths and higher ductilities than their rolled counterparts. Interestingly, the yield strengths of all the

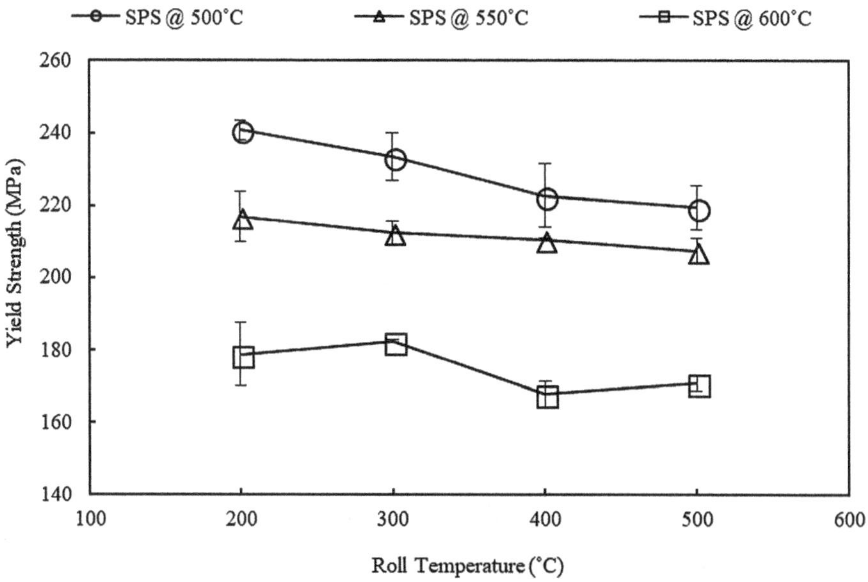

Fig. 12 Comparison of the yield strengths of SPS-processed specimens rolled at temperatures of 200 °C, 300 °C, 400 °C, and 500 °C

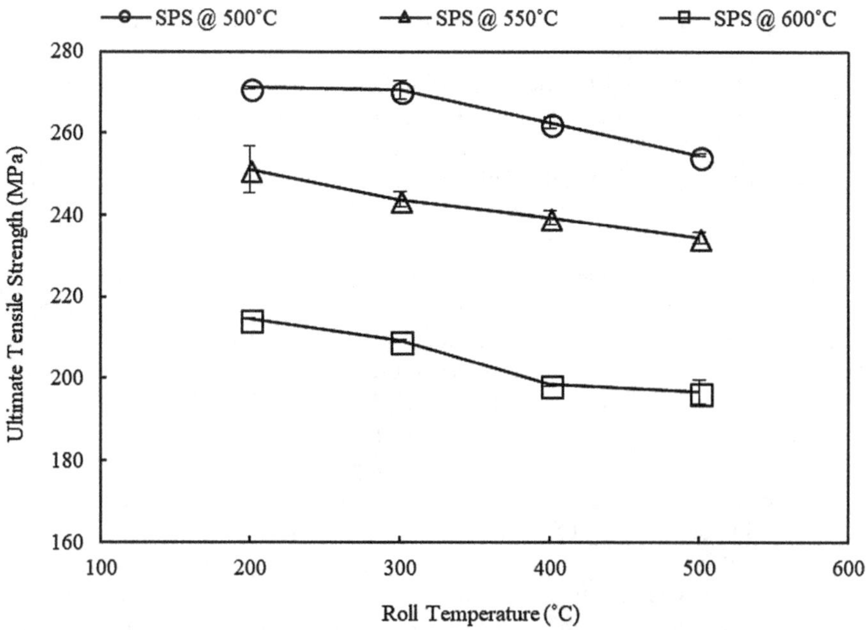

Fig. 13 Comparison of the tensile strengths of SPS-processed specimens at roll temperatures of 200 °C, 300 °C, 400 °C, and 500 °C

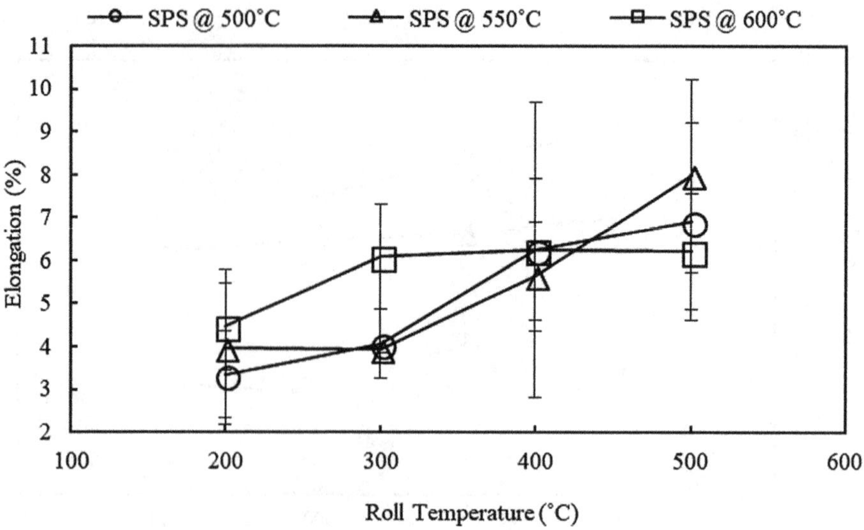

Fig. 14 Comparison of the tensile properties of SPS-processed specimens at roll temperatures of 200 °C, 300 °C, 400 °C, and 500 °C

sintered preforms rolled at 500 °C constituted the lowest yield strengths measured for the different rolling temperatures although, comparatively, these values remained greater than the yield strengths of their as-sintered counterparts. In this sense, the yield strengths measured for the products sintered at 500 °C, 550 °C, and 600 °C were 195 MPa, 163 MPa, and 106 MPa, respectively. Once these same materials were rolled at 500 °C, these values then increased to 219 MPa, 207 MPa, and 171 MPa. The magnitude and trend of these values confirmed that the rolling process had significantly influenced the tensile strength of the sintered preforms.

One of the targets of this work was to identify SPS + ASR parameters that would produce a material with the best combination of yield strength and elongation. To investigate this, the elongation to failure was plotted against the yield strength (Fig. 15). All products originally sintered at 600 °C exhibited an inferior combination of tensile properties. On the contrary, those sintered at 500 °C offered the best balance of properties as shown by the encircled region in Fig. 15.

Analyses were then completed using EBSD and TEM techniques to examine microstructures of the rolled products in greater detail. Previous TEM analyses confirmed that the $Al_3(Sc,Zr)$ precipitates had coarsened from a nominal size of 5 nm to 30 nm as the SPS temperature was increased from 500 to 600 °C. Meanwhile, it is important to note that the highest ASR temperature did not exceed 500 °C, the lowest sintering temperature considered. Hence, it would be counterintuitive to expect the size of the precipitates to increase appreciably during hot ASR. This was corroborated by the TEM micrographs presented in Fig. 16 for specimens sintered at 500 °C/rolled at 200 °C and 600 °C/rolled at 500 °C. These micrographs clearly substantiated that the $Al_3(Sc,Zr)$ precipitates had not coarsened during rolling.

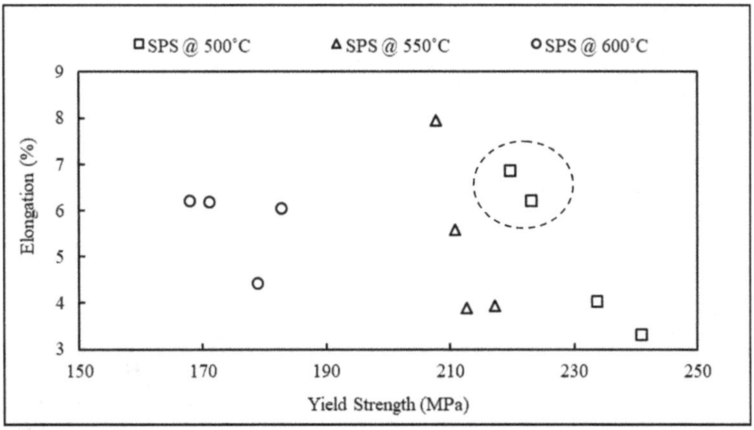

Fig. 15 Comparison of the yield strengths for the SPS-processed specimens rolled at different temperatures

Examples of EBSD findings are given in Figs. 17 and 18. These images implied that dynamic recrystallization was operative and that this effect occurred to a greater extent in the specimen sintered/rolled at 500 °C/200 °C compared to its 600 °C/200 °C counterpart. As stated earlier, the coherent precipitates which prevailed in the 500 °C sintered preform should have pinned grain boundaries and in turn stifled recrystallization. It is important to note that the preservation of a non-recrystallized structure would have also contributed to the measured gains in yield/tensile strengths.

To better understand the trends in tensile properties, a continued analysis of the EBSD data was done in conjunction with the TEM results. It can be recounted that the yield strength obtained for the specimen sintered/rolled at 500 °C/200 °C was ~26% higher than that of the specimen sintered/rolled at 600 °C/200 °C. Nevertheless, the yield strength decreased with increasing roll temperature in both instances. It has already been established in this study that the size of $Al_3(Sc,Zr)$ precipitates was not affected by the hot ASR process, regardless of the temperature employed. This implied that the $Al_3(Sc,Zr)$ precipitates could not have been exclusively responsible for the trends in tensile properties; a more plausible feature of relevance was the LAGB concentrations. In this sense, per Fig. 5 it was determined that the starting billets sintered at 500 °C and 600 °C were effectively identical in terms of their respective concentrations of LAGBs (48%) and HAGBs (52%). This situation was then changed appreciably after ASR as evident in Fig. 19 which compares the fraction of LAGBs with the yield strengths of the rolled products. Statistical analyses of the grain boundary misorientation angles showed that the specimen sintered at 500 °C/rolled at 200 °C contained the highest fraction of LAGB (62%) compared to only 39% for the specimen sintered at 600 °C/rolled at 200 °C. In addition, the fraction of LAGBs measured in the specimens SPS processed at 500 °C/rolled at 200 °C and 500 °C was higher than counterparts sintered at 600 °C/rolled at 200 °C and 500 °C. Furthermore, it can be recounted that the precipitate size was not

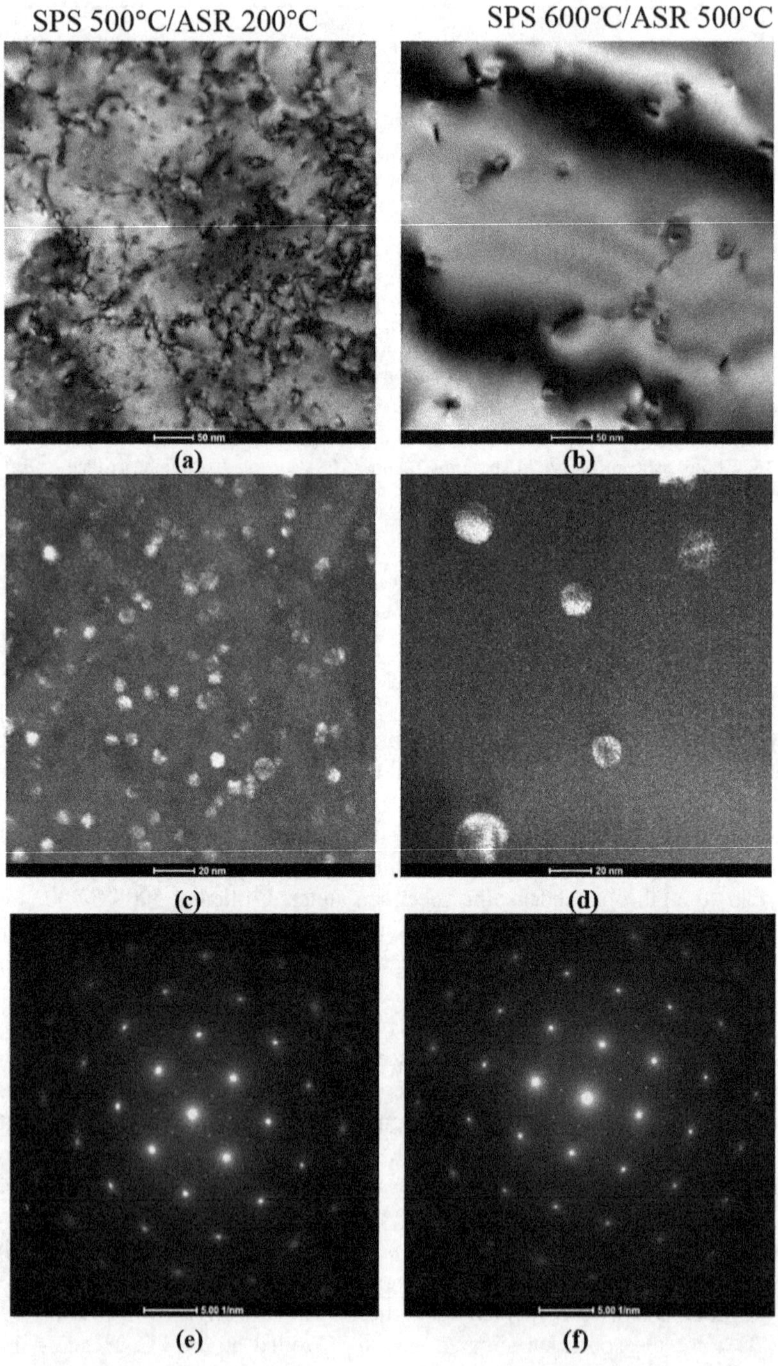

Fig. 16 TEM results acquired from SPS specimens consolidated via SPS at 500 °C + ASR at 200 °C (**a**, **c**, **e**) and SPS at 600 °C + ASR at 500 °C (**b**, **d**, **f**). Bright-field images (**a** and **b**), dark-field images (**c** and **d**), and selected area diffraction patterns recorded with the beam oriented closely parallel to the <100> zones axis (**e** and **f**)

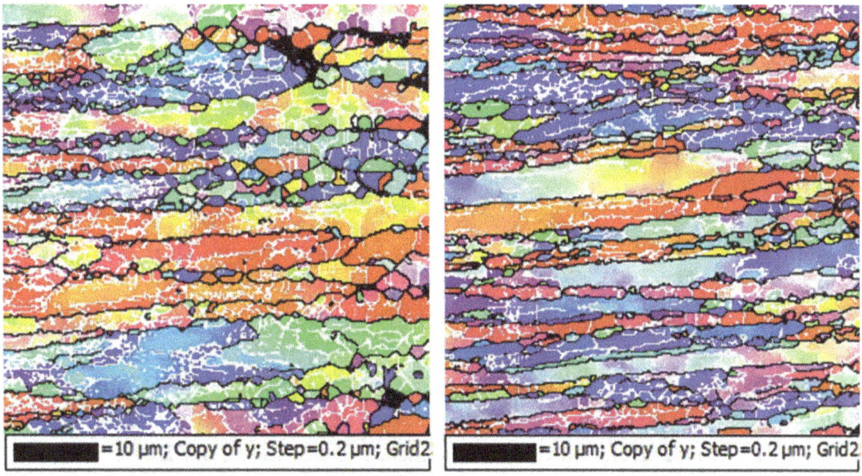

(a) (b)

Fig. 17 EBSD maps of specimens sintered at 500 °C and subsequently hot rolled at (**a**) 200 °C and (**b**) 500 °C. Black lines indicate HAGBs; white lines indicate LAGBs

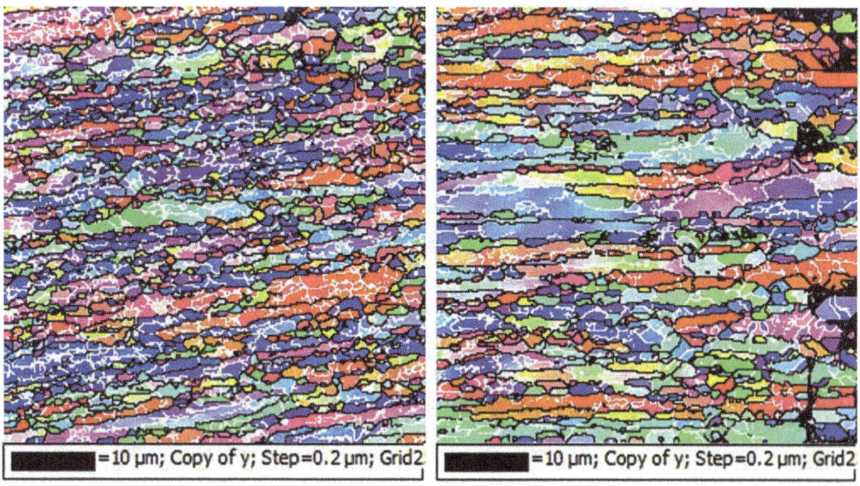

(a) (b)

Fig. 18 EBSD maps of specimens sintered at 600 °C and subsequently hot rolled at (**a**) 200 °C and (**b**) 500 °C. Black lines indicate HAGBs; white lines indicate LAGBs

influenced significantly by the rolling process, confirming that the precipitates were not primarily responsible for the trends in mechanical properties of the rolled products. Collectively, the fraction of LAGB decreased with increasing roll temperature, and yield strength correlated well with the evolution of the fraction of LAGBs. This led to the conclusion that the key mechanism responsible for the increased

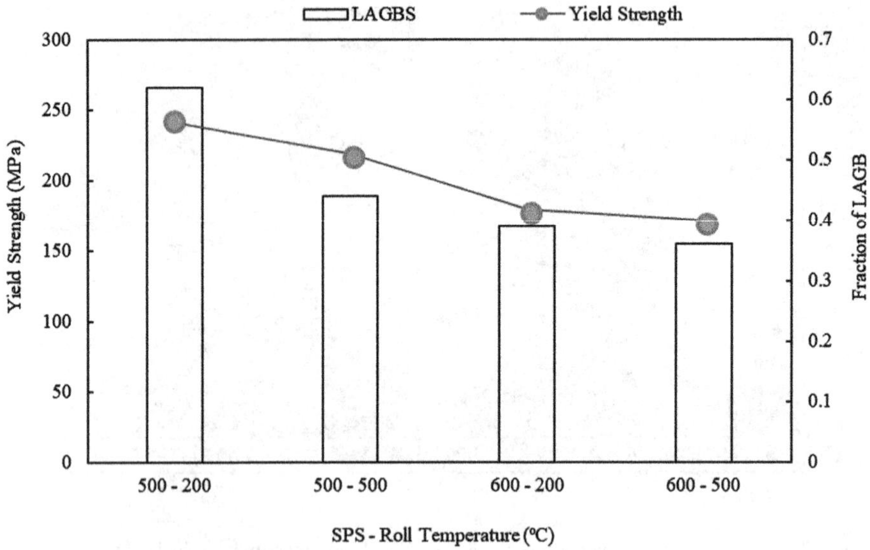

Fig. 19 Trends in yield strength and the fraction of LAGBs for different SPS + ASR process combinations

strengths instilled through hot ASR was subgrain strengthening. This is consistent with the works of other researchers who have reported improved strengths in aluminum alloys due to this effect (Zakharov 2003; Radmilović et al. 2006).

5 Conclusions

1. The Al-0.3Sc-0.2Zr powder was successfully consolidated through SPS.
2. Heating associated with the SPS process invoked the formation of $Al_3(Sc,Zr)$ precipitates. This phase coarsened appreciably as the SPS temperature increased from 500 °C to 600 °C and was the primary driving force behind observed trends in tensile properties.
3. SPS-consolidated billets of the Al-0.3Sc-0.2Zr alloy were responsive to hot ASR. All rolling schedules imparted full densification as well as higher mechanical properties compared to their as-sintered preforms.
4. The hot ASR process had no appreciable influence on the coarsening of $Al_3(Sc, Zr)$ precipitates that preexisted in the sintered preforms.
5. Decreases in the rolling temperature yielded marginal improvements in the mechanical properties of the Al-0.3Sc-0.2Zr PM alloy.
6. Lower SPS and ASR temperatures promoted the formation of higher concentration of subgrains in the final product.

7. The specimen sintered at 500 °C and rolled at 200 °C exhibited the highest concentration of LAGBs and the most desirable combination of hardness and tensile properties.

Acknowledgments The authors would like to graciously acknowledge the financial support provided by Boeing Research and Technology (research contract 11-6392), the Natural Sciences and Engineering Research Council of Canada (NSERC) via collaborative research and development grant #451466, and GKN Powder Metallurgy. TEM assistance provided by Dr. Derek Oliver and Dr. William Caley at the Manitoba Institute for Materials (University of Manitoba) is graciously acknowledged as is the laboratory assistance provided by Mr. Randy Cooke, Mrs. Patricia Scallion, and Mr. Dean Grijm at Dalhousie University.

References

Capus J (2005) Metal powders: a global survey of production, applications and markets. 4th edn. Elsevier, Oxford

Fuller CB, Murray JL, Seidman DN (2005) Temporal evolution of the nanostructure of Al(Sc,Zr) alloys: part I – chemical compositions of al3(Sc1−xZrx) precipitates. Acta Mater 53:5401–5413

Jia Z, Røyset J, Solberg JK, Liu Q (2012) Formation of precipitates and recrystallization resistance in Al–Sc–Zr alloys. Trans Nonferrous Met Soc China 22:1866–1871

Kirchoff SD, Adkins JY, Griffith WM, Martorell IA (1986) Effective method for degassing evaluation of aluminum pm alloys. Rapidly solidified powder aluminum alloys. ASTM publication code number (PCN) 04-890000-04. 354–368

Kondoh K, Kimura A, Watanabe R (2000) Effect of Mg on sintering phenomenon of aluminum alloy powder. J Jpn Soc Powder Powder Metall 47:36–41

Quist WE, Lewis RE (1986) The need for rapidly solidified powder metallurgy aluminum alloys for aerospace applications. Rapidly solidified powder aluminum alloys. ASTM publication code number (PCN) 04-890000-04. 7–36

Radmilović V, Tolley A, Lee Z, Dahmen U (2006) Core-shell structures and precipitation kinetics of Al_3 (Sc, Zr) Li_2 intermetallic phase in Al-rich alloy. Metalurgija 12:309–314

Rostova TD, Davydov VG, Yelagin VI, Zakharov VV (2000) Effect of scandium on recrystallization of aluminum and its alloys. Mater Sci Forum 331–337:793–798

Simões FJP, De Sousa RJA, Grácio JJA, Barlat F, Yoon JW (2009) Effect of asymmetrical rolling and annealing on the mechanical response of an 1050-O sheet. Int J Mater Form 2:891–894

Song SX, Wang Z, Shi GP (2013) Heating mechanism of spark plasma sintering. Ceram Int 39:1393–1396

Sweet GA, Brochu M, Hexemer RLJ, Donaldson IW, Bishop DP (2014) Microstructure and mechanical properties of air atomized aluminum powder consolidated via spark plasma sintering. Mater Sci Eng A 608:273–282

Vlach M, Stulikova I, Smola B et al (2013) Precipitation in cold-rolled Al–Sc–Zr and Al–Mn–Sc–Zr alloys prepared by powder metallurgy. Mater Charact 86:59–68

Xie G, Ohashi O, Yoshioka T et al (2001) Effect of interface behavior between particles on properties of pure al powder compacts by spark plasma sintering. Mater Trans 42:1846–1849

Yamasaki M, Kawamura Y (2006) Changes in the surface characteristics of gas-atomized pure aluminum powder during vacuum degassing. Mater Trans 47:1902–1905

Zakharov VV (2003) Effect of scandium on the structure and properties of aluminum alloys. Metal Sci Heat Treat 45:246–253

Zuo F, Jiang J, Shan A, Fang J, Zhang X (2008) Shear deformation and grain refinement in pure al by asymmetric rolling. Trans Nonferrous Met Soc China 18:774–777

Spark Plasma Sintering Produced W-Fiber-Reinforced Tungsten Composites

Y. Mao, J. W. Coenen, J. Riesch, S. Sistla, C. Chen, Y. Wu, L. Raumann, R. Neu, C. Linsmeier, and C. Broeckmann

1 Introduction

Tungsten and tungsten-based materials are widely used refractory material in industry. Currently, it is also considered as the top candidate for plasma facing materials due to their advanced high-temperature properties. Tungsten material

Y. Mao (✉) · J. W. Coenen · L. Raumann · C. Linsmeier
Forschungszentrum Jülich GmbH, Institut für Energie- und Klimaforschung – Plasmaphysik, Jülich, Germany
e-mail: y.mao@fz-juelich.de; j.w.coenen@fz-juelich.de; l.raumann@fz-juelich.de; ch.linsmeier@fz-juelich.de

J. Riesch
Max-Planck-Institut für Plasmaphysik, Garching bei München, Germany
e-mail: johann.riesch@ipp.mpg.de

S. Sistla · C. Broeckmann
Institut für Werkstoffanwendungen im Maschinenbau (IWM), RWTH Aachen University, Aachen, Germany
e-mail: s.sistla@iwm.rwth-aachen.de; c.broeckmann@iwm.rwth-aachen.de

C. Chen
School of Materials Science and Engineering, Hefei University of Technology, Hefei, China
e-mail: chench@hfut.edu.cn

Y. Wu
School of Materials Science and Engineering, Hefei University of Technology, Hefei, China

China International S&T Cooperation Base for Advanced Energy and Environmental Materials, Hefei, China
e-mail: ycwu@hfut.edu.cn

R. Neu
Max-Planck-Institut für Plasmaphysik, Garching bei München, Germany

Technische Universität München, Garching, Germany
e-mail: rln@ipp.mpg.de

© Springer Nature Switzerland AG 2019

P. Cavaliere (ed.), *Spark Plasma Sintering of Materials*,
https://doi.org/10.1007/978-3-030-05327-7_9

combines the properties of a high melting point, high creep resistance, high-temperature strength, good thermal conductivity, low vapor pressure, and good erosion resistance (Smid et al. 1998). However, in contrast to these excellent high-temperature properties, the major concern when using tungsten is its brittleness at low temperatures. Like most body-centered cubic (bcc) metals, tungsten has a characteristic brittle to ductile transition at a high temperature. An exact value of this transition temperature cannot be given, as it depends strongly on the microstructure and testing condition of the material (Rupp and Weygand 2011; Giannattasio and Roberts 2007). When the application temperature of tungsten is lower than the ductile-to-brittle transition temperature (DBTT), tungsten will behave brittly and show low tolerance toward pre-cracking or surface damage.

It has been widely reported that tungsten material can show good ductility after high deformation machining process, like cold rolling (Reiser et al. 2012; Rieth et al. 2013). This ductility relies on the highly deformed grain structure. However, neutron embrittlement causes a potential issue when using this kind of highly deformed tungsten as plasma facing material. In the early stage of operation, the material may behave ductile. After some operation time, the material would be embrittled by the neutron irradiation and recrystallization. Again, we may face the issue of the cracking. Therefore, an extrinsic reinforcement principle to increase the tungsten fracture toughness like a fiber-reinforced ceramic material could be the potential solution. With this concept, the increased toughness relies not on the intrinsic properties of the material but on the interaction between each component of the composite. The overall effect could be that, with both brittle fiber and matrix, the material is still able to reveal good toughness. Still many studies and developments are required to prove whether this composite concept could be the potential solution for tungsten as the plasma facing material.

In recent years, W-fiber-reinforced tungsten (W_f/W) composites have been developed relying on an extrinsic toughing mechanisms (Riesch et al. 2014a, 2016c; Neu et al. 2016; Du et al. 2010a). Commercially available W-fibers (OSRAM GmbH) have been chosen to reinforce the tungsten matrix. It has been reported that the W-fibers show ductile behavior and high tensile strength caused by the as-drawn microstructure (Riesch et al. 2016a, c). Similar to ceramic fiber-reinforced composites (Chawla 1998; Stang and Shah 1986; Li et al. 1991), with the incorporation of the fibers, the tungsten composite can develop the so-called pseudo-ductile behavior. This mechanism allows the composite to obtain a certain tolerance toward cracking and damage compared to conventional tungsten (Stang and Shah 1986; Riesch et al. 2014a). In general, similar to the ceramic fiber-reinforced composites, a relatively weak interface between the fiber and the matrix is considered to be necessary to achieve pseudo-ductility (Czél and Wisnom 2013; Ming-Yuan and Hutchinson 1989; Mao et al. 2018).

Initially, the W_f/W composite production was solely based on a chemical vapor deposition (CVD) process (Gietl et al. 2017; Riesch et al. 2016b). In this process, the tungsten matrix is deposited on several layers of aligned long W-fibers to form a bulk material. The basic structure is shown in Fig. 1a. Extensive studies on W_f/W have been performed to understand the reinforcement mechanisms for the CVD

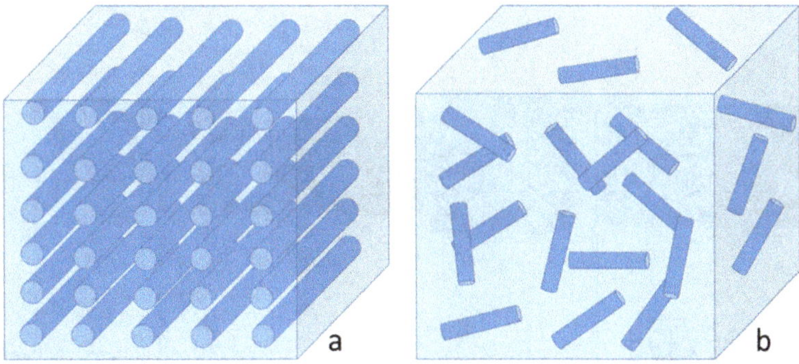

Fig. 1 Schematic drawing of the composite structure for: (**a**) CVD produced aligned long fiber W_f/W; (**b**) PM produced short random fiber W_f/W

produced W_f/W (Jasper et al. 2015, 2016; Riesch et al. 2013, 2014a, b; Du et al. 2010a, b, 2011; Gietl et al. 2017, 2018).

Industrially, powder metallurgical (PM) processes are the main paths to produce tungsten materials. Therefore, the development of a PM route is important for tungsten composites in view of large-scale production. As technically introducing aligned long fibers into the PM process is quite challenging, the approach for PM W_f/W is to use short randomly distributed fibers. The basic structure of the PM produced W_f/W is shown in Fig. 1b.

In comparison to the CVD production route, the PM process, as a more conventional industrial process for the manufacturing of tungsten-based material, has several benefits, such as substantial experience with bulk production, higher production rate, and an easier realization of alloyed materials. However, the fiber-matrix interface might be damaged due to the high temperatures and pressures needed during the consolidation process (Coenen et al. 2017; Mao et al. 2017b). Additionally, the high temperature during the PM process can cause the change of the as-drawn microstructure of the W-fiber and, hence, reduce the fiber ductility and strength (Riesch et al. 2016a, d; Zhao et al. 2017).

As one of the standard PM process, the spark plasma sintering (SPS) is chosen to produce the W_f/W material (Guillon et al. 2014; Kim et al. 2009; Allahar et al. 2011; Ohser-Wiedemann et al. 2012). SPS is a low-voltage, pulsed direct current (DC) activated, pressure-assisted sintering and synthesis technique. This method can be used to synthesize new compounds and/or to densify materials in a single step (Arzt et al. 1983; Orrù et al. 2009). The process is similar to hot pressing (HP), but the way the heat is produced and transmitted to the sintering material is different (Orrù et al. 2009). If the powder is electrically conductive, heat can be generated directly within the sample. Apart from the high temperature and pressure supplied, there are some special sintering mechanisms during SPS due to the heating principle applied, for example, Joule heating can cause local vaporization or melting of the powder surface to accelerate the sintering rate, and current flow through the specimen might enhance the diffusion process (electromigration) (Arzt et al. 1983; Orrù et al. 2009).

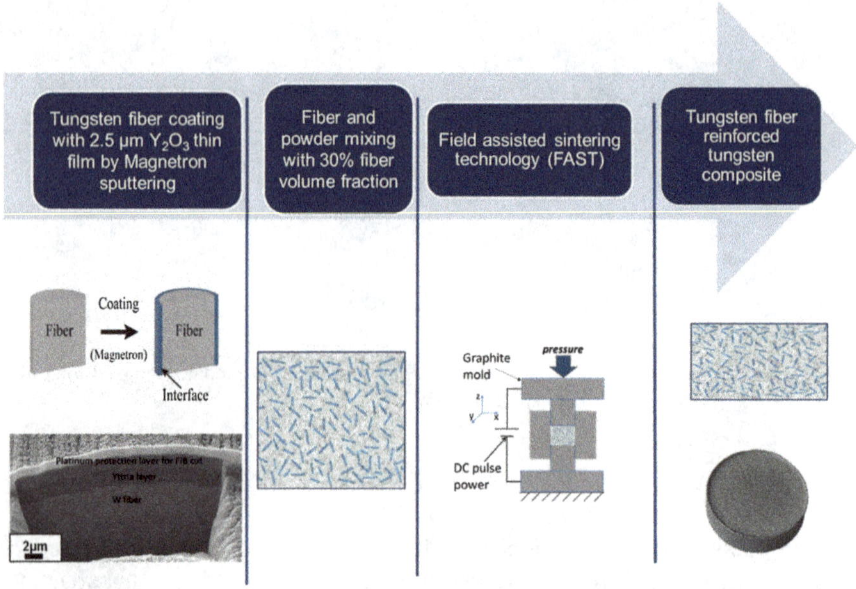

Fig. 2 Production route of W-fiber-reinforced tungsten composite

Consequently, SPS is a relatively quick process compared to the typical HP process. The schematic diagram of the SPS sintering mechanism is shown in Fig. 2 (third step). For tungsten sintering, the densification times by SPS are typically close to a few minutes (Kim et al. 2009; Allahar et al. 2011; Ohser-Wiedemann et al. 2012; Cho et al. 2004; Kellogg et al. 2008).

1.1 W-Fiber-Reinforced Self-Passivating Tungsten Alloy

For the applications of tungsten as the future fusion plasma wall material, one other potential issue is its poor oxidation resistance. In case of a severe accident scenario, the so-called loss-of-coolant accident with simultaneous air ingress, active cooling of the first wall will fail. At such conditions, the neutron-irradiated, radioactive tungsten can form a volatile tungsten oxide, which may be released into the environment.

To solve this problem, the so-called self-passivating "smart" advanced tungsten-based alloys are under development (Koch and Bolt 2007; Wegener et al. 2016, 2017; Litnovsky et al. 2017a, b) aiming at accommodating their properties to the exposure conditions. During the regular plasma operation, plasma ions and neutrals of high energy may sputter the surface of the smart alloy. The preferential sputtering will first remove the lighter candidate alloying elements chromium (Cr) and yttrium (Y) from the plasma-exposed surface thus providing almost pure tungsten surface facing the plasma. In case of an accident, the alloying elements remaining in the bulk of the smart alloy system will form their own dense oxides, protecting tungsten from sublimation.

However, this alloy material, similar to pure tungsten, is also very brittle. In order to increase the toughness of the smart alloy, smart W-alloy matrix material could be reinforced by W-fibers. The SPS route is chosen to realize a proof of principle (Coenen et al. 2017).

In this work, the short multi-fiber-reinforced tungsten composites are produced by SPS process (Coenen et al. 2017; Mao et al. 2017a, 2018). The bulk W_f/W composites with short multi-fibers are microstructurally and mechanically characterized. The goal of this work is to prove the suitability of the SPS process to produce tungsten-based composite material.

2 Experiment

2.1 W_f/W Production

The W_f/W manufacturing route is shown in Fig. 2. Pure tungsten powders (provided by OSRAM GmbH) and W-fibers are used as the raw material. The tungsten powders are the mixture of two sets of powders with finer (~5 μm average particle size) and coarser (>50 μm average particle size) particles. This combination is an optimized powder mixture to achieve higher density for sintering process based on information provided by OSRAM. The W-fibers are potassium-doped short W-fibers with 0.24 mm diameter and 2.4 mm length (provided by OSRAM GmbH). The W-fibers are produced via a drawing process and then manufactured into the required length. After the drawing process, the W-fibers have an extremely elongated grain structure. Relying on this microstructure, the W-fibers show ductile behavior and high tensile strength (~3000 MPa) (Riesch et al. 2016a, d). Potassium is insoluble in tungsten; the doping (approx. 60–75 ppm) is present in the form of nano-dispersed bubble rows along the elongated grains pinning the grain boundaries. It leads to a good high-temperature microstructural stability (Briant et al. 1993) compared to normal W-fibers.

As a relative weak interface is required between the fiber and matrix to achieve the desired pseudo-ductility (Mao et al. 2018), an yttrium oxide layer is coated on the W-fibers as the interface material as the first production step. The magnetron deposition parameters are described in (Mao et al. 2017b). The coating is facilitated by a four-step process with magnetron sputtering similar to (Mao et al. 2018). As an example, the SEM image of one of the coated fiber cross section prepared by focus ion beam (FIB) is shown in Fig. 2 (left). In this study, the yttria coating thickness is either 1 μm or 2.5 μm.

In a next step, the W-fibers were mixed with the tungsten powders by manual shaking in a vessel with a reasonable fiber volume fraction. The fiber volume fraction is 30% in this study. After the SPS process, no obvious fiber cluster was visible in the mixture and in the bulk material.

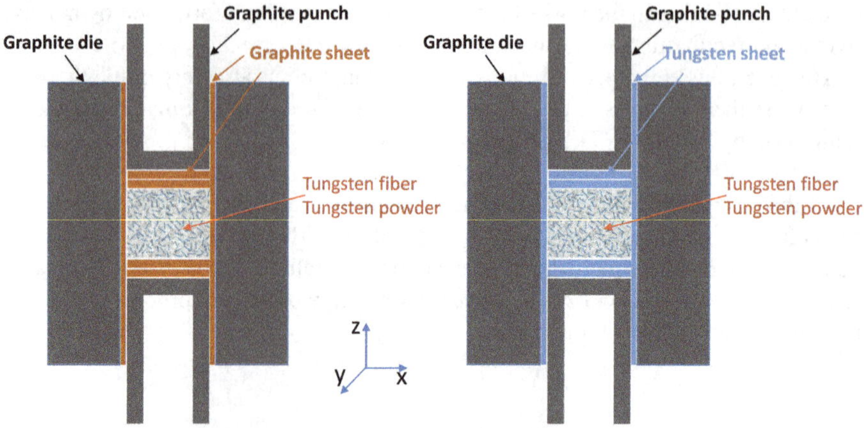

Fig. 3 Dies used during SPS. Left, die with C foil; right, die with W foil

The fiber/powder mixture is consolidated by SPS process as the next production step. A commercial SPS system (HP D 25–2) from "FCT Systeme GmbH" (max force, 250 KN; max temperature, 2200 °C; max heating rate, 400 K/min) is used in this study.

The molds for the SPS process are shown in Fig. 3. Two different mold structures are used:

1. Graphite foil is placed between graphite tool and the mixture (Fig. 3a), aiming to reduce the damage of the punch surface and to facilitate sample removal after the SPS process. This is a relative conventional setup.
2. Tungsten foils with 25 μm thickness are used to separate the powder-fiber mixture. The tungsten foil is applied here to reduce carbon contamination in the W-fiber during sintering, as discussed in (Coenen et al. 2018; Müller et al. 2018). Carbon embrittlement is one of the main challenges when using SPS to produce W_f/W material. Some detail of this content will be discussed in Sect. 3.3.

For both cases, an electrically insulating boron nitride coating inside the graphite die guarantees that the electrical current only flows through the powder/fiber mixture and not along the die. Hence, the sintering efficiency is increased.

During the SPS process, the sintering parameters are a constant heating rate of 200 °C/min, holding temperature of 1600–2000 °C (the temperature is measured at the bottom of the hollow punch), 4–6 min holding time, and a constant pressure of 60 MPa (Ohser-Wiedemann et al. 2012; Chanthapan et al. 2012; Cho et al. 2004). The sintering pressure is limited by the strength of the graphite die material (maximum ~70 MPa). The average cooling rate from holding temperature to 400 °C is ~375 °C/min. The sintering is performed in vacuum below 0.1 mbar. A typical sintering time-temperature curve is shown in Fig. 4. As result of the sintering, coin-shaped samples (20 mm or 40 mm diameter with ~5 mm height) are produced (examples in Fig. 5).

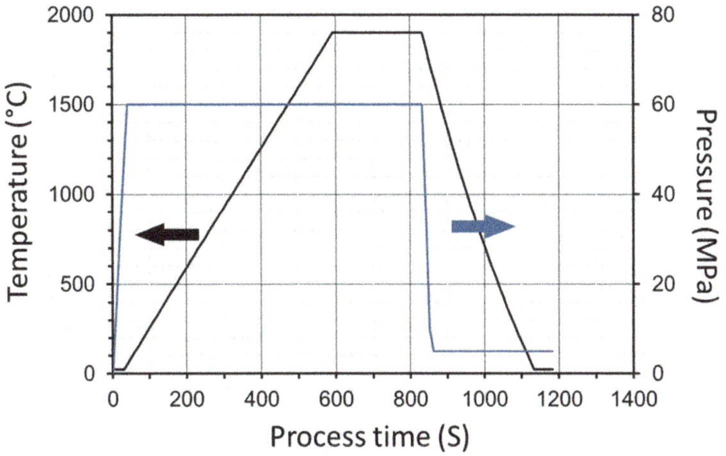

Fig. 4 Typical time-temperature curve of SPS process during the production

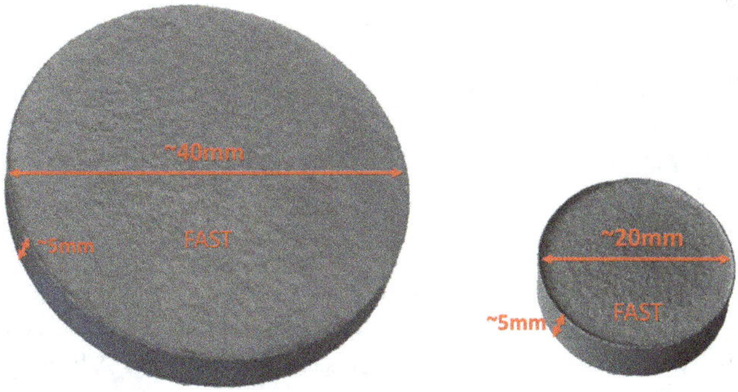

Fig. 5 Bulk W_f/W sample examples after SPS process

For the W-fiber-reinforced self-passivating W-alloy production (87.5 wt% W, 12 wt% Cr, and 0.5 wt% Y), powders are sintered together with 2.5 μm yttria- coated W-fibers.

2.2 Density, Microstructure, and Mechanical Testing

The mass density of the W_f/W samples after the consolidation is measured by the Archimedes principle. During the measurement, once the sample is dipped into the ethanol, the measured weight starts to change slowly and continuously. This is due to the existence of open pores where ethanol can fill in the sample. To get a valid measurement, the measured weight is recorded after the samples have been dipped into the ethanol for 10 s. In this way, not many open pores are filled by the ethanol, but the surface wetted well.

The microstructure and morphology of the composites are analyzed by a LEO 982 scanning electron microscope (SEM) after mechanical polishing. An energy-dispersive X-ray spectroscopy (EDX) system is used for qualitative surface composition measurements. The high-resolution TEM is conducted on Tecnai G20 transmission electron microscope, and the operated accelerating voltage is 200 KV.

A 3-point bending test on notched samples (at RT) is chosen to investigate the fracture behavior as given in (Mao et al. 2018). Sample dimensions are $18 \times 2 \times 4\ mm^3$ (length × width × thickness, manufactured by diamond wire cutting) with a ~2 mm deep notch. The pre-notch for the 3-point bending test is prepared by diamond wire cutting with wire diameter of 0.04 mm (GS-1000 wire cutting machine from company Sommer Präzisionstechnik) followed by manually razor blade polishing for 3 min. The sharpness of the notches are not identical, therefore, this is not a quantitative test. The absolute values of each measurement are not directly comparable. Important here is the shape of the force-displacement curves.

3 Results and Discussion

3.1 Microstructure Overview and Interface Challenge

After the SPS process, a typical microstructure of the as-fabricated W_f/W is given in Fig. 6, similar to the results published in (Mao et al. 2017a).

The cross sections shown are parallel and perpendicular to the z-axis (compression direction as shown in Fig. 3). No major defects can be observed. A homogeneous fiber distribution can be observed. Potential fiber distribution inhomogeneity (areas marked by dash lines in Fig. 6) could be caused by insufficient shaking during mixing. The samples tend to have a two-dimensional planar fiber distribution. Most fibers tend to orient parallel to x-y plane. It is caused by the uniaxial deformation during process.

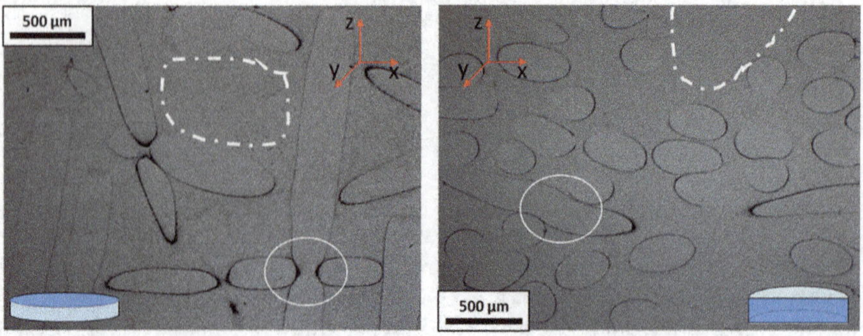

Fig. 6 Typical microstructure of the as-fabricated W_f/W

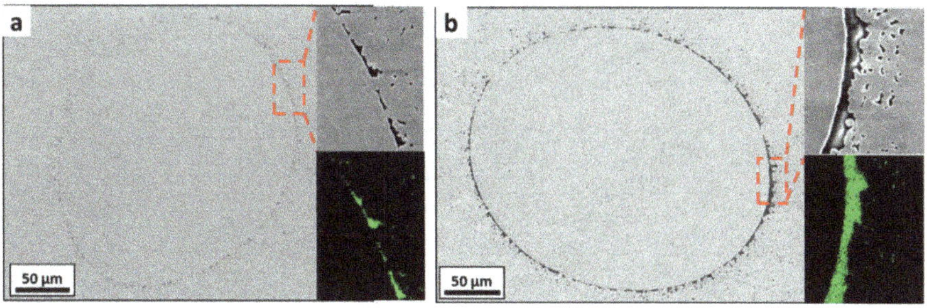

Fig. 7 SEM images of the fiber-matrix interface region of the SPS-produced W_f/W composites. An energy-dispersive X-ray spectroscopy mapping analysis was performed. The location of yttrium is shown in green

Additionally, fiber deformation during consolidation is visible, especially when fibers are in contact with each other, as marked by solid circles in Fig. 6.

During the SPS process, one of the issues is the yttria interface damage. The interface area for the samples with 1 µm and 2.5 µm coating thickness is shown in Fig. 7. An energy-dispersive x-ray spectroscopy (EDX) mapping analysis was performed to detect local interface chemical composition. The location of elemental yttrium signals is shown in green. The fiber-matrix interfaces are visible after sintering for the samples from both processes. In contrast to the dense and homogenous yttria layer in the as-coated state (Fig. 2 the first step), the yttria interface shows damage after the consolidation process. Especially the outer shell of the interface got penetrated by the tungsten powders owing to the external pressure and the elevated temperature during the SPS process (Mao et al. 2017b; Bonifacio et al. 2013, 2014; Coenen et al. 2017). Also an effect called dielectric breakdown effect might play a role (Bonifacio et al. 2013, 2014), as yttria is electrically isolating. For the 1 µm yttria interface W_f/W, it can be observed that the interface is impacted strongly by the tungsten powder particles. In some areas, the tungsten powder and W-fiber are directly joined together. Therefore, a mixed Yttria/tungsten interface is formed. In the case of 2.5 µm yttria interface, only the outer shell of the interface is damaged, while the interface itself remains intact. Hence, tungsten fiber and matrix are clearly separated by yttria. However, some yttria particles can be detected within the matrix near the interface region indicating the destruction and migration of the yttria layer during the SPS process. For the later on mechanical testing, W_f/W with 2.5 µm interface is used due to the necessity of an intact weak interface.

3.2 *Sintering Density*

The densities of the PM produced samples with various sintering parameters are shown in Table 1. The error bar in the table is the standard deviation from the measurement of multiple samples. From Table 1, it can be determined that with

Table 1 Sample sintering parameters and relative densities after sintering process

Sample	Production process	Sintering temperature (K)	Holding time (min)	Heating rate (K/min)	Sintering pressure (MPa)	Relative density
W	SPS with C foil	2073	6	100	50	91.4
W	SPS with C foil	2173	4	200	60	92.7 ± 0.4
W	SPS with C foil	2273	4	200	60	93.5
W	SPS with C foil	2173	8	200	60	94.0
W	SPS with C foil	2173	6	200	60	93.9
W_f/W	SPS with C foil	2073	6	100	50	92.6
W_f/W	SPS with C foil	2173	4	200	60	93.8 ± 0.5
W_f/W	SPS with C foil	2273	4	200	60	95.1
W_f/W	SPS with C foil	2173	4	200	60	93.7 ± 0.4
W_f/W	SPS with W foil	2173	4	200	60	91.7 ± 0.5

current tungsten powders, SPS-produced pure tungsten samples have ~91%–95% relative density. Generally speaking, with increased pressure and temperature, the density increases. The relative density of pure tungsten material is slightly lower than that of the W_f/W material. This is because 30% already dense fibers are introduced (Mao et al. 2017a). For the samples that are sintered with W foil, the relative density is lower than the samples sintered with C foil. This is due to the higher current resistance of the sample during sintering when using tungsten foil. As the contact material between powders and mold, W foil is not as soft and flexible as the carbon foil. During the powder filling step, some gaps could potentially remain between powders and W foil, which will also increase the current resistance. During the sintering, the measured current resistance when using W foil is ~50 mΩ which is around five times higher than for the graphite case (~10 mΩ). The corresponding sintering current is ~1 A and ~5 A for W foil and C foil processes, respectively. For the SPS process, even though the sintering temperature is the same, the sintering current also plays a vital role during the consolidation. Lower sintering current could result in lower sintering efficiency and lower sintering density.

Because a temperature of 2273 K is already very challenging for the applied facility, in addition the sintering efficiency is not evidently higher than at 2173 K sintering (Table 1); the samples for the mechanical tests are all sintered at 2173 K.

Table 2 Sample sintering parameters and relative density after SPS process with carbon reinforced carbon tool

Sample	Production process	Sintering temperature (K)	Holding time (min)	Heating rate (K/min)	Sintering pressure (MPa)	Relative density
W	SPS with C foil	2173	4	200	60	91.6
W	SPS with C foil	2173	4	200	80	93.9
W_f/W	SPS with C foil	2173	4	200	80	94.5
W	SPS with C foil	2173	4	200	90	Tool failure

3.2.1 Sample Production with Carbon Reinforced Carbon Tools

For the SPS process, the sintering pressure is limited by the graphite tool material. To get higher relative density with the currently used tungsten powders, the tool material with higher strength could be applied. Carbon reinforced carbon composite (CFC) is used as alternative for the tool material as it has much higher strength compared to graphite. The relative density of the samples that are sintered with CFC tool is shown in Table 2.

When using the CFC tool with the same parameter like the graphite tool case (compare the 2173 K, 4 min, 60 MPa sintering), the achieved relative density is slightly lower. This could be due to the lower electrical conductivity of the CFC material compared to graphite. During the sintering, the electrical resistance of the CFC tool is around 60 mΩ, while the resistance of the graphite tool is only around 10 mΩ. Even at the same sintering temperature, lower sintering current could result in lower sintering efficiency and lower sintering density. When the sintering pressure is increased to 80 MPa with the CFC tool, the achieved relative density is higher but still limited to lower than 95%. When the sintering pressure increased to 90 MPa, the CFC tool fails. The sintering pressure limitation is around 80 MPa in the performed experiments. For the process with CFC tools and W foil, the resistance becomes even higher (~100 mΩ) as the contact between powders and the W foil is worse as discussed before. For this case, the SPS facility cannot establish a stable sintering process, since the facility is initially adjusted to the low resistance graphite system.

3.2.2 Sample Production with Different Powder Particle Size

Apart from varying the sintering parameters, powder particle size is also a critical factor to get higher relative density in SPS process. Though it is not the focus in this study, some initial attempts have been made to try to find an optimized particle size. Pure tungsten material has been sintered by SPS with different powder particle sizes.

Table 3 Relative densities of the pure W samples produced by SPS using different powder particle sizes

Average powder particle size (μm)	Density (%)
12	87.5
8.6	91.1
8.6 pre-pressed	91.0
Mixture of 5, 8.6, and 12	90.1
5.6	91.5
4	91.1
5	92.8
<1	95.5

The 5 μm powders are the standard powders from OSRAM (the mixture of 5 μm and >50 μm particles). Pre-pressed in this figure means the powders got pre-pressed under 100 MPa then filled into the die. The ratio of the mixture sample in this table is 1/3, 1/3, and 1/3

Fig. 8 Light microscope image showing microstructure overview of the W_f/W produced with submicron tungsten powders

The sintering parameters are 2173 K, 60 MPa, and 4 min with graphite tool and foil. Relative densities after sintering are shown in Table 3.

Generally speaking, the relative density tends to increase with decreasing particle size. Pre-pressing and powder mixing without optimization for this case do not show significant influence on the sintering efficiency. The highest relative density is given by the submicron powders. However, it was found that the submicron powders will cause stronger fiber/matrix interface damage. When W_f/W is produced by submicron powders, an intact interface cannot be retained after sintering. As shown in Fig. 8, the yttria interface is no longer visible between the fiber and matrix. This could be due to the external pressure where the fine powders could cause local etching of the weak interface during sintering.

3.3 *Initial Mechanical Testing and W-Fiber Challenge*

After the sample production, the 3-point bending tests are performed to test whether the pseudo-ductile behavior can be achieved by the SPS-produced W_f/W composite. Pure tungsten sample that produced by SPS process with the same parameter is also tested as a reference. Figure 9 shows the typical force-displacement curves and the corresponding fracture sections for SPS-produced pure tungsten and W_f/W samples produced with graphite foil and W foil.

At RT, the pure tungsten sample shows pure linearly elastic loading until a severe load drop during failure without having any plastic deformation. Also, a clear brittle intergranular fracture surface is observed. Additionally, on the fracture surface, as the relative density is only ~93%, some remaining pores on grain boundaries are visible.

Both W_f/W samples give different force-displacement curves as shown in Fig. 9: after the first load drop, the load still increases; a massive load drop occurs after reaching the maximum load; afterward, the samples tend to have a stepwise or continuous load drop instead of a complete failure. Interface de-bonding and fiber pullout can be observed in both cases, as shown in the fracture surface in Fig. 9. Additionally, the overall fracture surface of W_f/W is more uneven compared to pure tungsten, pointing toward crack deflection.

The above results give a strong indication that the SPS-produced W_f/W composites are able to exhibit typical pseudo-ductile behavior similar to W_f/W produced by CVD (Riesch et al. 2014b): the first load drop signifies the matrix failure; after the matrix failure, the load still increases due to the crack bridging by the fibers; the massive load drop indicates unstable crack propagation; and then the stepwise load decreasing is likely supplied by frictional fiber pullout from the matrix. Here the fiber pullout, elastic fiber bridging, and crack deflection are likely the energy dissipation mechanisms which contribute to the elevated fracture resistance compared to pure tungsten (Stang and Shah 1986; Riesch et al. 2013).

Additionally, some differences between samples covered by graphite foil and W foil can be identified, with respect to the shape of the curves and the fiber fracture surface. For the curves, the main difference is visible after maximum loading: for the W foil covered sample, the load drop is more continuous (has more steps) and gentler. It can be assumed that this difference is caused by the different fiber properties after the SPS process. When looking at the fiber fracture surface, the W-fiber in the graphite-covered sample shows a brittle transgranular cleavage fracture with sharply defined river line patterns (Fig. 9 middle right). For the W foil protected sample fiber fracture surface, a clear knife edge-shaped pattern is visible together with some necking (Fig. 9 bottom right). This is a typical ductile fracture surface for W-fiber (Zhao et al. 2017; Riesch et al. 2016c).

When using PM processes to produce W_f/W, one of the major concerns is the degradation of the W-fiber properties. This degradation can be caused by elevated temperature causing recrystallization or the introduction of impurities during the high-temperature sintering processes. The recrystallization may cause the

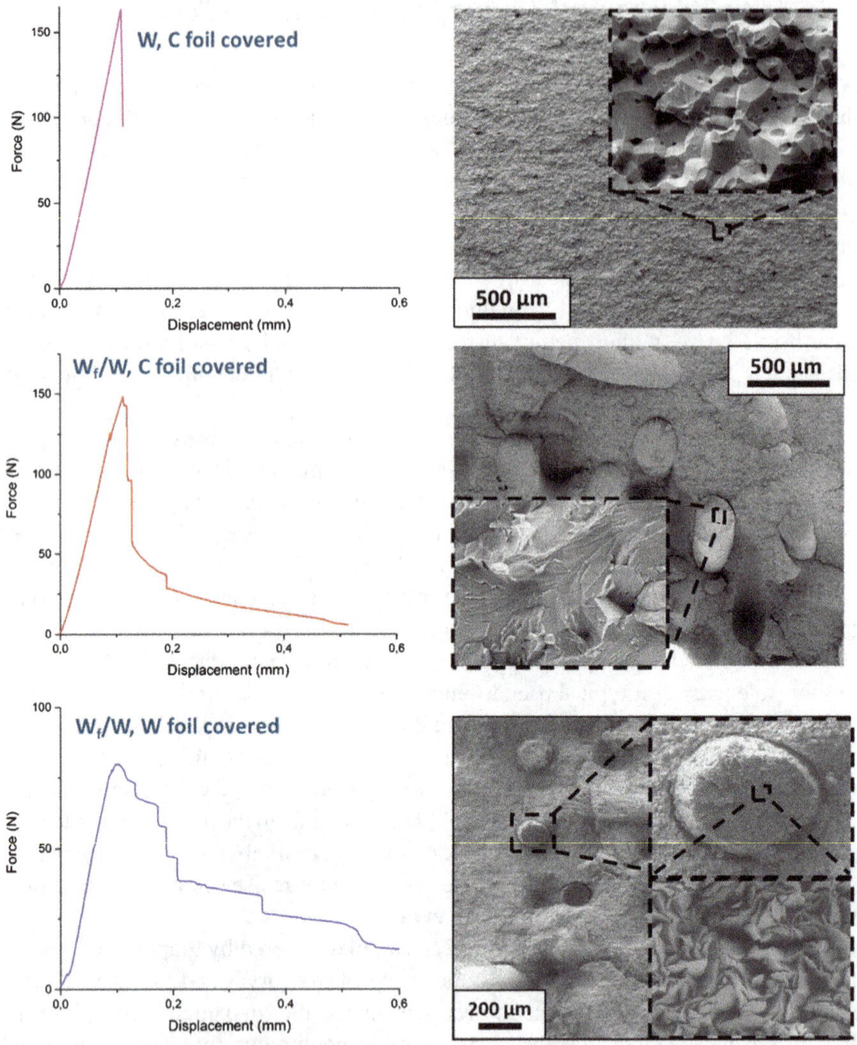

Fig. 9 Force-displacement curves of the pre-notched 3-point bending tests and SEM images of the corresponding fracture sections after the bending tests for pure tungsten, W_f/W produced with graphite foil, and W_f/W produced with W foil

degradation of the mechanical properties of the fiber. It has been reported in (Zhao et al. 2017) that the recrystallized W-fibers have a tensile strength around 900 MPa, while the initial fiber tensile strength is around 3000 MPa.

In order to minimize recrystallization and grain growth, potassium-doped fibers are already used in W_f/W. From previous studies, the potassium bubbles can reduce grain size after high-temperature annealing. Indeed, the W-fiber remains ductile to some extent after annealing at 2373 K for 30 min (Riesch et al. 2016c). However, even though the sintering temperatures of SPS processes are lower than 2373 K, the graphite-covered W_f/W still show brittle fibers after sintering. This can be distinguished by the fracture surface after mechanical testing (Fig. 9). It means the high temperature is not the dominant reason of W-fiber embrittlement. Rather the C contamination could play a key role here.

Earlier literature has reported that tungsten is rather sensitive to carbon contamination (Stephens 1964). Only very small amounts of carbon can significantly increase the DBTT temperature of tungsten. Some previous experiments have proven that the carbon has major influence on the W-fiber as well (Coenen et al. 2018). Based on the study by Müller et al. (2018), when doing annealing in a carbon-contaminated atmosphere, the W-fiber can be easily embrittled at even low temperature (lower than 1573 K). With the same annealing temperature, without carbon atmosphere, the fiber will remain ductile. During the SPS process, conventionally, graphite foils are used. However, in this case, carbon can diffuse into the sample during sintering, as they are in direct contact. To reduce the carbon contamination during the SPS process, W foil is used to separate the graphite tool material and the sample. In this way, the carbon embrittlement is reduced as one can see from the fracture section. This fiber property difference explains the different shape of the two bending curves. During the bending tests, after the maximum load, the fracture of the ductile fiber needs more energy and can allow higher displacement, so the slope of the load decrease is much smoother than in the brittle case.

To confirm this fiber embrittlement is caused by the carbon impurities, a transmission electron microscope (TEM) analysis on W-fibers after sintering is performed. Figure 10 shows the TEM results with deflection patterns for the sample produced with a graphite foil. It can be seen that many nano-sized carbides are (WC_{1-x}, W_2C) formed in the W-fiber. These particles cannot be detected in the W-fiber when the sample is protected by a tungsten foil during production. This is a clear proof that carbides can be formed during the SPS process in tungsten material if not protected.

To make sure this different fracture behavior is not caused by the grain structure difference, the W-fiber grain structure is investigated, as shown in Fig. 11. The fiber cross sections are prepared by focus ion beam (FIB) from the center of the fiber. And each section is parallel to the fibers longitudinal direction.

As shown in Fig. 11, the fiber grain size gets much larger after the high-temperature processes. The sample produced with graphite foil does however not have larger or different grain size based on this figure and the grain size measurement in (Coenen et al. 2018). This suggests again the conclusion that carbon impurities likely cause the tungsten fiber embrittlement. By using a tungsten foil, this fiber property degradation can be reduced.

3.4 W-Fiber-Reinforced Self-Passivating Alloy

For the SPS-produced W-fiber-reinforced self-passivating alloy, production parameters and sample density are shown in Table 1. The overview microstructure and the fiber/matrix interface details are shown in Fig. 12. The microstructure is quite similar

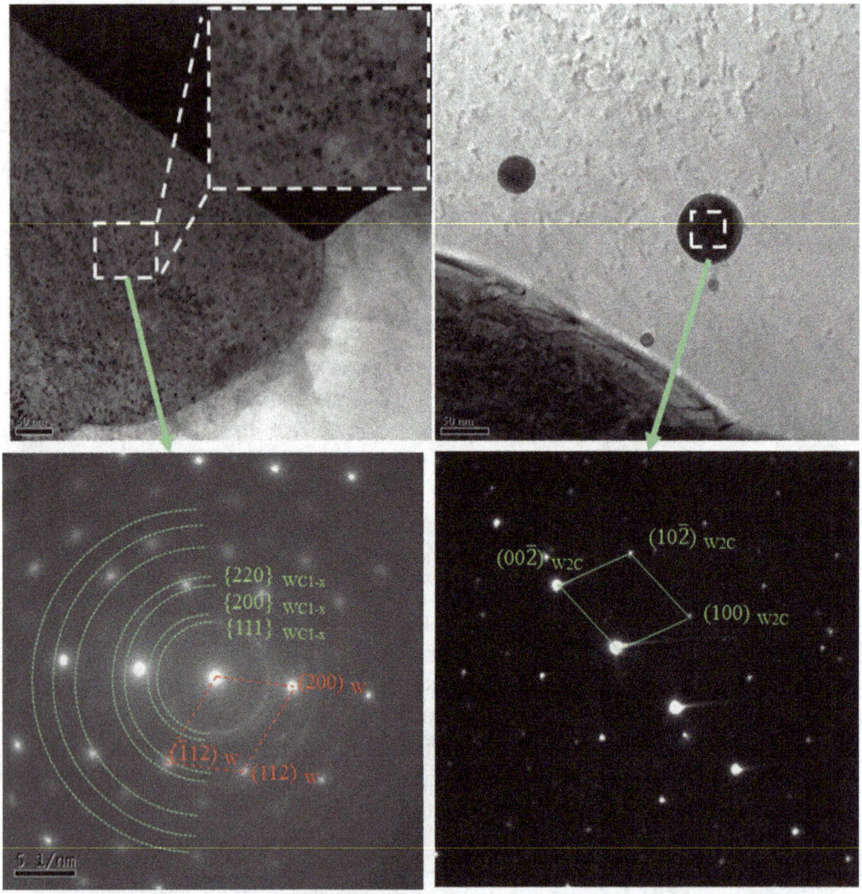

Fig. 10 TEM analysis on a W-fiber after sintering coved by graphite foil. Two types of nano-sized carbides can be detected in the tungsten grains: $WC_{1\text{-}x}$ and W_2C

to the conventional W_f/W material. The randomly distributed W-fibers and yttria interface between fiber and matrix are visible. However, it should be noticed that the interface between the fiber and matrix seems to be more intact with less outer shell damage. This is due to the lower sintering temperature during the SPS process.

Figure 13 gives the EDX line scan analysis across the yttria interface. A Y and O signal jump can be detected showing the yttria interface. Noticeably, the Cr signal can be detected even in the W-fiber. This indicates that during the sintering, the Cr alloy can penetrate through the yttria interface and diffuse into the pure W-fibers. But this could also be an effect of the sample polishing that some matrix material is scratched into the fiber surface area.

After the sintering, the sample density is measured. The results are shown in Table 4. The SPS-sintered smart alloy density can reach more than 98%. Compared

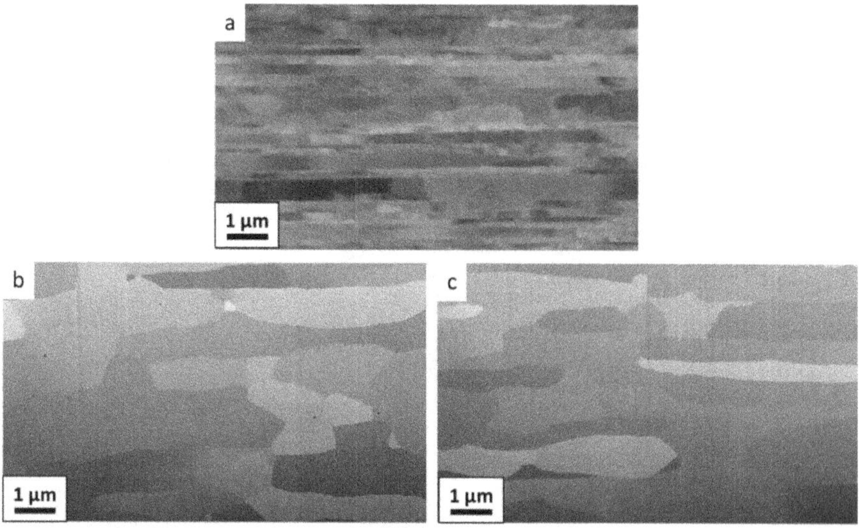

Fig. 11 Fiber microstructure after different sintering processes: (**a**) as-fabricated fiber, (**b**) fiber after SPS process with tungsten foil at 2173 K, (**c**) fiber after SPS process with graphite foil at 2173 K

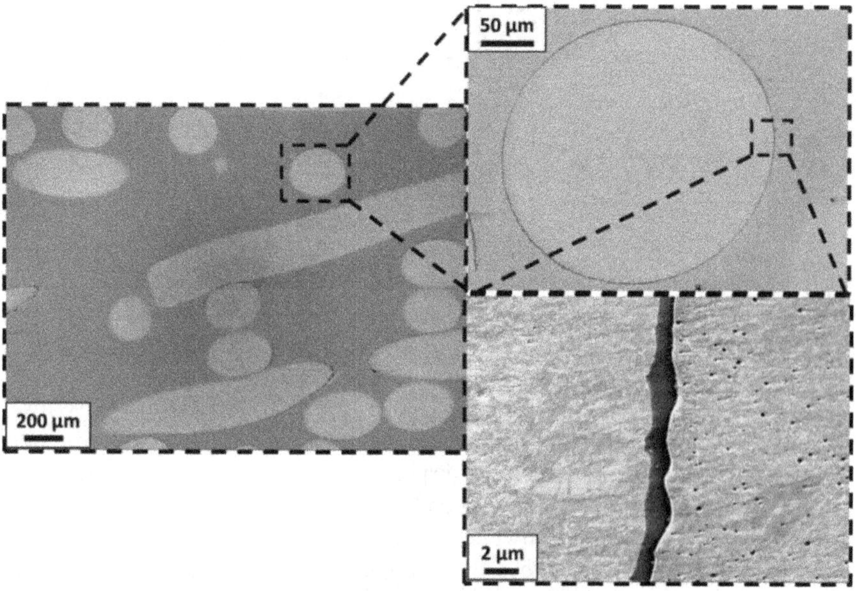

Fig. 12 Overview on the microstructure and the fiber/matrix interface detail of the W-fiber (30% fiber volume fraction) reinforced smart alloy with 2.5 μm Yttria fiber/matrix interface

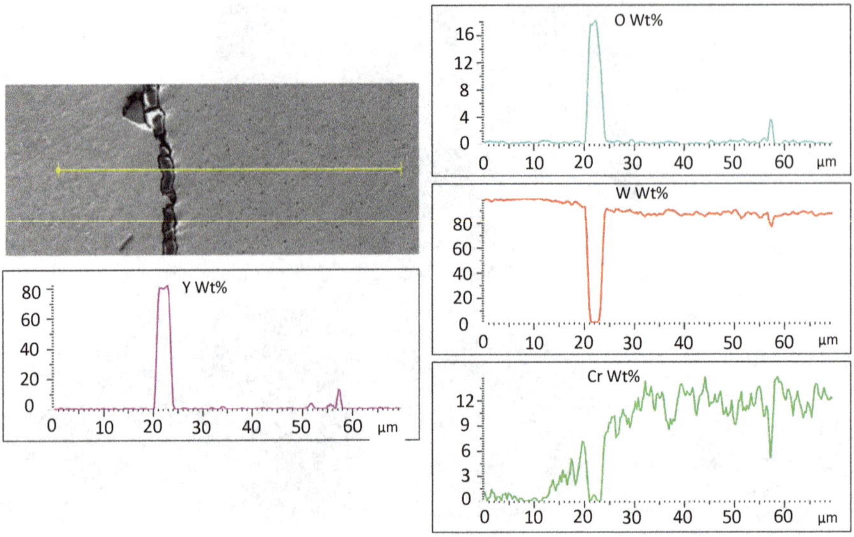

Fig. 13 EDX line scan analysis across the yttria interface for W-fiber-reinforced self-passivating alloy

Table 4 Sintering parameters and relative density of W-fiber-reinforced self-passivating alloy

Sample	Production process	Sintering temperature (K)	Holding time (min)	Heating rate (K/min)	Sintering pressure (MPa)	Relative density
W_f/W smart alloy	SPS with C foil	2073	4	200	60	98.6
Smart alloy	SPS with C foil	1873	4	200	60	98.9

to pure tungsten, the low melting point alloying elements, like Cr, can behave as a binding material; the sintering efficiency can thus be significantly increased (Litnovsky et al. 2017b). Even with lower sintering temperature, the relative density after sintering is higher than that of pure tungsten samples.

To investigate that this reinforcement principle also works also for W-fiber-reinforced W-alloy, the 3-point bending test on notched samples is performed similar to the pure W and W_f/W. The force-displacement curves and the corresponding fracture sections are shown in Figs. 14 and 15, respectively.

Similar to the pure W_f/W, after the elastic part, a load drop can be identified, indicating matrix cracking. Further load-bearing capability is established relying on the W-fiber. During the fiber pullout stage, the material cannot dissipate much energy. Here, a sudden load drop happens after reaching the maximum load. It can be assumed that this difference is caused by the different fiber-matrix interface strength compared to normal W_f/W. After the sintering, the interface is more intact

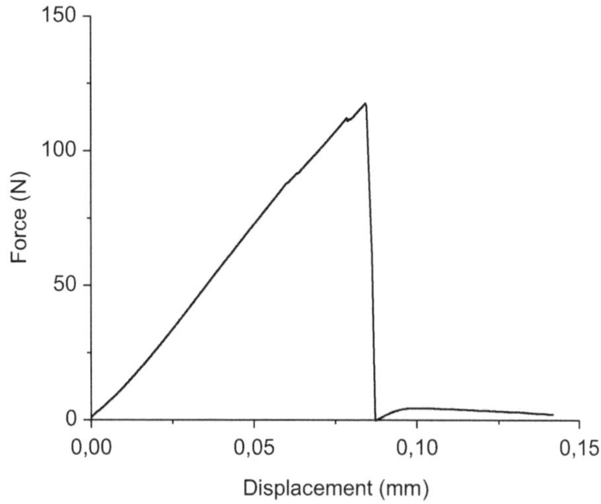

Fig. 14 Typical load-displacement curves of SPS-produced W-fiber-reinforced self-passivating smart alloy

Fig. 15 SEM image of the fracture sections after the pre-notched 3-point bending tests of SPS-produced W-fiber-reinforced self-passivating smart alloy

for W-fiber-reinforced self-passivating alloy. However, with a more damaged fiber/matrix interface, the mechanical bonding/friction between the fiber and matrix can be higher due to the residual stress. This leads to a higher interface strength compared to a clean yttria interface layer (Jasper 2017; Jasper et al. 2015; Schönen et al. 2018). During the pullout stage, a higher strength interface can dissipate more energy (fiber-matrix bonding and friction is higher), so one ends up with a more continuous load decrease compared to the alloy sample with a relatively low interface strength. From the fracture surface (Fig. 15), several exposed fiber ends are visible. This phenomenon indicates a very weak fiber-matrix interface. Before the matrix failure, the interface at the fiber ends is already failing due to a high shear stress state (Mao et al. 2018). This issue could potentially be solved by reducing the interface thickness to increase the interface strength.

To sum up, W-fiber-reinforced self-passivating tungsten alloy has been produced by the given SPS process with ~98% relative density. The microstructure and fracture behavior are quite similar to that of normal W_f/W material. The basic principle of utilizing alloys via the SPS routes for W_f/W production is established. In the future, the sintering parameters could be further optimized based on the specific alloy condition to facilitate better composite properties. However, one challenge that remains is that the production of smart alloy fibers is very difficult because of the brittleness of the smart alloy causing fiber failure during the drawing process of the fiber production.

4 Summary and Outlook

Multi short-fiber W_f/W can be produced by a PM-SPS process. The samples tend to have a two-dimensional planar fiber orientation, caused by the uniaxial deformation during the consolidation process. During the sintering process, the isolating yttria interface with lower melting point and low strength is damaged, especially when studying the outer shell, due to the impact of elevated temperature and external pressure. By increasing the coating thickness, this damage can be compensated, so that fiber and matrix are efficiently separated. After the standard sintering procedure, the W-fiber grain structure changes, and their strength decreases, obviously due to the carbon contamination during the SPS process. The carbon embrittlement can be reduced by using a tungsten foil during the SPS process. With current tungsten powders (mixture of 5 μm and >50 μm particles), the optimized sample density after sintering is ~94%. With initial mechanical testing, the SPS-produced W_f/W reveals a typical pseudo-ductile behavior. To establish the availability of using of W-alloys as matrix during the SPS process, W-fiber-reinforced self-passivating tungsten alloy has been produced by the given SPS process with ~98% relative density. The microstructure and fracture behavior are quite similar to that of normal W_f/W material.

Based on these results, some ideas can be considered to further improve the sintering density. This could be potentially realized by using different tungsten powders with finer particle size or using optimized powder mixture with various particle sizes On the other hand, die material or structure (Sokol et al. 2014) could be changed for higher sintering pressure which may also be helpful to increase the sintering efficiency.

Acknowledgments This work has been carried out within the framework of the EUROfusion Consortium and has received funding from the Euratom research and training program 2014–2018 under grant agreement No. 633053. The views and opinions expressed herein do not necessarily reflect those of the European Commission.

Reference

Allahar KN, Butt DP, Webb J, Charit I (2011) Electrochemical properties of spark plasma sintered tungsten. In: NACE – international corrosion conference series

Arzt E, Ashby MF, Easterling KE (1983) Practical applications of hotisostatic pressing diagrams: four case studies. Metall Trans A 14(1):211–221

Bonifacio CS, Holland TB, van Benthem K (2013) Evidence of surface cleaning during electric field assisted sintering. Scr Mater 69(11–12):769–772

Bonifacio CS, Holland TB, Van Benthem K (2014) Time-dependent dielectric breakdown of surface oxides during electric-field-assisted sintering. Acta Mater 63:140–149

Briant CL, Horacsek O, Horacsek K (1993) The effect of wire history on the coarsened substructure and secondary recrystallization of doped tungsten. Metall Trans A 24(4):843–851

Chanthapan S, Kulkarni A, Singh J, Haines C, Kapoor D (2012) Sintering of tungsten powder with and without tungsten carbide additive by field assisted sintering technology. Int J Refract Met Hard Mater 31:114–120

Chawla KK (1998) Ceramic matrix composites. In: Composite materials. Springer, New York, pp 212–251

Cho KC, Woodman RH, Klotz BR, Dowding RJ (2004) Plasma pressure compaction of tungsten powders. Mater Manuf Process 19(4):619–630

Coenen JW, Mao Y, Almanstotter J, Calvo A, Sistla S, Gietl H, Jasper B, Riesch J, Rieth M, Pintsuk G, Klein F, Litnovsky A, Mueller AV, Wegener T, You JH, Broeckmann C, Garcia-Rosales C, Neu R, Linsmeier C (2017) Advanced materials for a damage resilient divertor concept for DEMO: powder-metallurgical tungsten-fibre reinforced tungsten. Fusion Eng Des 124:964–968

Coenen JW, Mao Y, Sistla S, Riesch J, Hoeschen T, Broeckmann C, Neu R, Linsmeier C (2018) Improved pseudo-ductile behavior of powder metallurgical tungsten short fiber-reinforced tungsten (Wf/W). Nucl Mater Energy 15:214–219

Czél G, Wisnom MR (2013) Demonstration of pseudo-ductility in high performance glass/epoxy composites by hybridisation with thin-ply carbon prepreg. Compos A: Appl Sci Manuf 52:23–30

Du J, Höschen T, Rasinski M, Wurster S, Grosinger W, You JH (2010a) Feasibility study of a tungsten wire-reinforced tungsten matrix composite with ZrOx interfacial coatings. Compos Sci Technol 70(10):1482–1489

Du J, Höschen T, Rasinski M, You JH (2010b) Interfacial fracture behavior of tungsten wire/ tungsten matrix composites with copper-coated interfaces. Mater Sci Eng A 527(6):1623–1629

Du J, Höschen T, Rasinski M, You JH (2011) Shear debonding behavior of a carbon-coated interface in a tungsten fiber-reinforced tungsten matrix composite. J Nucl Mater 417 (1–3):472–476

Giannattasio A, Roberts SG (2007) Strain-rate dependence of the brittle-to-ductile transition temperature in tungsten. Philos Mag 87(17):2589–2598

Gietl H, Riesch J, Coenen JW, Hoschen T, Linsmeier C, Neu R (2017) Tensile deformation behavior of tungsten fibre-reinforced tungsten composite specimens in as-fabricated state. Fusion Eng Des 124:396–400

Gietl H, Müller A v, Coenen JW, Decius M, Ewert D, Höschen T, Huber P, Milwich M, Riesch J, Neu R (2018) Textile preforms for tungsten fibre-reinforced composites. J Compos Mater:002199831877114

Guillon O, Gonzalez-Julian J, Dargatz B, Kessel T, Schierning G, Räthel J, Herrmann M (2014) Field-assisted sintering technology/spark plasma sintering: mechanisms, materials, and technology developments. Adv Eng Mater 16(7):830–849

Jasper B (2017) Einzelfaserkomposite aus Pulvermetallurgischem Wolfram-faserverstaerktem Wolfram. PhD thesis, Ruhr-Universitaet Bochum

Jasper B, Coenen JW, Riesch J, Höschen T, Bram M, Linsmeier C (2015) Powder metallurgical tungsten fiber-reinforced tungsten. Mater Sci Forum 825–826:125–133

Jasper B, Schoenen S, Du J, Hoeschen T, Koch F, Linsmeier C, Neu R, Riesch J, Terra A, Coenen JW (2016) Behavior of tungsten fiber-reinforced tungsten based on single fiber push-out study. Nucl Mater Energy 9:416–421

Kellogg FR, Cho KC, Klotz BR, Klier EM, Dowding RJ (2008) Plasma Pressure Compaction (P2C) of nanoscale tungsten powder. Paper read at proceedings of the 7th International Conference on Tungsten, Refractory and Hardmaterials

Kim Y, Lee KH, Kim E-P, Cheong D-I, Hong SH (2009) Fabrication of high temperature oxides dispersion strengthened tungsten composites by spark plasma sintering process. Int J Refract Met Hard Mater 27(5):842–846

Koch F, Bolt H (2007) Self-passivating W-based alloys as plasma facing material for nuclear fusion. Phys Scr 2007(T128):100–105

Li VC, Wang Y, Backer S (1991) A micromechanical model of tension-softening and bridging toughening of short random fiber reinforced brittle matrix composites. J Mech Phys Solids 39 (5):607–625

Litnovsky A, Wegener T, Klein F, Linsmeier C, Rasinski M, Kreter A, Tan X, Schmitz J, Coenen JW, Mao Y, Gonzalez-Julian J, Bram M (2017a) New oxidation-resistant tungsten alloys for use in the nuclear fusion reactors. Phys Scr 2017(T170):014012

Litnovsky A, Wegener T, Klein F, Linsmeier C, Rasinski M, Kreter A, Unterberg B, Vogel M, Kraus S, Breuer U, Garcia-Rosales C, Calvo A, Ordas N (2017b) Smart alloys for a future fusion power plant: first studies under stationary plasma load and in accidental conditions. Nucl Mater Energy 12:1363–1367

Mao Y, Coenen JW, Riesch J, Sistla S, Almanstötter J, Jasper B, Terra A, Höschen T, Gietl H, Bram M, Gonzalez-Julian J, Linsmeier C, Broeckmann C (2017a) Development and characterization of powder metallurgically produced discontinuous tungsten fiber reinforced tungsten composites. Phys Scr 2017(T170):014005

Mao Y, Engels J, Houben A, Rasinski M, Steffens J, Terra A, Linsmeier C, Coenen JW (2017b) The influence of annealing on yttrium oxide thin film deposited by reactive magnetron sputtering: process and microstructure. Nucl Mater Energy 10:1–8

Mao Y, Coenen JW, Riesch J, Sistla S, Almanstotter J, Jasper B, Terra A, Hoschen T, Gietl H, Linsmeier C, Broeckmann C (2018) Influence of the interface strength on the mechanical properties of discontinuous tungsten fiber-reinforced tungsten composites produced by field assisted sintering technology. Compos Part A Appl Sci Manuf 107:342–353

Ming-Yuan H, Hutchinson JW (1989) Crack deflection at an interface between dissimilar elastic materials. Int J Solids Struct 25(9):1053–1067

Müller A v, Ilg M, Gietl H, Höschen T, Neu R, Pintsuk G, Riesch J, Siefken U, You JH (2018) The effects of heat treatment at temperatures of 1100°C to 1300°C on the tensile properties of high-strength drawn tungsten fibres. Nucl Mater Energy 16:163–167

Neu R, Riesch J, Coenen JW, Brinkmann J, Calvo A, Elgeti S, García-Rosales C, Greuner H, Hoeschen T, Holzner G, Klein F, Koch F, Linsmeier C, Litnovsky A, Wegener T, Wurster S, You JH (2016) Advanced tungsten materials for plasma-facing components of DEMO and fusion power plants. Fusion Eng Des 109–111:1046–1052

Ohser-Wiedemann R, Martin U, Müller A (2012) Fast sintering of mechanical alloyed Mo-W powders. Paper read at proceedings of the international Euro Powder Metallurgy Congress and Exhibition, Euro PM 2012

Orrù R, Licheri R, Locci AM, Cincotti A, Cao G (2009) Consolidation/synthesis of materials by electric current activated/assisted sintering. Mater Sci Eng R Rep 63(4–6):127–287

Reiser J, Rieth M, Dafferner B, Hoffmann A (2012) Tungsten foil laminate for structural divertor applications – basics and outlook. J Nucl Mater 423(1–3):1–8

Riesch J, Buffiere JY, Höschen T, di Michiel M, Scheel M, Linsmeier C, You JH (2013) In situ synchrotron tomography estimation of toughening effect by semi-ductile fibre reinforcement in a tungsten-fibre-reinforced tungsten composite system. Acta Mater 61(19):7060–7071

Riesch J, Hoschen T, Linsmeier C, Wurster S, You JH (2014a) Enhanced toughness and stable crack propagation in a novel tungsten fibre-reinforced tungsten composite produced by chemical vapour infiltration. Phys Scr T159:014031

Riesch J, Höschen T, Linsmeier C, Wurster S, You JH (2014b) Enhanced toughness and stable crack propagation in a novel tungsten fibre-reinforced tungsten composite produced by chemical vapour infiltration. Paper read at Physica Scripta

Riesch J, Almanstötter J, Coenen JW, Fuhr M, Gietl H, Han Y, Höschen T, Linsmeier C, Travitzky N, Zhao P, Neu R (2016a) Properties of drawn W wire used as high performance fibre in tungsten fibre-reinforced tungsten composite. IOP Conf Ser Mater Sci Eng 139:012043

Riesch J, Aumann M, Coenen JW, Gietl H, Holzner G, Höschen T, Huber P, Li M, Linsmeier C, Neu R (2016b) Chemically deposited tungsten fibre-reinforced tungsten – the way to a mock-up for divertor applications. Nucl Mater Energy 9:75–83

Riesch J, Han Y, Almanstotter J, Coenen JW, Hoschen T, Jasper B, Zhao P, Linsmeier C, Neu R (2016c) Development of tungsten fibre-reinforced tungsten composites towards their use in DEMO-potassium doped tungsten wire. Phys Scr 2016(T167):014006

Riesch J, Han Y, Almanstötter J, Coenen JW, Höschen T, Jasper B, Zhao P, Linsmeier C, Neu R (2016d) Development of tungsten fibre-reinforced tungsten composites towards their use in DEMO – potassium doped tungsten wire. Paper read at Physica Scripta

Rieth M, Dudarev SL, Gonzalez de Vicente SM, Aktaa J, Ahlgren T, Antusch S, Armstrong DEJ, Balden M, Baluc N, Barthe MF, Basuki WW, Battabyal M, Becquart CS, Blagoeva D, Boldyryeva H, Brinkmann J, Celino M, Ciupinski L, Correia JB, De Backer A, Domain C, Gaganidze E, García-Rosales C, Gibson J, Gilbert MR, Giusepponi S, Gludovatz B, Greuner H, Heinola K, Höschen T, Hoffmann A, Holstein N, Koch F, Krauss W, Li H, Lindig S, Linke J, Linsmeier C, López-Ruiz P, Maier H, Matejicek J, Mishra TP, Muhammed M, Muñoz A, Muzyk M, Nordlund K, Nguyen-Manh D, Opschoor J, Ordás N, Palacios T, Pintsuk G, Pippan R, Reiser J, Riesch J, Roberts SG, Romaner L, Rosiński M, Sanchez M, Schulmeyer W, Traxler H, Ureña A, van der Laan JG, Veleva L, Wahlberg S, Walter M, Weber T, Weitkamp T, Wurster S, Yar MA, You JH, Zivelonghi A (2013) Recent progress in research on tungsten materials for nuclear fusion applications in Europe. J Nucl Mater 432 (1–3):482–500

Rupp D, Weygand SM (2011) Loading rate dependence of the fracture toughness of polycrystalline tungsten. J Nucl Mater 417(1):477–480

Schönen S, Jasper B, Coenen JW, Du J, Höschen T, Riesch J, Natour G, Neu R, Linsmeier C (2018) Insight into single-fiber push-out test of tungsten fiber-reinforced tungsten. Compos Interfaces:1–20

Smid I, Akiba M, Vieider G, Plöchl L (1998) Development of tungsten armor and bonding to copper for plasma-interactive components. J Nucl Mater 258–263(Part 1):160–172

Sokol M, Kalabukhov S, Dariel MP, Frage N (2014) High-pressure spark plasma sintering (SPS) of transparent polycrystalline magnesium aluminate spinel (PMAS). J Eur Ceram Soc 34 (16):4305–4310

Stang H, Shah SP (1986) Failure of fibre-reinforced composites by pull-out fracture. J Mater Sci 21 (3):953–957

Stephens JR (1964) Effects of Interstitial Impurities on the Low-Temperature Tensile Properties of Tungsten, Technical report, National Aeronautics and Space Administration Cleveland OH Lewis Research Center, 1964. www.dtic.mil/dtic/tr/fulltext/u2/a396979.pdf

Wegener T, Klein F, Litnovsky A, Rasinski M, Brinkmann J, Koch F, Linsmeier C (2016) Development of yttrium-containing self-passivating tungsten alloys for future fusion power plants. Nucl Mater Energy 9:394–398

Wegener T, Klein F, Litnovsky A, Rasinski M, Brinkmann J, Koch F, Linsmeier C (2017) Development and analyses of self-passivating tungsten alloys for DEMO accidental conditions. Fusion Eng Des 124:183–186

Zhao P, Riesch J, Hoschen T, Almanstotter J, Balden M, Coenen JW, Himml R, Pantleon W, von Toussaint U, Neu R (2017) Microstructure, mechanical behaviour and fracture of pure tungsten wire after different heat treatments. Int J Refract Met Hard Mater 68:29–40

Part III
SPS of Hard Metals

Microstructural Evolution and Related Kinetics During Pulsed Electric Current Sintering of Tungsten

Jonathan A. Webb, Indrajit Charit, and Darryl P. Butt

1 Introduction

Tungsten is a refractory metal with the highest melting temperature (3667 K $\pm$ 19 K) of any known metals (Baum et al. 2002; Lassner and Schubert 1999; Storms 1967). Tungsten has been considered as a matrix material for refractory fission fuels where tungsten would encapsulate a ceramic nuclear fuel such as uranium dioxide (UO_2) (Farrell et al. 1967; Stephens 1963; Webb and Charit 2011, 2012; Argonne National Laboratory 1968; General Electric 1974a, b). Tungsten has also been considered as a plasma facing first wall liner for experimental fusion reactors (Kaufmann and Neu 2007; Mitteau et al. 2007). While in a general sense tungsten offers a superior high-temperature strength compared to most other metals, proper microstructural engineering can enhance the tensile and creep strength while decreasing the DBTT of pure tungsten (Luo et al. 1992; Klopp and Raffo 1964).

Appropriate processing techniques can optimize the microstructure by producing fine-grained tungsten with increased strength. One novel approach to fabricating tungsten-based materials is pulsed electric current sintering (PECS). While PECS methods are similar to hot (uniaxial) pressing (HP), a material can be consolidated at a significantly lower temperature in PECS than those by HP (Perkins 2011; Luke 2010). The decrease in sintering temperature has been shown to reduce the grain size of the sintered material, which would increase the strength of the tungsten specimens. However, some PECS studies with Al_2O_3, $LaNbO_4$, and ZrO_2 have suggested that the grain size is a function of relative density and is no different between HP and

J. A. Webb · I. Charit (✉)
Nuclear Engineering Program and Department of Chemical & Materials Engineering, University of Idaho, Moscow, ID, USA
e-mail: icharit@uidaho.edu

D. P. Butt
Department of Metallurgical Engineering, University of Utah, Salt Lake City, UT, USA
e-mail: darryl.butt@utah.edu

© Springer Nature Switzerland AG 2019

P. Cavaliere (ed.), *Spark Plasma Sintering of Materials*,
https://doi.org/10.1007/978-3-030-05327-7_10

PECS (Langer et al. 2009; Syvertsen et al. 2012; Bernard-Granger et al. 2010). Regardless, PECS has been shown to have a particle cleansing effect by which the oxide barriers on the surface of the particles are removed during early-stage sintering, leading to less oxide formation and a stronger material.

Two theories currently exist which might explain the very rapid low-temperature densification of metals and ceramic powders consolidated by PECS techniques. The first theory has come to be known as the *plasma theory* and states that the current flowing through the materials forms a plasma within the interparticle gaps, which removes the oxide barriers from the particles (Anderson et al. 1999; Xie et al. 2005). The suspected existence of this plasma is why the PECS technique is also frequently referred to as spark plasma sintering (SPS). If this plasma exists, it is also hypothesized that the plasma deposits material in the necking joints and allows for rapid densification (Munir et al. 2006; Song et al. 2006). However, the existence of a plasma during PECS has recently come into questions due to the lack of experimental validation and the lack of appropriate current and an inadequate spark-gap dimension (Hulbert et al. 2008, 2009). Another theory has been proposed to explain the rapid densification process during PECS and has become known as the *field theory* (Olevsky and Froyen 2006, 2009; Olevsky et al. 2007). The field theory model predicts that early-stage sintering in PECS is accompanied by a very high localized temperature at the interparticle contact points that are caused by the high current densities at those locations. The heat is predicted to diffuse around the particle periphery, causing the oxide barriers surrounding the particles to be volatilized. The highly localized temperatures cause a very brief period of plastic flow that crushes the particles together, skipping the first stage of sintering in which necking joints between particles traditionally evolve. The high temperatures also lead to enhanced densification process by increasing the diffusional and dislocation-based creep rates, which is known as the Ludwig-Soret effect (Olevsky and Froyen 2009).

The work presented herein attempts to characterize the effects of the PECS process on the sintered microstructure of tungsten. This work also attempts to characterize the mechanisms present during the PECS process that cause consolidation of the tungsten powders.

2 Analytical Framework

Both Coble and Nabarro-Herring creep are diffusion-based creep mechanisms where vacancies diffuse either through the grain boundaries (Coble creep) or the grain interior (Nabarro-Herring creep). The mathematical relationships that describe Coble and Nabarro-Herring creep as applicable in the sintering of porous materials are given in Eqs. 1 and 2, respectively (Coble 1970), where Ω is the atomic volume in units of m^{-3}, ρ represents density in percentage of the theoretical density, t is time, D_b and D_l represent the grain boundary and lattice diffusion coefficients, G represents the average grain size, k represents Boltzmann's constant, σ represents the applied pressure, γ represents surface energy, R represents the gas constant of

8.314 J/K.mol, and r represents the average pore radius of curvature. Finally, the variable X is a constant that is 47.5 during intermediate-stage sintering (before pore-boundary separation) and 7.5 during final-stage sintering (after pore-boundary separation) (Olevsky and Froyen 2006). The term c is also a constant that is generally considered to be 1 before the final stage of sintering and 2 during the final stage of sintering. It should also be noticed that the left-hand side of Eqs. 1 and 2 ($1/\rho \cdot d\rho/dt$) also represents a creep rate ($\dot{\varepsilon}$) in units of s^{-1}.

$$\dot{\varepsilon} = \frac{1}{\rho}\frac{d\rho}{dt} = X\frac{\Omega\, bD_b^0 \exp\left(-\frac{Q_b}{RT}\right)}{G^3 kT}\left(\frac{\sigma}{\rho} + c\frac{\gamma}{r}\right) \quad (1)$$

$$\dot{\varepsilon} = \frac{1}{\rho}\frac{d\rho}{dt} = \frac{40}{3}\frac{\Omega\, D_l^0 \exp\left(-\frac{Q_l}{RT}\right)}{G^2 kT}\left(\frac{\sigma}{\rho} + c\frac{\gamma}{r}\right) \quad (2)$$

Power-law creep (PLC) is a dislocation-based deformation mechanism, which can be divided into two stages. At high values of modulus-compensated stress ($\sim\sigma/E > 0.15$), deformation occurs by dislocation gliding with a stress exponent (n) around 7, while at lower values of modulus-compensated stress, deformation occurs by dislocation climb with a stress exponent of 5. Most metals have been found to have a power-law creep rate that is proportional to the inverse grain size up to ~100 μm at which point slip mechanisms become inhibited and the creep rate becomes independent of grain size above this grain size (Kassner and Perez-Prado 2009). However, tungsten seems to exhibit a different behavior in which the creep rate seems to be proportional to the square of the grain size (Klopp and Raffo 1964; Klopp et al. 1964; Klopp and Witzke 1966; King and Sell 1965). The point at which tungsten seems to transition from 5 to 7 PLC region is at a modulus-compensated stress level of ~0.16, and the relationship for 5 PLC and 7 PLC in polycrystalline tungsten has been empirically correlated by the authors of this study and is shown in Eq. 3. At σ/E values below 0.16, the values of A and n are $1.52 \times 10^{-12}\ s^{-1}$ and 4.22, respectively, whereas at σ/E values above 0.16, the values of A and n are $5.54 \times 10^{-10}\ s^{-1}$ and 7.44, respectively (Harrison and Ellison 1966; Conway and Flagella 1968; Robinson and Sherby 1969).

$$\dot{\varepsilon} = \frac{1}{\rho}\frac{d\rho}{dt} = A\left(\frac{G}{b}\right)^{1.75}\left(\frac{D_{\text{eff}} Eb}{kT}\right)\left(\frac{\sigma}{\rho E}\right)^n \quad (3)$$

Both dislocation-core diffusion and lattice diffusion are thought to be important for dislocation climb mechanisms that enable power-law creep. Dislocation-core diffusion is the dominant mechanism from 0.3 to $0.5T_m$, and lattice diffusion is the dominant climb mechanism from $0.5T_m$ and higher (Frost and Ashby 1983; Raj et al. 1995). The effective diffusion that accounts for both lattice and dislocation-core diffusion is shown in Eq. 4. It should be noted that the activation energy used for dislocation-core diffusion is almost the same as that of grain boundary diffusion, which is because the short-circuiting vacancies diffuse through dislocation cores.

$$D_{\mathrm{eff}} = D_{\mathrm{dc}}^{0}\exp\left(-\frac{Q_b}{RT}\right) + D_l^{0}\exp\left(-\frac{Q_l}{RT}\right) \tag{4}$$

The relationships for Coble, Nabarro-Herring, and power-law creep can be compared as a function of temperature to determine the regions of dominance for each type of creep on a plot of stress versus temperature. Equations 1, 2, 3, and 4 were solved with the input parameters from Table 1 to develop creep deformation

Table 1 Kinetic parameters for the development of creep deformation mechanism map

Parameter	Value	Reference
Ω	1.59×10^{-29} m^3	Frost and Ashby (1983), Raj et al. (1995)
b	2.74×10^{-4} m	Frost and Ashby (1983)
Q_b	385 kJ/Mol	Ashby (1974), German (1996)
D_b^0	1.09×10^{-3} m^2/s	Ashby (1974), German (1996)
Q_l	520 kJ/Mol	Krieder and Bruggeman (1967), Pawel and Lundy (1969)
D_l^0	8.00×10^{-6} m^2/s	
D_{dc}^0	9.22×10^{-9} m^2/s	Klopp and Witzke (1966), Coble (1970)

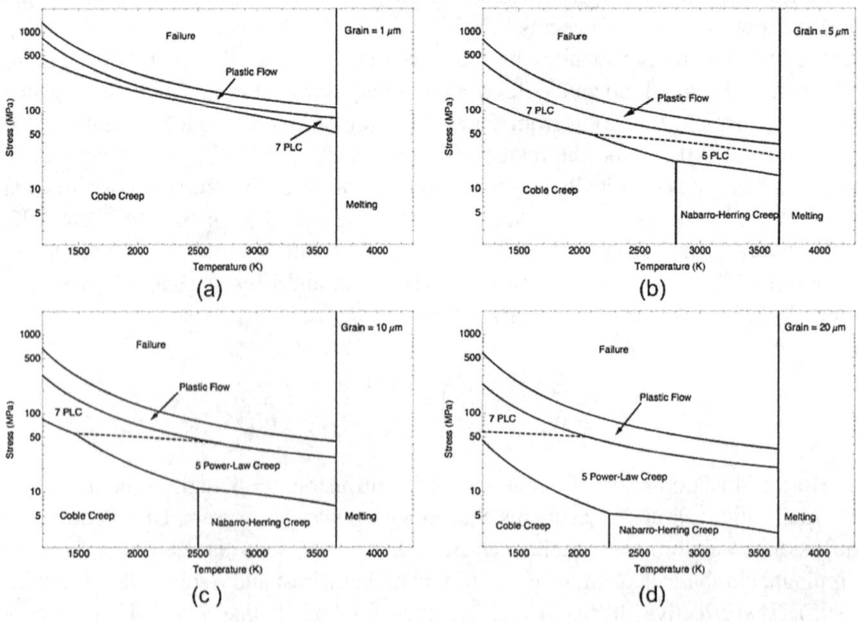

Fig. 1 Creep deformation mechanism maps for tungsten with grain sizes of **(a)** 1 μm, **(b)** 5 μm, **(c)** 10 μm, and **(d)** 20 μm

mechanisms for full-density tungsten at grain sizes of 1, 5, 10, and 20 μm which are shown in Fig. 1. While the data in Fig. 1 is for full-density and near-fully recrystallized tungsten, it can be used as a general indicator for the types of mechanisms that may be applicable to intermediate- and final-stage PECS assuming the use of tungsten feedstock in the correct grain sizes. Early-stage sintering may be accompanied by highly localized temperatures which make prior predictions of sintering mechanisms difficult via Eqs. 1, 2, 3, and 4.

3 Experimental Apparatus and Methods

A series of sintering experiments were conducted to compare the sintering kinetics and microstructures of tungsten particles consolidated via HP and PECS processes. Two separate batches of tungsten powders were acquired from Buffalo Tungsten Inc. in Depew, NY. Prior to each sintering experiment, the particles were reduced in a flowing 6%H_2–94%N_2 environment at 1123 K for 6 h. Each particle batch was analyzed via scanning electron microscopy (SEM) images and digital analysis of electron back-scatter diffraction (EBSD) scans for average particle and crystallite size prior to sintering. It was found that each particle batch consisted of mostly single crystals with few agglomerates; however, the average powder size among batches varied from 90 nm to 471 nm.

Three sets of PECS experiments were conducted on two different PECS units. Two sets of experiments were conducted on a Fuji Electronics Co. (formerly SPS Syntex) Dr. Sinter 515S unit housed at the Center for Advanced Energy Studies in Idaho Falls, ID, USA. Another series of pressure-based experiments were conducted in a Thermal Technology 25–10 PECS unit at the Thermal Technology LLC factory in Santa Rosa, CA, USA. Every PECS sintering experiment regardless of the unit was conducted with a 12 ms current pulse followed by 2 ms of no current.

Tungsten compacts consisting of 10 g of powder were consolidated in graphite dies that were also lined with graphite foil and in some cases graphite foil and niobium foil. A series of sintering experiments were also carried out with a Thermal Technology hot press, so comparisons could be made between HP and PECS techniques. Following PECS processing, the tungsten specimens were still not fully dense, which requires post-processing to annihilate the remaining porosity. A series of ten specimens were also manufactured in the Thermal Technology 25–10 PECS unit and were then further consolidated by hot isostatic pressing (HIP) at the American Isostatic Pressing facility in Columbus, OH, USA, in order to validate HIP techniques as a post-processing method.

Each die was also machined with a borehole at the midplane of the die that had a diameter of 1.6 mm. The borehole of each die was machined to extend within 4 mm of the inner annulus where the tungsten powders were poured. The borehole acted as a quasi-blackbody cavity to which a pyrometer was aimed to determine the sintering temperature. The pyrometers used in this study have calibrated accuracy to within 0.5% of the true temperature between measured temperatures of 873 K and 1873 K and 1.0% between 1873 K and 2273 K. The true temperatures in the specimen were estimated by one-dimensional heat transfer calculations with knowledge of the power input and the distance between the end of the pyrometer borehole and the specimen annulus (Perkins 2011; Pottlacher et al. 1993).

Following consolidation or post-processing, each specimen was ground using SiC grinding papers to a finish of 120 grit. After grinding, specimens were cut in half, one specimen was mounted and further polished to a 1200 grit with SiC grinding paper, and finally it was vibratory polished for 48 h in a 0.02 μm colloidal silica solution for microscopy analysis. The second half of the specimen was soaked in a vacuum pump oil and then held in a 7 kPa vacuum to close any open-pore structures. Following the vacuum treatment, the specimen was analyzed for density using the ASTM B-962-08 standard for Archimedes testing. Each specimen was analyzed via the EBSD technique to determine the average grain size, pore size, and area fraction of porosity.

4 Results and Discussion

4.1 Microstructural Evolution During PECS

Multiple specimens were produced using PECS to observe the microstructural characteristics and physical properties such as density and microhardness. The effects of powder particle size, sintering time, temperature, and applied stress on sintering kinetics and microstructure were assessed in this study.

4.1.1 Effects of Sintering Temperature on Microstructure

Two different sintering experiments were performed in the Dr. Sinter 515S PECS unit to determine the effects of sintering temperature and tungsten particle size on the consolidation of the tungsten powders. Seven tungsten specimens were sintered with an initial particle size of 471 nm at an applied stress of 45 MPa. Eleven more specimens were consolidated in the same PECS unit at 65 MPa. The specimens were heated at a rate of 100 K/min to a peak temperature where they were held at temperature for 2 min to allow for thermal equilibrium to be attained. The peak sintering temperatures ranged from 1098 K to 2096 K. The measured specimen densities and grain sizes are shown in Figs. 2 and 3, respectively. As shown in Fig. 2, the temperature-dependent densities of the specimens are nearly identical even though the particle sizes and applied stress were different. It is not until a sintering temperature of 1700 K that the density of the specimens with a particle size of 260 nm begins to have a higher density than those with 471 nm. The measurement uncertainties in the density were so small that they were not plotted in Fig. 2. Figure 3 shows that the specimens with the larger initial particle size have much larger grain sizes than the specimens with the smaller particle sizes.

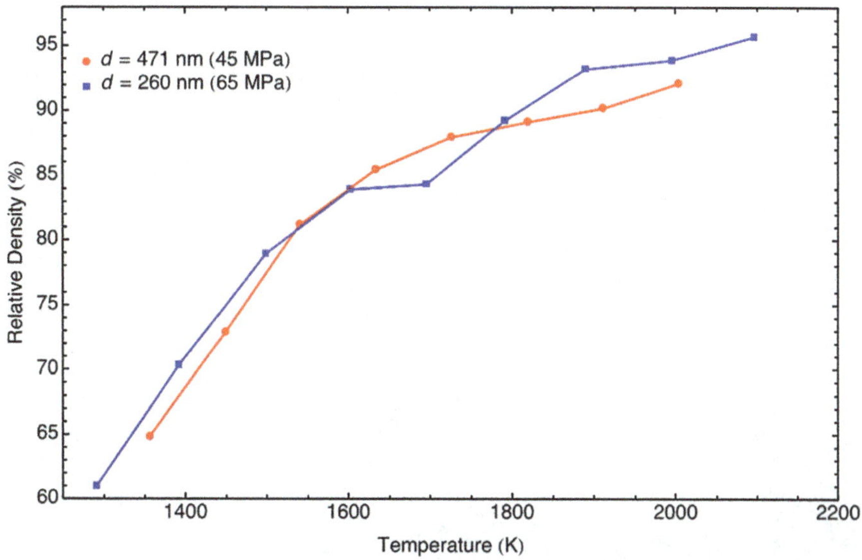

Fig. 2 Temperature-dependent density of specimens produced from tungsten powders with mean diameters of 260 nm and 471 nm at a heating rate of 100 K/min under applied stresses of 65 MPa and 45 MPa, respectively

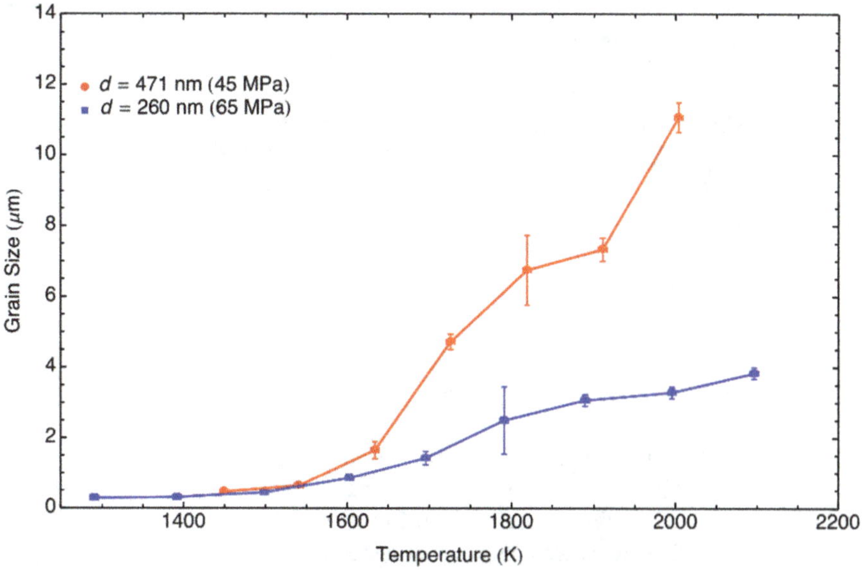

Fig. 3 Grain size as a function of temperature of specimens produced from tungsten particles with average diameters of 260 nm and 471 nm at a heating rate of 100 K/min under applied stresses of 65 MPa and 45 MPa, respectively

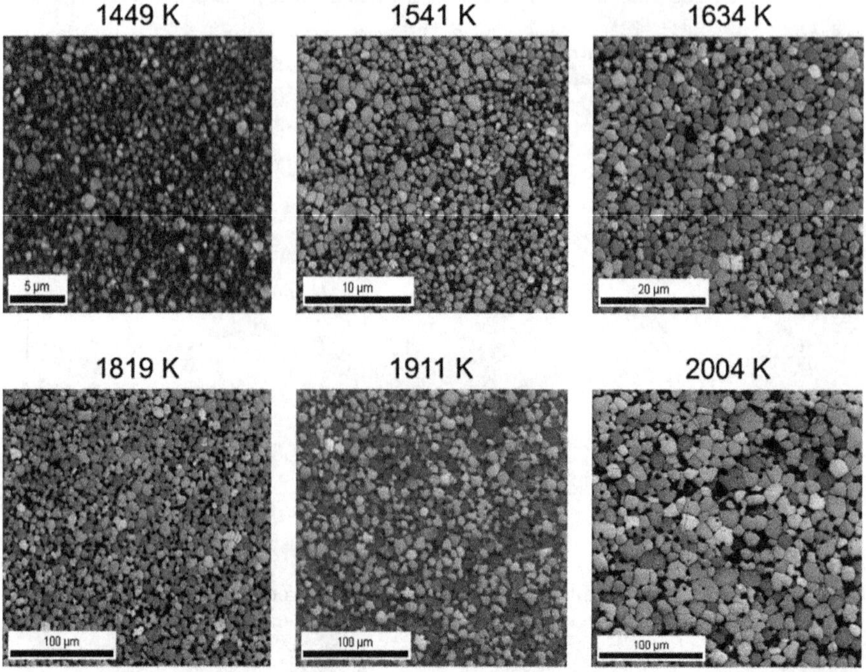

Fig. 4 Temperature-dependent microstructure produced from consolidation of 471 nm tungsten particles during PECS process with a heating rate of 100 K/min, applied stress of 45 MPa, and isothermal hold of 2 min

Figure 4 shows EBSD images as a function of temperature for specimens sintered from 471 nm tungsten particles between 1449 K and 2004 K at a heating rate of 100 K/min with an isothermal hold of 2 min. The first stages of sintering are characterized by particle coalescence and the beginnings of grain evolution. At about 1540 K, the interparticle gaps can be seen forming into cylindrical pore structures which are beginning to shrink and spheroidize by 2004 K.

4.1.2 Effects of Isothermal Hold Time

Tungsten powders with an initial crystallite size of 471 nm consolidated in the Dr. Sinter 515S PECS unit were isothermally sintered at temperatures of 1819 K, 1911 K, and 2004 K for times ranging from 5 to 30 min. The heating rate was 50 K/min, and the applied stress was 45 MPa. Another experiment was also conducted where tungsten particles with an initial grain size of 260 nm were isothermally sintered at 2096 K at soak times ranging from 2 to 26 min. The relative densities and average grain sizes of the specimens are shown in Figs. 5 and 6. The relative densities of the specimens fabricated at 1819 K, 1911 K, and 2004 K increase with temperature. The rate of densification is also shown to slightly increase as a

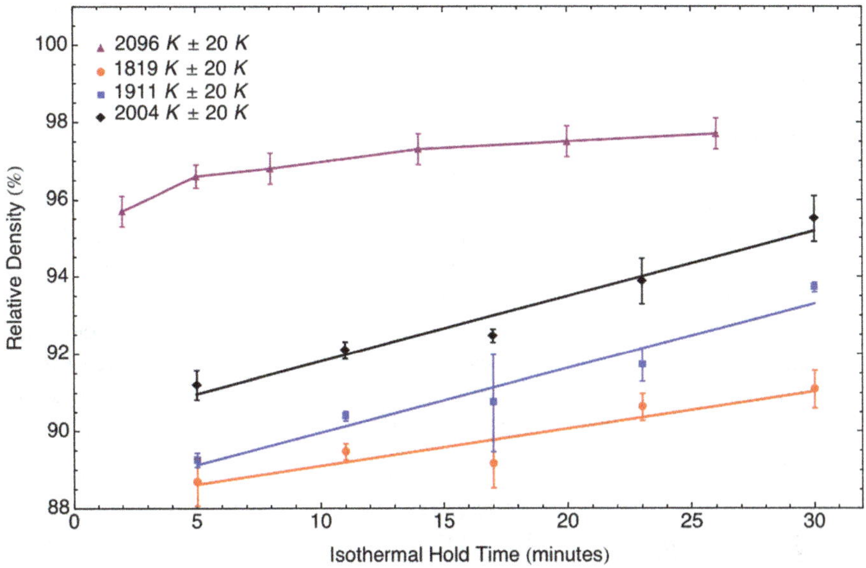

Fig. 5 Relative density of tungsten particles consolidated at temperatures of 1819 K, 1911 K, 2004 K, and 2096 K with isothermal hold times ranging from 2 to 30 min

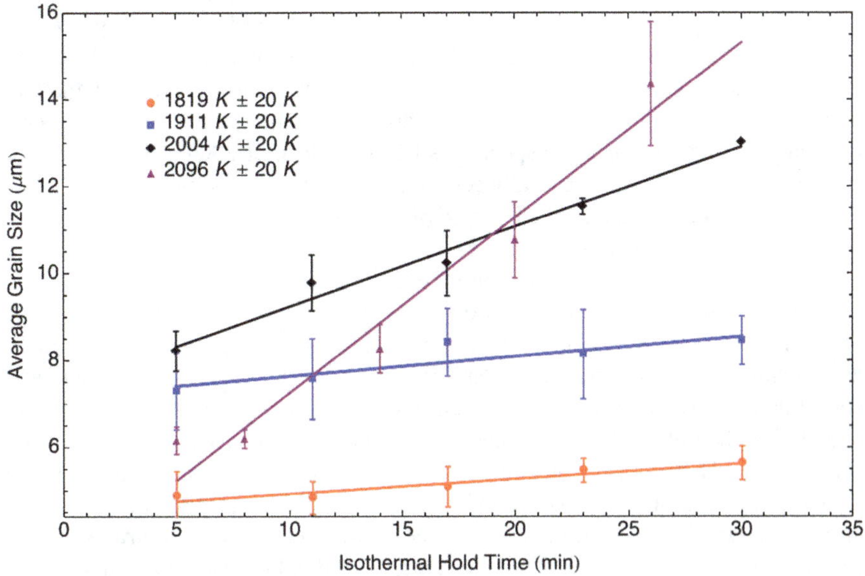

Fig. 6 Average grain size of tungsten specimens consolidated at temperatures of 1819 K, 1911 K, 2004 K, and 2096 K as a function of soak time. All specimens had an initial powder size of 471 nm, except the 2096 K specimens which had a powder size of 260 nm

function of sintering temperature up to 2004 K. At 2096 K the rate of consolidation seems to decrease after a mere 5 min at which point the sintering rate decreases. EBSD and SEM imaging revealed a large degree of separation between the grain boundaries and pores at 2096 K after 5 min, which leads to a decrease in the level of consolidation. However, prior to pore-boundary separation, the rate of sintering clearly increases with increasing sintering temperature.

Figure 6 shows the average grain size of the specimens as a function of soak time at each temperature. The grain sizes of the specimens sintered at 1819 K, 1911 K, and 2004 K all increase at approximately the same rate with the higher temperatures yielding the larger grain sizes. The smaller initial crystallite size of the specimens processed at 2096 K led to smaller initial grain sizes than the specimens processed at the other three temperatures; however, there is a large change in the rate of grain growth before and after a soak time of 5 min, caused by pore-boundary separation. The action of pore-boundary separation leads to unpinned grain growth and a decrease in sintering.

4.1.3 Effects of Stress

Five tungsten specimens with an initial crystallite size of 90 nm were consolidated in the Thermal Technology LLC 25–10 PECS unit. The specimens were heated at 100 K/min to 1073 K, where they underwent an isothermal soak for 2 min at which point the heating continued again at a rate of 100 K/min up to 1173 K. The process of heating and soaking was continued at 100 K/min with a 2 min hold every 100 K up to a peak temperature of 2004 K. The specimens underwent a series of isothermal holds to allow any potential volatiles to escape the sintering specimen before pore-boundary separation. Once the specimens had been held at 2004 K for 2 min, the specimens were allowed to naturally cool back to room temperature. The specimens were consolidated at an applied stress of 20, 40, 60, 70, and 80 MPa at the temperature conditions described above.

The stress clearly had an effect on the relative density as shown in Fig. 7. The specimen density increased as a function of applied stress up to a value of 96.8% at a stress of 80 MPa. The trend of density versus stress seems to be exponentially approaching 97% where pore-boundary separation is expected to be complete and consolidation essentially comes to a halt. Once the pores separate from the grain boundaries, the only method available to annihilate the porosity is lattice diffusion. The process of lattice diffusion aids in the migration of the vacancies from the pore structures to the grain boundaries, and it is also the mechanism for plastic flow which also helps annihilate porosity; however, these two mechanisms require very high pressures and/or temperatures. The graphite dies used in this study could not sustain the stresses required to obtain plastic flow in tungsten at these temperatures.

Figure 8 shows the average grain size of the specimens sintered 2004 K as a function of applied stress. The grain size seems to be relatively unaffected by the applied stress, except over the range of 20–40 MPa. It is likely that the increased stress causes a slight increase in the driving force for grain boundary diffusion that

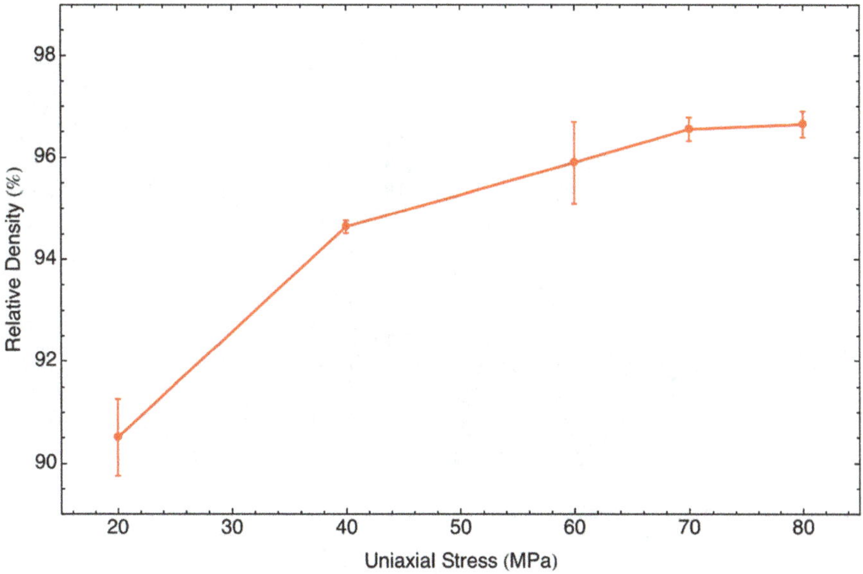

Fig. 7 Relative density of tungsten specimens produced from 90 nm tungsten particles at 2004 K as a function of applied uniaxial stress

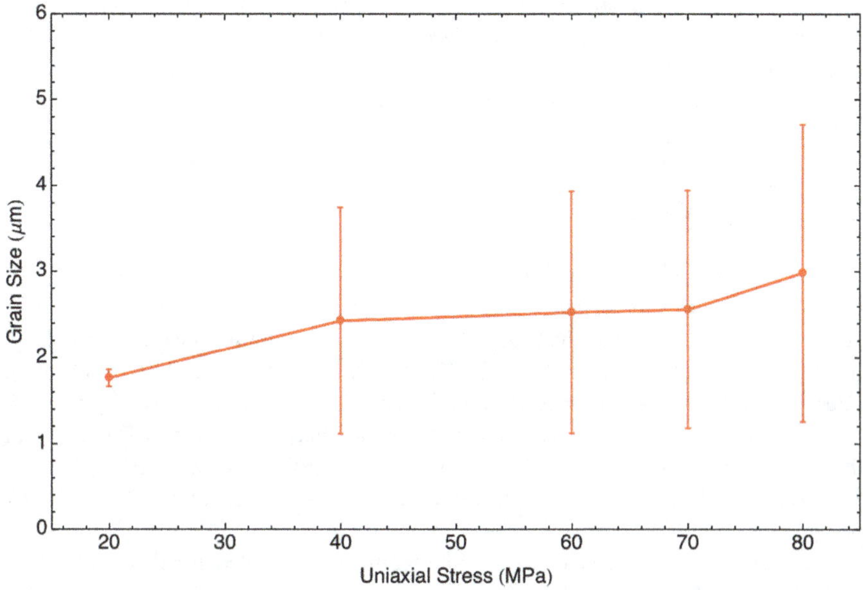

Fig. 8 Average grain size of tungsten specimens consolidated from 90 nm particles at 2004 K as a function of applied stress

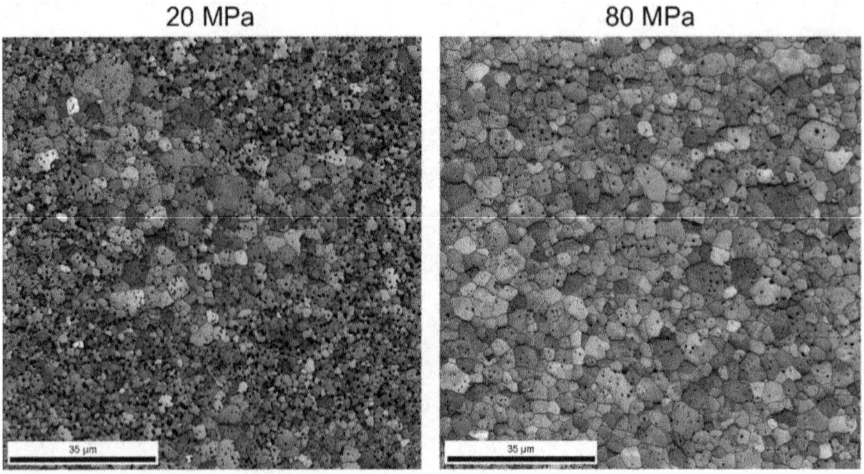

Fig. 9 EBSD images of tungsten specimens processed at 20 MPa and 80 MPa showing the evolution of porous microstructure

causes a slight increase in grain size; however, above that level of stress, pore-boundary separation seems to dominate any diffusional effects on grain growth. Figure 9 shows EBSD images of the specimens fabricated at 20 MPa and 80 MPa, clearly showing the separation of pores from the boundaries. The pores start off in cylindrical shapes, decrease in size and become spherical, and then collapse into the grain structure at which point densification begins to slow. The effects of stress seem to have little if any effect on the grain morphology.

4.1.4 Relationship Between Grain Growth and Density

The final grain sizes measured by EBSD techniques were normalized to the initial crystallite sizes of the particles, so that specimens with different initial crystallite sizes could be compared against each other. Four more specimens were consolidated by hot pressing at temperatures of 1573 K, 1673 K, 1773 K, and 1873 K and heating rates of 30 K/min. Each specimen underwent an isothermal soak at the peak temperature for 30 min and was sintered at an applied stress of 42 MPa. As with the specimens processed by PECS, the average grain size was measured and normalized by the 260 nm crystallite size (or in this case power size) of the feedstock powders.

The normalized grain sizes of all 67 specimens fabricated for this study were plotted in Fig. 10 as a function of density to see if a clear trend would emerge. Many important aspects of PECS can be extracted from Fig. 10. First, PECS consolidation methods follow a similar trend to other traditional sintering methods. PECS consolidation is initiated by pure densification and little grain growth up to an approximate relative density of 80%. Between 80% and 96% relative density, PECS

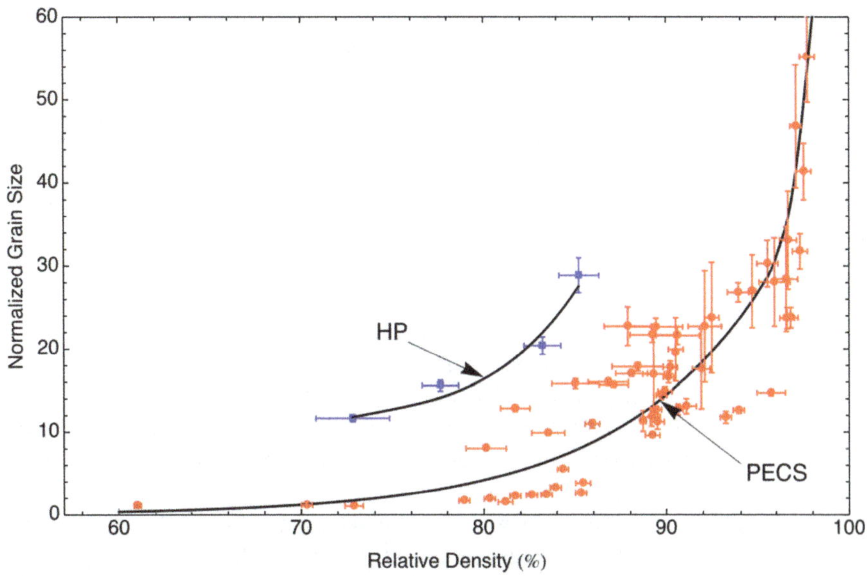

Fig. 10 Normalized grain sizes of tungsten specimens processed by PECS and HP methods as a function of relative density

consolidation is characterized by both grain growth and densification. Above a relative density of 96%, the rate of densification process is retarded due to pore-boundary separation, and grain growth begins to increase at a drastic rate. Most importantly, the rate of grain growth incurred during PECS consolidation is significantly lower than the grain growth obtained during HP indicating the mechanisms responsible for PECS densification can produce a finer-grained tungsten structure compared to other conventional sintering techniques. The fine microstructures produced by PECS can make it a promising technique for fabricating tungsten specimens for engineering applications. While one of the likely reasons for the finer microstructure of PECS-processed tungsten is due to the rapid heating rates of the PECS machine, specimens processed at a slower rate of 20 K/min still had a finer microstructure than the HIP specimens heated at 30 K/min.

4.2 *Kinetics of PECS Consolidation and Grain Growth*

Thus far we have demonstrated the ability to consolidate materials at lower temperatures using PECS methods than with conventional sintering techniques and thereby decrease the average grain size of the sintered material. The primary question remains: How does the PECS consolidation process achieve such rapid densification and minimal grain growth? Multiple sintering and grain growth kinetics experiments were designed to evaluate whether the mechanisms for sintering by PECS methods are diffusional in nature as they are with traditional sintering methods.

4.2.1 Non-isothermal Sintering Kinetics

Eleven specimens were consolidated from a tungsten feedstock with an average crystallite size of 260 nm, and eight more were consolidated from a different tungsten feedstock with an average crystallite size of 471 nm. The specimens were heated at a rate of 100 K/min to peak temperatures ranging from 1073 K to 2096 K at which point the specimens were held at temperature for 2 min before being allowed to naturally cool back to room temperature by radiation. The microstructures of these materials were discussed in Sect. 4.1, and the specimen densities and grain sizes were shown in Figs. 2 and 3, respectively. The time-dependent density of each specimen was fitted to a sigmoid function which was then differentiated as a function of time to yield time-dependent values of $d\rho/dt$.

The grain sizes and average pore radii were determined by EBSD. The average grain size, density, and pore radii were used in Eqs. 1, 2, and 3 to empirically determine the values of the diffusion coefficients at each temperature for Coble creep, Nabarro-Herring creep, and Power-law creep. The values of the natural log of the diffusion coefficients were plotted as a function of inverse temperature. The slope of the curves was multiplied by the gas constant R to determine the activation energy, and the pre-exponential frequency factor was determined from the y-intercept. With respect to power-law creep, the ratio of $\sigma/\rho E$ was tracked for each specimen. If the ratio was higher than 0.16, a power-law exponent of 7.44 was used, and if the value was less than 0.16, 4.22 was used along with the appropriate value of A. The radii of the pores in the specimens ranged from 0.3 to 1.4 μm, and it is also important to note that the pores were contained on the grain boundaries in all specimens; however, the higher temperature specimens exhibited the beginnings of pore-boundary separation.

If the mechanisms for densification in PECS are the same as for traditional sintering, we would expect the measured diffusion activation energies to match that of grain boundary diffusion or lattice diffusion (Olevsky and Froyen 2009; Olevsky et al. 2007). Figures 11 and 12 show the Arrhenius relationship between the diffusion coefficients and inverse temperature for all three plotted mechanisms with respect to the sintering of the 260 nm and 471 nm tungsten particles. The sintering temperatures for the experiments ranged from 1300 to 2200 K, with pore sizes that ranged from 0.3 to 0.5 μm and grain sizes that ranged from 0.3 to 12 μm. As seen in Figs. 11 and 12, no trend with a negative slope could be established for power-law creep; however, reasonable trends could be established for Coble and Nabarro-Herring creeps. The activation energies for diffusion related to Coble creep and Nabarro-Herring creep determined from Eqs. 1 and 2 for the 260 nm tungsten particles were 339 ± 36 kJ/mol and 203 ± 11 kJ/mol, respectively.

The uncertainties in the calculated activation energies are based on the average deviation of the points from the fitted trend. The average deviation was used to perturb the slope of the curve from which an average deviation in the activation

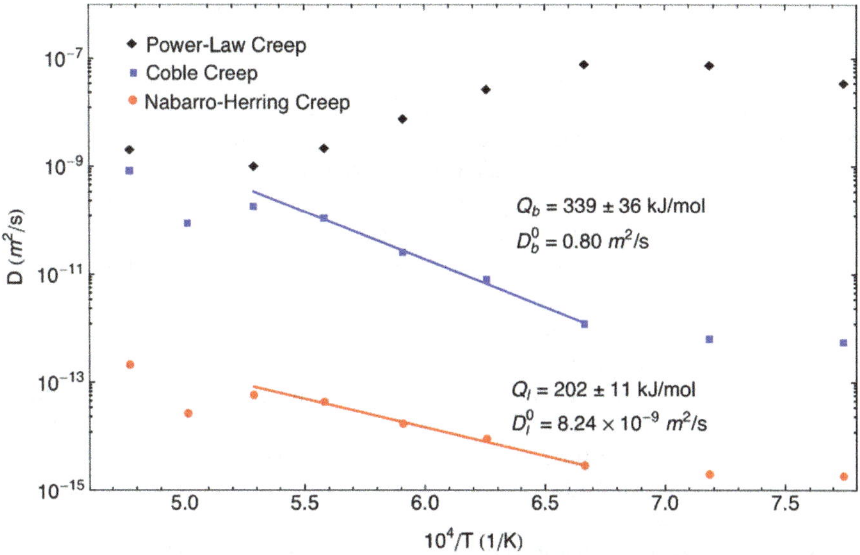

Fig. 11 Experimental diffusion coefficients for Coble creep, Nabarro-Herring creep, and power-law creep established for non-isothermal sintering of 260 nm tungsten particles

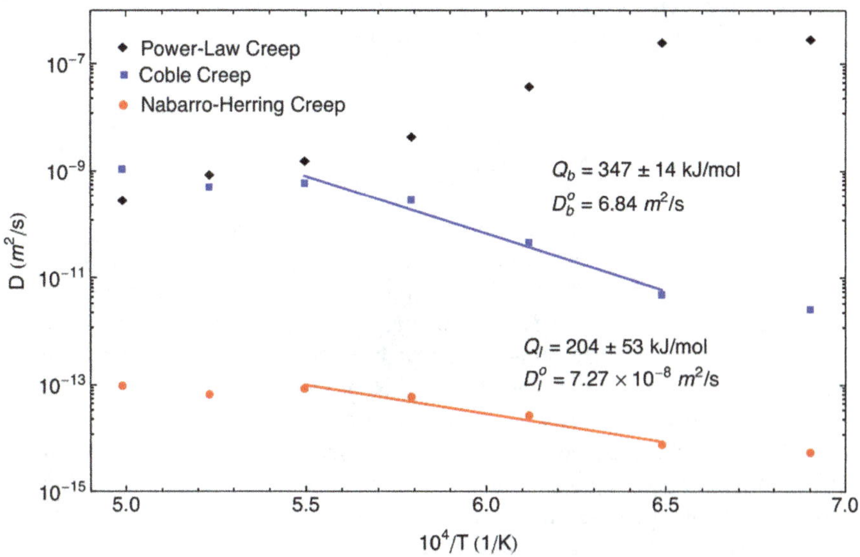

Fig. 12 Experimental diffusion coefficients for Coble creep, Nabarro-Herring creep, and power-law creep for non-isothermal sintering of tungsten particles with an average particle size of 471 nm

energy could be developed. The activation energy of 203 kJ/mol was less than half the expected activation energy for lattice diffusion; however, the energy of 339 kJ/mol is fairly close to that of grain boundary diffusion in tungsten (Mundy et al. 1978; Andelin et al. 1965). Experiments with the 471 nm tungsten powders yielded a grain boundary and lattice diffusion activation energy of 347 ± 14 kJ/mol and 204 ± 53 kJ/mol, respectively, which were developed from the Coble creep and Nabarro-Herring creep relationships. Thus the activation energies of 339 ± 36 kJ/mol and 347 ± 14 kJ/mol derived from the Coble creep relationship in Eq. 1 are very close to the accepted value for grain boundary diffusion in tungsten; however, the activation energies produced from the Nabarro-Herring creep relationship in Eq. 2 are less than half of the accepted value for lattice diffusion. While Coble creep does appear to be the densification mechanism for these two non-isothermal sintering experiments, the pre-exponential frequency factors were between two and three orders of magnitude higher than would be expected for traditional grain boundary diffusion at these temperatures. The pre-exponential frequency factors were determined to be 0.8 m^2/s and 6.84 m^2/s for the 260 nm and 471 nm tungsten particles, respectively.

The abovementioned experiments appear to show that Coble creep dominates the densification of these materials over the temperature range of 1450 K to 1820 K; however, Coble creep seems to taper off as a sintering mechanism at temperatures above 1820 K. It is possible that above 1820 K for the grain sizes within these specimens, the power-law creep begins to become dominant, but the event of pore-boundary separation makes it difficult to compare the microstructures of the different specimens in order to correlate the grain size, porosity, and density to the deformation mechanisms.

4.2.2 Isothermal Sintering Kinetics

A series of isothermal sintering experiments were performed to help remove the effects of pore-boundary separation from the kinetics results. In this set of experiments, tungsten powder batches with an average size of 471 nm were sintered at temperatures of 1819 K, 1911 K, and 2004 K with isothermal soak times of 5, 11, 17, 23, and 30 min and at an applied stress of 45 MPa. One more isothermal data set was added after the initial experiment where 260 nm tungsten particles were sintered at a temperature of 2096 K and with dwell times of 2–26 min. After a 5-min isothermal hold, the specimens with an initial crystallite size of 260 nm began to exhibit pore-boundary separation. In order to ensure that the data points with pore-boundary separation did not influence the results, only the specimens sintered with an isothermal soak of less than 5 min were used from that data set.

The specimen density values and grain sizes for these sintering experiments were discussed in Sect. 4.1 and are shown in Figs. 5 and 6. Instead of using the non-isothermal sintering time to develop a relationship for $d\rho/dt$, the isothermal sintering time was used to develop a value of $d\rho/dt$. The higher sintering temperatures were selected to help grow larger grains and push the sintering envelope into the predicted range for power-law creep. The grain size and porosities were determined from EBSD analysis and were used in Eqs. 1, 2, and 3 along with the density to empirically evaluate the diffusion coefficients for Coble creep, Nabarro-Herring creep, and power-law creep. As expected, based on the deformation mechanism maps in Fig. 1, the power-law creep did seem to be the dominant sintering mechanism at these elevated temperatures and grain sizes. Figure 13 shows the Arrhenius relationship between the experimentally derived diffusion coefficient for power-law creep versus inverse temperature. The activation energy was found to be 356 ± 40 kJ/mol, with a pre-exponential frequency factor of 4.20×10^{-11} m^2/s, which both match predictions for dislocation-core diffusion as a low-temperature dislocation climb mechanism for power-law creep (King and Sell 1965; Robinson and Sherby 1969).

The data from the isothermal sintering experiments were also used with Eqs. 1 and 2. Equation 1 (the Coble creep relationship) yielded an activation energy of 794 ± 42 kJ/mol, and the relationship for Nabarro-Herring creep yielded an activation energy of 671 ± 22 kJ/mol. The experimentally determined value for grain boundary diffusion activation energy was nearly twice as large as the accepted value of 385 kJ/mol and as such was rejected as a rate-limiting mechanism. The activation energy for Nabarro-Herring creep was actually at the upper end of accepted values for lattice diffusion, which ranges from 520 kJ/mol to 670 kJ/mol; however, the experimentally determined value of pre-exponential frequency factor was 3.99×10^7 m^2/s which is nearly 15 orders of magnitude higher, and as such it was rejected as the rate-limiting mechanism (German 1996; Krieder and Bruggeman 1967).

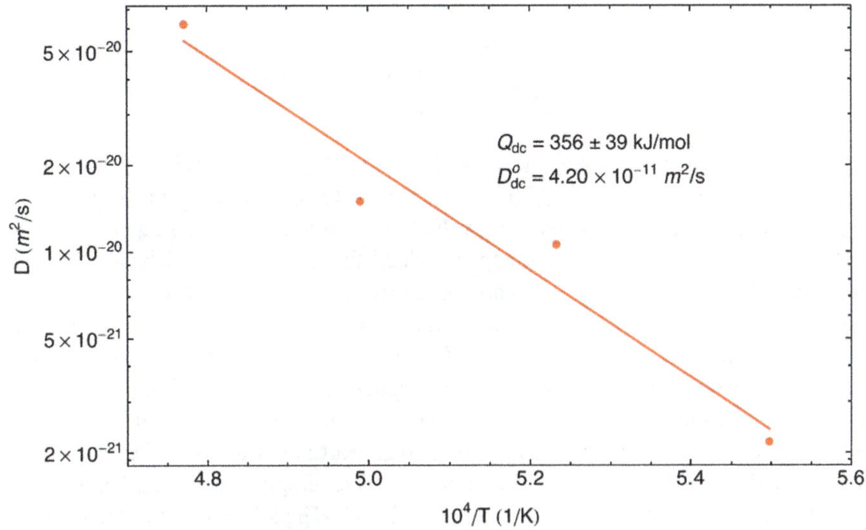

Fig. 13 Diffusion coefficients determined for high-temperature sintering of tungsten based on the power-law creep relationship

Based on the results of Eq. 3 which was empirically derived, it appears as if power-law creep is the rate-limiting mechanism for sintering of tungsten with the grain sizes shown in Fig. 5 at temperatures above 1819 K. The data was also analyzed by Eq. 5 which is a purely analytical relationship proposed by Helle and Taylor (Helle et al. 1985; Taylor et al. 1993), where ρ is the specimen instantaneous density; ρ_0 is the density of the specimens at the beginning of the isothermal hold, which is extrapolated based on the isothermal trend; D_{PLC} is the diffusion coefficient for power-law creep; μ is the shear modulus which is determined from Eq. 6 (Frost and Ashby 1983; Pottlacher et al. 1993); and B^p is a constant that has already been determined as 2/3 based on experimental evidence. The terms μ_0 and C are material constants which are 1.65×10^{11} Pa and $-$ 0.38, respectively, for tungsten, and the term T_m represents the melting temperature (3667 K).

$$\dot{\rho} = \frac{5.3}{\sqrt{3}}\left(\rho^{\frac{2}{3}}\rho_0^{\frac{1}{3}}\right)\left(\frac{\rho - \rho_0}{1 - \rho_0}\right)^{\frac{1}{2}}\frac{\mu D_{PLC}}{kT}\left(\frac{B^p\sigma}{3\rho\mu}\right)^n \tag{5}$$

$$\mu = \mu_0\left[1 + \left(\frac{T - 300}{T_m}\right)C\right] \tag{6}$$

Equation 5 can be rearranged to solve for D_{PLC} with knowledge of the isothermal densification rate ($\dot{\rho}$), the instantaneous density (ρ), and the applied stress (σ) for each data point in the isothermal sintering series shown in Figs. 5 and 6. However, prior to solving for the expected diffusion coefficients, the values of n at each temperature must be determined by developing a plot of

$$\log\left[\frac{\dot{\rho}}{\rho^{\frac{2}{3}}(\rho - \rho_0)^{\frac{1}{2}}}\right] \text{vs.} \log\left[\frac{B^p\sigma}{3\rho\mu}\right]$$

which can yield a constant slope line for all of the data points in each isothermal data set. The slope of the trend should be equal to the value of n for the data points at that temperature. More importantly, the value of n should yield a numerical value between 4 and 7 if power-law creep is indeed the dominant mechanism. Indeed, when the data points from Fig. 5 and the appropriate sintering rates were used, the data points all yielded values of n that ranged from 4.65 to 6.51 in a manner similar to what would be expected in the deformation mechanism maps based on the effective stress, sintering temperatures, and specimen grain sizes. Figure 14 shows the plot from which the values of n were extrapolated. The values of n determined in Fig. 14 were input into Eq. 5 which was then rearranged to solve for D_{PLC}. The values of D_{PLC} were plotted as a function of inverse temperature for each data point and then fitted to a trend which is shown in Fig. 15. The slope and y-intercept were used to determine the value of activation energy to be 338 ± 32 kJ/mol with a pre-exponential frequency factor of 3.14×10^{-14} m^2/s, which appears to confirm that power-law creep is the dominant mechanism for sintering at the grain sizes,

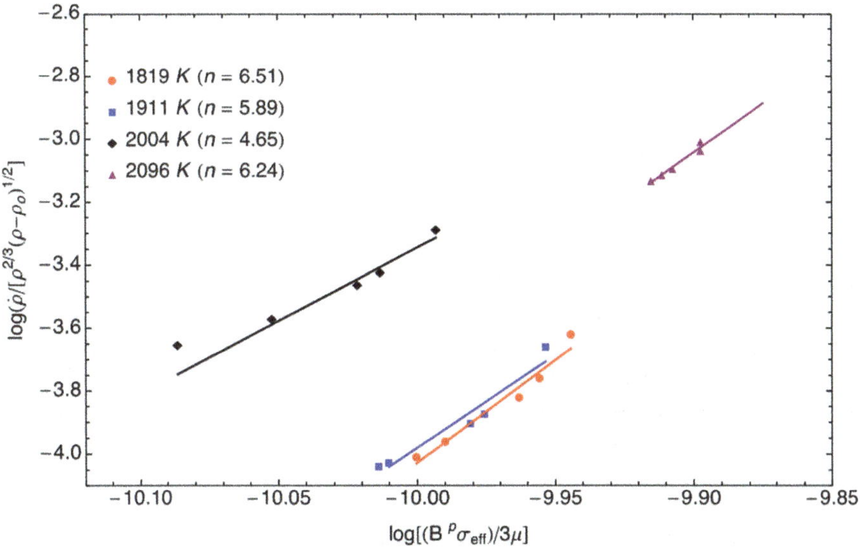

Fig. 14 Plot developed for experimental determination of power-law exponents (n) at each isothermal sintering temperature

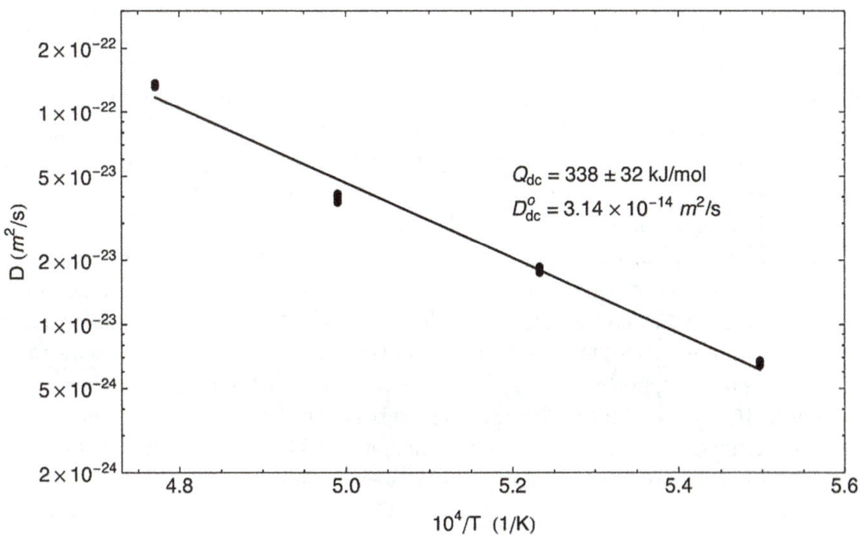

Fig. 15 Diffusion coefficient related to power-law creep as determined by the Helle method using the experimental data

temperatures, and stresses used in this sintering study. The data also lends evidence that supports the idea that dislocation-core diffusion is a low-temperature mechanism that allows for dislocation climb in tungsten. However, PECS is a complicated process, and to fully understand the essence of the rapid densification process, further robust models involving effect of the direct current would be necessary (Deng et al. 2017, 2018).

4.2.3 Grain Growth Mechanisms

The grain sizes shown in Fig. 3 from the temperature-dependent sintering of tungsten particles with initial crystallite sizes of 260 nm and 471 nm were analyzed in accordance with Eq. 7 to determine the activation energy and pre-exponential frequency factor for grain growth. Within Eq. 7 the term G_0 represents the initial grain size, G represents the instantaneous grain size, K represents a temperature-dependent grain growth rate coefficient, t represents the sintering time during the isothermal hold, and n represents a geometric constant that is usually equal to 2 or 3.

$$G^q - {G_0}^q = Kt \tag{7}$$

The value of q can be determined by plotting the value of $G^q - {G_0}^q$ versus t for all data points in an isothermal data set and varying the value of n until a slope of G_0 is achieved. These experiments have been attempted by Mistler (1968) on tungsten previously (Mistler 1968) and yielded values of 2 between 1723 K and 1873 K and values of 3 at 1923 K. These experiments were performed on the tungsten specimens sintered by PECS methods at 1819 K, 1911 K, 2004 K, and 2096 K and yielded values of 3.26, 2.92, 2.23, and 1.84, respectively. It seems as if the value of q decreases with increasing temperature; however, this is contrary to the results of Mistler. More importantly the purpose of using a value of 3 at higher temperatures is to correct for effects of pore-boundary separation at which point grains begin to grow much more rapidly. For the purposes of this study, the results obtained by Mistler were used; at temperatures below 1820 K, a value of 2 was used; and above 1820 K, a value of 3 was used. The value of K was determined from the data in Fig. 3, and it was noted that if a common value of q was used, there was a marked change in the slope of the curve at an approximate value of ~1800 K, which seems to validate Mistler's findings. Also, the initiation of pore-boundary separation was noticed at ~1800 K, which supports a change in the value of n at that temperature.

Figure 16 shows a plot of the relationship between the grain growth rate coefficient of tungsten versus the inverse temperature. The results indicate that the activation energy for the 471 nm particles were essentially the same before and after a value of 1800 K, yielding values of 357 ± 32 kJ/mol and 361 ± 28 kJ/mol, respectively, indicating that grain boundary diffusion is responsible for the grain growth. The grain growth activation energy for the 260 nm particles yielded an activation energy of 278 ± 54 kJ/mol at a temperature below 1800 K and an activation energy of 88 ± 33 kJ/mol above 1800 K. The data points for the 260 nm particles are consistent with surface diffusion, which may be driven by the high surface area to volume ratio of the particles (Bettler and Charbonnier 1960; Barbour et al. 1960; Ehrlich and Hudda 1966; Bowder and Singer 1969).

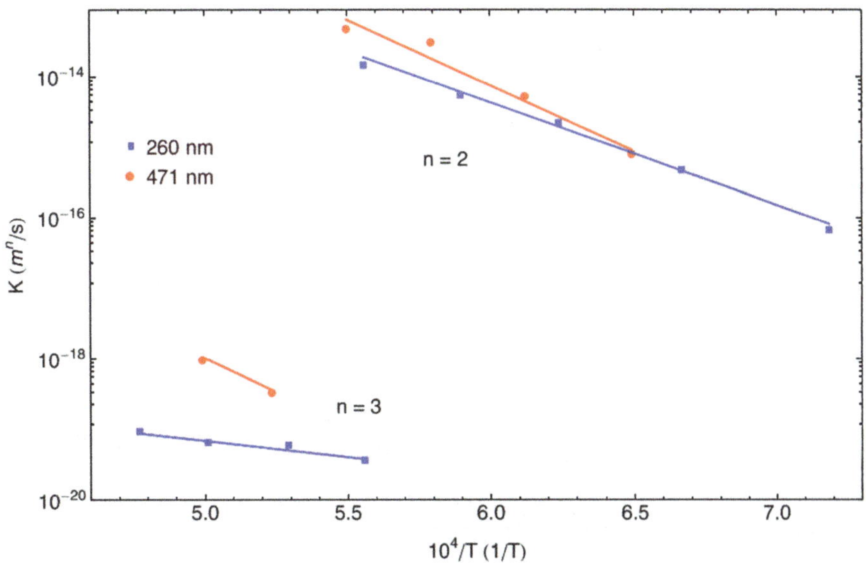

Fig. 16 Plot of the relationship between $(G^n - G_0{}^n)/t$ versus inverse temperature for 260 nm and 471 nm tungsten particles

4.3 Post-process HIP

While the PECS technique does offer the ability to rapidly consolidate materials at low temperatures with minimal grain growth, the inherent design of PECS makes it very difficult to obtain tungsten densities above 97% of the theoretical density. Post-process hot isostatic pressing (HIP) was explored as a possible post-process technique to fully densify materials with minimum grain growth. Two tungsten specimens were consolidated from 20 g of tungsten particles each with a particle feedstock consisting of 90 nm particles with the Thermal Technology LLC 25–10 PECS unit. The specimens were heated at a rate of 100 K/min and underwent an isothermal hold of 2 min every 100 K starting at a temperature of 1073 K, until they reached a peak temperature of 2000 K, where the specimens were soaked for 20 min.

Once the specimens were consolidated, they were removed from the graphite dies and polished for density measurements and initial microscopy analysis. SEM/EBSD methods were used to examine that the pores had completely collapsed and migrated into the grain structures. Each specimen was sectioned into five pie-shaped wedges which were then further consolidated in an American Isostatic Presses HIP unit; no canning was used since the pore structures had already fully collapsed. The specimens were heated to temperatures of 1473 K, 1673 K, 1873 K, and 1973 at a rate of 20 K/min in an argon atmosphere. The specimens were held at peak temperature for 10, 30, and 60 min at an applied isotropic stress of 310 MPa. Following consolidation, the specimens were removed, polished once again, and analyzed for density and average grain size. Prior to sintering, the average grain size of the particles was

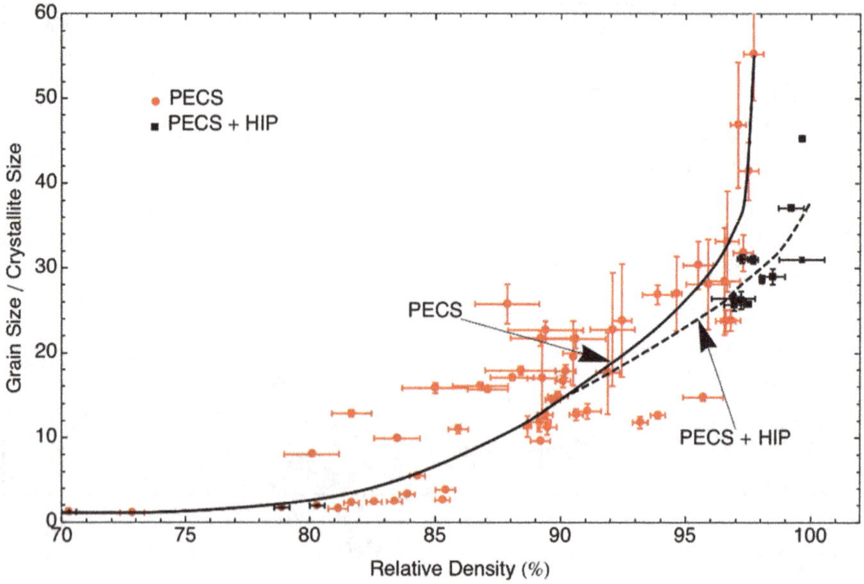

Fig. 17 Normalized grain size versus relative density for tungsten specimens processed by pure PECS and by PECS with post-process HIPing at 310 MPa. Here the grain size represents the grain size in the SPSed specimens, whereas crystallite size is the starting monocrystalline powder size

2.57 μm with an average relative density of ~96%. It was found that the specimens HIPed at 1873 K and 1973 K were able to achieve full density; however, the specimens sintered at 1973 K had excessive grain growth. The optimum HIP post-processing temperature that achieved full density with the minimum grain growth was found to be 1873 K with a 60-min isothermal hold. Figure 17 shows a plot of normalized grain size versus relative density for the specimens processed by only PECS methods and those processed by a combination of PECS and HIP methods. The specimens that underwent post-process HIP consolidation clearly have a lower density-dependent grain size during the final stage of sintering than those processed by pure PECS only. This study seems to confirm that the combination of initial PECS sintering followed by HIP post-processing can achieve a finer microstructure than other sintering techniques with full density, which may be very advantageous to high-temperature engineering applications.

5 Conclusions

Tungsten powder particles with an average size of 90 nm, 260 nm, and 471 nm were consolidated by PECS at heating rates that ranged from 20 K/min to 100 K/min and values of applied stress that ranged from 20 MPa to 80 MPa. PECS was shown to produce a much finer microstructure during the initial and intermediate stages of sintering than can be achieved with hot uniaxial pressing; however, achieving a relative density above 97% proved to be difficult to attain due to the limitations in the mechanical strength of the graphite dies. Post-process HIP was pursued as a post-processing technique to achieve full densification process and successfully demonstrated its ability to complete the densification process while not increasing the average grain size over that of the pure PECS method. Dynamic kinetics studies were pursued to determine the mechanisms for the rapid and low-temperature densification achieved during PECS. The kinetics studies indicated that the mechanisms for consolidation during PECS sintering appeared to be no different than traditional sintering methods and are based on diffusion- and dislocation-based forms of creep that are dependent on the sintering temperature, applied stress, and tungsten grain size. The grain growth of the particle batches sintered seemed to be based on surface diffusion and grain boundary diffusion, with larger grain sizes favoring grain boundary diffusion over surface diffusion. It is, however, possible that even larger grain structures may favor lattice diffusion. The study seems to confirm that PECS is capable of producing very fine tungsten microstructures that could be advantageous for high-temperature engineering applications where increased mechanical strength is required.

Acknowledgments The authors would like to thank Dr. Daniella Fredrick of Thermal Technology LLC for conducting some sintering experiments. JW would also like to thank Tammy Trowbridge of the Idaho National Laboratory for conducting the EBSD and SEM analysis presented in this work. Finally, the research was performed using funding received from the DOE Office of Nuclear Energy's Nuclear Energy University Programs (NEUP contract # 42246-57). The authors would also like to acknowledge the contributions of Dr. Mark Carroll and Cory Sparks during the execution of the project.

References

Andelin RL, Knight JD, Khan D (1965) Diffusion of tungsten and rhenium tracers in tungsten. Trans AIME 233:19–24

Anderson KR, Groza JR, Fendorf M, Echer CJ (1999) Surface oxide debonding in field assisted powder sintering. Mater Sci Eng A270:278–282

Argonne National Laboratory (1968) Nuclear Rocket Program (NTR) Terminal Report, ANL-7236

Ashby MF (1974) A first report on deformation mechanism maps. Acta Metall 22:275–289

Barbour JP, Charbonnier FM, Dolan WW, Dyke WP, Martin EE, Trolan JK (1960) Determination of the surface tension and surface migration constants for tungsten. Phys Rev 117(6):1452–1459

Baum EM, Knox HD, Miller TR (2002) Nuclides and isotopes: chart of the nuclides. Knolls Atomic Power Laboratory, Niskayuna

Bernard-Granger G, Addad A, Fantozzi G, Bonnefont G, Guizard C, Vernat D (2010) Spark plasma sintering of commercially available granulated zirconia powder: comparison with hot-pressing. Acta Mater 58:3390–3399

Bettler PC, Charbonnier FM (1960) Activation energy for the surface migration of tungsten in the presence of a high electric field. Phys Rev 119(1):85–93

Bowder FP, Singer KE (1969) Surface self-diffusion of tungsten. Nature 222(5197):977–979

Coble RL (1970) Diffusion models for hot pressing with surface energy and pressure effects as driving forces. J Appl Phys 41:4798–4807

Conway JB, Flagella PN (1968) Physical and mechanical properties of reactor materials. 7th Annual AEC Fuels and Materials Development Program Report, General Electric Company

Deng S, Yuan T, Li R, Zeng F, Liu G, Zhou X (2017) Spark plasma sintering of pure tungsten powder: densification kinetics amd grain growth. Powder Technol 310:264–271

Deng S, Yuan T, Li R, Zeng F, Liu G, Zhou X (2018) Direct current-enhanced densification kinetics during spark plasma sintering of tungsten powder. Scr Mater 143:25–29

Ehrlich G, Hudda FG (1966) Atomic view of surface self-diffusion: tungsten on tungsten. J Chem Phys 44(3):1039–1049

Farrell K, Schaffhauser AC, Stiegler JO (1967) Recrystallization, grain growth and ductile-to-brittle-transition in tungsten sheet. J Less-Common Met 13(2):141–155

Frost HJ, Ashby MF (1983) Deformation mechanism maps, plasticity and creep of metals and ceramics. Permagon Press, Oxford

General Electric Company (1974a) High temperature gas reactor program summary Report (1974) Vol. I: Summary

General Electric Company (1974b) 710 High temperature gas reactor program summary report (1974) Vol. VI: Fuel element drawings and manufacturing specifications

German RM (1996) Sintering theory and practice. Wiley, New York

Harrison B, Ellison E (1966) Creep and tensile properties of heavily drawn tungsten wire. Trans ASM 59:744–754

Helle AS, Easterling KE, Ashby MF (1985) Hot-isostatic pressing diagrams: new developments. Acta Metall 33(12):2163–2174

Hulbert DM, Anders A, Dudina DV, Anderson J, Jiang D, Unuvar C, Anselmi-Tamburini U, Lavernia EJ, Mukherjee A (2008) The absence of plasma in "spark plasma sintering". J Appl Phys 104:03305

Hulbert DM, Anders A, Anderson J, Lavernia EJ, Mukherjee AK (2009) A discussion of the absence of plasma in spark plasma sintering. Scr Mater 60:835–838

Kassner ME, Perez-Prado MT (2009) Fundamentals of creep in metals and alloys. Elsevier, Oxford

Kaufmann M, Neu R (2007) Tungsten as a first wall material in fusion devices. J Fusion Eng Des 82:521–527

King G, Sell H (1965) The effects of thoria on the elevated temperature tensile properties of recrystallized high purity tungsten. Trans AIME 233:1104–1113

Klopp W, Raffo P (1964) Effects of purity and structure on recrystallization, grain growth, ductility, tensile strength and creep properties of arc melted tungsten. National Aeronautics and Space Administration, NASA TN D-2503

Klopp W, Witzke W (1966) Mechanical properties and recrystallization behavior of electron-beam melted tungsten compared with arc-melted tungsten, National Aeronautics and Space Administration, NASA TN D 3232, 1966

Klopp W, Witzke W, Raffo P (1964) Effects of grain size on the tensile stress and creep properties of arc-melted and electron-beam melted tungsten at 2250 F to 4150 F. National Aeronautics and Space Administration, NASA TM X-54756, 1964

Krieder KG, Bruggeman G (1967) Grain boundary diffusion in tungsten. Trans AIME 239 (8):1222–1226

Langer J, Hoffmann MJ, Guillon O (2009) Direct comparison between hot pressing and electric field-assisted sintering of submicron alumina. Acta Mater 57:5454–5465

Lassner W, Schubert WD (1999) Tungsten properties, chemistry, technology of the element, alloys and chemical compounds. Kluwer Academic/Plenum Publishers, Norwell

Luke MT (2010) Microstructural evolution of nickel during spark plasma sintering. MS Thesis, Boise State University

Luo A, Jacobson DL, Shin KS (1992) Strength properties and deformation behavior of arc-melted and powder-metallurgy processed tungsten at high temperatures. J Refract Met Hard Mater 11:1–8

Mistler RE (1968) Comments on the paper grain growth during sintering of tungsten. J Appl Phys 39:4875

Mitteau R, Missiaen JM, Brustolin P, Ozer O, Durocher A, Russet C, Lungu CP, Courtois X, Dominicy C, Maier H, Grisolia C, Piazza G, Chuppuis P (2007) Recent developments towards the use of tungsten Armour material in plasma facing components. J Fusion Eng Des 82:1700–1705

Mundy JN, Rothman SJ, Lam NQ, Hoff HA, Nowicki LJ (1978) Self diffusion in tungsten. Phys Rev B 18:6566

Munir ZA, Anselmi-Tamburini U, Ohyanagi M (2006) The effect of electric field and pressure on the synthesis and consolidation of materials: a review of the spark plasma sintering method. J Mater Sci 41:763–777

Olevsky E, Froyen L (2006) Constitutive modeling of spark-plasma sintering of conductive materials. Scr Mater 55:1175–1178

Olevsky EA, Froyen L (2009) Impact of thermal diffusion on densification during SPS. J Am Ceram Soc 92(S1):S122–S132

Olevsky E, Kandukuri S, Froyen L (2007) Consolidation enhancement in spark plasma sintering: impact of high heating rates. J Appl Phys 102:114913

Pawel RE, Lundy TS (1969) Tracer diffusion in tungsten. Acta Metall 17(8):979–988

Perkins JB (2011) Spark plasma sintering of tungsten and tungsten-ceria: microstructures and kinetics. MS Thesis, Boise State University

Pottlacher G, Hixson RS, Melnitzky S, Kaschnitz E, Winkler MA, Jager H (1993) Thermophysical properties of POCO AXF-5Q graphite up to melting. Thermochim Acta 218:183–193

Raj S, Iskovitz I, Freed A (1995) Modeling of the role of dislocation substructure during class-M and exponential creep. National Aeronautics and Space Administration, NASA TM-106986

Robinson S, Sherby OD (1969) Mechanical behavior of polycrystalline tungsten at elevated temperature. Acta Metall 17(2):109–125

Song X, Liu X, Zhang J (2006) Neck formation and self-adjusting mechanism of neck growth of conducting powders in spark plasma sintering. J Am Ceram Soc 89(2):494–500

Stephens JR (1963) Effects of oxygen on mechanical properties of tungsten. National Aeronautics and Space Administration, NASA TN D-1581

Storms E (1967) The refractory carbides. Academic, Cambridge, MA

Syvertsen GE, Estournes C, Fjeld H, Haugsrud R, Einarsrud MA, Grande T (2012) Spark plasma sintering and hot pressing of hetero-doped $LaNbO_4$. J Am Ceram Soc 95(5):1563–1571

Taylor N, Dunand DC, Mortensen A (1993) Initial stage hot pressing of monosized Ti and 90% Ti-10% TiC powders. Acta Mater 41(3):955–965

Webb JA, Charit I (2011) Monte Carlo criticality analysis of simple geometries containing tungsten-rhenium alloys engrained with uranium dioxide and uranium mononitride. J Nucl Eng Des 241(8):2968–2973

Webb JA, Charit I (2012) Analytical determination of thermal conductivity of W-UO_2 and W-UN CERMET nuclear fuels. J Nucl Mater 427:87–94

Xie G, Ohashi O, Song M, Mitsuishi K, Furuya K (2005) Reduction mechanism of surface oxide films and characterization of formations on pulse electric-current sintered Al-Mg alloy powders. Appl Surf Sci 241:102–106

Spark Plasma Sintering of Metallic Glasses

Loïc Perrière, Yannick Champion, and Frédéric Bernard

1 Introduction

The amorphous nature of metallic glasses induces a unique set of chemical, physical, and mechanical properties. However, their lack of macroscopic plasticity and their limited casting dimensions hinder their industrial development at a large scale. Bulk metallic glasses are limited in composition, whereas a wide variety of amorphous alloys can be obtained by rapid solidification processes.

In this context, powder metallurgy is a relevant alternative route to produce large-sized and intricate-shaped samples. One of the main benefits is that micrometric powders can be prepared from almost any glass-forming alloy, even the ones exhibiting low glass-forming ability (GFA). Amorphous powders are indeed successfully manufactured since the 1970s by atomization techniques (Miller and Murphy 1979), and dense samples have been since produced, using sintering techniques such as hot pressing or hot extrusion (Cox et al. 2012) or severe deformation (Krämer et al. 2017), taking advantage of the viscous flow of MG in the supercooled liquid region. Mechanical alloying is also another widely studied amorphous powder processing route. Furthermore, the development of spark plasma sintering (SPS)/field-assisted sintering technique (FAST) gives a new perspective for MG processing. The main asset of SPS is to provide extremely rapid heating (up to

L. Perrière (✉)
ICMPE UMR7182 CNRS/UPEC, Thiais, France
e-mail: perriere@icmpe.cnrs.fr

Y. Champion
SIMaP UMR5266, CNRS/Grenoble INP/Université Grenoble Alpes, Saint Martin d'Hères, France
e-mail: yannick.champion@simap.grenoble-inp.fr

F. Bernard
ICB UMR6303 CNRS/Université de Bourgogne, Dijon, France
e-mail: fbernard@u-bourgogne.fr

© Springer Nature Switzerland AG 2019
P. Cavaliere (ed.), *Spark Plasma Sintering of Materials*,
https://doi.org/10.1007/978-3-030-05327-7_11

several hundreds of K.min^{-1}) and cooling, which is an invaluable advantage in order to limit the devitrification of MG during their processing. It consequently widens the spectrum of compositions that can be manufactured at a centimetric scale without losing their amorphous characteristics. It is, for instance, of prime importance for Fe-based alloys, which are notoriously difficult to manufacture in large dimensions, as a consequence of the specific chemical bonds in these systems. Moreover, as the other powder metallurgy routes, SPS favors the preparation of MG-based composites that can overcome the drawbacks of MG, as, for instance, their lack of macroscopic ductility.

In the following paragraphs, we will first briefly approach the specificities of metallic glasses and how SPS can be part of their manufacturing routes. We will then focus on the sintering of amorphous powders in order to produce dense amorphous parts. Composites based on metallic glasses, either produced in situ in the SPS apparatus (i.e., from the amorphous alloy partial devitrification or a chemical reaction) or ex situ (i.e., from prior powder mixing), will also be covered. Finally, we will bring up the specific sintering mechanisms that have been observed on these materials.

2 Background: A Brief Introduction to Metallic Glasses

2.1 Discovery, Evolution, and Spectrum of Use

Contrary to crystallized alloys, amorphous metallic alloys, or metallic glasses, do not present long-range order. They have been discovered in the 1960s (Klement et al. 1960) as Au-Si ribbons prepared by rapid solidification. The critical quenching rate, i.e., the cooling rate that is needed to obtain the amorphous phase, was comprised between 10^4 and 10^6 K.s^{-1} and the maximal thickness less than 0.1 mm. More complex alloys, containing three or more elements, were later discovered and can be cooled less rapidly into bulk parts (with a thickness >1 mm and critical cooling rate <10^3 K.s^{-1}). However, these few alloys were based on expensive metals such as Pt or Pd (Chen 1974). Since the beginning of the 1990s, numerous new compositions based on Mg, Zr, Fe, Ni, Cu, Ti, or Co were discovered and can be manufactured to dimensions up to tens of centimeters (Greer 2009). To obtain an amorphous alloy, crystallization must be avoided. It can be achieved by diminishing the nucleation of crystals during the solidification, or inhibiting the crystalline growth by making diffusion and, consequently, long-range ordering, difficult. This was translated into semiempirical rules to design amorphous alloys with high GFA (Inoue 2000): (i) at least three, generally 5, elements, (ii) atomic radius difference of at least 12%, (iii) composition near deep eutectic points in order to slow down atomic movements in the liquid state, and (iv) negative pair mixing enthalpy.

Recently, new combinatory metallurgy techniques have also been used to design amorphous alloys (Khan et al. 2017). As there is a wide spectrum of metallic glasses available, the domain of their applications is vast. MG combine high strength and toughness, making them good candidates for structural materials (Kruzic 2016). Their ability to store elastic energy and their high specific strength (Jiang et al. 2015) are indeed particularly interesting. The absence of crystalline defects and grain boundaries is an advantage for corrosion resistance over their crystalline counterparts (Wang 2016). Biomedical applications (Kumar et al. 2016), magnetic materials (Neamţu et al. 2014), and catalysis (Hu et al. 2016) are domains where MG are also considered appealing. Finally, their superplastic behavior allows nano-molding and microelectromechanical system fabrication (Kumar et al. 2009).

2.2 *Structural Properties*

Condensed matter generally possesses a crystalline structure at equilibrium which is at the origin of the Bragg diffraction peaks on X-ray diffraction (XRD) patterns. It is also at the origin of numerous properties: deformation, electrical and thermal conductivity, magnetism, etc. In a vitreous state, atomic positions are only short-range correlated and appear randomly distributed. Consequently, no Bragg peak will be visible on XRD patterns, and many properties are also affected. Short-range order nevertheless exists, and the probability to find specific atoms next to each other depends on their chemical nature (Blétry 2004).

The glass transition temperature, T_g, is not a constant but depends on the cooling rate: the faster the cooling, the higher the T_g, similar to a thermally activated phenomenon. T_g can be measured by differential scanning calorimetry (DSC), as shown in Fig. 1. Three characteristic values are then visible: the onset T_{g1}, or the deviation from the vitreous state curve, the inflection point T_{g2}, and the offset T_{g3}, the temperature from which the material behaves like a supercooled liquid. T_x corresponds to the onset of the first exothermic peak of crystallization.

From a thermodynamic viewpoint, a glass can be considered as a liquid whose configuration is not at a free energy minimum. The endothermic step associated with

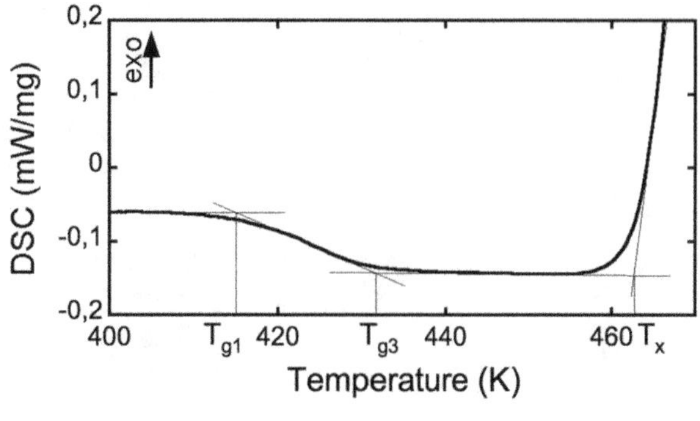

Fig. 1 Thermogram of a $Zr_{45}Cu_{45}Al_{10}$ metallic glass (20 k.min^{-1})

the glass transition corresponds to a relaxation of the atomic arrangement. MG are thus not at a thermodynamical equilibrium and will evolve upon any thermal cycle, even at $T < T_g$. This structural relaxation is controlled by topological short-range ordering (Mao et al. 1995) and chemical short-range ordering (Taub and Spaepen 1980) and is accompanied by a density increase (Suh and Dauskardt 2003). The structural relaxation has a complex role on the devitrification of MG, as it can slow the process down by decreasing the diffusion coefficients (Zumkley et al. 2001). However, one could argue that structural relaxation can also enhance crystallization due to the chemical ordering it induces.

Being out of equilibrium, MG have the tendency to lose their amorphous structure over a thermal treatment. Crystallization of MG can occur in various conditions, and the devitrification products can take a lot of different forms. It has been summarized as a phenomenon strongly dependent on the chemistry of the alloy (Louzguine-Luzgin 2014a). It has indeed been shown that nucleation of crystals in MG is heterogeneous and that nucleation sites likely preexist in the amorphous alloys (Louzguine-Luzgin 2014b). An interesting tool to predict crystallization of a specific MG during heating is the time-temperature-transformation diagram (see Fig. 2), or time-temperature-crystallinity diagram (Graeve et al. 2013).

One can also evaluate the GFA and stability of an amorphous alloy by using "glass-forming ability indicators" such as the reduced glass transition temperature, $T_{rg} = T_g/T_m$, or the width of the supercooled liquid region, $\Delta T = T_x - T_g$. Empirically, T_{rg} has been shown to be correlated to the critical casting diameter, while ΔT gives an insight on the temperature stability of the alloy and what forming time-temperature window the alloy accepts. If T_{rg} and ΔT are easily accessible, one should however notice that those indicators can only help to discriminate alloys among others. GFA prediction is not possible with such basic parameters and necessitates extensive modeling (Khan et al. 2017).

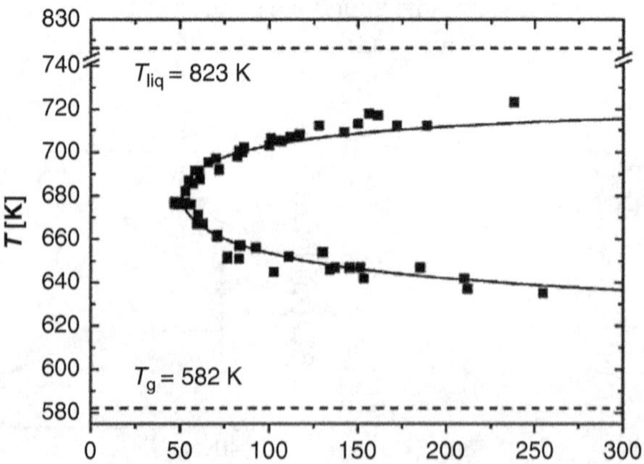

Fig. 2 Time-temperature-transformation diagram of a Pd-based MG (Löffler et al. 2000)

A MG does not possess a unique structure but rather exhibits various atomic arrangements depending on the cooling and annealing that can be rejuvenated by adapted thermomechanical treatments (Pan et al. 2018). This peculiar feature has a dramatic impact on properties. An emblematic example is for MG that can retain free volume (such as Zr-based) in rapid quenching: those glasses show plastic deformation though localized in shear bands in an unrelaxed state and fragile behavior in fully relaxed state. X-ray absorption observations coupled with Monte-Carlo and molecular dynamic simulations have shown that glasses are described by the arrangement of various icosahedral clusters (Cheng et al. 2009; Ma 2015). Various degrees of perfection can thus be identified for clusters, depending on composition and preparation, and are associated with the GFA. Another description consists in a mean field approach considering random distribution of atoms and free volumes, which is useful for analyzing mechanical properties, where flow and stress are accommodated by atoms and free volume rearrangement. Consequently, as a thermomechanical-coupled electric process, one easily anticipates that SPS would have strong impact on metallic glass properties.

2.3 Mechanical Properties

MG are isotropic materials and are less dense than their crystalline counterparts. This implies that the mean interatomic distances in MG are longer, which leads to ~30% lower values of E (Young's modulus) and ν (Poisson's ratio). Furthermore, the absence of crystalline lattice induces that the classical plastic deformation mechanisms such as dislocation movements are not present in MG. Consequently, at room temperature, their ultimate strength reaches elevated levels (up to 4 GPa for Fe-based MG), and the maximum elastic deformation is exceptionally high (around 2%). Macroscopically, MG appear fragile (fracture without macroscopic plastic strain). However, traces of plastic deformation localized in thin shear bands have been observed and extensively studied (Greer et al. 2013). Hence, at room temperature, the deformation is heterogeneous.

At higher temperature ($T > 0.8\ T_g$), the deformation regime gradually switches from heterogeneous to homogeneous, where all the atoms can participate in the deformation. In the homogeneous regime, shear bands disappear, and the MG can overcome more than 10^4% total strain (Inoue et al. 1997). Around T_g, the flow stress decreases dramatically as low as a few tens MPa and, as shown in Fig. 3, depends on both the temperature and the strain rate. This behavior has been described using diffusive displacements and free volume (Spaepen 1977) or shear transformation zones (Argon 1979). These two underlying plasticity mechanisms are thermally activated and describe the existence of the homogeneous deformation at high temperature and the heterogeneous deformation into shear bands at lower temperature. Yet, deformation of MG is still the subject of intensive research (Hufnagel et al. 2016).

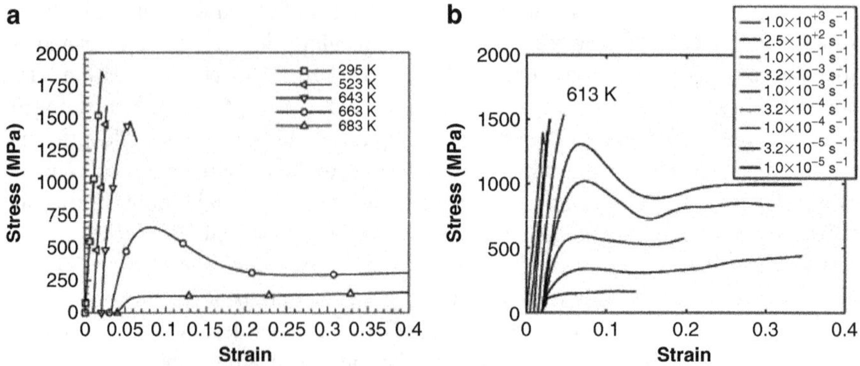

Fig. 3 Stress-strain curves as (**a**) a function of the temperature and (**b**) as a function of the strain rate (Lu et al. 2003; Johnson et al. 2002)

3 Sintering of Bulk Metallic Glasses

3.1 Sintering of Bulk Amorphous Parts

3.1.1 Al-Based Alloys

Studies on Al-based MG processed by SPS are reported in Table 1. Several experimental parameters are listed, such as powder type, die material, alloy characteristic temperatures, and sintering parameters (sintering temperature T_s, dwell time, heating rate, and applied stress). For the sake of comparison between the different chemical compositions, the ratio T_s/T_g is also indicated.

Al-based MG present a relatively low GFA and are thus mainly produced by mechanical alloying. Moreover, compared to other MG, their T_g is much lower, making them sensitive to devitrification. Several studies thus focused on studying the influence of SPS parameters on the microstructure. For instance, it was shown on two different alloys that increasing the applied stress not only permitted to increase the densification rate of the samples but also to reduce the nano-crystallization (Li et al. 2013c; Maurya et al. 2016b). The final microstructure is also strongly dependent on T_s, since an increase favors the formation of nanocrystalline phases (Sasaki et al. 2008; Deng et al. 2015; Maurya et al. 2016a). The role of the heating rate has also been investigated (Li et al. 2013b), and it was concluded that low heating rates were detrimental to the densification of the alloy, while increasing the risk of devitrification, thus leading to lower mechanical properties. Compressive strength values in the range of 800–900 MPa and around 1200 MPa for a $Al_{86}Ni_6Y_{4.5}Co_2La_{1.5}$ MG sintered at 5 and 40 K.min^{-1}, respectively, were reported (Li et al. 2013b).

Table 1 Al-based amorphous alloys

Composition	Powder[a]	Die[b]	T_g (K)	T_x (K)	T_s (K)	T_s/T_g	Dwell time (min)	Heating rate (K. min^{-1})[c]	Applied stress (MPa)	References
$Al_{82}La_{10}Fe_4Ni_4$	m	w*	560	650	617	1.10	3	10	400–600	Choi et al. (2007)
$Al_{85}Ni_{10}La_5$	a	w	541	546	473–753	0.87–1.39	–	–	150–440	Sasaki et al. (2008)
$Al_{86}Ni_6Y_{4.5}Co_2La_{1.5}$	a	w	544	554	521–581	0.96–1.07	1–20	5–40	200–600	Li et al. (2013a, b, c, d)
	m	w	?	472	673	–	15	100	25–400	Maurya and Laha (2016)
$Al_{86}Ni_6Y_6Co_2$	m	w	?	448	573–773	–	15	100	500	Maurya et al. (2016a)
$Al_{86}Ni_7Y_{4.5}La_{1.5}Co_1$	a	w	495	506	523–673	1.05–1.35	1.5	–	30	Deng et al. (2015)
$Al_{86}Ni_8Y_6$	m	w	492	537	523–773	1.06–1.57	2–20	100	100–500	Maurya et al. (2016b, c)
$Al_{88}Ni_6Ti_6$	m	w, g	749	793	723–773	0.97–1.03	10	–	75–450	Mula et al. (2010)

[a] *m* mechanically alloyed, *a* gas atomized, *r* milled melt-spun ribbon, *b* milled bulk ingot
[b] *w* tungsten carbide, *g* graphite, *s* steel, *k* kentanium. If not mentioned, graphite dies are innerly protected by a graphite foil, and ‡ indicates the use of a Ta foil instead. * points to an assumption on the die material, based on the experimental conditions detailed in the paper
[c] Values separated by a dash represent a range of values, whereas values separated by arrows indicate a change of the heating rate during a single densification

3.1.2 Fe-Based Alloys

The highest number of publications relative to SPS of MG is found for Fe-based alloys. This variety is due to the impact that industry could potentially benefit from "amorphous steels." Since the GFA of Fe-based MG is on the low side, the SPS processing of such alloys has rapidly been considered as very promising to obtain larger objects than by casting, with a first occurrence in 2002 (Shen et al. 2002). Table 2 gathers a number of studies that have been carried out on Fe-based MG.

Fe-based MG are often studied for their soft magnetic properties (references in italic in Table 2). One can, for instance, focus on a study of commercially available Finemet® ribbons (Gheiratmand et al. 2016). The authors obtain a powder from milling ribbons, which is further sintered to up to 97% density by SPS. A two-step densification mechanism is proposed, consisting in particle rearrangement followed by viscous flow. The measured magnetization and coercivity reach 122.9 emu.g^{-1} and 123 A.m^{-1}, respectively, and are attributed to the porosity and the presence of crystalline Fe(Si) particles.

Other studies emphasize the thermal behavior of Fe-based MG. Yamada et al. build a time-temperature-transformation diagram for a $Fe_{36}Co_{36}B_{19.2}Si_{4.8}Nb_4$ MG in SPS conditions (Yamada et al. 2009). The same kind of approach has been utilized to construct a "time-temperature-crystallinity" diagram for two Fe-based alloys (Graeve et al. 2013) in order to predict the amount and size of crystallized particles for specific sintering conditions. This can also be used as a guidance tool to identify optimal sintering conditions.

As it can be observed in Table 2, metalloid and nonmetallic elements, such as B, C, or P, are required to obtain Fe-based MG and are responsible for the properties of this family of amorphous alloys (strength up to 4 GPa but also brittleness and low GFA). However, they also easily react with metallic elements to form carbides or borides when the MG is heated near its crystallization temperature. Such phases are thus logically observed on SPSed Fe-based MG, as reported in several studies. $(Fe, Cr)_{23}C_6$ carbides were, for instance, observed on samples sintered at $1.06 \times T_g$ and further annealed at $1.14–1.26 \times T_g$ (Singh et al. 2012) or sintered near T_x ($T_s > 1.10 \times T_g$; Singh et al. 2013).

The role of the heating rate on the sintering of a $Fe_{48}Mo_{14}Cr_{15}Y_2C_{15}B_6$ MG was also analyzed (Paul and Harimkar 2017a, b). As shown in Fig. 4, which depicts the shrinkage as a function of temperature, a faster heating rate induces to lower-temperature densification. A second densification event is also observed, but it occurs at temperatures largely above T_x. The first densification event was further analyzed taking into account the viscosity of the alloy using Frenkel's equation. Since the viscosity coefficient follows an Arrhenius equation, it is possible, by plotting $\ln(d(\Delta L/L_0)/dT)$ versus $1/T$, to calculate the activation energy, Q, of the densification mechanism (viscous flow). The activation energy for the first sintering stage decreased with increasing heating rate, from 94 kJ.mol^{-1} at 25 K.min^{-1} to 55 kJ.mol^{-1} at 150 K.min^{-1}, explaining the lower-temperature densification.

Table 2 Fe-based amorphous alloys

Composition	Powder	Die	T_g (K)	T_x (K)	T_s (K)	T_s/T_g	Dwell time (min)	Heating rate (K. min^{-1})	Applied stress (MPa)	References
$Fe_{36}Co_{36}B_{19.2}Si_{4.8}Nb_4$	r	w	815	847	773–833	0.95–1.02	5	–	450–650	Prasanna et al. (2015)
	a	w	807	855	673–848	0.83–1.05	1–3	30–40	75–400	Yamada et al. (2009)
	a	w, g	807	855	753–833	0.93–1.03	1–3	33–42	75–400	Yodoshi et al. (2009)
$Fe_{48}Mo_{14}Cr_{15}Y_2C_{15}B_6$	a	g, w	848	926	748–1273	0.88–1.37	0–30	25–150	20–225	Graeve et al. (2013) Harimkar et al. (2009) Paul and Harimkar (2016, 2017a, b) Paul et al. (2017, 2018) Singh et al. (2010, 2012, 2013)
$Fe_{49.7}Cr_{17.7}Mn_{1.9}Mo_{7.4}W_{1.6}B_{15.2}C_{3.8}Si_{2.4}$	a	g	852	–	873–913	1.02–1.07	0–30	60	70	Graeve et al. (2013)
$Fe_{65}Co_{10}Ga_5P_{12}C_4B_4$	a	w	723	773	723–763	1.00–1.06	10	–	80–300	*Shen* et al. (2002, 2004) *Shen and Inoue* (2003)
$Fe_{65}Co_{10}Ga_5P_{12}C_4B_4$	r	w	806	854	763–813	0.95–1.01	1–15	–	300	*Ishihara* et al. (2003)
$Fe_{68.8}B_{22.9}Nb_{3.8}Y_{4.5}$	a	w	873	980	673–973	0.77–1.11	?	40	600	Lee et al. (2009a) Lee et al. (2009b)
$Fe_{75}Nb_3Si_{13}B_9$	m	w	804	854	723–883	0.9–1.10	10	30	500	Feng et al. (2016)
$Fe_{75}Si_{20}B_5$	m	g	583	748	723–1173	1.24–2.01	1–15	–	30	Neamțu et al. (2014)
$Fe_{76}Si_{9.6}B_{8.4}P_6$	a	w	783	831	720–780	0.92–0.99	–	–	600	*Li* et al. (2011a)

(continued)

Table 2 (continued)

Composition	Powder	Die	T_g (K)	T_x (K)	T_s (K)	T_s/T_g	Dwell time (min)	Heating rate (K. min^{-1})	Applied stress (MPa)	References
$Fe_{76}Si_9B_{10}P_5$	a	k	780	830	760	0.97	1–3	50–225	–	*Wang* et al. (2013a)
	a	w*	780	832	680–740	0.87–0.95	6	100	600	*Zhao* et al. (2009)
$Fe_{77}Ga_3P_{9.5}C_4B_4Si_{2.5}$	a	w*	743	771	723–743	0.97–1.00	–	–	600	*Shen* et al. (2005)
$Fe_{79.3}Mo_{4.5}P_{8.1}C_{6.75}B_{1.35}$	m	g‡	–	–	1073–1223	–	0–20	60–110	50	Yang et al. (2014a)
Finemet®	r	g	738	838	Max 833	< 1.13	–	30	40	Gheiratmand et al. (2016)

See Table 1 for notes on experimental details

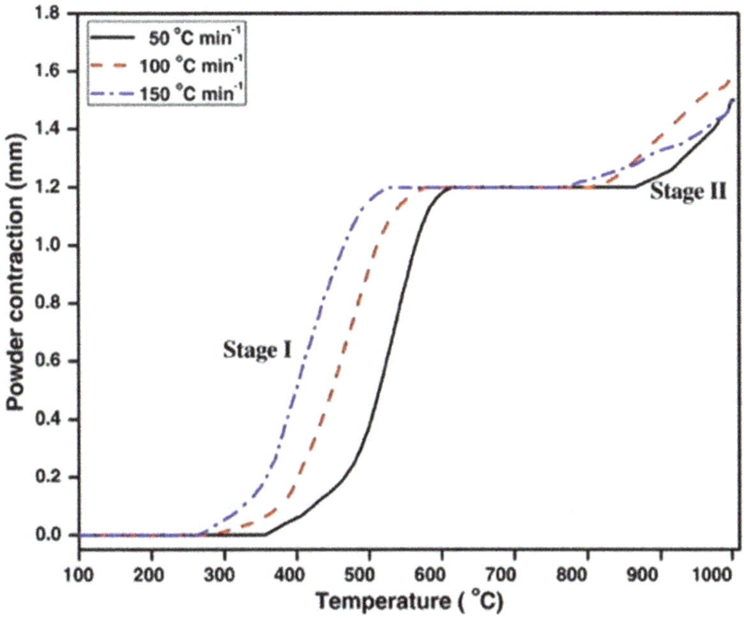

Fig. 4 Influence of the heating rate on the powder contraction during SPS. For reference, $T_g = 575$ K (Paul and Harimkar 2017a)

3.1.3 Ti-Based Alloys

Ti-based MG are also widely studied because of their potential in the biomedical domain and especially the biocompatibility of Ni-free alloys (Xie et al. 2012a). Studies relative to spark plasma sintering Ti-based MG are listed in Table 3.

One may notice a study of a $Ti_{40.6}Cu_{15.4}Ni_{8.5}Al_{5.5}W_{30}$ alloy prepared by mechanical alloying (El-Eskandarany et al. 2017). The study exhibits the importance of being in the supercooled liquid domain in order to attain full densification, since for $T_s < 700$ K (~0.9 × T_g), the relative density is as low as 65%. Optimal mechanical properties ($H_V \approx 12$ GPa) are obtained at 775 K (~1.05 × T_g) without apparent devitrification. However, depending on the MG, if T_s is increased, crystallization may start to occur at sintering temperatures as low as 1.01 × T_g (Li et al. 2012).

One can also highlight an analysis of the densification by viscosity (Chen et al. 2013a) using an approach similar to Paul and Harimkar's for Fe-based MG (see previous paragraph). The calculated activation energy of 340 $kJ.mol^{-1}$ is compatible with viscous flow-controlled densification. One may however keep in mind that this analysis implies a homogeneous temperature throughout the sample, which might not be the case, especially during the very first moments of the process.

Table 3 Ti-based amorphous alloys

Composition	Powder	Die	T_g (K)	T_x (K)	T_s (K)	T_s/T_g	Dwell time (min)	Heating rate (K. min^{-1})	Applied stress (MPa)	References
$Ti_{58}Cu_{22}Ni_{12}Al_8$	m	g	727	776	300–775	0.41–1.06	–	–	10	El-Eskandarany et al. (2017)
$Ti_{40.6}Cu_{15.4}Ni_{8.5}Al_{5.5}W_{30}$	m	g	737	827	300–775	0.41–1.05	–	–	10	El-Eskandarany et al. (2017)
$Ti_{40.6}Zr_{9.4}Cu_{37.5}Ni_{9.4}Al_{3.1}$	m	g[‡]	631	729	1050	1.66	5	20–140	30	Liu et al. (2015a)
$Ti_{45}Zr_{10}Cu_{31}Pd_{10}Sn_4$	a	w	663	717	623–717	0.94–1.08	10	–	600	Xie et al. (2012a)
$Ti_{48}Ni_{32}Cu_8Si_8Sn_4$	a	w	797	853	773–873	0.97–1.09	3–8	100 → 20	450	Chen et al. (2013a)
$Ti_{50}Cu_{20}Ni_{20}Al_{10}$	m	w	691	771	633–733	0.92–1.06	3	–	500	Nguyen et al. (2009)
$Ti_{50}Cu_{23}Ni_{20}Sn_7$	m	w	751	787	753–793	1.01–1.06	1–1.5	40–50	500	Li et al. (2009a, 2012)
$Ti_{42.5}Cu_{40}Zr_{7.5}Ni_5Sn_5$	a	w	696	749	673–713	0.97–1.02	1.5	40	500	Wang et al. (2010)

See Table 1 for notes on experimental details

3.1.4 Zr-Based Alloys

Contrary to Fe-based MG, Zr-based amorphous alloys are widely produced by casting. Some of them are even commercialized (Vitreloy alloys). They are thermally stable and often present a wide supercooled liquid region. The number of studies is thus lower than for other alloys, yet they can represent good candidates for studying the process or the structural evolution of the alloy during SPS. The different Zr-based alloys fabricated by SPS are gathered in Table 4.

The work of Xie et al. focuses on manufacturing porous materials (see paragraph 3.3.2), thus implying low sintering temperatures ($T_s < 0.9 \times T_g$). Since T_s is relatively low, no devitrification is observed. The evolution of mechanical properties, such as compressive strength (Xie et al. 2006, 2007a) or elasticity moduli (Xie et al. 2010a) with the porosity content, is analyzed, leading to the observation that the apparent elastic moduli decrease with increasing porosity. However, a low amount of porosity (<5%) can induce a small ductility ($\varepsilon_p \sim 1\%$) of the material, without reducing the compressive strength compared to an as-cast material.

The effect of the size of the amorphous powder was investigated on $Zr_{57}Cu_{20}Al_{10}Ni_8Ti_5$ (Nowak et al. 2011), indicating that greater consolidation is achieved for samples with small initial particle size. Further consolidation is also obtained if particles are selected to a thin size distribution. The consolidation is thus believed dependent on the particle size, which can be explained by a higher reactivity of fine powders, in relation to their larger surface curvature. One however emphasizes that the sintering rate is modulated by physical properties of the amorphous phase, in particular its viscosity. The size effect on consolidation efficiency is well known in conventional sintering. For SPS, it seemingly holds, but in the particular case of MG powders, it must be balanced with the devitrification amount. A compromise is then necessary in terms of particle size in order to obtain strong sintered glasses with limited crystallization.

3.1.5 Other Alloys

Finally, Co-, Cu-, Mg-, and Ni-based MG are gathered in Table 5. Similar to Fe-based MG, Co-based MG are widely prepared as ribbons, due to their interesting magnetic properties, and used in high-frequency transformers, for example. Yet, only one study to date has been identified on SPS fabrication of bulk parts from a Co-based MG (El-Eskandarany et al. 2005). In this study, the authors report a fully densified (99.6% density), high-strength ($\sigma_{max} = 2.4$ GPa, $E = 176$ GPa, $H_V \approx 8.7$ GPa) material, showing no sign of devitrification.

The work of Cardinal et al. is based on $Cu_{50}Zr_{45}Al_5$, which is only available to diameters up to 3 mm by casting (Cardinal et al. 2016). This emphasizes the interest of powder metallurgy and SPS to overcome the limitations of traditional processes. Devitrification is observed at 0.99 T_g and is complete for $T_s = 1.02 \times T_g$. However, an optimization of the sintering parameters shows that similar densification and

Table 4 Zr-based amorphous alloys

Composition	Powder	Die	T_g (K)	T_x (K)	T_s (K)	T_s/T_g	Dwell time (min)	Heating rate (K. min^{-1})	Applied stress (MPa)	References
$Zr_{55}Cu_{30}Al_{10}Ni_5$	a	g	685	774	573–623	0.83–0.91	10	50 → 5	20–40	Xie et al. (2006, 2007a 2010a)
$Zr_{57}Cu_{20}Al_{10}Ni_8Ti_5$	a	w, g	660	720	663	1.01	0–10	60 → 10	100 or 500	Nowak et al. (2011) Perrière et al. (2011, 2012)
$Zr_{65}Al_{10}Ni_{10}Cu_{15}$	a	w*	645	750	620–705	0.96–1.09	–	50 → 5	600	Shin et al. (2007, 2010)

See Table 1 for notes on experimental details

Table 5 Co-, Cu-, Mg-, and Ni-based alloys processed by SPS

Composition	Powder	Die	T_g (K)	T_x (K)	T_s (K)	T_s/T_g	Dwell time (min)	Heating rate (K.min^{-1})	Applied stress (MPa)	References
$Co_{65}Ti_{20}W_{15}$	m	–	786	878	843	1.07	3	300	20–40	El-Eskandarany et al. (2005)
$Cu_{46}Zr_{42}Al_7Y_5$	a	w	672	773	593–703	0.88–1.05	10	50 → 5	600	Chu et al. (2012a)
$Cu_{47.6}Ti_{32.7}Zr_{10.5}Ni_{6.1}Sn_{2.1}Si_1$	a	w, g	722	759	723–753	1.00–1.04	0.5–5	40	80–300	Kim et al. (2005a, 2008a)
$Cu_{50}Zr_{45}Al_5$	a	g, w	711	769	673–723	0.95–1.02	5–30	100 → 20	90	Cardinal et al. (2016)
$Cu_{54}Ni_6Zr_{22}Ti_{18}$	a	w*	715	756–767	743–843	1.04–1.18	1	–	280	Kim et al. (2005b, 2008b)
$Mg_{55}Cu_{30}Gd_{15}$	b	g	428	452	373–448	0.87–1.05	5	50 → 12.5	23.5	Méar et al. (2009)
$Mg_{55}Cu_{30}Y_{15}$	b	g	425	458	373–448	0.88–1.05	5	50 → 12.5	23.5	Méar et al. (2009)
$Mg_{65}Cu_{25}Gd_{10}$	a	s	421	475	423–443	1.00–1.05	3	340	500	Zheng et al. (2013)
$Mg_{75}Ni_{15}Si_{10}$	m	w*	839	890	843	1.01	1	–	100–300	Ozaki et al. (2004)
$Ni_{52.5}Nb_{10}Zr_{15}Ti_{15}Pt_{7.5}$	a	w	821	904	673–821	0.82–1.00	10	50 → 5	600	Xie et al. (2007b, 2009a)
$Ni_{59}Zr_{15}Ti_{13}Si_3Sn_2Nb_7Al_1$	a	w*	839	890	843	1.01	1	–	100–300	Lee et al. (2007a)

See Table 1 for notes on experimental details

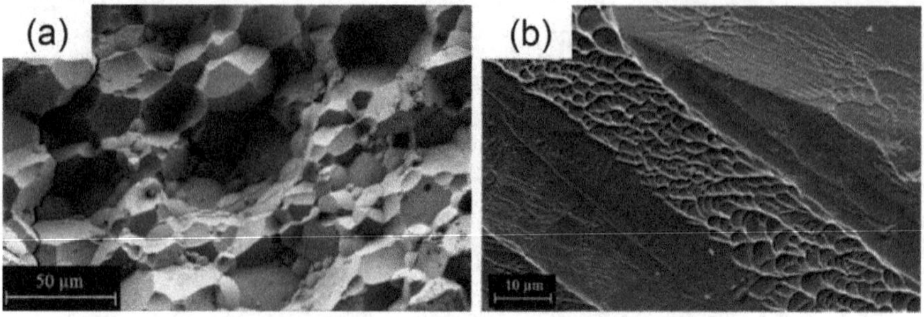

Fig. 5 SEM fractography images for $Cu_{50}Zr_{45}Al_5$ sintered in the same conditions (693 K for 5 min), except for the applied stress of (**a**) 90 MPa and (**b**) 1 GPa (Cardinal et al. 2016)

mechanical properties can be achieved in different conditions: 5 min at 693 K and 30 min at 683 K give the same density. More specifically, this study sheds light on the role of the applied stress, since a high compacting stress can modify the fracture mode: intergranular fracture is observed for a low applied stress, while ductile-type fracture is spotted for a pellet sintered under a higher stress (see Fig. 5).

Xie et al. studied the sintering of a Ni-based MG (Xie et al. 2007b, 2009a). Full densification is obtained for sintering cycles of 10 min and T_s as low as $0.91 \times T_g$, but XRD measurements show that the first signs of devitrification appear for $T_s = 0.97 \times T_g$. The optimal sintering temperature seems to be $T_s = 773$ K ($0.94 \times T_g$), since the pellet is fully dense (99.95%), is totally amorphous, and exhibits a compressive strength of 2.49 GPa. The compressive strength is maximal for $T_s = 0.97 \times T_g$ and decreases for higher T_s, indicating that the crystallized phases are detrimental for the mechanical properties.

3.2 *Sintering of Metallic Glass-Based Composites*

3.2.1 MG Matrix: MG Reinforcement

MG-based composites are widely studied, since the intrinsic macroscopic brittleness of bulk amorphous alloys is a major challenge to overcome. Powder metallurgy is a promising route to add a second phase that can add some degree of plastic deformation to a glass. The second phase can also be used to hinder the initiation of shear bands or deviate them during their propagation. In this case, the composite will present a delayed fracture. The first type of MG-based composites that is reported here is MG-MG composites or "dual-phase amorphous composites," and relative studies are listed in Table 6.

Composite materials combine properties from each phase that they associate. By mixing two MG powders, one can obtain a material exhibiting high strength and good magnetic properties (Xie et al. 2009b, 2011a). A Ni-based MG and a Fe-based MG, with similar particle sizes, were mixed in an equal volume ratio. A density of

Table 6 Ex situ MG-MG composites prepared by SPS

Alloy 1			Alloy 2								
Composition	Powder	T_g (K)	Composition	Powder	T_g (K)	fraction	T_s (K)	Dwell time (min)	Heating rate (K. min^{-1})	Applied stress (MPa)	References
$Cu_{54}Ni_6Zr_{22}Ti_{18}$	a	716	$Ni_{59}Zr_{15}Ti_{13}Nb_7Si_3Sn_2Al_1$	a	836	10 and 90 wt.%	743–843	1	–	280	Kim et al. (2005c, 2007a)
$Cu_{54}Ni_6Zr_{22}Ti_{18}$	a	716	$Zr_{62}Al_8Ni_{13}Cu_{17}$	a	660	10–30 vol. %	720	1	–	600	Lee et al. (2007b, c, d, 2010a) Lee (2007)
$Fe_{55}C_{10}B_5P_{10}Ni_{15}Mo_5$	a	719	$Fe_{81.4}Si_3B_{10}P_5Cu_{0.6}$	a	–	50–85 wt. %	733	10	–	–	Luan et al. (2016)
$Ni_{52.5}Nb_{10}Zr_{15}Ti_{15}Pt_{7.5}$	a	829	$Fe_{73}Si_7B_{17}Nb_3$	a	818	50 vol.%	748–798	10	50 → 5	600	Xie et al. (2009b, 2011a)

Please refer to Table 1 notes for details on powder types. All materials listed in this table have been or are assumed to have been sintered in a WC die

99% is reached for $T_s = 773$ K, which is lower than both T_g of the two MG. The composite displays a compressive stress of 2.6 GPa, comparable to the Ni-based MG sintered in the same conditions, and a saturation magnetization of 0.6 T originating from the random distribution of Fe-based MG particles.

Lee et al. associated a Cu-based MG with a Zr-based MG with various volume ratios (Lee et al. 2010a). The two amorphous powders are milled together and remain amorphous after milling and SPS. A third amorphous phase appears during the milling step and acts as an interface between the two initial MG, which is believed to transfer the load between the two phases. The addition of up to 30 vol.% of Zr-based MG allows the composite to attain ~3% plastic strain.

3.2.2 MG Matrix: Ceramic Reinforcement

Composites combining MG and a ceramic or inorganic reinforcement are collected in Table 7 and sorted by nature of the reinforcement. Several types of ceramics have been associated with MG: oxides such as alumina or zirconia, carbides, nitrides, and carbon in different allotropic forms. Except for diamond additions, the ceramic quantity is below 30 vol.%. In a large majority of the studies, the ceramic particle size is lower than a few microns.

Addition of a second ceramic phase, brittle by nature, is counterintuitively employed to increase the ductility of the MG. This has, for example, been observed with ZrO_2 reinforcement as it is displayed in Fig. 6 (Chu et al. 2015a). Depending on the size and quantity of the reinforcement, a few percents of plastic deformation can be achieved. The mismatch of elastic properties induces stress concentrations at the interfaces, which leads to the initiation of multiple shear bands in the glass, and ultimately dissipates a fraction of the mechanical energy before one shear band propagates and the composite fails. The presence of exogenous particles conducts to multiplication, deviation, and branching of the shear bands in the MG, as observed for $Ni_{52.5}Nb_{10}Zr_{15}Ti_{15}Pt_{7.5}$-SiC composites (Xie et al. 2010b). Fracture surfaces of MG-ceramic composites indeed exhibit a clear difference between monolithic MG (single plane at about 45° from the compression direction), since numerous ceramic particles can be observed on the rough fracture surface.

Magnetic properties of composites combining Fe-based MG with nanometric $Zn_{0.5}Ni_{0.5}Fe_2O_4$ (Liu et al. 2012, Zhao et al. 2016) have also been investigated. The obtained microstructure is micro-cellular, the MG being localized at the core, surrounded by the ceramic. This structure enhances the resistivity and decreases the eddy loss at high frequency. Both phases being ferromagnetic, the permeability attains 108.9 $H.m^{-1}$ (Liu et al. 2012). The sintering temperature (~800 K in both studies) is in the range of T_g of both MG, whereas $Zn_{0.5}Ni_{0.5}Fe_2O_4$ is usually sintered between 1273 and 1473 K. Yet, no porosity is observed and full densification seems reached. A complex interface between the two phases, containing numerous α-Fe nanocrystals, is observed (Zhao et al. 2016). This peculiar microstructure is believed to originate from local overheating between powder particles due to the resistive nature of $Zn_{0.5}Ni_{0.5}Fe_2O_4$.

Table 7 Ex situ MG-ceramic composites prepared by SPS, sorted by the nature of the second phase

MG composition	Powder	T_g (K)	Second phase	Fraction	T_s (K)	Dwell time (min)	Heating rate (K. min^{-1})	Applied stress (MPa)	References
$Zr_{55}Cu_{30}Al_{10}Ni_5$	a	685	Al_2O_3	10 vol.%	623	10	50 → 5	600	Xie et al. (2008a)
$Ti_{66}Nb_{13}Cu_8Ni_{6.8}Al_{6.2}$	m	715	B_4C	0.2–0.3 wt.%	1023	0	50–325	50	Li et al. (2011b)
$Cu_{54}Ni_6Zr_{22}Ti_{18}$	a	712	Diamond	30–50 vol.%	740–753	0.5–1	40	300	Kim et al. (2009b) Shin et al. (2009a)
$Zr_{65}Al_{10}Ni_{10}Cu_{15}$	a	–	Diamond	10–70 vol.%	–	–	–	–	Shin et al. (2009b)
$Ti_{42.5}Cu_{40}Zr_{7.5}Ni_5Sn_5$	a	696	Diamond	5 vol.%	713	1.5	40	500	Wang et al. (2013b)
$Ti_{66}Nb_{13}Cu_8Ni_{6.8}Al_{6.2}$	m	715	Graphite	0.1 wt.%	1023	0	50–325	50	Li et al. (2011b)
$Fe_{48}Cr_{15}Mo_{14}Y_2C_{15}B_6$	–	–	Graphite	1–4 wt.%	848	15	100	70	Ji et al. (2016)
$Ti_{45}Zr_{10}Cu_{31}Pd_{10}Sn_4$	a	663	Hydroxyapatite	2 wt.%	643	10	50 → 12.5	600	Xie et al. (2013)
$Ti_{40}Zr_{10}Cu_{36}Pd_{14}$	a	665	Hydroxyapatite	12–30 vol.% or 1–10 wt.%	643	10	50 → 10 → 5	600	Zhu et al. (2008, 2011)
$Zr_{55}Cu_{30}Al_{10}Ni_5$	a	685	Si_3N_4	10 vol.%	623	10	50 → 5	600	Xie et al. (2008a)
$Ni_{52.5}Nb_{10}Zr_{15}Ti_{15}Pt_{7.5}$	a	821	SiC	10 vol.%	773	10	50 → 5	600	Xie et al. (2008b, 2009c, 2010b)

(continued)

Table 7 (continued)

MG composition	Powder	T_g (K)	Second phase	Fraction	T_s (K)	Dwell time (min)	Heating rate (K. min^{-1})	Applied stress (MPa)	References
$Fe_{73}Si_7B_{17}Nb_3$	a	818	SiC	0.5–10 vol.%	773	10	–	600	Xie et al. (2012b)
$Al_{65}Cu_{20}Ti_{15}$	m	–	TiO_2	10 wt.%	473–873	10	150–200	50	Roy et al. (2007, 2008)
$Ti_{66}Nb_{13}Cu_8Ni_{6.8}Al_{6.2}$	m	715–746	WC	4.5–12.5 vol.%	1013–1173	6	50–850	50	Li et al. (2009b) Yang et al. (2012)
$Fe_{73.5}Cu_1Nb_3Si_{13.5}B_9$	a	–	$Zn_{0.5}Ni_{0.5}Fe_2O_4$	5 wt.%	823	3	130	600	Liu et al. (2012)
$Fe_{76}Si_9B_{10}P_5$	a	780	$Zn_{0.5}Ni_{0.5}Fe_2O_4$	10 wt.%	740–800	0.5	100	500	Zhao et al. (2016)
$Ni_{52.5}Nb_{10}Zr_{15}Ti_{15}Pt_{7.5}$	a	821	ZrO_2	10 vol.%	753	10	50 → 5	600	Xie et al. (2007c)
$Zr_{55}Cu_{30}Al_{10}Ni_5$	a	685	ZrO_2	5–20 vol.%	623	10	50 → 5	600	Chu et al. (2013, 2015a) Xie et al. (2007d, e)

Please refer to Table 1 notes for details on powder types. Except (Zhao et al. 2016) for which the sintering die is kentanium, all samples have been sintered either in graphite die (stress <100 MPa) or WC die (stress >100 MPa)

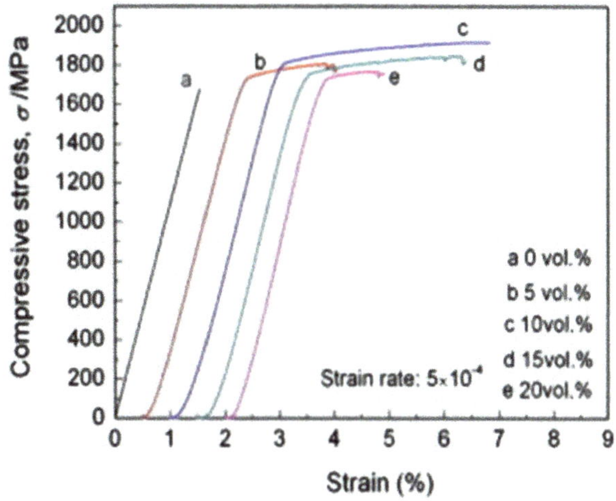

Fig. 6 Compressive stress-strain curves of $Zr_{55}Cu_{30}Al_{10}Ni_5$-$ZrO_2$ composites prepared by SPS. MG particle size is under 53 μm; ZrO_2 particle size is under 20 μm (Chu et al. 2015a)

Biocompatible composites combining Ti-based MG (without nickel nor beryllium) and hydroxyapatite have also been investigated. Hydroxyapatite is coated on medical implants to favor bone growth, but those coatings are subject to flaking, which reduce the lifespan of the implant. Co-sintering of the structural material and the functional one is thus a promising solution to develop new strategies of medical implant manufacturing. Sintering of such composites is carried out in the vicinity of T_g of the Ti-based MG (~0.97 × T_g), much higher than the temperature of decomposition of hydroxyapatite (>423 K). The added quantity (<3 wt.%) and the sintering conditions, especially the use of a high compacting stress, are determinant to obtain a dense material showing no sign of devitrification in the glass and no sign of hydroxyapatite decomposition, while retaining an acceptable level of mechanical strength (Zhu et al. 2011).

3.2.3 MG Matrix: Metallic Reinforcement

Table 8 aggregates the studies that have been carried out on composites containing a MG and a crystalline metallic phase. The vast majority of work focus on the addition of a pure metal, such as Al or Cu, or a ductile alloy to a metallic glass. The aim of such a strategy is clear: taking advantage of both the ductility of the second phase and the strength of the MG to obtain a unique combination of high strength (>1 GPa) with a ductile behavior. Contrary to the MG-ceramic composites, the quantity of the second phase is high, since volume ratio as high as 90 vol.% has been reported in the literature (see Table 8). The granulometry of the two phases is also closer to the exception of a study where a Zr-based MG was combined to nanostructured Fe (Cho et al. 2009).

The sintering conditions are mainly dictated by the thermal properties of the MG. T_s is thus chosen in the vicinity of T_g in order to benefit from the viscous flow of

Table 8 Ex situ MG-metal composites prepared by SPS, sorted by the nature of the second phase

MG composition	Powder	T_g (K)	Second phase	Fraction	T_s (K)	Dwell time (min)	Heating rate (K. min^{-1})	Applied stress (MPa)	References
$Zr_{57}Cu_{20}Al_{10}Ni_8Ti_5$	a	660	Al	10–90 vol.%	663	3	60 → 10	100	Perrière and Champion (2012) Sauvage et al. (2014)
$Cu_{47}Ti_{33}Zr_{11}Ni_6Sn_2Si_1$	a	722	Brass	20–40 vol.%	733	0.5	40	700	Son et al. (2009)
$Cu_{54}Ni_6Zr_{22}Ti_{18}$	a	716	Brass	10–30 vol.%	720	1	–	200–600	Lee et al. (2005, 2007d, e) Lee and Kim (2007)
$Ni_{59}Zr_{15}Ti_{13}Si_3Sn_2Nb_7A_{l1}$	a	837	Brass	10–30 vol.%	843	1	–	300	Lee et al. (2007e) Lee (2008)
$Ni_{59}Zr_{20}Ti_{16}Si_2Sn_3$	a	843	Brass	40 vol.%	843	–	–	300	Kim et al. (2013)
$Cu_{47}Ti_{33}Zr_{11}Ni_6Sn_2Si_1$	a	722	Cu	10–40 vol.%	733	0.5	40	300–700	Kim et al. (2007b, c) Nam et al. (2008) Son et al. (2009)
$Cu_{50}Zr_{45}Al_5$	a	714	Cu	5–40 vol.%	693	10	50 → 5	600	Xie et al. (2010c)
$Fe_{68.2}C_{5.9}Si_{3.5}B_{6.7}P_{9.6}Cr_{2.1}Mo_2Al_2$	a	790	Cu	5–15 wt.%	793–843	10	–	500	Kim et al. (2009a)
$Fe_{85\text{-}95}Si_{5\text{-}10}B_{1\text{–}5}$	Ribbon	–	Cu	15 vol.%	723	20	100	350	Harimkar et al. (2011)
$Ni_{52.5}Nb_{10}Zr_{15}Ti_{15}Pt_{7.5}$	a	821	Cu	1–40 vol.%	773	10	50 → 5	600	Xie et al. (2011b)
$Ni_{57}Zr_{20}Ti_{16}Si_2Sn_3$	a	821	Cu	10 wt.%	703–873	10	–	350	Kim et al. (2006, 2007d)
$Ni_{59}Zr_{20}Ti_{16}Si_2Sn_3$	a	843	Cu	40 vol.%	843	–	–	300	Kim et al. (2013)

$Zr_{62}Al_8Ni_{13}Cu_{17}$	a	660	$Fe_{84.9}Si_{9.5}Al_{5.6}$	20 vol.%	660–738	–	–	300	Cho et al. (2009)
$Al_{65}Cu_{16.5}Ti_{18.5}$	m	–	$Al_{0.6}CoCrFeNi$	50 vol.%	823	–	–	400	Tan et al. (2016)
$Cu_{54}Ni_6Zr_{22}Ti_{18}$	a	715	Ni	15–35 vol.%	715–756	–	–	–	Huh et al. (2007)
$Fe_{85\text{-}95}Si_{5\text{-}10}B_{1-5}$	Ribbon	–	Ni	15 vol.%	723	20	100	350	Harimkar et al. (2011)
$Fe_{49.7}Cr_{17.1}Mn_{1.9}Mo_{7.4}W_{1.6}B_{15.2}C_{3.8}Si_{2.4}$	a	–	Ta	< 40 vol.%	800–970	0–40	50–500	110	Kelly et al. (2016)
$Ti_{40}Zr_{10}Cu_{36}Pd_{14}$	a	665	Ti	5–20 wt.%	643	10	50→10→5	600	Zhu et al. (2013)
$Zr_{55}Cu_{30}Al_{10}Ni_5$	a	685	TiNb	5–20 vol.%	623	10	50 → 5	600	Chu et al. (2012b, 2013, 2015b)
$Ni_{59}Zr_{20}Ti_{16}Si_2Sn_3$	a	843	W	50 vol.%	850	4–7	40	300	Lee et al. (2015)
$Fe_{48}Cr_{15}Mo_{14}Y_2C_{15}B_6$	a	848	W	5–10 wt.%	798	15	150	–	Singh and Harimkar (2010)
$Fe_{49.7}Cr_{17.1}Mn_{1.9}Mo_{7.4}W_{1.6}B_{15.2}C_{3.8}Si_{2.4}$	a	–	W	< 40 vol.%	800–970	0–40	50–500	110	Kelly et al. (2016)
$Ni_{52.5}Nb_{10}Zr_{15}Ti_{15}Pt_{7.5}$	a	821	W	5–20 vol.%	773	10	50 → 5	600	Xie et al. (2008b, 2010d)
$Zr_{57}Cu_{20}Al_{10}Ni_8Ti_5$	a	660	W	50 vol.%	723	3	56 → 10	510	Nowak et al. (2009)
$Cu_{50}Zr_{45}Al_5$	a	711	Zr	5–50 vol.%	693	10	100 → 20	600	Cardinal et al. (2018)

Please refer to Table 1 notes for details on powder types. All studies have been carried out either with graphite die (stress <100 MPa) or WC die (stress >100 MPa) the amorphous alloy. The applied compacting stress is generally higher than for the monolithic MG. It is hypothesized to be necessary to attain full densification if the second phase particles do not deform or densify by themselves under the conditions required by the MG. However, in some cases, co-sintering of the MG and the reinforcement is possible. For instance, Zr-based MG overcome their T_g between 600 and 700 K, a range of temperature where aluminum can be sintered. This favorable domain has been taken into advantage in a study on $Zr_{57}Cu_{20}Al_{10}Ni_8Ti_5$-Al composites (Perrière and Champion 2012).

The mechanical behavior is often the center of interest of the studies on MG-metal composites, even though some composites made from copper as the reinforcement have been studied for their electrical properties. This is, for instance, the case of the study of the electrical conductivity on $Ni_{52.5}Nb_{10}Zr_{15}Ti_{15}Pt_{7.5}$-Cu composites (Xie et al. 2011b). Generally, the properties of a composite are defined by the amount of the second phase, the morphology of the reinforcement (shape, size, etc.), and the nature of the interface.

Modeling plays an important part in explaining such evolution of properties and predicting the behavior of composite materials. An example lies in the study of a $Cu_{50}Zr_{45}Al_5$-Zr composite (Cardinal et al. 2018). The mean interparticle spacing was calculated and connected to the size of the plastic zone in the metallic glass. This parameter obviously depends on the volume fraction and the size of the reinforcement particles. This simple model predicts whether a composite will exhibit some macroscopic plasticity or not. Indeed, this study shows that the interparticle distance has to be smaller than the plastic zone of the amorphous matrix in order to obtain macroscopic ductility. The toughness of the MG, whose square root is proportional to the size of the plastic zone, is also taken into account in order to predict an optimal size over quantity ratio in order to observe plastic deformation. Cardinal et al. conclude that the tougher the matrix, the easier the plasticizing effect is obtained. The same result is obtained when the particle size is reduced.

One may also mention the description of the mechanical behavior of a $Zr_{55}Cu_{30}Al_{10}Ni_5$-TiNb composite by finite element calculations (Chu et al. 2015b). This study sheds light on the stress concentrations arising around the TiNb particles, which favor shear-band initiation, ultimately leading to a more homogeneous macroscopic plastic deformation (in MG, plastic deformation is localized at room temperature; see Fig. 3a). This is confirmed by compression tests, since the MG shows a strictly elastic behavior, while the composites containing 20 vol.% of TiNb exhibit ~4% of plastic strain.

A recent study has been carried out on a composite combining an Al-based MG to a so-called "high entropy alloy" (HEA) $Al_{0.6}CoCrFeNi$ (Tan et al. 2016). HEA belong to a recent class of metallic alloys that is extensively studied for their promising mechanical properties. The glass is partially devitrified after sintering, due to the elevated $T_s > T_x$. However, the resulting microstructure (Fig. 7) exhibits wide interdiffusion layers between the two phase domains. The interdiffusion layer forms from the enhanced diffusion at the contact points between powder particles, which is a consequence of a potential temperature inhomogeneity. The resulting composite presents an extremely high strength (3.1 GPa, almost double than the

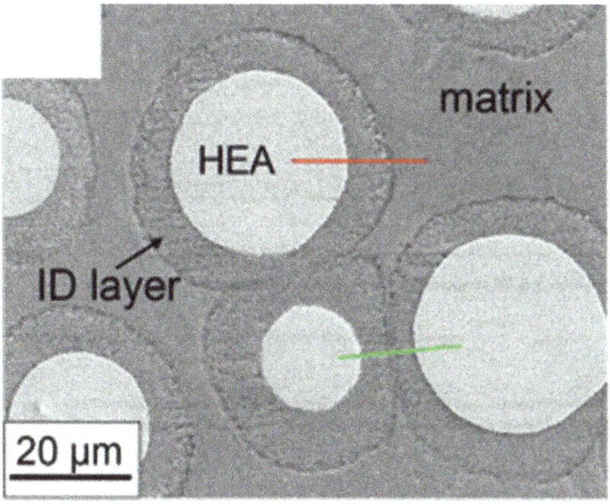

Fig. 7 Microstructure of a $Al_{65}Cu_{16.5}Ti_{18.5}$ (MG)-$Al_{13}Co_{21.75}Cr_{21.75}Fe_{21.75}Ni_{21.75}$ (HEA) composite. "matrix" denotes the partially devitrified MG (Tan et al. 2016)

monolithic MG), which is attributed to the crack propagation in the composite: cracks initiate in the amorphous domains but are deflected in the interdiffusion layers and arrested by the ductile HEA domains. However, the composite shows no plasticity, presumably due to a poor load transfer between the different ductile HEA particles.

Another direction for MG-metal composites is the addition of a heavy metal, essentially tungsten. Due to its high melting point, W stays inert during SPS, since the temperatures required for MG densification are quite low (roughly 700–1000 K from Table 8). T_s is thus close to T_g of the used MG, meaning that the densification of MG-W composites relies on the densification of the MG. Consequently, the size of the W particles becomes an important parameter, since any contact between W particles will be a potential weak point in the microstructure and must be avoided. Hence, one has to adjust the quantity of reinforcement according to the particle size and vice versa. For instance, a 20 µm MG powder and a 1–5 µm W powder were used to prepare a composite (Singh and Harimkar 2010) containing 10 wt.% of W for a relative density of 89%. Xie et al. could manufacture a composite from a Ni-based MG with 20 vol.% W attaining 98% of relative density using a 30 µm MG powder and 5 µm W powder (Xie et al. 2010d). Larger W contents (up to 40 vol.%) are reported using micrometric powders (Kelly et al. 2016), the final density (and thus the densification) depending on the granulometry of the W particles.

Finally, an alternative route to prepare MG-metal composite consists in functionalizing the amorphous powder before sintering. Kim et al. coated a Ni-based MG powder with Cu by spray-drying (Kim et al. 2007d). The sintered microstructure is cellular, with 50 µm Ni-MG particles surrounded by a < 1-µm-thick copper layer. The same strategy was adopted on a Cu-based MG coated with crystalline Ni (Huh et al. 2007). Another occurrence concerns a Fe-based powder coated by nanocrystalline Cu (Kim et al. 2009a). This approach leads to microstructures where the glass is dispersed into a metallic matrix, even though the volume fraction of the crystalline phase is very low. In a similar approach, the amorphous Fe-based melt-spun ribbons were coated with Cu or Ni, stacked, and sintered (Harimkar et al. 2011). A high compacting stress (350 MPa) is required to avoid debonding between the layers of the laminated microstructure.

3.2.4 MG Matrix: Polymer Reinforcement

MG also have been associated with a polymer, polytetrafluoroethylene (PTFE), and sintered by SPS (Zadorozhnyy et al. 2017) in order to improve physical and mechanical properties, PTFE being well known for its low friction coefficient. The choice of the amorphous phase ($Cu_{54}Pd_{28}P_{18}$) is driven by the thermal properties of the polymer, in order to avoid its degradation or melting. For $Cu_{54}Pd_{28}P_{18}$, $T_g = 519$ K and $T_x = 551$ K, whereas the melting point of PTFE is around 615 K. In this study, the melt-spun, milled MG and the polymer were mixed at a 10:1 weight ratio. The SPS was carried out in graphite molds under a compacting stress of 50 MPa, at 518 K ($0.99 \times T_g$) for 5 min at 50 K.min^{-1}. The repartition of the two phases appears rather homogeneous, and no porosity is visible. However, the authors emphasize on the appearance of crystalline phases, as well as copper oxide, at the boundaries between the particles which are detrimental to the physical properties.

3.2.5 Metallic Matrix: MG Reinforcement

Ex situ composites can also be prepared with a crystalline matrix and amorphous reinforcement. This strategy has been used to reinforce light metals. It opens opportunities to significantly improve the performances of metallic matrix composites, since MG exhibit high strength. For example, Ferre et al. added 4 vol.% of spherical $Zr_{57}Cu_{20}Al_{10}Ni_8Ti_5$ amorphous particles to pure aluminum and a 5083 Al alloy (Ferre et al. 2015) and investigated the damage resistance by X-ray tomography. Tan et al. also combined pure aluminum with 40–60 vol.% of a mechanically alloyed $Al_{65}Cu_{16.5}Ti_{18.5}$ MG (Tan et al. 2018). T_s was set above T_x, in order to partially crystallize the amorphous phase and promote interdiffusion between the Al particles and the amorphous ones. The obtained microstructure is capable to withstand plastic strain up to 4.3% and fracture stress higher than 1.7 GPa. Finally, Wang et al. reinforced a 7075 Al alloy with 15 vol.% of Zr-based MG fibers (Wang et al. 2016), leading to a yield strength increase from 168 MPa to 366 MPa. This composite also exhibited 25% plastic strain. This damage tolerance was explained by a 2–3 nm interdiffusion layer at the interfaces, which allowed to avoid interface failure and crack blunting.

This concept has also been utilized to prepare a W- $Hf_{44.5}Cu_{27}Ni_{13.5}Ti_5Al_{10}$ composite (Lee et al. 2010b) containing 50–70 vol.% of W for application as a kinetic energy penetrator. The two powders were milled together, in order to induce nano-structuration. The sintered microstructure is fully dense and composed of 200–500-nm-wide layers of W and MG. If the strength of the composite decreases compared to the monolithic MG, plastic deformation is slightly enhanced and can reach almost 1% for 70 vol.% W.

3.2.6 In Situ Composites by Partial Devitrification

For in situ composites, the reinforcements are created directly during sintering from partial and controlled devitrification of the MG. This intentional devitrification of the MG matrix induces a volume fraction of small crystallites that are hoped to strengthen the material. Usually, nanocrystals are induced in MG during a thermal treatment and aimed at deviating the shear bands by shear delocalization and delay fracture by deviating crack propagation (Qiao et al. 2016). The apparent ductility is thus increased, and the glass can withstand a small amount of plastic deformation.

The sintering temperatures for the fabrication of in situ composites by SPS tend to be higher than T_g or even higher than T_x. The dwell time and heating rate are set according to allow for devitrification. The chemical composition of the alloy is also important since it has an important influence on the GFA and the thermal stability of the alloy. Indeed, a poor glass former alloy, for instance, an Al- or Fe-based MG, will tend to crystallize more rapidly than a good glass former alloy such as a Pt-based MG. Table 9 gathers studies on partially crystallized MG processed by SPS. For instance, Chen et al. report the apparition of a volume fraction of 45% of 5-nm-wide β-Ti crystallites (Chen et al. 2013b), whereas Drescher et al. obtain up to 45 vol.% of 4–16 nm crystallites (Drescher et al. 2016). Singh et al. fabricated a Fe-based MG with ~20 nm $(Fe,Cr)_{23}(C,B)_6$ crystallites which allowed the hardness to increase from 1150 H_V for fully amorphous MG to 1375 H_V (Singh et al. 2016).

3.3 Other Sintered Parts Using Metallic Glasses

3.3.1 Fabrication of Ultrafine Grain Materials

Some studies have been carried out to fabricate fully crystallized, ultrafine grain materials. Full crystallization is achieved by sintering MG at high temperature (usually higher than T_x). By controlling the nucleation rate, it is possible to contain the final grain size of the sintered alloy under the micron scale, with the objective to obtain high strength value according to the Hall-Petch law. Suitable alloys need to exhibit low crystallization enthalpy, and apparition of brittle intermetallic phases must be avoided. In this respect, the composition of the amorphous alloys used for this objective is usually strongly concentrated in one element. This is, for instance, the case for Al-based metallic glasses, which are sintered at at least $T_s > T_x + 80$ K under high uniaxial load (150–500 MPa). $Al_{85}Ni_{10}Ce_5$ (Wang and Shen 2015), $Al_{85}Ni_{10}La_5$ (Sasaki et al. 2008), and $Al_{87}Ni_8La_5$ (Scudino et al. 2009) all devitrify

Table 9 In situ composites by partial devitrification of the MG matrix

Composition	Powder	Die	T_g (K)	T_x (K)	T_s (K)	T_s/T_g	Dwell time (min)	Heating rate (K.min^{-1})	Second phase(s)	References
$Al_{65}Cu_{16.5}Ti_{18.5}$	m	w	–	620	723	–	–	100 → 20	i	Tan et al. (2015)
$Al_{85}Y_8Ni_5Co_{1.5}Fe_{0.5}$	m	w*	525	575	553	1.05	3	–	Al	Kim et al. (2009c)
$Cu_{60}Hf_{25}Ti_{15}$	m	w*	690	781	753	1.09	3	10	–	Choi et al. (2006)
$Fe_{48}Cr_{15}Mo_{14}Y_2C_{15}B_6$	a	g	848	903	848	1.00	10–20	100–150	c, n	Singh and Harimkar (2010) Singh et al. (2014, 2016)
$Fe_{49.7}Cr_{17.1}Mn_{1.9}Mo_{7.4}W_{1.6}B_{15.2}C_{3.8}Si_{2.4}$	a	g	–	810	800–903	–	0–40	50–500	b, c, n	Kelly et al. (2016) Khanolkar et al. (2016)
$Ti_{66}Nb_{13}Cu_8Ni_{6.8}Al_{6.2}B_{0.79}C_{0.65}$	m	w	673	775	673–773	1.00–1.15	10	30	β-Ti	Chen et al. (2013b)
$Zr_{70}Cu_{24}Al_4Nb_2$	a	g	673–684	743	633–733	0.93–1.08	5	100	Zr + i	Drescher et al. (2016)

Please refer to Table 1 notes for details on powders and dies
Types of observed secondary phases: *b* borides, *c* carbides, *i* intermetallic phases, *n* nitrides

to form a matrix of fcc-Al, the Ni_3Al intermetallic phase, and an aluminum lanthanide. The grain size is contained under 200 nm, which allows to attain higher strength levels than classically manufactured Al alloys.

A similar procedure has been carried out on $Fe_{77.5}B_{18.5}Nd_4$ (Saito et al. 2004). When sintered at 873 K under 50 MPa, the material is composed of a mixture of Fe_3B (soft magnetic phase) and $Nd_2Fe_{14}B$ (hard magnetic phase). Li et al. prepared a bulk crystallized alloy from $Ni_{53}Nb_{20}Ti_{10}Zr_8Co_6Ta_3$ (Li et al. 2015) with a unique combination of compressive fracture strength higher than 3 GPa and hardness of 1000 Hv, due to the presence of several intermetallic phases in the microstructure.

The main MG family that has been used for full devitrification purposes remains the Ti-based alloys, which are gathered in Table 10. It is worth noting that T_s ranges from 1.37 to 1.89 × T_g. The devitrification leads to the crystallization of a matrix of titanium in its β form and several intermetallic phases. The sintering temperature T_s and heating rate are the main parameters the final grain size depends on. The grain size is kept submicronic, with values of 100 and 300 nm reported for $Ti_{66}Nb_{13}Cu_8Ni_{6.8}Al_{6.2}$ (Li et al. 2009c, 2010a). Moreover, this technique generates equiaxed microstructures with a homogeneous dispersion of strengthening intermetallic precipitates. Consequently, the mechanical properties of such materials are generally higher than β-Ti alloys and also combine high strength with large ductility. For instance, fracture strength of 2.4 GPa and fracture strain of 32.6% were reported for $Ti_{66}Nb_{13}Cu_8Ni_{6.8}Al_{6.2}$ sintered at 1173 K at 170 $K.min^{-1}$ for 5 min (Li et al. 2010b).

One other study worth mentioning concerns the semisolid processing of a Ti-based amorphous alloy by SPS (Liu et al. 2016b). Semisolid sintering on a $Ti_{62}Nb_{12.2}Fe_{13.6}Co_{6.4}Al_{5.8}$ amorphous alloy gives rise to a bimodal microstructure composed of a nanostructured β-Ti matrix and micrometric $(CoFe)Ti_2$ twins, exhibiting a yield strength of 1.8 GPa and 15.5% of plastic strain.

3.3.2 Metallic Glasses with Controlled Porosity

Porous MG are achieved by sintering at low temperature (lower than T_g) in order to limit the densification of the alloy. The compacting force is also kept as low as possible in order to avoid viscous deformation of the MG during the sintering process. For instance, one can highlight a study on $Ti_{42.5}Cu_{40}Zr_{10}Ni_5Sn_{2.5}$ (Wang et al. 2013c), which aimed to tune the electrical properties by varying the porosity content. The volume fraction of pores ranges from 15 to 26 vol.% and was obtained by sintering at different temperatures. T_s was comprised between 0.95 and 0.99 × T_g. A similar work was performed on a $Zr_{55}Cu_{30}Al_{10}Ni_5$ MG (Xie et al. 2006, 2007a, 2010a): the porosity content was controlled by varying T_s (in the range 0.84–0.91 × T_g) as well as the applied stress (20–40 MPa). Porosity varied from 5 to almost 40 vol.%, an increase of the sintering temperature or applied stress leading to a decrease of the porosity content. The apparent elastic constants of the porous MG obviously depend on the porosity content, but their values indicate that the deformation modes tend to evolve when the porosity content is the highest.

Table 10 Ti-based MG SPSed to produce ultrafine grain materials. All amorphous powders were mechanically alloyed

Composition	T_g (K)	T_x (K)	T_s (K)	T_s/T_g	Dwell time (min)	Heating rate (K. min^{-1})	Applied stress (MPa)	References
Eight different compositions	–	–	1023–1273	–	–	100	–	Liu et al. (2016a)
$Ti_{58}V_9Cu_{12.6}Ni_{10.7}Al_{9.7}$	744	807	1023–1310	1.37–1.76	0–5	41–273	50	Yang et al. (2013)
$Ti_{64}Nb_{12}Cu_{11.2}Ni_{9.6}Sn_{3.2}$	675	775	1123–1223	1.66–1.81	0–10	50–150	50	Yang et al. (2014b)
$Ti_{65}Nb_{23.33}Zr_5Ta_{1.67}Si_5$	–	820	1223	–	5	20	50	Li et al. (2016)
$Ti_{66}Nb_{13}Cu_8Ni_{6.8}Al_{6.2}$	715	799	1023–1173	1.43–1.64	0–20	50–850	–	Li et al. (2009b, 2010a, b)
$Ti_{66}Nb_{13}Fe_8Co_{6.8}Al_{6.2}$	674	751	1173–1273	1.74–1.89	–	100	50	Liu et al. (2015b)
$Ti_{66}Nb_{18}Cu_{6.4}Ni_{6.1}Al_{3.5}$	675	788	1023–1173	1.52–1.74	0–20	50–800	50	Chen et al. (2012)
$Ti_{66}V_{13}Cu_8Ni_{6.8}Al_{6.2}$	720	770	1023–1310	1.42–1.82	5–10	82–273	50	Yang et al. (2013)

Fig. 8 Porous MG obtained by co-sintering a $Ti_{45}Zr_{10}Cu_{31}Pd_{10}Sn_4$ MG with NaCl and subsequent dissolution of the second phase (Xie et al. 2014)

A different approach has been explored by Xie et al. (2014). A mixture of a Ti-based MG and NaCl was sintered at $0.97 \times T_g$ under a 600 MPa stress to reach full densification. After the sintering, NaCl was dissolved in order to reveal a porous microstructure (Fig. 8). This method leads to large volumes of pores (50–60 vol.% porosity), as well as large pores compared to the MG particles.

3.3.3 Metallic Glasses as a Binding Agent

Complementary to the works presented above, a Ti-based metallic glass, $Ti_{60.8}Al_{25}Nb_{10}V_3Mo_1B_{0.2}$, was used as a binding agent to sinter dense titanium boride TiB_2 (Yang et al. 2016). In this study, the ceramic powder was blended with 5 wt.% metallic glass and sintered between 1573 and 1773 K for 5 min, under a 30 MPa stress in a graphite die (Ta foil). Since $T_x = 833$ K, full devitrification of the MG is inevitable. However, the MG viscous flow still accelerates the densification at low temperature, which reaches 99+ %. Moreover, the crystallization of TiAl helps in improving the fracture toughness of the material.

4 Sintering Mechanisms

A lot of studies conclude that the main MG densification mechanism is the viscous flow, in relation with the experimental conditions ($T_s > T_g$ and high applied stress). However, due to the nature of the SPS process, T_s is just an average temperature that does not describe the temperature inside the amorphous powder during the sintering. FEM calculations indeed showed that the temperature in the center of the sample could be 10–20 K higher than at the edge (Wang et al. 2010), which is, furthermore, often not the measured T_s. This can be problematic for MG, since their physical and mechanical properties are sensitive to marginal temperature variations. Three different and complementary strategies (phenomenological, analytical, and numerical) have been developed in order to better understand the mechanisms of densification and consolidation.

4.1 Phenomenological and Analytical Approaches

Densification of MG by SPS is controlled by the viscous flow of the amorphous alloy. However, amorphous alloys evolve during sintering, encountering at least structural relaxation. Moreover, postmortem analysis of sintered specimens reveals non-systematic partial devitrification: with similar experimental conditions, devitrification is observed (Harimkar et al. 2009) or not (Xie et al. 2009b). As it is pointed out in paragraph 2.2, devitrification involves complex mechanisms. Despite its numerous advantages, SPS is also a process involving several physical properties of the material of interest and of the die material: electrical resistivity, calorific capacity, diffusion coefficient, mechanical properties, etc. Additionally, the temperature field resulting from Joule effect produced by the pulsed current in a powder is hypothetically highly heterogeneous on different scales (particles, pellet, powder, and die ensemble). All these parameters make the understanding of amorphous alloys sintering mechanisms extremely difficult to apprehend.

To approach specific densification mechanisms of SPS, a first densification sequence was proposed (Xie et al. 2003, 2004a, b), taking into account the breakout of the oxide layer on the surface of powder particles. It is divided into four main steps:

1. The oxide layer on the powder particles is broken by the electric field, and the subsequent contamination on the particle is removed through spark generation and sputtering effect.
2. Necks are formed between the powder particles.
3. The electric current is focused at the contact interfaces; thus, the temperature generated by Joule effect might be higher than the average temperature, especially during the early stages of sintering.
4. The temperature gradient in the particles and the electric field enhance atom migration, resulting in an accelerated neck growth.

Consequently, the reactivity of the amorphous powders should be enhanced, which permits SPS to be carried out at lower temperature and for shorter times than conventional sintering processes.

Another analysis was reported for $Zr_{57}Cu_{20}Al_{10}Ni_8Ti_5$ (Nowak et al. 2011) and is based on a microstructural analysis of the sintered pellets. Full densification was achieved, independently of the particle size and the SPS conditions detailed in Table 4. However, different levels of devitrification were evidenced by X-ray diffraction, even though T_s is close to T_g. Noticeably, high-temperature phases, including CuZr (stable above 1000 K), have been observed, oddly for the shortest

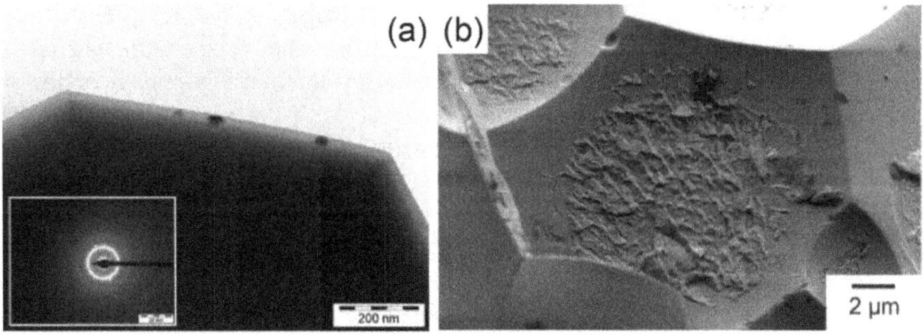

Fig. 9 (**a**) TEM evidence of nanocrystals at the particle neck formed during the SPS processing; (**b**) SEM image of a fracture surface, evidencing disparities on the interparticle surfaces (Nowak et al. 2011)

dwell times. It would indicate that high-temperature and less stable phases appear first at short annealing time and disappear after further annealing. Interestingly, the devitrification products were only observed in the vicinity of necks between particles using transmission electron microscopy (Fig. 9a). This would indicate that devitrification occurs at narrow particle necks, where electric current density and associated Joule dissipation effect are the largest (i.e., first stage of sintering).

After compression tests, fracture surfaces reveal trans-particle shear banding as well as "interparticle" rupture. "Interparticle" zones are rough in the center with a dimple-like structure characteristic of a ductile fracture and totally smooth at the periphery (Fig. 9b). This peculiar fracture surface has also been observed in another work (Xie et al. 2010a). Ductile fracture would mean that free volume is still present in the glass and that micro-shear banding might be initiated by crystalline particles present underneath. In contrast, the origin of the smooth zone is not straightforward, and the fracture surface suggests that the cohesion at particle contact is partial.

A new sequence for the SPS densification of this metallic glass powder was thus proposed. It is slightly different from the one proposed by Xie et al. and includes devitrification:

(i) Generation of heat at the initial particle necks is supposed to be larger than in the rest of the powder due to higher current densities in these areas. During this first stage of the sintering, necks are narrow, and temperature relaxation should be very fast due to the high thermal diffusivity of the MG. This might lead to the observed partial devitrification.
(ii) When necks are wide enough to limit current density, heating becomes more homogeneous.
(iii) The amorphous particles can thus viscously flow, leading to full densification.

Generally speaking, it is reminded that the first stage of sintering is dominant for future properties of the sintered parts. Unfortunately, it cannot be controlled by the SPS monitoring tools, since these events are localized at necks between particles and represent a negligible volume fraction of the material. Gradual increase of the macroscopic current intensity to moderate current density at the growing particle neck could be the most relevant process to avoid detrimental temperature and stress gradient as well as phase decomposition. This means that the SPS apparatus power should be controlled, not just the temperature cycle.

The originality of this work was to further analyze the thermal and electrical contributions to the sintering are considered. Those can be expressed in the general form of the first Fick law:

$$J = -D \cdot \mathrm{grad}(c) + \frac{Dc}{kT}(F_C + F_{\mathrm{th}} + F_i) \quad (1)$$

where J is the local atomic flux, D is the atomic diffusion coefficient, c is the atomic concentration, k is the Boltzmann constant, and T is the temperature. The mass flow forces are resulting from the neck topology that is surface curvature (Laplace force), F_C; the electric potential, F_i; and the thermal gradient, F_{th}. In a two-sphere model, the Laplace force, F_C, is perpendicular to the neck surface and is radial with respect to the neck boundary. F_i is derived from the electric current passing through the particles and is assumed normal to the neck boundary in a first approximation. With thermal gradient from the neck to the particle core, a thermo-transport force F_{th} parallel to F_i is resulting. As mentioned above, heat concentration in the vicinity of the neck occurs since the neck size is small compared to the particle size. In this case, much larger electric current density gets through the necks than through the core of the particles. Being related to the thermal gradient and the electric current, F_i and F_{th} should be assumed normal to the neck boundaries. As a consequence, the Laplace force, F_C, should be the main driving force in neck growth (sintering rate) since it is the only one that is radial to the neck. F_i and F_{th} are normal to the neck plane and should have minor contributions to the sintering rate. However, they might be involved in crystalline particle growth along the concentration gradient built up during devitrification.

The observation of devitrification suggests that a heterogeneous repartition of temperature takes place in the powder. Consequently, the thermo-transport resulting from the temperature gradients is far from easy to evaluate, since it emanates from the balance between heat generation at the neck by Joule effect and heat dissipation from the neck to particle center, these two events having different timescales. If one adds that SPS is operating by application of short time pulses of current and that metallic glasses are characterized by high thermal diffusivity, it follows that a simple evaluation of thermo-transport is difficult and necessitates modeling for a relevant approach. For its part, electro-transport may be effective during the early stages of the sintering and thus might be responsible for the nucleation of crystalline phases. Interestingly, this result concurs with a study on diffusion in Ag-Zn and Cu-Zn couples in the presence of an electric current (Trzaska 2015).

This approach has however to be balanced: the current density and the macroscopic current are difficult to estimate as they depend on the resistivity of the powder and die/punch material, tool geometry, and change during the process because of powder density and temperature variations.

4.2 Numerical Studies

Numerical modeling of a process such as SPS must be multi-physics, since in the various stages of sintering, the microstructure evolves with or without densification: particle adhesion, neck growth, grain polygonization and grain growth, phase transformation, and eventual devitrification have to be taken into consideration. The calculations also have to associate thermal, electrical, and mechanical simulations in order to describe the process. To the best knowledge of the authors, no study describes the entire SPS process. Several attempts have however been made to emulate the SPS process. In this context, one can highlight the work on $Mg_{65}Cu_{25}Gd_{10}$ (Zheng et al. 2013). The temperature distribution was calculated (see Fig. 10) for two idealized systems with varying density (52 and 63%). The electric current was adjusted to emulate realistic sintering temperatures, and the applied electric potential was 3 mV. One can notice that the average temperature is higher in the case of the highest packing density (around 485 K, against 422 K for the lowest density), which is supposedly a consequence of a lower resistivity and higher thermal conductivity of this configuration. More importantly, these simulations show a very small temperature difference between the necks and the particle core: 2 K in the lowest density configuration and 5 K in the highest one. This indicates that the temperature gradients should not be sufficient to cause thermo-transport at large scale. However, local high temperatures could promote surface activation and interface bonding.

A third densification sequence was proposed, taking into account the breakout of the oxide layer and the structural relaxation of the amorphous alloy in addition to viscous flow and mass transport. The underlying mechanism is that this localized overheat causes structural relaxation in the MG, which enhances the atomic mobility and consequently the atomic diffusion, thus leading to an accelerated densification of

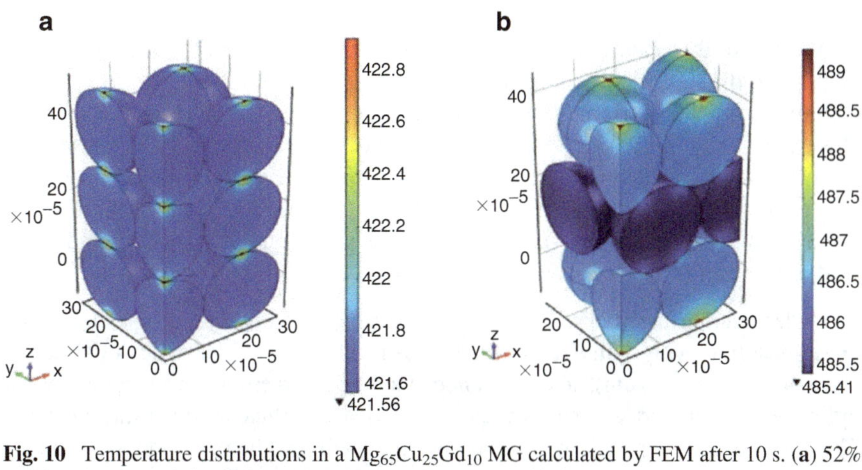

Fig. 10 Temperature distributions in a $Mg_{65}Cu_{25}Gd_{10}$ MG calculated by FEM after 10 s. (**a**) 52% packing density and (**b**) 63% packing density (Zheng et al. 2013)

the powder. This mechanism is nevertheless hindered by the presence of oxides, which limit the metal-metal contacts and thus the cohesion between the powder particles. It has to be noticed that the low amplitude of local thermal gradients has also been evidenced for crystalline TiAl alloys (Collard et al. 2017). The temperature difference between a neck and the core of the particle is strongly dependent on the diameter of the particle. Interestingly, for micronic powders, the temperature difference is only a few degrees. Even though this study has to be nuanced since it concerns crystalline alloys, it still highlights that overheating at particle contacts is not as important as previously admitted. Overheating presumably takes place, but its order of magnitude is rather low, and it should not promote thermo-transport as a principal mechanism of densification and/or devitrification. The MG sensibility to temperature should nevertheless balance this statement, since a difference of a few degrees can induce drastic structural evolutions, as exposed in paragraph 2.2.

5 Conclusions

As exposed in this chapter, lots of efforts have been devoted to study SPS of MG, including monolithic alloys and composites, and understand MG densification. The range of obtained properties is very wide and mainly spans from mechanical to magnetic properties. From this mass of work, one can evidence similarities in the SPS conditions that can be summarized as follows:

- The sintering temperature plays a direct role on the sintered density and the devitrification of the amorphous phase, which is evidently related to the MG thermal behavior.
- The applied stress also has an influence on the sintered density but compensates the effect of the temperature on the devitrification. In this respect, a majority of the studies have been carried out in WC dies, which allow applying elevated stress.
- The role of the heating rate is not as well understood, but an elevated heating rate tends to diminish the devitrification of the amorphous phase.
- Stress application has not been studied. From the 184 articles gathered in the various tables, only 33 give information about this parameter (19, stress applied before heating; 11, during heating; and 3, stress relieved for cooling), and only 1 explicitly studies the role of dynamic pressing (Kelly et al. 2016). It is however an important parameter, since a premature stress application can hinder the densification of the MG once viscous flow is activated.

Understanding the role of experimental parameters is necessary to obtain fully dense materials with optimized properties. It is also vital in the optics of sintering large samples (Ø > 10 cm), since this understanding can facilitate the development of optimized temperature and stress cycles, as well as an efficient power management. The engineering of specific multi-materials dies will also be of importance (Huang et al. 2017). For instance, one can anticipate that WC dies will not be sustainable for sintering large MG samples. Indeed, WC-specific gravity, combined to the powder mass, will make them difficult to handle. More importantly, its low electrical resistivity induces larger power consumptions than similar graphite die and punches.

Finally, studying MG sintering is an opportunity to learn more about SPS itself, since experimental observations point out some physical events that are unobservable in crystalline solids, such as the apparition of crystalline phases. One can imagine that experimental studies, coupled with numerical modeling, could lead to a better understanding of current repartition at the scale of the powder particles, which could be of interest for the whole SPS community.

Acknowledgments The authors also gratefully acknowledge the helpful comments and suggestions of Marc Blétry.

References

Argon AS (1979) Plastic deformation in metallic glasses. Acta Metall 27:47–58

Blétry M (2004) Déformation dans l'intervalle de transition vitreuse de verres métalliques. PhD thesis, Institut National Polytechnique de Grenoble

Cardinal S et al (2016) Influence of spark plasma sintering parameters on the mechanical properties of $Cu_{50}Zr_{45}Al_5$ bulk metallic glass obtained using metallic glass powder. Mater Sci Eng A 677:116–124

Cardinal S et al (2018) Manufacturing of Cu-based metallic glasses matrix composites by spark plasma sintering. Mater Sci Eng A 711:405–414

Chen HS (1974) Thermodynamic considerations on the formation and stability of metallic glasses. Acta Metall 22:1505–1505

Chen W et al (2012) Fabrication of ultrafine-grained $Ti_{66}Nb_{18}Cu_{6.4}Ni_{6.1}Al_{3.5}$ composites with high strength and distinct plasticity by spark plasma sintering and crystallization of amorphous phase. Mater Trans 53:531–536

Chen Q et al (2013a) Viscous flow during spark plasma sintering of Ti-based metallic glassy powders. J Alloys Compd 557:98–101

Chen Y et al (2013b) Ti-based bulk metallic glass matrix composites with in situ precipitated b-Ti phases fabricated by spark plasma sintering. J Non-Cryst Solids 359:15–20

Cheng YQ et al (2009) Atomic level structure in multicomponent bulk metallic glass. Phys Rev Lett 102:245501

Cho SM et al (2009) Fe-based nano-structured powder reinforced Zr-based bulk metallic glass composites by powder consolidation. Korean J Mater Res 19:504–509

Choi PP et al (2006) Production of Cu-Hf-Ti bulk glassy composites by mechanical alloying and spark-plasma sintering. Solid State Phenom 118:655–660

Choi PP et al (2007) Al-La-Ni-Fe bulk metallic glasses produced by mechanical alloying and spark-plasma sintering. Mater Sci Eng A 449-451:1119–1122

Chu ZH et al (2012a) Consolidation and mechanical properties of $Cu_{46}Zr_{42}Al_7Y_5$ metallic glass by spark plasma sintering. J Non Cryst Solids 358:1263–1267

Chu Z et al (2012b) Relationship between the reinforcement size and mechanical properties of Zr-based glassy matrix composites. Mater Trans 53:879–884

Chu ZH et al (2013) Correlation between the enhanced plasticity of glassy matrix composites and the intrinsic mechanical property of reinforcement. Mater Sci Eng A 560:40–46

Chu Z et al (2015a) The effect of size and volume fraction of the reinforcement on mechanical property and deformation mechanism of the bulk metallic glassy composite. J Alloys Compd 644:25–29

Chu Z et al (2015b) The study on interface and property of TiNb/Zr-based metallic glassy composite fabricated by SPS. J Non Cryst Solids 426:83–87

Collard C et al (2017) Theoretical and experimental investigations of local overheating at particle contacts in spark plasma sintering. Powder Technol 321:458–470

Cox ME et al (2012) Amorphous Hf-based foams with aligned, elongated pores. Mater Sci Eng A 533:124–127

Deng SS et al (2015) Spark plasma sintering of gas atomized AlNiLaCo amorphous powders. Adv Powder Technol 26:1696–1701

Drescher P et al (2016) Composites of amorphous and nanocrystalline Zr-Cu-Al-Nb bulk materials synthesized by spark plasma sintering. J Alloys Compd 667:109–114

El-Eskandarany MS et al (2005) Solid state synthesis of new glassy $Co_{65}Ti_{20}W_{15}$ alloy powders and subsequent densification into a fully dense bulk glass. J Mater Res 20:2845–2853

El-Eskandarany MS et al (2017) Fabrication of new metallic glassy $Ti_{40.6}Cu_{15.4}Ni_{8.5}Al_{5.5}W_{30}$ alloy powders by mechanical alloying and subsequent SPS consolidation. Adv Powder Technol 28:814–819

Feng YT et al (2016) Preparation of bulk $Fe_{75}Nb_3Si_{13}B_9$ amorphous and nanocrystalline alloys by spark plasma sintering. Ferroelectrics 505:190–199

Ferre A et al (2015) Damage characterization in aluminium matrix composites reinforced with amorphous metal inclusions. Mater Sci Technol 31:579–586

Gheiratmand T et al (2016) Fabrication of FINEMET bulk alloy from amorphous powders by spark plasma sintering. Powder Technol 289:163–168

Graeve OA et al (2013) The kinetics of devitrification of amorphous alloys: the time–temperature–crystallinity diagram describing the spark plasma sintering of Fe-based metallic glasses. Scr Mater 69:143–148

Greer AL (2009) Metallic glasses… on the threshold. Mater Today 12:14–22

Greer AL et al (2013) Shear bands in metallic glasses. Mater Sci Eng R 74:71–132

Harimkar SP et al (2009) Microstructure and properties of spark plasma sintered Fe-Cr-Mo-Y-B-C bulk metallic glass. J Non Cryst Solids 355:2179–2182

Harimkar SP et al (2011) Spark plasma sintering of amorphous-crystalline laminated composites. Mater Sci Eng A 528:1901–1905

Hu YC et al (2016) A highly efficient and self-stabilizing metallic-glass catalyst for electrochemical hydrogen generation. Adv Mater 28:10293–10297

Huang L et al (2017) Reducing electric current and energy consumption of spark plasma sintering via punch configuration design. Ceram Int 43:17225–17228

Hufnagel TC et al (2016) Deformation of metallic glasses: recent developments in theory, simulations, and experiments. Acta Mater 109:375–379

Huh MY et al (2007) Deformation behavior of amorphous composites containing crystalline nickel in the supercooled liquid region. Mater Sci Eng A 449-451:916–919

Inoue A (2000) Stabilization of metallic supercooled liquid and bulk amorphous alloys. Acta Mater 48:279–306

Inoue A et al (1997) High strain rate superplasticity of supercooled liquid for amorphous alloys. Mater Sci Forum 233-234:147–154

Ishihara S et al (2003) Consolidation of Fe-Co-Nd-Dy-B glassy powders by spark plasma sintering and magnetic properties of the consolidated powders. Mater Trans 44:138–143

Ji X et al (2016) Dry sliding wear behavior of spark plasma sintered Fe-based bulk metallic glass/graphite composites. Technologies. https://doi.org/10.3390/technologies4030027

Jiang JZ et al (2015) Low-density high-strength bulk metallic glasses and their composites: a review. Adv Eng Mater 17:761–780

Johnson WL et al (2002) Deformation and flow in bulk metallic glasses and deeply undercooled glass forming liquids – a self-consistent dynamic free volume model. Intermetallics 10:1039–1046

Kelly JP et al (2016) Designing in situ and ex situ bulk metallic glass composites via spark plasma sintering in the super cooled liquid state. Mater Des 93:26–38

Khan MM et al (2017) Recent advancements in bulk metallic glasses and their applications: a review. Crit Rev Sol State Mater Sci. https://doi.org/10.1080/10408436.2017.1358149

Khanolkar GR et al (2016) Shock wave response of iron-based in situ metallic glass matrix composites. Sci Rep 6:22568

Kim CK et al (2005a) Microstructure and mechanical properties of Cu-based bulk amorphous alloy billets fabricated by spark plasma sintering. Mater Sci Eng A 406:293–299

Kim TS et al (2005b) Consolidation of $Cu_{54}Ni_6Zr_{22}Ti_{18}$ bulk amorphous alloy powders. Mater Sci Eng A 402:228–233

Kim TS et al (2005c) Synthesis and properties of amorphous matrix composites using Cu-based/Ni-based amorphous alloys. J Korean Powder Metall Inst 12:406–412

Kim JC et al (2006) Spark plasma sintering of the ductile Cu-gas-atomized Ni bulk metallic glass composite powders. J Korean Powder Metall Inst 13:351–359

Kim TS et al (2007a) Synthesis of Cu-base/Ni-base amorphous powder composites. Mater Sci Eng A 449-451:804–808

Kim CK et al (2007b) Microstructure and mechanical properties of Cu-based amorphous alloy/copper composites consolidated by spark plasma sintering. Metall Mater Trans A 38A:2754–2762

Kim CK et al (2007c) Microstructure and mechanical properties of Cu-base amorphous alloy matrix composites consolidated by spark plasma sintering. Mater Sci Eng A 449-451:924–928

Kim YJ et al (2007d) Synthesis and densification of Cu-coated Ni-based amorphous composite powders. Mater Sci Eng A 449-451:1071–1074

Kim CK et al (2008a) Effects of consolidation temperature and pressure on microstructures and mechanical properties of Cu-based bulk amorphous alloys consolidated by spark plasma sintering. J Alloys Compd 453:108–114

Kim TS et al (2008b) Devitrification behavior and mechanical property of $Cu_{54}Ni_6Zr_{22}Ti_{18}$ glass powders. Mater Lett 62:323–326

Kim JC et al (2009a) Synthesis and densification of Cu added Fe-based BMG composite powders by gas atomization and electrical explosion of wire. J Alloys Compd 483:28–31

Kim TS et al (2009b) Wettability of Cu-based amorphous phases to the diamond particles during consolidation. J Alloys Compd 483:488–490

Kim JS et al (2009c) Synthesis of Al-Y-Ni-Co composites by mechanical alloying and consecutive spark-plasma sintering. J Alloys Compd 486:511–514

Kim SY et al (2013) Synthesis of Ni-based metallic glass composite fabricated by spark plasma sintering. J Korean Powder Metall Inst 20:33–36

Klement JW et al (1960) Non-crystalline structure in solidified gold-silicon alloys. Nature 187:869–870

Krämer L et al (2017) From powders to bulk metallic glass composites. Sci Rep 7:6551

Kruzic JJ (2016) Bulk metallic glasses as structural materials: a review. Adv Eng Mater 18:1308–1331

Kumar G et al (2009) Nanomoulding with amorphous metals. Nature 457:868–872

Kumar GP et al (2016) Deployment of a bulk metallic glass-based self-expandable stent in a patient-specific descending aorta. ACS Biomater Sci Eng 2:1951–1958

Lee JK (2007) Fabrication of metallic glass/metallic glass composites by spark plasma sintering. J Korean Powder Metall Inst 14:405–409

Lee JK (2008) Synthesis of Ni-based bulk metallic glass composites. J Korean Powder Metall Inst 15:297–301

Lee JK, Kim TS (2007) Consolidation and characterization of Cu-based bulk metallic glass composites. J Korean Powder Metall Inst 14:399–404

Lee JK et al (2005) Synthesis of Cu-based bulk metallic glass matrix composites by warm processing of gas atomized powders. Mater Sci Forum 475-479:3419–3422

Lee JK et al (2007a) Deformation behavior of Ni-based bulk metallic glass synthesized by spark plasma sintering. J Mater Process Technol 187-188:801–804

Lee JK et al (2007b) Plasticity in bulk metallic glass composites containing dual amorphous phases. Mater Sci Forum 539-543:2026–2030

Lee JK et al (2007c) Fabrication of bulk metallic glass composites by mechanical milling with subsequent spark plasma sintering process. J Korean Powder Metall Inst 14:197–201

Lee JK et al (2007d) Consolidation of bulk metallic glass composites. Mater Sci Forum 534-536:501–504

Lee JK et al (2007e) Consolidation of Cu- and Ni-based bulk metallic glass composites. J Alloys Compd 434-435:336–339

Lee S et al (2009a) Displacement behavior study of the shear stress effect on the early viscous flow nature of the Fe-B-Nb-Y metallic glassy powder in spark plasma sintering. Mater Trans 50:490–493

Lee S et al (2009b) Fabrication and soft-magnetic properties of Fe-B-Nb-Y glassy powder compacts by spark plasma sintering technique. Intermetallics 17:218–221

Lee JK et al (2010a) Synthesis of bulk amorphous composites with three amorphous phases by consolidation of milled amorphous powders. Intermetallics 18:2019–2023

Lee MH et al (2010b) Mechanical behavior of metallic glass reinforced nanostructured tungsten composites synthesized by spark plasma sintering. Intermetallics 18:2009–2013

Lee JK et al (2015) Effect of reinforcement phase on the mechanical property of tungsten nanocomposite synthesized by spark plasma sintering. Int J Refract Met Hard Mater 54:14–18

Li Q et al (2009a) $Ti_{50}Cu_{23}Ni_{20}Sn_7$ bulk metallic glasses prepared by mechanical alloying and spark-plasma sintering. Int J Mater Prod Technol 209:3285–3288

Li X et al (2009b) Microstructure and mechanical properties of SPSed (spark plasma sintered) $Ti_{66}Nb_{13}Cu_8Ni_{6.6}Al_{6.2}$ bulk alloys with and without WC addition. Mater Trans 50:1720–1724

Li YY et al (2009c) Ultrafine-grained $Ti_{66}Nb_{13}Cu_8Ni_{6.8}Al_{6.2}$ composites fabricated by spark plasma sintering and crystallization of amorphous phase. J Mater Res 24:2118–2122

Li Y et al (2010a) High plastic $Ti_{66}Nb_{13}Cu_8Ni_{6.8}Al_{6.2}$ composites with *in situ* β-Ti synthesized by spark plasma sintering of mechanically alloyed glassy powders. Mater Sci Forum 638-642:1642–1647

Li YY et al (2010b) Nucleation and growth mechanism of crystalline phase for fabrication of ultrafine-grained $Ti_{66}Nb_{13}Cu_8Ni_{6.8}Al_{6.2}$ composites by spark plasma sintering and crystallization of amorphous phase. Mater Sci Eng A 528:486–493

Li X et al (2011a) $Fe_{76}Si_{9.6}B_{8.4}P_6$ glassy powder soft-magnetic cores with low core loss prepared by spark-plasma sintering. Mater Sci Eng B 176:1247–1250

Li YY et al (2011b) Effect of minor B_4C or C on amorphous $Ti_{66}Nb_{13}Cu_8Ni_{6.8}Al_{6.2}$ alloy powders synthesized by mechanical alloying. J Inorg Organomet Polym 21:802–808

Li Q et al (2012) The crystal growth of $Ti_{50}Cu_{23}Ni_{20}Sn_7$ during the SPS process. Adv Mater Res 428:190–195

Li XP et al (2013a) Abnormal crystallization in $Al_{86}Ni_6Y_{4.5}Co_2La_{1.5}$ metallic glass induced by spark plasma sintering. Intermetallics 39:69–73

Li XP et al (2013b) The critical role of heating rate in enabling the removal of surface oxide films during spark plasma sintering of Al-based bulk metallic glass powder. J Non-Cryst Solids 375:95–98

Li XP et al (2013c) Applied pressure on altering the nano-crystallization behavior of $Al_{86}Ni_6Y_{4.5}Co_2La_{1.5}$ metallic glass powder during spark plasma sintering and its effect on powder consolidation. J Nanomater 2013:101508

Li XP et al (2013d) Fabrication of 10mm diameter fully dense $Al_{86}Ni_6Y_{4.5}Co_2La_{1.5}$ bulk metallic glass with high fracture strength. Mater Sci Eng A 568:155–159

Li W et al (2015) Crystallization kinetics and spark plasma sintering of amorphous $Ni_{53}Nb_{20}Ti_{10}Zr_8Co_6Ta_3$ powders prepared by mechanical alloying. Vacuum 14:93–100

Li YH et al (2016) Non-isothermal and isothermal crystallization kinetics and their effect on microstructure of sintered and crystallized TiNbZrTaSi bulk alloys. J Non-Cryst Solids 432:440–452

Liu T et al (2012) High frequency properties of $Fe_{73.5}Cu_1Nb_3Si_{13.5}B_9/Zn_{0.5}Ni_{0.5}Fe_2O_4$ soft magnetic composite with micro-cellular structure. Sci China Phys Mech Astron 55:2392–2396

Liu LH et al (2015a) Densification mechanism of Ti-based metallic glass powders during spark plasma sintering process. Intermetallics 66:1–7

Liu LH et al (2015b) Ultrafine grained Ti-based composites with ultrahigh strength and ductility achieved by equiaxing microstructure. Mater Des 79:1–5

Liu LH et al (2016a) Equiaxed Ti-based composites with high strength and large plasticity prepared by sintering and crystallizing amorphous powder. Mater Sci Eng A 650:171–182

Liu LH et al (2016b) A new insight into high strength $Ti_{62}Nb_{12.2}Fe_{13.6}Co_{6.4}Al_{5.8}$ alloys with bimodal microstructure fabricated by semi-solid sintering. Sci Rep 6:23467

Löffler JF et al (2000) Time-temperature-transformation diagram of bulk glass forming $Pd_{40}Cu_{30}Ni_{10}Pd_{20}$. Appl Phys Lett 2000:681–683

Louzguine-Luzgin DV (2014a) Vitrification and devitrification processes in metallic glasses. J Alloys Compd 586:S2–S8

Louzguine-Luzgin DV (2014b) Early stage crystallization kinetics in metallic glass-forming alloys. J Alloys Compd 586:216–219

Lu J et al (2003) Deformation behavior of the Zr-Ti-Cu-Ni-Be bulk metallic glass over a wide range of strain rates and temperatures. Acta Mater 51:3429–3443

Luan J et al (2016) Mechanically strong nanocrystalline Fe-Si-B-P-Cu soft magnetic powder cores utilizing magnetic metallic glass as a binder. AIP Adv 6:055934

Ma E (2015) Turning order in disorder. Nature 14:547–552

Mao M et al (1995) X-ray diffraction study of structural relaxation in metallic glasses. Phys Rev D 51:2798–2804

Maurya RS, Laha T (2016) Microstructure and phase evolution in spark-plasma-sintered $Al_{86}Ni_6Y_{4.5}Co_2La_{1.5}$ glassy alloy. Philos Mag Lett. https://doi.org/10.1080/09500839.2016.1215606

Maurya RS et al (2016a) Effect of sintering temperature on phase transformation during consolidation of mechanically alloyed $Al_{86}Ni_6Y_6Co_2$ amorphous powders by spark plasma sintering. J Non-Cryst Solids 453:1–7

Maurya RS et al (2016b) Effect of consolidation pressure on phase evolution during sintering of mechanically alloyed $Al_{86}Ni_8Y_6$ amorphous powders via spark plasma sintering. Mater Sci Eng A 649:48–56

Maurya RS et al (2016c) Quantitative $Al_{86}Ni_8Y_6$ bulk glassy alloy synthesized by consolidating mechanically alloyed amorphous powder via spark plasma sintering. Mater Des 93:96–103

Méar FO et al (2009) Spark plasma sintering of Mg-based amorphous ball-milled powders. Mater Trans 50:588–591

Miller SA, Murphy RJ (1979) A gas-water atomization process for producing amorphous powders. Scr Metall 13:673–676

Mula S et al (2010) Structure and mechanical properties of Al-Ni-Ti amorphous powder consolidated by pressure-less, pressure-assisted and spark plasma sintering. Mater Sci Eng A 527:3757–3763

Nam DH et al. (2008) Mechanical properties of Cu-based amorphous alloy matrix composites consolidated by spark plasma sintering. Proceedings of MN2008 MN2008-47048

Neamţu BV et al (2014) Preparation and soft magnetic properties of spark plasma sintered compacts based on Fe-Si-B glassy powder. J Alloys Compd 600:1–7

Nguyen HV et al (2009) Mechanical properties of bulk amorphous $Ti_{50}Cu_{20}Ni_{20}Al_{10}$ fabricated by high energy ball milling and spark plasma sintering. J Korean Powder Metall Inst 16:358–362

Nowak S et al (2009) Metallic glass $Zr_{57}Cu_{20}Al_{10}Ni_8Ti_5$-tungsten composites. Mater Manuf Process 24:1162–1167

Nowak S et al (2011) Approach of the spark plasma sintering mechanism in $Zr_{57}Cu_{20}Al_{10}Ni_8Ti_5$ metallic glass. J Alloys Compd 509:1011–1019

Ozaki K et al (2004) Preparing Mg-Ni-Si amorphous powders by mechanical alloying and consolidation by pulsed current sintering. Mater Sci Eng A 375-377:857–860

Pan J et al (2018) Extreme rejuvenation and softening in a bulk metallic glass. Nat Commun 9:560

Paul T, Harimkar SP (2016) Initial stage densification during spark plasma sintering of Fe-based amorphous alloy powder: analysis of viscous flow. J Appl Phys 120:134901

Paul T, Harimkar SP (2017a) Viscous flow activation energy adaptation by isochronal spark plasma sintering. Scr Mater 126:37–40

Paul T, Harimkar SP (2017b) Prediction of heating rate controlled viscous flow activation energy during spark plasma sintering of amorphous alloy powders. J Phys D Appl Phys 50:27LT01

Paul T et al (2017) Densification and crystallization in Fe-based bulk amorphous alloy spark plasma sintered in the supercooled liquid region. Adv Eng Mater. https://doi.org/10.1002/adem.201700224

Paul T et al (2018) Pressure controlled micro-viscous deformation assisted spark plasma sintering of Fe-based bulk amorphous alloy. J Alloys Compd 738:10–15

Perrière L, Champion Y (2012) Phases distribution dependent strength in metallic glass-aluminium composites prepared by spark plasma sintering. Mater Sci Eng A 548:112–117

Perrière L et al (2011) Spark plasma sintering of a Zr-based metallic glass. Adv Eng Mater 13:581–586

Perrière L et al (2012) Spark plasma sintering for metallic glasses processing. Rev Métal 109:5–10

Prasanna ML et al (2015) Microstructure and mechanical properties of $[(Fe_{0.5}Co_{0.5})_{0.75}B_{0.2}Si_{0.05}]_{96}Nb_4$ bulk metallic glass compacts. Trans Indian Inst Metals 68:1033–1037

Qiao J et al (2016) Metallic glass matrix composites. Mater Sci Eng R 100:1–69

Roy D et al (2007) Microstructure and mechanical properties of mechanically alloyed and spark plasma sintered amorphous-nanocrystalline $Al_{65}Cu_{20}Ti_{15}$ intermetallic matrix composite reinforced with TiO_2 nanoparticles. Intermetallics 15:1595–1605

Roy D et al (2008) Effect of sintering on microstructure and mechanical properties of-nano-TiO_2 dispersed $Al_{65}Cu_{20}Ti_{15}$ amorphous/nanocrystalline matrix composite. J Alloys Compd 460:320–325

Saito T et al (2004) Structures and magnetic properties of Nd-Fe-B bulk nanocomposite magnets produced by the spark plasma sintering method. J Mater Res 19:2730–2737

Sasaki TT et al (2008) Bulk nano-crystalline $Al_{85}Ni_{10}La_5$ alloy fabricated by spark plasma sintering of atomized amorphous powders. Mater Sci Eng A 490:343–350

Sauvage X et al (2014) Structure and properties of a nanoscaled composition modulated metallic glass. J Mater Sci. https://doi.org/10.1007/s10853-014-8279-z

Scudino S et al (2009) High-strength $Al_{87}Ni_8La_5$ bulk alloy produced by spark plasma sintering of gas atomized powders. J Mater Res 24:2909–2916

Shen B, Inoue A (2003) Fabrication of large-size Fe-based glassy cores with good soft magnetic properties by spark plasma sintering. J Mater Res 18:2115–2121

Shen B et al (2002) Preparation of $Fe_{65}Co_{10}Ga_5P_{12}C_4B_4$ bulk glassy alloy with good soft magnetic properties by spark-plasma sintering of glassy powder. Mater Trans 43:1961–1965

Shen B et al (2004) Bulk glassy soft-magnetic cores produced by spark-plasma sintering $Fe_{65}Co_{10}Ga_5P_{12}C_4B_4$ glassy powder. Mater Sci Eng A 375-377:666–670

Shen B et al (2005) Fabrication of Fe-based glassy cores with high saturation magnetization and good soft magnetic properties by spark plasma sintering. Mater Sci Forum 475-479:3397–3400

Shin S et al (2007) Application of PM to the consolidation of metallic glassy powder and its composites. J Korean Powder Metall Inst 14:348–353

Shin S et al (2009a) Binding and dispersion behavior of amorphous powders in diamond particles. J Alloys Compd 483:363–365

Shin S et al (2009b) Effects of phase fraction and metallic glass-diamond size ratio on the densification of metallic glass/diamond composite. J Korean Powder Metall Inst 16:173–179

Shin S et al (2010) The influence of spark plasma sintering temperature on the mechanical properties and corrosion resistance of $Zr_{65}Al_{10}Ni_{10}Cu_{15}$ metallic glass powder. Intermetallics 18:2005–2008

Singh A, Harimkar SP (2010) Spark plasma sintering of in situ and ex situ iron-based amorphous matrix composites. J Alloys Compd 497:121–126
Singh A et al (2010) Microstructure and tribological behavior of spark plasma sintered iron-based amorphous coatings. Mater Sci Eng A 527:5000–5007
Singh A et al (2012) Densification behavior and wear response of spark plasma sintered iron-based bulk amorphous alloys. Adv Eng Mater 14:400–407
Singh A et al (2013) Nanocrystallization in spark plasma sintered $Fe_{48}Cr_{15}Mo_{14}Y_2C_{15}B_6$ bulk amorphous alloy. J Appl Phys 114:054903
Singh AK et al (2014) Structural relaxation and nanocrystallization induced laser surface hardening of Fe-based bulk amorphous alloys. J Miner Met Mater 66:1080–1087
Singh A et al (2016) In situ nanocrystallization-induced hardening of amorphous alloy matrix composites consolidated by spark plasma sintering. J Miner Met Mater. https://doi.org/10.1007/s11837-016-1914-1
Son CY et al (2009) In situ microfracture observation of Cu-based amorphous alloy matrix composites containing copper or brass particles. Mater Sci Eng A 508:15–22
Spaepen F (1977) A microscopic mechanism for steady state inhomogeneous flow in metallic glasses. Acta Metall 25:407–415
Suh D, Dauskardt RH (2003) Effect of open-volume regions on relaxation time scales and fracture behavior of a Zr-Ti-Ni-Cu-Be bulk metallic glass. J Non Cryst Solids 317:181–186
Tan Z et al (2015) Strength-improved $Al_{65}Cu_{16.5}Ti_{18.5}$ amorphous/crystalline alloy synthesized by spark plasma sintering. Mater Sci Eng A 642:377–380
Tan Z et al (2016) High-entropy alloy particle reinforced Al-based amorphous alloy composite with ultrahigh strength prepared by spark plasma sintering. Mater Des 109:219–226
Tan Z et al (2018) A multiple grain size distributed Al-based composite consist of amorphous/nanocrystalline, submicron grain and micron grain fabricated through spark plasma sintering. J Alloys Compd 737:308–316
Taub AI, Spaepen F (1980) The kinetics of structural relaxation of a metallic glass. Acta Metall 28:1781–1788
Trzaska Z (2015) Etude des mécanismes activés par SPS dans un alliage TiAl et dans le système Ag-Zn. PhD thesis, Université Toulouse 3 Paul Sabatier
Wang S (2016) Corrosion resistance and electrocatalytic properties of metallic glasses. In: Movahedi B (ed) Metallic glasses – formation and properties. InTech, Vienna
Wang D, Shen J (2015) High strength aluminium alloys prepared by nanocrystallisation of amorphous powders using spark plasma sintering. Powder Metall 58:363–368
Wang DJ et al (2010) Temperature influence on sintering with concurrent crystallization behavior in Ti-based metallic glassy powders. Mater Sci Eng A 527:2662–2668
Wang M, Liu T, Zhao Z (2013a) Preparation of $Fe_{76}Si_9B_{10}P_5$ nanocrystalline magnetic cores by direct-nanocrystallizing method during spark-plasma sintering. Mater Sci Forum 745-746:740–745
Wang DJ et al (2013b) Mechanical behaviors of diamond reinforced Ti-based bulk metallic glassy composites prepared by spark plasma sintering. Mater Sci Eng A 560:841–846
Wang DJ et al (2013c) Mechanical, acoustic and electrical properties of porous Ti-based metallic glassy/nanocrystalline composites. Mater Des 44:69–73
Wang Z et al (2016) Microstructure and mechanical behavior of metallic glass fiber-reinforced Al alloy matrix composites. Sci Rep 6:24384
Xie G et al (2003) Behavior of oxide film at the interface between particles in sintered Al powders by pulse electric-current sintering. Metall Mater Trans A 34A:699–703
Xie G et al (2004a) Effect of Mg on the sintering of Al-Mg alloy powders by pulse electric-current sintering process. Mater Trans 45:904–909
Xie G et al (2004b) Reduction of surface oxide films in Al-Mg alloy powders by pulse electric current sintering. J Mater Res 19:815–819
Xie G et al (2006) Fabrication of porous Zr-Cu-Al-Ni bulk metallic glass by spark plasma sintering process. Scr Mater 55:687–690

Xie G et al (2007a) Microstructure and mechanical properties of porous $Zr_{55}Cu_{30}Al_{10}Ni_5$ bulk metallic glass fabricated by spark plasma sintering process. Mater Trans 48:1589–1594

Xie G et al (2007b) Nearly full density $Ni_{52.5}Nb_{10}Zr_{15}Ti_{15}Pt_{7.5}$ bulk metallic glass obtained by spark plasma sintering of gas atomized powders. Appl Phys Lett 90:241902

Xie G et al (2007c) Fabrication of $Ni_{52.5}Nb_{10}Zr_{15}Ti_{15}Pt_{7.5}$ bulk metallic glassy matrix composite containing dispersed ZrO2 particulates by spark plasma sintering. Mater Sci Forum 561-565:1291–1294

Xie G et al (2007d) Fabrication of ZrCuAlNi metallic glassy matrix composite containing ZrO_2 particles by spark plasma sintering process. Mater Trans 48:158–162

Xie G et al (2007e) Ceramic particulate reinforced Zr55Cu30Al10Ni5 metallic glassy matrix composite fabricated by spark plasma sintering. Mater Trans 48:1600–1604

Xie G et al (2008a) Microstructure and properties of ceramic particulate reinforced metallic glassy matrix composites fabricated by spark plasma sintering. Mater Sci Eng B 148:77–81

Xie G et al (2008b) Large-size ultrahigh strength Ni-based bulk metallic glassy matrix composites with enhanced ductility fabricated by spark plasma sintering. Appl Phys Lett 92:121907

Xie G et al (2009a) Densification of gas atomized Ni-based metallic glassy powders by spark plasma sintering. Mater Trans 50:1273–1278

Xie G et al (2009b) Dual phase metallic glassy composites with large-size and ultra-high strength fabricated by spark plasma sintering. Intermetallics 17:512–516

Xie G et al (2009c) Characterization of interface between the particles in NiNbZrTiPt metallic glassy matrix composite containing SiC fabricated by spark plasma sintering. J Alloys Compd 483:239–242

Xie G et al (2010a) Ultrasonic characteristics of porous $Zr_{55}Cu_{30}Al_{10}Ni_5$ bulk metallic glass fabricated by spark plasma sintering. Intermetallics 18:2014–2018

Xie G et al (2010b) Microstructure and mechanical properties of crystalline particulates dispersed Ni-based metallic glassy composites fabricated by spark plasma sintering. Intermetallics 18:851–858

Xie G et al (2010c) Cu particulate dispersed $Cu_{50}Zr_{45}Al_5$ bulk metallic glassy composite with enhanced electrical conductivity. Intermetallics 18:1973–1977

Xie G et al (2010d) Microstructure and mechanical properties of crystalline particulates dispersed Ni-based metallic glassy composites fabricated by spark plasma sintering. Intermetallics 18:851–858

Xie G et al (2011a) Formation and properties of two-phase bulk metallic glasses by spark plasma sintering. J Alloys Compd 509S:S214–S218

Xie G et al (2011b) Bulk metallic glassy composites with excellent electrical conductivity and enhanced plasticity fabricated by spark plasma sintering. Mater Sci Forum 675-677:197–200

Xie G et al (2012a) Ni-free Ti-based bulk metallic glass with potential for biomedical applications produced by spark plasma sintering. Intermetallics 29:99–103

Xie G et al (2012b) SiC dispersed Fe-based glassy composite cores produced by spark plasma sintering and their high frequency magnetic properties. Intermetallics 20:76–81

Xie G, Zhu S, Qin F (2013) Ti-based bulk metallic glass composites produced by spark plasma sintering. Mater Sci Forum 750:52–55

Xie G et al (2014) Corrosion behavior of porous Ni-free Ti-based bulk metallic glass produced by spark plasma sintering in Hank's solution. Intermetallics 44:55–59

Yamada R et al (2009) Consolidation of Fe-Co based metallic glassy powder by SPS method. J Jpn Inst Metals 73:299–305

Yang C et al (2012) Microstructure and mechanical properties of nanocrystalline WC-particle-reinforced Ti-based composites with nano/ultrafine-grained intermetallic matrix from spark plasma sintering and crystallization of amorphous phase. Int J Mater Res. 103:613–619

Yang C et al (2013) Effect of V content on microstructure and mechanical property of a TiVCuNiAl composite fabricated by spark plasma sintering. Mater Des 52:655–662

Yang C et al (2014a) Sinter-hardening with concurrent improved plasticity in iron alloys induced by spark plasma sintering. J Mater Res 29:981–988

Yang C et al (2014b) Intrinsic relationship between crystallization mechanism of metallic glass powder and microstructure of bulk alloys fabricated by powder consolidation and crystallization of amorphous phase. J Alloys Compd 586:542–548

Yang C et al (2016) Tough TiB_2-based ceramic composites using metallic glass powder as the sintering aid. Adv Eng Mater. https://doi.org/10.1002/adem.201600234

Yodoshi N et al (2009) Consolidation of $[(Fe_{0.5}Co_{0.5})_{0.75}Si_{0.05}B_{0.2}]_{96}Nb_4$ metallic glassy powder by SPS method. Mater Trans 50:2264–2269

Zadorozhnyy VY et al (2017) Investigation of structure and thermal properties in composite materials based on metallic glasses with small addition of polytetrafluoroethylene. J Alloys Compd 707:264–268

Zhao Z et al (2009) Preparation of bulk glassy $Fe_{76}Si_9B_{10}P_5$ as a soft magnetic material by spark plasma sintering. Mater Trans 50:487–489

Zhao Z et al (2016) Interfacial behavior of $Fe_{76}Si_9B_{10}P_5/Zn_{0.5}Ni_{0.5}Fe_2O_4$ amorphous soft magnetic composite during spark plasma sintering process. Prog Nat Sci: Mater Int 26:85–89

Zheng B et al (2013) Influence of the mechanically milled powder and high pressure on spark plasma sintering of Mg-Cu-Gd metallic glasses. Acta Mater 61:4414–4428

Zhu S et al (2008) Synthesis of Ti-based glassy alloy/hydroxyapatite composite by spark plasma sintering. Mater Trans 49:502–505

Zhu SL et al (2011) Effect of hydroxyapatite content on the microstructure, thermal and mechanical properties of Ti-based glassy alloy/hydroxyapatite composite prepared by spark plasma sintering. Intermetallics 19:572–576

Zhu S et al (2013) Ti particles dispersed Ti-based metallic glass matrix composite prepared by spark plasma sintering. Mater Trans 54:1335–1338

Zumkley T et al (2001) Effect of reversible structural relaxation on diffusion in a ZrTiCuNiBe bulk glass. Scr Mater 45:471–477

Liu LH et al (2016b) A new insight into high-strength $Ti_{62}Nb_{12.2}Fe_{13.6}Co_{6.4}Al_{5.8}$ alloys with bimodal microstructure fabricated by semi-solid sintering. Sci Rep 6:23467

Impact of High-Energy Mechanical Activation on Sintering Kinetics and Mechanical Properties of UFG Heavy Tungsten Alloys: SPS Versus Sintering in Hydrogen

Aleksey Nokhrin, Vladimir Chuvil'deev, Maksim Boldin, Gleb Baranov, Vladimir Belov, Eugeniy Lantcev, Nikolay Melekhin, Yu. V. Blagoveshchenskiy, Nataliya Isaeva, and Aleksandr Popov

1 Introduction

W-Ni-Fe tungsten heavy alloys with tungsten content of over 90 wt.% present an unparalleled combination of high density (17–19 g/cm^3), high strength (ultimate strength up to 1000 MPa), and elasticity (25–30% elongation to failure at room temperature) (Green et al. 1954). Due to this combination of physical and mechanical properties, researchers see potential in tungsten alloys both for real-life applications and for studying the effects that structural and phase states of interphase boundaries have on physical and mechanical properties of materials. Especially attractive is the possibility of improving mechanical qualities of tungsten heavy alloys for applications in high-strong structural elements.

Today, the most common technique that uses W-Ni-Fe alloys to produce three-dimensional objects is sintering in hydrogen at temperatures exceeding the lowest melting point among the components of the alloy, i.e., liquid phase sintering (Green et al. 1954; Krock and Shepard 1963). In W-Ni-Fe alloys, melting point is the lowest in γ-solid solution of Ni-Fe doped with tungsten (Green et al. 1954; Krock and

A. Nokhrin (✉) · V. Chuvil'deev · M. Boldin · E. Lantcev · N. Melekhin · A. Popov
Lobachevsky State University of Nizhny Novgorod – National Research University, Nizhny Novgorod, Russia
e-mail: nokhrin@nifti.unn.ru

G. Baranov · V. Belov
Russian Federal Nuclear Center – National Research Institute of Experimental Physics, Nizhny Novgorod, Russia

Y. V. Blagoveshchenskiy · N. Isaeva
A.A. Baikov Institute of Metallurgy and Materials Science, Russian Academy of Sciences, Moscow, Russia
e-mail: yuriblag@imet.ac.ru

© Springer Nature Switzerland AG 2019

P. Cavaliere (ed.), *Spark Plasma Sintering of Materials*,
https://doi.org/10.1007/978-3-030-05327-7_12

Shepard 1963). Research is already well advanced in the mechanisms of liquid phase sintering of W-Ni-Fe alloys, the impacts that phase composition and structural state of interphase boundaries have on the strength of W-Ni-Fe alloys, of sintering medium, of preliminary mechanical activation, original grain size on the kinetics of liquid phase sintering of W-Ni-Fe alloys, etc.

Tungsten alloys obtained by liquid phase sintering have a coarse-grained, nonuniform structure and, therefore, poor mechanical properties. Yield point of W-Ni-Fe alloys with tungsten content of 90–95 wt% is under 800–1000 MPa (Liu et al. 2012; Das et al. 2009). This seriously limits applicability of such materials in structural elements that have higher strength requirements. Additional deformation processing can increase the material's ultimate strength to ~1400–1560 MPa while preserving sufficiently high elasticity (Kiran et al. 2012; Gong et al. 2011).

In light of the above, high-rate solid-phase sintering that produces an ultrafine grain (UFG) structure is an especially promising avenue of research that may lead to further improvement of mechanical properties of tungsten alloys.

To achieve this, various methods of producing nano- and ultrafine W-Ni-Fe compositions are commonly used, mechanical activation (MA) being the most widely spread among those (Zhang et al. 2004; Jang et al. 2007). References (Jang et al. 2007; Akhtar 2008; Hu et al. 2011b) show that MA can lower optimal sintering temperature of tungsten alloys by 100–200 °C. For some cases, it has also been demonstrated that the dependence of tungsten alloy density on sintering temperature is non-monotonic (Hu et al. 2011a, b; Huang et al. 2003). However, the reasons that cause a peak in the dependence of tungsten alloy density on sintering temperature are yet to be determined.

One of the techniques that hold promise for obtaining high-density UFG materials is spark plasma sintering (SPS) (Tokita 2006, 2013; Chuvil'deev et al. 2015; Munir et al. 2006; Munir et al. 2011). This technique enables high heating rates, which are especially important for limiting grain growth and ensuring a uniform UFG structure. A review of previously obtained results indicates that SPS allows for producing tungsten heavy alloys with high ultimate strength at relatively low sintering temperatures (Krock and Shepard 1963; Liu et al. 2012; Das et al. 2009).

Still, a number of aspects regarding SPS of tungsten heavy alloys are underresearched: the influence of original size of tungsten particles on the kinetics of solid phase SPS, the reasons for fine-grained tungsten alloys sintering faster when SPS is used, the reasons behind a non-monotonic dependence of density on sintering temperature, etc.

This study focuses on SPS of mechanically activated W-Ni-Fe nanopowders and investigates the influence of tungsten particle size and MA modes on the kinetics of solid-phase sintering of UFG tungsten heavy alloys, as well as evaluates their mechanical properties.

2 Materials and Methods

This study investigated 95 W-3.5Ni-1.5Fe and 95 W-3.5Ni-1.0Fe-0.5Co alloys. The alloys were produced using W, Ni, Fe, and Co powders with average particle sizes of 4 μm, 20 μm, 11 μm, and 45 μm, respectively. To obtain original coarse-grained W-Ni-Fe and W-Ni-Fe-Co compositions, component powders were mixed in a double cone blender for 5 h at 46 rpm.

Powders for sintering were obtained by (Table 1):

- Low-energy MA of original coarse-grained powders (Series 1)
- Adding ultrafine W-Ni-Fe powders of a similar composition to the original coarse-grained W-Ni-Fe composition followed by low-energy MA (Series 2)
- High-energy MA of original coarse-grained powders (Series 3, 4).

The resulting powders were sintered using two techniques: (1) sintering pre-compacted powders in hydrogen (Series 1–3) and (2) SPS (Series 4).

Series 1. The powders in Series 1 were 95W-3.5Ni-1.0Fe-0.5Co alloy. The alloy was produced using tungsten powder with an average grain size of 4 μm, nickel powder with an average grain size of 20 μm, iron powder with an average grain size of 11 μm, and cobalt powder with a grain size under 45 μm. For chemical composition of powders, see Table 2.

Mechanical mixing of coarse-grained powders was performed in a standard double cone blender for 5 h at $V_{MA} = 46$ rpm. For the structure of the original W-Ni-Fe composition, see Fig. 1.

Low-energy MA (grinding ball acceleration 0.4 g, where g is the gravity acceleration) was performed in a MB0-10 ball mill (Russia). Mechanical activation time t_{MA} ranged from 0 to 10 h. Container rotational speed was $V_{MA} = 100$ rpm. In all

Table 1 Investigated samples and their preparation parameters

No.	Alloy (wt%)	Average particle size, μm		MA mode	Sintering mode
		Original state	After MA		
1	95 W-3.5Ni-1Fe-0.5Co	3.9	~1.0	Low-energy MA (0.4 g, $t_{MA} = 5$ h, $V_{MA} = 46$ rpm)	Sintering in hydrogen ($V_s = 5°$/min, $T_{sint} = 1200–1500$ °C)
2	95 W-3.5Ni-1Fe-0.5Co	0.12–0.20	~0.1–0.15	Low-energy MA (0.4 g, $t_{MA} = 72$ h, $V_{MA} = 100$ rpm)	Sintering in hydrogen ($V_s = 5°$/min, $T_{sint} = 1200–1500$ °C)
3	95 W-3.5Ni-1.5Fe	3.9	0.1	High-energy MA (60 g, $t_{MA} = 20$ min, $V_{MA} = 1450$ rpm)	Sintering in hydrogen ($V_s = 5°$/min, $T_{sint} = 1200–1500$ °C)
4	95 W-3.5Ni-1.5Fe	3.9	0.1	High-energy MA (60 g, $t_{MA} = 5–40$ min, $V_{MA} = 1450$ rpm)	SPS (vacuum 6 Pa, $V_s = 50–300°$/min, $T_{sint} = 900–1300$ °C)

Table 2 Chemical composition of original powders

Powder	chemical composition, wt%										
	O	Fe	C	S	P	Ni	Co	Si	Cu	Mo	Mn
W	0.08	0.02	0.01	–	0.005	0.01	–	0.005	0.01	0.045	0.002
Ni	–	0.0015	0.10	0.0006	0.001	99.89	0.0007	0.001	0.001	–	0.0003
Fe	0.2	99.623	0.048	0.004	–	–	–	0.01	–	–	–
Co	0.3	0.2	0.20	–	–	0.40	99.35	0.02	0.04	–	–

Fig. 1 95 W-3.5Ni-1.5Fe powder composition structure. Original state

Series, the material of the containers and ∅10 mm grinding balls was 95 W-3.5Ni-1.5Fe in order to eliminate abrasion contamination. Mass ratio of balls to powders was 10:1. MA was carried out under argon with addition of ethanol. After MA, moist powder blend was taken off the balls under argon followed by drying in a vacuum.

Prior to sintering in hydrogen, preliminary monoaxial compaction of the blend was carried out in a steel compression mold of ∅21 mm int. dia. under 150 MPa.

Sintering in hydrogen of samples of original diameter ∅20 mm was carried out in a two-stage mode: heating at 25°/min up to 950 °C, temperature maintained for 2 h, and heating up to sintering temperature at the rate of $V_s = 5°/min$. The temperature of 30-min sintering in hydrogen ranged from $T_{sint} = 1200–1500$ °C. Temperature readings in the heating chamber were taken by tungsten – rhenium thermal couples. Measurement accuracy was ±6 °C. Sintering was carried out on a corundum underlay.

Sintering in hydrogen was followed by annealing in a vacuum at 1000 °C for 1 h to eliminate alloy brittleness due to hydrogen.

Series 2. Series 2 samples were obtained by adding ultrafine powders of a similar composition (95%W-3.5%Ni-1.0%Fe-0.5%Co) to original coarse-grained 95%W-3.5%Ni-1.0%Fe-0.5%Co powders followed by low-energy MA of the resulting mixture and sintering it in hydrogen.

Ultrafine tungsten powders were produced by DC arc plasma synthesis (Chuvil'deev et al. 2017; Krasovskii et al. 2015). Specific surface of recovered tungsten powders was $S_s = 1.7–2.4\ m^2/g$, which corresponds to the particle size of 0.12–0.20 μm. Ultrafine Fe, Ni, and Co powders were obtained by metal hydroxide precipitation from salt, drying, and recovery at lower temperatures (300–350 °C, 2 h for Ni and Co, 750 °C, 4 h for Fe). The powders had the following particle size and specific surface: Ni powder, 0.4 μm ($S_s = 1.6\ m^2/g$); Fe powder, ~1 μm ($S_s = 0.7–0.9$- m^2/g); and Co powder, 0.08 μm ($S_s = 15\ m^2/g$).

Low-energy MA (grinding ball acceleration 0.4 g) was carried out in a MB0-10 ball mill for $t_{MA} = 72$ h at container rotational speed of $V_{MA} = 100$ rpm. Samples were sintered in hydrogen in a mode similar to that of Series 1.

Series 3. Nanopowder 95 W-3.5Ni-1.5Fe compositions were produced by MA in an APF-3 high-energy planetary mill (Russia) with grinding ball acceleration of 60 g (rotational speed $V_{Ma} = 1450$ rpm). MA time was $t_{Ma} = 20$ min. Series 3 samples were sintered in modes similar to those of Series 1.

Series 4. 95 W-3.5Ni-1.5Fe nanopowder composition for Series 4 was produced by high-energy MA in modes similar to those used for sample preparation in Series 3. MA times varied from 5 to 40 min. Series 4 samples were sintered by SPS on a Dr. Sinter model SPS-625 (SYNTEX INC., Japan) in a vacuum (6 Pa). Sample diameter was ∅20 mm; height was h = 3 mm. Sintering temperature ranged from 900 °C to 1300 °C. Temperature readings were taken with a Chino IR-AH infrared thermometer (Japan) focused on the outer surface of the graphite compression mold. Measurement accuracy was ±10 °C.

Samples were heated with 3.3-ms electrical pulses. Sintering speed (V_s) ranged from 50 to 300°/min. Sintering was performed under 50–70 MPa.

A dilatometer, which is supplied with the Dr. Sinter model SPS-625 unit, was used to measure effective shrinkage (L_{eff}, mm) in a continuous heating mode. To account for the impact on effective shrinkage L_{eff} produced by thermal expansion, occurring in the unit-sample system, additional thermal expansion measurements were taken on an empty compression mold. The obtained dependencies $L_0(T)$ were then subtracted from dependence $L_{eff}(T)$, and calculations were done for dependence of shrinkage speed (S) on SPS temperature (Fig. 2). Shrinkage accuracy was 0.01 mm.

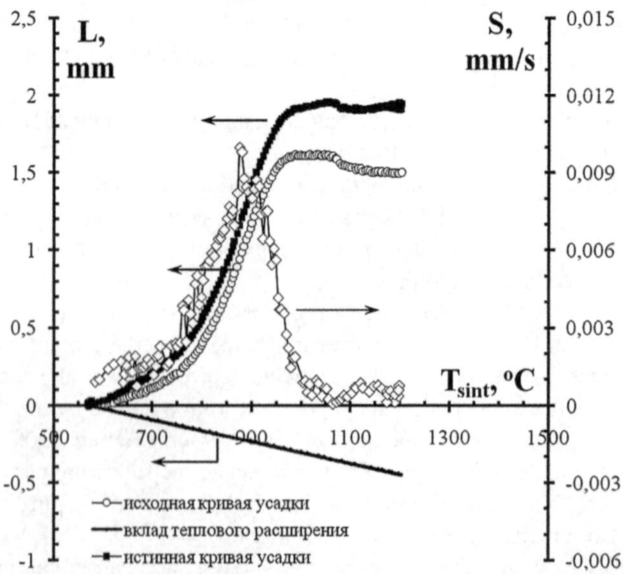

Fig. 2 Dependencies of shrinkage and shrinkage rate on heating temperature during SPS heating rate $V_s = 100°/min$ ($P = 50$ MPa). 95 W-3.5Ni-1.5Fe composition after high-energy MA ($t_{ma} = 20$ min)

Prior to SPS sintering, the powders were additionally annealed in hydrogen at 800 °C for 2 h. Hydrogen concentration in W-Ni-Fe powder compositions after annealing was under 0.03 wt%.

Specific powder surface (S) was measured by a Micromeritics TriStar 3000 analyzer. Chemical composition was analyzed by an Ultima 2 ICP atom-emitting spectrometer. XRD analysis was performed on a Shimadzu XRD-700 diffractometer (emission CuK_{α}, scanning rate from 0.5°/min (for investigation of X-ray peak widenings) to 2°/min (for overview images), increment from 0.02° to 0.5°). Phase volume percentage was determined by the Rietveld method.

The structure of sintered alloys was studied with a Leica IM DRM optical microscope and a Jeol JSM-6490 scanning electron microscope. Average grain size in the sintered ceramics was determined using the intercept method.

Density of the samples was measured by hydrostatic weighing in distilled water using a Sartorius CPA analytical balance. Density accuracy was ±0.01 g/cm^3.

Relaxation testing was used to study mechanical properties of the obtained samples, which allows for using compression tests to determine the macro-elastic limit (σ_o) and the yield stress (σ_y) (Nokhrin 2012). Macro-elastic limit and yield stress were measured on cylindrical samples of 3 mm dia. and 6 mm height. Measurement accuracy for σ_o and σ_y was ±20 MPa.

3 Experimental Results

3.1 Impact of Low-Energy MA on Optimal Sintering Temperature of 95 W-3.5Ni-1.0Fe-0.5Co Alloy (Series 1)

Tungsten particle size in 95 W-3.5Ni-1.0Fe-0.5Co alloy after low-energy mechanical activation is under 1 μm (Table 1). For XRD results for powder composition in its original state and after MA, see Fig. 3. The diffraction patterns of the powder composition in its original state (prior to MA) clearly show X-ray peaks that correlate to major phases (W, Ni, Fe). Mass fraction of the phases (W, Ni, Fe), calculated by the Rietveld method based on XRD results, corresponds to the masses of the mixed components within the margin of error (±0.5–1 wt%).

Figure 3b shows that low-energy MA results in slightly lower and wider peaks. The X-ray peaks that correspond to Ni and Fe virtually disappear, unlike the (111) nickel peak (Fig. 3b). The α-W peaks not only widen after MA but become asymmetric. Figure 4 shows that, with longer MA, the (110) α-W peak widens lopsidedly toward larger reflection angles.

Investigation of the 95 W-3.5Ni-1.0Fe-0.5Co alloy structure reveals that, in this alloy, active recrystallization occurs during sintering in hydrogen. Average grain size after sintering at 1500 °C depends on MA duration and ranges from 47 μm to 70 μm. MA lengthening from 2 to 10 h results in a more fine-grained structure

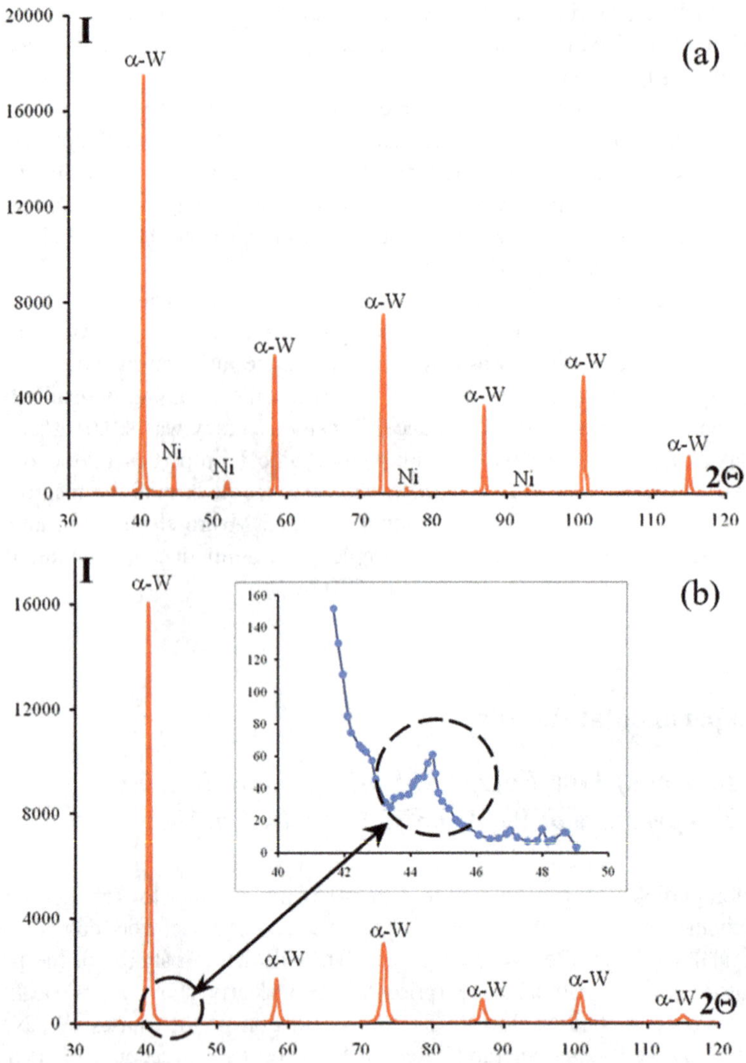

Fig. 3 XRD results for 95%W-3.5%Ni-1.0%Fe-0.5%Co powders (Series 1): (**a**) – original state, (**b**) – after low-energy MA for $t_{ma} = 10$ h. Scanning rate 2°/min. Increment 0.5°

(Fig. 5), which is caused by a known phenomenon of tungsten powder particles decreasing in size due to longer MA (Zhang et al. 2004; Jang et al. 2007).

Dependence of alloy density on sintering temperature is non-monotonic (Fig. 6a). At sintering temperatures between 1300 °C and optimal temperature T_1, density increases from ~17.00 to 18.05 g/cm^3 (from 93.9% to 99.7% of ρ_{th}); as the temperature continues to grow from T_1 to 1500 °C, the density declines to 16.5–17.8 g/cm^3, which corresponds to 91.1–98.3% of the alloy's theoretical density. Optimal sintering temperature T_1 largely depends on MA duration t_{Ma}; as t_{Ma} increases

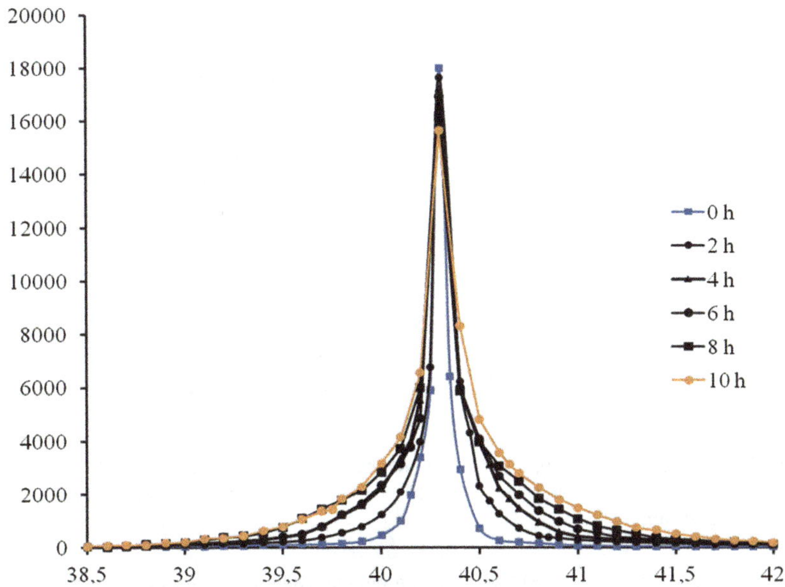

Fig. 4 X-ray diffraction pattern of (110) α-W peak after different exposure to low-energy MA of 95 W-3.5Ni-1.0Fe-0.5Co powder composition. Scanning rate 0.5°/min. Increment 0.1°

Fig. 5 Dependence of grain size on sintering temperature in hydrogen of 95%W-3.5%Ni-1.0%Fe-0.5%Co alloy (Series 1). The diagram shows MA time

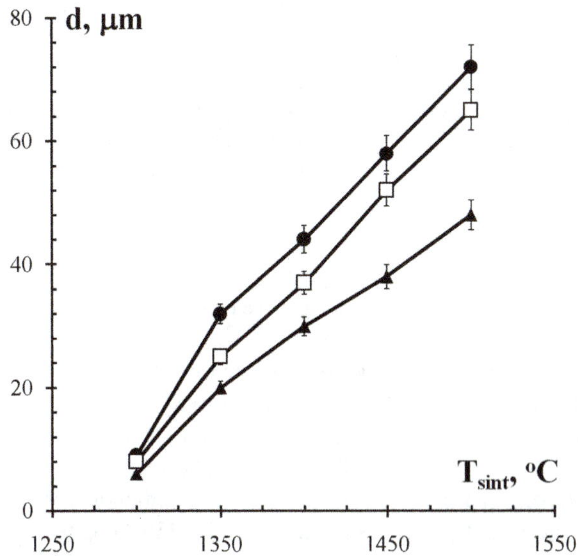

from 2 to 10 h, T_1 shifts from 1450 °C to 1500 °C toward 1300 °C (Table 3). It should be noted that raising t_{MA} from 2 to 10 h leads to a lower density in alloys sintered at $T_{sint} = 1500$ °C, from 18.02 down to 16.45 g/cm^3 (from 99.6% to 90.9% of ρ_{th}) (Fig. 6b).

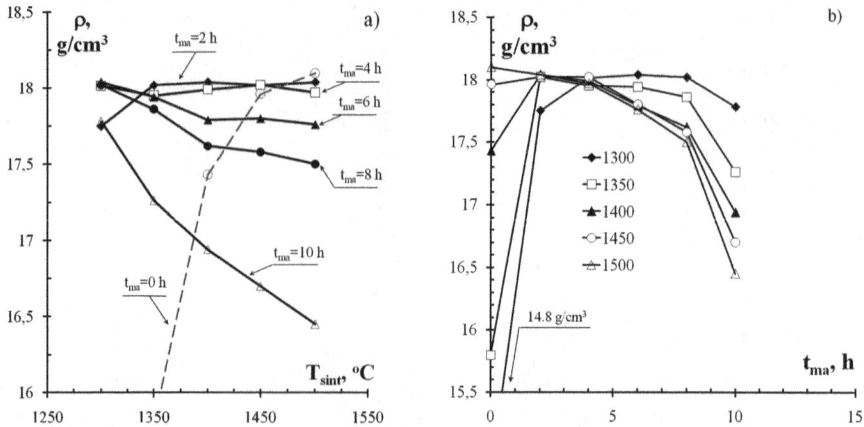

Fig. 6 Dependence of density on (**a**) sintering temperature in hydrogen and (**b**) MA time of 95% W-3.5%Ni-1.0%Fe-0.5%Co alloy (Series 1). The figures show (**a**) MA time for feedstock and (**b**) sintering temperature

Table 3 Impact of low-energy MA duration on optimal sintering temperature and 95 W-3.5Ni-1.0Fe-0.5Co alloy structure (Series 1)

		Alloy density at optimal sintering temperature $T_{sint} = T_1$			Mechanical properties		
t_{ma}, h	Optimal sintering temperature (T_1), °C	g/cm³	%	Average grain size d, µm	σ_o, MPa	σ_y, MPa	K_{HP}, MPa·m$^{1/2}$
0	1450–1500	18.08–18.10	100	50–70	250	510	2.1
2	1450	18.02–18.05	99.6–99.7	50	280	550	1.9
4	1400	18.00	99.5	36–38	300	550	1.5–1.6
6	1350	17.95–18.00	99.2–99.5	15–20	320	590	1.0–1.2
8	1300	17.97	99.3	5–10	740	930	0.4–0.6
10	≤1300*	17.78	98.2	3–5	780	1000	0.4–0.5

*The temperature corresponding to the maximum density for dependence $\rho(T_{sint})$ (see Fig. 6b)

However, the density decrease during the second stage of sintering (at $T_1 \leq T_{sint} \leq T_2$) is not linked to pore formation: detailed metallographic and electron-microscopy testing revealed no change in volume percentage of pores.

With longer MA, alloy strength shows a monotonic increase (Fig. 7a), whereas dependencies σ_o and σ_y on sintering temperatures are non-monotonic (Fig. 7b). During solid-phase sintering (T ≤ 1450 °C), temperature increase results in a monotonic decline in the yield point value caused by an onset of recrystallization.

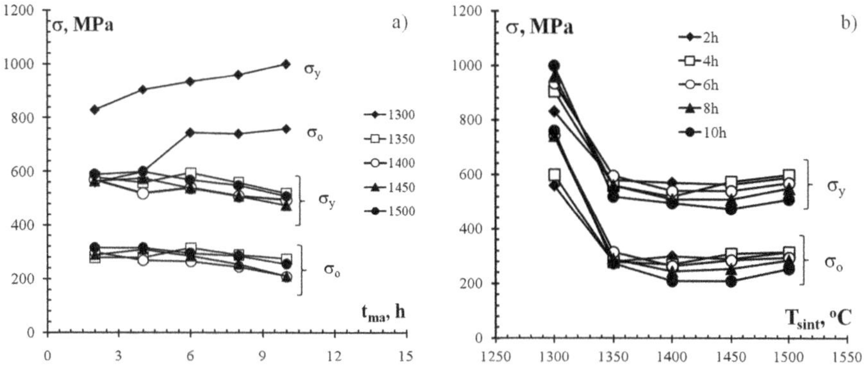

Fig. 7 Dependence of macro-elastic limit and yield point of 95 W-3.5Ni-1Fe-0.5Co alloy (Series 1) on (**a**) low-energy MA time and (**b**) sintering temperature in hydrogen. The figures show (**a**) sintering temperature and (**b**) MA time

3.2 Impact of Adding Ultrafine Powders on Optimal Sintering Temperature of 95 W-3.5Ni-1.0Fe-0.5Co Alloy (Series 2)

Figure 8 shows dependence of 95 W-3.5Ni-1.0Fe-0.5Co alloy density with varying content of ultrafine powders on sintering temperature in hydrogen. Dependence of 95 W-3.5Ni-1.0Fe-0.5Co alloy density with a 10% addition of ultrafine-powder mixture on sintering temperature presents three stages. During Stage 1, as sintering temperature T_{sint} increases from 1200 to T_1 = 1400 °C, alloy density undergoes a monotonic increase from 15.77 to 17.77 g/cm^3 (from 87.1% to 98.2% of the alloy's theoretical density ρ_{th}); during Stage 2, as sintering temperature increases to T_2 ~ 1450 °C, the density decreases to 17.53 g/cm^3 (down to 96.85%). As the temperature further increases to 1500 °C during Stage 3, the density goes up to 17.67 g/cm^3 (Table 4 and Fig. 8).

Temperature dependence of 95 W-3.5Ni-1.0Fe-0.5Co alloy density with a 20% addition of ultrafine-powder mixture has two stages. During Stage 1, as T_{sint} increases from 1200 °C to T_1 ~ 1400 °C, material density increases from 16.34 to 17.73 g/cm^3 (from 90.3% to 98.0%). As the temperature further increases to T_2 = 1500 °C during Stage 2, the density goes down to 17.32 g/cm^3 (down to 95.7% of ρ_{th}, see Table 4).

The dependence of alloy density with 100% ultrafine powder content on sintering temperature also has two stages. As T_{sint} increases from 1200 °C to optimal sintering temperature T_1 ~ 1300 °C, densification occurs, from 16.66 to 17.82 g/cm^3 (from 92.0% to 98.5% of the alloy's theoretical density). As T_{sint} further increases to T_2 ~ 1500 °C, the density declines to 17.04 g/cm^3 or 94.1% of ρ_{th}.

Noteworthy is that the obtained dependencies $\rho(T_{sint})$ for the powders of 95 W-3.5Ni-1.0Fe-0.5Co alloy differ from the dependence obtained from sintering coarse-grained mechanically activated powders (Fig. 8). (Coarse-grained W-Ni-Fe alloys (Green et al. 1954; Krock and Shepard 1963) present regular, monotonic

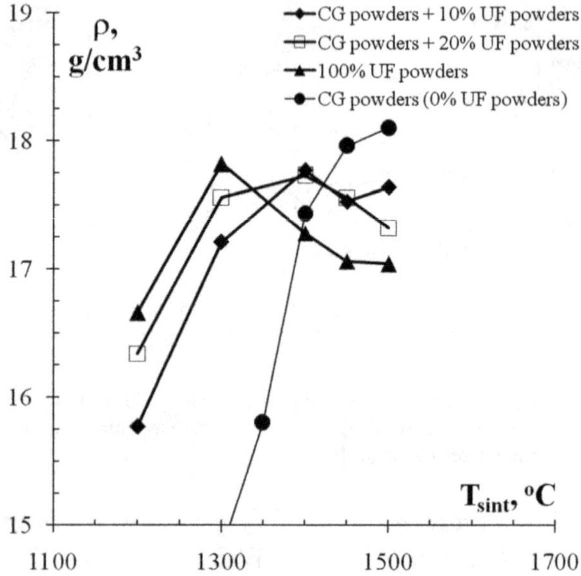

Fig. 8 Dependence of density of 95%W-3.5%Ni-1.0%Fe-0.5%Co alloy with varying content of ultrafine powders (Series 2) on sintering temperature in hydrogen

Table 4 Effect of additions (volume percentage) of ultrafine 95 W-3.5Ni-1.0Fe-0.5Co powder mixture on optimal sintering temperature, density and mechanical properties of 95 W-3.5Ni-1.0Fe-0.5Co alloy obtained by low-energy MA, and sintering in hydrogen (Series 2)

Content of ultrafine powders in alloy, %	Regular sintering temperatures		Alloy density at optimal sintering temperature $T_{sint} = T_1$		Average grain size	Mechanical properties		
	T_1, °C	T_2, °C	g/cm^3	%	d, μm	σ_o, MPa	σ_y, MPa	K_{HP}, MPa·m$^{1/2}$
0	1450	1500	18.02	99.6	22	185	635	2.1
10	1400	1450	17.77	98.2	15	230	655	1.6
20	1350–1400	1450	17.73	98.0	10	240	665	1.3
100	1300	1450	17.82	98.5	4	635	905	0.5

dependencies of density on sintering temperature). As with Series 1 samples, investigation of structure revealed no increase in the volume percentage of pores during the second stage of sintering at $T_1 \leq T_{sint} \leq T_2$, whereas the density decrease in 100% ultrafine-powder alloys from 17.82 to 17.04 g/cm^3 corresponds to a pore volume percentage that is higher by ~5%.

A closer look at the structure shows that adding ultrafine-powder into 95 W-3.5Ni-1.0Fe-0.5Co alloy results in smaller grain size. Average grain size in alloys without ultrafine powders after sintering at optimal temperature T_1 is 22 μm. In alloys with 10% and 100% ultrafine-powder content sintered at optimal temperature T_1, average grain size is ~15 μm and ~4 μm, respectively (Table 4).

Mechanical testing of 95 W-3.5Ni-1.0Fe-0.5Co alloy with different UFG content indicates that increasing the volume percentage of ultrafine powders from 10% to 100% results in a higher macro-elastic limit (σ_o), up from 230 MPa to 635 MPa, and higher yield stress (σ_y), up from 655 MPa to 905 MPa (Table 4).

3.3 Impact of High-Energy MA on Optimal Sintering Temperature of 95 W-3.5Ni-1.5Fe Alloys (Series 3)

After MA, average α-W grain size is under 100 nm (Table 1 and Fig. 9a). Particles form easily degradable conglomerates of ~1 μm on average (Fig. 9a).

Diffraction patterns show that longer MA times lead to weaker and wider X-ray peaks that correspond to α-W (Fig. 10a, b). It is important to note that longer MA times result in an asymmetric widening of (110) α-W peak toward larger reflection angles (Fig. 11). This suggests formation of a supersaturated solid solution of nickel and iron in the surface layer of α-W. In sintered tungsten alloys, the (110) α-W peak has no asymmetry.

For dependence of average grain size and alloy density on sintering temperature, see Fig. 12a. The data show that, when heating temperature exceeds optimal sintering temperature T_1, grain growth accelerates. In alloys sintered at T = 1100 °C and 1450 °C, average grain size is 1.5 μm and 22 μm, respectively. This is larger than tungsten particle size in its original state by an order of magnitude and more (see Figs. 8b and 12a).

Dependence of density on sintering temperature in Series 3 alloys has two stages; it is similar to dependencies $\rho(T_{sint})$ for alloys in Series 1 and 2. Increasing sintering temperature from 1100 °C to T_1 ~ 1300 °C leads to densification from 15.86 to 18.02 g/cm^3 (from 87.6% to 99.6% of ρ_{th}), whereas further temperature increase to 1450–1500 °C results in a lower density of 17.37–17.48 g/cm^3 (96.0–96.6% of ρ_{th}). It can be seen that the optimal sintering temperature T_1 in hydrogen for mechanically activated powders is 1300 °C, which is 150–200 °C lower than that of coarse-grained powders (~1450–1500 °C in hydrogen).

Temperature dependence curves for the macro-elastic limit (σ_o) and the yield stress (σ_y) of 95 W-3.5Ni-1.5Fe alloy are shown in Fig. 12b. The figure shows that dependencies $\sigma_o(T_{sint})$ and $\sigma_y(T_{sint})$ have two stages. Mechanical qualities of the material are the highest (σ_o = 1030 MPa, σ_y = 1320 MPa) in alloys sintered at T_{sint} = 1250 °C. Further temperature increase results in lower strength due to recrystallization.

Figure 13 shows dependencies of volume percentage of pores on temperatures of sintering mechanically activated powders in Series 3. Higher sintering temperatures lead to a monotonic decrease in the volume percentage of pores from ~20–25% at T_{sint} = 1150 °C down to zero at ~1300–1350 °C. Between 1350 °C and 1450 °C,

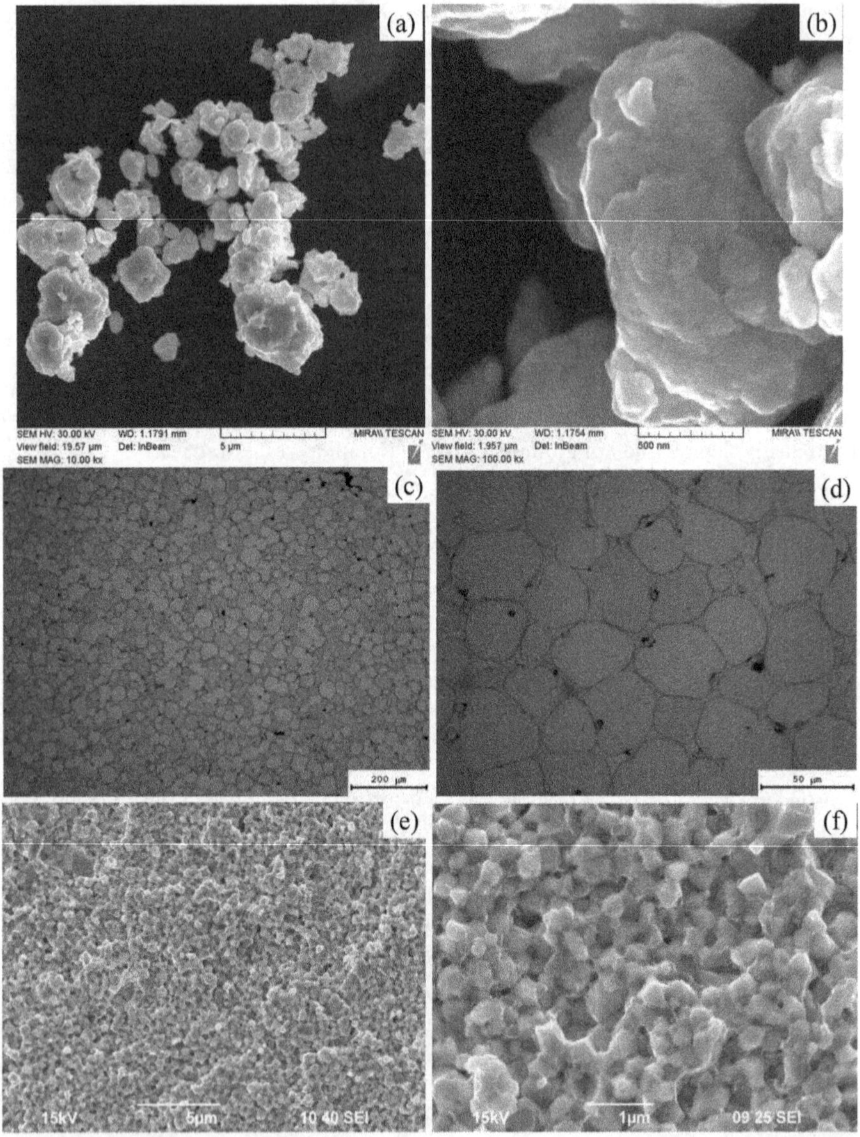

Fig. 9 Conglomerate structure in 95%W-3.5%Ni-1.5%Fe powders after high-energy MA for t_{MA} = 20 min (**a**, **b**), after sintering in hydrogen at 1450 °C (**c**, **d**) and SPS at 1100 °C (**e**, **f**)

where the alloy loses density (see above), no increase in the volume percentage of pores was revealed by metallographic and SEM testing. Alloys sintered at T_{sint} = 1350 °C have a high-density structure with pore volume percentage remaining under 0.2%.

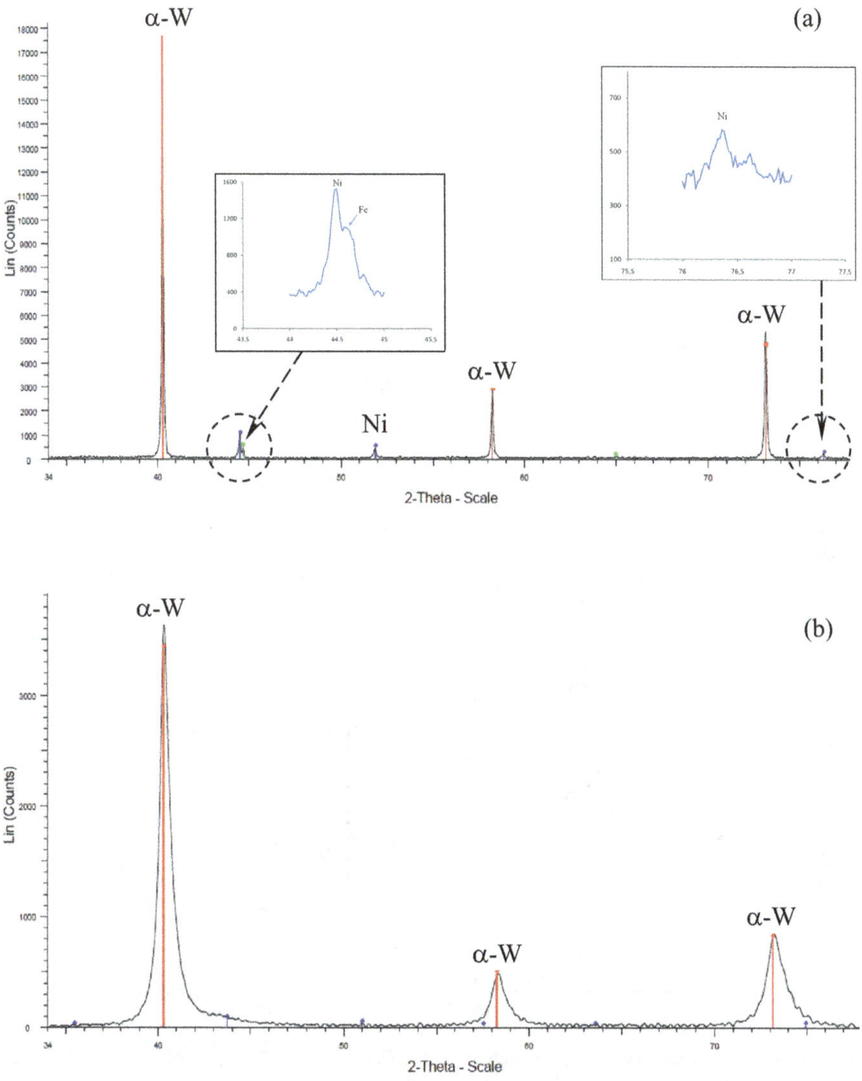

Fig. 10 XRD results for 95 W-3.5Ni-1.5Fe powders: (**a**) – original state; (**b**) – after high-energy MA (t_{Ma} = 20 min). Scanning rate 2°/min. Increment 0.02°

3.4 Impact of High-Energy MA on Optimal Sintering Temperature of 95 W-3.5Ni-1.5Fe Alloy (Series 4)

Figure 14 shows dependencies of shrinkage and shrinkage rates during SPS for Series 4 nanopowders obtained by high-power MA (t_{ma} = 20 min).

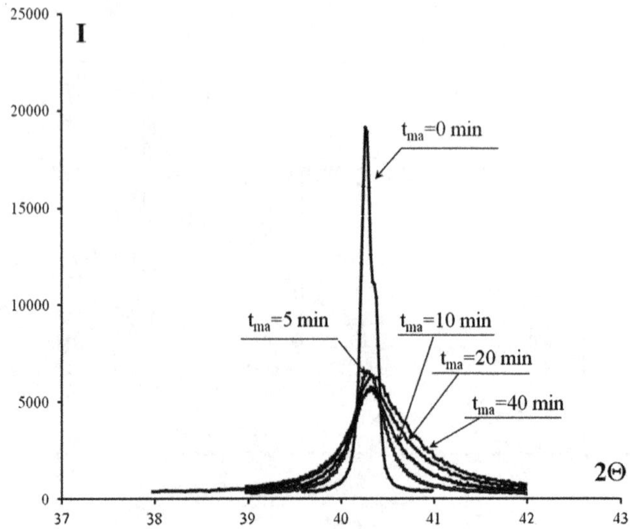

Fig. 11 X-ray diffraction pattern of (110) α-W peak after different exposure to high-energy MA of 95 W-3.5Ni-1.5Fe powder composition. Scanning rate 0.5°/min. Increment 0.02°

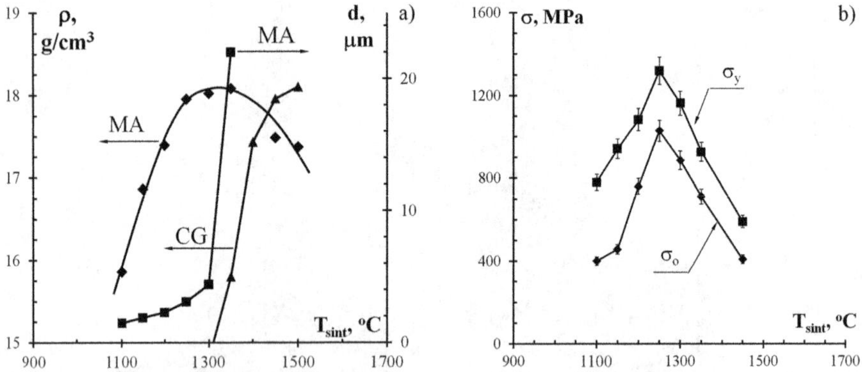

Fig. 12 Dependence of (**a**) density, grain size, and (**b**) mechanical properties on temperature of sintering in hydrogen of 95 W-3.5Ni-1.5Fe alloy after high-energy MA

Generally, the results indicate that the optimal SPS temperature for 95 W-3.5Ni-1.5Fe alloy ($t_{ma} = 20$ min) at the heating rate of $V_s = 100$ °C/min is $T_1 \sim 1100$ °C. Slower heating rates result in optimal SPS temperature drifting toward higher values; the optimal SPS temperature at 50 °C/min is $T_1 = 1200$ °C. For coarse-grained powders, the optimal SPS temperature is ~1300 °C (Fig.14a, b). Figure 15a reveals that dependence of density on SPS temperature for mechanically activated powders is non-monotonic and has a maximum, similar to Series 1–3. For coarse-grained powders, dependence of density on SPS temperature is monotonic (Fig. 15a).

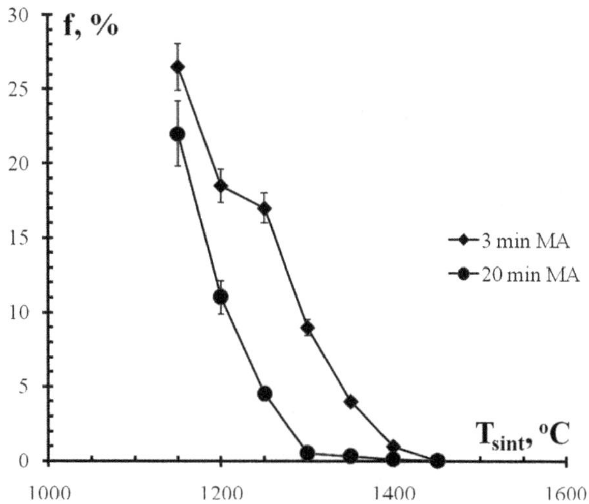

Fig. 13 Dependence of volume percentage of pores on free-sintering temperature of mechanically activated 95 W-3.5Ni-1.5Fe nanopowders after MA for 3 and 20 min

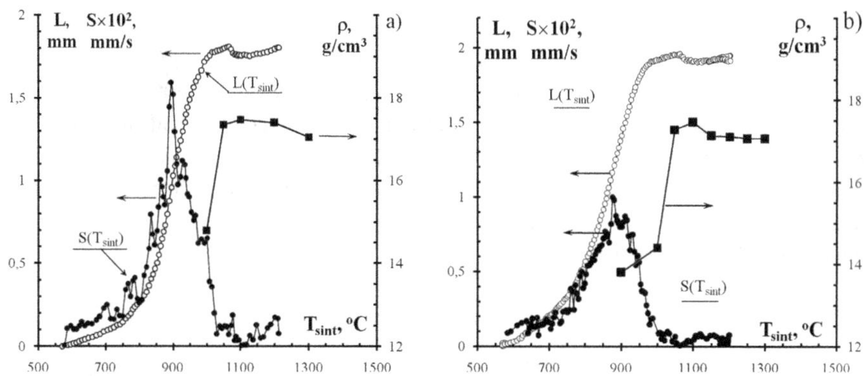

Fig. 14 Dependencies of shrinkage and shrinkage rate on SPS temperature for 95 W-3.5Ni-1.5Fe nanopowders obtained by high-energy MA ($t_{ma} = 20$ min): (**a**) – sintering at heating rate of 50°/min; (**b**) – sintering at heating rate of 100°/min

It should be emphasized that, in mechanically activated nanopowders, maximum densities during sintering in hydrogen and SPS are seen during solid-phase sintering unlike regular sintering of coarse-grained powders where maximum densities are achieved only during liquid-phase sintering at over 1450–1500 °C. Structural studies of the alloy reveal that at $T_1 = 1100$ °C, an UFG structure forms with a grain size of ~500 nm (Fig. 9c). Average grain size in 95 W-3.5Ni-1.5Fe alloy ($t_{ma} = 20$ min) after SPS at 1300 °C ($V_s = 100$ °C/min) is under 3–3.5 μm (Fig. 15b).

Figure 15b shows dependence curves of mechanical properties of Series 4 alloy ($t_{ma} = 20$ min) on SPS temperature at the heating rate of 100 °C/min. Dependencies $\sigma_o(T_{sint})$ and $\sigma_y(T_{sint})$ have two phases. The alloy achieves its best mechanical properties ($\sigma_o = 2250$ MPa, $\sigma_y = 2500$ MPa) after heating at 100 °C/min to $T_{sint} = 925$–950 °C. However, alloy plasticity after sintering in this mode is low

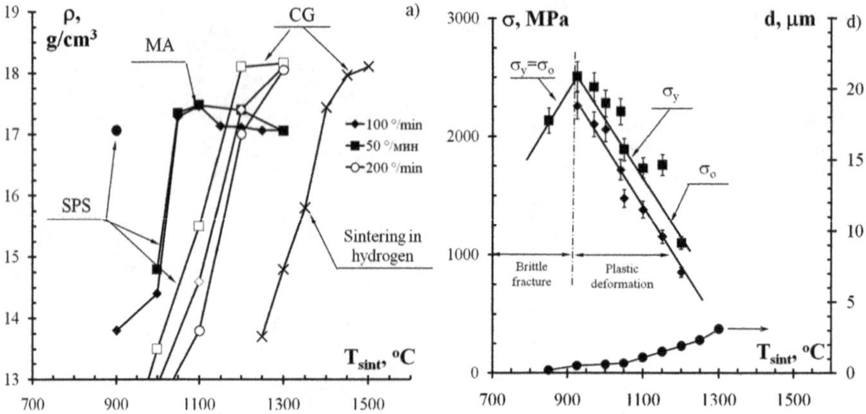

Fig. 15 Dependence of (**a**) density, mechanical properties, and (**b**) grain size on temperature of SPS 95%W-3.5%Ni-1.5%Fe alloy after high-energy MA. In (**a**), solid markers indicate mechanically activated powders, and hollow markers indicate coarse-grained powders. (**b**) shows data for the heating rate of 100°/min

under 0.2–0.5% for compression. At higher SPS temperatures (over 1050 °C), obtained samples show weaker strength ($\sigma_o \leq 1500$ MPa, $\sigma_y \leq 1900$ MPa), but do not fail after ~10% sinking strain. Comparing the experimental results with published data (Table 5), the yield strength of the obtained alloys is ~1.5–2 higher than that of regular coarse-grained alloys. It can be concluded, therefore, that the density and mechanical properties of tungsten alloys have a non-monotonic dependence on SPS temperature.

4 Discussion

Generalizing the experimental results, three points that require further elaboration should be raised first: (1) lower optimal sintering temperature in mechanically activated nanopowders, (2) higher sintering intensity in nanopowders under high heating rates, and (3) non-monotonic dependence of density on sintering temperature of mechanically activated powders.

For the purposes of analyzing the results, it should be noted that – as has been established (Larikov and Yurchenko 1987) – the intensity of diffusion-based mass transfer (I) is proportional to the diffusion coefficient D, which depends exponentially on diffusion activation energy Q ($D = D_0\exp(-Q/kT)$) and to the concentration gradient of the diffusing substance (C): $I \sim -D \cdot \mathrm{grad}C$, where D_0 is pre-exponential multiplying factor and k is the Boltzmann constant.

Table 5 Comparison of data on mechanical properties of regular coarse-grained and fine-grained W-Ni-Fe alloys and ultrafine-grained 95 W-3.5Ni-1.5Fe alloy obtained by sintering in hydrogen and SPS in this paper

Alloy		ρ, g/cm^3	σ_o, MPa	σ_y, MPa	σ_b, MPa	δ, %*
Coarse-grained 90 W-7Ni-3Fe alloy (Baranov 2010)		16.9–17.1	450–550	650–700	950–1000*	15–20
Coarse-grained 95 W-3.5Ni-1.5Fe alloy (Baranov 2010)		18.0–18.1	550–600	680–700	930–950*	10–15
Coarse-grained 97.5 W-1.5Ni-0.5Fe alloy (Baranov 2010)		18.5–18.7	–	700–730	700–750*	1
Coarse-grained 90 W-7Ni-3Fe alloy (Green et al. 1954)		17.01–17.09	–	358–452	560–604*	20.4–20.9
Coarse-grained 92 W-Ni-Fe alloy (Krock and Shepard 1963)		–	400–450	700–800	–	25–29
Fine-grained 88 W-Ni-Fe alloy (Lee et al. 2007)		16.60	–	–	894*	30
Fine-grained 93 W-Ni-Fe alloy (Lee et al. 2007)		17.52	–	–	996*	23
Fine-grained 95 W-3.5Ni-1.5Fe alloy (Lee et al. 2007)		17.98	–	–	916*	11
Fine-grained 94 W-Ni-Fe alloy (Humail et al. 2007)		–	–	600–650	900*	23
Fine-grained 93 W-(Ni:Fe at 7:3) alloy (German and Chur 1984)		–	–	–	1006*	22
Fine-grained 95 W-(Ni:Fe at 7:3) alloy (German and Chur 1984)		–	–	–	995*	20
Fine-grained 90 W-7Ni-3Fe alloy (SPS) (Xiang et al. 2013a, b)		93.8–96.12%**	–	–	1020*	–
Fine-grained 93 W-5.6Ni-1.4Fe alloy (SPS)	Li et al. (2013)	94.8–98.7%**	–	–	780–1062	–
	Li et al. (2014)	~95%**	–	–	835–1270	–
Fine-grained 90 W-7Ni-3Fe alloy (SPS) (Xiang et al. 2012)		96.07%**	–	–	1335***	–
Fine-grained W-2Mo-7Ni-3Fe alloy (SPS)	Xiang et al. (2013a, b)	96.42%**	–	–	922*	–
	Ding et al. (2012b)	< 90%**	–	–	390*	–
Fine-grained 95 W-3.5Ni-1.5Fe alloy (sintering in hydrogen) (*this study*)		18.02	1030 $K_{HP} = 0.5$ MPa·m$^{1/2}$ (d = 3.1 μm)	1320	–	≥10 (upsetting)

(continued)

Table 5 (continued)

Alloy	ρ, g/cm^3	σ_o, MPa	σ_y, MPa	σ_b, MPa	δ, %*
UFG 95 W-3.5Ni-1.5Fe alloy (SPS) (*this study*)	18.05	1475	1885	–	≥10 (upsetting)
		$K_{HP} = 0.6$ MPa·m$^{1/2}$ (d = 1.9 μm)			
	17.90	2250	2500	–	≤0.2 (upsetting)
		$K_{HP} = 0.2$ MPa·m$^{1/2}$ (d = 0.5 μm)			

σ_b is ultimate strength
δ is relative elongation
*Tensile testing results
**Of ρ_{th} (the reference gives no absolute values for density)
***Compression strength

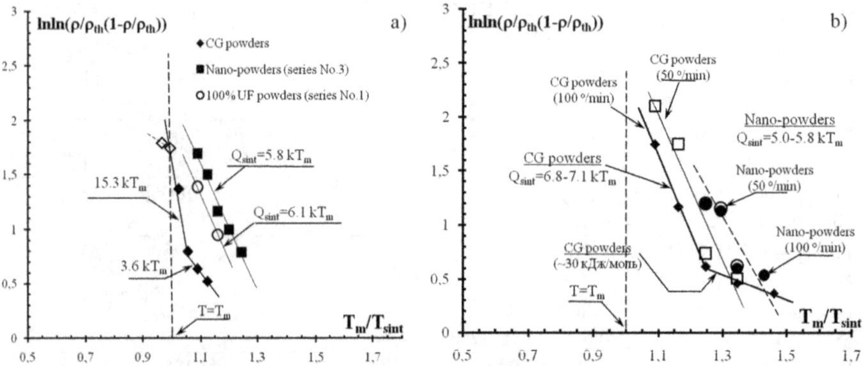

Fig. 16 Dependence of density on sintering temperature of tungsten alloys in coordinates $\ln(\ln(\rho/\rho_{th}[1-\rho/\rho_{th}])) - T_m/T_{sint}$: (**a**) sintering in hydrogen; (**b**) SPS

4.1 Calculation of Sintering Activation Energy. Accelerating Effect of MA on Tungsten Alloy Sintering

Primary data in sintering nano- and ultrafine powders are dependencies of density on sintering temperature $\rho/\rho_{th}(T_{sint})$. These dependencies have two regular stages: intensive densification (Stage I) and slow densification (Stage II) where the density of metallic materials changes ever slightly (from ~98% to their theoretical density ρ_{th}) and intensive grain growth occurs (Sōmiya and Moriyoshi 1990; Rahaman 2003).

According to the method found in references (Potanina et al. 2016; Chuvil'deev et al. 2015, 2017), sintering activation energy for nano- and fine-grained materials can be determined by the angle of dependence $\rho/\rho_{th}(T_{sint})$ represented in double

Table 6 Optimal sintering temperature (numerator) and sintering activation values (denominator) of 95%W-3.5%Ni-1.5%Fe tungsten alloy (t_{ma} = 20 min)

	Sintering in hydrogen	SPS
Coarse-grained powders	1500 °C	1250–1300 °C
	11.5–11.9 kT_m (163–169 kJ/mol)	7.9–8.8 kT_m* (112–125 kJ/mol)
Mechanically activated nanopowders	1300 °C	1100–1200 °C
	6.8–7.0 kT_m (97–100 kJ/mol)	8.3 kT_m (118 kJ/mol)** 6.2 kT_m (88 kJ/mol)***

*For heating rates of 50–200 °C/min
**For V_s = 50 °C/min
***For V_s = 100 °C/min

logarithmic coordinates of reversed temperature T_m/T_{sint}: $\ln(\ln[(\rho/\rho_{th})/(1-\rho/\rho_{th})])$ – T_m/T_{sint}. (Here, T_m = 1738 K is the melting point of the γ-phase (Green et al. 1954; Krock and Shepard 1963), and ρ_{th} is the alloy's theoretical density.) For the obtained dependencies, see Fig. 16. For calculated activation energies, see Table 6.

Activation energy in hydrogen for coarse-grained powders is Q_{sint} = 163–169 kJ/mol ~11.5–11.9 kT_m. This value is close to the activation energy of grain-boundary diffusion Q_b in tungsten in the γ-phase, which is, according to Frost and Ashby 1982, ~(0.5–0.6)Q_v, where Q_v is activation energy of volume diffusion. In case of ^{181}W diffusion in nickel, Q_v is ~268 kJ/mol (Larikov and Yurchenko 1987). In case of volume diffusion of tungsten in Ni-W and in γ-Fe-W, Q_v is 295–306 kJ/mol and 268 kJ/mol, respectively (Park et al. 2006; Seith 1955). (Available data on sintering activation energies of W-Ni-Fe tungsten alloys give dramatically different values; Q_{sint} for 88%W-8.4%Ni-3.6%Fe alloy is 250 kJ/mol according to Blaine et al. (2006), whereas reference (Park et al. 2013) gives Q_{sint} = 367 kJ/mol for 88%W-8.4%Ni-3.6%Fe alloy and 480 kJ/mol for 95%W-3%Ni-2%Fe.)

SPS activation energy for coarse-grained powders (7.9–8.8 kT_m ~ 112–125 kJ/mol) is lower than that of sintering such powders in hydrogen. We believe that the difference in activation energies of sintering in hydrogen and SPS for coarse-grained powders is due to application of pressure during SPS (Tokita 2013). Sintering activation energy in hydrogen for nanopowders obtained by high-energy MA is ~6.8–7.0 kT_m (97–100 kJ/mol). SPS activation energy for mechanically activated nanopowders depends on heating rate; as V_s goes up from 50 °C/min to 100 °C/min, sintering activation energy decreases from 8.3 kT_m to 6.2 kT_m. (Table 6). (The obtained results are in close agreement with the data in reference (Hu et al. 2013), which also shows that SPS activation energy of 93 W-5.6Ni-1.4Fe alloys greatly depends on the heating rate: at $V_s < 30$ °C/min Q_{sint} = 454 kJ/mol, at $V_s > 200$ °C/min Q_{sint} = 200 kJ/mol).

In our view, optimal temperature of UFG tungsten alloy sintering decreases because of low sintering activation energy Q_{sint}, which is low due to the low activation energy of grain-boundary diffusion Q_b in mechanically activated powders.

Let us take a closer look at the roots of this phenomenon.

According to the theory of non-equilibrium grain boundaries in metals and alloys (Chuvil'deev 2004), the activation energy of grain-boundary diffusion goes down due to increasing free grain boundary area, which is proportional to the density of orientation mismatch dislocations that can be found in grain boundaries under intensive plastic deformation. Grinding of γ-phase particles with face-centered cubic lattice is difficult because of high particle plasticity. Relaxation of MA energy then occurs through formation of grain boundaries in γ-phase particles, with a high density of orientation mismatch dislocations (Chuvil'deev 2004). The result is that the diffusion permeability of γ-phase grain boundaries after MA is much higher than that of γ-phase equilibrium grain boundaries formed by sintering of individual γ-phase particles. Higher diffusion permeability in grain boundaries of "deformation origin" leads to a higher intensity tungsten atom diffusion through the γ-phase, which lowers the optimal sintering temperature of UFG tungsten alloys.

Let us turn to the causes of the second effect described above, namely, increased sintering intensity in mechanical activated W-Ni-Fe nanopowders with high heating rates. We believe that faster sintering is caused by change in the gradient of tungsten concentration grad(C_w) between α-W particles and γ-phase, which is a solid solution of tungsten in nickel (Green et al. 1954).

As has been detailed above (110), α-W X-ray peak distorts asymmetrically (toward larger reflection angles) during MA. This effect is not observed during MA of pure α-W powders. Therefore, the (110) α-W peak asymmetry is caused by deformation-based dissolution of a Ni and Fe atoms from the γ-phase in the surface layer of α-W particles.

Uneven concentration of Ni and Fe atoms in the surface area of α-W particles after MA leads to a diffusion-regulated migration of Ni and Fe atoms from the α-W lattice into the γ-phase when heated during sintering (see also (Han et al. 2012)). This results in higher local concentration of Ni and Fe atoms in the γ-phase close to the interphase boundary (α-W) – (γ-phase). As a result, there is an increase in the gradient of tungsten atoms concentration grad(C_w) between the α-W particles and γ-phase, which intensifies diffusion-based mass transfer through the γ-phase and speeds up sintering.

4.2 Impact of MA on Dependence of Density on Sintering Temperature

The effect of deformation-induced "dissolution" of nickel atoms in tungsten may be the reason why the dependence of density on sintering temperature in tungsten alloys is in non-monotonic. Density decrease during tungsten alloy sintering was described in references (Ding et al. 2012a, b; Hu et al. 2011a, b; Li et al. 2010; Ryu et al. 1997; Xiang et al. 2013a, b); it is also observed in all cases of sintering mechanically activated powders. As has been noted above, significant density decrease – by over

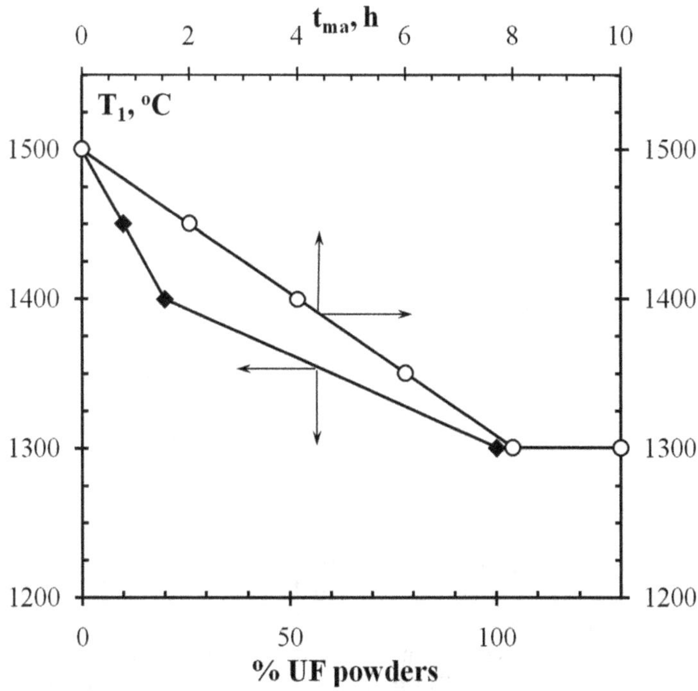

Fig. 17 Dependence of optimal free-sintering temperature T_1 of 95 W-3.5Ni-1Fe-0.5Co alloy on volume percent of ultrafine powders in the MIX and MA time

5–7% – during Stage 2 of sintering is yet to be explained since it is not related to higher pore volume percentage.

The intensity of density decreases during Stage 2 of sintering $\rho(T_{sint})$, and the optimal sintering temperature T_1 depends on parameters of original microstructure (grain size, duration, and intensity of MA) and on heating modes during sintering. Addition of ultrafine powders and grinding of powders during MA make the optimal sintering temperature T_1 shift toward lower values (see Fig. 17).

As has been demonstrated above, there is an asymmetric (toward larger reflection angles) distortion of (110) α-W peak during MA. It is caused by deformation-induced dissolution of nickel and iron atoms on the surface layer of α-W particles. Evidently, this will increase the density of α-W particles.

When heated, the material will be decreasing its internal energy by diffusion-based mass transfer of excessive nickel and iron atoms away from the strongly deformed α-W body-centered cubic lattice, which lessens Ni and Fe supersaturation in α-W and decreases density (Fig. 18a). Further densification of the alloy as temperature continues to rise is caused by regular acceleration at higher sintering temperatures of diffusion-based mass transfer, which is exponential $D = D_0\exp(-Q/kT)$.

Longer MA and faster grinding ball operation increase the solution supersaturation. This decreases the temperature at which nickel atoms start leaving tungsten and

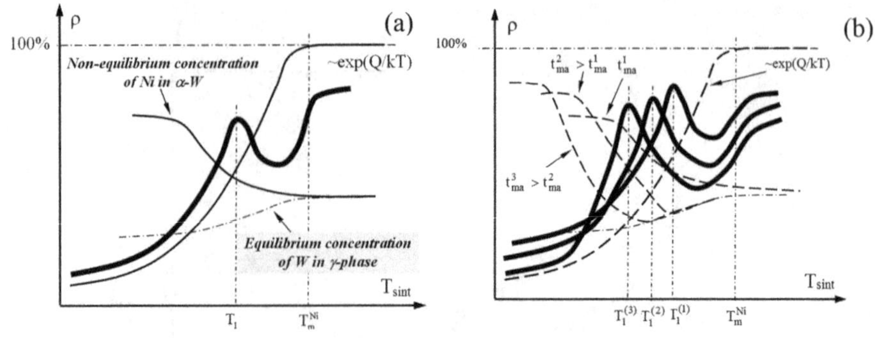

Fig. 18 Schematic curve of (**a**) dependence of density on sintering temperature of nano- and ultrafine 95%W-Ni-Fe alloys and (**b**) impact of MA on density decrease in alloys during the second stage of sintering: $t_{ma}^{(1)}$, $t_{ma}^{(2)}$, $t_{ma}^{(3)}$ – regular MA times; $T_1^{(1)}$, $T_1^{(2)}$, $T_1^{(3)}$ – optimal sintering temperatures corresponding to times $t_{ma}^{(1)}$, $t_{ma}^{(2)}$ and $t_{ma}^{(3)}$, respectively. Schematic

intensifies the process. For this reason, alloy density change will be more intensive during Stage 2 of sintering, and, by extension, the optimal temperature of sintering for mechanically activated tungsten alloys will be lower (see Fig. 18b) – as has been observed in the experiments.

Consequently, density decrease in mechanically activated tungsten alloys during sintering Stage 2 ($T_1 < T_{sint} < T_2$) is caused by intensive, diffusion-based loss of nickel atoms that are leaving tungsten particles (see also Han et al. 2012).

4.3 Impact of MA on Mechanical Properties of Tungsten Alloys

Investigation of sintered samples shows that dependence of yield point σ_y on grain size d in coordinates σ_y-$1/d^{1/2}$ can be interpolated as a straight line (see Fig. 19a). This means that the Hall-Petch relationship applies to the investigated UFG alloys[1]:

$$\sigma_y = \sigma_0 + K_{HP}/\sqrt{d}, \tag{1}$$

where K_{HP} is the Hall-Petch coefficient and d is α-W particle size.

[1]This is a rather unexpected finding since it is commonly assumed that the yield stress and ultimate strength of tungsten alloys are largely determined by the strength and volume percentage of the γ-phase. We believe that this result suggests that the volume percentage and particle size of the γ-phase in mechanically activated UFG alloys are smaller than in regular, sintered coarse-grained alloys. The effect can be caused by a large surface area of grain boundaries in UFG alloys, which leads to a proportional decrease in the size (thickness) of γ-phase particles that are located between α-W particles. Another possible factor is "dissolution" of Ni and Fe atoms within the γ-phase, in the α-W lattice during MA.

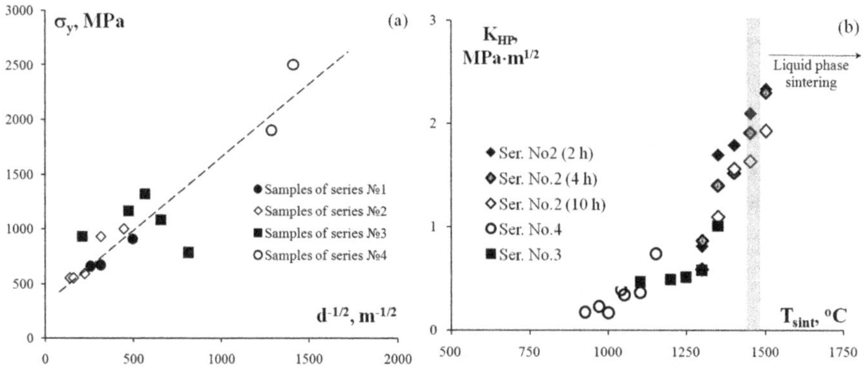

Fig. 19 Dependence of (**a**) yield point on grain size and (**b**) Hall-Petch coefficient K_{HP} on sintering temperature for mechanically activated tungsten alloys (**a**) shows generalized data on alloys obtained by sintering at optimal temperature T_1

K_{HP} describes the impact of grain boundaries on alloy strength and depends on the structural state of interphase and inter-grain boundaries. K_{HP} can assume a wide range of values and can, i.e., depend on the concentration of alloying elements at grain boundaries.

Equation (1) can be used to calculate K_{HP} for the investigated alloys. For calculation results, see Tables 3, 4, and 5. Results indicate that, as sintering temperature increases, coefficient K_{HP} also increases (Fig. 19b), whereas maximum values $K_{HP} > 2$ MPa·m$^{1/2}$ are observed in the samples obtained by liquid-phase sintering at 1450–1500 °C (Tables 1 and 2).

We believe that the Hall-Petch coefficient K_{HP} increases as sintering temperature grows because of higher tungsten atom concentration in the γ-phase. According to references (Lyakishev 1996; Povarova et al. 2002), as heating temperature increases, tungsten concentration in the γ-phase displays monotonic growth and reaches 20–22 wt% at the melting point of the γ-phase (over 1450 °C) (Fig. 19b). Higher tungsten concentration in the γ-phase results in a stronger γ-phase. In our case thin interlayers of this phase, found between α-W particles, serve as the interphase (α-γ) boundary.

5 Conclusions

1. During MA of W-Ni-Fe powder compositions, an asymmetric (toward larger reflection angles) distortion of the (110) α-W peak is observed in X-ray diffraction patterns. With longer MA, both the width and asymmetry of the (110) α-W peak increase. Asymmetry in the (110) α-W peak is caused by formation of a supersaturated solid solution of nickel atoms in the surface layer of α-W particles.

2. Experiments were carried out to study the structure of UFG W-Ni-Fe tungsten alloys obtained by solid-phase sintering in hydrogen and SPS. The dependence of alloy density on sintering temperatures has been determined to be non-monotonic with a maximum corresponding to the optimal sintering temperature. It has been demonstrated that dependence of alloy density on sintering temperature is non-monotonic due to change in nickel concentration in the surface layer of α-W particles, which has been formed earlier, during MA.
3. Addition of ultrafine powders, longer MA, and faster operation of grinding balls have been shown to lower the optimal sintering temperature and sintering activation energy. The optimal sintering temperature of mechanically activated nanopowders in hydrogen is lower than that of coarse-grained alloys by 400 °C. The optimal SPS temperature of mechanically activated nanopowders in hydrogen is lower than that of coarse-grained alloys by ~200–250 °C.
   The reason for a lower optimal sintering temperature is established to be a decreased activation energy of grain-boundary diffusion and formation of a non-equilibrium solid solution of nickel and iron in the surface layer of tungsten particles during high-energy MA.
4. The impact of MA modes and SPS temperature and speed modes on specific features in formation of mechanical properties and structure in various tungsten alloys has been studied. High-energy MA and high-speed SPS were used to obtain UFG W-Ni-Fe tungsten alloys with high strength, ~2 times higher than that of coarse-grained tungsten alloys obtained by conventional liquid-phase sintering.
5. The Hall-Petch relationship applies to the investigated alloys. It has been shown that, as sintering temperature increases, the Hall-Petch coefficient increases from 0.3–0.6 $MPa{\cdot}m^{1/2}$ to 2.0–2.4 $MPa{\cdot}m^{1/2}$. Minimum Hall-Petch coefficient values are found in samples obtained by SPS; maximum values – in samples sintered in hydrogen at 1450–1500 °C, which corresponds to the melting point of the γ-phase.

Acknowledgments This study has been funded by the Ministry of Education and Science of Russia (Grant No. 11.5944.2017/6.7).

The authors recognize N.Yu. Tabachkova of the National University of Science and Technology "MISiS" (Moscow) for help in analyzing XRD data.

References

Akhtar F (2008) An investigation on the solid state sintering of mechanically alloyed nano-structured 90W-Ni-Fe tungsten heavy alloy. Int J Refract Met Hard Mater 26:145–151. https://doi.org/10.1016/j.ijrmhm.2007.05.011

Baranov GV (2010) Developments and study of nano- and fine-grained tungsten heavy alloys with high mechanical properties. PhD dissertation, Alekseev Nizhny Novgorod Technical University (in Russian)

Blaine DC, Park SJ, Suri P, German RM (2006) Application of work-of-sintering concepts in powder metals. Metall Mater Trans A 37:2827–2835. https://doi.org/10.1007/BF02586115

Chuvil'deev VN (2004) Nonequilibrium grain boundaries in metals. Theory and applications. Fizmatlit, Moscow (in Russian)
Chuvil'deev VN, Boldin MS, Dyatlova YG, Rumyantsev VI, Ordan'yan SS (2015) A comparative study of hot pressing and spark plasma sintering of Al_2O_3-ZrO_2-Ti(C,N) powders. Inorg Mater 51:1047–1053. https://doi.org/10.1134/S0020168515090034
Chuvil'deev VN, Blagoveshchenskiy YV, Nokhrin AV, Boldin MS, Sakharov NV, Isaeva NV, Shotin SV, Belkin OA, Popov AA, Smirnova ES, Lantsev EA (2017) Spark plasma sintering of tungsten carbide nanopowders obtained through DC arc plasma synthesis. J Alloys Compd 708:547–561. https://doi.org/10.1016/j.jallcom.2017.03.035
Das J, Kiran UR, Chakraborty A, Prasad NE (2009) Hardness and tensile properties of tungsten based heavy alloys prepared by liquid phase sintering technique. Int J Refract Met Hard Mater 27:577–583. https://doi.org/10.1016/j.ijrmhm.2008.08.003
Ding L, Xiang DP, Li YY, Li C, Li JB (2012a) Effects of sintering temperature on fine-grained tungsten heavy alloy produced by high-energy ball milling assisted spark plasma sintering. Int J Refract Met Hard Mater 33:65–69. https://doi.org/10.1016/j.ijrmhm.2012.02.017
Ding L, Xiang DP, Li YY, Zhao YW, Li JB (2012b) Phase, microstructure and properties evolution of fine-grained W-Mo-Ni-Fe alloy during spark plasma sintering. Mater Des 37:8–12. https://doi.org/10.1016/j.matdes.2011.12.010
Frost HJ, Ashby MF (1982) Deformation-mechanism maps. Pergamon Press, London
German RM, Chur KS (1984) Sintering atmosphere effects on the ductility of W-Ni-Fe heavy metals. Metall Trans A 15:747–754. https://doi.org/10.1007/BF02644206
Gong X, Fan JL, Ding F, Song M, Huang BY, Tian JM (2011) Microstructure and highly enhanced mechanical properties of fine-grained tungsten heavy alloy after one-pass rapid hot extrusion. Mater Sci Eng A 528:3646–3652. https://doi.org/10.1016/j.msea.2011.01.070
Green EC, Jones D, Pitkin WR (1954) Developments in high-density alloys. Symp Powder Metall 58:253–256
Han Y, Fan J, Liu T, Cheng H, Tian J (2012) The effect of trace nickel additive and ball milling treatment on the near-full densification behavior of ultrafine tungsten powder. Int J Refract Met Hard Mater 34:18–26. https://doi.org/10.1016/j.ijrmhm.2012.02.014
Hu K, Li X, Yang C, Li Y (2011a) Densification and microstructure evolution during SPS consolidation process in W-Ni-Fe system. Trans Nonferrous Metals Soc China 21:493–501. https://doi.org/10.1016/S1003-6326(11)60742-5
Hu K, Zheng D, Li Y (2011b) SPS densification behavior of W-5.6Ni-1.4Fe heavy alloy powders. Rare Metals 30:581–587. https://doi.org/10.1007/s12598-011-0351-z
Hu K, Li X, Qu S, Li Y (2013) Effect of heating rate on densification and grain growth during spark plasma sintering of 93W-5.6Ni-1.4Fe heavy alloys. Metall Mater Trans A 44:4323–4336. https://doi.org/10.1007/s11661-013-1789-5
Huang B, Fan J, Liang S, Qu X (2003) The rheological and sintering behavior of W-Ni-Fe nano-structured crystalline powder. J Mater Process Technol 137:177–182. https://doi.org/10.1016/S0924-0136(02)01090-7
Humail IS, Akhtra F, Askari SJ, Tufail M, Qu X (2007) Tensile behavior change depending on the varying tungsten content of W-Ni-Fe alloys. Int J Refract Met Hard Mater 25:380–385. https://doi.org/10.1016/j.ijrmhm.2006.12.003
Jang JSC, Fwu JC, Chang LJ, Chen GJ, Hsu CT (2007) Study on the solid-phase sintering of the nano-structured heavy tungsten alloy powder. J Alloys Compd 434–435:367–370. https://doi.org/10.1016/j.jallcom.2006.08.215
Kiran UR, Rao AS, Sankaranarayana M, Nandy TK (2012) Swaging and heat treatment studies on sintered 90W-6Ni-2Fe-2Co tungsten heavy alloy. Int J Refract Met Hard Mater 33:113–121. https://doi.org/10.1016/j.ijrmhm.2012.03.003
Krasovskii PV, Malinovskaya OS, Samokhin AV, Blagoveshenskiy YV, Kazakov VA, Ashmarin AA (2015) XPS study of surface chemistry of tungsten carbide nanopowders produced through DC thermal plasma/hydrogen annealing processes. Appl Surf Sci 339:46–54. https://doi.org/10.1016/j.apsusc.2015.02.152

Krock R, Shepard H (1963) Mechanical behavior of the two-phase composite tungsten-nickel-iron. Trans Metall Soc AIME 227:1127–1134

Larikov LN, Yurchenko YF (1987) Diffusion in metals and alloys. Handbook. Naukova Dumka, Kiev (in Russian)

Lee K, Cha SI, Ryu HJ, Hong SH (2007) Effect of oxide dispersoids addition on mechanical properties of tungsten heavy alloy fabricated by mechanical alloying processes. Mater Sci Eng A 452–453:55–60. https://doi.org/10.1016/j.msea.2006.10.155

Li X, Xin H, Hu K, Li Y (2010) Microstructure and properties of ultra-fine tungsten heavy alloys prepared by mechanical alloying and electric current activated sintering. Trans Nonferrous Metals Soc China 20:443–449. https://doi.org/10.1016/S1003-6326(09)60160-6

Li Y, Hu K, Li X, Qu S (2013) Fine-grained 93W-5.6Ni-1.4Fe heavy alloys with enhanced performance prepared by spark plasma sintering. Mater Sci Eng A 573:245–252. https://doi.org/10.1016/j.msea.2013.02.069

Li X, Hu K, Qu S, Yang C (2014) 93W-5.6Ni-1.4Fe heavy alloys with enhanced performance prepared by cyclic spark plasma sintering. Mater Sci Eng A 599:233–241. https://doi.org/10.1016/j.msea.2014.01.089

Liu W, Ma Y, Zhang J (2012) Properties and microstructural evolution of W-Ni-Fe alloy via microwave sintering. Int J Refract Met Hard Mater 35:138–142. https://doi.org/10.1016/j.ijrmhm.2012.05.004

Lyakishev NP (1996) Phase diagrams of binary metallic systems. Mashinostroenie, Moscow (in Russian)

Munir ZA, Anselmi-Tamburini U, Ohyanagi M (2006) The effect of electric field and pressure on the synthesis and consolidation materials: A review of the spark plasma sintering method. J Mater Sci 41(3):763–777. https://doi.org/10.1007/s10853-006-6555-2

Munir ZA, Quach DV, Ohyanagi M (2011) Electric current activation of sintering: A review of the pulsed electric current sintering process. J Am Ceram Soc 94(1):1–19. https://doi.org/10.1111/j.1551-2916.2010.04210.x

Nokhrin AV (2012) Peculiarities of change of strength properties at annealing of submicrocrystalline metals and alloys received by equal-channel angular pressing method. Deformation and Fracture of Materials 11:23–31 (in Russian)

Park SM, Martin JM, Guo JF, Johnson JL, German RM (2006) Densification behavior of tungsten heavy alloy based on master sintering curve concept. Metall Mater Trans A 37:2837–2848. https://doi.org/10.1007/BF0258611

Park SJ, Johnson JL, Wu Y, Kwon YS, Lee S, German RM (2013) Analysis of the effect of solubility on the densification behavior of tungsten heavy alloys using the master sintering curve approach. Int J Refract Met Hard Mater 37:52–59. https://doi.org/10.1016/j.ijrmhm.2012.10.016

Potanina E, Golovkina L, Orlova A, Nokhrin A, Boldin M, Sakharov N (2016) Lanthanide (Nd, Gd) compounds with garnet and monazite structures. Powders synthesis by "wet" chemistry to sintering ceramics by spark plasma sintering. J Nucl Mater 473:93–98. https://doi.org/10.1016/j.jnucmat.2016.02.014

Povarova KB, Makarov PV, Zavarzina EK, Ratner AD, Volkov KV (2002) VNZH-90-type heavy alloys. I. Effect of alloying and the conditions of fabricating tungsten powders on their structure and the properties of sintered alloys. Russ Metall 4:39–48 (in Russian)

Rahaman MN (2003) Ceramic processing and sintering. 2nd edition, Marcel Dekker Inc., New York 876 p

Ryu HJ, Hong SH, Baek WH (1997) Mechanical alloying process of 93W-5.6Ni-1.4Fe tungsten heavy alloy. J Mater Process Technol 63:292–297. https://doi.org/10.1016/S0924-0136(96)02638-6

Seith W (1955) Diffusion in Mettallen. Platzwechselreaktionen. Springer, Berlin/Göttigen/Heidelberg

Sõmiya Sh, Moriyoshi Yu (Eds.) (1990) Sintering Key Papers. Elsevier Science Publishing Co Inc., London & New York, 801 p

Tokita M (2006) Industrial applications of advanced spark plasma sintering. Am Ceram Soc Bull 85:32–34

Tokita M (2013) Spark plasma sintering (SPS) method, systems, and applications (Chapter 11.2.3). In: Somiya S (ed) Handbook of advanced ceramics, 2nd edn. Academic Press. https://doi.org/10.1016/B978-0-12-385469-8.00060-5

Xiang DP, Ding L, Li YY, Li JB, Li XQ, Li C (2012) Microstructure and mechanical properties of fine-grained W-7Ni-3Fe heavy alloy by spark plasma sintering. Mater Sci Eng A 551:95–99. https://doi.org/10.1016/j.msea.2012.04.099

Xiang DP, Ding L, Li YY, Chen XY, Zhang TM (2013a) Fabricating fine-grained tungsten heavy alloy by spark plasma sintering of low-energy ball-milled W-2Mo-7Ni-3Fe powders. Mater Sci Eng A 578:18–23. https://doi.org/10.1016/j.msea.2013.04.065

Xiang DP, Ding L, Li YY, Chen GB, Zhao YW (2013b) Preparation of fine-grained tungsten heavy alloys by spark plasma sintered W-7Ni-3Fe composite powders with different ball milling time. J Alloys Compd 562:19–24. https://doi.org/10.1016/j.jallcom.2013.02.014

Zhang ZW, Zhou JE, Xi SQ, Ran G, Li PL (2004) Phase transformation and thermal stability of mechanically alloyed W-Ni-Fe composite materials. Mater Sci Eng A 379:148–153. https://doi.org/10.1016/j.msea.2004.02.039

Part IV
SPS of Ceramics and Diamond-Based Materials

Spark Plasma Sintering of Ultrahigh Temperature Ceramics

T. N. Maity, N. K. Gopinath, Krishanu Biswas, and Bikramjit Basu

1 Introduction

Since the 1960s, extensive research has been focused on the investigation of the eminent class of high-temperature materials based on the group IV and group V transition metal borides (MB_2), carbides (MC), and their composites (MB_2-MB_2, MB_2-MC, MB_2-SiC, etc., where M = Ti, Zr, Hf, and Nb), which are considered to be promising candidates for thermal protection system for the airframe of hypersonic vehicles, especially for leading edge, nose cones, cowl, struts, etc., operating beyond 2000 °C in oxidizing environment. This family of ceramics materials, known as ultrahigh-temperature ceramics, has unique combination of exceptional properties including extremely high melting point (>3000 °C), high hardness (>20 GPa), high elastic modulus (>400 GPa) (due to the presence of strong covalent bonds between transition metal and boron and carbon), chemical inertness, and remarkable oxidation resistance along with higher electrical and thermal conductivity due to the presence of varying degree of metallic bond. Among the UHTCs family, boride-based materials possess higher thermal conductivity as compared to the carbide-based materials providing them better thermal shock resistance and making them suitable for high-temperature structural applications. These unique combinations of ceramic-like and metal-like characteristics enable UHTCs to survive extremely high-operating temperature, combined thermomechanical loads, chemical reactivity, and other environmental conditions during service. These are beyond the capabilities of

T. N. Maity · K. Biswas (✉)
Department of Materials Science and Engineering, Indian Institute of Technology Kanpur, Kanpur, India
e-mail: kbiswas@iitk.ac.in

N. K. Gopinath
Centre of Excellence in Hypersonics, Indian Institute of Science, Bangalore, India

B. Basu
Materials Research Centre, Indian Institute of Science, Bangalore, India

© Springer Nature Switzerland AG 2019

P. Cavaliere (ed.), *Spark Plasma Sintering of Materials*,
https://doi.org/10.1007/978-3-030-05327-7_13

other existing materials used for structural applications. Among the UHTC materials, boride-based (ZrB_2, NbB_2, and TiB_2) and carbide-based (ZrC, TiC, NbC) UHTCs have achieved wider attention in the last few decades because of the combination of high-temperature structural stability and good oxidation and corrosion resistance.

However, the effective use of UHTCs requires proper densification of these materials available in powder form. The densification of transition metal borides and carbide-based UHTCs has always been a challenge due to the high melting point, low self-diffusion coefficient, and lower vapor pressure of the constituent elements. The poor sinterability of UHTCs is primarily due to their high binding energy and directional properties of strong covalent bond along with occasional strong ionic bond. As the remarkable mechanical properties are of intense concern for developing UHTCs, tailoring of microstructure during sintering is necessary. This can be performed by optimizing appropriate sintering parameters, viz., staging of sintering scheme, heating cycle, holding duration, applied pressure, etc., as well as the choice of sintering technique. To develop UHTCs with fine-scale microstructures and minimal grain coarsening, spark plasma sintering (SPS) technique is the best choice over conventional sintering technique. SPS is a well-known sintering technique, also known as FAST (field-assisted sintering technique), PECS (pulse electric current sintering), etc. Spark plasma sintering (SPS) is considered to be a novel compaction method which provides dense UHTCs at a relatively lower sintering temperature as compared to the conventional sintering techniques such as pressureless sintering, hot pressing, hot isostatic pressing, high-pressure sintering, etc. (Mukhopadhyay et al. 2013). The novelty in SPS technique lies in simultaneous application of DC current and uniaxial pressure along with high heating rate (100–600 °C/min) which leads to achieving fully dense fine-grained microstructure of UHTCs within a relatively shorter time. Based on the experimental investigations on SPS of UHTCs (ZrB_2, HfB_2, TiB_2, TaB_2, ZrB_2-TiB_2, SiC, ZrB_2-TaB_2, HfB_2-SiC etc.) with explicitly designed heating cycles, sintering parameters and sinter-aid, it has been proven that by suitably controlling the interplay among various diffusion mechanisms (grain boundary diffusion, surface diffusion, and pipe diffusion) and applied uniaxial pressure-imposed creep at high sintering temperature, furnishes better efficiency of the SPS. In case of spark plasma sintering of non-oxide ceramic-based UHTCs, the type and amount of second phase, which act as sinter-aid, have profound impact on the development of fine-scale microstructure and consequently result in phenomenal mechanical properties.

To realize the potential of UHTCs for high-temperature structural applications, a proper understanding of correlations among the processing microstructure properties is required. The main focus of this chapter is to provide insights on powder metallurgical processing involving SPS, microstructure, and properties of selected transition metal boride and carbide-based UHTCs investigated and reported in the literature. We shall illustrate the influence of tailoring SPS heating cycle, sintering parameters, diffusion mechanisms, and sinter-aid on sinter density, microstructural evolution, and mechanical properties. In addition, usage of SPS technique to obtain multicomponent entropy stabilized borides will be elaborated and discussed. Toward the conclusion, application of UHTC relevant to hypersonic aviation application is detailed with emphasis on translation of concepts to technologies, pertaining to thermal protection at sharp leading edges of hypersonic space vehicles.

This chapter will be classified into four sections: boride UHTCs (with/without sinter-aid), carbide UHTCs (with/without sinter-aid), and boride-boride UHTCs, boride-carbide UHTCs, and entropy-stabilized UHTC borides. We shall discuss each different classes of UHTCs in the subsequent sections, stressing on the processing using SPS.

2 Classification of Transition Metal (TM)-Based UHTCs

Based on the chemical composition, the transition metal-based UHTCs can be broadly classified into four main categories:

I. Monolithic borides
II. Boride-based composites
III. Monolithic carbide
IV. Carbide-based composites

Monolithic boride includes TM boride of Ti, Zr, Hf, Nb, and Ta, whereas, boride-based UHTCs includes the transition metal boride as mentioned, containing appropriate sinter-aid and other borides, carbides, etc. Similarly, monolithic carbides include the carbide of Ti, Zr, Hf, Nb, and Ta, and the carbide-based UHTCs are the metal carbides containing sinter-aid and other boride, carbide, etc. It is well reported that addition of sinter-aid to monolithic UHTCs plays an important role to obtain fully densified sinter compact with fine-scale microstructure with minimal grain coarsening. Selection of additives or second phase other than sinter-aid has a significant impact in improving the thermomechanical properties as well as the oxidation resistance.

2.1 *Monolithic Borides (MB_2)*

This section will illustrate the synthesis of monolithic transition metal boride (ZrB_2, NbB_2, and TiB_2) powders, followed by the consolidation of the powders using SPS. The phase assemblage, microstructural features, and the mechanical properties of the sintered monolithic borides will be discussed in detail.

2.1.1 Synthesis of MB_2 Powders

The various processing routes to synthesize different transition metal boride powders have been reported by many research groups (Mckenna 1936; Maeda et al. 1994;

Fahrenholtz et al. 2016; Wu et al. 2013; Balci et al. 2016a, b; Zhao et al. 1995; Ran et al. 2010; Peshev et al. 1968; Jung et al. 2015; Guo et al. 2011; Mukhopadhyay et al. 2013). The diboride powder fabrication/mixing route can be categorized into the following groups.

Carbothermal Reduction

Synthesis of MB_2 powders through carbothermal reduction is carried out using transition metal oxides (MO_2), boron oxide (B_2O_3), and carbon, where metal oxide acts as a metal source and combination of B_2O_3 and carbon is used as a boron source (R1). McKenna implemented the carbothermal reduction method to fabricate ZrB_2 powder by heating ZrO_2, with excess B_2O_3 and minimal amount of carbon as reactants at about 2000 °C (McKenna 1936). This is one of the first literature reports on carbothermal reduction process.

$$ZrO_2(s) + B_2O_3(l) + 5C(s) \rightarrow ZrB_2(s) + 5\,CO(g) \tag{R1}$$

Later, Maeda et al. synthesized NbB_2 by carbothermal reduction of Nb_2O_5 with B_2O_3 and cornstarch powder (source of carbon) (R2). The mixtures of powders were placed in a molybdenum boat and kept in the hot zone (1350–1550 °C) of the horizontal reactor for about 10 min. To reduce the partial pressure of CO gas, argon gas was purged at a rate of 12 ml/min (STP) (Maeda et al. 1994). Argon gas could reduce the diffusivity of product gas (CO) and enhance the rate of forward reaction to produce NbB_2. The diffusivity of CO in argon atmosphere is four times slower than the helium and hydrogen gas atmosphere (Dewan et al. 2009). Therefore, high flow rate of argon was used to reduce the CO diffusivity significantly. According to the literature (Maeda et al. 1994), the purity and particle size of the NbB_2 powder depend on many factors, such as calcination temperature of the carbonaceous materials, particle size of the precursor powders (Nb_2O_5, B_2O_3, and carbon after calcination), and temperature of reduction reaction. The calcination of cornstarch at 700 °C can activate the carbonaceous material present in the precursor composition. The smaller the particle size of the precursor powders, the longer the precursor particle could stay in the hot zone of the reactor, resulting in a better yield of NbB_2 powder. Carbothermal reduction is advantageous to obtain fine powder of NbB_2 with less amount (2 wt.%) of residual carbon (Maeda et al. 1994).

$$Nb_2O_5(s) + 2B_2O_3(l) + 11C(s) \rightarrow NbB_2(s) + 11\,CO(s) \tag{R2}$$

Mechanochemical Synthesis/Mechanical Alloying

Mechanochemical synthesis of transition metal diborides uses the elemental powders of transition metals and amorphous boron. Wu et al. described the synthesis of

ZrB_2 from the elemental powders of zirconium and amorphous boron (R3) via mechanical alloying. Zirconium to boron molar ratio of 1:2 was used to obtain stoichiometric ZrB_2. The milling operation was carried out using tungsten carbide milling media, with a ball to powder weight ratio of 10:1, at 250 rpm in high-energy planetary ball (Wu et al. 2013).

$$Zr + 2B \rightarrow ZrB_2 \tag{R3}$$

The size and shape of the transition metal powders profoundly influence the mechanochemical synthesis process. During mechanical alloying of elemental boron and transition metals in attrition or planetary ball mill, boron particle (brittle in nature) size is reduced with milling time; in contrast, the milling of transition metals (ductile in nature) results in the formation of flakes/thin platelet due to strong impact and shear force during milling operations. Therefore, large amount of materials from the milling media would be added to the parent diborides, which will deteriorate the mechanochemical reactions, sinter density, and eventually the mechanical properties of the resulting diboride. To reduce milling time and impurity level, the brittle compound such as transition metal hydrides or chlorides can be used as starting materials instead of ductile transition metals (R4). Fahrenholtz et al. adopted this technique to prepare ZrB_2 using zirconium hydride and elemental boron, where zirconium hydride spontaneously transformed into metallic zirconia during heating (Fahrenholtz et al. 2016).

$$ZrH_2(s) + 2B\ (s) \rightarrow ZrB_2(s) + H_2(g) \tag{R4}$$

This reaction is feasible at temperature lower than 975 °C due to low partial pressure of hydrogen (< 101 kPa). Ran et al. and Guo et al. separately explored that dissociation of zirconium hydride occurred at the temperature of 600–1300 °C, depending on the decomposition condition and powder particle size of the starting powders (Jung et al. 2015; Fahrenholtz 2003; Guo et al. 2011; Ran et al. 2010).

Mechanochemical Synthesis and Selective Leaching Using Strong Acid Solution

Mechanical synthesis (reaction during mechanical milling) has widely been used for synthesis of borides. Balci et al. synthesized NbB_2 powder from Nb_2O_5, B_2O_3, and Mg (R5) powders by mechanical alloying using high-energy ball mill with a ball to powder ratio of 10:1 and milling speed of 1200 rpm. It is reported that both reduction and replacement reactions of Nb_2O_5 and B_2O_3 can occur simultaneously in presence of Mg, which ignite the reactant powder particles spontaneously without application of external heat and hence reduce the total time of reaction. Mg also reduces Nb_2O_5 and B_2O_3 simultaneously to produce elemental Nb and B, respectively, along with large amount of heat energy as the reduction reaction is exothermic in nature and enhance the yield of NbB_2. The as-synthesized NbB_2 powder contains very less amount of residual Mg and MgO, which can be easily removed by acid leaching resulting in a highly pure ultrafine powder of NbB_2. The leaching was carried out in an ultrasonicator for 15 min using 4 mol/L HCl solution with solid to liquid ratio of 1 g/10 cc (Balci et al. 2016a, b).

$$Nb_2O_5(s) + 2B_2O_3(l) + 11Mg\ (s) \rightarrow 2NbB_2(s) + 11MgO\ (s) \tag{R5}$$

Borothermic Reduction

Another important technique to obtain boride is the borothermic reduction, in which the metal oxide is reduced by elemental boron or boron carbide (B_4C) with/without carbon. Mukhopadhyay et al. synthesized TiB_2 powder by borothermic reaction (R6) of commercially available TiO_2 powder and in-house synthesized B_4C powder (Mukhopadhyay et al. 2013). The mean particle diameter (D_{50}) and specific surface area of as-prepared TiB_2 powders were about 1.1 μm and 1.36 m^2/g, respectively, with traces of oxygen (0.5 wt.%) and carbon (0.6 wt.%) (Mukhopadhyay et al. 2013).

$$7TiO_2(s) + 5B_4C(s) \rightarrow 7TiB_2(s) + 3B_2O_3(l) + 5\ CO\ (g) \tag{R6}$$

Similarly, Zhao et al. synthesized ZrB_2 by reduction of ZrO_2 with B_4C and carbon (R7). The volatile B_2O_3 was formed at intermediate stage of the reaction, resulting in sudden loss of boron which was compensated by addition of excess boron in the reaction (Zhao et al. 1995). In similar lines, Ran et al. prepared ZrB_2 powder from ZrO_2, amorphous boron, and H_3BO_3, by borothermic reaction using multidirectional mixer by wet milling in ethanol medium for 24 h using zirconia balls and polyethylene vials. As B_2O_3 formation reduces the boron concentration during reaction, H_3BO_3 powder was added to maintain the boron concentration in the reaction (Ran et al. 2010).

$$2ZrO_2(s) + B_4C\ (s) + 3C\ (s) \rightarrow 2ZrB_2(s) + 4\ CO(g) \tag{R7}$$

Peshev et al. synthesized NbB_2 using niobium oxide (Nb_2O_5) and elemental boron (R8). The formation of NbB_2 by the reaction of Nb_2O_5 and boron was observed in the temperature range of 1000–1750 °C, but the complete stoichiometric NbB_2 was obtained by heat treatment of reactant powders at 1600 °C for 1 h (Peshev et al. 1968).

$$Nb_2O_5(s) + 9B\ (s) \rightarrow 2\ NbB_2(s) + 5BO\ (g) \tag{R8}$$

It is worthwhile to mention that fine transition metal boride powder with high purity can be synthesized through the route (R5) (Balci et al. 2016a, b), and the same reaction may be carried out to obtain other transition metal borides (TiB_2, ZrB_2, etc.).

However, large scale production (in kg) of transition metal borides, especially TiB_2 can be synthesized using borothermic reduction. But it is not possible to produce nanosize powder adopting borothermic reduction (Basu and Balani 2011).

Gu et al. (Basu and Balani 2011; Gu et al. 2003) reported the synthesis of nanocrystalline powder using the solvothermal reaction of elemental boron (B) and titanium tetrachloride ($TiCl_4$) in the presence of elemental sodium at 400 °C in benzene medium, wherein benzene helps in controlling the reaction kinetics as well as the particle size of the product powder (R9).

$$TiCl_4(s) + 2B\ (s) + 4\ Na\ (s) \rightarrow TiB_2(s) + 4\ NaCl\ (s) \qquad (R9)$$

Likewise, Chen et al. synthesized TiB_2 nanocrystalline powder with particle size in the range of 10–20 nm, by reaction between $TiCl_4$ and $NaBH_4$ (R10) at different temperatures (500–700 °C) (Chen et al. 2004; Basu and Balani 2011).

$$TiCl_4(s) + 2NaBH_4(s) = TiB_2(s) + 2NaCl\ (s) + 2HCl\ (g) + 3H_2(g) \qquad (R10)$$

Other than the reaction synthesis methods, fine powders of NbB_2, TiB_2, and ZrB_2 can be prepared by wet milling using high-energy ball milling or attrition milling of the commercially available powders by selecting appropriate milling parameters.

From the above discussion, it is important to mention that both laboratory scale and commercial scale powder processing routes are available to synthesize submicron to nanocrystalline powders of transition metal borides. As far the sintering is concerned, finer powder containing less numbers of particles agglomerates are suitable to get highly dense sintered product, as narrow particle size distribution, helps in particle packing and rearrangement at the initial stage of sintering. According to Herring's scaling law, sintering temperature reduces by 3–4 orders of magnitude with decrease in the particle size by an order of magnitude. The powder containing agglomerates reduces the sinterability and initiates micro- and macrocracks in the sintered samples. In spite of this, the purity of the powder plays a vital role in sinterability of the powder. Boride powders generally contain impurities such as oxygen, nitrogen, and carbon. While oxygen has detrimental effects in densification of powder, the presence of less amount of carbon is advantageous, because it removes the surface oxide film present in the boride powders by reduction reaction at the initial stage of sintering, especially when sintering is done using spark plasma sintering technique (Basu and Balani 2011).

2.1.2 Spark Plasma Sintering of MB_2 Powders

Sintering of monolithic UHTCs, especially transition metal borides such as ZrB_2, TiB_2, and NbB_2, is inherently arduous due to the three main characteristics of these materials, viz., (a) presence of surface oxides (MO_2 and B_2O_3) in the starting powders, (b) high melting point due to predominantly covalent bonding as well as low self-diffusion coefficient, and (c) relatively higher vapor pressure of transition metals and boron. To achieve the optimum sinter density (>99%ρ_{th}), it is required to sinter the monolithic borides at temperature greater than $0.7T_m$, where T_m is melting point of the boride in Kelvin (Basu and Balani 2011; Telle et al. 1988). In this context, choice of sintering technique along with sintering parameters plays a significant role toward achieving full sinter density of boride UHTCs. Optimum sinter density can be achieved by hot pressing of the diboride powders, but the longer processing time (several hours) leads to exaggerated grain growth, which can degrade the mechanical properties of the obtained materials. Mukhopadhyay et al. consolidated TiB_2 with final holding at 1400 °C for 10 min under the mechanical pressure of 40 MPa and reported sinter density of ~98%ρ_{th} when spark plasma sintering technique is used. But same sinter density had been obtained by at 1800 °C for more than 1 h of sintering when conventional hot pressing technique is used (Mukhopadhyay et al. 2009, 2013; Murthy et al. 2006).

To avoid this limitation, many research groups developed UHTCs adopting spark plasma sintering technique (Licheri et al. 2008; Mukhopadhyay et al. 2013). As the starting powders of borides contain thin oxygen-rich layer on the particle surface, sinterability of the powders is reduced significantly. Baik and Bechers explored the adverse effect of oxygen layer on the surface of powder particle to be sintered on the consolidation of TiB_2, which can be introduced during powder processing and handling. To obtain optimum sinter density while suppressing exaggerated grain growth, it has been recommended that the oxygen content should be less than 0.5 wt.% or some amount of reducing agent should be added to the starting powder to eliminate the TiO_x below 1600 °C (Basu and Balani 2011; Baik and Bechers 1987). In this context, SPS technique is much more advantageous as compared to the conventional techniques, because at the initial stage of sintering, the supplied pulsed DC current initiates spark discharge between the particle to particle interface, thereby removing the surface oxide film and activating the powder particles, which simultaneously enhance the densification kinetics. In the presence of electric field during SPS, the migration of grain boundary significantly reduces the closed porosity and improves the sinter density. In SPS process, densification is dominated over the grain coarsening due to higher heating rate, lower holding time, and relatively lower sintering temperature. UHTCs synthesized adopting SPS usually have higher sinter density with fine-scale microstructure, which helps in achieving desired mechanical properties. In the following, we shall discuss a few salient research work carried out using SPS (using single stage, two stage, and multistage) to achieve optimum sinter density. The effect of multistage sintering over the two and three stage sintering has been discussed in details in the latter section of this chapter.

Zhang et al. reported the consolidation of commercial TiB_2 powder, with particle size in the range of 1–2 μm, without incorporating any additives. The consolidation was carried out at six different sintering temperatures ranging from 1200 °C to 1800 °C with a final holding at maximum temperature for 5 min, at a heating rate of 150 °C/min under 50 MPa uniaxial pressure. The sinter density increases from 78.2% ρ_{th} to 97.6% ρ_{th} when the sintering temperature increases from 1500 °C to 1800 °C (Zhang et al. 2010). However, the sinter density below 1500 °C is not reported. Ipekci et al. reported the sintering of TiB_2 powder synthesized using borothermic process with particle size of ~0.54 μm. The sintering was performed with final holding at 1500 °C for 5 min, at a lower heating rate of 50 °C/min under 50 MPa uniaxial pressure and which resulted in sinter density of 95% ρ_{th} (İpekçi et al. 2017). Mukhopadhyay et al. reported the consolidation of commercial TiB_2 powder synthesized using borothermic process with mean particle size of 1.1 μm, without incorporating sinter-aid. The sintering was carried out at temperature ranging from 1400 °C to 1500 °C with a final holding at maximum temperature for 10 min, at a heating rate of 500 °C/min under 40 MPa pressure. The sinter density decreases from 97.6% ρ_{th} to 96% ρ_{th} when the sintering temperature increases from 1400 °C to 1500 °C (Mukhopadhyay et al. 2013), while and dense monolithic TiB_2 (97.5% ρ_{th}) could be obtained using hot pressing at 1800 °C for 1 h. However, the decrease in sinter density with increasing the sintering temperature is attributed to the formation of TiB phase, which affect the sintering parameters encountered during spark plasma sintering process (Mukhopadhyay et al. 2013; Murthy et al. 2006; Mukhopadhyay et al. 2009). Such investigations clearly indicate the advantages of SPS, in lowering the sintering temperature as well as the time of sintering process.

In an interesting work by Sairam et al. (2014), the consolidation of monolithic NbB_2 by reactive synthesis using spark plasma sintering of mixed powder of elemental niobium and amorphous boron milled for 2 h in tungsten carbide lined motorized pestle is reported. The samples sintered using two-stage sintering with final holding at 1900 °C for 15 min with a heating rate of 100 °C/min under 30 MPa, 50 MPa, and 70 MPa pressure showed sinter density of 91%ρ_{th}, 97.1%ρ_{th}, and 97.7%ρ_{th}, respectively. Significant improvement of sinter density (~6%) is observed when the applied pressure increases from 30 MPa to 50 MPa, but only a marginal rise of 0.6% is observed when the pressure increases from 50 to 70 MPa as shown in Fig. 1. Figure 1 also shows that the sinter density is slightly increased when holding time at initial stage of sintering, i.e., at 1200 °C, increases from 1 min to 15 min, with the same conditions at final holding, i.e., 1900 °C, for 15 min (Sairam et al. 2014).

Analogously, Maity et al. reported the consolidation of NbB_2 powder with/without ZrO_2 second phase with final holding at 1750 °C for 2 min under uniaxial pressure of 50 MPa with heating rate of 100 °C/min using multistage spark plasma sintering. The commercial NbB_2 powder was milled in acetone for 16 h using zirconia milling media with ball to powder ratio of 20:1 at 200 rpm. The sinter density of 83% ρ_{th} was obtained for NbB_2 powder without ZrO_2, while maximum sinter density of 99.6% ρ_{th} was achieved when submicron ZrO_2 was added to NbB_2 from milling media. Staging in sintering scheme resulted in sinter density of 96.4% ρ_{th} when single-stage SPS scheme was used, while 98.3% ρ_{th} was achieved for three-stage SPS scheme, with the same sintering temperature and total time of sintering. From this discussion, it can be concluded that careful addition of sinter-aid (ZrO_2) and proper selection of sintering scheme have a profound effect toward achieving fully dense NbB_2 sample. The increase in densification is reasoned to ZrO_2 addition which fills the free space available in the coarser NbB_2 particle, which restricts the grain growth during sintering and helps in achieving the fully dense NbB_2 compact, in addition to reducing the sintering temperature.

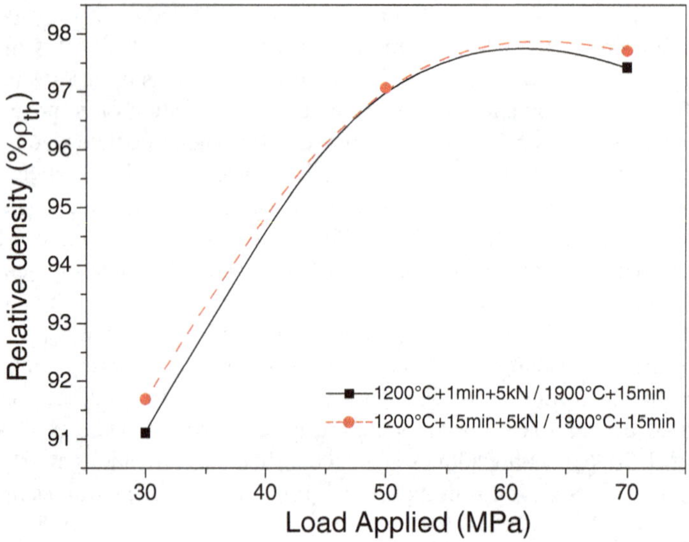

Fig. 1 Sinter density of niobium diboride (NbB_2) as a function of applied mechanical pressure during reaction spark plasma sintering of mixed powder of niobium and boron with final holding at 1900 °C for 15 min holding period (Sairam et al. 2014)

Balci et al. reported the consolidation of NbB_2 powder (~200 nm) synthesized adopting combined method of mechanochemical synthesis and acid leaching, without adding second phase. The consolidation was performed with final holding at 1500 °C for 15 min, at lower heating rate of 100 °C/min under 60 MPa uniaxial pressure resulting in sinter density of ~84% ρ_{th} (Balci et al. 2016a, b). As NbB_2 has high melting point (3060 °C), the achievement of lower sinter density is due to the lower sintering temperature used in this experiment. Balci et al. also carried out sintering of same powder using cold pressing (CP) and pressureless (PS) sintering at 1500 °C for 4 h, but the sinter density of ~84% ρ_{th} was obtained. To improve the sinter density, Co additives (2.5–5 wt.%) have been used, thereby obtaining maximum sinter density of ~98% ρ_{th}, using CP-PS technique. The improvement in sinter density with addition of Co phase is attributed to the liquid-phase sintering mechanism and higher wettability of boride phase in cobalt.

In another set of experiments, Solvas et al. described the fabrication of ZrB_2 from commercial powder with particle size (D_{50}) of ~2.5 μm, in which sintering was performed with final holding at 2000 °C for 6 min, at heating rate of 100 °C/min under 50 MPa uniaxial applied pressure, resulting in sinter density of ~96%ρ_{th}. The maximum sinter density (>99% ρ_{th}) was obtained at a sintering temperature of 1950 °C using the sinter-aid, SiC and La_2O_3. The increase in sinter density is attributed to the fine particle size of sinter-aid and domination of densification mechanism over the grain coarsening in presence of sinter-aid (Solvas et al. 2013). Sintering at lower temperature than the monolithic one is attributed to the low melting point sinter-aid.

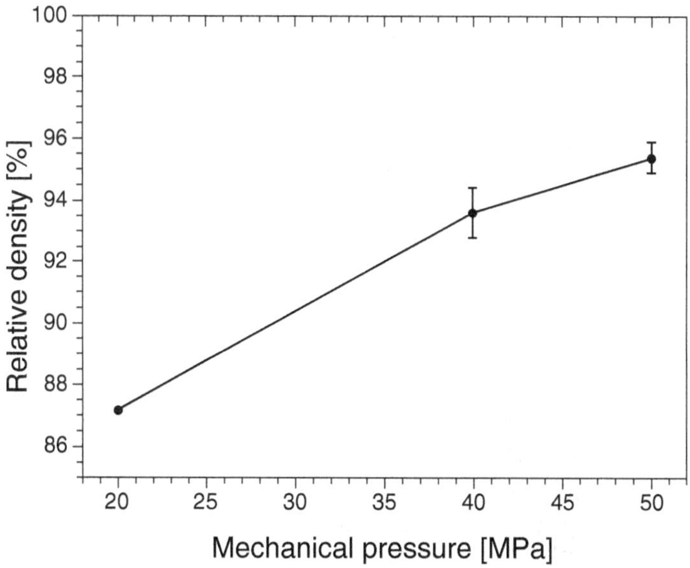

Fig. 2 Effect of the applied pressure on sinter density of bulk zirconium diboride (ZrB_2) obtained by reactive spark plasma sintering of mixed powder of zirconium and boron (Licheri et al. 2015)

Similarly, Licheri et al. demonstrated the synthesis of bulk monolithic ZrB_2 by reactive spark plasma sintering of elemental Zr and B at 1900 °C for 3 min at a heating rate of 500 °C/min. The elemental powder was mixed in shaker mill for 20 min using zirconia balls and plastic vials. The sinter density increased enormously from 87% ρ_{th} to 94% ρ_{th} with increasing applied pressure from 20 MPa to 40 MPa (Fig. 2), but only marginal rise in sinter density (1%) is observed when the pressure increases from 40 to 50 MPa. The achievement of poor sinter density even at 1900 °C under high pressure is ascribed to relatively lower sintering time and high heating rate, which leads to negligible grain coarsening. Sometimes slight grain growth is beneficial toward obtaining a fully dense sintered microstructure (Basu and Balani 2011; Licheri et al. 2015).

3 Phase Evolution and Microstructure of Sintered MB_2

The sinter density and mechanical properties of sintered compact of any materials largely depend on the phase evolution, reaction product phase, phase distribution, grain size, and grain size distribution in the microstructure. The phase evolution and microstructural studies play a key role in materials science to assess the performance of sintered specimen. The correlation between property and microstructure has been investigated by several researchers.

Zhang et al. reported XRD pattern (not shown) of sintered TiB_2, which reveals slight increase in intensity of peaks of monolithic TiB_2 with increasing the sintering temperature up to 1700 °C. However, a distinguishable increment of peak intensity is observed at 1800 °C. As composition of the materials remains the same, increase of intensities at 1800 °C is credited to the rapid rate grain coarsening at this temperature (Zhang et al. 2010). Figure 3 shows the microstructure of monolithic TiB_2, sintered at different sintering temperatures such as (a) 1200 °C, (b) 1400 °C, (c) 1500 °C, (d) 1600 °C, (e) 1700 °C, and (f) 1800 °C. Based on the phase evolution observed in the sintered microstructures (Fig. 3), sintering process can be divided into three stages:

I. Activation of particles
II. Particle to particle contact and neck formation
III. Rapid densification and grain growth

At the initial stage of sintering, slight change of shape and size of the TiB_2 particles is observed due to its low self-diffusion coefficient (Fig. 3a). However, when sintering temperature increases to 1400°C, neck formation is observed at the particle contact points due to removal of the oxide layer and enhancement of atomic

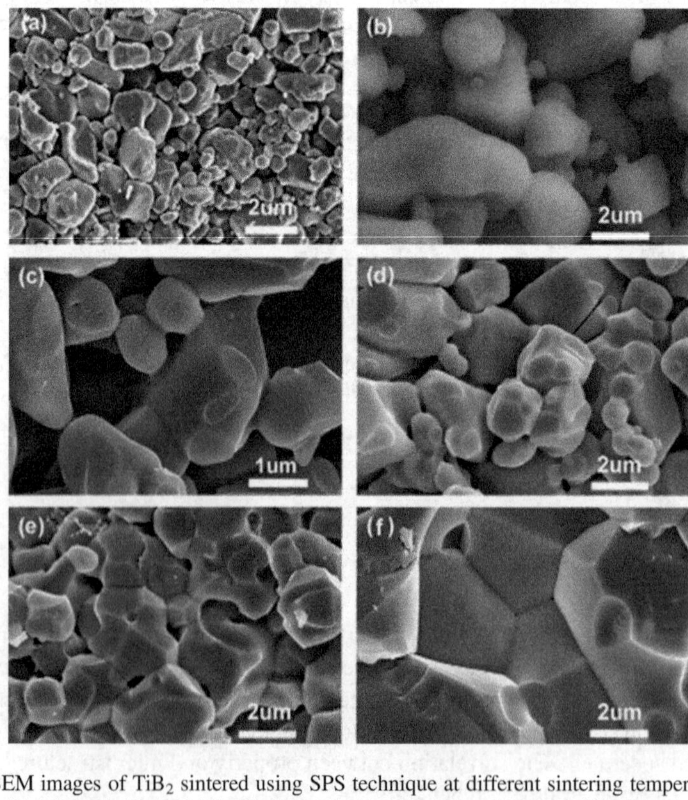

Fig. 3 SEM images of TiB_2 sintered using SPS technique at different sintering temperatures: (**a**) 1200 °C, (**b**) 1400 °C, (**c**) 1500 °C, (**d**) 1600 °C, (**e**) 1700 °C, and (**f**) 1800 °C (Zhang et al. 2010)

diffusion as shown in Fig. 3b. Surface activation of particles occurred at this stage. At high sintering temperature, the neck growth and particle to particle connection is improved due to evaporation and condensation (Fig. 3c–e). At 1800 °C, rapid enhancement of densification and grain growth happened due to high sintering temperature. Figure 3f shows the dense and coarse-grained TiB_2 microstructure (Zhang et al. 2010).

Sairam et al. reported NbB_2 synthesis by reactive spark plasma sintering (RSPS) using elemental powder as mentioned in the previous section. Formation of single-phase NbB_2 at 1200 °C is observed when holding times are 1 min and 15 min. It is interesting to note that single-phase stochiometric NbB_2 is synthesized adopting RSPS with a dwell time of 1 min, which is a very short duration as compared to the conventional sintering technique. Alternatively, Matsudaira et al. reported the synthesis of NbB_2 by heat treatment of elemental powder of niobium and boron at 1000 °C for 1 h (Matsudaira et al. 1989). The reason for formation of NbB_2 at very short time (1 min) can be explained using the novelty of SPS technique. In SPS, the pulsed DC current and uniaxial pressure can be applied at same time during processing. In this case, the uniaxial pressure in the range of 10 MPa to 20 MPa was applied during sintering, which helps in increasing the particle to particle contacts and chemical reaction to form NbB_2. Passage of large pulsed DC in the range of 700 A to 2000 A through green powder compact is reported to show a local rise in temperature exceeding the reaction temperature at the particle contacts (Sairam et al. 2014). This can significantly increase the diffusivity of niobium and boron, thereby accelerating the formation of NbB_2 within 1 min. In contrast, the conventional synthesis routes need much longer time at similar processing temperatures (Sairam et al. 2014). Figure 4 shows the secondary electron SEM image of fractured NbB_2 consolidated using two-stage SPS technique, which reveals a dense microstructure of NbB_2 with less number of pores (black contrast). However, predominant grain growth is observed owing to holding at 1900 °C for 15 min. The crack propagation through grains indicates transgranular mode of fracture.

The XRD patterns of the sintered compact prepared from NbB_2 powders containing ZrO_2 with final holding at 1750 °C for 2 min under uniaxial pressure of

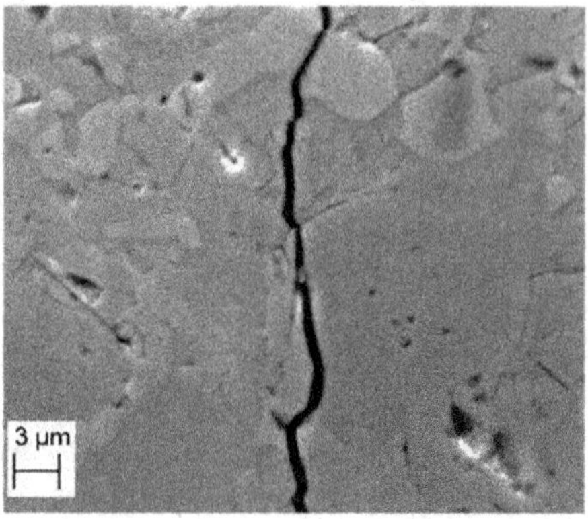

Fig. 4 Scanning electron micrograph of monolithic NbB_2 sintered with final holding at 1900 °C for 15 min using two-stage spark sintering technique, showing the transgranular crack (Sairam et al. 2014)

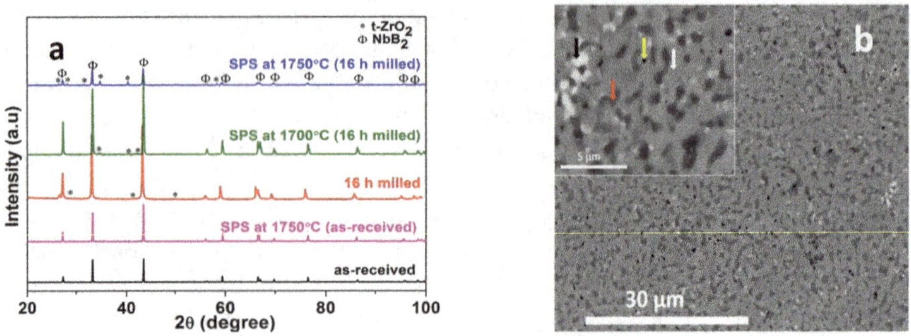

Fig. 5 (**a**) XRD pattern of 16 h milled NbB_2 powders as well as the as-received NbB_2 powder sintered at 1700 °C and 1750 °C using multistage spark plasmsintering; (**b**) BSE-SEM images of bulk NbB_2 specimen (prepared using 16 h milled NbB_2 containing ZrO_2) sintered at 1750 °C using four-stage spark plasma sintering technique, showing distribution of different phases (marked by different color arrow) in the microstructure (Maity et al. 2018)

50 MPa and heating rate of 100 °C/min using four-stage spark plasma sintering are shown in Fig. 5a. The detailed analysis exhibits the presence of NbB_2 along with the tetragonal zirconia (t-ZrO_2). It is evident that t-ZrO_2 is retained after sintering at 1750 °C. This has been confirmed by thorough analysis of the peaks present in the XRD patterns. It can be clearly identified that the XRD pattern of the as-received NbB_2 powder and the sintered pellet is similar without measurable peak shifts.

Figure 5b exhibits the BSE-SEM image of the NbB_2 specimen sintered at 1750 °C with the inset showing higher magnification micrograph. It is evident that there exist two different phases with gray contrast (marked by red arrow) and bright contrast (marked by black arrow). The small black regions in the microstructure (marked by yellow arrow) correspond to pores in the sintered compact. The compositional analyses using EDS indicate that the gray areas (white arrows) are NbB_2, whereas the white contrast is due to zirconia. The SEM image exhibits the uniform distribution of ZrO_2 phase over the NbB_2 phase with minimal grain growth (Maity et al. 2018).

The XRD pattern shown in Fig. 6a reveals the peaks of only zirconium phase at time = τ1, with absence of boron peaks, which evidence the amorphous nature of boron. At time = τ2, the XRD pattern shows the peaks of ZrB_2 phase without the presence of elemental zirconium and boron, which indicates the complete conversion of elemental constituent into ZrB_2 under high heating rate of 500 °C/min with final holding at 1900 °C for 3 min. (Licheri et al. 2015).

Figure 6b shows a secondary electron SEM image of the fracture surface of sintered ZrB_2 obtained by reactive SPS of Zr and B at 1900 °C for 3 min at a heating rate of 500 °C/min under 50 MPa applied pressure synthesis. The SEM image exhibits the presence of few micron-sized residual closed porosity along with some isolated pores. The formation of isolated pores is reasoned to the removal of significant amount of gaseous substance formed during combustion. This aspect also can be explained considering the presence of 1.1 wt.% B_2O_3 and 2 wt.% ZrO_2 in the

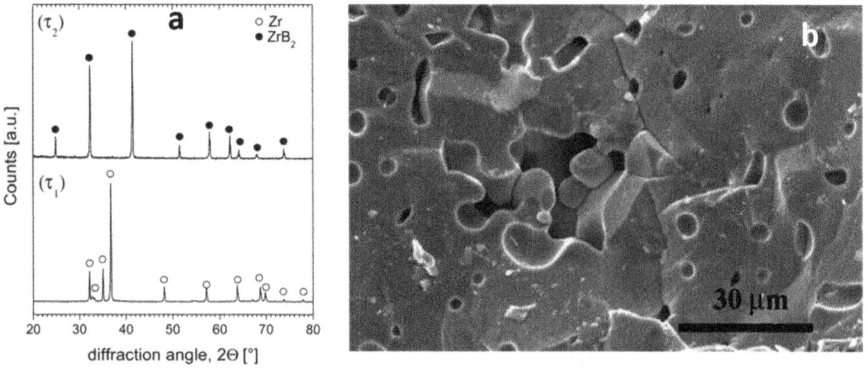

Fig. 6 (**a**) X-ray diffraction patterns of sintered ZrB_2 obtained at two different time intervals (τ1and τ2) by reactive SPS of elemental Zr and B at 1900 °C for 3 min at a heating rate of 500 °C/min under 20 MPa applied pressure; (**b**) secondary electron SEM image of the fracture surfaces of sintered ZrB_2 obtained by reactive spark plasma sintering of elemental Zr and B at 1900 °C for 3 min at a heating rate of 500 °C/min under 50 MPa applied pressure (Licheri et al. 2015)

starting zirconium and boron powder. Moreover, rapid gas evolution (causes sudden increase of gas pressure) during reaction synthesis causes an adverse effect involving significant loss of powder, up to 30% of the total weight of the reactants, before reaction. As the reaction product does not contain any kind of reactants, stoichiometry of the final product was maintained (Licheri et al. 2015).

3.1 Properties of Sintered MB_2

Table 1a contains the summary of the properties obtained from bulk monolithic NbB_2 as well as NbB_2-based materials. It can be observed that bulk NbB_2 containing 2 wt.% Co prepared using cold pressing followed by pressureless sintering at 1500 °C for 4 h has higher elastic modulus (Balci et al. 2016a, b) as compared to other NbB_2-based ceramics. The decrease in elastic modulus value with increasing the weight fraction of sinter-aid is due to the lower elastic modulus of Co and $NbCo_2$ phase formed during sintering stage, as well as the ZrO_2 sinter-aid. But the presence of cobalt decreases the high-temperature strength of the sintered compact due to the lower melting point of Co. Also, NbB_2-ZrO_2 composite (Maity et al. 2018) has higher fracture toughness as compared to bulk NbB_2 prepared by Sairam et al. (2014). The higher fracture toughness of NbB_2 containing ZrO_2 phase may be attributed to the crack deflection at the NbB_2 grain boundary and stress-induced transformation of tetragonal zirconia to monoclinic zirconia (phase transformation toughening) as the huge amount of stress generate at the crack tip when external load was applied to generate crack at the indent corner (Basu et al. 2002). However, the lower elastic modulus of the sintered NbB_2-ZrO_2 composite is attributed to the development of microcracks in the sintered specimen due to phase transformation

Table 1 Comparison of the relative density and mechanical properties of the as-sintered specimen processed via various sintering techniques: (a) monolithic NbB_2 and NbB_2-based ceramics; (b) monolithic ZrB_2 and ZrB_2-based ceramics

(a) Monolithic NbB_2 and NbB_2-based specimen						
Sample	**Sintering techniques and parameters**	**Sinter density (% ρ_{th})**	**Elastic modulus (GPa)**	**Hardness (GPa)**	**Fracture toughness (K_{IC}) ($MPa.m^{1/2}$)**	**References**
NbB_2	FS-SPS at 1750 °C, 2 min under 50 MPa	82.8	171	3.3 ± 0.9	NM[a]	Maity et al. (2018)
NbB_2-ZrO_2	FS-SPS at 1750 °C, 2 min under 50 MPa	99.6	496.2	21.7 ± 0.3	5.2 ± 0.4	
NbB_2	TS-SPS at 1900 °C, 15 min under 70 MPa	97.7	539	20.3 ± 1.7	4 ± 0.6	Sairam et al. (2014)
NbB_2	CP and SS-PS at 1500 °C, 4 h	81.1	88.7	1.9 ± 0.3	NA	Balci et al. (2016a, b)
NbB_2+ 2 wt.% Co	CP and SS-PS at 1500 °C, 4 h	93.7	552	24.8 ± 2.1	4.05 ± 0.6	
NbB_2+ 5 wt.% Co	CP and SS-PS at 1500 °C, 4 h	97.6	508	27.7 ± 1.9	4.74 ± 0.2	
NbB_2	SS-SPS at 1500 °C, 15 min under 60 MPa	83.8	243	13.7 ± 0.7	2.07 ± 0.1	
(b) Monolithic ZrB_2 and ZrB_2-based specimen						
Sample	**Sintering parameters (°C/min/MPa)**	**Sinter density (% ρth)**	**Flexural strength (MPa)**	**Hardness (GPa)**	**Fracture toughness (K_{IC}) ($MPa.m^{1/2}$)**	**References**
ZrB_2	2000/6/50	95.8 ± 0.7	450 ± 40	16.5 ± 0.9	3.6 ± 0.3	Solvas et al. (2013)
ZrB_2– 20 vol.% SiC	1950/4/50	99.1 ± 0.7	700 ± 90	21.1 ± 0.6	6.4 ± 0.6	
ZrB_2– 20 vol.% SiC –2 wt. % La_2O_3	1950/3/50	99.6 ± 0.5	600 ± 70	19.3 ± 0.6	5.2 ± 0.5	

FS four stage, *SS* single stage, *SPS* spark plasma sintering, *CP* cold pressing, *PS*[a] pressureless sintering, *NA* not available, *True density#* true density measured using Archimedes principle of tetragonal ZrO_2 to monoclinic ZrO_2 involving 3–7% volume expansion and 16–18% shear straining during cooling stage of sintering process (Maity et al. 2018; Basu et al. 2002; Basu 2005).

Some of the mechanical properties of ZrB_2-based ceramics are tabulated in Table 1b, which shows increase in room temperature flexural strength of ZrB_2 from 450 to 700 MPa with addition of second phase (SiC and La_2O_3). This can be attributed to the second-phase pinning and achievement of fine-grain microstructure with increasing number of additives, which restrict the grain growth of ZrB_2. It is well known that the defects cause failure in materials and the defect proportion increases with decreasing the grain size of the parent phase (Guo et al. 2011; Solvas et al. 2013). The larger the grain size, the higher is the amount of defects, which subsequently lowers the strength. The hardness increase with addition of SiC due to higher sinter density compared to monolithic ZrB_2 and the high hardness value of the SiC. But incorporation of La_2O_3 decreases the hardness of ZrB_2–20 vol.% SiC composite due to lower hardness of La_2O_3. The increase in fracture toughness with addition of SiC is ascribed to crack deflection at the hard SiC phase boundary as well as the tensile stresses along the SiC grain boundaries generated owing to differing thermal expansion coefficient (CTE of SiC 4.7×10^{-6} and CTE of ZrB_2 is 6×10^{-6}). The presence of tensile stress at grain boundaries promotes the effective stress transfer and enhances crack deflection during propagation of crack. The addition of La_2O_3 results in reduction of fracture toughness which is attributed to its high thermal expansion coefficient (CTE of La_2O_3 is $8–10 \times 10^{-6}$) among the mentioned components. This leads to rapid shrinkage during cooling that provides free space for SiC grain to effectively release the strain energy (Solvas et al. 2013).

Table 2 shows the mechanical properties along with the sinter density of monolithic TiB_2 processed using various sintering technique including spark plasma sintering. It is evident that elastic modulus and hardness value of the spark plasma-sintered TiB_2 decrease with increasing the sintering temperature. This is due to a decrease in sinter density resulting from increase in the volume fraction of TiB phase from 14% to 18%. As the TiB phase possesses lower elastic modulus (~370 GPa) and hardness (16 GPa) than the TiB_2 phase, the overall elastic modulus and hardness decrease accordingly. It is important to mention from Table 2 that the hot-pressed TiB_2 has superior mechanical properties as compared to the spark plasma-sintered TiB_2 mainly due to the presence of TiB phase in the spark plasma-sintered sample, which degrades the mechanical properties significantly. The indentation fracture toughness of the spark plasma-sintered TiB_2 varies in the range of 5.4–5.8 MPa. $m^{1/2}$. The fracture toughness of spark plasma-sintered TiB_2 is considerably higher than TiB_2 sintered using hot pressing and hot isostatic pressing. The superior fracture toughness value of SPSed TiB_2 is due to the pullout of TiB agglomerates during crack propagation, which might accumulate significant amount of energy in the crack tip (stress concentrator). The better electrical properties of hot-pressed sample over spark plasma-sintered sample signify the full sinter density of the former one than the latter one (Mukhopadhyay et al. 2013).

Table 2 Comparison of relative density, sintering techniques, and properties of monolithic TiB_2, with final holding time of 10 min for SPS $\geq$1 h for other sintering routes

Sintering technique	Sintering temperature (°C)	Sinter density (%ρ_{th})	Elastic modulus (GPa)	Hardness (GPa)	Fracture toughness (K_{IC}) (MPa.m$^{1/2}$)	Electrical conductivity (MΩ^{-1} cm^{-1})	References
SPS	1400	97.6	428.1	18.3 ± 0.6	5.8 ± 0.8	0.072	Mukhopadhyay et al. (2013)
SPS	1500	96.1	397.9	17.4 ± 0.3	5.4 ± 0.4	NA	
HP	1800	97.5	497	25–26.0	4.6 ± 0.5	0.066	Murthy et al. (2009), Mukhopadhyay et al. (2009)
HP	1800	97.0	NA	26.0	NA	NA	Telle et al. (1988)
HP	1800	99.5	512	35.0	NA	0.1	Königshofer et al. (2005)
PS	2150	93.1	NA	NA	5.4	NA	Wang et al. (2002)
HIP	1500	97.6	NA	26.0	3.5	NA	Muraoka et al. (1996)
HPS	2227	97.1	NA	21.7	2.8	NA	Bhaumik et al. (2000)

3.2 Monolithic Carbides

Monolithic transition metal carbide-based UHTCs such as TiC, ZrC, NbC, etc. have been drawing wider attention due to its high melting point, high hardness, corrosion resistance, fatigue resistance, etc. (Abderrazak et al. 2011; Sun et al. 2013; Sciti et al. 2008a; Shen and Nygren 2005). But the transition metal carbides have poor sinterability due to their strong covalent bonding and low diffusivity and high vapor pressure of the constituents. To overcome this limitation, powders of monolithic carbides are consolidated adopting novel sintering technique, i.e., sparks plasma sintering. The phase assemblage and microstructural and mechanical properties of the sintered monolithic carbides will be discussed in this section.

3.3 Consolidation of MC Powders Using Spark Plasma Sintering

In this section, the consolidation of monolithic transition metal carbide such as TiC, ZrC, NbC, etc. powders using spark plasma sintering technique will be detailed. Abderazak et al. synthesized monolithic TiC by consolidating the nanostructured powders of TiC prepared by mechanical alloying technique. The elemental powders of titanium and carbon are mechanically milled for 2–20 h using steel milling media, maintaining ball to powder ratio of 67:1. With increase in milling time 2–20 h, the crystallite size of TiC powders varies from 28 to 6 nm, respectively. The consolidation of the milled powders was carried out using the three-stage spark plasma sintering with final holding at 1650 °C for 6 min 15 sec. The sinter density of the compact of titanium carbides was found to increase from 92.8% ρ_{th} to 95.1% ρ_{th}, when the mechanical alloying was carried for 2–20 h, respectively. The first stage of spark plasma sintering leads to cold compaction, particle rearrangement, and removal of surface oxide film to activate the powder particle for further sintering, due to simultaneous application of DC current and uniaxial mechanical pressure. The second stage involves rapid densification by grain boundary diffusion through widespread neck formation, and the final stage leads to the formation of closed pores, followed by densification and grain growth by lattice diffusion, pipe diffusion, and grain boundary diffusion. Increase in sinter density with decreasing particle size due to increasing milling time is attributed to the high surface area and surface energy of smaller particles as compared to the larger one. Also, with increasing the milling time, dislocation density get increased, which helps in increasing the mass transport via pipe diffusion during final stage of sintering and improves the sinter density (Maity et al. 2018; Abderrazak et al. 2011).

Sun et al. reported the consolidation of ZrC powder with particle size of 1 μm, synthesized by carbothermal reduction of commercial zirconia powder and carbon. The consolidation of the powders was carried out using spark plasma sintering by reactive synthesis route at three different sintering temperatures such as 1700 °C (dwell time of 5 min, applied pressure of 5 MPa, and sinter density of 49.6%ρ_{th} were obtained), 1850 °C (dwell time of 10 min, applied pressure of 60 MPa, and sinter density of 85.3%ρ_{th} were obtained), and 1900 °C (dwell time of 10 min, applied pressure of 60 MPa, and sinter density of 96.1%ρ_{th} were obtained) at heating rate of 100 °C/min. At the sintering temperature of 1650 °C, complete reaction and formation of ZrC is observed. The maximum sinter density of 96.1% ρ_{th} was obtained at 1900 °C, which indicates that ZrC needs sintering temperature higher than 1900 °C. Likewise, Sciti et al. (2008a, b) reported the sinter density of 99% ρ_{th} by sintering of elemental Zr and C powders at 2100 °C by adopting reactive spark plasma sintering technique. From the above discussion, it can be concluded that the sintering temperature has profound effect toward achieving high sinter density of ZrC (Sun et al. 2013).

Sciti et al. (2008a, b) also investigated the consolidation behavior of ZrC with $MoSi_2$ (0–9 vol.%) as sinter-aid. The sintering of monolithic ZrC powder was carried out with final holding at 2100 °C for 3 min under the mechanical pressure of 65 MPa, and sinter density of ~99%ρ_{th} was achieved. The sintering of ZrC powder containing 1 vol% $MoSi_2$, 3 vol.% $MoSi_2$, and 9 vol.% $MoSi_2$was performed with final holding at 1950 °C, 1900 °C, and 1900 °C, respectively, for 3 min under the mechanical pressure of 100 MPa, and sinter densities of 98% ρ_{th}, 99.7%ρ_{th}, and 99% ρ_{th} were obtained, respectively. The presence of carbon in the starting ZrC powder enhances the sinter kinetics by cleaning the surface oxide film from the starting ZrC powder by carbothermal reduction process at initial stage of spark plasma sintering.

From Fig. 7a, it can be concluded that increasing the vol.% of $MoSi_2$ decreases the maximum sintering temperature required to achieve near theoretical density of the sintered compact. Monolithic ZrC could be sintered up to 99%ρ_{th} with final holding at 2100 °C (actual temperature more than 2300 °C) for 3 min, and sinter density more than 97–98%ρ_{th} was achieved at sintering temperature in between 1900 °C and 1950 °C for ZrC-1 vol.% $MoSi_2$ and ZrC-3 vol.% $MoSi_2$. Sinter density

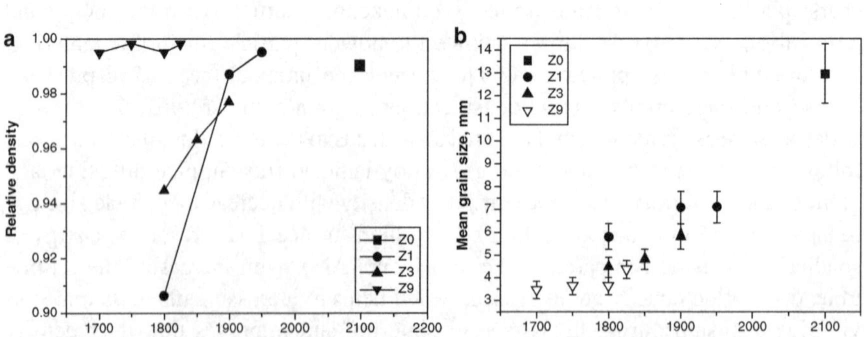

Fig. 7 (**a**) Relative density and (**b**) grain size of sintered specimen of monolithic ZrC and ZrC-$MoSi_2$ composite ($MoSi_2$ contain vary from 1 to 9 vol.%) as a function of sintering temperature (Sciti et al. 2008a, b)

in excess of 99% ρ_{th} was obtained at about 1750 °C (actual temperature more than 1900 °C) using 9 vol.% $MoSi_2$ (Sciti et al. 2008a, b). The change in mean grain size as a function of sintering temperature for ZrC-$MoSi_2$ composites can be seen in Fig. 7b, which clearly indicates depletion of mean grain size from 13 μm to 3 μm, with increasing the volume percent of $MoSi_2$ from 0% to 9% (Sciti et al. 2008a, b). The incorporation of $MoSi_2$ as a sinter-aid in ZrC caused an improvement in volume shrinkage rate, which can be assigned to the change in densification mechanism from solid state for monolithic carbide to liquid-phase sintering for ZrC-$MoSi_2$ composite. The optimum shrinkage is believed to have occurred due to change in diffusion mechanism from surface diffusion for solid-state sintering to grain boundary diffusion for liquid-phase sintering (Shen and Nygren 2005). This maximum shifted toward lower sinter densities with increasing the volume percent of $MoSi_2$, which indicates that liquid-phase formation enhances the rate of grain boundary diffusion at lower densities, thus restricting the grain coarsening caused by surface diffusion. Therefore, lower sintering temperature and earlier activation of grain boundary diffusion help in achieving sintered compact with finer microstructure (Sciti et al. 2008a, b; Shen and Nygren 2005).

3.4 *Phase Evolution and Microstructure of Sintered MC*

Abderrazak et al. investigated the XRD pattern (not shown) of the sintered TiC prepared using mechanically alloyed nanocrystalline powder, which revealed highly intense and sharp crystalline peaks of TiC after sintering. This is ascribed to the better recrystallization and increase in grain size after SPS. The crystallite size was estimated adopting the Scherer formula (Abderrazak et al. 2011):

$$D = \frac{0.89\lambda}{B\text{Cos}\theta} \tag{1}$$

where D denotes the crystallite size in angstrom (Å), λ is the wavelength, B is the full width at half maxima (FWHM), and θ is Bragg's angle. The mean crystallite size was calculated using the reflection from (111), (200), and (220) peaks. The grain size of sintered TiC, prepared using 20 h milled powder, is 91 nm, whereas the grain size of 120 nm is obtained using 2 h mechanically alloyed TiC powders. It can be concluded that grain size of the final sintered sample strongly depends on the particle size of the starting powder (Abderrazak et al. 2011).

The morphology of the sintered TiC fabricated using 2–20 h mechanically alloyed powders is represented in Fig. 8. The sintered TiC obtained from 2 h MA powder reveals smaller size pores with significant sinter density. The larger size pores progressively appeared with relatively lower sinter density when the mechanical alloying time increases from 10 to 20 h (Abderrazak et al. 2011).

Phase stability of bulk ZrC sintered at three different temperatures such as 1750 °C (dwell time of 5 min), 1850 °C (dwell time of 10 min), and 1900 °C

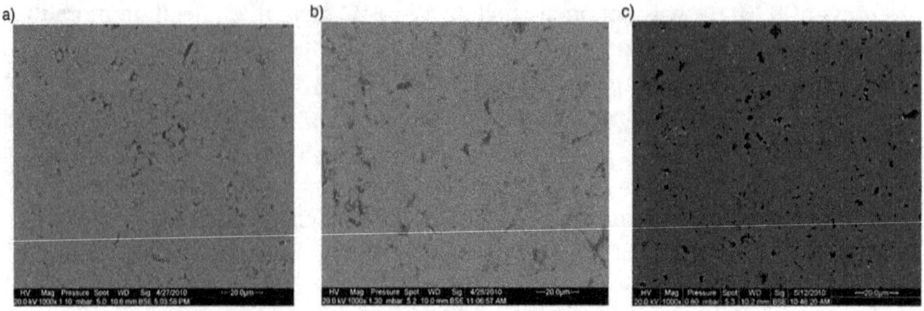

Fig. 8 BSE-SEM image of sintered TiC consolidated using spark sintering of mechanically alloyed powders (**a**) 2 h, (**b**) 10 h and (**c**) 20 h. The black contrast in the image indicates the pores (Abderrazak et al. 2011)

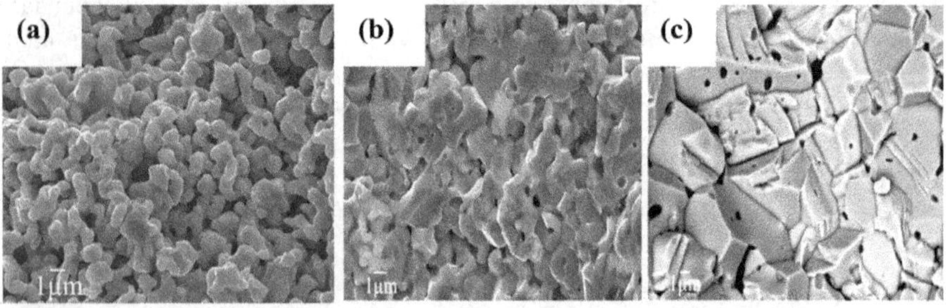

Fig. 9 BSE-SEM image bulk ZrC prepared by combined process of carbothermal reduction and reactive spark plasma sintering of ZrO_2 and carbon powders, (**a**) sintered with final holding at 1750 °C for 5 min, (**b**) fractured surface of ZrC sintered at 1850 °C for 10 min, and (**c**) fractured surface of ZrC sintered at 1900 °C for 10 min (Sun et al. 2013)

(dwell time-10 min) has been studied by Sun et al. All the XRD patterns (not shown) reveal intense peaks of monoclinic ZrC with wide range of homogeneity without peak shifting. This indicates that the stoichiometric ratio of Zr and C was maintained at 1:1 after the reactive spark plasma sintering at different temperatures and time (Sun et al. 2013). The BSE-SEM micrographs of the bulk ZrC are shown in Fig. 9. The SEM image of the sample sintered at 1750 °C for 5 min resembled green compact of ZrC powders as shown in Fig. 9a. The microstructure of fractured surface of bulk ZrC sintered at 1850 °C reveals a fine-scale microstructure (see Fig. 9b) with smaller grain size (~2 μm) and less numbers of residual closed pores (black contrast). ZrC sintered at 1900 °C as shown in Fig. 9c exhibits uniform microstructure with grain size below 10 μm, which is much larger than the samples sintered at 1850 °C. It is worthwhile to mention that in situ synthesis of bulk ZrC from ZrO_2 and C powder by carbothermal reduction followed by reactive spark plasma sintering favors the formation of dense ZrC with uniform microstructure and smaller grain size at a relatively lower sintering temperature as compared to the conventional technique.

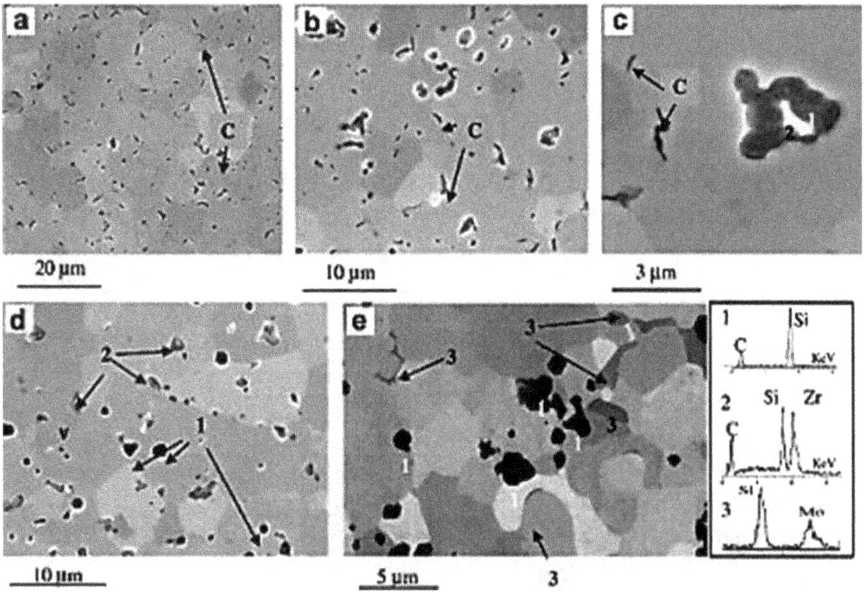

Fig. 10 BSE-SEM image of monolithic and ZrC-$MoSi_2$ composite with EDS spectra: (**a**) ZrC-0 vol.% $MoSi_2$, (**b**, **c**) ZrC-1 vol.% $MoSi_2$, (**d**) ZrC-3 vol.% $MoSi_2$, and (**e**) ZrC-9 vol.% $MoSi_2$. Number 1, 2, and 3 in EDS spectra denote SiC, ZrSi, and $MoSi_2$, respectively (Sciti et al. 2008a, b)

Most importantly, some of the adverse effect caused by oxide impurity could be avoided using this combined method of carbothermal reaction and reactive SPS (Sun et al. 2013).

Sciti et al. demonstrated the synthesis of bulk monolithic and ZrC containing 1, 3, and 9 vol % $MoSi_2$ as a sinter-aid. The XRD patterns (not shown) of sintered ZrC containing 0–3 vol.% of $MoSi_2$ contained peaks only due to bulk ZrC, whereas the XRD pattern of sintered ZrC containing 9 vol.% $MoSi_2$ contained high-intensity peaks due to ZrC and some low-intensity peaks due to $MoSi_2$ and reaction product, i.e., SiC (Sciti et al. 2008a, b). The microstructural investigations of monolithic ZrC and ZrC containing 1–9 vol.% $MoSi_2$ carried out using SEM are shown in Fig. 10. Monolithic ZrC shows dense microstructure with the presence of residual carbon (from starting ZrC powder) in the microstructure (marked by C as shown in Fig. 10a), while the ZrC containing 1 vol.% $MoSi_2$ (see Fig. 10b, c) sintered at 1950 °C exhibits the presence of carbon along with the reaction product (SiC and ZrSi) phases in the microstructure. Formation of reaction product phases, i.e., ZrSi and SiC, was also seen in ZrC-3 vol.% $MoSi_2$ sintered at 1900 °C (see Fig. 10d) and ZrC-9 vol.% $MoSi_2$ sintered at 1700 °C (Fig. 10e). The volume fraction of SiC was found to increase from 2% to 5% with increase in the volume fraction $MoSi_2$ (Sciti et al. 2008a, b).

3.5 Properties of MC

From Fig. 11, it can be concluded that the sintered TiC prepared from 20 h milled powder has high hardness (2700 HV, i.e., ~26.5 GPa) due to low diffraction crystallite size (DCS) of 90 nm and high relative density (95.1%ρ_{th}). Basu and Balani reported that the hardness value of sintered monolithic ceramics are directly related to their sintered density, which clearly emphasizes that the higher the sinter density, the higher would be the hardness (Abderrazak et al. 2011; Basu and Balani 2011). Similarly, Eskandarany (2000) reported the hardness of 3200 HV (~31.4 GPa) of fully dense SPSed TiC using mechanically alloyed TiC powder. Also, Cheng et al. (2012) reported the Vickers hardness of ~30.3 GPa of bulk TiC with sinter density of 99.9% ρ_{th} obtained by spark plasma sintering of commercially available TiC powder with final holding at 1600 °C for 5 min under 50 MPa applied pressure. This discussion strengthens the statement "increasing hardness of monolithic ceramics with increasing sinter density owing to increase in sintering temperature" (Abderrazak et al. 2011).

Table 3 shows the relative density, grain size, and mechanical properties of monolithic ZrC as well as the ZrC containing 1–9 vol.% of $MoSi_2$. The hardness measured using nanoindentation method varies in the range of 25–27 GPa. This slight variation is attributed to the formation of SiC and ZrSi in ZrC-containing $MoSi_2$ phase. The similar value of hardness of monolithic ZrC was reported by Pierson (1996). Minimal increase in hardness is observed in ZrC containing 9 vol.% $MoSi_2$ due to grain refinement in presence of second phase. The higher hardness value measured using nanoindentation with respect to the hardness value obtained from Vickers hardness measurement is reasoned to the difference in contact area

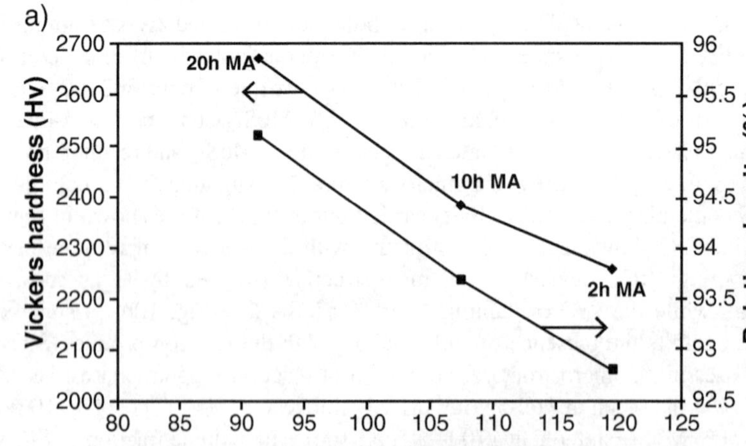

Fig. 11 Variation of Vickers hardness (measured applying 0.2 kg load with dwell time of 7 sec) of sintered TiC as a function of diffraction crystallite size (DCS) after spark plasma sintering with final holding at 1650 °C for 6 min 15 sec (Abderrazak et al. 2011)

Table 3 Relative density, grain size, and mechanical properties of monolithic ZrC as well as the ZrC-based ceramics (Sciti et al. 2008a, b)

Sample composition	Sintering parameters (°C/min/MPa)	Sinter density (%ρ_{th})	Nano hardness (GPa)	Elastic modulus (GPa)	$HV_{1.0}$ (GPa)	Indentation fracture toughness (MPa. $m^{1/2}$)	Flexural strength (MPa)
ZrC	2100/3/65	98	25.2 ± 1.4	464 ± 22	17.9 ± 0.6	Not available	407 ± 38
ZrC +1 vol.% $MoSi_2$	1950/3/100	98	24.9 ± 1.0	420 ± 10	18.8 ± 0.3	2.1 ± 0.2	495 ± 61
ZrC + 3 vol.% $MoSi_2$	1900/3/100	99.7	25.1 ± 2.3	444 ± 17	18.4 ± 0.7	3.2 ± 0.4	Not available
ZrC + 9 vol.% $MoSi_2$	1700/3/100	99	26.8 ± 1.3	467 ± 22	20.0 ± 0.5	3.3 ± 0.4	591 ± 48

considered to calculate Vickers and Berkovich hardness (Cripps 2004) as well as the indentation size effect (Basu and Balani 2011). The decrease in elastic modulus from 464 GPa to 420 GPa with addition of 1 vol.% $MoSi_2$ is due to the lower elastic modulus of $MoSi_2$. But with increasing the volume fraction from 3 vol.% to 9 vol.%, the elastic modulus increases from 420 GPa to 467 GPa because of the increase in SiC volume fraction, which enhances the elastic modulus. This results in slightly higher value of elastic modulus for ZrC-9 vol.% $MoSi_2$ as compared to that of monolithic ZrC. Also, the increase of fracture toughness with increasing volume fraction $MoSi_2$ in ZrC is due to the increase in phase fraction of SiC, which deflects the propagating crack thereby enhancing the fracture toughness. Another reason for the enhancement of fracture toughness is mismatching of thermal expansion coefficient among the ZrC, $MoSi_2$, SiC, and ZrSi phases, which initiate compressive stress in ZrC and promotes to increase the fracture toughness by closing the crack tip present in the vicinity of stress region (Sciti et al. 2008a, b). The flexural strength continuously increases with increasing $MoSi_2$ volume fraction in ZrC, mainly due to the enhancement of the fracture toughness and uniform fine-grain size, along with increment of SiC phase fraction. The estimated values of flexural strength are much higher than the reported value for other ZrC-based ceramics (220–320 MPa) available in the literature (Haga and Scott 1988; Sciti et al. 2008a, b).

4 Mixed UHTCs

In this section, we primarily focus on the sintering behavior and microstructure and mechanical properties of UHTC borides in presence of other borides, sinter-aids, and carbides.

4.1 *Boride-Based UHTCs*

It is well known that the monolithic transition metal diborides have lower sinterability due to their high melting point (>3000 °C), lower self-diffusion coefficient, high vapor pressure of constituent transition metal and boron in diboride system, as well as the presence of inherent surface oxide layer (transition metal oxide: MO_2 and occasionally B_2O_3). Sintering of monolithic diboride without sinter-aid or secondary phase may lead to coarse-grained microstructure, which would degrade the mechanical properties as well as the performance of the ceramic. To overcome this issue, different sinter-aid with/without other additives, for the improvement of specific properties desired for particular applications, is reported. For example, Basu and Balani (2011) reported that incorporation of SiC in ZrB_2 system results in significant enhancement in oxidation resistance of ZrB_2, which is attributed to the formation of protective SiO_2 layer on the surface of the ZrB_2-SiC composite at higher temperature. Also, Zhang et al. reported that a higher sinter density (>99%

ρ_{th}) at lower sintering temperature of TiB_2 could be achieved by adding 2.5 wt.% Ti as a sinter-aid in TiB_2 (Zhang et al. 2012).

4.2 *Processing, Microstructure, and Properties of Boride-Based UHTCs*

BSE-SEM images of the sintered ZrB_2-$MoSi_2$ composites are shown in Fig. 12. The samples sintered using PS reveal 1% residual porosity, whereas HP and SPS samples exhibit almost pore-free microstructure. Phase analysis of the sintered samples using XRD (image not shown) confirmed the presence of ZrO_2 in HP1 and SiC in HP3 including the ZrB_2 and $MoSi_2$ phase. But no crystalline peaks of ZrO_2 were observed in the starting powders. The ZrB_2 grain and $MoSi_2$ appeared in a globular and irregular shape, respectively. As the $MoSi_2$ grain is ductile in nature, it fills the pore region between the ZrB_2 grains. The mean grain size is seen to be in the range of 1.4–2.5 μm and can be found in Table 4. PSed sample exhibits clean grain boundary,

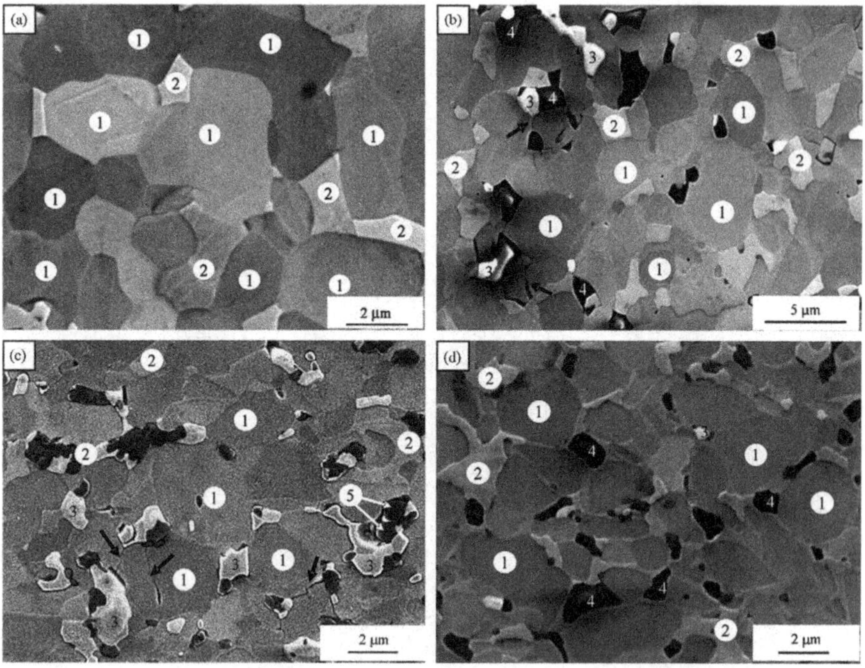

Fig. 12 BSE-SEM image of polished surfaces of sintered ZrB_2-based composite: (**a**) pressureless sintered (PS), (**b**) hot pressed 1 (HP1), (**c**) hot pressed 3 (HP3), (**d**) spark plasma sintered (SPS). The numerical number (1) indicates ZrB_2 phase, (2) $MoSi_2$ phase, (3) ZrO_2 phase, (4) SiO_2 phase, and (5) SiC phase. The black arrow indicates microcracking, initiated due to difference in thermal expansion coefficient of the different phases (Sciti et al. 2006)

Table 4 Relative density, grain size, and mechanical properties of ZrB_2-based ceramics (Sciti et al. 2006)

Sample composition (vol.%)	Sintering technique	Sintering parameters (°C/min/MPa)	Sinter density (%ρ_{th})	grain size (μm)	E (GPa)	$HV_{1.0}$ (GPa)	K_{IC} (MPa. $m^{1/2}$)	σ at room temperature (MPa)	σ at 1200 °C (MPa)	σ at 1500 °C (MPa)	$CTE_{25-1300\ °C}$ ($10^{-6}\ K^{-1}$)
ZrB_2 + 20 $MoSi_2$	Pressureless sintering	1850/30	99.1	~2.5	489 ± 4	16.0 ± 0.4	2.3 ± 0.2	531 ± 46	655 ± 17	500 ± 58	8.2
ZrB_2 + 15 $MoSi_2$	Hot pressing 1	1750/20/30	97.7	~1.8	NA	14.9 ± 0.5	NA	704 ± 98	NA	333 ± 3	NA
ZrB_2 + 20 $MoSi_2$	Hot pressing 2	1800/5/30	98.1	~1.8	NA	15.0 ± 0.4	NA	552 ± 336	NA	388 ± 23	8.1
ZrB_2 + 15 $MoSi_2$ + 2.3 C	Hot pressing 3	1780/12/30	99.2	~1.5	NA	17.0 ± 0.8	NA	498 ± 382	NA	479 ± 13	NA
ZrB_2 + 15 $MoSi_2$	Spark plasma sintering	1750/7/30	98.1	~1.4	479 ± 4	16.2 ± 0.5	2.6 ± 0.3	643 ± 97	632 ± 5	357 ± 48	7.9

NA not available, σ flexural strength, E elastic modulus, HV vickers hardness, K_{IC} indentation fracture toughness but some MoB phase was identified in the XRD pattern. Hot-pressed samples exhibit the nonuniform distribution of silica pockets throughout the microstructure with 2–3 vol.% ZrO_2 agglomerates, which was detected by SEM-EDX analysis (not provided here). As silica accommodate itself at the pore region and triple junctions, degradation in the mechanical properties is expected, especially the flexural strength. To reduce the silica content in the hot-pressed samples, 2.3 vol.% reducing agent (C) was intentionally added to reduce and replace the silica pockets with SiC. The formation of SiC is highly likely due to the lower heating rate and higher sintering temperature in hot pressing.

It is important to mention that both transgranular and intergranular microcracks were observed in hot-pressed sample as shown in Fig. 12b and c. The formation of microcracks in the hot-pressed samples is attributed to the martensitic phase transformation of tetragonal zirconia to monoclinic zirconia during cooling stage of sintering, which involves 3–7% volume contraction and 16–18% shear strain generation (Platt et al. 2014; Basu 2005; Maity et al. 2018). It is evident that sinter density of 98.1% ρ_{th} with smallest grain size (~1.4 μm) could be obtained using spark plasma sintering, but higher sinter density more than 99% ρ_{th} could be achieved by adopting PS and HP, however, with larger grain size, as mentioned in Table 4. This clearly indicates that with increasing the volume fraction of $MoSi_2$, sinter density increases due to the ductile nature of $MoSi_2$ above 1000 °C (Jeng and Lavernia 1994). Also, $MoSi_2$ restricts the grain growth during sintering by filling the pore region between the matrix grains. Therefore, it can be concluded that $MoSi_2$ is an effective sinter-aid and grain growth prohibitor to obtain a fully dense and fine-grain ZrB_2-based composite. Table 4 also shows the highest hardness for HP3 due to its highest sinter density among the all composites mentioned, while the elastic modulus of PSed sample is higher than the SPSed sample due to higher sinter density and absence of silica pockets in the PSed sample. This indicates the highest value of PSed sample is due to its pure and homogenous microstructure. The fracture toughness of the composites is comparable with a monolithic ZrB_2 phase (Monteverde et al. 2002), i.e., 2.8 MPa. $m^{1/2}$. This indicates that there is no activation of any kind of toughening mechanisms in presence of $MoSi_2$ at room temperature. Also, due to similar thermal expansion coefficient of ZrB_2 and $MoSi_2$, there is no generation of compressive stress. From Table 4, the large difference in flexural strength of HP2 and HP3 samples is noticed. As large defects were not explored in HP2 and HP3, it is believed that low flexural strength is due to coalescence of microcracks as shown in Fig. 12b and c. The reason of microcracks generation has been briefed in the earlier part of this section. In case of SPSed sample, no significant change in room temperature flexural strength is observed at high temperature. But enhancement of flexural strength of PSed sample is observed at high temperature, which is due to the oxidation of $MoSi_2$ leading to the formation of silica glass through the reaction

$$5MoSi_2 + 7O_2 = Mo_5Si_3 + 7\ SiO_2 \qquad \text{(R11)}$$

At 1500 °C, substantial loss of flexural properties of HP1, HP2, and SPSed sample is observed (see Table 4), which is due to the softening of silica pockets at the grain boundary, which reduce the grain boundary strength. But PS and HP3 sample almost retained the room temperature flexural strength (~500 MPa and ~479 MPa, respectively), which clearly indicates that the secondary phase does not instigate plasticity or crack propagation. Particularly, HP3 retained the flexural strength at 1500 °C due to the formation of SiC phase at high temperature by carbothermal reduction of silica by carbon (Sciti et al. 2006).

Effect of $TiSi_2$ sinter-aid and multistage sintering scheme on densification, microstructure, and mechanical properties of the ZrB_2-SiC composite consolidated using spark plasma sintering technique has been studied by Gupta et al. (2012). To compare the effect of multistage sintering scheme over two-stage and single-stage sintering scheme (sintering schedule is provided in Fig. 13), consolidation of ZrB_2-SiC is containing 2.5 wt.% $TiSi_2$ as a sinter-aid with final holding at 1500 °C for net holding time of 10 min and heating rate of 200 °C/min under the applied pressure of 50 MPa.

The increase of sinter density with increasing weight fraction of $TiSi_2$ in ZrB_2-SiC is clearly emphasized in Table 5, which is attributed to the plastic deformation and activation of transient liquid-phase sintering by $TiSi_2$ as the ductile to brittle transition temperature of $TiSi_2$ is 800 °C (Raju et al. 2009). Fully dense composite is obtained for ZrB_2-SiC composite containing 5 wt.% $TiSi_2$. The microstructure of ZrB_2-SiC-2.5 wt.%$TiSi_2$ composite sintered through various sintering schemes (single-stage, two-stage, and multistage sintering schemes) is shown in Fig. 14. It can be observed that there is no secondary-phase segregation at the grain boundary of the parent phases. Figure 14c clearly exhibits the strong bonding between ZrB_2 and SiC grains, where the SiC grains are uniformly dispersed in between the ZrB_2 grains. The presence of irregular-shaped grains in the microstructure clearly indicates the occurrence of liquid-phase sintering, which was also confirmed by the fine-scale microstructural analysis using TEM. The fine-scale microstructural analysis (image not shown) confirms the presence of all the grains of starting powders without any reaction product phase segregation at the grain boundary. Also, the microstructure presents uniform dispersion of equiaxed nanocrystalline SiC grains in between the submicron ZrB_2 grains and throughout the microstructure. The presence

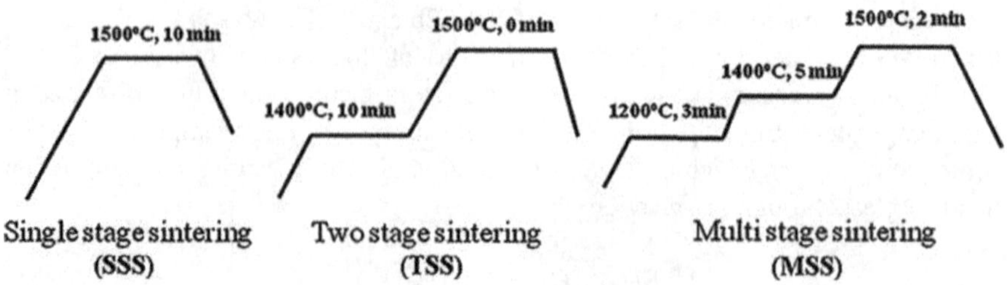

Fig. 13 Schematic representation of the heating profile followed to sinter ZrB_2–18SiC–2.5$TiSi_2$ composite using SPS technique (Gupta et al. 2012)

Table 5 Relative density, sintering parameters, and mechanical properties of (a) spark plasma-sintered ZrB_2-based UHTCs (Gupta et al. 2012); (b) comparison of ZrB_2-based UHTCs prepared using various techniques (Gupta et al. 2012)

(a) Spark plasma-sintered ZrB_2-based UHTCs

Sample details	SPS parameters (T/P/HR/Ht) (°C/MPa/°C min^{-1}/min)	RD (% ρ_{th})	*HV* (GPa)	K_{IC} (MPa. $m^{1/2}$)	FS (MPa)	References
ZrB_2–18 wt.% SiC-0 wt.% $TiSi_2$	1600/50/145/10 (SSS)	97.8	17.1 ± 0.8	3.17 ± 0.7	373 ± 40	Gupta et al. (2012)
ZrB_2–18 wt.% SiC-2.5 wt.% $TiSi_2$	1600/50/145/10 (SSS)	98.2	17.9 ± 0.4	4.04 ± 1.2	NA	
ZrB_2–18 wt.% SiC-5$TiSi_2$	1600/50/145/10 (SSS)	~100	20.3 ± 0.5	5.07 ± 0.3	497 ± 22	
ZrB_2–18 SiC-2.5$TiSi_2$	1500/30/200/10 (SSS)	99.2	25.1 ± 1.9	3.5 ± 0.3	391 ± 17	
ZrB_2–18 SiC-5$TiSi_2$	1500/30/200/0 (TSS)	96.9	23.2 ± 2.1	3.2 ± 0.1	385.2 ± 17	
ZrB_2–18 SiC-5$TiSi_2$	1500/30/200/2 (MSS)	99.9	27 ± 1.6	20.3 ± 0.5	455.1 ± 37	

(b) Comparison of ZrB_2-based UHTCs prepared using various technique

Sample composition (wt.%)	Sintering technique and parameters	RD (% ρ_{th})	*HV* (GPa)	K_{IC} (MPa. $m^{1/2}$)	References
ZrB_2	PS, 215 °C, 9 h	98	14.5	Not available	Chamberlain et al. (2006b)
ZrB_2–20$MoSi_2$	PS, 1850 °C, 30 min	99.1	16.1	2.3	Sciti et al. (2006)
ZrB_2–30SiC	RHP, 1800 °C	99	27	2.1	Chamberlain et al. (2006a)
ZrB_2–15$TaSi_2$	HP, 1900 °C	99	17.8	3.8	Sciti et al. (2008a, b)
ZrB_2–18.5SiC–3.7Si_3N_4–1Al_2O_3–0.5Y_2O_3	HP, 1760 °C, 10 min	98	14.2	4.6	Monteverde et al. (2003)
ZrB_2–SiC	SPS, 1400 °C,	98.3	17.3	5.3	Zhao et al. (2008)

(continued)

Table 5 (continued)

	30 MPa, 12 min				
ZrB_2–40ZrC–12SiC	SPS, 1800 °C, 20 MPa, 10 min	99.5	16.9	5.9	Licheri et al. (2008)
ZrB_2–60SiC	SPS, 2100 °C, 20 MPa, 6 min	99	26.8	3.5	Akin et al. (2009)
ZrB_2–6Cu	SPS, 1500 °C, 15 min	95	19.1	7.4	Venkateswaran et al. (2006)
ZrB_2–18SiC–5$TiSi_2$	SPS, 1600 °C, 10 min, 50 MPa	99.9	20.3	5.07	Gupta et al. (2012)
ZrB_2–18SiC–2.5$TiSi_2$	MSS SPS, 1500 °C, 2 min	99.9	27	4.7	

SPS spark plasma sintering, *SSS* single-stage sintering, *TSS* two-stage sintering, *MSS* multistage sintering

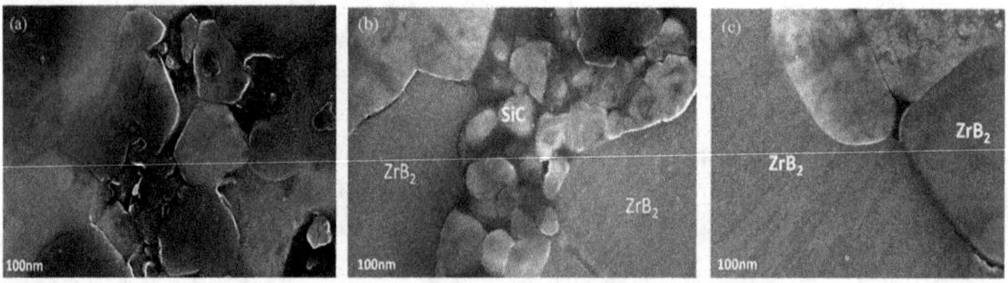

Fig. 14 BSE-SEM image of spark plasma-sintered $ZrB_2$18 wt.% and SiC 2.5 wt.% $TiSi_2$ composite obtained via (**a**) two-stage sintering (**b** and **c**) multistage sintering (Gupta et al. 2012)

of $TiSi_2$ phase in the ZrB_2/ZrB_2 and ZrB_2/SiC grain boundaries as well as in the triple junctions was observed. The morphology of the $TiSi_2$ along with anisotropic ZrB_2 grain with faceted microstructure confirms the occurrence of liquid-phase sintering. Table 5a indicates the mechanical properties of all ZrB_2-SiC-$TiSi_2$ composites. From Table 5a, it can be commented that the sinter density, hardness, and fracture toughness increase with increasing the $TiSi_2$ content. The increase in hardness with increasing $TiSi_2$ content is due to improved sinter density, as the hardness is directly related to the sinter density. The enhancement of fracture toughness with $TiSi_2$ content is imputed to the toughening by crack deflection and crack bridging near the SiC phase boundary. Superiority of multistage sintering over two-stage and single-stage sintering is observed in terms of better densification and higher hardness as shown in Table 5a. Increase in hardness for multistage sintering is due to uniform fine-grain microstructure and higher sinter density. Similar effect of multistage sintering over the two- and single-stage sintering was reported by Reddy et al. (2010a, b, c). The flexural strength of the composite varies between 352 and 450 MPa, with highest value occurring for the sample sintered using multistage sintering scheme. In order to emphasize the effectiveness of ZrB_2-SiC composite containing 2.5–5 wt.% $TiSi_2$, the properties are compared with the other reported ZrB_2-based materials processed by spark plasma sintering (SPS), hot pressing (HP), and reactive hot pressing (RHP). From the Table 5b, it can be observed that to obtain fully dense ZrB_2-based specimens through conventional sintering route, considerably higher sintering temperature (>1700 °C) is required. Moreover, the Vickers hardness and fracture toughness values of ZrB_2-based materials vary in the range of 14.2–27 GPa and 2.3–7.4 MPa. $m^{1/2}$, respectively. An investigation reported by Chamberlain et al. (2006a) stated that ZrB_2–30 SiC composite fabricated by RHP exhibits high hardness but poor fracture toughness, which is contrary to the results obtained by Gupta et al. (2012). However, it can be concluded from the Table 5b that Gupta et al. obtained the best combination of mechanical properties among the ZrB_2-based materials listed in Table 5b. Detailed investigations on high-temperature properties of ZrB_2-based ceramics are reported by Basu and Balani (2011).

Microstructure and mechanical properties of TiB_2 consolidated with 2.5 wt.% Ti as a sinter-aid at various sintering temperatures are investigated by Zhang et al. (2012). The final holding temperatures of 1350 °C, 1450 °C, 1550 °C, and 1650 °C for 5 min under the uniaxial pressure of 50 MPa with a heating rate of 150 °C/min are used. Sinter density of 62.6% ρ_{th} was obtained at the sintering temperature of 1350 °C, which indicated the presence of large fraction of pores. The sinter density, grain size, and mechanical properties of the monolithic TiB_2 sintered with final holding at 1800 °C for 5 min under the uniaxial pressure of 50 MPa (Zhang et al. 2012) are studied, which revealed exaggerated grain growth (grain size >10 μm) in the monolithic TiB_2. However, TiB_2–2.5 wt.% Ti sintered at 1650 °C shows equiaxed fine-grain microstructure with grain size of 2.5 μm (Fig. 15). Subramanian et al. (2007) also observed rapid grain coarsening in monolithic TiB_2 system with increasing sintering temperature beyond 1700 °C. From Figs. 15 and 16, it is evident that the intergranular mode of fracture with cleavage steps is observed in monolithic

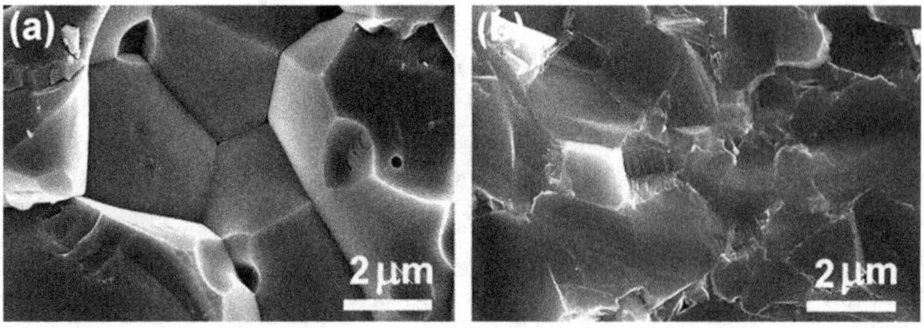

Fig. 15 SEM images of fracture surface of (**a**) monolithic TiB_2 spark plasma sintered at 1800 °C and (**b**) the TiB_2–2.5 wt.% Ti spark plasma sintered at 1650 °C (Zhang et al. 2012)

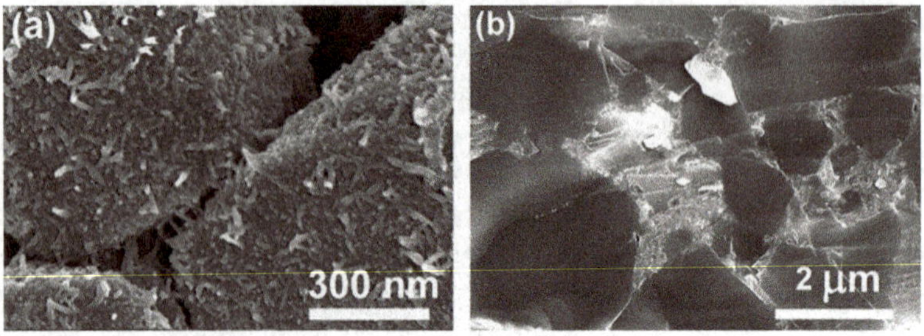

Fig. 16 High magnification SEM images of fracture surface of TiB_2–2.5 wt.% Ti composite (**a**) sintered at 1450 °C (**b**) sintered at 1650 °C using SPS (Zhang et al. 2012)

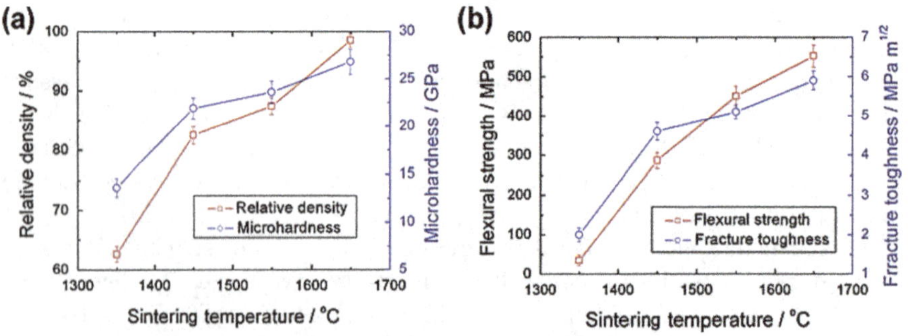

Fig. 17 Effect of sintering temperature on sinter density and the mechanical properties of the TiB_2–2.5 wt.% Ti composite (Zhang et al. 2012)

TiB_2 due to lower bond strength at TiB_2/TiB_2 grain boundary. In case of TiB_2–2.5 wt.% Ti composite, transgranular mode of fracture is evidenced which is reasoned to the enhancement of the bond strength at TiB_2/TiB_2 grain boundary as TiB phase (needle or rod like shape as shown in Fig. 16) formation occurred around the TiB_2 grain in presence of sinter-aid (Ti). It was noted that the sinter density and mechanical properties (microhardness, flexural strength, and fracture toughness) increase with increasing the sintering temperature, and maximum sinter density and mechanical properties were obtained at the sintering temperature of 1650 °C (as shown in Fig. 17). However, for better mechanical properties at room temperature as well as high temperature, a fully dense (> 97% ρ_{th}) monolithic TiB_2 is required (Raju and Basu 2007). But to obtain such sintered compact of monolithic TiB_2, higher sintering temperature (1800 °C) and spark plasma sintering technique are essential (see Table 6). But, higher sinter density (99.1% ρ_{th}) with excellent mechanical properties of TiB_2 could be obtained using 2.5 wt.% Ti as a sinter-aid at temperature of 1650 °C (Zhang et al. 2012).

The sinter density and mechanical properties of TiB_2 fabricated using various sinter-aids and sintering techniques are summarized in Table 6. It is apparent that TiB_2–2.5 wt.% Ti sintered at 1650 °C using SPS has better sinter density and

Table 6 Relative density and mechanical properties of sintered monolithic TiB_2 ceramic as well as TiB_2-based composites prepared using various sintering methods and schemes (Zhang et al. 2012)

Sample details (wt.%)	Processing details (°C, min, MPa)	Relative density (%ρ_{th})	Vickers hardness (*HV*) (GPa)	Fracture toughness (MPa. $m^{1/2}$)	Flexural strength (MPa)	References
TiB_2–2.5% $MoSi_2$	HP, 1700, 60, 30	99	27.6	–	391 ± 31	Raju et al. (2009)
Monolithic TiB_2	HP, 1650, 60, 30	94.5	18.3 ± 1.2	3.8 ± 0.6	365 ± 89	Raju and Basu (2007)
TiB_2–2.5% $TiSi_2$	HP, 1650, 60, 30	98.8	24.8 ± 0.9	4.3 ± 0.3	380 ± 74	Raju and Basu (2007)
TiB_2–5% CrB_2	HP, 1750, 60, 35	94.8	23.7 ± 1.96	4.3 ± 0.4	–	Murthy et al. (2009)
Monolithic TiB_2	SPS, 1800, 5, 50	97.6	29.6 ± 2.5	5.2 ± 0.4	538 ± 45	Zhang et al. (2010)
TiB_2–0.5% Cr + 0.5% Fe	PS, 1800, 120, 0	97.6	27.0 ± 2.5	6.2 ± 0.9	506 ± 82	Kang et al. (2001)
TiB_2–2.5% Ti	SPS, 1650, 5, 50	99.1	26.8 ± 1.6	5.9 ± 0.3	558 ± 39	Zhang et al. (2012)

mechanical properties as compared to the other TiB_2-based composites mentioned in the Table 6 processed using hot pressing/pressureless sintering (HP/PS) technique. The discrepancy in the obtained mechanical properties is attributed to the coarse-grain microstructure arising from long soaking time in HP/PS at higher sintering temperature. However, it can be concluded that use of Ti sinter-aid and SPS technique paves an effective way to obtain highly dense TiB_2-based composites with fine-grain microstructure and excellent mechanical properties (Zhang et al. 2012).

4.3 MB_2 with MB_2

This section focuses on the fabrication of mixed MB_2-MB_2 such as ZrB_2-TiB_2, ZrB_2-CrB_2, and ZrB_2-TaB_2 powder followed by compaction using spark plasma sintering and finally the microstructure and properties of bulk MB_2-MB_2 composite. Addition of one transition metal boride to other may lead to improvement in densification, and resisting the grain growth restriction of the parent phase, leading to enrichment of the mechanical properties of mixed boride system. It is well known that monolithic metal borides are difficult to sinter due to their high melting point above 3000 °C, low self-diffusion coefficient, and high vapor pressure of the constituents elements present in the metal boride system (Basu and Balani 2011). Addition of sinter-aid of MB_2 type may aid to overcome these difficulties. First, we shall discuss the synthesis of mixed borides, followed by spark plasma sintering of the mixed boride system.

4.4 Synthesis of MB_2-MB_2 Powder

This section explores the synthesis of fine-mixed boride powders of ZrB_2-TiB_2 by mechanical alloying of elemental powders using zirconia milling media. Ping et al. synthesized nanocrystalline mixed powder of (Ping et al. 2016) ZrB_2-TiB_2 by mechanical alloying using elemental powders of Zr, Ti, and amorphous boron with irregular shape and particle size in the range of 1–50 μm. To obtain the desired stoichiometric mixed boride of ZrB_2-TiB_2, the elemental powders were mixed with molar ratio of 1:1:4 (Zr/Ti/B). Mechanical alloying was carried out in planetary ball mill using zirconia milling media with ball to powder ratio of 10:1 and rotational speed of 500 rpm, under argon atmosphere. The mechanical alloying was carried out for different times starting from 10 to 120 h. Appearance of mixed ZrB_2-TiB_2 phase after 40 h of milling without any individual elements is confirmed by XRD pattern (not shown here). The description of the mechanism of formation of mixed boride from the elemental powders by mechanical alloying can be categorized into three stages:

(1) At the activation period, i.e., below 20 h of milling, welding of relatively softer Zr and Ti particles occurs due to collision among the balls, vials, and powder particles, which promotes the formation of layered structure where boron powder particles are trapped in-between the Zr and Ti powder particles. With increasing milling time beyond 20 h, long duration impact, compression, and shear result in extensive plastic deformation and strain hardening, which further induce a significant refinement of particle size of Zr and Ti along with the boron, which are uniformly distributed in the metal matrix (Ping et al. 2016). Mechanical alloying for extended period leads to the formation of substantial amount of lattice defects such as vacant lattice sites, dislocations, and stacking faults, along with large number of grain boundaries. These types of lattice defects enhance the interfacial area between the constituents, providing more diffusional paths, which decrease the activation energy for diffusional mass transport. The powder mixtures are subjected to severe plastic deformation, which generates significant amount of structural defects at the first reaction stage (milling time below 20 h). (2) Slight increase of temperature is observed, possibly due to the frictional heating at the second reaction stage (milling time below 20–40 h). Once the reaction is initiated in a localized region, it spreads expeditiously throughout the entire mixture until the reaction is completed. A sharp increase in temperature is attributed to the self-activated reaction induced by high-energy ball milling. (3) Further continuing the milling process beyond 40 h leads to continuous decrease of particle size of ZrB_2-TiB_2 due to the

absence of ductile components in the mixture. Finally after 120 h of milling, ZrB_2-TiB_2 composite powder with particle size <20 nm is obtained.

Synthesis of mix powder of TiB_2-CrB_2 by mixing the 95 vol.% TiB_2 and 5 vol.% of CrB_2 has been done by Sulima et al. using a tubular mixer for 8 h (Sulima et al. 2016).

4.5 SPS of MB_2-MB_2

Karthiselva and Bakshi (2016) demonstrated the synthesis of ZrB_2-TiB_2 composite by mechanical mixing of commercially available elemental powders of Zr with particle size less than 30 μm and Ti with particle size less than 44 μm and amorphous boron, using planetary ball mill followed by reactive spark plasma sintering. The elemental powders were stoichiometrically mixed and wet milled in toluene for 8 h using WC milling media. Zr, Ti, and B were mixed in various proportions to develop ZrB_2-based composite containing different volume percent of TiB_2 as an additive. The compositional details are provided in Table 7. Consolidation is performed using reactive spark plasma sintering with final holding at 1400 °C for 10 min under the mechanical pressure of 50 MPa and heating rate of 50 °C/min. Monolithic ZrB_2, TiB_2, and composites obtained after sintering are described in Table 7 (Karthiselva and Bakshi 2016).

Figure 18a shows the nature of displacement of different ZrB_2-TiB_2 composite, as function of sintering time, which clearly depicts four distinct stages. In stage I, gradual shrinkage is observed, which may be due to the rearrangement of the particle. Stage II indicates rapid displacement of specimen, which might be due to the exothermic reaction between the milled powders. This stage is indicated by sudden increase in displacement over the temperature range of 550–700 °C. During this stage of reactive spark plasma sintering, a bright flash of light was observed at the bottom and top of the die-punch combination, which is attributed to the exothermic reaction of the reactant powders resulting in rapid displacement. The sample displacement in the present context includes the displacement contribution from die-punches and spacers, but it is hard to estimate the displacement contribution from the die-punches and spacers. It is worthwhile to mention that consolidation using reactive spark plasma sintering facilitates a large displacement of the samples or full densification as compared to the normal spark plasma sintering. The simultaneous application of direct current (DC) and mechanical pressure play a significant role in amplifying the sintering kinetics. In stage III, further densification occurs by plastic flow mechanism which is observed only in the ZrB_2-TiB_2 composite, as no plastic flow mechanism is reported for the monolithic ZrB and TiB system. In stage IV, densification happens due to final holding at high temperature, by grain boundary, pipe, and lattice diffusion mechanisms. For the monolithic ZrB and TiB system, the densification at stage III and IV is caused primarily due to grain boundary, pipe, and lattice diffusion (Karthiselva and Bakshi 2016; Reddy et al. 2010a, b, c). In another set of investigations, Sulima et al. (2016) reported the synthesis of bulk TiB_2

Table 7 Sinter density, grain size, and mechanical properties of reaction spark plasma-sintered ZrB_2 system with/without TiB_2 and TiB_2 system with/without CrB_2

Composition	Sintering conditions (°C/min/MPa/(°C/min))	Sinter density (% ρ_{th})	Grain size (μm)	Hardness (GPa)	Elastic modulus (GPa)	Indentation toughness (MPa. $m^{1/2}$)	References
ZrB_2	2000/10/50/50	97	1.2 ± 0.3	29 ± 2^a	495 ± 43	3.2 ± 0.6	Karthiselva and Bakshi (2016)
ZrB_2–25 vol.% TiB_2	2000/10/50/50	96	0.7 ± 0.3	33 ± 4^a	496 ± 47	3.5 ± 0.5	Karthiselva and Bakshi (2016)
ZrB_2–50 vol.% TiB_2	2000/10/50/50	97	1.5 ± 0.5	34 ± 5^a	523 ± 52	3.9 ± 0.6	Karthiselva and Bakshi (2016)
ZrB_2–75 vol.% TiB_2	2000/10/50/50	98	1.5 ± 0.4	33 ± 6^a	568 ± 44	3.7 ± 0.6	Karthiselva and Bakshi (2016)
TiB_2	2000/10/50/50	98	0.5 ± 0.2	34 ± 4^a	658 ± 32	3.4 ± 0.5	Karthiselva and Bakshi (2016)
TiB_2	2000/10/35/200	80	NA	NM	282 ± 7	NA	Sulima et al. (2016)
TiB_2–5 vol.% CrB_2	2000/10/35/200	88	NA	NM	453 ± 13	NA	Sulima et al. (2016)
TiB_2	2200/10/35/200	87	NA	NM	340 ± 9	NA	Sulima et al. (2016)
TiB_2–5 vol.% CrB_2	2200/10/35/200	95	NA	16.1^b	477 ± 14	NA	Sulima et al. (2016)

NA not available, *NM* not measured, [a]indicates hardness measured by nano-indentation method, [b]indicates Vickers hardness with/without CrB_2 additive. The consolidation of the commercially available monolithic TiB_2 powders along with mixer of TiB_2–5 vol.% CrB_2 powders, mixed for 8 h using Tubular mixer, with final holding at 2000 °C and 2200 °C for 10 min with a heating rate of 200 °C/min under mechanical pressure of 35 MPa is conducted using spark plasma sintering. The sinter densities of the specimens are provided in Table 7, which shows considerably lower 80–87% ρ_{th} for monolithic TiB_2 as compared to the monolithic TiB_2 (Karthiselva and Bakshi 2016) prepared by Karthiselva et al. at 2000 °C by reactive spark plasma sintering of 8 h ball milled powder. But the maximum sinter density (95% $\rho_{th)}$ was obtained for TiB_2–5 vol.% CrB_2 consolidated at 2200 °C for 10 min. The incorporation of CrB_2 in TiB_2 was observed to enhance the sinter density as well as impede the grain coarsening during sintering process.

ZrB_2-TaB_2-type mixed boride systems have been examined by Demirskyi and Vasylkiv (2017) with final holding at 2300 °C for 6 min under the 26 kN mechanical pressure using three-stage spark plasma sintering technique, resulting in sinter density of 99.2% ρ_{th}. The composite powder was synthesized by wet mixing of the commercially available ZrB_2 and TaB_2 powders in alcohol medium, followed by drying at 100 °C and screening through 60, 400, and 1250 mesh screens. Tantalum foil instead of graphite foil was inserted on the top and bottom of the powder bed as well as inside the graphite die for easy removal of the sample after sintering as well as to avoid the carbon diffusion from the graphite foil or the die-punch set up at higher temperature (>2000 °C) (Demirskyi and Vasylkiv 2017).

Figure 18b reveals the data obtained during sintering of ZrB_2–44% TaB_2, which helps in understanding the sintering behavior. Three distinct peaks in the electric current curve are observed which corresponds to three holding temperatures. The first, second, and third peaks indicate the temperature of 900 °C, 1730 °C, and 2300 °C, respectively. For the primary two peaks, 1 min pause time is provided to stabilize the current value. The third peak does not arise immediately as the temperature reaches to 2300 °C but only after holding for 15 sec. The characteristics of the

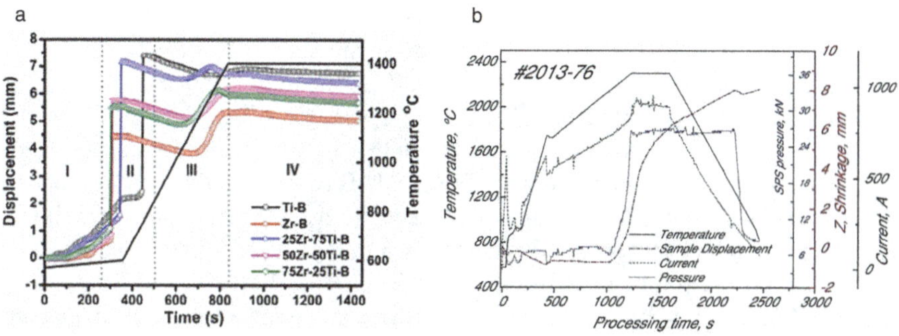

Fig. 18 (**a**) Displacement versus time obtained during reaction spark plasma sintering of ZrB_2 with/without TiB_2 (Karthiselva and Bakshi 2016); (**b**) data recorded during sintering of ZrB_2-TaB_2 ceramic composite with final holding at 2300 °C for 6 min under 26 kN mechanical pressure. Three different heating rates were used, i.e., ~450 °C/min up to 900 °C, ~200 °C/min up to 1730 °C, and ~50 °C/min to reach final sintering temperature of 2300 °C (Demirskyi and Vasylkiv 2017)

electrical current curve can be explained using the loading scheme used. The maximum pressure was applied above 2150 °C to reduce the ZrB_2/TaB_2 particle contact to diminish the possibility of the formation of (Zr, Ta) B_2 solid solution and to reduce the grain coarsening process. As the mechanical pressure reaches a maximum value (26 kN) at 2300 °C, rapid shrinkage occurs, indicating rapid densification, resulting in large current consumption by the specimen, which may cause overheating. In order to avoid overheating, maximum limit of current was set up by controlling the SPS equipment manually.

4.6 Properties of Bulk MB_2-MB_2

4.6.1 Mechanical Properties of MB_2-MB_2 Composites

Karthiselva et al. synthesized ZrB_2-TiB_2 composite by varying the volume percent of TiB_2 by spark plasma sintering technique and reported the mechanical properties of sintered ZrB_2-TiB_2 composite as provide in Table 7. Nano-hardness and elastic modulus of the sintered composite were measured using nanoindenter by applying 8 mN load, and estimation of the hardness value was carried out adopting Oliver-Pharr model (Oliver and Pharr 1992). Short crack toughness or indentation toughness was measured using Vickers hardness tester by applying 5 kg load with 10 sec pause time. The indentation fracture toughness was evaluated using the Antis equation (Antis et al. 1981).

$$K_{\mathrm{IC}} = 0.016\left(\frac{E}{HV}\right)^{0.5}\frac{P}{C^{1.5}} \tag{2}$$

where K_{IC} is the fracture toughness, E is elastic modulus, HV is the Vickers hardness, P is the load applied during indentation to generate crack at the indent diagonals, and c is the crack length. From Table 7 and Fig. 19, it is evident that indentation fracture toughness and grain size of ZrB_2-TiB_2 composites are maximum for the composition of 50Zr-50Ti-B as compared to the monolithic ZrB_2 and TiB_2 samples, which clearly indicates the profound influence of secondary-phase TiB_2/ZrB_2 in enhancing the mechanical properties (Karthiselva and Bakshi 2016). The significant enhancement of fracture toughness was observed for 50Zr-50Ti-B as compared to other compositions, which is attributed to the toughening mechanisms of the ceramic composite by crack branching and crack deflections (as shown in Fig. 20). Different coefficient of thermal expansion (CTE) of ZrB_2 ($\alpha_a = 6.66$, $\alpha_c = 6.93 \times 10^{-6}\ \mathrm{K}^{-1}$) and TiB_2 ($\alpha_a = 6.36$, $\alpha_c = 9.30 \times 10^{-6}\ \mathrm{K}^{-1}$) in the composites can cause significant improvement of fracture toughness by residual stress toughening mechanism. Residual stress produce microcracks during cooling stage of reaction spark plasma sintering process, which dissolve the strain energy of the existing large crack and generate compressive stress near the long crack tip and resist the propagation of long crack and increase the fracture toughness (Karthiselva and Bakshi 2016; Zhang et al. 2012).

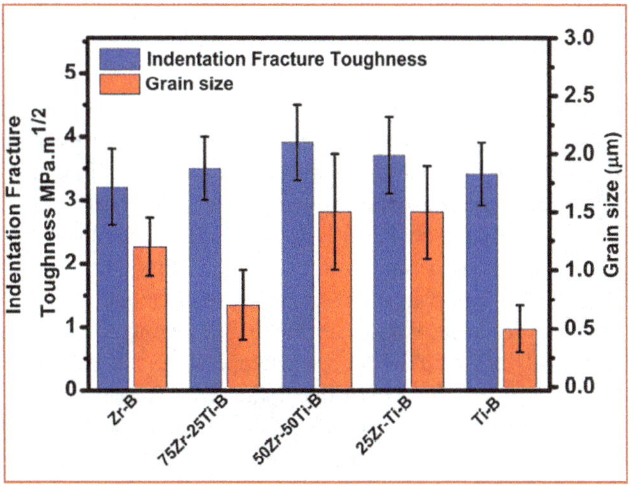

Fig. 19 Effect of TiB_2 addition on fracture toughness and grain size of ZrB_2 (Karthiselva and Bakshi 2016)

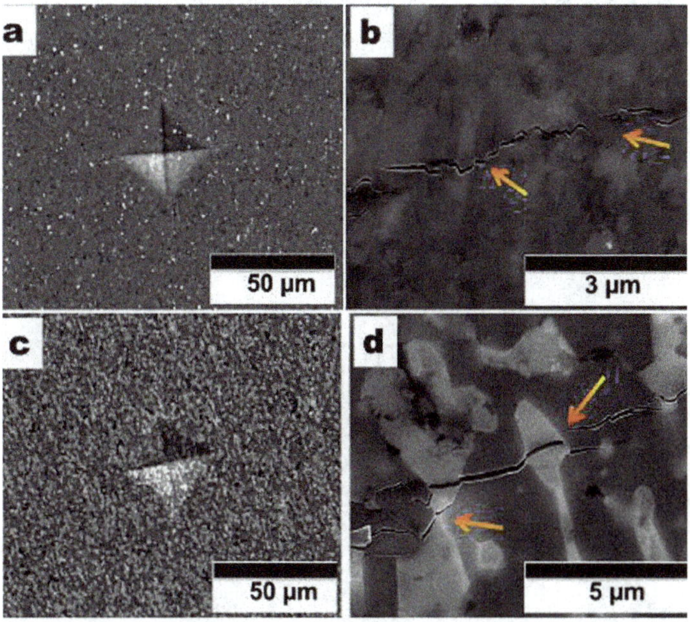

Fig. 20 SEM image of the indents containing cracks at the indent corner of TiB_2 (**a**, **b**) and ZrB_2–50 vol.% TiB_2 (**c**, **d**) samples showing the toughening mechanism by crack deflection and crack branching (Karthiselva and Bakshi 2016)

Improvement in other mechanical properties can be attributed to the formation of solid solution of ZrB_2-TiB_2 of varying extent. According to the principle of energy, when a solid solution forms between ZrB_2 and TiB_2, the Ti^+ and Zr^+ substitute each other, which cause an increase in internal energy due to difference in ionic radius of the two ions. As the solid solution reaction starts at the interphase region, this can lead to restricted grain boundary movement resulting in a finer grain size and improved mechanical properties (Karthiselva and Bakshi 2016).

Figure 21a and b (Karthiselva and Bakshi 2016) reveals the stress-strain plot obtained under compressive loading of SPSed samples (monolithic ZrB_2, TiB_2, and ZrB_2 with 0–75 vol.% TiB_2 composites) at room temperature and 800 °C, respectively. Both the figures clearly indicate the higher compressive strength of the composites, especially 50Zr-50Ti-B at room temperature and 800 °C, as compared to the monolithic ZrB_2 and TiB_2. This is due to the formation of solid solution, which may cause significant enhancement in mechanical properties, though 50Zr-50Ti-B shows some plasticity at room temperature (see Fig. 21a). From Fig. 21b, it is evident that no significant change of compressive strength is observed between ZrB_2–50 vol.% TiB_2 (indicated by 50ZR-50Ti-B) and monolithic ZrB_2 (indicated by Zr-B) until the stress value reached to the machine safety limit, i.e., 6 kN load.

From Table 7, it is evident that both elastic modulus and hardness increases with increase in the volume fraction of TiB_2 in ZrB_2, with maximum hardness observed for ZrB_2–50 vol.% TiB_2 composite and maximum elastic modulus value found for ZrB_2–75 vol.% TiB_2 composite. The pop in phenomenon is observed in ZrB_2–50 vol.% TiB_2 during the hardness testing (plot is not shown). The possible reasons for pop in phenomenon are as follows:

I. Phase transformation
II. Multiplication of dislocations under the indenter during indentation loading
III. Reaching the stress value to a critical resolved shear stress at a specific load

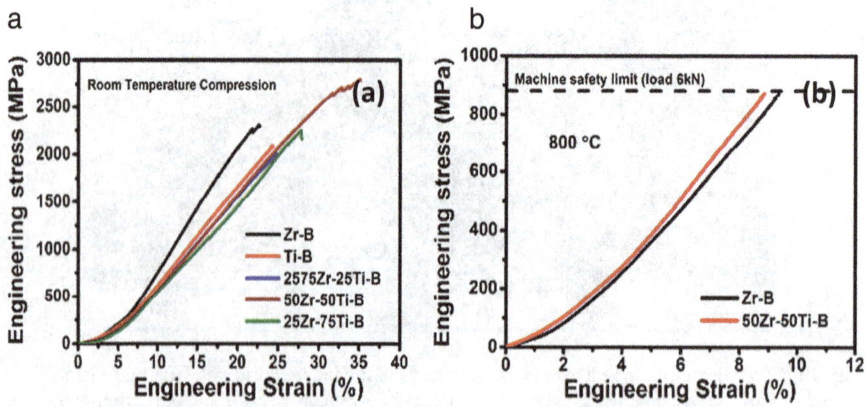

Fig. 21 (**a**) Stress-strain plot obtained under compression of (**a**) spark plasma-sintered samples (monolithic ZrB_2, TiB_2, and ZrB_2 with 0–75 vol.% TiB_2 composites) at room temperature and (**b**) spark plasma sintered monolithic ZrB_2 and ZrB_2 containing 50 vol.% TiB_2 composites at 800 °C (Karthiselva and Bakshi 2016)

It is demonstrated that this type of phenomenon occurs due to the movement of the indenter inside the grain, which clearly indicates that ZrB_2–50 vol.% TiB_2 shows some level of plasticity at room temperature. In spite of this, further detail investigation is required to explore the slip or twin system activation under the indenter.

Sulima et al. reported the sinter density along with mechanical properties (elastic modulus) of monolithic TiB_2 and TiB_2–5 vol.% CrB_2 composite as provided in Table 7. The monolithic TiB_2 sintered at 2000 °C and 2200 °C show poor relative density, thus resulting in lower elastic modulus as compared TiB_2–5 vol.% CrB_2 fabricated under same route and conditions. The hardness was not calculated due to the presence of large volume fraction of pores, which may bring about significant distortion of the indent. The addition of 5 vol.% CrB_2 in TiB_2 matrix give rise to significant improvement in the sinter density as well as the mechanical properties, with the latter being approximately 40–50% more than the monolithic TiB_2 (Sulima et al. 2016).

5 MB_2 with MC

It is well known that monolithic UHTCs, especially the transition metal borides (TiB_2, ZrB_2, etc.), are difficult to sinter due to their high melting point (>3000 °C), strong covalent boding, low self-diffusion coefficient, and high vapor pressure of the transition metals and boron contained in the borides. Monolithic borides have lower fracture toughness as compared to the boride-based UHTCs. Therefore, to improve the sinterability as well as the mechanical properties, in particular, the fracture toughness of monolithic diboride, transition metal carbide with/without sinter-aid is incorporated as additional phases in monolithic boride system. In this section, processing, microstructure, and mechanical properties of TiB_2 and ZrB_2 with carbides additives are explored in detail.

5.1 Spark Plasma Sintering, Microstructure, and Properties of MB_2-MC

Demirskyi et al. demonstrated the fabrication of bulk TiB_2-TaC composite by mechanical mixing of commercially available TiB_2 powder (d_{av} = 1–2.4 μm) with 10 wt.% TaC powder [1.5– 2 μm], through wet milling in alcohol medium followed by drying at 100 °C and screening through 60 mesh sieve. Sintering of then as-mixed TiB_2–10 wt.% TaC and individual monolithic powders is carried out with final holding at 2000 °C for 10 min under the applied pressure of 40 MPa, with heating rate of 50 °C/min in argon atmosphere using the two-stage spark plasma sintering (Demirskyi et al. 2016).

Table 8 The relative density, grain size, and the flexural strength of the sintered composite

Composition (wt.%)	Relative density (%ρ_{th})	Grain size of sintered samples (μm)	Flexural strength (MPa)			Reference
			RT	600 °C	1600 °C	
TiB_2	~ 96.0	6–12	370 ± 15	390	200 ± 7	Demirskyi et al. (2016)
TiB_2–10TaC	> 98	2–7	533 ± 13	590	480 ± 15	
TaC	>92.0	8–20	569 ± 16	520	190 ± 12	

The sinter density, grain size, and flexural strength of TiB_2-TaC are provided in Table 8. Flexural strength measurement was carried out by 3-point bending method at room temperature as well as at high temperature, up to 1600 °C. It is clear that TiB_2-TaC composite has higher sinter density with significantly lower grain size, as compared to the monolithic TiB_2 and TaC fabricated under same conditions. The flexural strength of TiB_2-TaC composite increases with increasing the test temperature up to 600 °C, while rapid reduction in strength is observed for TaC samples at 600 °C. TiB_2 shows improvement in flexural strength. It can also be noted that room temperature flexural strength of TiB_2-TaC composite is much higher (~160 MPa) than the monolithic TiB_2 but the slightly lower (~30 MPa) than the monolithic TaC. Also, at 1600 °C, the composite shows significantly higher flexural strength than both the monolithic TiB_2 and TaC. The higher flexural strength of the composite at 1600 °C is attributed to the formation of high-temperature (Ti, Ta)(C, B) solid solution. Significant reduction of flexural strength of the monolithic TiB_2 and TaC at 1600 °C occurs due to the plastic deformation caused by high-temperature creep (Demirskyi et al. 2016).

Fractographical analysis of fractured surface of sintered monolithic TiB_2, TaC, and TiB_2-TaC composite (Fig. 22) after the room temperature bending test shows transgranular fracture in larger grain size region, intragranular fracture for smaller grain size region, and some cleavage fractures at the grain boundary regions (image not shown) of TiB_2. TaC reveals the presence of some sharp-edged pores and spherical pores, with pore size up to 500 nm (along with impurity phase near the pore), cleavage fracture, and surface marking in the fractures regions of TaC (image not shown). Figure 22 exhibits a typical core-rim structure mostly in TaTiC phase region (white contrast) (solid solution phase of TaTiC is formed by the high-temperature reaction between TiB_2 and TaC (Demirskyi et al. 2015)) surrounded by TiB_2 phase along with cleavage and surface markings in all the phases present in the composite microstructure (Demirskyi et al. 2016).

High-temperature fractographical analysis is shown in Fig. 23 for TiB_2-TaC after the high-temperature (1600 °C) bending test. Intergranular fracture, grain pull-out are observed to be the dominant modes of fracture in all the samples. For TiB_2 samples, high-temperature etching occurred as concluded from the matte grain surface (image not shown). Also, some ultrafine cleavage steps in the vicinity of the sharp-edge pores are identified. The intragranular fracture is also observed in TaC samples at 1600 °C with very less number of sharp-edged pore (image not shown). The solid solution phase of TaTiC phase appears bright in the fracture

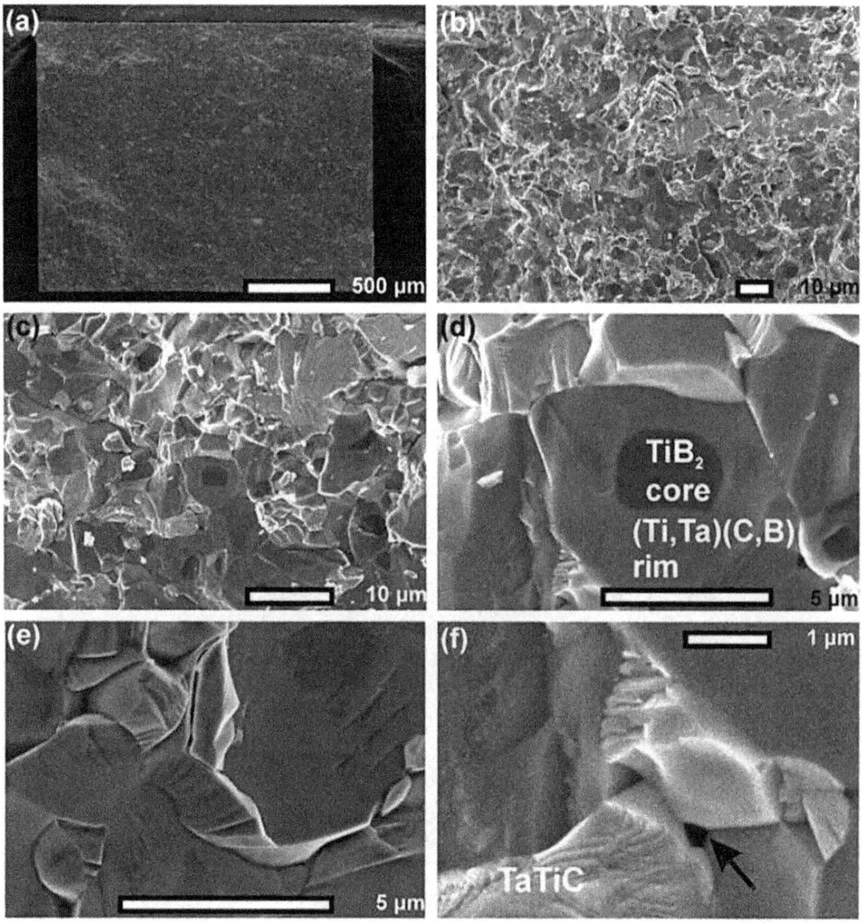

Fig. 22 SE-SEM image: (**a–f**) the fractured surface of TiB -TaC composite at room temperature captured in different magnifications; (**d–f**) showing core rim structure, cleavage fractures and surface marking along with spherical and sharp edged pore (Demirskyi et al. 2016)

microstructure. Some cleavage fracture is observed at the TaTiC grain boundaries and the size of the voids formed due to grain pulling (Fig. 23). Moreover, some smaller size sharp-edge pores, spherical pores, fine surface markings, and ultrafine cleavage steps can only be distinguished at high magnifications (Demirskyi et al. 2015, 2016).

Demirskyi et al. also demonstrated the fabrication of TiB_2-NbC composite by mechanical mixing (by wet milling in alcohol medium followed by drying at 100 °C and screening through 60 mesh sieve) of commercially available TiB_2 powder (d_{av} = 1–2.4 μm) with (10, 30, and 50 wt.%) NbC powder [1.5–3 μm], which were used as the starting materials. Two stage reactive spark plasma sintering of

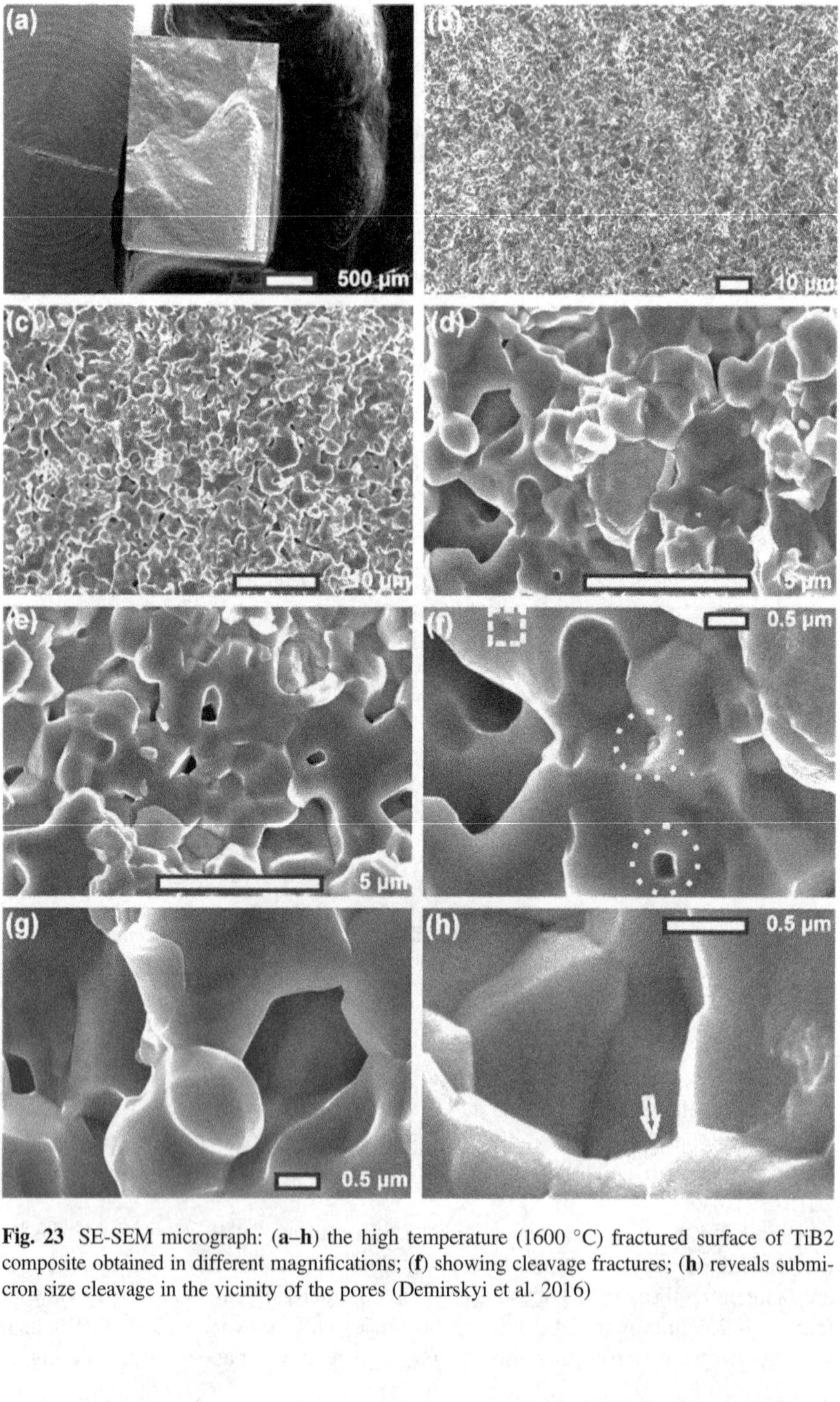

Fig. 23 SE-SEM micrograph: (**a–h**) the high temperature (1600 °C) fractured surface of TiB2 composite obtained in different magnifications; (**f**) showing cleavage fractures; (**h**) reveals submicron size cleavage in the vicinity of the pores (Demirskyi et al. 2016)

TiB_2-(10, 30, and 50 wt. %) NbC and individual monolithic powders have been conducted with final holding at 1800 °C and 2000 °C for 5 min under the applied pressure of 40 MPa, with heating rate of 50 °C/min in argon atmosphere (Demirskyi et al. 2015). The effect of dwell time on densification and microstructure of the composites is investigated by choosing five holding times (10 sec, 1 min, 5 min, 10 min, and 40 min). The plot of relative density as a function of final dwell time of TiB_2-NbC composites along with the monolithic specimen sintered at 2000 °C under 40 MPa pressure in argon atmosphere is shown in Fig. 24. It can be found that relative density increases linearly with increasing dwell at 2000 °C. Most curves are not linear. The maximum sinter density is found in TiB_2–10 wt.% NbC composite due to formation of solid solution phase of (Nb,Ti)C and $(Ti,Nb)B_2$. Lower sinter density is observed in monolithic TiB_2, due to rapid grain coarsening above 1900 °C (Baumgartner and Steiger 1984). But in the case of TiB_2-NbC composite, the densification and chemical reaction happen simultaneously, which reduce the grain growth significantly, as most of the heat energy will be used in the chemical reaction as well as in the densification, rather than grain growth (Vasylkiv et al. 2012; Demirskyi et al. 2015).

Hardness measurement of all the sintered samples was carried out using Vickers hardness tester by applying 1 kg load. The fracture toughness of the sintered specimens was measured by indentation method in a hardness tester by applying 10–50 kg load to generate crack at the indent corners. It is well known that to generate cracks at the indent corners, higher load is most essential as compared to the mechanical load required to measure the Vickers hardness without initiating crack during the hardness measurement because adequate energy will be absorbed to generate crack at the indent corner rather than the formation of indent. The fracture toughness was evaluated from the following equation (Lawn and Fuller 1975; Lawn 1993; Tanaka 1987; Basu and Balani 2011; Demirskyi et al. 2015):

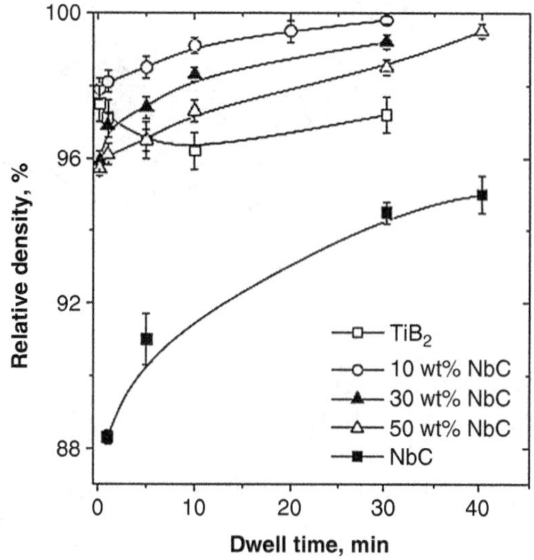

Fig. 24 Relative density as a function of dwell time of TiB_2-NbC composite sintered at 2000 °C under 40 MPa pressure

$$K_{IC} = \frac{0.073 \times 10^{-9} p}{C^{3/2}} \tag{3}$$

where K_{IC} is the fracture toughness, P is the applied load required to generate crack at the indent corners, and C is the half crack length. Figure 25 reveals that the grain size of solid solution phases of (Nb,Ti)C and (Ti,Nb)B_2 formed during RSPS is similar to the TiB_2 (dark contrast region) and NbC grain (see Fig. 25b). It is worthwhile to mention that some unreacted TiB_2 phase has lower grain size which is even smaller that initial powder particle size (Fig. 25b, d), which clearly indicate dissolution of TiB_2 during the RSPS of the TiB_2-NbC composites (Vasylkiv et al. 2012; Demirskyi et al. 2015).

However, the grain size of newly formed solid solution phases (Nb,Ti)C and (Ti, Nb)B_2 increases in pore-free composites, which clearly indicates that the grain growth and densification occur simultaneously during the final stage of RSPS of TiB_2-NbC composite. It is observed that some finer size (Nb,Ti)C precipitates (50–300 nm) at the grain boundary region as well as at triple junction of the grains at 1800 °C and 2000 °C. The size of the precipitates increases to 500–600 nm above the sintering temperature of 2000 °C. Similar kind of precipitates formation was reported for the TaC-TiB_2 composite (Vasylkiv et al. 2012; Demirskyi et al. 2015). As the solid solution of carbides, (Nb,Ti)C and borides, (Ti, Nb)B_2 have higher hardness as compared to the individual phases, the overall mechanical properties of the composites are improved significantly.

The sinter density, mean grain size, and the mechanical properties of the individual phases of TiB_2-NbC are summarized in Table 9. Hardness (measured by applying 0.1 kg load) of (Ti,Nb)B_2 is significantly higher than the individual TiB_2 and NbC phases as well as the (Nb,Ti)C phases. The hardness values of the

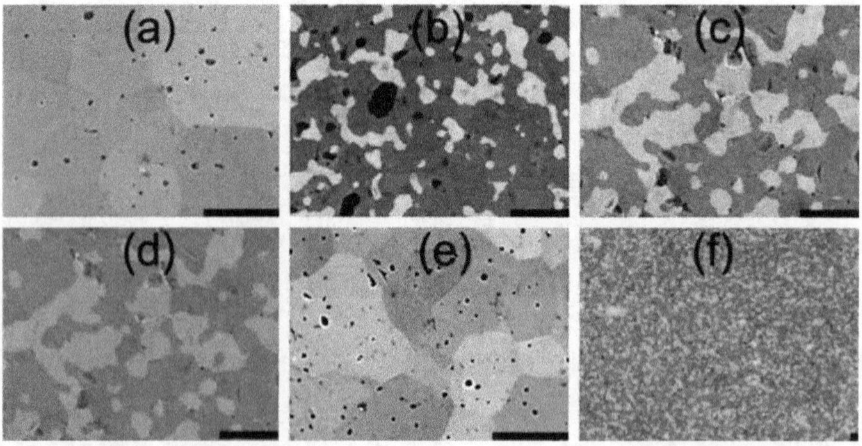

Fig. 25 BSE-SEM images of sintered TiB_2-NbC composite sintered with final holding at 2000 °C for 10 min under 40 MPa pressure (**a**) TiB_2; (**b**) TiB_2–10 wt.% NbC; (**c**, **f**) TiB_2–30 wt.% NbC; (**d**) TiB_2–50 wt.% NbC; (**e**) NbC. The scale bar is 10 μm

Table 9 The relative density and mechanical properties of the individual phases present in TiB_2-NbC composite consolidated using RSPS at 1800 °C and 2000 °C (Demirskyi et al. 2015)

Sample composition (wt.%)	Relative density (%ρ_{th})	Grain size of sintered samples (μm)	Major phases	Lattice parameter[a] a (Å)	Lattice parametera[a], c (Å)	Microhardness[b] (GPa)	HV^c (GPa)	K_{IC} (MPa m^{-2}
TiB_2	> 97	4–12	TiB_2, TiB_2	3.036 3.036	3.238 3.238	29.5 ± 1.7 28.7 ± 1.5	24.5 ± 2.5 26.8 ± 1.4	5.5 ± 0.3 7.6 ± 0.3
10 wt% NbC	> 98	2–4	(Nb,Ti)C, $(Ti,Nb)B_2$	4.3755 3.0629	– 3.2902	28.8 ± 1.8 32.1 ± 1.1	– –	– –
30 wt% NbC	> 98	4–6	(Nb,Ti)C, $(Ti,Nb)B_2$	4.4016 3.0719	– 3.2732	28.8 ± 1.8 31.3 ± 0.6	25.9 ± 1.7	6.9 ± 0.7
50 wt% NbC	97	4–6	(Nb,Ti)C, $(Ti,Nb)B_2$	4.4392 3.074	– 3.2996	28.8 ± 1.8 24.3 ± 1.1	22.7 ± 1.5	6.3 ± 1.1
NbC	95	5–15	NbC	4.4702	–	22.1 ± 1.8	19.7 ± 1.9	2.9 ± 0.2

[a]Error is within 0.004 Å; [b]indentation was performed with load of 0.1 kg; [c]indentation was performed with load of 1 kg

composites measured by applying 0.1 kg load are slightly higher than the hardness values measured applying 1 kg load, which is reasoned to the indentation size effect (Basu and Balani 2011). From, Table 9 it is evident that TiB_2–10 wt.% NbC composite has highest hardness due to the finer grain size as compared to the other composites as well as the monolithic TiB_2 and NbC.

A similar kind of study was reported earlier for the reactive spark plasma sintering of TaC-TiB_2 composite (Vasylkiv et al. 2012) (as provided in Table 10) (Mukhopadhyay et al. 2013; Chernysheva et al. 1979; Yu et al. 1979; Andrievskii et al. 1974; Kamiya et al. 1989; Woydt and Mohrbacher 2015; Demirskyi et al. 2015). From Table 10, it is evident that TiB_2-NbC composites shows significantly higher fracture toughness as compared to other ceramic system mentioned in the Table 10. The higher fracture of TiB_2-NbC is attributed to the crack deflection and crack bridging at the grain boundaries of the hard solid solution phases, namely, (Nb, Ti)C and (Ti, Nb)B_2, as shown in Fig. 26. Moreover, Wen et al. reported the stress field toughening in TiB_2-TiC system is due to the discrepancy in thermal expansion coefficient of TiB_2 and TiC phases present in the composite. As a result, a substantial amount of compressive stress as well as tensile stress would be generated at the TiB_2/TiC grain boundaries, which result in more energy being absorbed by the pre-existing crack (Wen et al. 2001).

Tsuchida and Yamamoto reported the fabrication of ZrB_2-ZrC composites by sintering of ZrB_2-ZrC powder mixture prepared by mechanically activated self-propagating high-temperature synthesis (MA-SHS) of elemental powder mixtures of Zr/B/C = 2/2/1. The mixture was first ball milled for 45 min in planetary ball mill using WC balls of 12 mm diameter in agate vials for surface activation of the powder particles. Then the mixture was exposed to air atmosphere for self-ignition to form ZrB_2-ZrC composite powder. Two-stage spark plasma sintering with final holding at 1800 °C for 5–10 min under the applied pressure of 40 MPa, with a heating rate of 40 °C/min, in vacuum environment of 6–10 Pa is used. To compare the densification behavior and mechanical properties, the following powder samples (sintered by SPS under same conditions) are considered: (I) mechanically activated power mixtures of Zr/B/C = 2/2/1 exposed to air being left for 24 h in agate vial to avoid self-ignition; (II) power mixtures of Zr/B/C = 2/2/1, ball milled for a few minutes for mixing but without surface activation; and (III) mechanical mixing of commercially available power of ZrB_2 and ZrC for 45 min in planetary ball mill using WC balls and agate vials (Tsuchida and Yamamoto 2007).

Figure 27 displays the BSE-SEM images of the ZrB_2-ZrC composites, sintered from different mixture of elemental powders (Zr, B, C) such as powder mixture prepared by MA-SHS method, the MA method without surface activation, the mixed powder of Zr, B, C, and mixed powder of the commercially available ZrB_2 and ZrC powder with final holding at 1800 °C for 5–10 min under 6–10 Pa (vacuum) using two-stage spark plasma sintering technique. From Fig. 27, it is evident that there exist two different contrast phases (bright contrast and gray contrast) other than pores (appeared black in the microstructure). From EPMA analysis, it is confirmed that the gray contrast is corresponding to ZrB_2 phase, and the bright contrast is corresponding to ZrC phase. Among sintered ZrB_2-ZrC

Table 10 Processing techniques, relative density, grain size, and mechanical properties of TiB_2-NbC composite. The data for the other ceramic systems are provided for comparison with the TiB_2-NbC composite (Demirskyi et al. 2015)

Sample composition (wt.%)	Processing details method/ sintering temperature (°C)/ Dwell time (min)	Relative density (%ρ_{th})	Grain size of sintered samples (μm)	Hardness (GPa)	Indentation fracture toughness (K_{IC}) (MPa.m$^{1/2}$)	References
TiB_2	SPS/1400/10	97.6	3	18.3 ± 0.6	5.8 ± 0.8	Mukhopadhyay et al. (2013)
TiC–B	CS and arc-melting	–	–	42.6 ± 0.18	–	Chernysheva et al. (1979)
TiC–B	CS	~ 99	3	24.0 ± 1.0	–	Yu et al. (1979)
TiC	HP	96–98	2	27.5 ± 1.0	–	Andrievskii et al. (1974)
NbC	HP/1600/60	99.3	9	19.1 ± 0.6	3.81 ± 0.05	Kamiya et al. (1989)
NbC	SPS/2000/12	–	–	16.5 ± 0.5	–	Woydt and Mohrbacher 2015
TiB_2	SPS/2000/10	>97	6–11	24.5 ± 2.5	5.5 ± 0.3	Demirskyi et al. (2015)
TiB_2–10 wt% TaC	SPS/2000/5	>98	2–4	28.8 ± 1.8	5.8 ± 0.4	Demirskyi et al. (2015)
TiB_2–10 wt% NbC	SPS/2000/5	>97	2–4	26.6 ± 1.4	5.9 ± 0.4	Demirskyi et al. (2015)
TiB_2–10 wt% NbC	SPS/2000/10	>98	2–5	26.9 ± 1.8	7.8 ± 0.2	
TiB_2–10 wt% NbC	SPS/2000/20	>99	3–6	27.2 ± 1.2	6.1 ± 0.3	
TiB_2–30 wt% NbC	SPS/2000/5	>98	4–6	25.9 ± 1.7	6.9 ± 0.7	
TiB_2–50 wt% NbC	SPS/2000/5	>97	4–6	22.7 (±1.5)	6.1 (±1.1)	
NbC	SPS/2000/40	95	5–15	19.7 (±1.9)	2.9 (±0.2)	

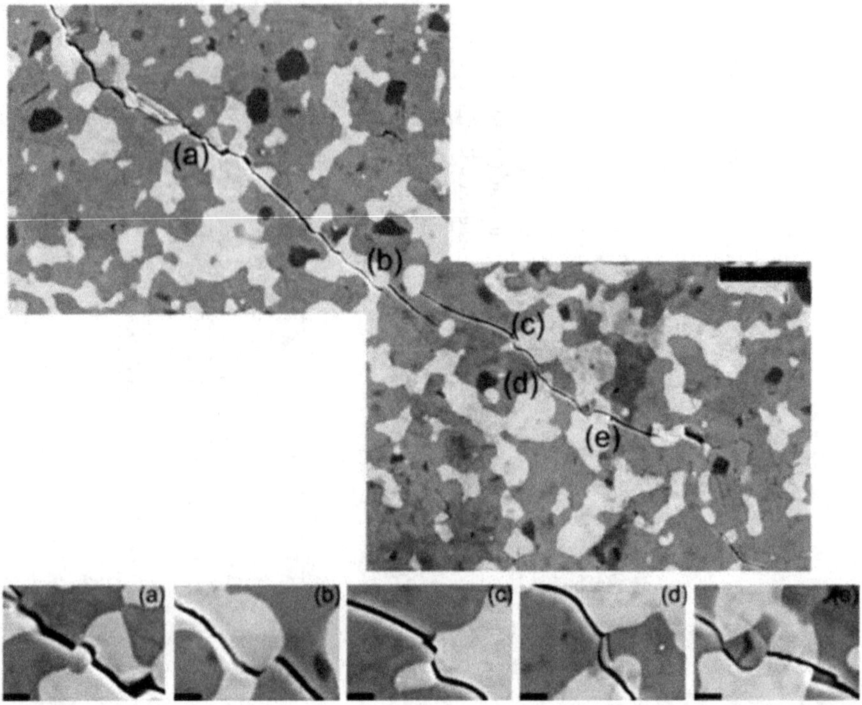

Fig. 26 Propagation of cracks generated during indentation loading of bulk TiB_2–30 wt.% NbC composite sintered with final holding at 2000 °C for 5 min. The scale bar for the top two images is 10 μm. Insets (bottom images) scale bar is 1 μm. (**a–e**) the different spots in the micrograph along with magnified view showing various crack propagation pathway

composite, the compact fabricated from MA-SHS powder reveals a uniform fine-scale microstructure (as shown in Fig. 27a) with grain size of ZrB_2 and ZrC below 5 μm, which is similar to the starting powder particle size. In contrast, other sintered composites as shown in Fig. 27b–d exhibit the agglomeration of coarser grain in the microstructure. More specifically, the sintered ZrB_2-ZrC composite prepared using the mixture of commercially available ZrB_2 and ZrC powders exhibits a merely heterogeneous microstructure with grain size more than 10 μm as shown in in Fig. 27d. Sinter density and mechanical properties (Vickers hardness and fracture toughness) of the ZrB_2-ZrC composite are provided in Table 11. The highest sinter density (98.5% ρ_{th}) and Vickers hardness (17.8 GPa) with moderate fracture toughness (3.8 $MPa.m^{1/2}$) were obtained from sintered ZrB_2-ZrC composite prepared using the MA-SHS powder mixture. But the highest fracture toughness (4.5 $MPa.m^{1/2}$) was obtained from ZrB_2-ZrC based on MA powder mixture. Hardness of the sintered specimens strongly depends on the sinter density, and sinter density of the compact highly depends on the particle size, surface area, defect structure, etc. of the starting powders. The composite with sinter density of 97.5% ρ_{th} shows highest fracture toughness value, which is attributed to the presence of the agglomerated coarser grain, along with large volume of pores in the sintered

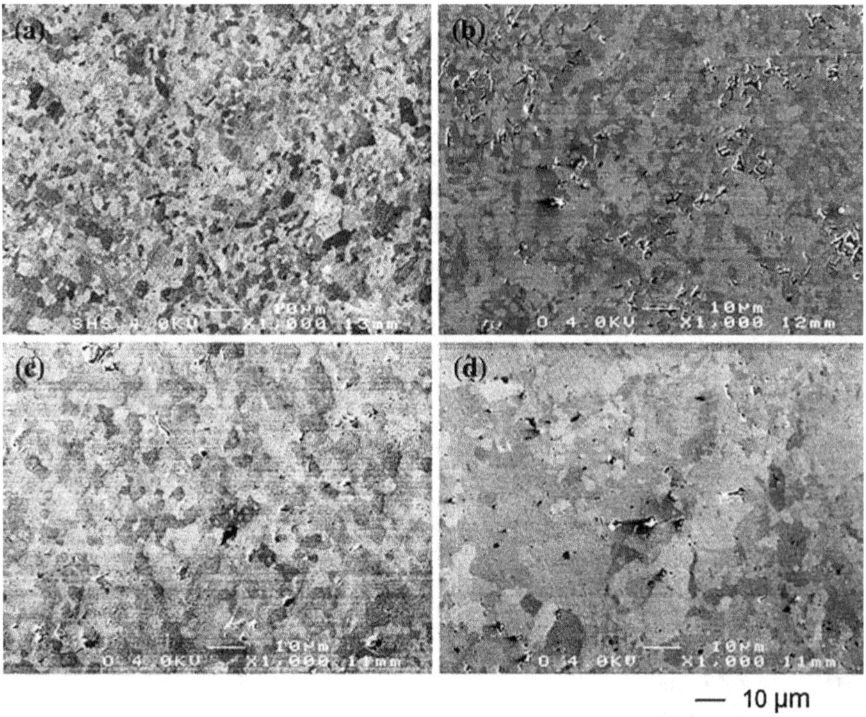

Fig. 27 BSE-SEM images of ZrB_2-ZrC composite sintered with final holding at 1800 °C for 5–10 min under vacuum using two-stage spark plasma sintering, from different mixed powders such as (**a**) MA-SHS powder, (**b**) MA powder, (**c**) mixing powder, and (**d**) commercial powder (Tsuchida and Yamamoto 2007)

microstructure. The agglomerated coarser grain may enhance the fracture toughness by crack deflection mechanism, and pores may increase the fracture toughness by crack bridging mechanisms. Figure 28 shows the BSE-SEM images of the sintered composite prepared from the MA-SHS powder mixture of Zr/B/C = 4/2/3–5/8/1. Depending on the Zr/B/C ratio, the difference in the distribution and agglomeration of ZrB_2 (gray contrast phase) and ZrC (bright contrast phase) grains can be observed. It is evident that grain growth and porosity increase with increasing the boron concentration. In fact, the sinter density of 97.5% ρ_{th} is obtained from the spark plasma-sintered composite, prepared from powder of Zr/B/C = 5/8/1 (i.e., ZrB_2/ZrC = 4/1). Similar trend also can be observed in the sintered composite by increasing the carbon content. Consequently, the mechanical properties strongly depend on the boron and carbon concentration in the sintered composite as well as the sinter density of the sample. Typically, ZrC-rich sintered composite obtained from the Zr/B/C = 4/2/3 and 3/2/2 powder mixtures revealed higher Vickers hardness and lower fracture toughness, while ZrB_2-rich compacts behave in reverse manner. The ZrB_2-rich sintered composite prepared from three powder mixtures of Zr/B/C = 3/4/1, 4/6/1, and 5/8/1 exhibits the lower Vickers hardness as well as

Table 11 Relative density and mechanical properties of ZrB_2-ZrC composites prepared from (a) MA-SHS, MA, mixing, and commercial powder mixtures and (b) MA-SHS powder mixtures of Zr/B/C = 4/2/3–5/8/1

Sample details	Vickers hardness, *HV* (GPa)	Fracture toughness, K_C (MPa m1/2)[a]	Relative density (% ρ_{th})	Reference
MA-SHS powder compact of Zr/B/C = 2/2/1	17.8	3.8	98.5	Tsuchida and Yamamoto (2007)
MA powder compact of Zr/B/C = 2/2/1	13.8	4.5	97.2	
Mixing powder compact of Zr/B/C = 2/2/1	13.7	3.8	96.9	
Commercial powder compact of ZrB2/ZrC = 1/1	16.6	3.4	97.6	
MA-SHS powder compact of Zr/B/C = 4/2/3	17.5	3.0	97.6	
MA-SHS powder compact ofZr/B/C = 3/2/2	17.4	2.9	98.8	
MA-SHS powder compact ofZr/B/C = 3/4/1	14.4	4.8	98.2	
MA-SHS powder compact ofZr/B/C = 4/6/1	14.5	5.1	98.4	
MA-SHS powder compact ofZr/B/C = 5/8/1	13.6	4.2	97.5	

[a]Estimated using the equation $K_C = 0.026(E^{1/2}P^{1/2}a/C^{3/2})$, where *P* is applied load during indentation, a is the half-length of the indent, *C* is the half crack length, and *E* is the elastic modulus of the sintered composites. (*E* = 435 GPa was used as the mean of 470 GPa for ZrB_2 and 400 GPa for ZrC) [JIS R1607 1995]

higher fracture toughness. For the comparison, commercially available monolithic ZrB_2 and ZrC powders were sintered using two-stage spark plasma sintering under the same conditions, and the sinter density of 84.7% ρ_{th} and 88.3% ρ_{th} was achieved, respectively. The sintered composite of ZrB_2-ZrC obtained from MA-SHS powder mixture of Zr/B/C = 2/2/1 showed significantly higher values of sintered density as well as mechanical properties (Tsuchida and Yamamoto 2007).

6 Entropy-Stabilized UHTCs

In spite of tremendous progress that has been made in the discipline of materials science and engineering in last few decades, there still remain technological difficulties toward design of more sophisticated and specialized class of alloys having homogeneous and stable microstructure with superior properties. Conventional alloys consist of one or two principal elements along with other minor elements. Recently Zhang et al. reported a new generation of alloy called "high-entropy alloys

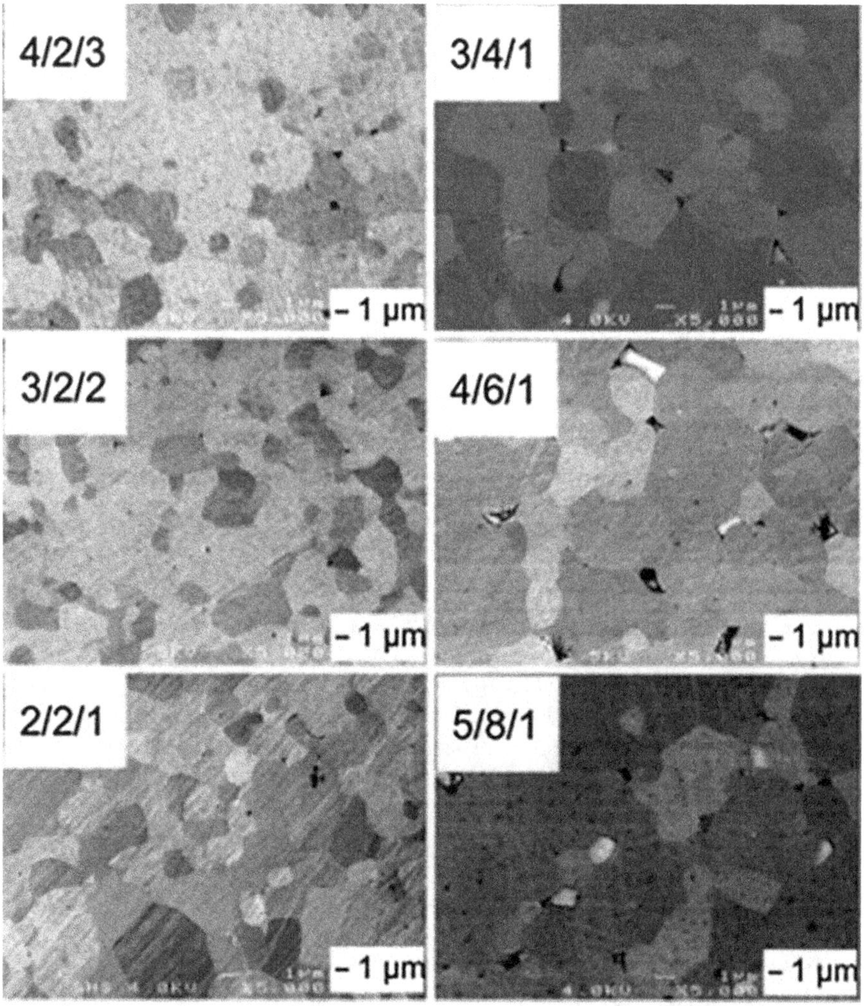

Fig. 28 BSE-SEM images of ZrB_2-ZrC composite sintered with final holding at 1800 °C for 5–10 min under vacuum using two-stage spark plasma sintering, from MA-SHS powder mixture of Zr/B/C = 4/2/3–5/8/1 (Tsuchida and Yamamoto 2007)

(HEAs)" consisting of five or more than five metallic elements. HEAs are differentiated from conventional alloys as the former has at least five principal elements, instead of two or more in the latter. HEAs or multicomponent alloys from its outset have allured lots of interest due to the formation of highly stable single-phase solid solutions from more than four principal elements with equimolar or near equimolar composition with significant configurational entropy or high entropy of mixing. The formation of nanocrystalline HEAs has made them even more interesting due to their fundamental and technological importance (Zhang et al. 2014).

Most recently, the synthesis and properties of high-entropy alloys (HEAs) have been investigated by many materials research groups (Zhang et al. 2014; Tsai and Yeh 2014; Tazuddin and Biswas 2017). The development of HEA involves mixing of at least five metallic elements of similar atomic radius and crystal structures with equimolar or near-equimolar compositions. In HEA, the configuration entropy could be significantly amplified as compared to the binary or ternary alloy to stabilize the HEA solid solution phase. Typically, mixing of five or more metallic elements with equimolar composition can be carried out to obtain optimum configurational entropy or entropy of mixing, i.e.

$$\Delta S_{\text{mix}} = R \ln N \tag{4}$$

where N defines the number of constituents and R denotes universal gas constant (Zhang et al. 2014; Tsai and Yeh 2014). A system can be defined as a thermodynamically stable only when its Gibbs free energy is minimum. However, the Gibbs free energy is the dictator of thermodynamic stability of any system under consideration at a particular pressure. The thermodynamical stability of a material is defined as,

$$G = H - TS \tag{5}$$

where G denotes the Gibbs free energy, H denotes enthalpy, T denotes temperature, and S denotes entropy of the system. Therefore, high-entropy materials with large configurational entropy can be more stable at high temperature. This key factor motivated the researchers to develop and study the phase stability of high-entropy transition metal diborides and carbides, which is a new class of ultrahigh-temperature ceramics, known as entropy-stabilized or high-entropy ultrahigh-temperature ceramics.

Extensive investigation on crystalline entropy-stabilized materials reported to date has been carried out for metallic high-entropy alloys of mostly BCC (body-centered cubic), FCC (face-centered cubic), and occasionally HCP (hexagonal closed pack) structure (Zhang et al. 2014; Tsai and Yeh 2014). Very limited studies have been reported so far on entropy-stabilized ceramics with complex non-cubic system. Very recently, Rost et al. reported that the configurational irregularities can be tailored compositionally into mixed oxide by positioning different types of metallic cations into a single sublattice, which promotes the formation of a novel and newly developed entropy-stabilized oxide ceramics containing five oxides $(Mg_{0.2}Co_{0.2}Ni_{0.2}Cu_{0.2}Zn_{0.2})O$ of FCC structure (single-phase rocksalt structure) (Rost et al. 2015). Berardan et al. (2016) reported similar investigation on the development of high-entropy oxides and its derivative, i.e., $(Mg_{0.2}Co_{0.2}Ni_{0.2}Cu_{0.2}Zn_{0.2})_{1\text{-}x\text{-}y}Ga_yA_xO$ (where A is Na, K, or Li), which have large dielectric constant and super ionic conductivities. It is worthwhile to mention that the idea of developing entropy-stabilized ceramics has originated from Rost et al., and these oxide ceramics along with its derivatives are the first crystalline entropy-stabilized ceramics reported so far. Gild et al. extended the investigation of

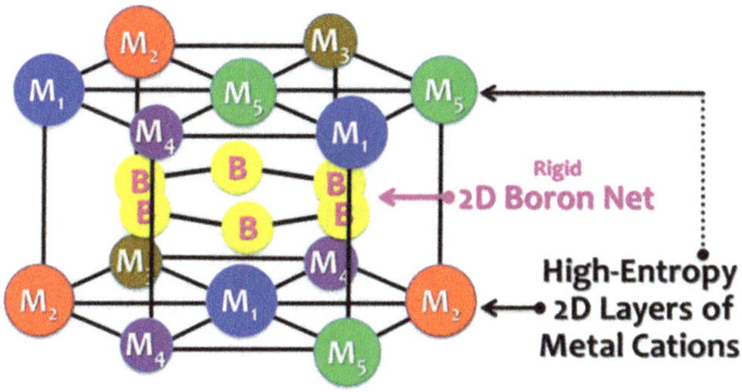

Fig. 29 Schematic depiction of atomic structure of entropy-stabilized transition metal diborides. Here, M_1, M_2, M_3, M_4, and M_5 represent various transition metals (chosen from Hf, Zr, Ti, Ta, Nb, Mo, and W). (Gild et al. 2016)

entropy-stabilized ceramics and developed the new class of high-entropy boride UHTCs by mechanical alloying of five different transition metal borides (selected from HfB_2, ZrB_2, TaB_2, NbB_2, W_2B_5 (to substitute WB_2, which is commercially unavailable), TaB_2, CrB_2, and MoB_2) (equimolar ratio) followed by spark plasma sintering. These metal borides are the first entropy-stabilized boride system reported to date, which reveals a distinctive layered HCP structure with alternating rigid 2D boron layer (consists of six boron atoms) and the high-entropy metal cations layers (consists of five different metal cations) as shown schematically in Fig. 29, which is essentially a unique class of quasi-2D entropy-stabilized ultrahigh-temperature ceramics materials.

6.1 Processing Challenge

Commercially available transition metal diboride powders of HfB_2, ZrB_2, TaB_2, NbB_2, W_2B_5 (to substitute WB_2, which is commercially unavailable), TaB_2, CrB_2, and MoB_2 were used as starting materials to fabricate seven equimolar entropy-stabilized metal diborides (ESDBs) by selecting and mixing five metal borides from the set of individual boride systems mentioned. Each ESDBs contains five different diborides of equimolar compositions. The mechanical alloying of each composition consisting of five borides was carried out via high-energy ball milling in Spex mill for 6 h using tungsten carbide milling media. The consolidation of the as-milled powders was carried out using spark plasma sintering with final holding at 2000 °C for 5 min under the mechanical pressure of 30 MPa with a heating rate of 100 °C/min. A molybdenum foil of 25 μm thickness was lined inside the die-punch combination, which prevent the reaction of specimen with the graphite at high temperature. The experimental and theoretical density along with the sinter density (% ρ_{th}) of various entropy-stabilized metal borides synthesized by Gild et al. is provided in Table 12, which shows that all the seven entropy-stabilized borides

Table 12 Comparison of relative density and Vickers hardness of as-sintered high-entropy diborides with the sintered individual diboride participated in high-entropy diboride formation except MoB_2, consolidated using spark plasma sintering with same sintering parameters (Gild et al. 2016)

Sample ID	Relative density (% ρ_{th})	Vickers hardness (GPa)
HEB #1	92.4	17.5 ± 1.2
HEB #2	92.4	19.1 ± 1.8
HEB #3	92.3	21.9 ± 1.7
HEB #4	92.2	22.5 ± 1.7
HEB #5	92.1	23.7 ± 1.7
HEB #7	92.2	19.9 ± 2.6
HfB_2	80	18.6 ± 1.3
ZrB_2	83.7	15.4 ± 1.1
TaB_2	94.8	17.5 ± 0.8
NbB_2	96.4	15.6 ± 0.9
TiB_2	91	25 ± 1.8
CrB_2	94.8	14.5 ± 0.7

could be sintered up to ~92% ρ_{th} using the abovementioned sintering conditions. To compare the sinter density and mechanical properties of the developed entropy-stabilized metal borides, the individual metal borides also have been consolidated by the spark plasma sintering technique using the same sintering parameters as shown in Table 12.

6.2 Microstructures

The cross-sectional SEM images along with the EDX mapping of seven HEBs are shown in Fig. 30, which clearly represent the homogeneous distribution of the individual borides in the entropy-stabilized metal borides. Also, uniform distribution of secondary oxide phase (Hf, Zr)O_2 with variable concentration in different specimens of HEBs can be noticed. The additional phase, i.e., $(Ti_{1.6}W_{2.4})B_4$, is observed only in HEB#6, and negligible amount of tungsten is observed in HEB#1, HEB#5, and HEB#7 specimens, which is the contamination phase coming out from WC milling media used in high-energy ball milling. EDX maps of HEB#1, HEB#3, and HEB#5 specimens exhibit the localization of Nb along with Zr and Mo clusters in the same region of the microstructure. It is quite surprising, considering the fact that NbB_2 forms uniform solid solutions with the other borides (Post et al. 1954). Typically, localization of Nb might be due to the kinetic effect, and this localized Nb can be homogenized by annealing for a long time or at a significantly high temperature below the melting point of the HEBs. The compositional homogeneity in HEBs is quite satisfactory as compared to the high-entropy alloy made up of refractory metals of BCC structure by casting, which promotes the formation of dendrites with significant compositional segregation (Gild et al. 2016).

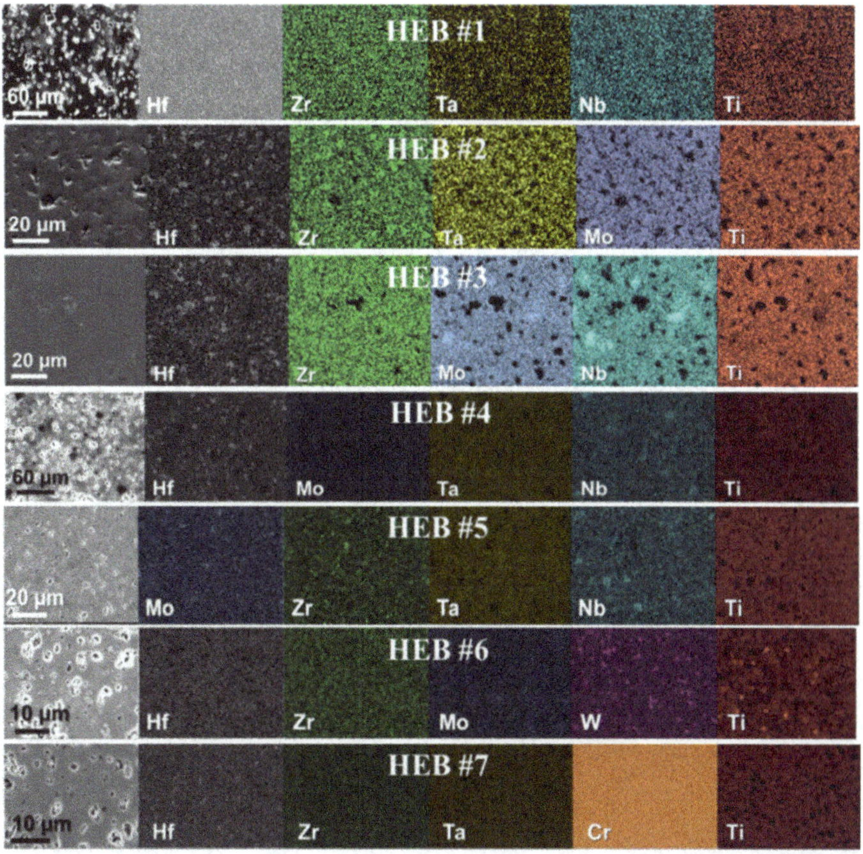

Fig. 30 Cross-sectional SEM micrographs along with SEM-EDX compositional maps of the sintered HEB (high-entropy borides) displaying homogeneous distribution of all the transition metals in the sintered microstructure that participate in the high-entropy diborides formation (Gild et al. 2016)

Fine-scale microstructural analysis has been carried out by Gild et al. using scanning transmission electron microscopy (STEM) high-angle annular dark field (HAADF) to confirm the uniformity in the HEB solid solution and most importantly the formation of two-dimensional layered structure mentioned in Fig. 29. The STEM-ABF and STEM-HAADF images (not shown) explore the atomic configuration in $[01\bar{1}0]$ zone axis of hexagonal lattice. The interplanar spacing between two (0001) planes is ~3.445 Å which is close to 3.316 Å estimated from XRD pattern. The same atomic arrangement and homogeneity could be found in other locations as well as in different specimens (image not shown). A careful image analysis explored that the standard deviation in the measurement of lattice spacing between two (0001) planes is about 0.6% of average value of lattice parameter along c or the estimated variations from STEM-HAADF and STEM-ABF images (~0.02 Å), which clearly indicates the presence of clean 2D high-entropy metal layers without any segregated metal layer in different (0001) basal planes, where these 2D high-entropy metal layers are distinctly separated by the rigid 2D boron layers. Therefore, the entropy-stabilized metal borides developed by Gild et al. can be known as quasi-two-dimensional-layered high-entropy ceramics.

Nanoscale Compositional Mapping Compositional homogeneity of HEB#2 was also confirmed by STEM-EDS mapping. The maps show the uniform distribution of all the elements (Hf, Zr, Ta, Mo, and Ti) present in the HEB#2, $(Hf_{0.2}Zr_{0.2}Ta_{0.2}Mo_{0.2}Ti_{0.2})B_2$. No segregation was observed throughout the scanned microstructure. Since the EDS maps were also captured with the electron beam parallel to zone axis of $[01\bar{1}0]$, it also confirmed that no elemental segregations are observed along the c axis in (0001) basal planes. Thus, this developed boride is indeed a quasi-2D high-entropy ceramics (Gild et al. 2016).

6.3 Properties

Gild et al. carried out measurement of hardness (see Table 12) and oxidation resistance (as shown in Fig. 31) of six entropy-stabilized diborides (HEB#1 to HEB#7) among the seven developed HEBs along with six individual diborides (HfB_2, TaB_2, ZrB_2, TiB_2, NbB_2, and CrB_2) fabricated adopting the same high-energy ball milling for 6 h using WC milling media followed by spark plasma sintering with final holding at 2000 °C for 5 min under 30 MPa pressure with heating rate of 100 °C/min. Only CrB_2 was sintered at a lower temperature (1800 °C for 5 min) due to its lower melting point (~2200 °C). MoB_2 was not studied due to its decomposition behavior at lower temperature below 1500 °C. Hardness measurements were carried out applying 200 kgf/mm^2 load with a dwell time of 15 sec. The density and hardness values are uniform throughout the region of the specimen (HEB#1-HEB#5) in all six individual diboride ceramics, but the hardness values

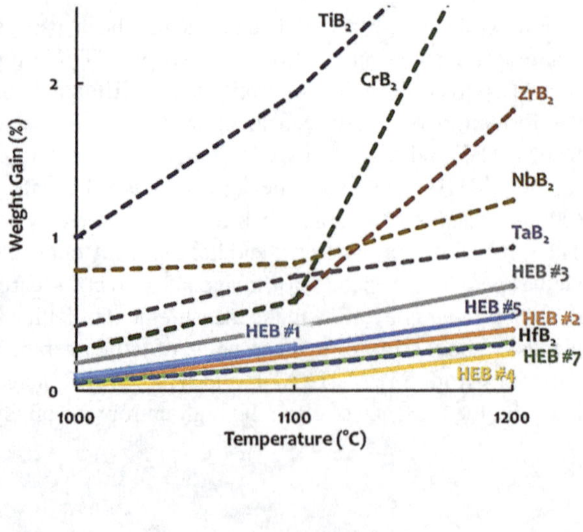

Fig. 31 Relative oxidation behavior of different entropy-stabilized metal diborides along with the individual metal diborides prepared by the same technique (SPS) with the same sintering parameters

varied in region to region (center of the specimen to edge of the specimen, HEB#6) due to density gradient and average values are provided in Table 12.

The oxidation studies were carried out in tube furnace under air atmosphere. The specimens (HEB#1 to HEB#7) were sequentially annealed for 1 h at different temperatures starting from 800 °C to 1500 °C with 100 °C interval with a heating rate of 10 °C/min. After annealing, the samples were furnace cooled with a cooling rate of ~100 °C/min and removed from the furnace, and weight was measured. The weight of the samples annealed at lower temperature below 1300 °C was measured directly. The weight of the samples annealed at higher temperature above 1300 °C was measured along with alumina crucible because at higher temperature alumina reacted with the samples. The precise weight was obtained from the samples oxidized in the range of 1000–1200°C and for the oxidizing temperature in the range of 1400–1500 °C, where weight changes were significantly large to allow to be measured accurately in crucibles. Except these two temperatures, the weight changes were typically on the same order of magnitude as the measurement errors. Due to this reason the data obtained from temperature above 1200 °C was not reported by Gild et al. (2016).

Table 12 indicates that hardness values of six as-prepared single-phase HEBs (high-entropy metal diborides) along with the individual diborides system follow the identical route. As MoB_2 decompose below 1500 °C, the hardness was not measured. Table 12 shows that the measured hardness values of HEBs are greater than the individual metal diboride system except TiB_2, but no significant increment of the hardness value is observed in entropy stabilized metal diborides due to the presence of pore (8%) and oxide impurities. However, it could be expected that high sinter density (>99% ρ_{th}) can lead to significant increase of hardness value of the HEBs than the measured hardness values (Gild et al. 2016).

Figure 31 shows the relative oxidation behavior of different entropy-stabilized metal diborides along with the individual metal diborides prepared by the same technique (SPS) with the same sintering parameters. Percentage weight gain versus oxidation resistance plot obtained for all the specimens (as mentioned earlier), during annealing at different temperatures, such as 1000 °C, 1100 °C, and 1200 °C under air atmosphere, is portrayed. HEB#1 has been taken as a reference material because it does not form volatile oxide in the temperature range of 1000–1200 °C and HEB#1 provide better performance in oxidizing environment as compared to the individual diboride system excepts HfB_2. Among all the specimens, TiB_2 exhibits less resistance to oxidation (Gild et al. 2016).

Gild et al. reported significant oxidation stability of high-entropy diborides mentioned in this section, especially HEB#1, HEB#2, HEB#3, and HEB#7, even at 1500 °C, while most of the monolithic diborides except HfB_2 oxidized significantly at 1500 °C. For example, TiB_2 sample, which is a well-known classical UHTC today, transformed into powder at 1500 °C (image not shown). Among the developed entropy-stabilized metal diborides, MoB_2 explored captivating oxidation resistance in diverse environment due to presence of volatile MoO_3 (Gild et al. 2016).

7 Case Study of ZrB_2-SiC-Based UHTCs for Hypersonic Space Vehicle Applications

In this section, let us focus on a specific application of UHTCs pertaining to thermal protection at sharp leading edges of hypersonic space vehicles using ZrB_2-SiC-based ceramics. A hypersonic vehicle flying at Mach 7 in the altitude of 30 km would result in stagnation temperature of 2200 °C, when sharp leading edges are employed, which is a vital aerodynamic design requirement for maneuverable hypersonic vehicle. ZrB_2-based ceramics are candidate materials for such extreme temperature structural applications, owing to high melting point, high hardness, enhanced oxidation, and corrosion resistance (Purwar et al. 2016). Interestingly, the superior thermal conductivity of ZrB_2-based ceramics enables heat dissipation from localized hot spots, which occur in hypersonic vehicle leading edges. Moreover, their high electrical conductivity results in easy machining of the ceramic using noncontact-type electric discharge machining (EDM), which can achieve complex 3D shapes as demanded by aerodynamic design. Figure 32 depicts the translation of concept to technology, toward development of UHTC for hypersonic application, which includes ceramic processing, performance qualifying assessments of the developed ceramics in relevant environment, and computational design of ceramic-based leading edges. UHTC processing involves sintering of the starting powders with suitable sinter-aids and sintering conditions to obtain dense ceramic compact with fine-scale microstructure. Understanding of processing-microstructure-property correlations is performed by conducting a variety of characterization including microstructure, phase stability, and mechanical and thermal properties of interest. The

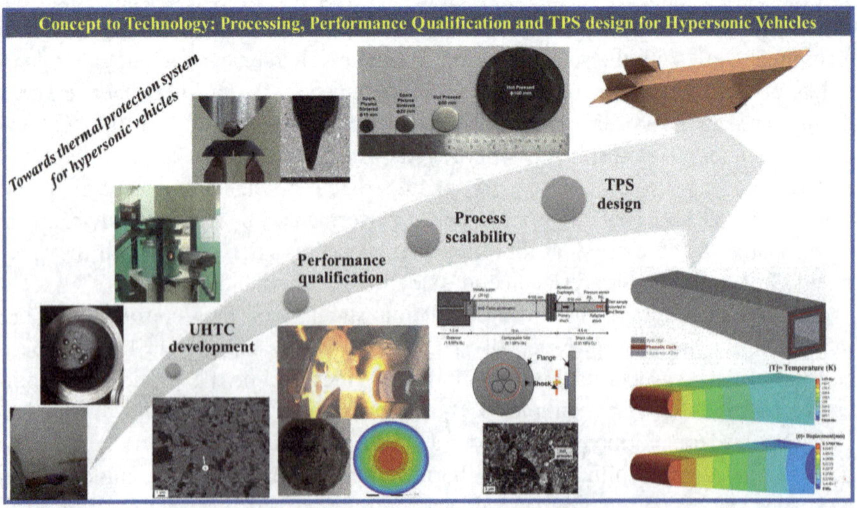

Fig. 32 Various stages in translation of concept to technology in the development of UHTCs pertaining to thermal protection at leading edges of hypersonic vehicle

sintering conditions and processing parameters are carefully tailored to obtain the desired property combinations as required for hypersonic flight mission.

As a proof of concept, the developed ceramics are subjected to rigorous testing in relevant hypersonic environment which involves testing in arc jet, shock tube facility, erosion rig, etc. At this stage, it is also required to evolve integrated framework of experimental investigation with computational studies, which will aid in enhanced understanding of certain important physics/phenomena such as thermo-oxidative stability, shock-material interaction, thermo-erosive stability, etc. thereby improving the material design process (Purwar et al. 2017). Moreover, such framework would result in reduction of the number of experiments involving parametric changes, which could significantly improve the design cycle time and reduce cost involved in ground experimentation. As an outcome of such integrated computational-experimental approach, the material combinations which sustain the simulated extreme hypersonic environment conditions are identified and chosen for the design of leading edges for hypersonic vehicles. However, such application-oriented development necessitates the demonstration of dimensional scalability to climb the technology readiness level (TRL) scale as required for any aerospace vehicles. This enables transition from typical lab-scale dimensions to near-flight vehicle-scale dimensions of the developed ceramic. But achieving large-scale dimensions with uniform densification and property distribution poses considerable challenges in optimization of sintering scheme and corresponding process parameters.

Once the process scale-up is proven, the TPS design for realistic vehicle dimension can be undertaken. Primarily, the thermal protection systems consist of UHTC-based top layer attached to the metallic airframe via an intermediate layer. The design and analysis involve optimization of various sizing parameters of the leading edge based on the coupled thermo-structural analysis, supplemented with computational fluid dynamic analysis. Predominantly, the analysis involves minimization of wall temperature, total deformation, and Von-Mises stress in the leading edge segment by proper selection of material combinations and thickness of ceramic top layer, intermediate layer, and metallic substrate (Purwar and Basu 2017), such that all values are well within the design limits involving factor of safety. At the end of the design iteration process, the detailed design for the TPS is evolved which can be implemented in the overall design of hypersonic vehicle. Thus, one can realize the technology for development of ceramic-based leading edge of hypersonic space vehicle, from fundamental science and concepts related to UHTC and hypersonic aerothermodynamics.

However, to accomplish successful hypersonic flight test, the computational design of TPS concepts alone is not sufficient, as TRL 9 has to be reached. This calls for an integrated perspective in TPS material development along with vehicle design, which is elaborated in detail in Fig. 33. As previously discussed, the translation in concept to technology results in ceramic processing, performance qualification, process scale-up, and TPS design using computational framework.

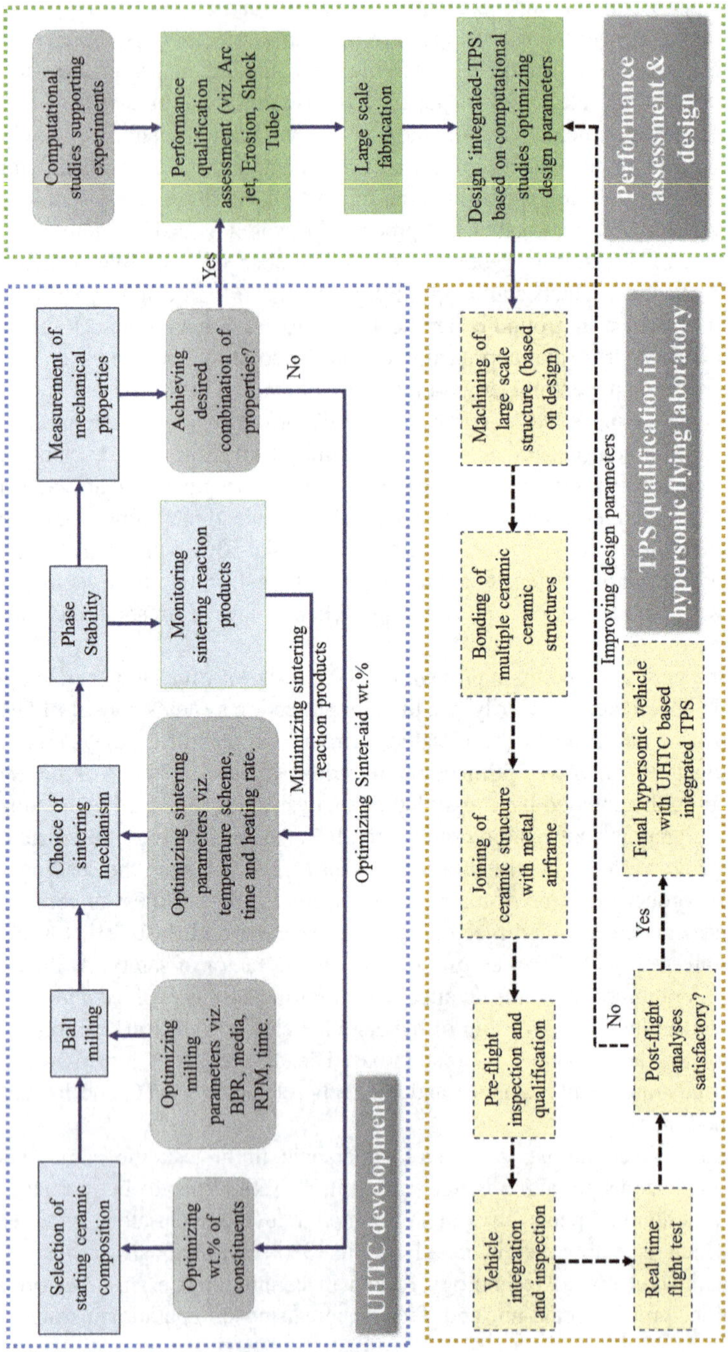

Fig. 33 Detailed flow chart depicting various stages involved in the fabrication and testing of UHTC for hypersonic vehicle leading edges

Following this, the large-scale ceramic composites obtained through dimensional scalability need to be machined as per the TPS design. Here, one can use the advantageous electrical conductivity property of ZrB_2-based ceramics for machining using EDM. The crucial part in TPS realization lies in bonding of multiple machined ceramics to form the flight vehicle structure (as even the dimensionally scaled-up ceramic may also not be produced as a single piece) and joining of the UHTC-based TPS with the rest of the metallic airframe, considering thermal compliance. Once this is successfully achieved with necessary coupon-level testing and validation, the TPS structure is ready to be part of the hypersonic vehicle airframe structure. As it is well known that there are no ground test facilities that exists in the world which can simulate all hypersonic flight conditions, viz., Mach number, Reynolds number, stagnation enthalpy, stagnation pressure, test duration, flow shear, etc. (Lu and Marren 2002), it is vital to conduct hypersonic flight test to assess the material behavior in real-time flight environment. Such hypersonic flight test for material qualifications can be found in HiFIRE series of flight experimentation, specifically in HiFIRE-5 FinEx (Bohrk et al. 2011; Dolvin 2016) as depicted in Fig. 34. Thus, the developed TPS structure is subjected to laborious preflight qualification tests and detailed inspection, followed by integration with the flight scale hypersonic vehicle and instrumenting with required sensors. The flight experimentation is conducted in the desired flight envelope of Mach number and dynamic pressure, with real-time transmission of telemetric data. Postflight test involves recovery of the flight vehicle and particularly the UHTC component to assess the material response to hypersonic flow conditions. Once the material behavior and response are found to be satisfactory, TRL 9 is declared achieved, by which time the manufacturing readiness level (MRL) would also have matured for enabling pilot line serial/batch production of TPS for hypersonic vehicle with quality control aspects included. Thus, specific application steers the development of UHTCs and enables translation from concept to technology and eventually aids in successful flight hardware realization.

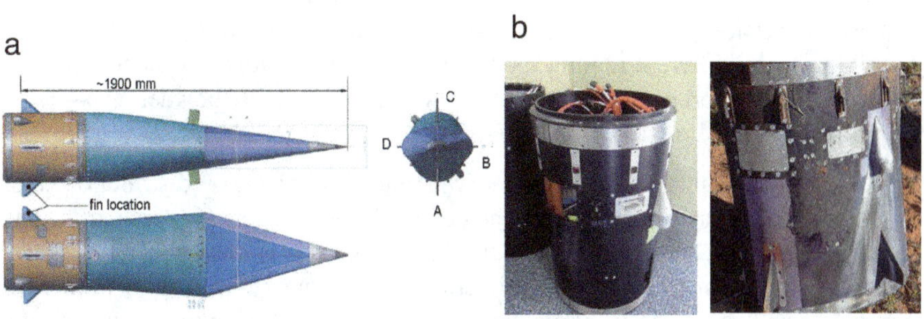

Fig. 34 Flight testing as means for qualifying UHTC for hypersonic vehicle: (**a**) HiFIRE-5 flight carrying piggyback FinEx experiment; (**b**) preflight and postflight conditions (Bohrk et al. 2011; Dolvin 2016)

8 Concluding Remarks and Future Perspectives

Extensive engineering applications concerned with the exposure to severe oxidative and corrosive environment at extremely high temperature beyond 2000 °C require the utilization of ultrahigh-temperature materials, particularly the borides, carbides, and mixed boride-carbides of transition metals due to their high melting points exceeding 3000 °C along with unique combination of physical and mechanical properties. In spite of this, there are some major issues regarding processing of UHTCs, such as lower sinterability due to their high melting point, low self-diffusion coefficient, and high vapor pressure of the constituent elements present in the UHTC system. The presence of undesirable oxygen containing impurities in as-received UHTC powders promote the formation of coarse-grained microstructure and, hence, degrade the mechanical properties (flexural strength, fracture toughness, etc.). Against this backdrop, the current chapter discussed the various powder processing techniques to synthesize ultrafine pure UHTCs powders followed by densification using reactive/nonreactive spark plasma sintering routes utilized to fabricate various bulk UHTCs with/without containing sinter-aid or other additives. In addition, the mechanical properties of developed UHTCs are critically analyzed with the issues associated with these materials. Particularly, the following can be summarized:

I. Laboratory-scale and commercial-scale synthesis routes are available to fabricate submicron to nanocrystalline ultrahigh-temperature ceramics powder based on transition metal borides, carbides, and their mixture. For example:

- Borothermic reduction and carbothermal reduction technique can be used to synthesize boride powders of Ti, Zr, and Nb.
- Mechanical alloying of elemental powders such as Ti, Zr and Nb, and amorphous boron (B) or carbon (C) using high energy ball milling can be carried out to synthesize borides and carbide powders of Ti, Zr and Nb respectively.
- Mechanical mixing followed by particle size reduction of different UHTCs borides, carbides, mixture of boride-boride, and boride-carbide powder by ball milling using different milling media and medium (acetone, methanol, etc.) by selecting appropriate milling parameters such as ball to powder mass ratio, milling time, and rotational speed.

II. As regards sintering, finer powder (with narrow particle size distribution helps in particle packing and rearrangement at the initial stage of sintering) containing less number of particle agglomerates is suitable to get highly dense sintered product. According to Herring's scaling law, sintering temperature reduces by 3–4 orders of magnitude with decrease in particle size by an order of magnitude. The powder containing agglomerates reduces the sinterability and initiates micro- and macrocracks in the sintered samples during cooling stage of sintering. In spite of this, the purity of the starting powder is the deciding factor of sinterability of the UHTCs powders. Boride powders generally contain impurity such as oxygen, nitrogen, and carbon. Oxygen has detrimental effect in densification of powder.

But presence of less amount of carbon is advantageous, because it removes the surface oxide film present in the boride powders by reduction reaction at the initial stage of sintering, especially when sintering is performed in spark plasma sintering machined due to application of pulsed DC current and uniaxial pressure at same time.

III. It is evident from the discussion in this chapter that immense research works have been carried out so far by various research group to utilize several additives (ZrO_2 in NbB_2, $MoSi_2$ in ZrB_2, $TiSi_2$ in TiB_2, etc.) to obtain fully dense and uniform fine-grain sintered microstructure of bulk UHTCs such as NbB_2, ZrB_2, TiB_2, ZrC, and TiC with superior mechanical properties using spark plasma sintering adopting multistage sintering scheme. Multistage spark plasma sintering is advantageous as compared to single-stage spark plasma sintering in order to obtain near theoretical density of the ceramics with desired microstructure, which has a significant impact on the mechanical properties. This aspect is established with multiple ceramics like NbB_2 and TiB_2. The rationale for holding the powder compact at different stages is to facilitate grain boundary and lattice diffusion in stepwise manner. In a single-stage SPS, all these diffusional mass transport processes take place concurrently. If the pores are progressively removed in different stages, then grain boundary diffusion/ lattice diffusion would take place in dense compact which would allow removal of residual porosity during holding at a higher holding temperature. The various sintering reaction with/without second-phase ($TiSi_2$, $MoSi_2$ with TiB_2, and ZrB_2, respectively) materials play a significant role toward achieving fully dense sintered UHTCs with superior mechanical properties including improvement in fracture toughness due to new product phase formation.

IV. Spark plasma sintering (using multistage heating profile) is much more advantageous as compared to the different conventional sintering technique due to lower sintering temperature, high heating rate, pulsed DC current, and uniaxial pressure application involved with it. Lower sintering temperature and higher heating rate help in achieving fine-grain sintered microstructure with negligible grain growth. Optimization of multistage sintering scheme can help in obtaining fully dense UHTCs along with uniform fine-grain microstructure with improved mechanical properties.

V. Fracture toughness of boride- and carbide-based UHTCs could be improved by the following ways:

- Incorporation of second phase which can react with the matrix and form hard new product phase like SiC (generated due to reaction between ZrC and $MoSi_2$) at the final stage of sintering and toughen the developed composite by crack deflection near the hard product phase boundary.
- By incorporating second phases in boride or carbide system which have different thermal expansion coefficient than the parent phase can toughen the developed materials by generating compressive stress near the crack tip and restrict the crack extension.
- By addition of ZrO_2 or other phase which has polymorphism involving change in volume due to phase transformation can improve the fracture by developing

microcracks in the sinter microstructure. The microcracks will generate significant compressive stress in the sintered microstructure and hence restrict the crack extension of the existing large cracks and toughen the ceramics.

VI. Future Prospects of Borides, i.e., High-Entropy UHTCs

Future generation UHTCs for the extensive applications in spacecraft vehicles or reentry vehicles require considerable enhancement in material properties especially high-temperature structural stability and oxidation resistance leading to increased reliability, safety, and intellectual design to allow the developed materials discussed in this chapter to meet future demands. UHTCs based on metal borides containing SiC are potential candidate materials for sharp leading edges application on reentry vehicle or hypersonic vehicle. The developmental cycle for application-specific UHTC development is elucidated, which brings out various critical stages in material design for hypersonic space applications. Emphasis on performance qualification assessments with integrated computational-experimental approach and process scalability signifies the need understanding of interesting physics/phenomena and maturation of technology readiness level in order to translate the concepts to technology. Moreover, the need for flight testing as a means to qualify materials is demonstrated, which would enable successful realization of long-range hypersonic vehicles.

References

Abderrazak H, Schoenstein F, Abdellaoui M, Jouini N (2011) Spark plasma sintering consolidation of nanostructured TiC prepared by mechanical alloying. Int J Refract Met Hard Mater 29:170–176

Akin MH, Sahin FC, Yucel O, Goller G, Goto T (2009) Microstructure and densification of ZrB_2–SiC composites prepared by spark plasma sintering. J Eur Ceram Soc 29:2379–2385

Andrievskii RA, Lanin AG, Rymashevskii GA (1974) Strength of refractory compounds. Metallurgiya, Moscow [in Russian]

Antis GR, Chantikul P, Lawn BR, Marshall DB (1981) A critical evaluation of indentation techniques for measuring fracture toughness: I, direct crack measurements. J Am Ceram Soc 64:533–538

Baik S, Becher PF (1987) Effect of oxygen contamination on densification of TiB_2. J Am Ceram Soc 70:527–530

Balci Ö, Ağaoğulları D, Muhaffel F, Öveçoğlu ML, Çimenoğlu H, Duman I (2016a) Effect of sintering techniques on the microstructure and mechanical properties of niobium borides. J Eur Ceram Soc 36:3113–3123

Balci O, Ağaoğulları D, Öveçoğlu ML, Duman I (2016b) Synthesis of niobium borides by powder metallurgy methods using Nb_2O_5, B_2O_3 and Mg blends. Trans Nonferrous Metals Soc China 26:747–758

Basu B (2005) Toughening of yttria-stabilized tetragonal zirconia ceramics. Int Mater Rev 50:239–256

Basu B, Balani K (2011) Advanced structural ceramics. Wiley, USA, American Ceramic Society, Hoboken

Basu B, Vleugels J, Biest VDO (2002) Development of ZrO_2-ZrB_2 composites. J Alloys Compd 334:200–204

Baumgartner HR, Steiger RA (1984) Sintering and properties of titanium diboride made from powder synthesized in a plasma-arc heater. J Am Ceram Soc 67:207–212

Berardan D, Franger S, Dragoe D, Meena AK, Dragoe N (2016) Colossal dielectric constant in high entropy oxides. Phys Status Solidi RRL 10:328–333

Bhaumik SK, Divakar C, Singh AK, Upadhyaya GS (2000) Synthesis and sintering of TiB_2 and TiB_2–TiC composite under high pressure. Mater Sci Eng A 279:275–281

Bohrk H, Lohle S, Fuchs U, Elsaber H, Weihs H (2011) FinEx-Fin Experiment On HIFIRE-5, 7th Symposium on Aerothermodynamics for Space Vehicles, Brugge

Chamberlain AL, Fahrenholtz WG, Hilmas GE (2006a) Low-temperature densification of zirconium diboride ceramics by reactive hot pressing. J Am Ceram Soc 89:3638–3645

Chamberlain AL, Fahrenholtz WG, Hilmas GE (2006b) Pressureless sintering of zirconium diboride. J Am Ceram Soc 89:450–456

Chen L, Gu Y, Qian Y, Shi L, Yang Z, Ma J (2004) A facile one-step route to nanocrystalline TiB2 powders. Mater Res Bull 39:609–613

Cheng L, Xie Z, Liu G, Liu W, Xue W (2012) Densification and mechanical properties of TiC by SPS-effects of holding time, sintering temperature and pressure condition. J Eur Ceram Soc 32:3399–3406

Chernysheva OF, Sinel'nikova VS, Kosolapova TY (1979) Structure and properties of alloys of the system TiC–B. Inorg Mater (USSR) 15:614–617

Cripps ACF (2004) Nanoindentation, 2nd edn. Springer, New York

Demirskyi D, Sakka Y, Vasylkiv O (2015) High-temperature reactive spark plasma consolidation of TiB -NbC ceramic composites. Ceram Int 41:10828–10834

Demirskyi D, Vasylkiv O (2017) Materials flexural strength behavior of a ZrB_2-TaB_2 composite consolidated by non-reactive spark plasma sintering at 2300°C. Int J Refract Met Hard Mater 66:31–35

Demirskyi D, Nishimura T, Sakka Y, Vasylkiv O (2016) High-strength TiB_2-TaC ceramic composites prepared using reactive spark plasma consolidation. Ceram Int 42:1298–1306

Dewan MAR, Zhang G, Ostrovski O (2009) Carbothermal reduction of titania in different gas atmospheres. Metall Mater Trans B Process Metall Mater Process Sci 40B:62–69

Dolvin D (2016) High speed flight research insight briefing, USAF AFOSR Industry Program Review, 2016-0746,0747 & 0748.

El-Eskandarany MS (2000) Structure and properties of nanocrystalline TiC full-density bulk alloy consolidated from mechanically reacted powders. J Alloys Compd 305:225–238

Fahrenholtz WG (2003) Reactive hot pressing of Al_2O_3–Ni composites. J Mater Sci 38:3073

Fahrenholtz WG, Binner J, Zou J (2016) Synthesis of ultra-refractory transition metal diboride compounds. J Mater Res 31:2757–2772

Gild J, Zhang Y, Harrington T, Jiang S, Hu T, Quinn MC, Mellor WM, Zhou N et al (2016) High-entropy metal Diborides: a new class of high-entropy materials and a new type of ultrahigh temperature ceramics. Sci Rep 6:2–11

Gu Y, Qian Y, Chen L, Zhou FA (2003) Mild solvothermal route to nanocrystalline titanium diboride. J Alloys Compd 352:325–327

Guo S, Hu C, Kagawa Y (2011) Mechanochemical processing of nanocrystalline zirconium diboride powder. J Am Ceram Soc 94:3643

Gupta N, Mukhopadhyay A, Pavani K, Basu B (2012) Spark plasma sintering of novel ZrB_2-SiC-$TiSi_2$ composites with better mechanical properties. Mater Sci Eng A 534:111–118

Haga EM, Scott WD (1988) Sintering and mechanical properties of ZrC-ZrO_2 composites. J Mater Sci 23:2865–2870

İpekçi M, Acar S, Elmadağlı M, Hennicke J, Balcı O, Somer O (2017) Production of TiB_2 by SHS and HCl leaching at different temperatures: characterization and investigation of sintering behavior by SPS. Ceram Int 43(2):2039–2045

Jeng YL, Lavernia EJ (1994) Processing of molybdenum disilicide. J Mater Sci 29:2557–2571

Jung HJ, Sohn Y, Sung HG, Hyun HS, Shin WG (2015) Physicochemical properties of ball milled boron particles: dry versus wet ball milling process. Powder Technol 269:548

Kamiya A, Nakano K, Okuda H (1989) Fabrication and properties of SiC whisker reinforced NbC composites. J Cerma Soc Jpn 97:947–953

Kang SH, Kim DJ, Kang ES, Baek SS (2001) Pressureless sintering and properties of titanium diboride ceramics containing chromium and iron. J Am Ceram Soc 84:893–895

Karthiselva S, Bakshi S (2016) Reactive spark plasma sintering and mechanical properties of zirconium Diboride–titanium Diboride ultrahigh temperature ceramic solid solutions. Technologies 4:30

Königshofer R, Fürnsinn S, Steinkellner P, Lengauer W, Haas R, Rabitsch K, Scheerer M (2005) Solid-state properties of hot-pressed TiB2ceramics. Int J Refract Met Hard Mater 23:350–357

Lawn BR (1993) Fracture of brittle solids. Cambridge University press, New York, pp 263–276

Lawn BR, Fuller ER (1975) Equilibrium penny-like cracks in indentation fracture. J Mater Sci 10:2016–2024

Licheri RO, Musa C, Cao G (2008) Combination of SHS and SPS techniques for fabrication of fully dense ZrB_2-ZrC-SiC composites. Mater Lett 62:432–435

Licheri R, Musa C, Orrù R, Cao G (2015) Influence of the heating rate on the in situ synthesis and consolidation of ZrB_2 by reactive spark plasma sintering. J Eur Ceram Soc 35:1129–1137

Lu FK, Marren D (2002) Principles of hypersonic test facility development. Prog Astronaut Aeronaut 198:17–28

Maeda H, Yoshikawa Y, Kusakabe K, Morooka S (1994) Synthesis of ultrafine NbB_2 powder by rapid carbothermal reduction in a vertical reactor. J Alloys Compd 215:127–134

Maity TN, Biswas K, Basu B (2018) Critical role of ZrO_2 on densification and microstructure development in spark plasma sintered NbB_2. Acta Mater. 152:215–228

Matsudaira T, Itoh H, Naka S, Hamamoto H (1989) Synthesis of niobium boride powder by solid state reaction between niobium and amorphous boron. J Less-Common Met 155(2):207–214

McKenna PM (1936) Tantalum carbide its relation to other hard refractory compounds. Ind Eng Chem 28(7):767–772

Monteverde F, Bellosi A, Guicciardi S (2002) Processing and properties of zirconium diboride based composites. J Eur Ceram Soc 22:279–288

Monteverde F, Guicciardi S, Bellosi A (2003) Advances in microstructure and mechanical properties of zirconium diboride based ceramics. Mater Sci Eng A 346:31–319

Mukhopadhyay A, Raju GB, Basu B, Suri AK (2009) Correlation between phase evolution, mechanical properties and instrumented indentation response of TiB_2-based ceramics. J Eur Ceram Soc 29:505–516

Mukhopadhyay A, Venkateswaran T, Basu B (2013) Spark plasma sintering may lead to phase instability and inferior mechanical properties: a case study with TiB_2. Scr Mater 69:159–164

Muraoka Y, Yoshinaka M, Hirota K, Yamaguchi O (1996) Hot isostatic pressing of TiB_2-ZrO_2 (2 Mol% Y_2O_3) composite powders. Mater Res Bull 31:787–792

Murthy TSRC, Sonber JK, Subramanian C, Fotedar RK, MR G, Suri AK (2009) Effect of CrB_2 addition on densification, properties and oxidation resistance of TiB_2. Int J Refract Met Hard Mater 27:976–984

Oliver WC, Pharr GM (1992) An improved technique for determining hardness and elastic modulus using load and displacement sensing indentation experiments. J Mater Res 7:1564–1583

Peshev P, Leyarovska L, Bliznakov G (1968) On the borothermic preparation of some vanadium, niobium and tantalum borides. J Less-Common Met 15:260–267

Pierson HO (1996) Handbook of refractory carbides and nitrides. William Andrew Publishing/ Noyes, Westwood, p 68

Ping L, Shijie D, Anzhuo Y, Shixuan S, Zhixiong X, Zhong Z, Wei Y (2016) ZrB_2-TiB_2 nanocomposite powder prepared by mechanical alloying. Rare Met Mater Eng 45:1381–1385

Platt P, Frankel P, Gass M, Howells R, Preuss M (2014) Finite element analysis of the tetragonal to monoclinic phase transformation during oxidation of zirconium alloys. J Nucl Mater 454:290–297

Post B, Glaser FW, Moskowitz D (1954) Transition metal diborides. Acta Metall 2:20–25

Purwar A, Basu (2017) Thermo-structural design of ZrB_2-SiC based thermal protection system for hypersonic space vehicles. J Am Ceram Soc 100(4):1618–1633

Purwar A, Mukherjee R, Ravikumar K, Ariharan S, Gopinath NK, Basu B (2016) Development of ZrB_2-SiC-Ti by multi stage spark plasma sintering at 1600°C. J Cerma Soc Jpn 124(4):393–402

Purwar A, Venkateshwaran T, Basu B (2017) Experimental and computational analysis of thermo-oxidative-structural stability of ZrB_2-SiC-Ti composite during arc-jet testing. J Am Ceram Soc 100(10):4860–4873

Raju GB, Basu B (2007) Densification, sintering reactions, and properties of titanium diboride with titanium disilicide as a sintering aid. J Am Ceram Soc 90:3415–3423

Raju GB, Biswas K, Mukhopadhyay A, Basu B (2009) Densification and high temperature mechanical properties of hot pressed TiB_2-(0–10wt. %) $MoSi_2$ composites. Scr Mater 61:674–677

Ran S, Biest DVO, Vleugels J (2010) ZrB_2 powders synthesis by borothermal reduction. J Am Ceram Soc 93:1586–1590

Reddy KM, Kumar N, Basu B (2010a) Inhibition of grain growth during the final stage of multi-stage spark plasma sintering of oxide ceramics. Scr Mater 63:585–588

Reddy KM, Kumar N, Basu B (2010b) Innovative multi-stage spark plasma sintering to obtain strong and tough ultrafine-grained ceramics. Scr Mater 62:435–438

Reddy KM, Mukhopadhyay A, Basu B (2010c) Microstructure-mechanical-tribological property correlation of multistage spark plasma sintered tetragonal ZrO_2. J Eur Ceram Soc 30:3363–3375

Rost CM, Sachet E, Borman T, Moballegh A, Dickey EC, Hou D, Jones JL, Curtarolo S, Maria JP (2015) Entropy-stabilized oxides. Nature Comm 6:8485

Sairam K, Sonber JK, Murthy TSRC, Subramanian C, Fotedar RK, Hubli RC (2014) Reaction spark plasma sintering of niobium diboride. Int J Refract Met Hard Mater 43:259–262

Sciti D, Monteverde F, Guicciardi S, Pezzotti G, Bellosi A (2006) Microstructure and mechanical properties of ZrB_2-$MoSi_2$ ceramic composites produced by different sintering techniques. Mater Sci Eng A 434:303–309

Sciti D, Guicciardi D, Nygren M (2008a) Spark plasma sintering and mechanical behaviour of ZrC-based composites. Scr Mater 59:638–641

Sciti D, Silvestroni L, Celotti G, Melandri C, Guicciardi S (2008b) Sintering and mechanical properties of ZrB_2-$TaSi_2$ and HfB_2-$TaSi_2$ ceramic composites. J Am Ceram Soc 91:3285–3291

Shen Z, Nygren M (2005) Microstructural prototyping of ceramics by kinetic engineering: applications of spark plasma sintering. Chem Rec 5:173–184

Solvas EZ, Jayaseelan DD, Lin HT, Brown P, Lee WE (2013) Mechanical properties of ZrB_2 and HfB_2 based ultra-high temperature ceramics fabricated by spark plasma sintering. J Eur Ceram Soc 33:1373–1386

Subramanian C, Murthy TSRC, Suri AK (2007) Synthesis and consolidation of titanium diboride. Int J Refract Met Hard Mater 25:345–350

Sulima I, Jaworska L, Pałka P, Hyjek P, Kurtyka P (2016) Influence of sintering temperature and CrB_2 addition on properties of titanium diboride produced by spark plasma sintering. Polish Soc Compos Mater 16:30–36

Sun SK, Zhang GJ, Wu WW, Liu JX, Suzuki T, Sakka Y (2013) Reactive spark plasma sintering of ZrC and HfC ceramics with fine microstructures. Scr Mater 69:139–142

Tanaka K (1987) Elastic/plastic indentation hardness and indentation fracture toughness: the inclusion core model. J Mater Sci 22:1501–1508

Tazuddin GNP, Biswas K (2017) In the quest of multicomponent multi-principle high entropy alloys. J Alloys Compd 697:434–442

Telle R, Meyer S, Petzow G, Franz ED (1988) Sintering behaviour and phase reactions of TiB_2 with ZrO_2 additives. Mater Sci Eng A 105/106:125–129

Tsai MH, Yeh JW (2014) High-entropy alloys: a critical review. Math Res Lett 2:107–123

Murthy TSR, Basu B, Balasubramaniam R, Suri AK, Subramanian C, Fotedar RK (2006) Processing and properties of TiB_2 with $MoSi_2$ sinter-additive: a first report. J Am Ceram Soc 89:131–138

Tsuchida T, Yamamoto S (2007) Spark plasma sintering of ZrB_2-ZrC powder mixtures synthesized by MA-SHS in air. J Mater Sci 42:772–778

Vasylkiv O, Borodianska H, Badica P, Grasso S, Sakka Y, Tok A, Su LT, Bosman M, Ma J (2012) High hardness B_aC_b-(B_xO_y/BN) composites with 3D mesh-like fine grain-boundary structure by reactive spark plasma sintering. J Nanosci Nanotechnol 12:959–965

Venkateswaran BB, Raju GB, Kim DY (2006) Densification and properties of transition metal borides-based cermets via spark plasma sintering. J Eur Ceram Soc 26:2431–2440

Wang W, Fu Z, Wang H, Yuan R (2002) Influence of hot pressing sintering temperature and time on microstructure and mechanical properties of TiB_2 ceramics. J Eur Ceram Soc 22:1045–1049

Wen G, Li SB, Zhang BS, Guo ZX (2001) Reaction synthesis of TiB_2–TiC composites with enhanced toughness. Acta Mater 49:1463–1470

Woydt M, Mohrbacher H (2015) The use of niobium carbide (NbC) as cutting tools and for wear resistant tribosystems. Int J Refract Met Hard Mater 49:212–218

Wu WW, Zhang GJ, Sakka Y (2013) Nanocrystalline ZrB_2 powders prepared by mechanical alloying. J Asian Ceramic Soc 1:304–307

Yu G, Tkachenko SS, Ordan'yan DZ, Yurchenko VK, Yulyugin GA, Bovkun VI, Unrod (1979) Strength and antifriction properties of alloys of the systems MIVC–MIVB2 over a wide range of component concentrations. Inorg Mater (USSR) 15:704–708

Zhang ZH, Shen XB, Wang FC, Lee SK, Wang L (2010) Densification behavior and mechanical properties of the spark plasma sintered monolithic TiB_2 ceramics. Mater Sci Eng A 527:5947–5951

Zhang ZH, Shen XB, Wang FC, Lee SK, Fan QB, Cao MS (2012) Low-temperature densification of TiB_2 ceramic by the spark plasma sintering process with Ti as a sintering aid. Scr Mater 66:167–170

Zhang Y, Zuo TT, Tang Z, Gao MC, Dahmen KA, Liaw PK, Lu ZP (2014) Microstructures and properties of high-entropy alloys. Prog Mater Sci 61:1–93

Zhao H, He Y, Jin Z (1995) Preparation of zirconium boride powder. J Am Ceram Soc 78:2534–2536

Zhao Y, Wang LJ, Zhang GJ, Jiang W, Chen LD (2008) Effect of holding time and pressure on properties of ZrB_2–SiC composite fabricated by the spark plasma sintering reactive synthesis method. Int J Refract Met Hard Mater 27:177–180

Spark Plasma Sintering of Diamond- and Nanodiamond-Metal Composites

Dina V. Dudina, Boris B. Bokhonov, Arina V. Ukhina, Vyacheslav I. Mali, and Alexander G. Anisimov

1 Introduction

Diamond particles are very attractive fillers in the design of composites due to a high hardness, a high thermal conductivity, and a low coefficient of thermal expansion of diamond. Micrometer-sized synthetic diamonds have become available on the market in large quantities (Choudhary and Bellare 2000). In composites, diamond particles with sizes from several tens to several hundreds of micrometers play the role of fillers and are used to impart high conductivity or abrasive properties/high cutting performance. Nanodiamonds present a promising reinforcement material for metal matrix composites (Blum and Molian 2009; Nunes et al. 2010; Popov 2013). Furthermore, nanodiamonds can be used to design porous graphitic materials when

D. V. Dudina (✉)
Lavrentyev Institute of Hydrodynamics, Siberian Branch of the Russian Academy of Sciences (LIH SB RAS), Novosibirsk, Russia

Novosibirsk State Technical University, Novosibirsk, Russia

Institute of Solid State Chemistry and Mechanochemistry, Siberian Branch of the Russian Academy of Sciences (ISSCM SB RAS), Novosibirsk, Russia

B. B. Bokhonov
Institute of Solid State Chemistry and Mechanochemistry, Siberian Branch of the Russian Academy of Sciences (ISSCM SB RAS), Novosibirsk, Russia

Novosibirsk State University, Novosibirsk, Russia

A. V. Ukhina
Institute of Solid State Chemistry and Mechanochemistry, Siberian Branch of the Russian Academy of Sciences (ISSCM SB RAS), Novosibirsk, Russia

V. I. Mali · A. G. Anisimov
Lavrentyev Institute of Hydrodynamics, Siberian Branch of the Russian Academy of Sciences (LIH SB RAS), Novosibirsk, Russia

© Springer Nature Switzerland AG 2019

P. Cavaliere (ed.), *Spark Plasma Sintering of Materials*,
https://doi.org/10.1007/978-3-030-05327-7_14

undergoing consolidation accompanied by graphitization without a catalyst or in the presence of a metal catalyst (Ukhina et al. 2015, 2017b).

Diamond-metal composites with micrometer-sized diamond particles can be produced by different methods, such as electrodeposition (Cho et al. 2016), infiltration (Abyzov et al. 2011), conventional sintering in vacuum (Sidorenko et al. 2013; Galashov et al. 2016a), hot pressing (Schubert et al. 2008b; Lee et al. 2011; Tan et al. 2013a, b), hot isostatic pressing (Galashov et al. 2016b), compaction of the products of self-propagating high-temperature synthesis conducted in mixtures composed of powders reacting via an exothermic reaction and forming a binder and diamond particles as an inert additive (Padyukov and Levashov 1993), electric pulse sintering using a pulse from a capacitor discharge (Tillmann et al. 2009, 2010), pulse plasma sintering (Schubert et al. 2008a; Ciupiński et al. 2009, 2017), hot pressing with direct resistance heating (Schubert et al. 2008b), and spark plasma sintering (SPS) (Chu et al. 2009, 2010a, b; Schmidt et al. 2010; Gao et al. 2010; Mizuuchi et al. 2011; Tan et al. 2016; Bokhonov et al. 2015; Che et al. 2015; Ukhina et al. 2017a, b). Nanodiamond-metal mixtures have been cold sprayed (Woo et al. 2013) and detonation sprayed (Dudina 2017) to produce composite coatings. For obtaining bulk materials from nanodiamond-metal mixtures, conventional sintering (Kaftelen and Öveçoğlu 2011), laser sintering (Blum and Molian 2009), hot pressing (Reinert et al. 2015), hot rolling (Livramento et al. 2008), and SPS (Hanada et al. 2007; Nunes et al. 2010, 2011; Zhang et al. 2017) have been attempted.

During consolidation of diamond- and nanodiamond-containing composites, one or several goals are to be reached depending on the target property of the sintered material: (1) preserving the metastable diamond phase in the bulk material; (2) partially graphitizing the diamond phase; and (3) making diamond react with other components of the composite material to form carbide layers or separate carbide particles.

The SPS method has recently made significant advancements in materials science and materials processing thanks to its capability of fast and efficient densification of a great variety of materials and the suitability of the equipment for conducting solid-state syntheses and syntheses, in which liquid phases can form within limited volumes – locally at the contacts between particles (Orrù et al. 2009; Hulbert et al. 2009; Garay 2010; Munir et al. 2011; Dudina and Mukherjee 2013). In this chapter, studies on SPS of micrometer-sized diamond-metal composites will be reviewed for materials developed for heat sink and cutting tool applications. The potential of SPS for fabrication of metal matrix composites with nanodiamond-derived reinforcing phases will be shown. The analysis of the available literature data as well as the authors' own research experience will be presented to show the possibilities offered by SPS for the production and design of diamond- and nanodiamond-metal composites. Future perspectives of using SPS for consolidation, synthesis, and design of these materials will be outlined.

2 SPS of Diamond-Metal Composites for Heat Sink Applications

In composites for heat sink applications, a high thermal conductivity and a reduced coefficient of thermal expansion relative to the binder metal in the monolithic form are the target properties. In order to increase the wettability of the surface of diamond by certain metals and reduce the interface-associated porosity in sintered composites, coatings are deposited on diamond particles (Tan et al. 2013b; Bai et al. 2014; Ukhina et al. 2017a), or alloying additives are introduced into the metal matrices (Sidorenko et al. 2013; Rape et al. 2013). These additives are usually carbide-forming elements, which ultimately react with diamond to form a carbide layer or carbide islands on the surface of diamond. An interesting approach to the formation of coatings on diamond in situ during sintering was used by Zhang et al. (2014). The presence of Si as an alloying element in an aluminum matrix led to the formation of a eutectic melt during SPS of diamond-Al-Si mixtures and precipitation of the eutectic Si on the diamond surface upon cooling, which is important for reducing a mismatch between the coefficients of thermal expansion of diamond and aluminum in the sintered composites. Approaches to improving the quality of the diamond filler/metal interfaces described above can be used during processing of composites by different sintering methods, including SPS, which is often selected as a fast method to densify the diamond-metal powder mixtures. Several examples of the preparation of diamond-metal composites are discussed below.

Rape et al. spark plasma sintered diamond particles with powders of CuCrZr and CuAgZr alloys and obtained composites having thermal conductivity values exceeding that of pure copper, which was not possible to achieve in diamond-unalloyed Cu composites (Rape et al. 2013). An increased thermal conductivity showed that interfaces between the alloy and the particles of the diamond filler were of better quality as compared with those observed in the diamond-unalloyed copper composites.

SPS was successfully used for the preparation of diamond-copper composites with diamonds coated with Cr prior to sintering (Ren et al. 2011): the 70 vol.% diamond-Cu composition showed a thermal conductivity of 657 $Wm^{-1}K^{-1}$.

A comparison of the structure and properties of spark plasma sintered diamond-Cu composites containing diamonds with coatings of different compositions has been made in a study conducted by our group (Ukhina et al. 2017a). Composites with uncoated diamonds were not dense enough to show high thermal conductivity (Table 1). The larger were the diamond crystals taken for manufacturing the composites, the lower was the relative density reached. Figure 1 shows the fracture surfaces of composites containing diamonds, on which tungsten and nickel were deposited by rotary chemical vapor deposition. The mixtures were subjected to SPS at 920 °C and 40 MPa for 3 min.

The X-ray diffraction phase analysis showed that a W-containing coating was composed of WC, W_2C, and W. The structures of the filler particle/copper interface in composites shown in Fig. 1a, b differ from each other. The W-containing coating

Table 1 Characteristics of 50 vol.% diamond-Cu composites spark plasma sintered at 920 °C and 40 MPa for 3 min and hot pressed at the same temperature for 10 min. The applied pressure was 40 MPa in both SPS and hot pressing (Ukhina et al. 2017a)

Consolidation method	Diamond particle size, μm	Composition of the coating on diamond crystals	Relative density, %	Thermal conductivity, $Wm^{-1}K^{-1}$
SPS	50	Uncoated	87	150
SPS	100	Uncoated	83	150
SPS	200	Uncoated	77	138
SPS	200	WC, W_2C, W	77	245
SPS	50	WC, W_2C, W	91	340
Hot pressing	50	WC, W_2C, W	85	150
SPS	100	Ni+small amounts of graphite	83	98

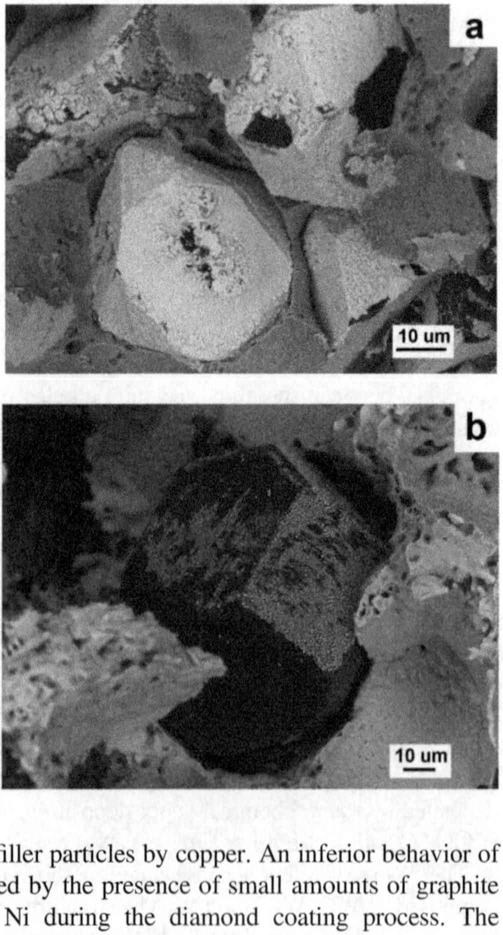

Fig. 1 Fracture surface of 50 vol.% diamond-Cu composites after SPS at 920 °C and 40 MPa for 3 min: diamond crystals having (**a**) WC-W_2C-W coating, (**b**) Ni-based coating. (Ukhina et al. 2017a)

did improve the wettability of the filler particles by copper. An inferior behavior of the Ni-based coatings was explained by the presence of small amounts of graphite deposited together with metallic Ni during the diamond coating process. The difference in the behavior of the coatings on diamond particles toward copper was reflected in the relative density and thermal conductivity of the composites (Table 1), with composites obtained from W-coated diamonds showing a higher relative density and a higher thermal conductivity. As can be inferred from comparing the relative density and thermal conductivity values of compacts produced by SPS and hot pressing (Table 1), hot pressing with a short holding time was less efficient than SPS.

Schubert et al. suggested that morphological features of the carbide phases forming on the diamond surface in the presence of carbide-forming elements should depend on the consolidation method of diamond-metal mixtures (Schubert et al. 2008b). In their opinion, direct and, therefore, rapid heating of diamond-metal compacts by electric current during hot pressing should lead to the formation of numerous nuclei of carbides at the diamond/metal interface and coatings with a smoother and finer structure in comparison with those forming during slow heating. Ciupiński et al. conducted a detailed analysis of thermal conductivity of compacts produced by different consolidation methods and defined the fabrication technique as one of the most important factors determining the formation and quality of the diamond/metal interface (Ciupiński et al. 2017). Experimentally, properties of composites obtained by SPS and vacuum hot pressing were compared by Tan et al. (2016). It was found that both SPS and vacuum hot pressing allow obtaining compacts with a high relative density. Vacuum hot pressing was conducted at 650 °C, while SPS was performed at 540–560 °C. Taking into account a difference in the measured and real temperatures of the sample known in the SPS practice, with real temperatures being about 100 °C higher, the real temperatures of the samples in these two methods could be assumed to be close to each other. However, the thermal conductivity of composites obtained by vacuum hot pressing was higher than that of composites processed by SPS, although the relative densities of the samples had close values. Indeed, a 40 vol.% diamond-Al composite spark plasma sintered at 560 °C had a relative density of 96.7%, while a hot-pressed sample of the same composition was 97.0% dense. The thermal conductivity of the spark plasma sintered material was only 330 $Wm^{-1}K^{-1}$, which is much lower than the value exhibited by the hot pressed material – 475 $Wm^{-1}K^{-1}$. It should be noted that the holding time was only 5 min in the SPS processing and 90 min in vacuum hot pressing. Interestingly, spark plasma sintered compacts showed a higher content of the Al_4C_3 phase forming as a result of a chemical reaction of aluminum with diamond. Tan et al. emphasized the importance of distinguishing between diffusion-bonded and reaction-bonded diamond/metal interfaces, the former being preferable for establishing low thermal resistances and easier to achieve by conventional hot pressing than by SPS. An inferior thermal conductivity of composites produced by SPS was explained by the temperature gradients developing at both micro- and macroscales during the process. Diamond particles located near high-temperature regions forming in the material under passing electric current (regions of necks forming between aluminum particles) were suggested to be more likely to develop the reaction-type bonding to the matrix.

3 SPS of Diamond-Metal Composites for Cutting Tool Applications

In the area of diamond-containing composites for cutting and grinding tools, conventional preparation methods are based on either a two-step process comprising pre-forming and free sintering or a single-step process of hot pressing, in which compaction and sintering occur during a single technological operation. In order to reduce the time required for consolidation of diamond-metal mixtures, electric current-assisted methods have been attempted (Tillmann et al. 2010). However, conventional sintering remains to offer a higher production capacity and a lower cost in comparison with other sintering methods (Kahraman et al. 2017).

Selvi et al. used conventional sintering and SPS for consolidating diamond-(Co-Ni-Cu-Sn) composites to compare the efficiencies of these two techniques (Selvi et al. 2013). The soaking time at the maximum temperature during SPS (850 °C) was only 5 min, while conventional sintering was carried out at 950–1150 °C and lasted 1–4 h. In the SPS compacts, alloying between Co and other metals contained in the binder proceeded to a smaller extent than in conventionally sintered compacts due to a shorter high-temperature exposure in the SPS process. Compacts produced by SPS showed a higher relative density, a higher hardness and a higher compressive strength than those that were conventionally sintered. When the behavior of compacts having close values of relative density (exceeding 98%) was compared, it was also found that the SPS-processed diamond-(Co-Ni-Cu-Sn) composites exhibit a slightly higher wear resistance compared with those obtained by conventional sintering (Bulut et al. 2016). For a better cutting performance, a lower wear resistance of conventionally sintered compacts is believed to be beneficial.

SPS was successful in producing diamond-steel-Cu composites to be used as wire saw beads (Schmidt et al. 2010). An interesting microstructural observation was made: in the fracture surfaces of SPS compacts, the diamond particle pullout sites were crack-free, while those in conventionally sintered compacts showed extensive cracking, which was likely to be related to the extensive formation of Fe_3C in the latter. Multi-cavity SPS dies were designed to increase the productivity of the process. Extensive degradation of diamond took place in conventionally sintered diamond-Fe-Cu composites (Tillmann et al. 2010), while SPS allowed bringing degradation to a minimum.

Diamond-WC-Co materials are designed to increase wear resistance in comparison with conventional cemented carbides and machinability in comparison with materials with a diamond filler only. When diamond particles are used as fillers in cemented carbides, higher temperatures are required for consolidation as compared with diamond-metal composites. In such situations, fast heating and sintering offered by electric current-assisted techniques is crucial for preventing extensive graphitization of the diamond phase, especially when the diamond crystals happen to be in contact with a metallic melt (Moriguchi et al. 2007). In a similar manner, fast consolidation by SPS was selected for preserving diamond while making diamond-SiC-Si-Al composites with a volume fraction of diamond particles of 55 vol.% by reactive sintering (Zhu et al. 2012). In this process, a soaking time of 5 min at a temperature close to the melting point of silicon allowed reaching a relative density of 99.6%.

4 Application of SPS for Depositing Coatings on Diamond Particles

An interesting application of SPS for the processing of diamond-metal mixtures has recently emerged in the area of surface modification of particles. Treatment of diamond-metal powder mixtures in a SPS chamber has been suggested as a method of depositing coatings on diamond particles (Chang et al. 2017a, b). Using this technique, Ti- and Mo-containing coatings were obtained on diamond grits when the latter were subjected to treatment with the corresponding metal powders by electric current in a SPS die (Fig. 2). During treatment of diamond-Ti mixtures, it was found that the content of titanium in the mixture affects the intensity of the coating process and sintering of the powder particles together. A mixture containing 20 wt.% of Ti sintered into a single piece during the SPS treatment. At a Ti content of 5 wt.%, the

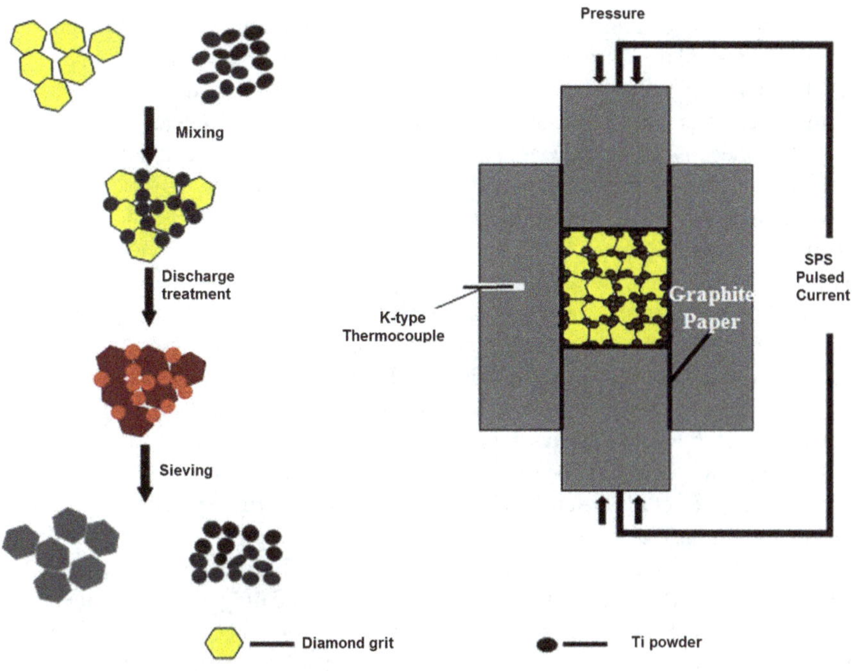

Fig. 2 Flow chart for preparing Ti-coated diamond grits by treatment in a SPS die. (Reprinted from Chang et al. (2017b), Copyright (2017), with permission from Elsevier)

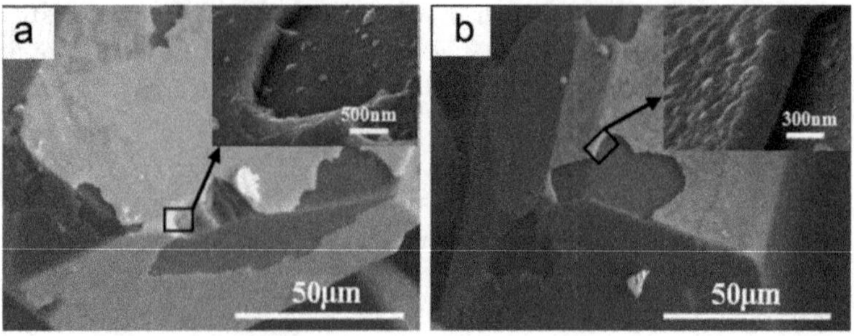

Fig. 3 Morphology of the Ti-containing coatings on diamond: treatment at 800 °C for 90 min (**a**), at 850 °C for 30 min (**b**). (Reprinted from Chang et al. (2017b), Copyright (2017), with permission from Elsevier)

product of treatment was a powder, in which the diamonds changed their color from yellow (the color of the raw diamonds) to black. A change in color indicated deposition of titanium on their surfaces. At a Ti content of 2.5 wt%, the color changes of the diamonds were less dramatic. Examples of the morphology of Ti-containing coatings on diamond crystals are shown in Fig. 3. In a similar process, a Mo-Mo_2C gradient coating 750 nm thick was obtained on diamond crystals by the SPS treatment of a diamond-Mo powder mixture containing 9 wt.% of Mo at 870 °C for 60 min.

For depositing coatings on diamond particles that are further used as components of composites, diamond-rich mixtures should be prepared in order to avoid sintering and formation of agglomerates. Experiments with layers of a titanium powder and diamond crystals placed successively into the die were conducted to study the formation mechanism of the coatings during the SPS treatment. As diamond crystals that were not in direct contact with titanium particles were also coated with Ti-containing layers, it was concluded that the transport of Ti to the surface of diamond occurred through the gaseous phase. On the diamond surface, Ti reacted with carbon to form TiC coatings. Consequently, the formation of coatings on diamond particles during treatment in a SPS die together with powders of carbide-forming metals follows the evaporation-condensation-interaction mechanism.

In our group, the SPS method was used to study the phase transformation and morphological evolution during the interaction between diamond and nickel at elevated temperatures (Bokhonov et al. 2015). In order to conduct morphological studies, partially densified 50 vol.% diamond-Ni compacts were obtained by SPS at 900 °C. On the fracture surface, nickel was found in the form of patches on diamond facets (Fig. 4): the diamond crystals recovered from the compacts were covered with an island-like coating. No Ni layer with a constant thickness was found on the surface of diamond crystals. The patches of nickel formed only in places of the direct contact of the diamond surface with nickel particles and indicated cohesive fracture of diamond-nickel joints formed during SPS. The absence of a uniform Ni coating can be related to limited mass transport through the gaseous phase, as the experiments were conducted under an applied pressure of 40 MPa and on a composition

Fig. 4 Fracture surface of the 50 vol.% diamond-Ni compact after SPS at 900 °C and 40 MPa for 5 min. (Bokhonov et al. 2015)

having a relatively high content of nickel (a ductile component) – conditions that do not favor extensive evaporation of nickel during SPS.

It was shown that tungsten carbide WC coatings on diamond particles can be deposited by treatment of a diamond-W-Cu powder mixture in a vacuum furnace at 1293–1373 K (Bai et al. 2013). The role of copper added to tungsten was to ensure the formation of a uniform carbide coating. Future research is needed to evaluate a possible influence of inert additives on the coating process of diamond crystals conducted with the help of the SPS treatment.

5 SPS of Nanodiamond-Metal Composites: Targeting Strengthening of Metals or Controlled Graphitization of Nanodiamonds

If the metal matrix is a carbide-forming element, nanodiamonds added to it react with it – fully or partially – during sintering, which leads to the formation of carbide particles distributed in the residual metal matrix. Mixtures of nanodiamonds with a titanium powder were consolidated by SPS by Zhang et al. (2017). Interaction of titanium with carbon led to the formation of solid solutions of carbon in titanium and titanium carbide. After soaking for 10 min at 900 °C, the composites still contained nanodiamond particles along with newly formed titanium carbide. Thanks to fast sintering accompanied by partial reaction, the consolidated titanium matrix composites were strengthened by the nanodiamond particles and the titanium carbide nanoparticles through Orowan mechanism, in addition to solution strengthening.

Results of hot extrusion were compared with those of SPS for nanodiamond-Cu powder mixtures (Nunes et al. 2010). Composite particles (agglomerates) were obtained by ball milling of the mixture of a nanodiamond powder and a copper powder. In terms of relative density, hot extrusion was more efficient in consolidating the composite powder. The hot extruded composites showed a uniform microstructure, in which boundaries between the initial powder agglomerates were indistinguishable. During SPS, the samples experienced a specific action of electric current passing through them. Sintering was conducted at 600 °C and 800 °C; a higher temperature allowed producing a higher relative density. Hot extrusion was performed at 600 °C. Despite the measured temperatures being lower than the melting point of copper, melting of the metal at the inter-agglomerate contacts did occur during SPS, being more significant at a measured temperature of 800 °C. It should be noted that similar observations were reported for other systems: evidence of melting of the boundary regions between particles during SPS was found for Ti_3SiC_2-Cu agglomerates (Dudina et al. 2013) and Ag-Fe agglomerates (Bokhonov and Dudina 2013) obtained by ball milling. Local melting is directly related to the passage of electric current through the sample. In systems with poor wettability of the solid by the melt, it leads to a change in the mutual distribution of the phases in the vicinity of the boundary regions. Melting was indeed detrimental for the relative density of the nanodiamond-Cu compacts, as it caused redistribution of the phases: at the former interparticle boundaries, the nanodiamond-depleted regions showed a decreased microhardness relative to regions with a remaining composite structure (Nunes et al. 2010). The microhardness of the center of the agglomerates was higher than that of the extruded sample and close to that of the milled powder. The boundary nanodiamond-depleted regions showed the lowest microhardness. Figure 5 shows an example of the microstructure of a sintered material, in which melting of the interparticle boundary regions occurred during

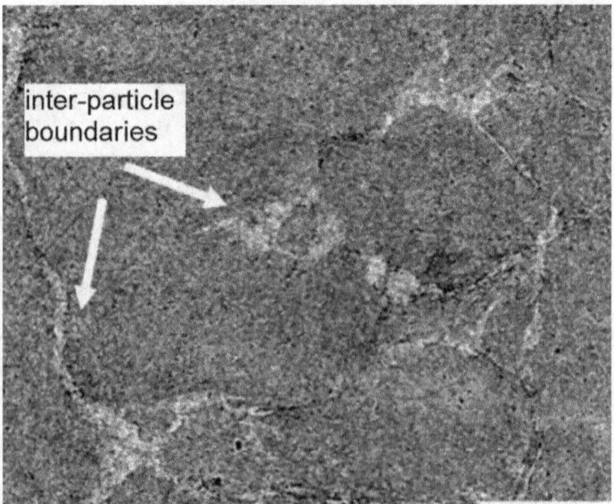

Fig. 5 Scanning electron microscopy image (back-scattered electron image) of nanodiamond-Cu(Cr) composite spark plasma sintered from a mechanically milled powder mixture at 800 °C. (Reprinted from Nunes et al. (2011), Copyright (2011), with permission from Elsevier)

SPS (Nunes et al. 2011). The image shows a nanodiamond-Cu(Cr) composite spark plasma sintered from a mechanically milled powder mixture.

SPS was successfully applied to WC-Co/nanodiamond composites (Nieto et al. 2017). Similar to the case of composites with micrometer-sized diamonds added to cemented carbides, SPS was chosen to enable consolidation of powders while avoiding graphitization of nanodiamonds. Although some porosity was retained in the sintered WC-Co/nanodiamond composites, some compositions showed increased wear resistance achieved through the complex influence of nanodiamonds on the microstructure and chemical reactions during sintering. Dissolution of WC in cobalt was inhibited by the presence of nanodiamonds.

Nickel catalyzes graphitization of nanodiamonds during annealing of powder mixtures (Nunes et al. 2012). If nanodiamonds are sintered with a metal that catalyzes their graphitization and their content in the compact is such that a network of graphite grains can be formed, porous materials can be obtained after the metal catalyst has been selectively dissolved. This approach was tested by Ukhina et al. (2017b) for mixtures containing nanodiamonds at a concentration ensuring the formation of 50 vol.% graphite-Ni composites upon complete graphitization. The morphology of the nanodiamond particles used in that study is shown in Fig. 6a. The nanodiamond-Ni mixtures were prepared by mechanical milling. Figure 7 shows XRD patterns of carbon materials obtained by selective dissolution of nickel from compacts spark plasma sintered from nanodiamond-Ni powder mixtures at 800 °C for 3 and 10 min and at 1000 °C for 3 min. The reflections of graphite are present in the patterns of all three samples. The diamond phase is detectable only in the compact sintered at 800 °C for 3 min. Raman spectra (Fig. 8) and the calculated I_D/I_G intensity ratios (Table 2) show that, as the sintering temperature increases from 800 °C to 1000 °C for the samples held for 3 min, the graphitization degree of nanodiamonds increases significantly. Figure 6b, c presents the fracture surfaces of porous compacts obtained by selective dissolution of nickel from the sintered compacts. Both scanning and transmission electron microscopy analyses have shown that the porous carbon material obtained from the composite sintered at 800 °C consists of smaller crystallites than that obtained at 1000 °C. The specific surface area of the initial nanodiamond powder was measured to be 360 m^2g^{-1} (Ukhina et al. 2015). SPS of nanodiamonds in the presence of nickel followed by removal of the metal by selective dissolution produces porous carbon materials with a significantly reduced specific surface area relative to the initial nanodiamond powder: the porous material obtained from the compact sintered at 1000 °C showed a specific surface area of only 20 m^2g^{-1}. When the same nanodiamond powder was spark plasma sintered without any metal additions, sintering resulted in graphitization and the formation of porous materials having either a slightly decreased or a significantly increased specific surface area. The trend in the specific surface area change observed in nanodiamond-Ni compacts is explained by a high catalytic activity of nickel in the graphitization process of nanodiamonds and growth of graphite platelets.

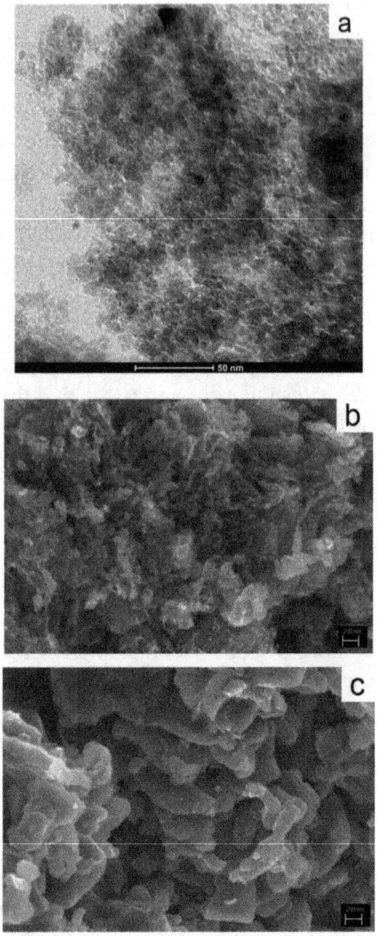

Fig. 6 Transmission electron microscopy image of the initial nanodiamond powder (**a**) and scanning electron microscopy images of the fracture surface of carbon materials obtained by selective dissolution of nickel from compacts spark plasma sintered from nanodiamond-Ni mixtures at 800 °C (**b**) and 1000 °C (**c**). The holding time at the maximum temperature during SPS was 3 min. (**a**) is reprinted from Ukhina et al. (2015), Copyright (2015), with permission from Elsevier; (**b-c**) is reprinted from Ukhina et al. (2017b), Copyright (2018), with permission of John Wiley & Sons)

6 Conclusions

Over the past few years, SPS has been widely used for the preparation of diamond-metal composites. The idea behind the use of SPS is to shorten the processing time of composites from the viewpoint of saving time and energy. For diamond-metal composites aimed to be used as heat sink materials, the structure of diamond/metal interface is of primary importance. Results of some studies speak in favor of hot pressing instead of SPS to avoid thermal gradients at different length scales and achieve interfaces with low thermal resistances and a high thermal conductivity of the composite as a whole. For composites to be used in making cutting and grinding tools, hardness, strength, and mechanical integrity are the key features, and, since they are ensured by the SPS processing, full benefit can be taken of the possibility to dramatically shorten the duration of the consolidation process by using SPS instead

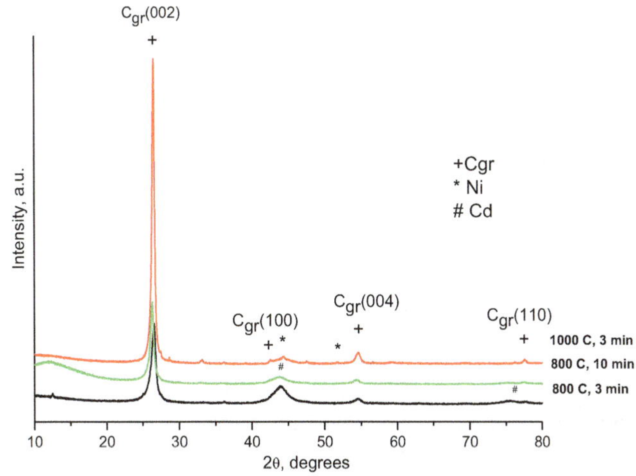

Fig. 7 XRD patterns of carbon-based materials obtained by selective dissolution of nickel from compacts spark plasma sintered from nanodiamond-Ni mixtures (Reprinted from Ukhina et al. (2017b), Copyright (2018), with permission of John Wiley & Sons)

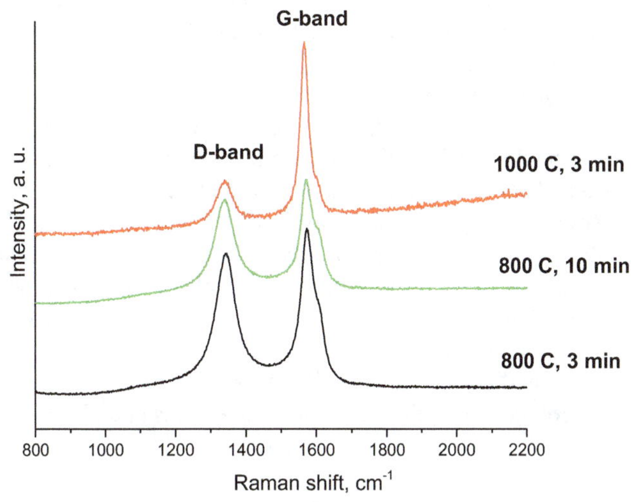

Fig. 8 Raman spectra of carbon-based materials obtained by selective dissolution of nickel from compacts spark plasma sintered from nanodiamond-Ni mixtures (Reprinted from Ukhina et al. (2017b), Copyright (2018), with permission of John Wiley and Sons)

of conventional methods. Interesting opportunities may lie in the combination of SPS with post-sintering annealing: densification is to be achieved at the first step, and additional property tailoring of composites can be made at the second step.

A new and promising direction in using SPS for the processing of diamond-metal composites is depositing coatings on micrometer-sized diamond particles. The

Table 2 Fabrication conditions and characteristics of the porous carbon materials obtained by SPS of nanodiamond-Ni mixtures followed by selective dissolution of the metal: sintering temperature and holding time, apparent density of the sintered metal-carbon compacts, and specific surface area and I_D/I_G ratio of Raman spectra of the porous carbon materials (Ukhina et al. 2017b)

SPS temperature, °C	Holding time, min	Apparent density, g·cm^{-3}	I_D/I_G	Specific surface area, m^2/g
800	3	4.94	0.86	240
800	10	4.98	0.95	180
1000	3	5.94	0.21	20

formation of coatings is based on evaporation-condensation-interaction mechanism. Evaporation is thought to be related to the formation of local high-temperature regions inside the powder samples subjected to pulsed electric current treatment. This emerging area of utilization of the SPS technique has a potential of being applied to a wider variety of coating materials and substrates (particles) to be coated.

In the fabrication of nanodiamond-metal composites, because of an extremely small size of nanodiamond particles, exposure to elevated temperatures during consolidation should be shortened as much as possible to avoid graphitization of the diamond phase. SPS has been used to conduct fast consolidation of powder mixtures up to high relative densities under conditions, in which the metastable phase of diamond could be fully or partially retained. It was shown that, in composites based on metals that have poor wettability to diamond, local melting during SPS can result in the phase redistribution (formation of nanodiamond-enriched and nanodiamond-depleted regions) and, therefore, should be avoided by adjusting the pressure and morphology of nanodiamond-metal composite particles (agglomerates). SPS has also been used for conducting graphitization of nanodiamonds both in the presence of metal catalysts and without any catalytic additives to produce bulk porous materials with controlled graphitization degrees and structural characteristics. In the future research in the area of SPS processing of nanodiamond-metal composites, the influence of the size, shape, inner structure, and surface characteristics of composite particles obtained by mechanical milling on the structural development of sintered composites and porous carbon materials should be studied in more detail to gain reliable control tools over the materials architectures at different length scales.

References

Abyzov AM, Kidalov SV, Shakhov FM (2011) High thermal conductivity composites consisting of diamond filler with tungsten coating and copper (silver) matrix. J Mater Sci 46:1424–1438

Bai H, Ma N, Lang J, Zhu C, Ma Y (2013) Thermal conductivity of Cu/diamond composites prepared by a new pretreatment of diamond powder. Compos Part B 52:182–186

Bai H, Dai D, Yu JH, Nishimura K, Sasaoka S, Jiang N (2014) Architecting boron nanostructure on the diamond particle surface. Appl Surf Sci 292:790–794

Blum R, Molian P (2009) Liquid-phase sintering of nanodiamond composite coatings on aluminum A319 using a focused laser beam. Surf Coat Technol 204:1–14

Bokhonov BB, Dudina DV (2013) Recrystallisation-accompanied phase separation in Ag–Fe and Ag–Ni nanocomposites: a route to structure tailoring of nanoporous silver. RSC Adv 3:12655–12661

Bokhonov BB, Ukhina AV, Dudina DV, Gerasimov KB, Anisimov AG, Mali VI (2015) Towards a better understanding of nickel/diamond interactions: the interface formation at low temperature. RSC Adv 5:51799–51806

Bulut B, Tazegul O, Baydogan M, Kayali ES (2016) The comparison of the sintering methods for diamond cutting tools. J Achiev Mater Manuf Eng 76:30–35

Chang R, Zang J, Wang Y, Yu Y, Lu J, Xu X (2017a) Preparation of the gradient Mo layers on diamond grits by spark plasma sintering and their effect on Fe-based matrix diamond composites. J Alloys Compd 695:70–75

Chang R, Zang J, Wang Y, Yu Y, Lu J, Xu X (2017b) Study of Ti-coated diamond grits prepared by spark plasma coating. Diam Relat Mater 77:72–78

Che QL, Zhang JJ, Chen XK, Ji YQ, Li YW, Wang LX, Cao SZ, Guo L, Wang Z, Wang SW, Zhang ZK, Jiang YG (2015) Spark plasma sintering of titanium-coated diamond and copper–titanium powder to enhance thermal conductivity of diamond/copper composites. Mater Sci Semicond Process 33:67–75

Cho HJ, Tam J, Kovylina M, Kim YJ, Erb U (2016) Thermal conductivity of bulk nanocrystalline nickel-diamond composites produced by electrodeposition. J Alloys Compd 687:570–578

Choudhary D, Bellare J (2000) Manufacture of gem quality diamonds: a review. Ceram Int 26:73–85

Chu K, Jia C, Liang X, Chen H, Gao W (2009) Effect of particle size on the microstructure and thermal conductivity of Al/diamond composites prepared by spark plasma sintering. Rare Metals 28:646–650

Chu K, Jia C, Liang X, Chen H (2010a) Effect of sintering temperature on the microstructure and thermal conductivity of Al/diamond composites prepared by spark plasma sintering. Int J Miner Metall Mater 17:234–240

Chu K, Liu Z, Jia C, Chen H, Liang X, Gao W, Tian W, Guo H (2010b) Thermal conductivity of SPS consolidated Cu/diamond composites with Cr-coated diamond particles. J Alloys Compd 490:453–458

Ciupiński Ł, Siemiaszko D, Rosiński M, Michalski A, Kurzydłowski KJ (2009) Heat sink materials processing by pulse plasma sintering. Adv Mater Res 59:120–124

Ciupiński Ł, Kruszewski MJ, Grzonka J, Chmielewski M, Zielińsk R, Moszczyńska D, Michalski A (2017) Design of interfacial Cr_3C_2 carbide layer via optimization of sintering parameters used to fabricate copper/diamond composites for thermal management applications. Mater Des 120:170–185

Dudina DV (2017) Phase and microstructure formation of composite materials and coatings obtained under conditions of non-equilibrium consolidation and pulse treatment. Doctoral thesis (Dr Sci), Novosibirsk State Technical University, Russia (in Russian)

Dudina DV, Mukherjee AK (2013) Reactive spark plasma sintering: successes and challenges of nanomaterial synthesis. J Nanomater 2013:625218, 12 p

Dudina DV, Mali VI, Anisimov AG, Bulina NV, Korchagin MA, Lomovsky OI, Bataev IA, Bataev VA (2013) Ti_3SiC_2-Cu composites by mechanical milling and spark plasma sintering: possible microstructure formation scenarios. Met Mater Int 19:1235–1241

Galashov EN, Yusuf AA, Mandrik EM, Atuchin VV (2016a) Preparation and thermo-physical parameters of diamond/W, Cu heat- conducting composite substrates. Int J Adv Manuf Technol 86:475–478

Galashov EN, Yusuf AA, Mandrik EM (2016b) Cu/synthetic and impact-diamond composite heat-conducting substrates. J Phys Conf Ser 690:012043

Gao W, Jia C, Jia X, Liang X, Chu K, Zhang L, Huang H, Liu M (2010) Effect of processing parameters on the microstructure and thermal conductivity of diamond/Ag composites fabricated by spark plasma sintering. Rare Metals 29:625–629

Garay JE (2010) Current-activated, pressure-assisted densification of materials. Annu Rev Mater Res 40:445–468

Hanada K, Yamamoto K, Taguchi T, Ōsawa E, Inakuma M, Livramento V, Correia JB, Shohoji N (2007) Further studies on copper nanocomposite with dispersed single-digit-nanodiamond particles. Diam Relat Mater 16:2054–2057

Hulbert DM, Jiang D, Dudina DV, Mukherjee AK (2009) The synthesis and consolidation of hard materials by spark plasma sintering. Int J Refract Met Hard Mater 27(2):367–375

Kaftelen H, Öveçoğlu ML (2011) Microstructural characterization and wear properties of ultra-dispersed nanodiamond (UDD) reinforced Al matrix composites fabricated by ball-milling and sintering. J Compos Mater 46:1521–1534

Kahraman Y, Bulut B, Sabri Kayali E (2017) Conventional sintering behavior of matrix materials used for diamond beads. Mater Test 59:647–652

Lee MT, Fu MH, Wu JL, Chung CY, Lin SJ (2011) Thermal properties of diamond/Ag composites fabricated by electroless silver plating. Diam Relat Mater 20:130–133

Livramento V, Correia JB, Nunes D, Carvalho PA, Fernandes H, Silva C, Hanada K, Shohoji N, Osawa E (2008) Novel approach to plasma facing materials in nuclear fusion reactors. AIP Conf Proc 996:166

Mizuuchi K, Inoue K, Agari Y, Sugioka M, Tanaka M, Takeuchi T, Kawahara M, Makino Y (2011) Thermal conductivity of diamond particle dispersed aluminum matrix composites fabricated in solid-liquid co-existent state by SPS. Compos Part B 42:1029–1034

Moriguchi H, Tsuduki K, Ikegaya A, Miyamoto Y, Morisada Y (2007) Sintering behavior and properties of diamond/cemented carbides. Int J Refract Met Hard Mater 25:237–243

Munir ZA, Quach D, Ohyanagi M (2011) Electric current activation of sintering: a review of the pulsed electric current sintering process. J Am Ceram Soc 94(1):1–19

Nieto A, Jiang L, Kim J, Kim DE, Schoenung JM (2017) Synthesis and multi scale tribological behavior of WC-Co/nanodiamond nanocomposites. Sci Rep 7:7060

Nunes D, Livramento V, Correia JB, Hanada K, Carvalho PA, Mateus R, Shohoji N, Fernandes H, Silva C, Alves E, Osawa E (2010) Consolidation of Cu-n diamond nanocomposites: hot extrusion vs spark plasma sintering. Mater Sci Forum 636–637:682–687

Nunes D, Correia JB, Carvalho PA, Shohoji N, Fernandes H, Silva C, Alves LC, Hanada K, Ōsawa E (2011) Production of Cu/diamond composites for first-wall heat sinks. Fusion Eng Des 86:2589–2592

Nunes D, Vilarigues M, Correia JB, Carvalho PA (2012) Nickel–carbon nanocomposites: synthesis, structural changes and strengthening mechanisms. Acta Mater 60:737–747

Orrù R, Licheri R, Locci AM, Cincotti A, Cao G (2009) Consolidation/synthesis of materials by electric current activated/assisted sintering. Mater Sci Eng R 63(4–6):127–287

Padyukov KL, Levashov EA (1993) Self-propagating high-temperature synthesis: a new method for the production of diamond-containing materials. Diam Relat Mater 2:207–210

Popov VA (2013) Metal matrix composites with non-agglomerated nanodiamond reinforcing particles. In: Wang X (ed) Nanocomposites. Nova Science Publishers Inc, New York

Rape A, Liu X, Kulkarni A, Singh J (2013) Alloy development for highly conductive thermal management materials using copper-diamond composites fabricated by field assisted sintering technology. J Mater Sci 48:1262–1267

Reinert L, Zeiger M, Suárez S, Presser V, Mücklich F (2015) Dispersion analysis of carbon nanotubes, carbon onions, and nanodiamonds for their application as reinforcement phase in nickel metal matrix composites. RSC Adv 5:95149–95159

Ren S, Shen X, Guo C, Liu N, Zang J, He X, Qu X (2011) Effect of coating on the microstructure and thermal conductivities of diamond–Cu composites prepared by powder metallurgy. Compos Sci Technol 71:1550–1555

Schmidt J, Knote A, Armbrüster M, Weißgärber T, Kieback B (2010) Spark plasma sintering of diamond impregnated wire saw beads. In: Proceedings of the World Powder Metallurgy Congress and Exhibition, World PM2010 – Diamond Tools 3

Schubert T, Ciupiński Ł, Zieliński W, Michalski A, Weißgärber T, Kieback B (2008a) Interfacial characterization of Cu/diamond composites prepared by powder metallurgy for heat sink applications. Scr Mater 58:263–266

Schubert T, Trindade B, Weißgärber T, Kieback B (2008b) Interfacial design of Cu-based composites prepared by powder metallurgy for heat sink applications. Mater Sci Eng A 475:39–44

Selvi E, Topaloglu F, Tazegul O, Kayali ES (2013) Conventional sintering of diamond cutting tool used in natural stone cutting. AIP Conf Proc 1569:427–432

Sidorenko DA, Zaitsev AA, Kirichenko AN, Levashov EA, Kurbatkina VV, Loginov PA, Rupasov SI, Andreev VA (2013) Interaction of diamond grains with nanosized alloying agents in metal – matrix composites as studied by Raman spectroscopy. Diam Relat Mater 38:59–62

Tan Z, Li Z, Fan G, Kai X, Ji G, Zhang L, Zhang D (2013a) Fabrication of diamond/aluminum composites by vacuum hot pressing: process optimization and thermal properties. Compos Part B 47(2013):173–180

Tan ZQ, Li ZQ, Fan GL, Guo Q, Kai XZ, Ji G, Zhang LT, Zhang D (2013b) Enhanced thermal conductivity in diamond/aluminum composites with a tungsten interface nanolayer. Mater Des 47:160–166

Tan Z, Ji G, Addad A, Li Z, Silvain JF, Zhang D (2016) Tailoring interfacial bonding states of highly thermal performance diamond/Al composites: spark plasma sintering vs. vacuum hot pressing. Compos Part A 91:9–19

Tillmann W, Kronholz C, Ferreira M (2009) Novel current induced short-time sintering processes for the production of diamond tools. In: Proceedings of the Euro International Powder Metallurgy Congress and Exhibition, Euro PM 2009 1

Tillmann W, Kronholz C, Ferreira M, Knote A, Theisen W, Schütte P, Schmidt J (2010) Comparison of different metal matrix systems for diamond tools fabricated by new current induced short-time sintering processes. In: Proceedings of the World Powder Metallurgy Congress and Exhibition, World PM2010 – Diamond Tools 3

Ukhina AV, Dudina DV, Anisimov AG, Mali VI, Bulina NV, Bataev IA, Skovorodin IN, Bokhonov BB (2015) Porous electrically conductive materials produced by spark plasma sintering and hot pressing of nanodiamonds. Ceram Int 41:12459–12463

Ukhina A, Bokhonov B, Samoshkin D, Stankus S, Dudina D, Galashov E, Katsui H, Goto T, Kato H (2017a) Morphological features of W- and Ni-containing coatings on diamond crystals and properties of diamond-copper composites obtained by spark plasma sintering. Mater Today Proc 4:11396–11401

Ukhina AV, Bokhonov BB, Dudina DV, Yubuta K, Kato H (2017b) Structural characterization of carbon-based materials obtained by spark plasma sintering of non-graphitic carbon with nickel and iron as catalysts and space holders. In: Bansal NP, Castro RHR, Jenkins M, Bandyopadhyay A, Bose S, Bhalla A, Singh JP, Mahmoud MM, Pickrell G, Johnson S (ed) Ceramic transactions, V. 261, Processing, properties, and design of advanced ceramics and composites II – Proceedings of 2016 Materials Science and Technology (MS&T'16), Salt Lake City, October 24–27, 2016, pp 117–126

Woo DJ, Sneed B, Peerally F, Heer FC, Brewer LN, Hooper JP, Osswald S (2013) Synthesis of nanodiamond-reinforced aluminum metal composite powders and coatings using high-energy ball milling and cold spray. Carbon 63:404–415

Zhang Y, Li J, Zhao L, Zhang H, Wang X (2014) Effect of metalloid silicon addition on densification, microstructure and thermal – physical properties of Al/diamond composites consolidated by spark plasma sintering. Mater Des 63:838–847

Zhang F, Liu S, Zhao P, Liu T, Sun J (2017) Titanium/nanodiamond nanocomposites: effect of nanodiamond on microstructure and mechanical properties of titanium. Mater Des 131:144–155

Zhu C, Ma N, Jin Y, Bai H, Ma Y, Lang J (2012) Thermal properties of Si(Al)/diamond composites prepared by in situ reactive sintering. Mater Des 41:208–213

Spark Plasma Sintering of Fine-Grain Ceramic–Metal Composites Based on Garnet-Structure Oxide $Y_{2.5}Nd_{0.5}Al_5O_{12}$ with Mo, W, and Ni

L. S. Golovkina, A. I. Orlova, A. V. Nokhrin, M. S. Boldin, E. A. Lantsev, and V. N. Chuvil'deev

1 Introduction

One of the main problems of modern nuclear energy is the management of plutonium and minor actinides (MA) stocks accumulated over many years. The best way to deal with this situation would be to implement a closed nuclear fuel cycle and to create an inert matrix fuel (IMF) for plutonium burning and MA transmutation. In this regard, the crystalline compounds, which could be analogues of natural minerals, and processes for the synthesis of ceramics based on such compounds with high thermal, chemical, and radiation stability seem to be promising enough.

Currently, many materials are regarded as IMF, including metals (Delage et al. 2011; Gotcu-Freis et al. 2011; O'Brien et al. 2009; O'Brien and Jerred 2013), nitrides (Konings and Haas 2002; Ryu et al. 2006), carbides (Ryu et al. 2006; Gosset and Provot 2001) and oxides of metals (Delage et al. 2011; Gotcu-Freis et al. 2011; Konings and Haas 2002; Raison and Haire 2003), and also crystalline compounds with the structure of natural minerals such as fluorite (Gotcu-Freis et al. 2011; O'Brien et al. 2009; O'Brien and Jerred 2013; Holliday et al. 2009; Raison and Haire 2003), monazite, garnet (Chauvin et al. 1999b; Neeft et al. 1999, 2003), spinel (Neeft et al. 2003; Konings et al. 1999; Chauvin et al. 1999a), kosnarite (Gregg et al. 2013), etc.

In recent years, interest in the process of spark plasma sintering for the production of ceramic materials for nuclear technologies (nonradioactive simulants and real radioactive materials) increases: Nd_2O_3–Nb (Williams et al. 2013); CeO_2–W (O'Brien et al. 2009); UO_2–W (O'Brien and Jerred 2013); Dy_2O_3–ZrC, TiC, ZrN, and TiN (Ryu et al. 2006); AmO_2–Cm_2O_3–ZrO_2 (Raison and Haire 2003);

L. S. Golovkina (✉) · A. I. Orlova · A. V. Nokhrin · M. S. Boldin · E. A. Lantsev
V. N. Chuvil'deev
Lobachevsky State University of Nizhny Novgorod, Nizhny Novgorod, Russia

© Springer Nature Switzerland AG 2019

P. Cavaliere (ed.), *Spark Plasma Sintering of Materials*,
https://doi.org/10.1007/978-3-030-05327-7_15

$CaZr_4(PO_4)_6$–Ni and $BaZr_4(PO_4)_6$–Ni (Gregg et al. 2013); $Ca_{0.5}Zr_2(PO_4)_3$–Ni (Gregg et al. 2013); $NaFeNb(PO_4)_3$ (Orlova et al. 2012); and $Y_3Al_5O_{12}$ (Suárez et al. 2009; Golovkina et al. 2013).

2 Materials and Techniques

This study focused on powders based on complex garnet-type oxide $Y_{2.5}Nd_{0.5}Al_5O_{12}$ – x vol. % Mo, W, and Ni (x = 0, 5, 10, 20).

To obtain a ceramic–metal composite, garnet powder $Y_{2.5}Nd_{0.5}Al_5O_{12}$ was placed into water solutions containing $Y_{2.5}Nd_{0.5}Al_5O_{12}$, $(NH_4)_4W_5O_{17}$, and $(NH_4)_6Mo_7O_{24}$, $NiCl_2$. The resulting composite powders were subjected to recovery annealing in a Nabertherm RS 120/750/13 furnace under hydrogen for 3 h at 700 °C (for W and Ni) and at 1000 °C (for Mo).

The obtained powder was ground in a Pulverisette 6 planetary ball mono mill for 30 min. To minimize abrasion contamination, zirconium dioxide grinding balls (10 mm dia.) were used for grinding in isopropyl alcohol. Mass ratio of powder/balls/alcohol was 2:4:1. Rotational speed was 300 rpm. After grinding, powder annealing was carried out for 8 h at 70 °C in a Nabertherm RHTC 80-230/15 circulating-air furnace.

Ceramics were obtained by SPS using a Dr. Sinter model SPS-625 unit in a vacuum (6 Pa). Heating rate varied from 50 °C/min to 300 °C/min. Applied monoaxial pressure varied from 40 MPa to 100 MPa. Powders were loaded into a graphite compression mold of 12 mm int. dia. and heated by applying millisecond-long strong pulses (up to 3 kA). Temperature readings were taken with a Chino IR-AH infrared thermometer focused on the surface of the graphite compression mold. Measurement accuracy was ±10 °C; pressure stability was 1 MPa.

A dilatometer was used to measure effective shrinkage (L_{eff}) during sintering. To analyze the correlation between true shrinkage and heating temperature, additional thermal expansion measurements were taken on an empty compression mold. True

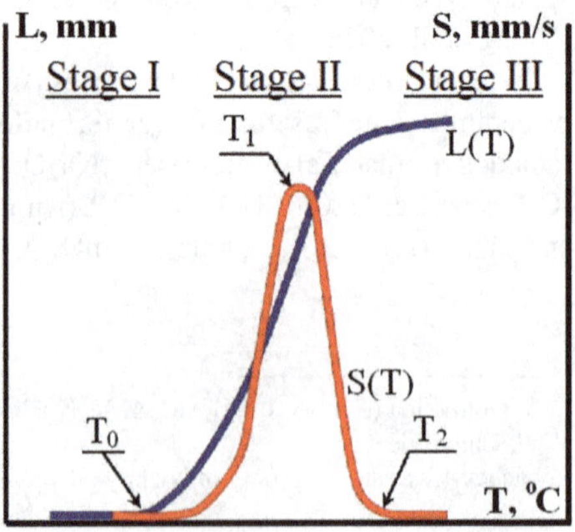

Fig. 1 Dependencies of shrinkage L and shrinkage rate S on temperature

shrinkage (L) was calculated by subtracting thermal expansion value ΔL from L_{eff}. (Figures below use conventional terminology, where effective shrinkage L_{eff} is marked as "experimental," thermal expansion as "baseline," and true shrinkage as "corrected.") Shrinkage measurement accuracy was ±0.01 mm. Dependencies L (T) were used to determine the temperature at which shrinkage begins (T_0) (see Fig. 1). Analysis of dependencies between shrinkage rate and temperature S(T) was used to determine T_1, at which the shrinkage rate peaks S_{max}, and T_2, which corresponds to cessation of active shrinkage (see Fig. 1).

Density of obtained ceramics was measured by hydrostatic weighing in distilled water using a Sartorius CPA 225D analytical balance. Accuracy was ±0.001 g/cm^3. Microstructure and elemental composition of samples were studied with a Jeol JSM-6490 scanning electron microscope (SEM) with an Oxford Instruments INCA 350 energy dispersive spectrometer. Sintered samples were studied at fracture surfaces and polished surfaces (pre-polished to roughness under 1 μm).

3 Experimental Results

Energy dispersive microanalysis (EDX) shows that compositions $Y_{2.5}Nd_{0.5}Al_5O_{12}$ with the theoretical content 10 vol.%W and 20 vol.%W contain 40 wt.% (~9.8 vol. %) and 72 wt.% (~17.6 vol.%) of tungsten, respectively; the compositions with the theoretical content of 10 vol.% and 20 vol.% Mo contain 24 wt.% (~11 vol.%) and 47 wt.% (~21.6 vol.%) molybdenum, respectively. Microstructural analysis of powder samples with tungsten and molybdenum shows inclusions of metal microparticles on the surfaces of garnet grains (Fig. 2): tungsten (Samples 1, 2) and

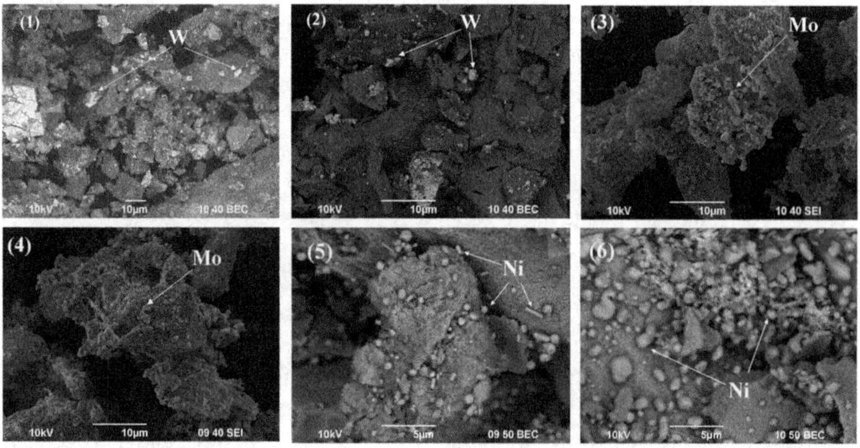

Fig. 2 Backscattered electron micrographs of powders: $Y_{2.5}Nd_{0.5}Al_5O_{12}$ – 10%W (1), $Y_{2.5}Nd_{0.5}Al_5O_{12}$ – 20%W(2), $Y_{2.5}Nd_{0.5}Al_5O_{12}$ – 10%Mo (3), $Y_{2.5}Nd_{0.5}Al_5O_{12}$ – 20%Mo(4), $Y_{2.5}Nd_{0.5}Al_5O_{12}$ – 5%Ni (5), $Y_{2.5}Nd_{0.5}Al_5O_{12}$ – 10%Ni(6)

molybdenum (Samples 3, 4). The size of garnet particles is under 0.5 μm; at the same time, they form easily degradable conglomerates of up to 30 μm. The size of metal particles ranges from under 100 nm to 1 μm.

EDS microanalysis of the composition of original powders $Y_{2.5}Nd_{0.5}Al_5O_{12}$–10% Ni (Fig. 2, Sample 6) reveals that nickel in the original composition is distributed nonuniformly, with nickel content varying in different areas from 15.6 wt.% (~8.8 vol.%) to 58.6 wt.% (~42.6 vol.%). It should be noted that nickel distribution in composite $Y_{2.5}Nd_{0.5}Al_5O_{12}$ – 5%Ni is significantly more uniform and evenly spread (Fig. 2, Sample 5). SEM images show that the average size of garnet particles is under 0.5 μm. The particles form easily breakable conglomerates of up to 10 μm. The average size of lighter-colored nickel particles, which have both equiaxed and elongated shapes, is under 1 μm.

For conditions of composites sintering, see Fig. 3 and Table 1. As Table 1 demonstrates, the density of the sintered composites is ~92–99% of the theoretical density. The highest density (99.0–99.1%) is observed in the molybdenum composites.

Increasing tungsten and molybdenum content results in higher shrinkage rates and values. In pure ceramic $Y_{2.5}Nd_{0.5}Al_5O_{12}$, maximum shrinkage rate at 100 °C/min heating rate under 70 MPa is $\Delta L_{max} = 2.2$ mm; in composites with 10%W and 20%W, maximum shrinkage rates are 3.87 mm and 4.05 mm, respectively; and in composites with 10%Mo and 20%Mo, maximum shrinkage rates are – 6.15 and 6.86 mm, respectively. In pure ceramic $Y_{2.5}Nd_{0.5}Al_5O_{12}$, the optimal temperature of high-speed sintering, which corresponds to cessation of active shrinkage (T_2), is 1380–1390 °C (see Fig. 3a), or ~0.75 $T_{m(Gr)}$ (here, $T_{m(Gr)} = 2243$ K is melting point of yttrium–aluminum garnet (Samsonov 1969; Andrievskiy and Spivak 1989)). In composites with 10%W and 20%W, the optimal sintering temperature (T_2) is 1310–1320 °C and 1290–1300 °C (see Fig. 3b, c), respectively (at a pressure of 55 MPa), or ~0.37–0.39 $T_{m(W)}$ (here, tungsten melting point is $T_{m(W)} = 3683$ K (Frost and Ashby 1982)). In composites with 10%Mo and 20%Mo, the optimal sintering temperature is 1370–1380 °C and 1300–1320 °C respectively (see Fig. 3d, e), or ~0.50–0.53 $T_{m(Mo)}$ (here: molybdenum melting point is $T_{m(Mo)} = 2883$ K (Frost and Ashby 1982)).

For regular temperature values at the beginning of active shrinkage T_0, temperatures correlating to maximum shrinkage rates T_1, and temperatures correlating to cessation of active shrinkage T_2 in ceramic $Y_{2.5}Nd_{0.5}Al_5O_{12}$, see Table 2.

Analysis of the shrinkage dependencies on the heating temperature for composites with nickel shows that nickel has a considerable effect on maximum shrinkage value L_{max}: with addition of 5% Ni maximum shrinkage in ceramics rises from 2.2 mm to 3.7 mm. Further increasing the nickel content up to 10%, L_{max} drops to 2.5 mm, and relative density of the ceramic decreases from 98.8% down to 92.1% (see Table 1). It can be seen that introduction of 5% Ni results in shrinkage starting at a lower temperature (T_0), down from 980–1000 °C to 840–870 °C, and in a decrease in the temperature correlating to the maximum shrinkage rate (T_1), down from 1335–1345 °C to 1220–1250 °C. For the composite with 10% Ni, regular temperatures T_0 and T_1 are 830–840 °C and 1360–1370 °C, respectively. The temperature

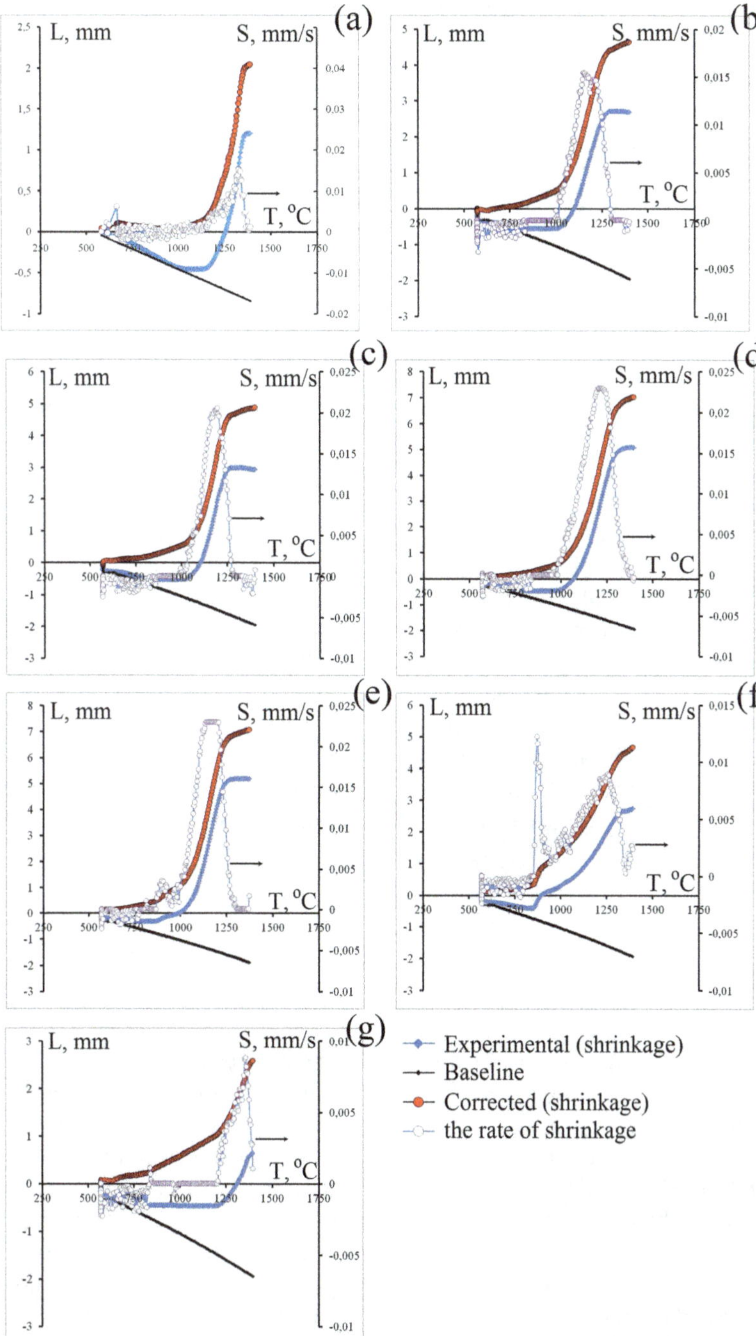

Fig. 3 Dependencies of shrinkage values (L) and shrinkage rates (S) of $Y_{2.5}Nd_{0.5}Al_5O_{12}$ (**a**), $Y_{2.5}Nd_{0.5}Al_5O_{12}$ – 10% W (**b**), $Y_{2.5}Nd_{0.5}Al_5O_{12}$ – 20% W (**c**), $Y_{2.5}Nd_{0.5}Al_5O_{12}$ – 10% Mo (**d**), $Y_{2.5}Nd_{0.5}Al_5O_{12}$ – 20% Mo (**e**), $Y_{2.5}Nd_{0.5}Al_5O_{12}$ – 5% Ni (**f**), $Y_{2.5}Nd_{0.5}Al_5O_{12}$ – 10% Ni (**g**)

Table 1 Sintering conditions for garnet-structure ceramics and composites based on them

	P_{max}, MPa	T_{sint}, °C	t_{sint}, min	Density g/cm^3	%
$Y_{2.5}Nd_{0.5}Al_5O_{12}$	70	1390	8	4.674	98.6
$Y_{2.5}Nd_{0.5}Al_5O_{12}$ – 5% Ni	70	1400	7	4.901	98.8
$Y_{2.5}Nd_{0.5}Al_5O_{12}$ – 10% Ni	70	1400	6	4.873	92.1
$Y_{2.5}Nd_{0.5}Al_5O_{12}$ – 10% W	55	1315	7	5.870	97.5
$Y_{2.5}Nd_{0.5}Al_5O_{12}$ – 20% W	55	1295	6	6.254	97.6
$Y_{2.5}Nd_{0.5}Al_5O_{12}$ – 10% Mo	55	1400	9	5.213	99.0
$Y_{2.5}Nd_{0.5}Al_5O_{12}$ – 20% Mo	55	1395	9	5.684	99.1

at which active shrinkage stops (T_2) in ceramic $Y_{2.5}Nd_{0.5}Al_5O_{12}$ is 1380–1390 °C; in composites with 5% and 10% Ni, it is 1360 °C and 1410–1420 °C, respectively (see Fig. 3f, g and Table 2).

We see that the temperature range within which sintering begins in composites with 10% Ni and 20% Ni has drifted toward lower temperatures relative to regular sintering temperatures of ceramic $Y_{2.5}Nd_{0.5}Al_5O_{12}$. The temperatures at which active sintering stops (T_2), though, are rather similar, falling within the range from 1360 °C to 1410–1420 °C (see Table 2 and Fig. 3g).

We also note that in the case of "garnet–tungsten" and "garnet–molybdenum" composites, no distinctly pronounced stage of the plastic flow of the material is observed, which, in our opinion, may indicate, on the one hand, a more even distribution of refractory tungsten and molybdenum in the ceramic matrix (in comparison with the distribution of nickel) and, on the other hand, the partial oxidation of the surface of tungsten and molybdenum particles during the synthesis (cladding of garnet particles), which will also impede plastic deformation of the particles.

4 Discussion

Obtained experimental results show that the dependencies of shrinkage on temperature L(T) have three stages: a stage of slow change in shrinkage at lower heating temperatures (Stage I), a stage of rapid shrinkage change at "medium" heating temperatures (Stage II), and a stage of slower shrinkage change at higher sintering temperatures (Stage III) (see Fig. 1). According to Rahaman (2003) and Somiya and Moriyyoshi (1990), the increase in density during Stage I is the result of particles slipping relative to each other due to the applied pressure and temperature; the increase in density during Stage II is due to diffusion-based "dissolution" of pores; and the increase in density during Stage III is due to grain growth (Chuvil'deev et al. 2017).

To compare the sintering kinetics, we have transformed dependencies L/L_0-T into density increase diagrams ρ/ρ_{th}-T, where ρ_{th} is theoretical density. The procedure for

Table 2 Spark plasma sintering conditions for ceramics and composites

Material	Regular SPS temperatures			L_{max}, mm	S_{max}, mm/s	Sintering activation energy, Q_s	
	T_0, °C	T_1, °C	T_2, °C			kT_m	kJ/mol
$Y_{2.5}Nd_{0.5}Al_5O_{12}$	980–1000	1335–1345	1380–1390	2.2	1.5 ± 10^{-2}	35.0 ± 2.0	630 ± 35
$Y_{2.5}Nd_{0.5}Al_5O_{12}$ – 5%Ni	840–870	1220–1250	1360	3.7	1.1 ± 10^{-2}	30.5 ± 2.5	550 ± 45
$Y_{2.5}Nd_{0.5}Al_5O_{12}$ – 10%Ni	830–840	1360–1370	1410–1420	2.5	1.3 ± 10^{-2}	32.0 ± 2.5	575 ± 45
$Y_{2.5}Nd_{0.5}Al_5O_{12}$ – 10% W	990–1000	1190–1210	1310–1320	3.87	1.5 ± 10^{-2}	8 ± 0.5	430 ± 4
$Y_{2.5}Nd_{0.5}Al_5O_{12}$ – 20% W	1020–1030	1180–1200	1290–1300	4.05	2.1 ± 10^{-2}	6 ± 0.5	350 ± 4
$Y_{2.5}Nd_{0.5}Al_5O_{12}$ – 10% Mo	950–960	1215–1235	1370–1380	6.15	2.3 ± 10^{-2}	13 ± 1.0	320 ± 5
$Y_{2.5}Nd_{0.5}Al_5O_{12}$ – 20% Mo	860–870	1170–1190	1300–1320	6.86	2.3 ± 10^{-2}	9 ± 0.5	200 ± 5

transforming density increase curves L/L_0-T into sintering diagrams ρ/ρ_{th}-T during SPS is detailed in references (Potanina et al. 2016; Chuvil'deev et al. 2015a, b). The accuracy of calculations of dependence ρ/ρ_{th}-T was checked by comparing densities corresponding to maximum shrinkage against value ρ_{max} measured in the experiment. According to the procedure described in references (Potanina et al. 2016; Chuvil'deev et al. 2015a, b), SPS activation energy (Q_s, in kT_m, where k is the Boltzmann constant and T_m is melting point) can be determined by the angle of experimentally obtained dependence $\rho/\rho_{th}(T)$, represented in a double logarithmic coordinate system $\ln(\ln[(\rho/\rho_{th})/(1\text{-}\rho/\rho_{th})]) - T_m/T$.

Our analysis shows that, during Stage II of density increase (Stage II in Fig. 1), dependencies L(T) in coordinates $\ln(\ln[(\rho/v_{th})/(1\text{-}\rho/\rho_{th})]) - T_m/T$ are straight lines. For ceramic $Y_{2.5}Nd_{0.5}Al_5O_{12}$, sintering activation energy Q_s is 630 ± 35 kJ/mol (~35 ± 2 kT_m). These estimates of Q_s are close to the activation energy of neodymium self-diffusion in the garnet crystal lattice (567 ± 15 kJ/mol according to reference (Hollingsworth et al. 2009), and 623 ± 70 kJ/mol according to reference (Hollingsworth et al. 2011)). (The activation energy of oxygen diffusion in the garnet crystal lattice lies between 325 kJ/mol and 519 kJ/mol (Haneda et al. 1984; Beke 1999); the activation energy for yttrium in the lattice of Y_2O_3 is between 184 kJ/mol (Mrowec 1988; Chuvil'deev and Smirnova 2017) and 290 kJ/mol (Samsonov 1969; Andrievskiy and Spivak 1989).) It should also be noted that regular temperatures of active shrinkage cessation T_2 = 1380–1420 °C are ~0.74–0.75 $T_{m(Gr)}$, where the major diffusion mechanism in most metals and ceramics is diffusion in the crystal lattice (Rahaman 2003; Somiya and Moriyyoshi 1990; Frost and Ashby 1982; Chuvil'deev and Smirnova 2017; Pelleg 2016).

Table 2 indicates that sintering activation energies Q_s for garnet–tungsten and garnet–molybdenum composites are low, significantly lower than that of pure garnet. Obtained estimates for SPS activation energy of composites are close to the activation energy of grain-boundary diffusion in corresponding metals: in tungsten (385 kJ/mol (Frost and Ashby 1982)) for garnet–tungsten composites (350–430 kJ/mol) and in molybdenum (263 kJ/mol (Frost and Ashby 1982)) for garnet–molybdenum composites (200–320 kJ/mol). SPS results indicate that sintering activation energy decreases as tungsten and molybdenum content increases to 20% (see Table 1). The reason is that a nearly "continuous mesh" of the metal phase forms around most garnet particles as the metal phase content reaches around 20 vol.% (see Fig. 4a, b).

A comparison of the calculated values of the activation energy for composites with nickel (for $Y_{2.5}Nd_{0.5}Al_5O_{12}$-5%Ni, Q_s = 174 ± 35 kJ/mol, and for $Y_{2.5}Nd_{0.5}Al_5O_{12}$-10%Ni, Q_s = 810 ± 40 kJ/mol) shows that the value of Q_s in a composite with 5% Ni is close to the activation energy of diffusion in nickel (the activation energy of bulk diffusion in nickel (Q_v = 284 kJ/mol (Frost and Ashby 1982)) and the activation energy of grain-boundary diffusion in nickel (Q_b = 115 kJ/mole (Frost and Ashby 1982)) and, consequently, the sintering process of the given composite is in fact carried out due to the diffusion mass transfer through the nickel phase.

Fig. 4 Microstructure of composites $Y_{2.5}Nd_{0.5}Al_5O_{12}$ – 20% W (**a**) and $Y_{2.5}Nd_{0.5}Al_5O_{12}$ – 20% Mo (**b**), $Y_{2.5}Nd_{0.5}Al_5O_{12}$ – 10% Ni (**c**) obtained by SPS. Lighter particles are metals

As already noted above, the uneven distribution of nickel in the composite $Y_{2.5}Nd_{0.5}Al_5O_{12}$-10% Ni leads to the fact that individual sections of the powder have an excessive (more than 40%) nickel content. It is obvious that in this case in the powder composition, there will be micro-regions of the powder, in which nickel will be absent. This will lead to the fact that the sintering process in these regions will be limited by the same kinetic conditions as in the case of sintering pure ceramics $Y_{2.5}Nd_{0.5}Al_5O_{12}$, while the diffusion and kinetic characteristics of the "neighboring" sections of the ceramics containing an excessive amount of nickel (in Fig. 4c such a region is marked by a dashed line) will correspond to the diffusion and kinetic characteristics of nickel. The superposition of the diffusion characteristics of such an inhomogeneously sintered ceramic provides, in a composite with 10% Ni, the "intermediate" activation energy values Qs = Q_s = 810 ± 40 kJ/mol, which are average (taking into account the weight fractions proportional to the volume fraction of enriched and nickel-depleted sites in the structure of ceramics) between the diffusion characteristics of nickel (115–285 kJ/mol) and pure ceramics (1152 ± 85 kJ/mol).

It should also be noted that garnet–tungsten and garnet–molybdenum composites display no distinctive stage of intensive plastic flow in the metal phase, which was observed in garnet–nickel composites during sintering (Golovkina et al. 2017). This indicates a more even distribution of tungsten and molybdenum particles within the ceramic matrix of garnet.

5 Conclusions

Systematic studies of the kinetics of sintering of composites based on garnet $Y_{2.5}Nd_{0.5}Al_5O_{12}$ with additives of 5, 10, and 20 vol.% metals (nickel, tungsten, molybdenum) deposited on the surface of synthesized garnet nanoparticles by precipitation of salts of the corresponding metals were carried out. The values of the relative density (they amounted to values from 92% to 99%) of sintered ceramics and composites were determined.

It is shown that the dependence of shrinkage of powders on the heating temperature by the SPS method has a characteristic two-stage character: the stage of densification of nanoparticles due to their rearrangement and plastic flow under the applied stress and the direct sintering stage (compaction of powders due to diffusion "collapse" of the pores).

It is shown that the temperature interval of the first stage and the maximum shrinkage value in the first stage essentially depend on the content and nature of the distribution of the metal phase in the powder composition. It was found that the nonuniform distribution of the metallic phase (in this case, nickel) leads to a significant expansion of the temperature interval of the first stage of compaction and to the formation of an inhomogeneous grain structure in which the regions of well-sintered material alternate with regions containing an increased number of pores and cracks.

Using the previously developed procedure for analyzing the temperature dependence of shrinkage, the activation energy of sintering Q_s was determined in the second stage of powder consolidation in the continuous heating mode.

It has been established that the mechanism responsible for compaction of ceramic composites in the second stage of sintering is the process of diffusion mass transfer through the grain boundaries (ceramic–metal interface), the activation energy of the sintering process Q_s with good accuracy corresponds to the activation energy of grain-boundary diffusion in metals that are a material of cladding layer for ceramic garnet nanoparticles.

At the same time, for a number of composites (containing, as a rule, an increased volume fraction of the metallic phase), an appreciable decrease in the activation energy of sintering Q_s is found, the value of which is 20–30% lower than the equilibrium activation energy of grain-boundary diffusion Q_b. It is suggested that the reason for the decrease in the activation energy of the sintering process can be the rapid growth of grains in ceramics under heating conditions.

Acknowledgments This work is supported by Russian Science Foundation (grant №16-13-10464).

References

Andrievskiy AR, Spivak II (1989) Strength of high-melting compounds and materials on their base. Metallurgiya, Chelyabinsk (in Russian)

Beke DL (1999) Diffusion in non-metallic solids (part 1). Landolt-Börnstein – Group III Condensed Mater. http://link.springer.com/book/10.1007/b59654

Chauvin N, Albiol T, Mazoyer R et al (1999a) In-pile studies of inert matrices with emphasis on magnesia and magnesium aluminate spinel. J Nucl Mater 274:91–97. https://doi.org/10.1016/S0022-3115(99)00080-X

Chauvin N, Konings RJM, Matzke H (1999b) Optimization of inert matrix fuel concepts for americium transmutation. J Nucl Mater 274:105–111. https://doi.org/10.1016/S0022-3115(99)00078-1

Chuvil'deev VN, Smirnova ES (2017) Phenomenological theory of bulk diffusion in metals oxides. Phys Solid State 58:1487–1499. https://doi.org/10.1134/S1063783416070118

Chuvil'deev VN, Boldin MS, Dyatlova YG et al (2015a) A comparative study of the hot pressing and spark plasma sintering of Al_2O_3-ZrO_2-Ti(C,N) powders. Inorg Mater 51:1047–1053. https://doi.org/10.1134/S0020168515090034

Chuvil'deev VN, Nokhrin AV, Sakharov NV et al (2015b) Sparking plasma sintering of tungsten carbide nanopowders. Nanotechnol Russ 10:434–448. https://doi.org/10.1134/S1995078015030040

Chuvil'deev VN, Blagoveshchenskiy YV, Nokhrin AV et al (2017) Spark plasma sintering of tungsten carbide nanopowders obtained through DC arc plasma synthesis. J Alloys Compd 708:547–561. https://doi.org/10.1016/j.jallcom.2017.03.035

Delage F, Belin R, Chen X-N et al (2011) ADS fuel developments in Europe: results from the EUROTRANS integrated project. Energy Procedia 7:303–313. https://doi.org/10.1016/j.egypro.2011.06.039

Frost HJ, Ashby MF (1982) Deformation-mechanism maps: the plasticity and creep of metals and ceramics. Pergamon Press, Oxford

Golovkina LS, Orlova AI, Nokhrin AV et al (2013) Ceramics based on yttrium aluminum garnet containing Nd and Sm obtained by spark plasma sintering. Adv Ceram Sci Eng (ACSE) 2:161–165

Golovkina LS, Orlova AI, Boldin MS et al (2017) Development of composite ceramic materials with improved thermal conductivity and plasticity based on garnet-type oxide. J Nucl Mater 489:158–163. https://doi.org/10.1016/j.jnucmat.2017.03.031

Gosset D, Provot B (2001) Boron carbide as a potential inert matrix: an evaluation. Prog Nucl Energy 38:263–266

Gotcu-Freis P, Hiernaut J-P, Colle J-Y, Nästrén C, Carretero AF, Konings RJM (2011) Vaporisation of candidate nuclear fuels and targets for transmutation of minor actinides. J Nucl Mater 411:119–125. https://doi.org/10.1016/j.jnucmat.2011.01.039

Gregg DJ, Karatchevtseva I, Triani G, Lumpkin GR, Vance ER (2013) The thermophysical properties of calcium and barium zirconium phosphate. J Nucl Mater 441:203–210. https://doi.org/10.1016/j.jnucmat.2013.05.075

Haneda H, Miyazawa Y, Shirasaki S (1984) Oxygen diffusion in single crystal yttrium aluminum garnet. J Cryst Growth 68:581–588. https://doi.org/10.1016/0022-0248(84)90465-2

Holliday K, Hartmann T, Pouneau F et al (2009) Synthesis and characterization of zirconia–magnesia inert matrix fuel: uranium homolog studies. J Nucl Mater 393:224–229. https://doi.org/10.1016/j.jnucmat.2009.06.007

Hollingsworth JP, Kuntz JD, Soules TF (2009) Neodymium ion diffusion during sintering of Nd:YAG transparent ceramics. J Phys D Appl Phys 42:052001:1–5. https://doi.org/10.1088/0022-3727/42/5/052001

Hollingsworth JP, Kuntz JD, Ryerson FJ, Soules TF (2011) Nd diffusion in YAG ceramics. Opt Mater 33:592–595. https://doi.org/10.1016/j.optmat.2010.10.048

Konings RJM, Haas D (2002) Fuels and targets for transmutation. C R Phys 3:1013–1022

Konings RJM, Bakker K, Boshoven JG, Hein H, Huntelaar ME, Van der Laan RR (1999) Transmutation of actinides in inert-matrix fuels: fabrication studies and modelling of fuel behaviour. J Nucl Mater 274:84–90. https://doi.org/10.1016/S0022-3115(99)00043-4

Mrowec S (1988) On the defect structure and diffusion kinetics in transition metal sulphides and oxides. React Solids 5:241–268. https://doi.org/10.1016/0168-7336(88)80025–1

Neeft EAC, Konings RJM, Bakkera K et al (1999) Neutron irradiation of pollycristalline yttrium aluminate garnet, magnesium aluminate spinel and A-alumina. J Nucl Mater 274:78–83. https://doi.org/10.1016/S0022-3115(99)00079-3

Neeft EAC, Bakker K, Schram RPC et al (2003) The EFTTRA-T3 irradiation experiment on inert matrix fuels. J Nucl Mater 320:106–116. https://doi.org/10.1016/S0022-3115(03)00176-4

O'Brien RC, Jerred ND (2013) Spark plasma sintering of W-UO_2 cermets. J Nucl Mater 433:50–54. https://doi.org/10.1016/j.jnucmat.2012.08.044

O'Brien RC, Ambrosi RM, Bannister NP et al (2009) Spark plasma sintering of simulated radioisotope materials within tungsten cermets. J Nucl Mater 393:108–113. https://doi.org/10.1016/j.jnucmat.2009.05.012

Orlova AI, Koryttseva AK, Kanunov AE et al (2012) Fabrication of $NaZr_2(PO_4)_3$-type ceramic materials by spark plasma sintering. Inorg Mater 48:313–317. https://doi.org/10.1134/S002016851202015X

Pelleg J (2016) Diffusion in ceramics. Series "solid mechanics and its applications". Springer, Cham. https://doi.org/10.1007/978-3-319-18437-1

Potanina E, Golovkina L, Orlova A et al (2016) Lanthanide (Nd, Gd) compounds with garnet and monazite structures. Powders synthesis by "wet" chemistry to sintering ceramics by spark plasma sintering. J Nucl Mater 473:93–98. https://doi.org/10.1016/j.jnucmat.2016.02.014

Rahaman MN (2003) Ceramic processing and sintering, 2nd edn. Marcel Dekker Inc, New York

Raison PE, Haire RG (2003) Structural investigation of the pseudo-ternary system AmO_2–Cm_2O_3–ZrO_2 as potential materials for transmutation. J Nucl Mater 320:31–35. https://doi.org/10.1016/S0022-3115(03)00165-X

Ryu HJ, Lee YW, Cha SI, Hong SH (2006) Sintering behaviour and microstructures of carbides and nitrides for the inert matrix fuel by spark plasma sintering. J Nucl Mater 352:341–348. https://doi.org/10.1016/j.jnucmat.2006.02.089

Samsonov GV (1969) The oxide handbook. Metallurgiya, Moscow (in Russian)

Somiya S, Moriyyoshi Y (1990) Sintering key papers. Elsevier Applied Science, London/New York

Suárez M, Fernández A, Menéndez JL, Torrecillas R (2009) Transparent yttrium aluminium garnet obtained by spark plasma sintering of lyophilized gels. J Nanomater. https://doi.org/10.1155/2009/138490

Williams HR, Ning H, Reece MJ, Ambrosi RM, Bannister NP, Stephenson K (2013) Metal matrix composite fuel for space radioisotope energy sources. J Nucl Mater 433:116–123. https://doi.org/10.1016/j.jnucmat.2012.09.030

Part V
SPS of Thermoelectric and Magnetic Materials

Consolidation and Behavior of FeCoV Soft Magnetic Materials via Spark Plasma Sintering

Alexander D. Dupuy, Baolong Zheng, Yizhang Zhou, Jean-Pierre Delplanque, Todd C. Monson, Enrique J. Lavernia, and Julie M. Schoenung

1 Introduction

Soft magnetic materials are widely used in a variety of electromagnetic technologies such as magnetic bearings and rotors. It is often desirable for soft magnetic materials to have very low coercivities due to the demands of their applications requiring that the magnetic materials be easily magnetized and demagnetized with minimal AC core losses. Many soft magnet applications also demand sustained exposure to high temperatures, requiring that soft magnetic materials have high Curie temperatures and temperature-independent magnetic properties. One material system that meets these needs is the compound FeCo. FeCo has a high Curie temperature (~980 °C) and a low coercive field (<10 Oe) (Sundar and Deevi 2005). Unfortunately, FeCo is extremely brittle and not suitable for many demanding soft magnetic applications. One solution to the brittleness of FeCo is to add 2 wt% V. The addition of V greatly improves the ductility of FeCo alloys while having a negligible effect on the magnetic properties (Sourmail 2005). FeCoV alloys have become widely used for a variety of soft magnetic applications due to their desirable properties.

Improvements to the magnetic and mechanical properties of these alloys can be achieved through nanostructuring. One method of directly synthesizing nanocrystalline FeCoV powders is through mechanical alloying, which involves using high

A. D. Dupuy (✉) · B. Zheng · Y. Zhou · E. J. Lavernia · J. M. Schoenung
University of California, Irvine, CA, USA
e-mail: dupuya@uci.edu; baolongz@uci.edu; yizhangz@uci.edu; lavernia@uci.edu; julie.schoenung@uci.edu

Jean-PierreDelplanque
University of California, Davis, CA, USA
e-mail: delplanque@ucdavis.edu

T. C. Monson
Sandia National Laboratories, Albuquerque, NM, USA
e-mail: tmonson@sandia.gov

© Springer Nature Switzerland AG 2019

P. Cavaliere (ed.), *Spark Plasma Sintering of Materials*,
https://doi.org/10.1007/978-3-030-05327-7_16

energy milling to break down, cold weld, and deform a mixture of powder (Suryanarayana 2001). Single phase nanocrystalline FeCoV powders have been successfully synthesized through mechanical alloying before (Behvandi et al. 2010). These mechanically alloyed FeCoV powders have excellent magnetic properties, with coercivities <70 Oe. Traditional metallurgical processing techniques often result in FeCoV parts with reduced mechanical properties due to grain growth. Processing techniques that involve deformation, such as cold rolling, result in materials with large amounts of residual stress, which will reduce magnetic properties (Shokrollahi 2009). Optimization of the mechanical and magnetic properties of FeCoV will require a processing technique that can retain nanocrystalline microstructures while avoiding the development of residual stress.

Spark plasma sintering (SPS) has garnered considerable attention for its ability to retain nanocrystalline microstructures and create materials with interesting functional properties (Maglia et al. 2013). This is in part due to its ability to process materials at lower temperatures and in shorter amounts of time compared to conventional techniques, making SPS a true nonequilibrium processing technique. Decades of research has led to SPS finding success in nearly every category of material and property (Orrù et al. 2009). Hall-Petch strengthening effects make retention of nanostructures in metals and alloys particularly important (Yip 1998; Meyers et al. 2006). SPS is uniquely positioned to take advantage of grain boundary strengthening due to its ability to process materials to high density with minimal grain growth.

Previous work on FeCo has shown that it is possible to retain nanocrystalline grain sizes and very low coercivities (<20 Oe) by consolidating mechanically alloyed powder using SPS (Kim et al. 2000). Despite this and the abundance of materials successfully consolidated by SPS, no work has been done on using SPS to consolidate FeCoV. In this study we show that it is possible to create FeCoV magnets with excellent properties using mechanical alloying and SPS. Thus, we explore the effect of mechanical alloying and SPS conditions on the microstructure of the powders and consolidated samples. These microstructural features will then be linked to the mechanical and magnetic properties of our samples.

2 Methods

Mechanically alloyed FeCoV powders were synthesized from elemental powders. Fe, Co, and V elemental powders were sourced from Advanced Powder Solutions (Houston, TX, USA) and were prepared with a stoichiometry of 49 wt% Fe, 49 wt% Co, and 2 wt% V. The elemental powders were premixed without any process control agent. The powder mixture was mechanically milled and alloyed using a SPEX 8000D (SPEX Sample Prep, Metuchen, NJ, USA) high energy ball mill. 316 L stainless steel jars and balls were used for milling. 100 g of 6.35 mm balls were used with a ball-to-powder weight ratio of 5:1. The milling experiments were performed at room temperature in an argon atmosphere. Powders were milled for

2–16 h. Direct current plasma emission spectroscopy was performed by Luvak, Inc. to confirm the chemical composition of the final powders. The effect of ball-to-powder weight ratio during mechanical alloying was assessed by milling for 8 h with a ratio of 5:1, 10:1, and 20:1. Phase identification for both powder and bulk samples was performed using a Scintag XDS 2000 XRD (Thermo Fisher Scientific, Waltham, MA USA) over the 2θ range of 30–90°. Crystallite size was determined from the powder XRD data using an integral breadth analysis method (Jiang et al. 1999)

Powders that were mechanically alloyed for 8 h were consolidated using a Fuji SPS model 825S (Fuji Electronic Industrial Co., Ltd., Kawasaki, Japan). Consolidation was performed at temperatures of 650–850 °C with a hold time of 2 min and a heating rate of 100–200 °C/min. Temperature was measured using a K-type thermocouple inserted into a hole drilled halfway through the die. An applied pressure of 100 MPa was used throughout the experiment. All experiments were performed under vacuum. All samples utilized for this study were 10 mm in diameter. A summary of the SPS samples and their processing conditions can be found in Table 1.

SPS-consolidated samples were polished metallographically, and their densities were measured using the Archimedes method. Grain size was determined from the XRD data using an integral breadth analysis method (Jiang et al. 1999). Microstructure analysis was performed on the powder and consolidated samples using a Phillips/FEI XL-30 SEM (Hillsborough, OR, USA) and a JEOL JEM-2500SE TEM (Tokyo, Japan). Micrographs were taken of both polished and fractured surfaces. Thin foils for TEM observations were prepared via mechanical grinding and polishing to a thickness of about 30 μm, followed by ion milling to an electron transparent thickness. This was carried out using a Gatan precision ion polishing system (PIPS) 691 (Pleasanton, CA, USA) with an accelerating voltage of 4 kV from both sides until perforation occurred. Hardness was measured on the consolidated samples using a Buehler Micromet 2004 Micro Hardness Tester (Lake Bluff, IL, USA).

Magnetic measurements were performed on powder and bulk samples using an alternating gradient magnetometer (AGM) and a vibrating sample magnetometer (VSM) (Lake Shore Cryotronic, Westerville, OH, USA). Hysteresis loops of mass normalized magnetization M (emu/g) vs field H (kOe) were collected by sweeping the applied field from −15 to 15 kOe. Coercivity, H_C [Oe], remanent magnetization M_R [emu/g], and saturation magnetization M_S [emu/g] were extracted from these hysteresis loops.

Table 1 List of SPS sample processing conditions used in this study

Milling time (h)	SPS temperature (°C)	Heating rate (°C/min)	Pressure (MPa)	Hold time (min)
8	650	100	100	2
8	700	200	100	2
8	700	100	100	2
8	800	100	100	2

3 Results

3.1 Mechanically Alloyed Powders

Elemental powders were used for the mechanical alloying of the FeCoV powders in this study. XRD confirms that the starting powders contain no extraneous phases (seen in Fig. 1). Direct current plasma emission spectroscopy measurements (not shown) indicate that the largest impurities in the final FeCoV powders are Mn (0.0023%) and Cr (0.018%). SEM microscopy of the elemental powders indicates that all three powders have different morphologies. The Fe powders have large flat agglomerates, as seen in Fig. 2. Both Co (Fig. 3) and V (Fig. 4) have much finer particles with a more spherical morphology. V powders, however, are more agglomerated than the Co powders.

The evolution of the FeCoV powder from mixed constituents to single-phase powder can be seen in Fig. 5. XRD measurements of the starting mixture (0 h) display peaks from the three starting constituents. However, a single-phase BCC powder is achieved after only 2 h of milling. Although further milling does not alter the phase state of the powders, it does decrease the grain size, as seen in Fig. 6. According to the XRD results, the grain size decreases linearly from 28 nm after 6 h of milling to 8 nm after 16 h of milling.

After 2 h of mechanical alloying, the FeCoV powders consist primarily of large flat particles and agglomerates, as seen in Fig. 7. The morphology of the powder becomes more spherical after 8 h of milling. Figure 8 shows that the powders milled for 8 h have a greatly reduced particle size. Additionally, the particles have a more homogenous size and shape. Individual particles are ~5 μm in size, and cold weld

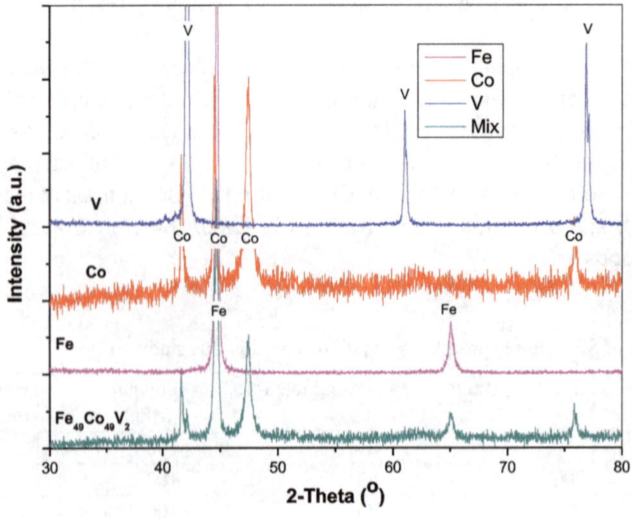

Fig. 1 XRD patterns of the elemental powders used to make FeCoV

Fig. 2 SEM micrograph of the elemental Fe powder used to make the FeCoV mechanically alloyed powders

Fig. 3 SEM micrograph of the elemental Co powder used to make the FeCoV mechanically alloyed powders

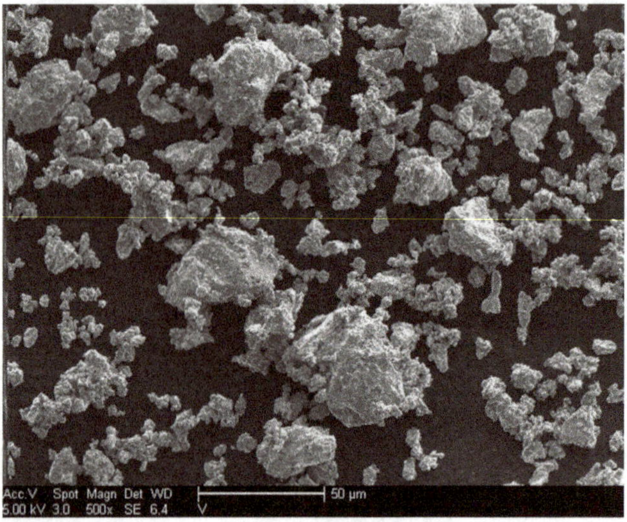

Fig. 4 SEM micrograph of the elemental V powder used to make the FeCoV mechanically alloyed powders

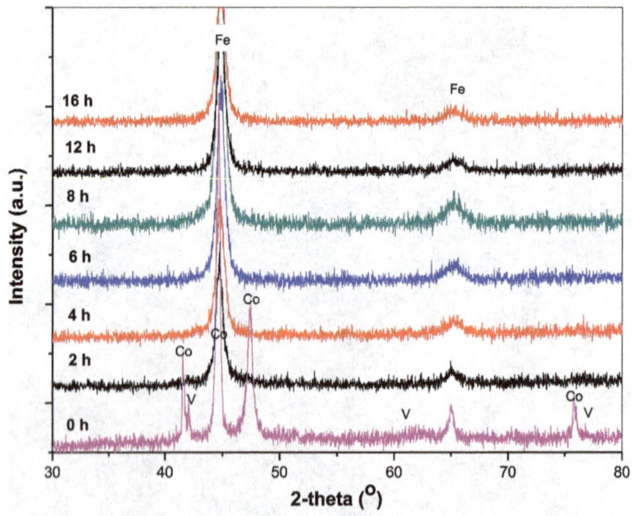

Fig. 5 XRD patterns for the mechanically alloyed FeCoV powders after milling for 0–16 h

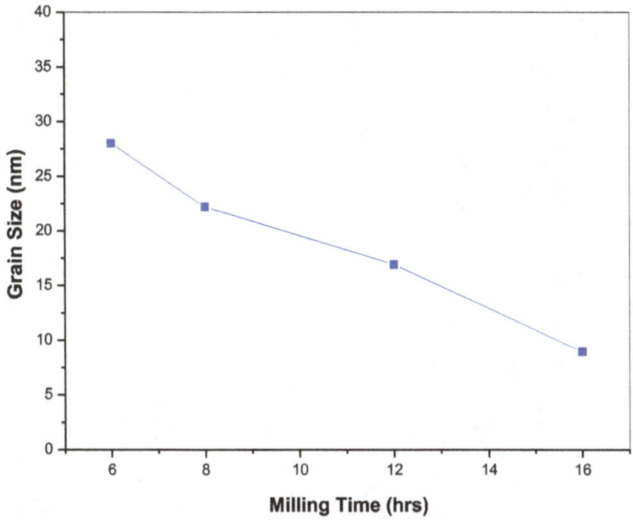

Fig. 6 Grain size vs milling time for mechanically alloyed FeCoV powders. Grain size is calculated from XRD

Fig. 7 SEM micrograph of the FeCoV powder after 2 h of mechanical alloying

Fig. 8 SEM micrograph of the FeCoV powder after 8 h of mechanical alloying

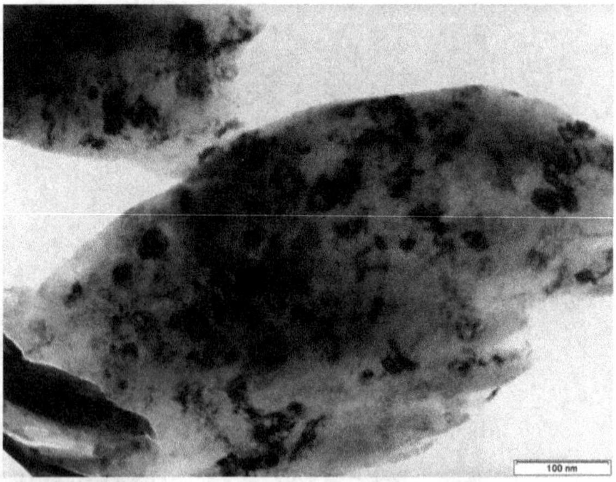

Fig. 9 TEM micrograph of the FeCoV powder after 8 h of mechanical alloying

into larger aggregates that are 20–40 μm in size. The particle morphology does not change significantly after this, even after 16 h of milling.

Direct observation of nanostructuring in the FeCoV mechanically alloyed powders can be seen using bright field TEM. Figures 9 and 10 are TEM micrographs of the SPEX milled powders after 8 h and 16 h, respectively. Analysis of the TEM micrographs indicate that the powders have a grain size of 23 nm after 8 h of milling and 14 nm after 16 h of milling. These results confirm the grain sizes calculated from XRD.

Fig. 10 TEM micrograph of the FeCoV powder after 16 h of mechanical alloying

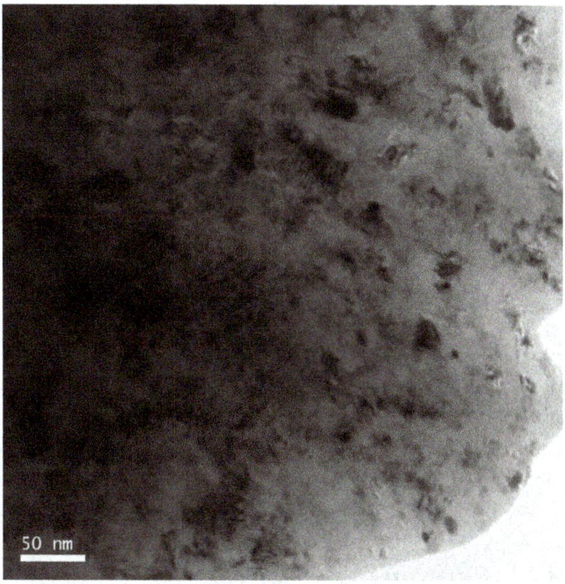

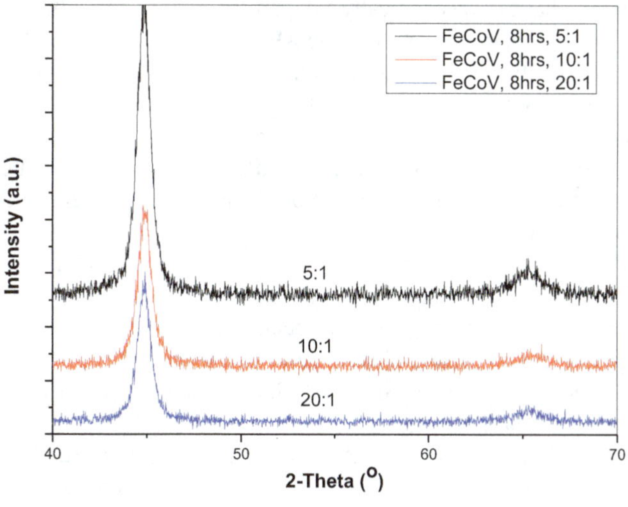

Fig. 11 XRD patterns for the mechanically alloyed FeCoV powders milled for 8 h with a ball to powder weight ratio of 5:1, 10:1, and 20:1

Increasing ball-to-powder weight ratio is shown to decrease the XRD peak intensity, as seen in Fig. 11. This reduction in peak intensity is likely due to an increase in strain and a decrease in grain size in the higher ratio powders. A higher ball-to-powder ratio means a greater frequency of collisions during milling, leading to a more efficient milling process. Additionally, the XRD patterns show no

Table 2 Relevant magnetic properties of the mechanically alloyed FeCoV powders

Milling time (h)	Coercive field, H_C (Oe)	Saturation magnetization, M_S (emu/g)	M_r/M_S (%)
8	62.8	201.2	4.65
12	42.7	213.0	3.92
16	63.0	208.3	4.45

extraneous peaks. The higher energy milling runs did not change the structure of the powder, only the grain size and strain. The remainder of this study will focus on powders that are mechanically alloyed using a 5:1 ball-to-powder weight ratio.

The FeCoV mechanically alloyed powders display a familiar soft magnetic hysteresis response (not shown). The coercive field and saturation magnetization of the three powders are very similar. However, small changes in the magnetic properties can be seen with increasing milling time. A summary of the magnetic properties for the three powders can be seen in Table 2.

3.2 Bulk Sample Properties

FeCoV powders that are mechanically alloyed for 8 h maintain their single-phase state after SPS consolidation, as seen in the XRD patterns in Fig. 12. The XRD peaks are visibly sharper after SPS consolidation compared to the mechanically alloyed powder. Higher SPS temperatures lead to further sharpening of the peaks. This change in peak shape can be attributed to grain growth and strain relaxation. A calculated grain size of 52 nm is achieved after consolidation at 700 °C with a 100 °C/min heating rate. This is a little more than double the calculated grain size found in the 8 h powder (23 nm).

SEM and TEM microscopy indicate that the SPS consolidated samples have a bimodal grain structure. This bimodal structure manifests as coarse grains surrounded by a matrix of fine grains, as seen in the fracture surface SEM micrograph in Fig. 13. These coarse grains range in size from 200 to 1000 nm. Figure 14 shows TEM microscopy of an SPS consolidated sample, confirming that these coarse grains are not nanostructured.

Increasing SPS temperature increases the relative density of the final bulk samples. This can be qualitatively visualized from micrographs of polished surfaces. The FeCoV sample consolidated at 700 °C shows obvious porosity on the surface (Fig. 15), while the 800 °C sample shows significantly less surface porosity (Fig. 16). This trend is confirmed by the Archimedes density measurements found in Fig. 17. The highest relative density in this study was 95.8%, achieved from sintering the 8 h mechanically alloyed powder at 800 °C and 100 °C/min. Changing the heating rate from 100 °C/min to 200 °C/min has no meaningful impact on the final density when sintering at 700 °C. It can be seen that the hardness of the SPS

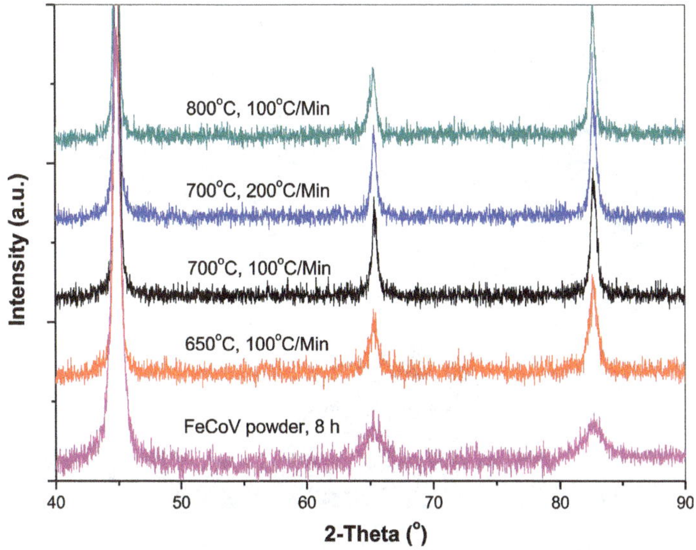

Fig. 12 XRD patterns for FeCoV samples consolidated using SPS from powders that are mechanically alloyed for 8 h

Fig. 13 SEM micrograph of a FeCoV SPS consolidated sample using the 8 h mechanically alloyed powder. The sample was consolidated at 700 °C and 100 °C/min. The bimodal microstructure displays both coarse-grained (CG) and fine-grained (FG) structures

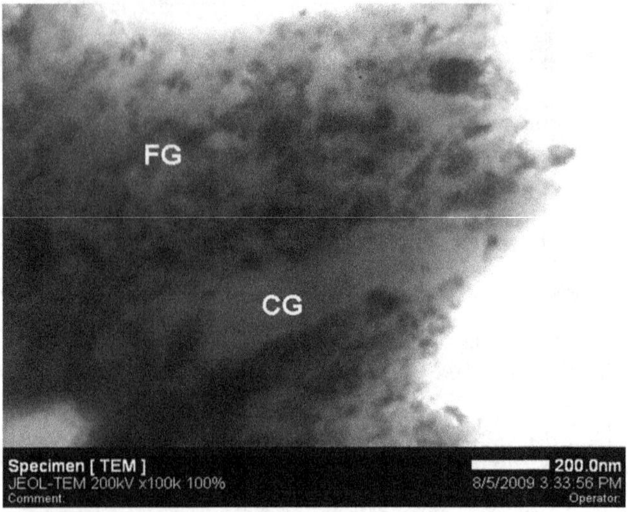

Fig. 14 TEM micrograph of a FeCoV SPS consolidated sample using the 8 h mechanically alloyed powder. The sample was consolidated at 700 °C and 100 °C/min. The bimodal microstructure displays both coarse-grained (CG) and fine-grained (FG) structures

Fig. 15 SEM micrograph of a polished surface from a FeCoV SPS consolidated sample using the 8 h mechanically alloyed powder. The sample was consolidated at 700 °C and 100 °C/min

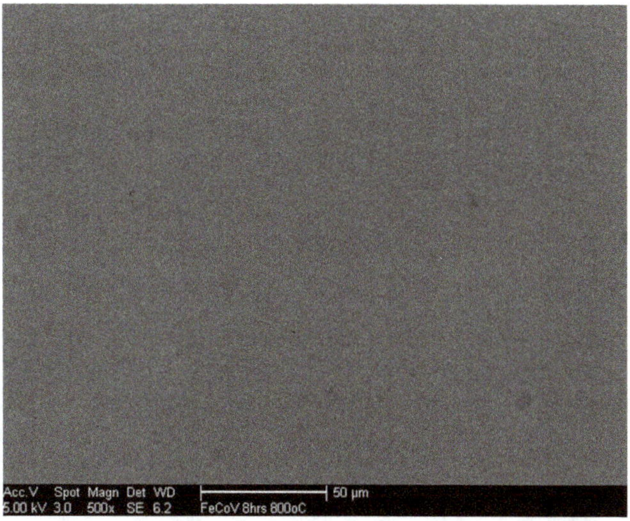

Fig. 16 SEM micrograph of a polished surface from a FeCoV SPS consolidated sample using the 8 h mechanically alloyed powder. The sample was consolidated at 800 °C and 100 °C/min

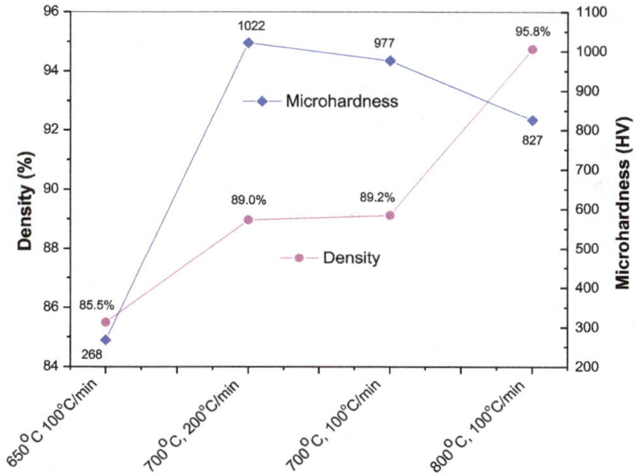

Fig. 17 Relative density and hardness vs sintering conditions for the FeCoV samples sintered using powder that was mechanically alloyed for 8 h

sample consolidated at 800 °C (827 HV) is lower than the sample consolidated at 700 °C (977 HV), despite the higher temperature sample having a higher density. The maximum hardness (1022 HV) is achieved from processing at 700 °C and 200 °C/min.

Table 3 Relevant magnetic properties of the SPS consolidated FeCoV samples made from powders mechanically alloyed for 8 h

SPS processing temperature (°C)	Heating rate (°C/min)	Coercive field, H_C (Oe)	Saturation magnetization, M_S (emu/g)
650	100	16.5	199.7
700	200	16.8	226.8
700	100	16.5	228.9
800	100	11.7	211.4

The SPSed FeCoV samples have lower coercive field values than the 8 h powders that they are made from, as summarized in Table 3. A further decrease in coercivity can be seen after SPS consolidation at 800 °C compared to the 700 °C and 650 °C samples. This can be attributed to grain growth from the increased SPS temperatures. The saturation magnetization is largely unaffected by SPS processing conditions.

4 Discussion

4.1 Mechanically Alloyed Powders

The mechanically alloyed powders are shown to be single phase after only 2 h. At this stage the Co and V peaks disappear, leaving behind only a BCC solid solution structure. This rapid solid-state synthesis occurs because of the high energies associated with SPEX milling, which lead to repeated fracturing and cold welding of the powders. The individual powder components are intimately mixed from the milling process. When the components are in intimate contact, the continued high energy impacts will result in diffusion and solid solution formation (Suryanarayana 2001).

The evolution of this alloying process can be seen in Fig. 5. As the milling time is increased, the primary Fe BCC peak shifts, indicating that the other elements are gradually being incorporated into this phase. This change is subtle and can be more easily visualized by looking at the change in lattice parameter, as seen in Fig. 18. A significant decrease in the lattice parameter occurs after 2 h of milling, which is when the majority of the alloying occurs. The decrease in lattice parameter is more subtle after 2 h, showing the continued incorporation of Co and V into the BCC structure.

It is known that FeCo forms a BCC solid solution (Moumeni et al. 2005). This structure forms during mechanical alloying by Co diffusing into Fe and sitting directly on Fe sites. Mechanical alloying therefore allows Fe and Co to be mixed at the atomic level. Despite the low solubility of V in FeCo (Chen 1961), V can also be readily incorporated into the FeCo structure through mechanical alloying (Behvandi et al.). Extensive milling times can produce the Co_3V intermetallic phase, which will reduce the magnetic properties. Our FeCoV mechanically alloyed

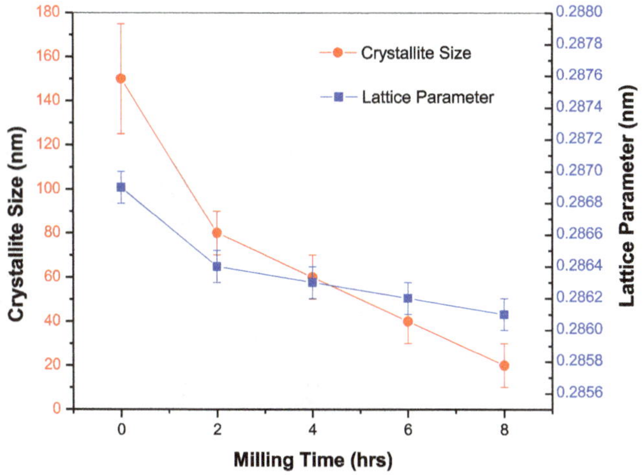

Fig. 18 Calculated crystallite size and lattice parameter for the mechanically alloyed FeCoV powders

powders do not show any Co_3V peaks in XRD, indicating that any intermetallic phase is in quantities smaller than the resolution of XRD. The site occupancy of V in FeCo is more controversial. Experimental evidence is contradictory, with some work indicating that V sits on the Fe site (Persiano and Rawlings 1987), while other measurements show that V sits on the Co site (Williams et al. 1990). Both of these studies also contradict simulations, which indicate that V has a mixed occupancy (Aykol et al. 2010).

Mechanical alloying reduces the grain size linearly starting from 8 h of milling. The smallest grain size in this study is 14 nm, achieved from mechanical alloying for 16 h. Grain size refinement during milling is often attributed to the formation of deformation-induced shear bands caused by the localized plastic deformation. Further milling results in these shear bands converting into small angle grain boundaries due to dislocation annihilation and recombination (Fecht 1995).

Significant changes occur in the particle morphology during mechanical alloying. At 2 h of milling, cold welding of the powder results in mostly flattened particles. After 8 h of milling, some flat particles remain, but the majority of particles are smaller and more equiaxed. This loss of flattened particles occurs because cold welding no longer dominates particle fracture at this stage of mechanical alloying. The particle morphology does not significantly change after this, even after 16 h of milling. A homogenous particle morphology is obtained after 8 h due the mechanical alloying process reaching an equilibrium between fracturing and cold welding (Chitsazan et al. 2011b).

Fig. 19 Photograph of a 50 mm and a 10 mm FeCoV SPS sample made at 700 °C

4.2 Bulk Sample Properties

SPS samples made from the 8 h mechanically alloyed powders were successfully consolidated to high density while still maintaining nanocrystalline grain sizes. Additionally, all of the bulk samples maintain the single-phase state from the mechanically alloyed powders. SPS enables the consolidation of FeCoV samples at low temperatures (650–800 °C) and high heating rates (100–200 °C/min in this study), allowing for microstructural equilibrium to be avoided in our samples. These processing conditions result in FeCoV samples that retain the phase state and much of the original microstructure of the mechanically alloyed powder. Although all analysis in this study was performed on samples with a diameter of 10 mm, samples as large as 50 mm could be successfully sintered. A photograph showing both a 10 mm sample and a 50 mm sample is provided in Fig. 19.

As anticipated, higher densities are achieved at higher SPS consolidation temperatures. Unexpectedly, however, this trend does hold for hardness. Higher SPS temperatures produce lower hardness values despite the increase in density. The mechanism behind this result becomes clear when examining the effect of SPS heating rate. When sintering at 700 °C, 200 °C/min and 100 °C/min heating rates produce similar densities. Processing at 200 °C/min yields a sample with a higher hardness (1022 HV) than one processed at 100 °C/min (977 HV). Faster heating rates will result in smaller grain sizes due to less time being spent at high temperature. This indicates that our FeCoV samples exhibit grain boundary strengthening, which has been seen in previous studies on FeCoV (Duckham et al. 2003). Due to its nanocrystalline microstructure, our FeCoV samples are harder than coarse-grained FeCo (~200–300 Hv) (Albaaji et al. 2017) and cold-rolled FeCoV (~300–450 HV) (Hosoda et al. 2000).

Our bulk SPSed FeCoV samples display a bimodal grain structure consisting of a nanostructured matrix surrounding coarse grains. This kind of microstructure has been seen before and is commonly attributed to localized heating at particle contact points due to joule heating (Xiong et al. 2012). Current will pass through conductive samples during SPS, leading to variations in the local temperature as the microstructure develops. In the early stages of sintering, the narrow contact points between particles can experience higher temperatures than the body of the particle. Even brief exposure to high temperatures can result in rapid grain growth.

4.3 Magnetic Properties

The mechanically alloyed FeCoV powders display high saturation magnetization and low coercivities, which are in line with previous studies on mechanically alloyed FeCoV (Chitsazan et al. 2011a). At 12 h of milling, there is an increase in magnetization and a decrease in coercivity compared to milling for 8 h. This trend is reversed after milling for 16 h. Such a change in properties during mechanical alloying is not unexpected and has been seen in previous studies (Chitsazan et al. 2011a). The decrease in coercivity and increase in magnetization after 12 h can be attributed to the completion of the mechanical alloying process. After 12 h the Co and V have been completely incorporated into the BCC lattice. The increase in coercivity and decrease in magnetization after 16 h can be attributed to an increase in dislocation density and the formation of small amounts of Co_3V intermetallic. These defects are known to pin domain walls, decreasing the magnetic properties (Shokrollahi 2009).

The SPS consolidated FeCoV samples have lower coercivities than the mechanically alloyed powders they are made from. This can be attributed to grain growth (Yu et al. 1999) and a decrease in dislocation density during SPS. Consolidation at 800 °C results in a lower coercivity compared to the samples consolidated at 650 °C and 700 °C, likely due to the increased grain growth and decreased residual stress. The coercivity values of our SPS samples are lower than conventionally processed FeCoV alloys (Sundar and Deevi 2005). However, our coercivity values are higher than many heat-treated FeCoV alloys, indicating that additional optimization of the microstructure may be necessary to maximize magnetic properties.

5 Conclusions

We prepared nanostructured FeCoV powders through mechanical alloying. Only 2 h of mechanical alloying was necessary to produce single-phase FeCoV powders. 8 h of milling yielded powders with a homogenous particle shape and size as well as <25 nm grain sizes. Mechanical alloying duration is shown to slightly modify the low magnetic coercivity and high magnetic saturation of the powders. SPS consolidation of these mechanically alloyed powders produces samples with a bimodal grain structure consisting primarily of the original nanocrystalline microstructure along with coarse grains. A high hardness of 1022 HV is obtained from sintering at

700 °C with a heating rate of 200 °C/min. Slower heating rates and higher temperatures result in lower hardness values despite increasing the density of the samples, indicating that the hardness is primary controlled by grain boundary strengthening. Lower coercivities are obtained in the SPS consolidated samples compared to the original mechanically alloyed powders due to grain growth and reduced residual stress. These findings show that the magnetic and mechanical properties of FeCoV are strongly dependent on microstructure. SPS is the ideal tool for controlling the final microstructure of FeCoV and optimizing its properties.

Acknowledgments This work is supported by the AFOSR contract #FA955009C0025. Thanks to Professor Kai Liu and Dr. Randy K. Dumas of the Department of Physics at the University of California, Davis, for the measurement of magnetic properties.

Sandia National Laboratories is a multimission laboratory managed and operated by National Technology and Engineering Solutions of Sandia LLC, a wholly owned subsidiary of Honeywell International Inc. for the US Department of Energy's National Nuclear Security Administration under contract DE-NA0003525. This paper describes objective technical results and analysis. Any subjective views or opinions that might be expressed in the paper do not necessarily represent the views of the US Department of Energy or the United States Government.

References

Albaaji AJ, Castle EG, Reece MJ, Hall JP, Evans SL (2017) Enhancement in the elongation, yield strength and magnetic properties of intermetallic FeCo alloy using spark plasma sintering. J Mater Sci 52:13284–13295. https://doi.org/10.1007/s10853-017-1435-5

Aykol M, Mekhrabov AO, Akdeniz MV (2010) Effect of vanadium on atomic ordering characteristics and anti-phase boundary energies of B2–FeCo alloys. Intermetallics 18:893–899. https://doi.org/10.1016/j.intermet.2009.12.029

Behvandi A, Shokrollahi H, Chitsazan B, Ghaffari M (2010) Magnetic and structural studies of mechanically alloyed nanostructured Fe49Co49V2 powders. J Magn Magn Mater 322:3932–3937. https://doi.org/10.1016/j.jmmm.2010.08.025

Chen CW (1961) Metallurgy and magnetic properties of an Fe-Co-V alloy. J Appl Phys 32:S348–S355. https://doi.org/10.1063/1.2000465

Chitsazan B, Shokrollahi H, Behvandi A, Ghaffari M (2011a) Magnetic, structural and micro-structural properties of mechanically alloyed nano-structured Fe48Co48V4 powder containing inter-metallic Co3V. J Magn Magn Mater 323:1128–1133. https://doi.org/10.1016/j.jmmm.2010.11.079

Chitsazan B, Shokrollahi H, Behvandi A, Mirzaee O (2011b) Characterization and magnetic coercivity of nanostructured (Fe50Co50)100-XVX=0,2,4 powders containing a small amount of Co3V intermetallic obtained by mechanical alloying. Powder Technol 214:105–110. https://doi.org/10.1016/j.powtec.2011.07.042

Duckham A, Zhang D, Liang D, Luzin V, Cammarata R, Leheny R, Chien C, Weihs T (2003) Temperature dependent mechanical properties of ultra-fine grained FeCo–2V. Acta Mater 51:4083–4093. https://doi.org/10.1016/S1359-6454(03)00228-3

Fecht H-J (1995) Nanostructure formation by mechanical attrition. Nanostructur Mater 6:33–42. https://doi.org/10.1016/0965-9773(95)00027-5

Hosoda H, Miyazaki S, Inoue K, Fukui T, Mizuuchi K, Mishima Y, Suzuki T (2000) Cold rolling of B2 intermetallics. J Alloys Compd 302:266–273. https://doi.org/10.1016/S0925-8388(00)00583-1

Jiang HG, Rühle M, Lavernia EJ (1999) On the applicability of the x-ray diffraction line profile analysis in extracting grain size and microstrain in nanocrystalline materials. J Mater Res 14:549–559. https://doi.org/10.1557/JMR.1999.0079

Kim YD, Chung JY, Kim J, Jeon H (2000) Formation of nanocrystalline Fe–Co powders produced by mechanical alloying. Mater Sci Eng A 291:17–21. https://doi.org/10.1016/S0921-5093(00)00982-5

Maglia F, Tredici IG, Anselmi-Tamburini U (2013) Densification and properties of bulk nanocrystalline functional ceramics with grain size below 50 nm. J Eur Ceram Soc 33:1045–1066. https://doi.org/10.1016/j.jeurceramsoc.2012.12.004

Meyers MA, Mishra A, Benson DJ (2006) Mechanical properties of nanocrystalline materials. Prog Mater Sci 51:427–556. https://doi.org/10.1016/j.pmatsci.2005.08.003

Moumeni H, Alleg S, Greneche JM (2005) Structural properties of Fe50Co50 nanostructured powder prepared by mechanical alloying. J Alloys Compd 386:12–19. https://doi.org/10.1016/j.jallcom.2004.05.017

Orrù R, Licheri R, Locci AM, Cincotti A, Cao G (2009) Consolidation/synthesis of materials by electric current activated/assisted sintering. Mater Sci Eng R Rep 63:127–287. https://doi.org/10.1016/j.mser.2008.09.003

Persiano AIC, Rawlings RD (1987) A Mössbauer Investigation of Equiatomic FeCo with Vanadium and Niobium Additions. Phys Status Solidi 103:547–556. https://doi.org/10.1002/pssa.2211030227

Shokrollahi H (2009) The magnetic and structural properties of the most important alloys of iron produced by mechanical alloying. Mater Des 30:3374–3387. https://doi.org/10.1016/j.matdes.2009.03.035

Sourmail T (2005) Near equiatomic FeCo alloys: constitution, mechanical and magnetic properties. Prog Mater Sci 50:816–880. https://doi.org/10.1016/J.PMATSCI.2005.04.001

Sundar RS, Deevi SC (2005) Soft magnetic FeCo alloys: alloy development, processing, and properties. Int Mater Rev 50:157–192. https://doi.org/10.1179/174328005X14339

Suryanarayana C (2001) Mechanical alloying and milling. Prog Mater Sci 46:1–184. https://doi.org/10.1016/S0079-6425(99)00010-9

Williams A, Kwei GH, Ortiz AT, Karnowski M, Warburton WK (1990) Combined neutron and x-ray powder diffraction study of Fe0.50Co0.48V0.02. J Mater Res 5:1197–1200. https://doi.org/10.1557/JMR.1990.1197

Xiong Y, Liu D, Li Y, Zheng B, Haines C, Paras J, Martin D, Kapoor D, Lavernia EJ, Schoenung JM (2012) Spark plasma sintering of cryomilled nanocrystalline Al alloy – part I: microstructure evolution. Metall Mater Trans A 43:327–339. https://doi.org/10.1007/s11661-011-0933-3

Yip S (1998) Nanocrystals: the strongest size. Nature 391:532–533. https://doi.org/10.1038/35254

Yu RH, Basu S, Zhang Y, Parvizi-Majidi A, Xiao JQ (1999) Pinning effect of the grain boundaries on magnetic domain wall in FeCo-based magnetic alloys. J Appl Phys 85:6655–6659. https://doi.org/10.1063/1.370175

Thermoelectric Materials Synthesized by Spark Plasma Sintering (SPS) for Clean Energy Generation

Sunanda Mitra and Tanmoy Maiti

1 Introduction

The ever-increasing burning of fossil fuels along with the emission of anthropogenic carbon dioxide lays a negative impact to the climatic situation and to meeting the global energy demand. So, the paramount interest is to establish useful measures for the elevating energy crisis and the menacing global warming. On the way to conquer such issues, we need to develop alternative green energy sources such as wind, solar, biomass, etc. On the other hand, in our daily life, the tremendous amount of unused waste heat is generated from engines, home appliances, automobiles, manufacturing plants, power plants, nuclear decay, etc. as shown schematically in Fig. 1. So, it would be extremely beneficial to develop a viable technology, which can directly convert this waste heat into electricity. In this context, thermoelectricity becomes an important regime of research, as it offers a unique power generation solution via direct conversion of thermal energy into electrical energy.

The three principle phenomena, on which the thermoelectric devices work, were developed by Seebeck (1823), Peltier (1834), and Thomson (1851a, b) in the nineteenth century. Clean energy generations using TE devices particularly use Seebeck effect, which is defined as the direct conversion of temperature difference into electrical energy.

S. Mitra · T. Maiti (✉)
Plasmonics and Perovskites Laboratory, Department of Materials Science and Engineering, Indian Institute of Technology Kanpur, Kanpur, Uttar Pradesh, India
e-mail: tmaiti@iitk.ac.in

© Springer Nature Switzerland AG 2019

P. Cavaliere (ed.), *Spark Plasma Sintering of Materials*,
https://doi.org/10.1007/978-3-030-05327-7_17

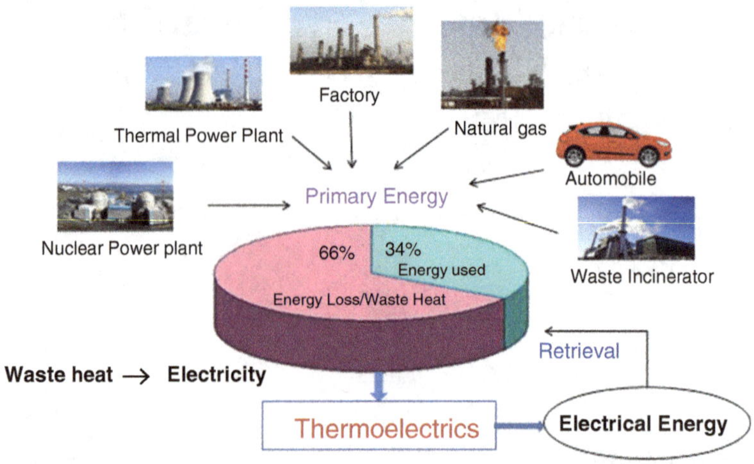

Fig. 1 Waste heat recovery

2 Applications of Thermoelectric Materials

Thermoelectric covers a wide range of application in the field of automotive industries, defense, healthcare, power plants, and space applications. In power plants and manufacturing industries, huge amount of heat is wasted, which can be scavenged to produce electricity using thermoelectric power generator. In automobiles thermoelectric are used as automotive thermoelectric generators (ATEGs) for increasing fuel efficiency. In addition, microscale power generators made by integrated circuit (IC) technology are used for energy conversion in micro-systems and wearable electronics.

A reliable power source is the prime concern for all the interstellar projects including a variety of space probes and flights sent to the Moon, Mars, or the outer planets of the Solar System. Environmental radiation in space such as solar radiation, earth infrared radiation or albedo radiation, and internal operational thermal loads produce heat flows inside the structure of satellites (Messerschmidt et al. 1997). Thermoelectrics had been used since the 1960s in radioisotope thermoelectric generators (RTGs) to supply space systems to deep space with electrical power (Hyder et al. 2003; Lange and Carroll 2008). In RTGs a radioactive decay of a radioisotope produces radiation that is decelerated by a graphene core. Therefore, the core is heated up (ca. 1200 K). This heat source is coupled by TE generators to radiator. RTGs had been used by NASA in various interstellar projects such as Pioneer and Voyager and the robotic rover Curiosity sent to Mars (Harland 2010; Bennett 2006). Furthermore, modern iteration of space-bound RTGs, named multimission radioisotope thermoelectric generators (MMRTGs), had been developed by NASA in which PbTe-based TE generators were used to obtain ~125 watts of electrical energy from Pu-238 fuel pellets over the course of 14 years (Ritz and Peterson 2004; Misra 2006).

3 Thermoelectric Module and Figure of Merit

TE cooling and power generation devices are mainly based on the Peltier and the Seebeck effect, respectively. Such devices usually comprise of the similar architecture and built up with a primary unit known as TE module. Each TE module contains n- and p-type legs lying between the two ceramic plates, which provide electrical insulation and mechanical resistance. Further, electrical contact pads are used to connect several n- and p-type legs in the main structure as shown in Fig. 2a. Likewise, in the device, n-type and p-type modules are connected in such a manner that heat flows in parallel and electric current flows in series. Both the n and p branches have a prominent role toward obtaining the effective TE effects: they possess Seebeck coefficients with opposite sign.

At first, upon exposure to heat source, the charge carriers such as electrons and holes in their respective legs move away from the hot end to cold end of a TE generator. Such flow of charge carriers gives rise to an electric current. That means, current will flow in a closed loop through the legs and an external load, when proper electrical contacts are made in the resultant structure. Thus, by utilizing the same module, it is possible to make the entire structure function as a power generator and a refrigerator as shown in Fig. 2b.

The coefficient of performance (COP) and efficiency (η) of TE device are important parameters to evaluate the potential of TE devices.

$$COP_{\max} = \frac{Q_c}{P} = \frac{T_h}{T_h - T_c}\left[\frac{\sqrt{1 + Z_{pn}T_M} - \left(\frac{T_h}{T_c}\right)}{1 + \sqrt{1 + Z_{pn}T_M}}\right] \tag{1}$$

On the other hand, the η can be expressed by the relation (2), i.e., the relation of the output electrical power (P_0) to the input thermal power at the hot side (Q_h).

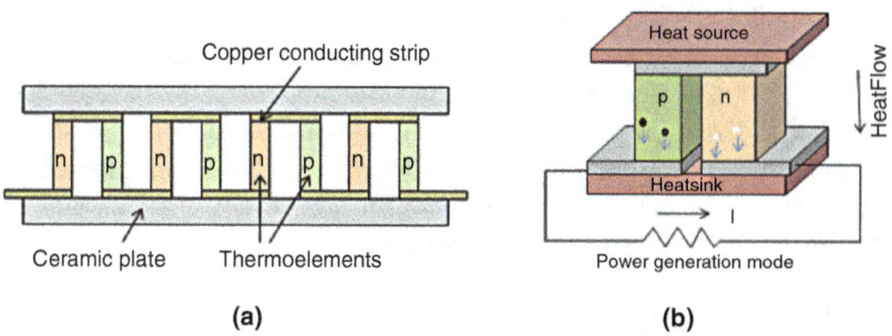

Fig. 2 (**a**) Schematic drawing of TE module and (**b**) the structure that function as both the power generation and refrigeration

$$\eta_{\max} = \frac{P_0}{Q_h} = \frac{T_h - T_c}{T_h}\left[\frac{\sqrt{1 + Z_{pn}T_M} - 1}{\sqrt{1 + Z_{pn}T_M + \frac{T_c}{T_h}}}\right] \tag{2}$$

These two equations show that they solely depend on the temperature at the hot (T_h) and cold (T_c) junctions and the dimensionless merit factor is $Z_{pn}T_M$. The figure of merit value for a TE couple is expressed as:

$$Z_{pn} = \frac{\left(s_p - s_n\right)^2 T}{\left(\rho_n k_n\right)^{1/2} + \left(\rho_p k_p\right)^{1/2}} \tag{3}$$

where, S, ρ, and k represent the Seebeck coefficient, electrical resistivity, and thermal conductivity of the respective materials n and p. Equation (3) shows that the Z_{pn} factor is incorporated with physical transport properties. Thus, to obtain maximum efficiency, materials should be chosen so to optimize this factor. However, in practice, it is rather easier to focus on each of the branches of the TE couple individually, in order to optimize them separately. So, for this, the dimensionless figure of merit (Altenkrich 1909) is given as:

$$ZT = \frac{S^2\sigma T}{k} = \frac{S^2 T}{\rho k} \tag{4}$$

where σ and T represent the electrical conductivity and the absolute temperature, respectively.

4 Challenges in Enhancing *ZT*

Over the long period, the motive of the researcher had been to achieve the increased value of *ZT*. In the previous section, we have noticed that the relation *ZT* combined with the parameters S, σ, and k. These parameters are interlinked with each other in such a manner that it is difficult to optimize them separately (Zheng 2008). In addition, material's properties such as electronic structure, carrier concentration, and crystal structure have an effect on these parameters. But, their interconnection puts a restriction over the choice of material for the TE development and applications. Figure 3 shows the schematic of interlink among these properties with respect to the carrier concentration.

The selection of material plays a crucial role in TE. Materials can be classified as metals, semiconductors, and insulators depending on the carrier concentration and electronic structure. Figure 3a shows that at low carrier concentration for insulators, S value is high, whereas σ and κ are low, thus making them poor thermal and electrical conductors. On the other hand, at high concentration, metals possess high

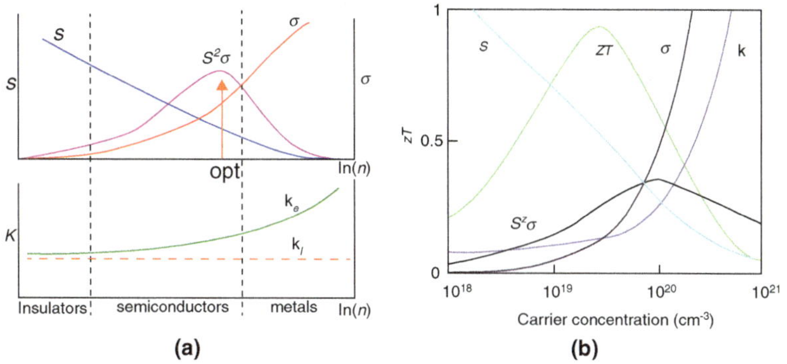

Fig. 3 (**a**) and (**b**); classification of material and schematic drawing of variation of their properties *ZT*, *s*, *k*, *σ*, and $s^2\sigma$ vs carrier concentration n. (Zheng 2008; Szczech et al. 2010)

value of σ and κ with reduced value of *S*. Thus, a trade-off should exist between these two materials: metals and insulators. In this context, semiconductors have occupied such place as shown in Fig. 3a. It exhibits relatively high σ, lower value of κ, and large value of $S^2\sigma$ that may contribute to an optimum *ZT* value. The optimum value lies in the mid region as shown in Fig. 3. Till now, semiconductors are proved to be the best candidate for a good TE material (Zhang et al. 2015). Besides the graphical interpretation, these parameters are also related mathematically, as summarized below.

So, the optimization of *ZT* is necessary in order to attain a compromise between such situations. That can be possible in either way: by lowering the *k* value or by increasing the $S^2\sigma$. Besides this graphical interpretation, these parameters are also related mathematically, as summarized below.

At first, the *S* value and n are connected with each other as per Eq. (5) (i.e., Mott-Jones relation). This relation works for the metals as well as the semiconductors.

$$S = \frac{8\pi^2 {K_B}^2}{3eh^2} m^* T \left(\frac{\pi}{3n}\right)^{2/3} \tag{5}$$

where K_B is the Boltzmann constant, *n* is the carrier concentration, *h* is the Planck's constant, and m^* is the effective mass of the carrier. According to this relation, with increase in the *n* value, *S* value decreased. The second relation which connects σ with n is given in Eq. (6). It shows that σ value increases with increases in n.

$$\sigma = \frac{1}{\rho} = ne\mu = \frac{ne^2}{m^*}\tau \tag{6}$$

where ρ is the resistivity, τ is the scattering rate, and $\mu = \frac{e\tau}{m^*}$ is the mobility of charge carrier.

Moreover, Eqs. (5) and (6) show that the effective mass m^* related to S and σ in a reverse way. The high value of m^* will raise the thermopower while on the other hand that lowers the σ value. The drop in σ value is attributed to the heavy carriers, which move with lower velocities and thus reduce the mobility. Therefore, a compromise should be made between the high μ and higher value of m^*.

The third factor is the thermal conductivity κ. The κ can be expressed by the relation (7). It consists of two parts κ_ℓ and κ_e.

$$\kappa = \kappa_\ell + \kappa_e \tag{7}$$

where κ_ℓ is the heat transport by lattice vibrations, i.e., phonons, and κ_e is the heat transport through electrons. κ_e is related to σ through Wiedemann-Franz law (Kittel 2005), as

$$\kappa_e = L_0 \sigma T = ne\mu L_0 T \tag{8}$$

where L_O is the Lorentz number expressed as $L_0 = \frac{\pi^2}{3}\left(\frac{K_B}{e}\right)^2$ and generally its value taken as $2.4 \times 10^{-8}\ \mathrm{J^2K^{-2}C^2}$. Further, the ZT value can be expressed by the relation (9) which combines both κ_ℓ and κ_e:

$$ZT = \frac{S^2/L}{1 + \frac{k_\ell}{k_e}} \tag{9}$$

As we see above that κ_e is directly proportional to σ, it is necessary to minimize the lattice part of k, i.e., κ_ℓ, so as to obtain the high ZT value. The last is the lattice part of k, i.e., κ_ℓ; decrease in its value will enhance the ZT. According to the classical kinetic theory for any heat transport, κ_ℓ can be expressed by the relation (10).

$$\kappa_\ell = \frac{1}{3}\left(C_v \vartheta_s L_{ph}\right) \tag{10}$$

where L_{ph} is the mean free path of phonons, C_v is the specific heat at constant volume, and υ_s is the velocity of sound. Generally, at high temperature, the κ_ℓ is determined only by the factor mean free path of phonon, as the specific heat and sound velocity are found to be independent of temperature.

5 Strategies for Designing Novel Thermoelectric Material: High *ZT*

At the beginning of TE research, Bi_2Te_3, PbTe, and SiGe were the attractive conventional materials among the researchers. But, these materials possess low-energy conversion efficiency in spite of their potential applications. Moreover, they also contain toxic, expensive, and scarce elements, which hinder from their wide-scale commercialization. So, the prime idea was to enhance the TE efficiency. Scientists adopted two different strategies in TE materials: (i) by improving the properties of the conventional materials and (ii) by designing the new material. During the period of improving the TE efficiency, two ideas came across in order to attain high *ZT*: the phonon glass electron crystal (PGEC) proposed by Slack (Slack 1995) and the nanostructuring by Dresselhaus and Hicks (Dresselhaus et al. 2007; Hicks et al. 1996). Based on these ideas, several groups of materials were discovered, e.g., clathrates, skutterudite, super lattices, nanowires, nanocomposites, etc., shown in Fig. 4.

In order to implement these strategies, several processing routes were adapted to obtain the high *ZT*, via solid solution, alloying, doping, and nanostructuring. All these ideas include synthesis of material by various means to obtain the first outcome, for instance, an ingot. The selection of synthesis route is a crucial point in determining the TE properties. Here are some examples of synthesis processes: melt and growth, arc melting, hydrothermal or solvothermal synthesis, mechanical alloying, single crystal growth, and others. Many of the processes involve ball milling or nano-milling too in order to reduce the grain size which makes the densification faster with small grain size. Further, the particle size and the density of the material under consideration have an impact on the thermal conductivity. The processing methods like HP (hot pressing) or SPS (spark plasma sintering) play a crucial role in the densification of the samples. It also helps to obtain the bar or disc sample out of an ingot, which is the required shape for the fabrication of a TE module. In case of thin films, restriction comes due to their expensive nature and poor scalability. In addition, their structures are not very stable in medium or high temperature and thus become a demerit for the applications. So, nowadays, another route, i.e., forming composites, has been adapted to overcome such issues. Nanocomposites are one of the less expensive alternatives to the high-cost

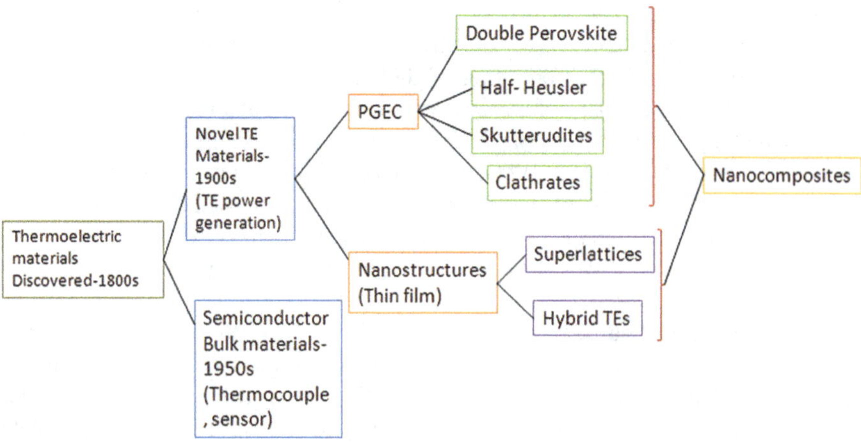

Fig. 4 A schematic flowcharts of TE materials

superlattices. Here, in this type of material, fine powders are normally prepared by milling or grinding, followed by sintering using HP or SPS. Further, the reduction in dimension helps in introducing several interfaces into the material, so as to reduce the thermal conductivity. For the device application, it is important to make the material into desired shapes and that can be scaled up for commercial purpose. So far, we have understood that densification plays one of the major roles in determining the thermoelectric efficiency. These days, researchers mostly prefer the sintering method SPS for TE materials over any other consolidation techniques, due to its own favorable characteristics, discussed in the following section.

6 Benefits of Spark Plasma Sintering in Thermoelectric

Over the years, several strategies were adapted to obtain the high figure of merit *ZT* value. As discussed above, the ultimate goal is to do the optimization of various parameters, i.e., S and σ should be high while κ should be reduced. To achieve such goal, nowadays, materials are synthesized with advanced sintering technique which can fulfill our growing requirements. Recently, the sintering technique, spark plasma sintering (SPS), gets much attention over other available consolidation methods in thermoelectric. It plays a crucial role toward the betterment of the TE properties, which we have mentioned in the following section.

As discussed in the previous chapters, spark plasma sintering (SPS) is a rapid powder consolidation technique. In this sintering system, DC pulse has been used to generate the spark that discharges inside the gaps of the powder. In comparison with other available consolidation methods, SPS is a fast and uniform sintering method. It provides material with low grain growth depending on the conditions applied. Crystal structure, grain orientation, size of the particle, microstructure, and densification, all these factors are important to consider in order to examine the characteristics of thermoelectric materials. For example, normally, in case of solid-state synthesis, the compounds are prepared by mixing the adequate amount of starting precursors followed by the formation of the pellets. Then, the pellets are sealed inside the silica glass tube for the heat treatments. On completion of the heating process, the ingots were taken out of the glass tube. It is further ground into powder form suitable for the final consolidation. In most of the cases, nowadays, SPS has been used for densification of the materials. After being pressed inside a die setup, disc-shape pellets are obtained which is suitable for the measurement purpose, as shown in Fig. 5. After that, to carry out the transport property measurements, the sintered dense pellets were cut into the proper shapes.

The sintering method SPS helps in various ways. The optimization of temperature, pressure, and atmosphere (N/Ar) in SPS plays an important role in determining the single phase in case of solid solution. At first, after synthesis, X-ray powder diffraction has been carried out to do the structural analysis of the material. The data

Fig. 5 A transformation of an ingot into a dense pellet using SPS

obtained were then analyzed to identify the phase of the material. So far, it has been observed that the main phase well match with the database. In some cases, little traces of secondary phases are observed along with the main phase. The secondary phases may arise due to several reasons, depending on the type of precursors used, temperature and duration of heating treatment, etc. Such phases have an adverse effect on the material as they can decrease the materials performance. In this context, optimization of SPS parameters helps to get rid of such unwanted phases or impurities in the ultimate output.

Secondly, the densification is another factor in determining the materials property. So far, we know that it is difficult to obtain the high *ZT* value in case of bulk materials. In such context, the idea of nanostructurization was developed so as to improve the efficiency. This concept contains the idea of reducing the particle size and dimensionality, which in turn helps in regulating the materials properties. Moreover, to keep the balance between the mean free path of the charge carriers and phonons, these days microstructural engineering emerges as another way to maintain it and helps in enhancing the *ZT* value. Again, *ZT* can be increased, if there exists a difference in the mean free path between phonons and charge carriers, as the k value reduces rapidly as compared to the rate of electrical conductivity. Thus, if it is possible to make the dimension of the material toward nanometer length, then this new variable length scale will control the material properties and leads to the improvement of *ZT*.

Besides, the grain size of the material and their orientation are another crucial parameter for the material. The alignment of grains has to be in the preferred direction, so that the electrical properties of the material may not decrease much. Such orientation of grains can be obtained with the process like as SPS. Further, the scattering of phonons occurs more effectively due to the increase of grain boundaries as a result of reduction in grain size. Therefore, the lattice thermal conductivity reduces via the interfacial scattering process and results in an improved *ZT* value. Moreover, in case of bulk nanostructure materials, the advantage of using SPS is that it preserves the nanoprecipitates or the nano-inclusions and does not allow these to enter into the matrix.

7 Synthesis of TE Materials Using SPS

Although discovered since long ago, the TE materials are still actively researched by the scientific community to improve their properties. Normally, based on the temperature range, materials are divided accordingly. Some TE materials operate at low temperatures whereas others at intermediate and high temperatures. The *ZT* values of several TE materials were shown below in Fig. 6 with respect to their suitable temperature range for applications.

Over decades, chalcogenide becomes an attractive group of materials among the researchers. In this class of material, semiconducting nature do exists in the wide range of compounds. Bi_2Te_3 and PbTe become the state-of-the-art materials in this group. In chalcogenide materials, SPS plays the effective role in the way, which we already have discussed in the earlier section. Chalcogenide materials contain chalcogen in it such as Se, S, or Te, which are volatile in nature especially at high temperature. For the synthesis, it is important to seal the starting precursors inside the silica quartz tube. During the heat treatment, high stress can occur inside the tube due to vapor pressure, and as a result, it can break the tube due to the considerable difference in thermal expansion between the samples inside and the quartz, which becomes more severe when materials undergo phase transition. So, in this situation, it can be troublesome to obtain a proper ingot. While doing SPS, the advantage is that controlled pressure can be applied simultaneously with increasing temperature resulting in lower processing temperature in achieving dense pellets with improved characteristics.

The materials such as SnSe, SnS, SnTe, PbTe, PbSe, PbS, GeSe, GeS, GeTe, $SnSe_2$, and SnS_2 belong to IV–VI chalcogenide family, which has shown good TE performances. The p-type PbTe-SrTe (4 mol%) system showed a high *ZT* value of 2.2 at 915 K while going from the nanoscale structure to mesoscale structure. Here, in this system, the mesoscale grains and nanoscale precipitates were obtained by using SPS that further facilitated the scattering of additional heat-carrying phonons

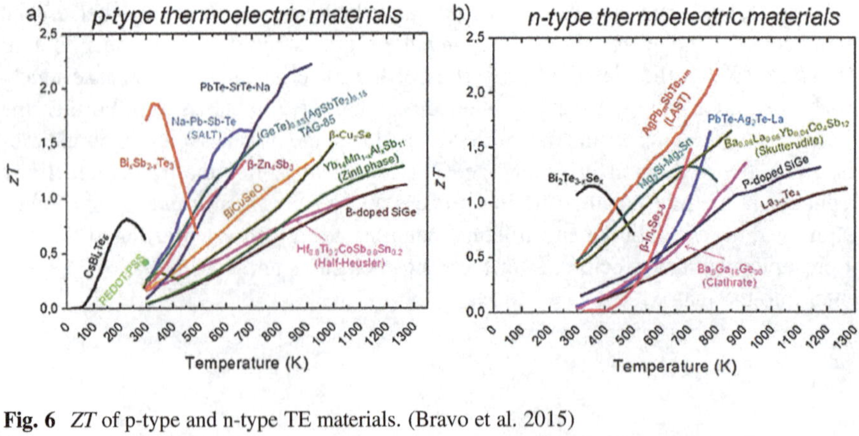

Fig. 6 *ZT* of p-type and n-type TE materials. (Bravo et al. 2015)

and reduced the lattice value of thermal conductivity (Biswas et al. 2008). Another interesting work made with bismuth antimony telluride ($Bi_{0.5}Sb_{1.5}Te_3$) showed an enhanced *ZT* value of 1.86 ± 0.15 at 320 K. Here, in this system, liquid-phase compacting process was applied in order to create periodic dislocations at the low-energy grain boundaries. This resulted in a reduced k_ℓ value of 0.33 $Wm^{-1} K^{-1}$ at 320 K. During the SPS cycle, at high temperature and pressure, the excess Te was liquefied and expelled out of the surface of graphite die as the applied SPS temperature 480 °C was higher than the melting point of the Te (450°C), shown in Fig. 7. The microstructure study of this as-sintered sample showed the presence of eutectic phase, which was rich in contained Te and $Bi_{0.5}Sb_{1.5}Te_3$ nanoparticles (Kim et al. 2015).

By co-doping (Na, K) in polycrystalline SnSe, the lattice thermal conductivity of the material was achieved as low as 0.29 $Wm^{-1} K^{-1}$ at 773 K, whereas *ZT* increased from 0.3 for 1% Na-doped SnSe to 1.2 for 1% (Na, K)-co-doped SnSe (Ge et al. 2017). In this co-doping, enhancement in TE properties arose due to multi-valley effects, enhanced solid solubility of Na and K sample, and as well as the presence of structural defects. The introduction of the PbSe phase in SnSe, i.e., (SnSe +1% PbSe) material, showed a record-high *ZT* ~1.7 at 873 K in polycrystalline SnSe (Tang et al. 2016). Here, low k_ℓ value occurred due to all-scale hierarchical architectures. Further, with the composite system 0.5 mol% SnTe/$Ag_{0.005}Sn_{0.995}Se$, a largest *ZT* value of 1.6 ± 0.2 was achieved at 875 K indicating that incorporation of nano-phase SnTe was effective in enhancing thermoelectric performance (Guo et al. 2016). Here, this distribution of nano-phase in the matrix was possible since fast sintering technique, SPS, was used. A record peak *ZT* of 1.36 ± 0.12 in polycrystalline $Sn_{0.98}Se$ macro-sized plates was obtained due to its high power factor of 6.95 $\mu Wcm^{-1} K^{-2}$ and ultralow thermal conductivity of 0.42 $Wm^{-1} K^{-1}$ at 823 K (Shi et al. 2018). In case of pristine SnSe synthesized by mechanical alloying, the SnSe powders had high surface volume and thus absorbed oxygen easily. And here, Sn mainly reacted with oxygen and formed SnO_2 on the grain boundaries. Whereas after K doping in SnSe, K reacted with oxygen instead of Sn, as K was more reactive than Sn and formed K_2O at the grain boundaries. But during the SPS process, K_2O

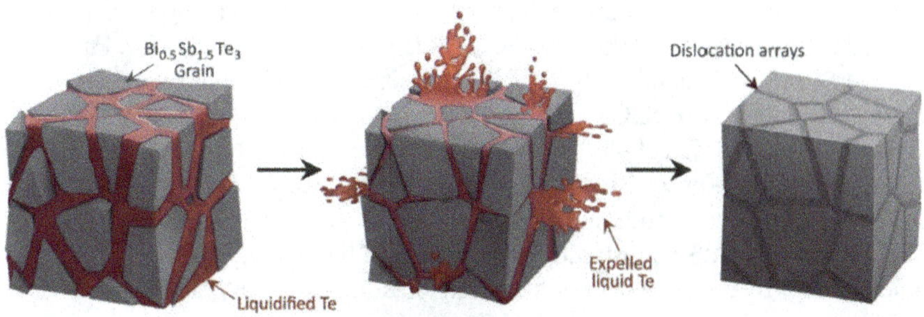

Fig. 7 Schematic illustration of showing the generation of dislocation arrays during the liquid-phase compaction process. (Figure is taken from Kim et al. 2015)

decomposed and oxygen was out of the matrix. And further, this K doping resulted in the formation of nanoprecipitates. This precipitates contained more Sn and K than in the matrix. Thus, the absence of Sn oxides at grain boundaries and the presence of nano-precipitates in the SnSe caused by K doping contributed to the low k_ℓ of ≈ 0.20 $Wm^{-1}\ K^{-1}$ resulting *ZT* of $\approx$1.1 at 773 K (Chen et al. 2016).

Tin sulfide (SnS), a direct bandgap IV–VI semiconductor compound, attracted particular attention due to its unique characteristics. Due to its low cost, less toxic, and readily abundance nature, this compound offered a potential place as promising thermoelectric materials in future commercial applications. It exhibits both the p- and n-type conduction with an energy bandgap of about 1.3 eV. SnS bulk materials prepared in combination with the mechanical alloying and SPS methods showed high Seebeck coefficient (above 500 μV/K at high temperature) and a low thermal conductivity (below 1.0 W/m/K). In such sample, highest *ZT* value of 0.16 at 823 K was achieved (Tan et al. 2014a, b, c). By Na doping in polycrystalline SnS, the carrier concentration obtained up to ~10^{19} cm^{-3} led to the highest *ZT* of 0.65 in polycrystalline SnS (Zhou et al. 2017). Ag doping in SnS samples exhibited high Seebeck coefficient of > + 400 μVK^{-1} and thermal conductivity dropped below 0.5 $Wm^{-1}\ K^{-1}$ at 873 K leading to a high *ZT* of 0.6. Earlier, it was reported that the thermal conductivity of Ag-doped SnSe system increased due to the formation of metallic secondary phase ($AgSnSe_2$), as a result of hot pressing. SPS has the potential to improve the sinter ability of the compounds while keeping the original microstructure, and it's possible to avoid secondary phase formation in the bulk samples too (Li et al. 2018a, b). The SnS compound and solid solutions of SnS-SnSe were fabricated which showed anisotropic thermoelectric performance along the parallel and perpendicular to the pressing directions. For SnS, along the parallel, the *ZT* value was 0.19 at 823 K, which was greater than along the perpendicular direction ($ZT = 0.16$). Moreover, *ZT* values of $SnS_{0.5}Se_{0.5}$ and $SnS_{0.2}Se_{0.8}$ were higher than those of the SnS compound. The peak *ZT* value of 0.82 was obtained for $SnS_{0.2}Se_{0.8}$ at 823 K, which was again more than four times higher than that of SnS (Han et al. 2015). The increase of carrier mobility and the reduction of k_ℓ with the increase in Se content was the main reason behind the higher *ZT*.

Tin telluride (SnTe) is a lead-free IV–VI narrow bandgap semiconductor that exhibited the same crystal structure and a similar electronic band structure with PbTe. Alloying with Cd atoms enabled a form of valence band engineering, which resulted in *ZT* values of ~0.96 at 823 K in p-type samples of $SnCd_{0.03}Te$. Further, in this work, the endotaxial CdS or ZnS nanoscale precipitates reduce the lattice thermal conductivity of $SnCd_{0.03}Te$ and such endotaxially nanostructure done with CdS and ZnS led to ZTs of ~1.3 and ~1.1 at 873 K, respectively (Tan et al. 2014a, b, c). Another work showed a significant enhancement of the TE performance of p-type SnTe over a broad temperature with a peak *ZT* value of ~1.4 at 923 K through In/Cd co-doping and a CdS nanostructuring approach. This doping reduced the thermal conductivities by effective phonon scattering obtained due to atomic point defects by doping, nanoscale precipitations by CdS nanostructuring, and mesoscale grains by using fast sintering method, SPS (Tan et al. 2015a, b, c, d). The detail *ZT* values of several compositions are presented in Fig. 8.

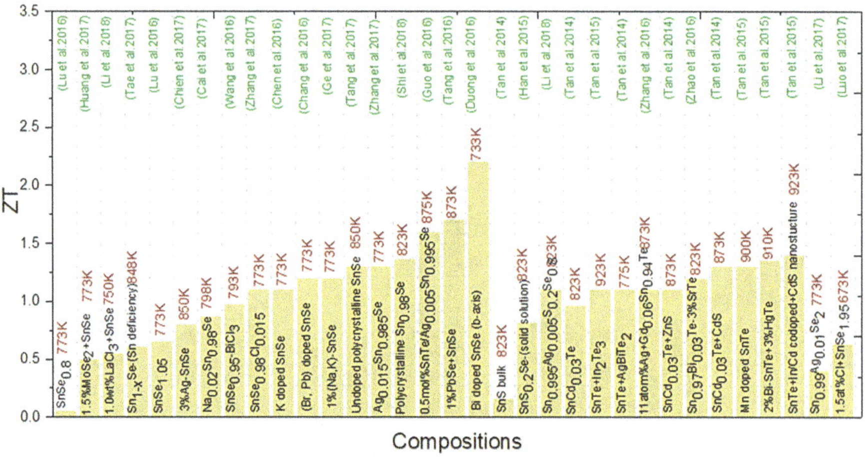

Fig. 8 A *ZT* plot of compositions belongs to the IV–VI chalcogenide group

At present, ternary chalcogenides (I-V-VI$_2$) are one of the attractive materials due to their potential applications in the field of TE, photovoltaic, and photothermal areas. They are normally based on the ternary semiconducting compounds. The most common examples belonging to this class of compounds are $AgSbSe_2$, $AgSbTe_2$, $AgSbS_2$, $AgBiS_2$, and $AgBiSe_2$ (Morelli et al. 2008). The commonly known AgSb-based materials are, namely, $AgSbTe_2$ and $AgSbSe_2$. They are favorable due to their very low lattice thermal conductivity value <1 W/mK. Recently, efforts were made to enhance the TE performance of these materials by tuning their electrical and thermal properties through doping or substitution. Such strategies are centered on the SPS method in order to make dense materials to obtain low k value resulting in an improved *ZT* value. For $AgSbTe_2$, the TE performance was improved by Na doping on p-type $AgSbTe_2$; *ZT* of 1.50 was reached at 570 K for $Ag_{0.99}Na_{0.01}SbTe_{2.02}$ sample (Du et al. 2011). Here, Na doping created a strong effect on the electron band structure, due to which the samples exhibit higher hole effective mass and enhanced Seebeck coefficient. Further, by adding little excess Te in such Na-doped samples, the hole concentration and electrical conductivity increased, and on the other side, low lattice thermal conductivity obtained at high temperature by strong scattering of phonons. Further, Zn-doped $AgSbTe_2$ sample showed a high *ZT* value of 1.9 for the composition p-type $AgSb_{0.96}Zn_{0.04}Te_2$ at 585 K. This was the highest measured value in this temperature range (450–650 K) (Roychowdhury et al. 2017). The thermal and mechanical stability had been increased by such doping, as it had suppressed the occurrence of intrinsic Ag_2Te impurity phases in $AgSbTe_2$. Here, in this material, low k_ℓ value was obtained due to the solid solution point defect phonon scattering and an increased electrical conductivity was observed due to enhanced carrier concentration. The detail *ZT* values of several compositions are presented in Fig. 9.

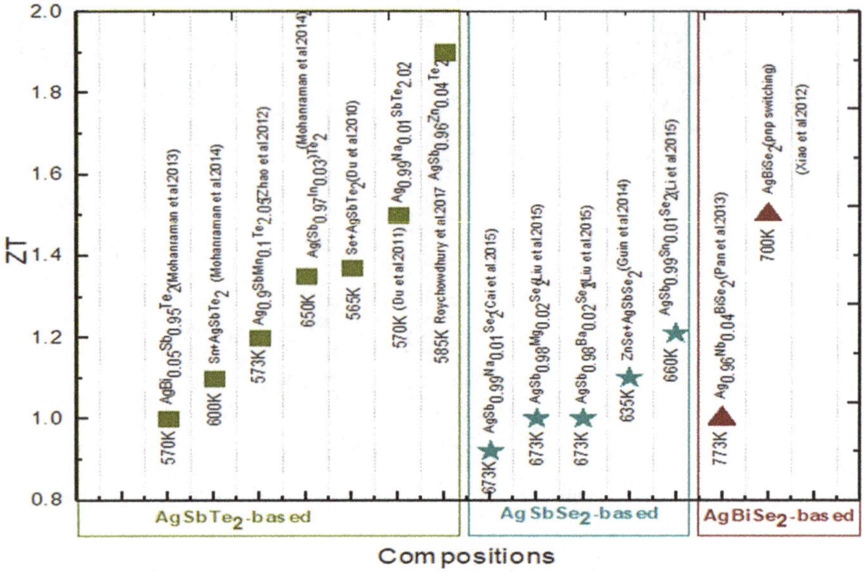

Fig. 9 A *ZT* plot of compositions belongs to the I-V-VI_2 chalcogenides

At present, most of the conventional Cu-based TE materials become one of the hottest topics of research due to the readily available copper element. Currently, several achievements with favorable *ZT* were obtained with Cu-based compounds which are comparable with other state-of-the-art TE materials, as per example, Cu_2SnSe_3, $CuGaTe_2$, Cu_3SbSe_4, $CuInTe_2$, $Cu_2ZnSnSe_4$, Cu_2S, $Cu_2SeCuCrSe_2$, etc. Here, in the following section, we have highlighted some of the Cu-based chalcogenide materials with reasonable *ZT* values, where densification was carried out using SPS.

Copper-based binary chalcogenides usually represent with a nominal composition, $Cu_{2-x}X$ (where X stands for chalcogens S, Se, or Te). Despite the complexity in its crystal structure, it was proved to be an ideal TE material, as it exhibited good electrical properties with low lattice thermal conductivity value enabling a high *ZT* value. The system like $Cu_{2-x}Se$ achieved a high *ZT* of 1.5 at 1000 K. Moreover, here the "liquid-like" behavior of copper ions around a crystalline sub-lattice of Se resulted in an intrinsically very low k_ℓ value (Liu et al. 2012a, b). Further, another work showed a high *ZT* of 2.1 at 973 K for Cu_2Se, where the thermal conductivity dropped to a very low value of the order of $0.34 Wm^{-1} K^{-1}$. This decreased in value primarily originated from the enhanced low-to-high wavelength phonon scattering by several defects, such as microstructural defects, nanocrystalline grain boundaries due to very small crystallite size, and low dimensions residual porosity created due to SPS (Gahtori et al. 2015). Moreover, lithium doping in $Cu_{2-x}Se$ showed a maximum $ZT > 1.4$ for $Li_{0.09}Cu_{1.9}Se$, which was higher than the undoped samples. Here, in this system, along with the improved thermoelectric performances, suppression of bipolar conduction was achieved due to an enlarged bandgap (Aydemir et al. 2017). Moreover, with increasing the Cu deficiency, the $Cu_{2-x}Se$ nanoplates showed a reduced *ZT*, such as 1.4 at 850 K for $Cu_{1.98}Se$ (Yang et al. 2016), whereas β-phase Cu_2Se nanostructures showed an ultralow lattice thermal conductivity as low as ~0.2 $Wm^{-1} K^{-1}$ which resulted in a record-high *ZT* of 1.82 at 850 K. Here, in this system, the formation of high-density small-angle grain boundaries in sintered Cu_2Se samples caused additional phonon scattering, which further reduced the lattice thermal conductivity (Yang et al. 2015). A *ZT* of 1.18 at 850 K was attained for $CuInTe_2$ sample due to a low k_ℓ value via strong phonon Umklapp process (Liu et al. 2012a, b). The *ZT* values of several compositions are presented in a bar chart shown in Fig. 10.

BiCu*Ch*O oxychalcogenides are considered as more suitable thermoelectric material for high-temperature applications. The most preferable choice of the chalcogen had been selenium over tellurium and sulfur. BiCuSeO crystallizes in a layered ZrCuSiAs structure with space group P4/nmm. Due to its layered structure, BiCuSeO showed anisotropic transport behaviors. Due to low mobility and carrier concentration, it exhibited poor electronic transport properties. But, these materials were developed due to their reduced value of intrinsic thermal conductivity arising from the weak bonding between the layers. Various strategies had been adapted to obtain enhanced *ZT*. Here, in the following section, we mention few examples of BiCuSeO-based materials along with their *ZT* values. By Cd-doping at Bi site in BiCuSeO, power factor increased up to 45% due to high electrical conductivity through bond anharmonicity and moderate Seebeck value. Further, this doping created the phonon-scattering centers, which had an effect in the local geometric and electronic structure. As a result, with 5% Cd-doping, *ZT* of 0.98 at 923 K was achieved (Farooq et al. 2016a, b). The heavy atom Ag substitution at Cu site in BiCuSeO enhanced the *S* value as well as reduced the *k* value by phonon scattering which was obtained due to the growth in number of grain boundaries, together with the appearance of nano-porosity by this doping. This resulted in an improved *ZT* value of 0.64, which was

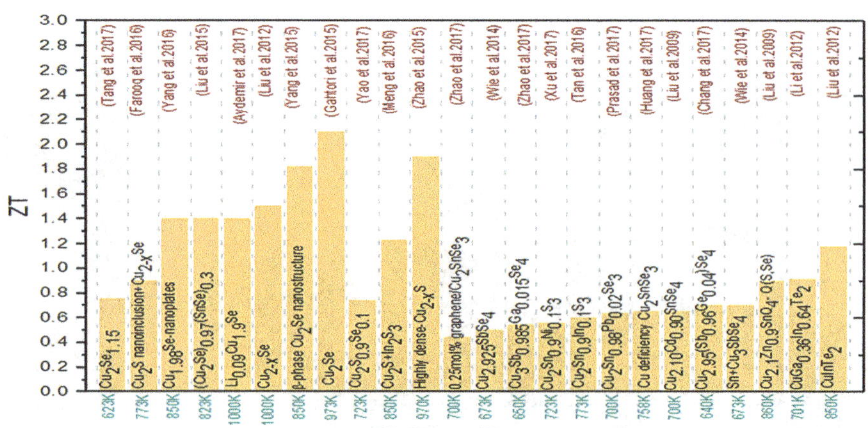

Fig. 10 A *ZT* plot of Cu-based chalcogenides

16% more than that for pure BiCuSeO at 923 K (Farooq et al. 2017). Ag was doped at Cu site effectively. Further, a high *ZT* value of ~1.1 at 923 K in the BiCuSeO system was achieved via heavily doping with Ba (Li et al. 2012a, b). Here, by this doping, low value of lattice thermal conductivity was obtained due to the presence of point defects scattering in the doped samples. Moreover, thermal and chemical stabilities up to 923 K were observed in BiCuSeO system. Recently, a highly enhanced *ZT* of BiCuSeO was obtained by partial substitution of Bi^{3+} by Ca^{2+} and Pb^{2+}. The microstructure of the doped samples was investigated after SPS. The result showed the presence of CaO_2 nano-clusters and nano-inclusions of Bi element in the system. This further reduced the lattice part of thermal conductivity due to the acoustic phonon scattering by the nano-clusters. As a result, the total thermal conductivity was very low. Moreover, this co-doping not only reduced the *k* value as low as 0.5 $Wm^{-1}\ K^{-1}$ at 873 K through dual atomic point-defect scattering but also showed a gain in power factor of $\approx 1 \times 10^{-3}$ $Wcm^{-1}\ K^{-2}$ from 300 to 873 K. A high *ZT* value of $\approx$1.5 at 873 K for the system $Bi_{0.88}Ca_{0.06}Pb_{0.06}CuSeO$ was obtained, making it an appropriate material for high-temperature applications (Liu et al. 2016). Another approach had been applied for the improvement in thermoelectric performance in a BiCuSeO system, i.e., through modulation doping by improving the carrier mobility. This led to the redistribution of the carrier and facilitated the electrical transport. Combining with the high power factor ranging from 5 to 10 $\mu Wcm^{-1}\ K^{-2}$ and with extremely low thermal conductivity value of ~0.25 $Wm^{-1}\ K^{-1}$at 923 K, a high $ZT \approx 1.4$ was achieved in a BiCuSeO system, making it suitable for medium temperature applications (Pei et al. 2014). The *ZT* values of several compositions are presented in a bar chart shown in Fig. 11.

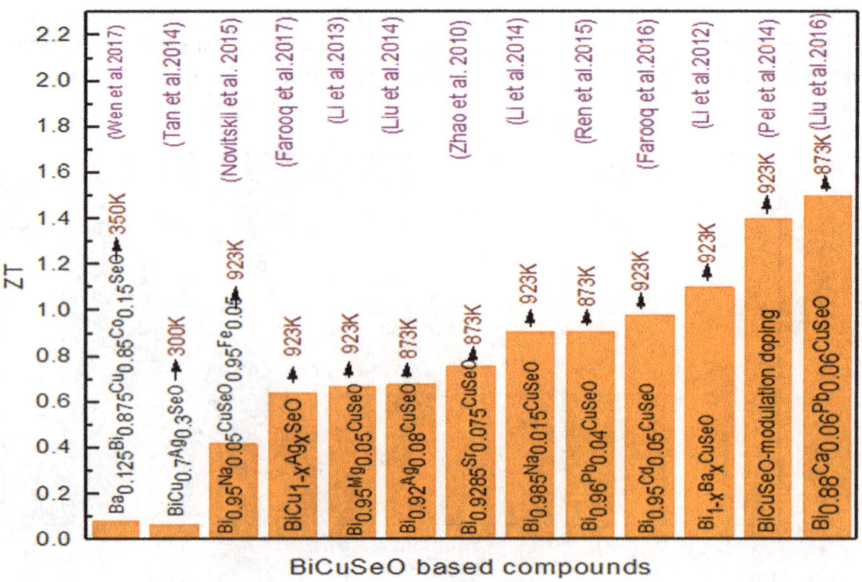

Fig. 11 A *ZT* plot of oxychalcogenide-based compounds

Clathrates exhibit cage-like structure, which is an open framework, composed of tetrahedrally coordinated atoms (Al, Ga, Si, Ge, Sn) that enclose relatively large metal atoms in it. The cage-like structure enables the rattling mechanism, which can lower the thermal conductivity value. Here, scattering of lattice phonons occurs due to the presence of guest atoms, which initiate the rattling activity in the structure and thus suppress the lattice thermal conductivity. The polycrystalline samples of $Sr_8Ga_xGe_{46-x}$ clathrates showed the maximum *ZT* value of 0.62 at 800 K by changing the amount of Ga in the composition (Fujita et al. 2006). Zn-doped $Ba_8Ga_{16}Zn_xGe_{30-x}$ clathrates showed a decrease in *k* value with increase in Zn content, and here in this system, scattering mechanism changed from ionized impurity to acoustical one with increase in temperature. Aluminum-doped type I clathrates $Ba_8Ga_{16}Al_xGe_{30-x}$ showed a maximum *ZT* of 0.61 at 760 K for the composition $Ba_8Ga_{16}Al_{3.0}Ge_{27.0}$. With this doping, k_ℓ value reduced very low because of mass fluctuation between the Al atom and other atoms present in the structure (Deng et al. 2008). The type I clathrate compound $Yb_{0.5}Ba_{7.5}Ga_{16}Ge_{30}$ double filled with ytterbium and barium showed a *ZT* of 1.1 at 950 K. Here, in this system, the substitution of Yb atom led to a significant reduction in the k_ℓ value (Tang et al. 2008). The type II clathrate $K_8Ba_{16}Ga_{40}Sn_{96}$ showed a *ZT* value of 0.93 at 637 K and this was more thermally stable at high temperatures (Koda et al. 2014). Further, the improved TE properties by optimizing the carrier concentration and high *ZT* of 1.19 was obtained at 630 K (Kishimoto et al. 2015).

Skutterudite is another promising TE compound whose general crystal structure is described by the formula $\square_2Co_8Sb_{24}/\square_2X_8Y_{24}$. The open square indicates the presence of two voids, which can be filled by the rare earth atoms to help in the reduction of thermal conductivity. Here are few examples of skutterudite material with *ZT* using SPS. By Se and Te co-doping in $CoSb_3$, *ZT* value of 1.29 was achieved at 780 K for $CoSb_{2.8}Te_{0.15}Se_{0.05}$. Such doping suppresses the thermal conductivity with significant increase in carrier concentration due to extra amount of Te in the composition (Dong et al. 2015). The *ZT* of 1.15 at 840 K is obtained for n-type $Ca_xCo_4Sb_{12}$ skutterudite with the nominal composition $Ca_{0.31}Co_4Sb_{12}$. Here, with increase in the filling fraction of Ca in the material, power factor enhanced and the k_ℓ value had reduced due to the effective phonon scattering from the rattling modes of filling atom (Li et al. 2016). Al-filled $Yb_{0.25}Co_4Sb_{12}$ showed a *ZT* of 1.36 at 850 K for the composite $Al_{0.1}Yb_{0.25}Co_4Sb_{12}$. Here, the microstructure of the sintered samples was investigated. Here, less number of Al was found inside the grains but was effectively present besides Sb in the grain boundaries. So, the aggregated AlSb nano-inclusions at the grain boundaries introduced point defects in the matrix instead of entering into the voids and thus enhanced the interfacial phonon scattering and resulted in a reduced value of *k* (Elsheikh et al. 2017). Moreover, *ZT* of 1.25 at 800 K is obtained for triple-filled skutterudite $Ba_{0.09}Yb_{0.07}In_{0.06}Co_4Sb_{11.97}$. Here, due to the presence of In and InSb nano-inclusions, *k* value had significantly reduced attributed to the enhanced phonon scattering (Ballikaya et al. 2011). By Te and S co-doping in $Co_4Sb_{11.3}Te_{0.7-x}S_x$ compound, *ZT* of 1.1 at 800 K has been achieved for $Co_4Sb_{11.3}Te_{0.63}S_{0.07}$. Here, enhancement in the TE performance occurred due to the sulfur defects, which in turn increased the scattering mechanism and thus reduced the *k* value (Duan et al. 2014).

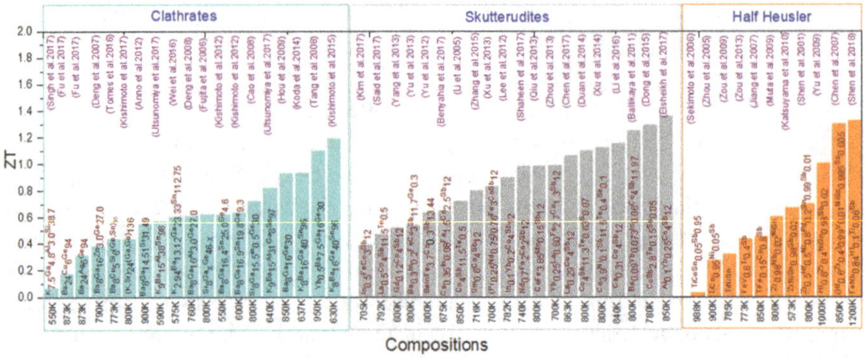

Fig. 12 A *ZT* plot of clathrates, skutterudites, and half-Heusler-based compounds

Half-Heusler (HH) intermetallic alloys also become profound as potential TE materials for high-temperature applications owing to their high-temperature stability. A *ZT* of 1.0 was obtained at 1000 K for the HH compound $Hf_{0.6}Zr_{0.4}NiSn_{0.98}Sb_{0.02}$ (Yu et al. 2009). For n -type $(Hf_{0.6}Zr_{0.4})_{0.99}V_{0.01}NiSn_{0.995}Sb_{0.005}$ alloy, *ZT* of 1.3 was obtained near 850 K (Chen et al. 2017a, b). The presence of Vanadium resonant states in this system resulted in enhanced Seebeck coefficient and reduced thermal conductivity value. A highest *ZT* of 1.32 at 1200 K was obtained in case of Hf-Ti dual-doping FeNbSb-based HH compound for the composition $FeNb_{0.84}Hf_{0.1}Ti_{0.06}Sb$ (Shen et al. 2018). Here, in this system, the dual doping not only reduced the k_ℓ value attributed to the point defect scattering but also simultaneously enhanced the power factor. The *ZT* values of several compositions are presented in a bar chart shown in Fig. 12.

8 Conclusion

In thermoelectric research, the fast sintering technique, SPS, is an attractive process among the researcher. As discussed in this chapter, it is evident that SPS plays a very important role in achieving high *ZT* values in thermoelectric materials. The nanoscale grain size obtained from SPS resulting in increase of grain boundaries causes the phonons to scatter more effectively. Therefore, the lattice thermal conductivity reduces via the interfacial scattering process and results in an improved *ZT* value via SPS. Further, in case of bulk nanostructure materials, the advantage of using SPS is that it preserves the nanoprecipitates or the nano-inclusions and does not allow these to dissolve into the matrix because of the fast sintering process. The SPS processing covers a wide range of TE materials as mentioned above. However, some adverse effect of SPS in the material was observed in a few cases, such as excess Te was expelled out of the graphite die, the absence of Sn oxides at grain boundaries, etc. Thus, nowadays, researchers are more focused toward the development of composites TE material for future use.

References

Altenkrich E (1909) Physikalische Zeitschrift. 10:560

Aydemir U, Qiu P, Dongmin S, Jan-hendrik P, Stoumpos CC, Hanus R, Anne M, Shi X, Chen L, Kanatzidis MG, Snyder GJ (2017) Enhanced stability and thermoelectric fi gure-of-merit in copper selenide by lithium doping. Mater Todays Phys 1:7–13

Ballikaya S, Wang G, Sun KAI (2011) Thermoelectric properties of triple-filled $Ba_xYb_yIn_zCo_4Sb_{12}$ skutterudites. J Electron Mater 40:570–576

Bennett GL (2006) Space nuclear power: opening the final frontier. Am Ins Aero Astro AIAA:4191

Biswas K, He J, Blum ID, Wu C, Hogan T, Seidman D, Dravid V, Kanatzidis MG (2008) High performance bulk thermoelectrics with all-scale hierarchical architechtures. Nat Mater 7:105–114

Chen Y, Ge Z, Yin M, Feng D, Huang X (2016) Understanding of the extremely low thermal conductivity in high-performance polycrystalline SnSe through potassium d. Adv Funct Mater 26:6836–6845

Chen C, Zhang L, Li J, Yu F, Yu D, Tian D, Xu B (2017a) Enhanced thermoelectric performance of lanthanum filled $CoSb_3$ synthesized under high pressure. J Alloys Compd 699:751–755

Chen L, Liu Y, He J, Tritt TM, Poon SJ (2017b) High thermoelectric figure of merit by resonant dopant in half-Heusler alloys. AIP Adv 7:065208 (1–6)

Deng S, Tang X, Li P, Zhang Q (2008) High temperature thermoelectric transport properties of p-type $Ba_8Ga_{16}Al_xGe_{30-x}$ type-I clathrates with high performance. J Appl Phys 103:073503 (1–6)

Dong J, Yang K, Xu B, Zhang L, Zhang Q, Tian Y (2015) Structure and thermoelectric properties of Se- and Se/Te-doped $CoSb_3$ skutterudites synthesized by high-pressure technique. J Alloys Compd 647:295–302

Du BL, Li H, Tang XF (2011) Enhanced thermoelectric performance in Na-doped p-type nonstoichiometric $AgSbTe_2$ compound. J Alloys Compd 509:2039–2043

Duan B, Zhai P, Xu C (2014) Thermoelectric performance of tellurium and sulfur double-substituted skutterudite materials. J Mater Sci 49:4445–4452

Dresselhaus MS, Chen G, Tang MY, Yang R, Lee H, Wang D, Ren Z, Fleurial JP, Gogna P (2007) New directions for low-dimensional thermoelectric materials. Adv Mater 19:1043–1053

Elsheikh MH, Faizul M, Sabri M, Said SM (2017) Rapid preparation of bulk $Al_xYb_{0.25}Co_4Sb_{12}$ ($x = 0, 0.1, 0.2, 0.3$) skutterudite thermoelectric materials with high figure of merit $ZT = 1.36$. J Mater Sci 52:5324–5332

Farooq MU, Butt S, Gao K, Sun X, Pang X, Khan SU (2016a) Enhanced thermoelectric efficiency of $Cu_{2-x}Se$–Cu_2S composite by incorporating Cu_2S nanoparticles. Ceram Int 42:8395–8401

Farooq MU, Butt S, Gao K, Zhu Y, Sun X, Pang X, Khan SU, Mohmed F, Mahmood A, Mahmood N, Xu W (2016b) Cd-doping a facile approach for better thermoelectric transport properties of BiCuSeO oxyselenides. RSC Adv 6:33789–33797

Farooq MU, Butt S, Gao K, Pang XL, Sun X, Mohmed F, Ahmad A, Mahmood N (2017) Improved thermoelectric performance of BiCuSeO by Ag substitution at Cu site. J Alloys Compd 691:572–577

Fujita I, Kishimoto K, Sato M, Anno H, Koyanagi H (2006) Thermoelectric properties of sintered clathrate compounds $Sr_8Ga_xGe_{46-x}$ with various carrier concentrations. J Appl Phys 99:093707 (1–7)

Gahtori B, Bathula S, Tyagi K, Budhani RC, Dhar A (2015) Giant enhancement in thermoelectric performance of copper selenide by incorporation of different nanoscale dimensional defect features. Nano Energy 13:36–46

Ge Z, Song D, Chong X, Zheng F, Jin L, Qian X, Zheng L, Dunin-borkowski RE, Qin P, Feng J, Zhao L (2017) Boosting the thermoelectric performance of (Na, K) -codoped polycrystalline SnSe by synergistic tailoring of the band structure and atomic-scale defect phonon scattering. J Am Chem Soc 139:9714–9720

Guo H, Xin H, Qin X, Zhang J, Li D, Li Y, Song C (2016) Enhanced thermoelectric performance of highly oriented polycrystalline SnSe based composites incorporated with SnTe nanoinclusions. J Alloys Compd 689:87–93

Han Y, Zhao J, Zhou M, Jiang X, Leng H, Li L (2015) Thermoelectric performance of SnS and SnS – SnSe solid solution. J Mater Chem A 3:4555–4559

Harland D (2010) Apollo 12 – On the ocean of storms. Praxis, p 269. https://doi.org/10.1007/978-1-4419-7607-9

Hicks LD, Harman TC, Sun X, Dresselhaus MS (1996) Experimental study of the effect of quantum-well structures on the thermoelectric figure of merit. Phys Rev B 53:10493–10496

Hyder AK, Wiley RL, Halpert G, Flood DJ, Sabripour S (2003) Spacecraft power technologies. Imperial College Press, London

Kim S, Lee K, Mun H, Yang D, Shin W, Li X, Lee Y, Snyder G (2015) Dense dislocation arrays embedded in grain boundaries for high performance bulk thermoelectrics. Sci Res Rep 348:109–113

Kishimoto K, Koda S, Akai K, Koyanagi T (2015) Thermoelectric properties of sintered type-II clathrates $(K,Ba)_{24}(Ga,Sn)_{136}$ with various carrier concentrations. J Appl Phys 118:125103 (1–7)

Kittel C (2005) Introduction to Solid state physics. Chemical Inductry Press, Beijing

Koda S, Kishimoto K, Akai K, Asada H, Koyanagi T (2014) Thermoelectric and transport properties of sintered n-type $K_8Ba_{16}Ga_{40}Sn_{96}$ with type-II clathrate structure. J Appl Phys 116:023710 (1–9)

Lange RG, Carroll WP (2008) Review of recent advances of radioisotope power systems. Energy Convers Manag 49:393–401

Li XY, Chen LD, Fan JF, Zhang WB, Kawahara T, Hirai T (2005) Thermoelectric properties of Te-doped $CoSb_3$ by spark plasma sintering. J Appl Phys 98:083702 (1–6)

Li Y, Meng Q, Deng Y, Zhou H, Gao Y, Li Y, Yang J, Cui J (2012a) High thermoelectric performance of solid solutions $CuGa_{1-x}In_xTe_2$. Appl Phys Lett 231903(1–4):100

Li J, Sui J, Pei Y, Barreteau C, Berardan D, Dragoe N, Cai W, He J, Zhao LD (2012b) A high thermoelectric figure of merit ZT > 1 in Ba heavily doped BiCuSeO oxyselenides. Energy Environ Sci 5:8543–8547

Li X, Zhang Q, Kang Y, Chen C, Zhang L, Yu D, Tian Y, Xu B (2016) High pressure synthesized Ca- filled $CoSb_3$ skutterudites with enhanced thermoelectric properties. J Alloys Compd 677:61–65

Li F, Wang W, Ge Z, Zheng Z, Luo J, Fan P, Li B (2018a) Enhanced thermoelectric properties of polycrystalline SnSe via $LaCl_3$ doping. Mater 11:203 (1–10)

Li Z, Sun F, Tang H, Dong J, Li J (2018b) Enhanced thermoelectric properties of p-type $SnS_{0.2}Se_{0.8}$ solid soultion doped with Ag. J Alloys Compd 745:172–178

Liu H, Shi X, Xu F, Zhang L, Zhang W, Chen L, Li Q, Uher C, Day T, Snyder GJ (2012a) Copper ion liquid-like thermoelectrics. Nat Mater 11:422–425

Liu R, Xi L, Shi X, Zhang W, Chen L (2012b) Ternary compound $CuInTe_2$: a promising thermoelectric material with diamond like structure. Chem Commun 48:3818–3820

Liu YC, Zheng YH, Zhan B, Chen K, Butt S, Zhang B, Lin YH (2014) Influence of Ag doping on thermoelectric properties of BiCuSeO. J Eur Ceram Soc 35:845–849

Liu Y, Zhao LD, Zhu Y, Liu Y, Li F, Yu M, Liu DB, Xu W, Lin YH, Nan CW (2016) Synergistically optimizing electrical and thermal transport properties of BiCuSeO via a dual-doping approach. Adv Energy Mater 6:1502423 (1–9)

Messerschmidt E, Bertnard, Pohelmann R, Raumstationen F (1997) Systeme und Nutzung. Springer, New York

Misra AK (2006) Overview of NASA program on development of radioisotope power systems with high specific power. Am Inst Aero Astro: 4187

Morelli DT, Jovovic V, Heremans JP (2008) Intrinsically minimal thermal conductivity in cubic I-V-VI_2 semiconductors. Phys Rev Lett 101:035901

Pei YL, Wu H, Wu D, Zheng F, He J (2014) High thermoelectric performance realized in a bicuseo system by improving carrier mobility through 3D modulation doping. J Am Chem Soc 136:13902–13908

Peltier JC (1834) Ann Phys Chim LV1, 371

Ritz F, Peterson CE (2004) Multi-mission radioisotope thermoelectric generator (MMRTG) program overview. Proc IEEE Aerospace Conf 5:2950–2957

Roychowdhury S, Panigrahi R, Perumal S, Biswas K (2017) Ultrahigh thermoelectric figure of merit and enhanced mechanical stability of p-type $AgSb_{1-x}Zn_xTe_2$. ACS Energy Lett 2 (2):349–356

Rull-Bravo M, Moure A, Fernándezb JF, Mańtin-González M (2015) Skutterudites as thermoelectric materials: revisited. RSC Adv 5:41653

Seebeck T (1822–1823) Magnetic polarization of metals and minerals 265

Shen J, Fu C, Liu Y, Zhao X, Zhu T (2018) Enhancing thermoelectric performance of FeNbSb half-Heusler compound by Hf-Ti dual-doping. Energy Storage Mater 10:69–74

Shi X, Chen Z, Liu W, Yang L, Hong M, Moshwan R (2018) Achieving high figure of merit in p-type polycrystalline $Sn_{0.98}Se$ via self- doping and anisotropy-strengthening. Energy Storage Mater 10:130–138

Slack G (1995) New materials and performance limits for thermoelectric cooling. In: Rowe DM (ed) CRC Handbook of Thermoelectrics. Taylor and Francis, London, p 407

Szczech JR, Higgings JM, Jin S (2010) Enhancement of thermoelectric properties in nanoscale and nanostructured materials. J Mater Chem 21:4037–4055

Tan G, Shi F, Sun H, Zhao LD, Uher C, Dravid VP, Kanatzidis MG (2014a) SnTe-$AgBiTe_2$ as an efficient thermoelectric material with low thermal conductivity. J Mater Chem A 2:20849–20854

Tan G, Zhao L, Shi F, Doak JW, Lo S, Sun H, Wolverton C, Dravid VP, Uher C, Kanatzidis MG (2014b) High thermoelectric performance of p-type SnTe via a synergistic band engineering and nanostructuring approach. J Am Chem Soc 136:7006–7017

Tan SG, Lei H, Shao DF, Lv HY, Lu WJ, Huang YN, Liu Y, Yuan B, Zu L, Kan XC, Song WH, Sun YP (2014c) Enhanced low temperature thermoelectric performance of Ag-doped BiCuSeO. Appl Phys Lett 105:082109 (1–5)

Tan G, Zeier WG, Shi F, Wang P, Je G, Dravid VP, Kanatzidis MG (2015a) High thermoelectric performance SnTe−In_2Te_3 solid solutions enabled by resonant levels and strong vacancy phonon scattering. Chem Mater 27:7801–7811

Tan G, Shi F, Doak W, Sun H, Zhao L (2015b) Extraordinary role of Hg in enhancing the thermoelectric performance of p-type SnTe. Energy Environ Sci 8:267–277

Tan G, Shi F, Hao S, Chi H, Zhao L, Uher C, Wolverton C, Dravid VP, Kanatzidis MG (2015c) Codoping in SnTe: enhancement of thermoelectric performance through synergy of resonance levels and band convergence. J Am Chem Soc 137:5100–5112

Tan G, Shi F, Hao S, Chi H, Bailey TP, Zhao L, Uher C, Wolverton C, Dravid VP, Kanatzidis MG (2015d) Valence band modification and high thermoelectric performance in SnTe heavily alloyed with MnTe. J Am Chem Soc 137:11507–11516

Tang X, Li P, Deng S, Zhang Q (2008) High temperature thermoelectric transport properties of double-atom-filled clathrate compounds $Yb_xBa_{8-x}Ga_{16}Ge_{30}$. J Appl Phys 104:013706 (1–7)

Tang G, Wei W, Zhang J, Li Y, Wang X, Xu G, Chang C, Wang Z, Du Y, Zhao L (2016) Realizing high figure of merit in phase-separated polycrystalline $Sn_{1-x}Pb_xSe$. J Am Chem Soc 138:13647–13654

Thomson W (1851a) Mathematical and Physical Papers 1:175

Thomson W (1851b) Proceedings of the Royal Society of Edinburgh 91

Yang L, Chen Z, Han G, Hong M, Zou Y (2015) High-performance thermoelectric Cu_2Se nanoplates through nanostructure engineering. Nano Energy 16:367–374

Yang L, Chen Z, Han G, Hong M, Zou J (2016) Impacts of Cu deficiency on the thermoelectric properties of $Cu_{2-X}Se$ nanoplates. Acta Mater 113:140–146

Yu C, Zhu T, Shi R, Zhang Y, Zhao X, He J (2009) High-performance half-Heusler thermoelectric materials $Hf_{1-x}Zr_xNiSn_{1-y}Sb_y$ prepared by levitation melting and spark plasma sintering. Acta Mater 57:2757–2764

Zhang Q, Chen C, Kang Y, Li X, Zhang L, Yu D, Tian Y, Xu B (2015) Structural and thermoelectric characterizations of samarium filled $CoSb_3$ skutterudites. Mater Lett 143:41–43

Zheng J (2008) Recent advances on thermoelectric materials. Front Phys China 3(3):269–279

Zhou B, Li S, Li W, Li J, Zhang X, Lin S, Chen Z, Pei Y (2017) Thermoelectric properties of SnS with Na-doping. ACS Appl Mater Interfaces 9:34033–34041

Part VI
SPS of High-Entropy Alloys

Spark Plasma Sintering of High Entropy Alloys

Sephira Riva, Stephen G. R. Brown, Nicholas P. Lavery, Adam Tudball, and Kirill V. Yusenko

1 Introduction

High-Entropy Alloys (HEAs) are multi-principal component systems defined according to the concentration of their constituent elements, ranging between 5 and 35 at.%. Following their discovery in 2004 by Yeh (Yeh et al. 2004) and Cantor (Cantor et al. 2004), HEAs have attracted much research interest due to their formidable array of compositional complexity and microstructural variation, and ultimately due to their outstanding mechanical properties. HEAs have been reported as having high compression yield (Senkov et al. 2011), fracture strength (Zhang et al. 2012), ductility (Laplanche et al. 2016; Chen et al. 2016) and toughness (Yu and Zhang 2016; Gludovatz et al. 2016), as well as extreme corrosion (Tsau and Lee 2016; Zheng et al. 2015), wear (Poulia et al. 2016) and fatigue (Hemphill et al. 2012) resistance and electro-catalytic features (Yusenko et al. 2017). These features have often been linked to the crystallization of HEAs in simple solid solutions, either as a single phase or as systems in which multiple solid solutions and/or intermetallic compounds coexist. The relative simplicity of HEAs despite their chemical complexity has been ascribed to three core effects: the high entropy, the lattice distortion and the sluggish diffusion effect (Miracle and Senkov 2017).

S. Riva · S. G. R. Brown · N. P. Lavery
College of Engineering, Swansea University, Wales, UK
e-mail: s.g.r.brown@swansea.ac.uk; n.p.lavery@swansea.ac.uk

A. Tudball
Kennametal Manufacturing (UK) Ltd., Wales, UK
e-mail: adam.tudball@kennametal.com

K. V. Yusenko (✉)
BAM Federal Institute aBAM Federal Institute of Materials Research and Testing, Berlin, Germany

Institute of Solid State Chemistry, Ekaterinburg, Russia
e-mail: kirill.yusenko@bam.de

© Springer Nature Switzerland AG 2019

P. Cavaliere (ed.), *Spark Plasma Sintering of Materials*,
https://doi.org/10.1007/978-3-030-05327-7_18

The high entropy effect envisages that the increase in configurational entropy in multi-principal component alloys might favour the formation of simple solid solution over competing intermetallic compounds. According to the lattice distortion effect, an HEA's crystal lattice is expected to be highly strained, with atoms displaced with respect to their average lattice site. Finally, diffusion in HEAs has been reported as being sluggish, due to the formation of nanocrystals and amorphous phases upon solidification (Tsai et al. 2013).

The existence and volume fraction of secondary phases in HEAs is strongly connected to synthetic conditions and post-synthetic mechanical and/or heat treatment. The manufacturing of bulk polycrystalline HEA has thus become a popular avenue of research, aiming to optimize the control of the process locally. The most popular processing routes are casting and mechanical alloying (MA) followed by sintering. MA is normally a dry, high-energy ball milling technique and has been employed to produce a variety of powder materials, ranging from amorphous to metastable phases, to intermetallics, to nanocomposites. Gas atomization of the HEA melt has also been widely reported as a method to prepare large quantities of homogeneous and single-phase powders prior to sintering. It has been proved that different processing routes lead to the formation of different phases in HEAs: unlike casting, mechanical alloying relies on the formation of defects, acting as energy storages along grain boundaries and providing extra driving force for the formation of a solid solution (Mohanty et al. 2015). However, the powder metallurgy route appears to have clear advantages with respect to casting technology. In fact, even if segregation was to happen due to the complex nature of the alloy, it would lead to de-mixing within the small dimensions of individual particles. The alloys would thus be easily homogenized during sintering (Eissman et al. 2017).

Eventually, the first additive manufactured HEA, FeCoCrNi, has been reported as a new benchmark for powder metallurgy (Brif et al. 2015).

2 Sintering of High-Entropy Alloys by SPS

Highly-performing bulk HEAs can be achieved via MA and SPS, which can rapidly consolidate the HEA powders to high density within a short soaking time. However, the advantageous speed of SPS is lost by the long time required by mechanical alloying. It is thus surprising that the preparation of a HEA by spark plasma sintering of elemental powder mixture has been reported only once: this methodology is simple and offers an inexpensive and efficient way for producing bulk HEAs based of first-row metals (Zhang et al. 2016).

On the other hand, the high temperatures involved make the sintering of refractive metals and alloys challenging. In studying the spark-plasma sintering of the ball-milled $W_{0.3}(TaTiVCr)_{0.7}$ HEA, Waseem and Ryu noticed a prominent effect of the sintering temperature on particle inter-diffusion (Waseem and Ryu 2017). An increase in relative density was observed with increasing pressure (Yusenko et al. 2018) and temperature. The temperature effect is related to the formation of a

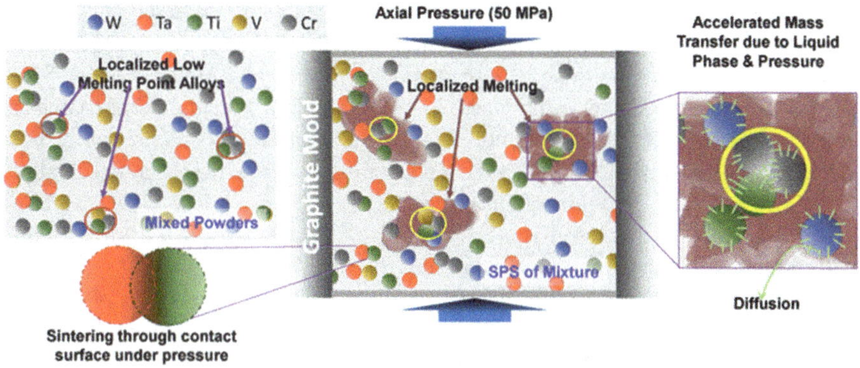

Fig. 1 Illustration of the sintering mechanism of the $W_xTaTiVCr$ HEA. (Reproduced with permission from Waseem and Ryu 2017)

liquid phase following the melting of low melting point constituents of the alloy and has been reported for W-Ni-X powders (X = Fe, Cu, Co or Cr) and WC-Co (Muddle 1984; German et al. 2009). The sintering process is illustrated in Fig. 1. The liquid phase guarantees fast diffusion and the dissolution of solid particles. By seeping through particles, the liquid phase causes wetting due to the mutual solubility of constituents (Kingery 1959).

2.1 *Phase Formation and Evolution*

The $Al_xCoCrFeNiTi_y$ and $Al_xCo_yCrCu_zFeNi$ systems have been widely investigated, since their phase constitution and mechanical properties can be tuned by adjusting Al, Co, Cu and Ti additions – but these elements have very different melting points and are not suitable for casting scale-up. Commonly observed effects upon MA are crystallite refinement, the increment of lattice strain and of dislocation density, and the dissolution of low-melting point elements into the lattice of high-melting point ones.

Crystallite refinement upon MA has been reported for $Al_{0.7}Co_{0.3}CrFeNi$ due to the circulation of crushed material and agglomeration typical for the process (Chen et al. 2013b). The crystallite refinement increases grain boundary friction, whereas lattice strain can be associated with the size mismatch between the constituents. The authors suggested the dissolution of all other elements in Cr and Fe with increasing milling time to be due to the *bcc* crystal structure of the two metals, which could accommodate other elements without much expansion, as well as to the higher melting point of Cr.

An investigation of the $AlCoCrFeNiTi_{0.5}$ highlighted that the major peaks of the constituent elements, still present after blending, disappear after the mechanical alloying procedure with the formation of a single *bcc* phase. The XRD patterns of the corresponding powders broaden, which suggests either lattice distortion due to the induced strains, or excessive grain refinement due to the repeated cold deformation of the powders. After SPS compaction at 1100 °C, however, the *bcc* phase transforms into a mixture of several phases (Moravcik et al. 2016).

Similarly, the mechanical alloying of $Al_{0.6}CoCrFeNiTi_{0.4}$ led to the appearance of broad *bcc* and *fcc* peaks. The authors highlighted that the crystalline size of both phases is significantly reduced with increasing milling time while the lattice strain of the phases increases gradually (Fu et al. 2013a).

An exception to this trend is represented by $Co_{1.5}CrFeNi_{1.5}Ti_{0.5}$. In fact, after MA, the alloy consists mainly of a mixture of two *fcc* phases of cell parameters close to Ni – the element with lowest melting point in the alloy (Moravcik et al. 2017). A throughout investigation of the dissolution of first raw metals (Al, Co, Cr and Fe) in Ni or Cu upon MA has been reported by Praveen and co-workers (Praveen et al. 2012).

As MA is a non-equilibrium process facilitating the formation of metastable supersaturated solutions, the high temperature associated with SPS often results in phase evolution. Atomic rearrangement can be an effect of the annihilation of the defects introduced in the structure during MA – but, unlike annealing, the short-term sintering exposure and intense, localized pulsed electric current of the SPS process might lead to non-equilibrium phases. Wang et al. reported a significant increase in the alloy's grain size (from about 10 nm in the MA powders to hundreds of nanometres in the bulk alloys) upon sintering. This suggest a dramatic decrease of grain boundaries after SPS, resulting in the release of mixing enthalpy from grain boundaries. The authors thus recommend that mixing enthalpy and free energy be considered as affecting phase evolution in SPS (Wang et al. 2016a).

The investigation of $Co_{0.5}CrFeNiTi_{0.5}$, for example, highlighted that phase evolution occurred during SPS, with the original *fcc* + *bcc* phase being substituted by a less strained *fcc* phase and the more stable σ-phase (Fu et al. 2013b). A summary of phase evolution occurring in HEAs fabricated by MA and subsequent SPS is summarized in Table 1.

The effect of sintering temperature on the phase transformations in HEAs have been investigated by Mohanty et al. on AlCoCuNiZn (Mohanty et al. 2014). Upon increasing sintering temperature, a supersaturated *fcc*1 solid solution transforms into a $L1_2$ and an *fcc*2 phase. *fcc*2 grows with temperature until it dominates the XRD profile. Conversely, Wang et al. highlighted the different phase evolution pathway occurring in the MA powder upon annealing or SPS. Annealing of the AlCoCrFeNiMn HEA at 800 °C resulted in the precipitation of the Mn_3Co_7 intermetallic, which does not appear in the sintered sample (Wang et al. 2014). Similar results have been reported in (Ji et al. 2014).

The influence of sintering conditions on microstructure and mechanical properties has been studied taking as example the $Al_{0.6}CoNiFe$ and $Al_{0.6}CoNiFeTi_{0.4}$ HEAs. The diverse sintering routes lead to different phase volume fractions, strength, Vicker's hardness and plasticity (Fu et al. 2014a, b).

Results showing that field-assisted techniques can effectively retain the initial microstructure in the mechanically alloyed powders after sintering motivated further investigations of spark-plasma sintered HEAs powders (Praveen et al. 2012).

Table 1 Summary of the synthetic conditions and phase evolution of HEAs fabricated by MA and SPS

Alloy	Milling conditions (ball to powder weight ratio; milling rate; milling time)	Phase (s) after MA	SPS conditions (temperature, pressure, holding time)	Phase (s) after SPS	Reference
AlBFeNiSi	15:1 350 rpm 240 h	Amorphous	1080 °C 50 MPa 10 min	*bcc* *fcc*1 *fcc*2	Wang et al. (2016b)
AlCoCuNiZn	20:1 200 rpm 20 h	*fcc*1	600–1000 °C 50 MPa 6 min	*fcc*2 $L1_2$	Mohanty et al. (2014)
AlCoCrCuFe	10:1 – 15 h	*bcc* *fcc*	900 °C 50 MPa 15 min	B2 *fcc* σ	Praveen et al. (2012)
	20:1 200 rpm 60 h	*fcc*1 *bcc*	1000 °C 50 MPa 6 min	*fcc*2 σ *fcc*3	Mohanty et al. (2017)
AlCoCrCuFeNi	5:1 – 60 h	*fcc*	750 °C 80 MPa 10 min	*fcc* B2	Ganji et al. (2017)
AlCoCrFeNi	15:1 250 rpm 60 h	*bcc*	900 °C 50 MPa 10 min	*bcc* *fcc*	Ji et al. (2014)
$Al_{0.6}CoCrFeNi$	10:1 300 rpm 42 h	*bcc* *fcc*	1000 °C 30 MPa 8 min	*fcc* *bcc*	Fu et al. (2015a)
$Al_{0.75}CoCrFeNi$	10:1 300 rpm 50 h	*bcc* *fcc*	1000 °C 30 MPa 8 min	*fcc* *bcc*	Chen et al. (2015)
$Al_{0.7}Co_{0.3}CrFeNi$	10:1 300 rpm 45 h	*bcc*	1000 °C 30 MPa 8 min	*bcc* B2 *fcc*	Chen et al. (2013b)
$Al_{0.75}CoCrFeNiTi_{0.25}$	10:1 300 rpm 50 h	*bcc* *fcc*	1000 °C 30 MPa 8 min	*fcc* *bcc*	Chen et al. (2015)
AlCoCrFeTi	10:1 250 rpm 10 h	*bcc*	900 °C 45 MPa 15 min	*bcc* *fcc*	Raphael et al. (2017)
AlCoFeNiTi	10:1 300 rpm 49 h	*bcc* *fcc*	1000 °C 30 MPa 8 min	*fcc* *bcc* Al_3Ti	Fu et al. (2015b)
	15:1 250 rpm 30 h	*bcc* *fcc*	800 °C 30 MPa 10 min	*bcc* *fcc* B2	Zhang et al. (2010)
$Al_{0.5}CoFeNiTi_{0.5}$	10:1 300 rpm 42 h	*fcc*	1000 °C 30 MPa 1 min	*fcc* Unknown	Fu et al. (2014b)

(continued)

Table 1 (continued)

Alloy	Milling conditions (ball to powder weight ratio; milling rate; milling time)	Phase (s) after MA	SPS conditions (temperature, pressure, holding time)	Phase (s) after SPS	Reference
$Al_{0.5}CoCrFeNiTi_{0.5}$	10:1 300 rpm 42 h	*fcc* *bcc*	1000 °C 30 MPa 1 min	*fcc* *bcc* Unknown	Fu et al. (2014b)
$AlCoCrFeNiTi_{0.5}$	10:1 250 rpm 24 h	*bcc*	1100 °C 60 MPa 8 min	B2 *fcc* σ TiC	Moravcik et al. (2016)
$Al_{0.6}CoCrFeNiTi_{0.4}$	10:1 300 rpm 49 h	*bcc* *fcc*	1000 °C 30 MPa 8 min	*fcc* (traces of *bcc*)	Fu et al. (2013b)
	10:1 300 rpm 49 h	*fcc*	1000 °C 30 MPa 6 min	*fcc* *bcc*	Chen et al. (2013a)
AlCoCrFeNiMn	15:1 250 rpm 60 h	*bcc* *fcc*	800 °C 50 MPa 10 min	*bcc* *fcc*	Wang et al. (2014)
$Al_{0.3}CoCu_{0.7}FeNi$	10:1 300 rpm 49 h	*fcc* (WC)	1000 °C 30 MPa –	*fcc* (WC)	Fu et al. (2016)
$AlCrCuFeMg_{0.5}$	10:1 - 20 h	*bcc*	700 °C 50 MPa 5 min	AlFe-type *bcc* Cu_2Mg σ	Maulik et al. (2016)
AlCrCuFeMg	10:1 – 20 h	*bcc*1 *bcc*2	700 °C 50 MPa 5 min	AlFe-type *bcc* Cu_2Mg σ Mg_2Cu	Maulik et al. (2016)
$AlCrCuFeMg_{1.7}$	10:1 – 20 h	*bcc*1 *bcc*2	700 °C 50 MPa 5 min	AlFe-type *bcc* Cu_2Mg σ Mg_2Cu	Maulik et al. (2016)
AlCuNiTiZn	15:1 350 rpm 200 h	amorphous	950 °C 90 MPa 10 min	*bcc* *fcc*1 *fcc*2 $AlNi_2Zr$	Ge et al. (2017)

(continued)

Table 1 (continued)

Alloy	Milling conditions (ball to powder weight ratio; milling rate; milling time)	Phase (s) after MA	SPS conditions (temperature, pressure, holding time)	Phase (s) after SPS	Reference
CoCrCuFeNi	10:1 – 15 h	*fcc* *bcc*	900 °C 50 MPa 15 min	*fcc*1 *fcc*2 σ	Praveen et al. (2012)
	10:1 300 rpm 70 h	*fcc* *bcc*	950 °C 50 MPa 6 min	*fcc*1 *fcc*2 σ	Wang et al. (2016a)
	10:1 300 rpm 15 h	*fcc* *bcc*	1000 °C – 15 min	*fcc*1 *fcc*2 σ (Cr_2O_3)	Praveen et al. (2013a)
$CoCr_{0.2}CuFeNi_{1.8}$	10:1 300 rpm 70 h	*fcc* *bcc*	950 °C 50 MPa 6 min	*fcc*1 *fcc*2 σ	Wang et al. (2016c)
$CoCr_{0.5}CuFeNi_{1.5}$	10:1 300 rpm 70 h	*fcc* *bcc*	950 °C 50 MPa 6 min	*fcc*1 *fcc*2 σ	Wang et al. (2016c)
$CoCr_{0.8}CuFeNi_{1.2}$	10:1 300 rpm 70 h	*fcc* *bcc*	950 °C 50 MPa 6 min	*fcc*1 *fcc*2 σ	Wang et al. (2016c)
CoCrFeNiMn	15:1 250 rpm 60 h	*bcc* *fcc*	800 °C 50 MPa 10 min	*fcc*	Ji et al. (2015a)
	Gas atomized alloy	–	1000 °C 30 MPa 8 min	*fcc*	Wang et al. (2016c)
$Co_{0.5}CrFeNiTi_{0.5}$	10:1 300 rpm 49 h	*fcc*	1000 °C 30 MPa 8 min	*fcc* σ	Fu et al. (2013b)
$Co_{1.5}CrFeNi_{1.5}Ti_{0.5}$	10:1 250 rpm 30 h	*fcc*1 *fcc*2 *bcc*	1150 °C 30 MPa 20 min	*fcc*	Moravcik et al. (2017)
CrCuFeTiZn	10:1 200 rpm 20 h	*fcc* *bcc*	900 °C 50 MPa –	*fcc* *bcc*	Mridha et al. (2013)
$CrCu_{1.5}FeTiZn_{0.5}$	10:1 200 rpm 20 h	*fcc* *bcc*	900 °C 50 MPa –	*fcc* *bcc*	Mridha et al. (2013)

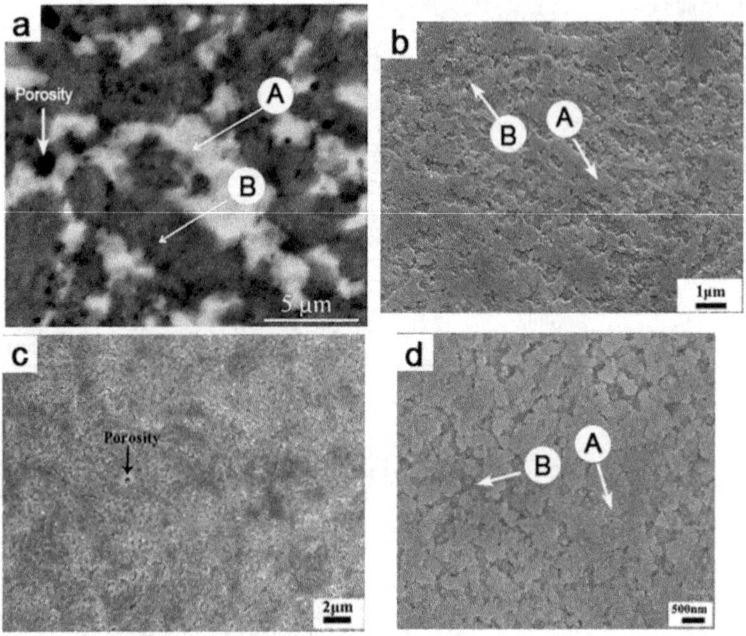

Fig. 2 SEM images of AlCoCrCuFe (**a**), $Al_{0.5}CoFeNiTi_{0.5}$ (**b**) and $Al_{0.6}CoFeNiTi_{0.4}$ (low magnified image in (**c**) and high magnified image in (**d**) after SPS. The regions labelled A and B correspond to *fcc* and *bcc* phases respectively. (Reproduced from Praveen et al. 2012; Fu et al. 2014b; Chen et al. 2013a)

2.2 Microstructural Variation

HEAs present a wide array of microstructural variation, which is retained in the sintered samples. The surface morphology of bulk alloys often comprises irregular areas, potholes, ultrafine porosity and/or nano-precipitates (see Fig. 2a, c).

Nevertheless, Fu et al. noticed that HEAs samples sintered by SPS display lower porosity than their hot pressed (HP) counterparts, because the application of current during SPS increases point defect concentration, reducing the activation energy for their migration and thus enhancing defect mobility (Fu et al. 2015a). By comparing the bulk $Al_{0.6}CrFeNi$ with the $Al_{0.6}CoCrFeNi$ alloy (the latter containing a higher percentage of *fcc* phase), the authors also demonstrated that the easily deformable *fcc* phase aids the densification process (Fu et al. 2015a). Figure 3 represents the sintering process of (*a*) $Al_{0.6}CoCrFeNi$ and (*b*) $Al_{0.6}CrFeNi$ respectively. In both cases, the three representative mechanically alloyed particles exhibit metastable *bcc* and *fcc* supersaturated solid solutions. During pressure-assisted sintering, the *fcc* phase can be deformed so as to incorporate the more brittle *bcc* phase.

The same effect was reported in other duplex-structured HEAs, thus leading to dendritic-like structures as the one displayed in Fig. 2.

TEM bright field images and corresponding SAED patterns have also been frequently used to identify phases with very close cell parameters (Fu et al.

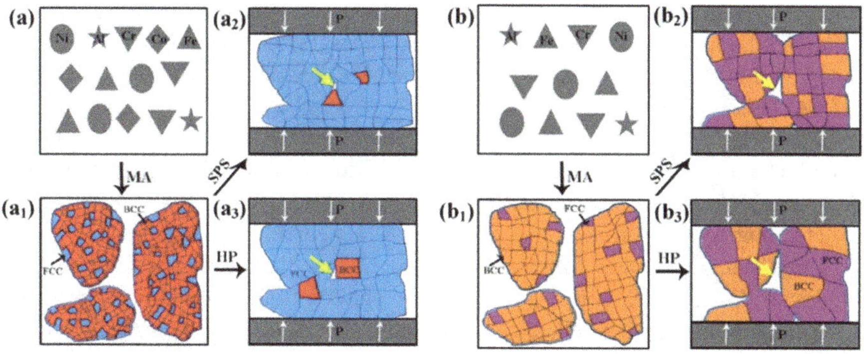

Fig. 3 Microstructure and phase evolution of (**a**) $Al_{0.6}CoCrFeNi$ and (**b**) $Al_{0.6}CrFeNi$ from elemental powders, through mechanical alloying, to pressure-assisted sintering (SPS or HP). The yellow arrow points to residual porosity after sintering. (Reproduced with permission from Fu et al. 2015a)

2013a, b; Ji et al. 2015a, b; Chen et al. 2013a, b; Zhang et al. 2010; Fu et al. 2014a, b; Wang et al. 2016a, b, c; Chen et al. 2015; Wang et al. 2014; Ji et al. 2014; Fu et al. 2015a, b; Mohanty et al. 2017; Mridha et al. 2013; Chen et al. 2013a, b; Mohanty et al. 2014; Ganji et al. 2017; Maulik et al. 2016; Fu et al. 2016). Nanoscale twins have been found in the *fcc* phases of AlCoCrCuFeNi (Ganji et al. 2017), $Al_{0.3}CoCu_{0.7}FeNi$ (Fu et al. 2016), $Al_{0.6}CoCrFeNiTi_{0.4}$ (Fu et al. 2013a, b), $Al_{0.6}CoNiFeTi_{0.4}$ (Chen et al. 2013a, b), $Al_{0.5}CoFeNiTi_{0.5}$, $Al_{0.5}CoCrFeNiTi_{0.5}$ (Fu et al. 2014a, b), $CoCr_{0.8}CuFeNi_{1.2}$ (Wang et al. 2016a, b, c), $Al_{0.75}CoCrFeNi$ (Chen et al. 2015), AlCoFeNiTi (Fu et al. 2015a, b), $Co_{1.5}CrFeNi_{1.5}Ti_{0.5}$ (Moravcik et al. 2017), AlCoCrFeNi (Mohanty et al. 2017), $Cu_{1.5}CrFeTiZn_{0.5}$ (Mridha et al. 2013). Deformation twinning can occur during MA or during SPS. The corresponding two formation mechanisms have been widely discussed in (Fu et al. 2013b).

Moravcik et al. observed a high density of twins inside the *fcc* solid solution grains using EBSD orientation maps (Moravcik et al. 2017). By considering the *fcc* structure to have low stacking fault energy, any eventual plastic deformation in the material could be realized via deformation twinning, to be summed to the prevalent dislocation slip mechanism. Finally, despite the uniaxial pressure of the sintering mechanism, the authors found no evident preferential orientation of the *fcc* phase grains (Moravcik et al. 2017).

2.3 Properties of Spark Plasma Sintered High-Entropy Alloys

2.3.1 Mechanical Properties

As previously mentioned, microstructural investigation of the alloys often indicates the presence of nano-precipitates and eutectoid-like decompositions. These features affect the mechanical properties of the alloy, such as hardness, elastic modulus at different length scales and fracture toughness. SPS'ed. HEAs have been frequently tested for compressive properties and Vicker's hardness. An overview of the results is reported in Table 2: due to the inhomogeneity of the samples, the values reported refer to bulk properties.

In general, both hardness and compressive properties of the sintered alloys are superior to those of HEAs prepared by casting, due to solid solution strengthening and nano-scale grains (Ji et al. 2015a).

The room temperature compressive stress-strain curves are presented in Fig. 4.

Three main behaviours can be observed, corresponding to diverse fracture morphologies. In $Al_{0.6}CoCrFeNiTi_{0.4}$ slip planes and point-like adhesion are observed, with dimples on the fracture surface. The fracture mechanism is thus reported to be intergranular and plastic (Fu et al. 2013a). In comparison, the $Al_{0.5}CoCrFeNiTi_{0.5}$ HEA reported in (Fu et al. 2014a) shows extremely limited compression plasticity. Finally, the $CoCr_xCuFeNi_{1-x}$ series displays remarkable plasticity and a quasi-cleavage fracture mode (Wang et al. 2016a). The differences in compressive behaviour arise from the diverse crystal structures of the alloy, the presence of nanoscale phases, grain size following sintering and solid solution strengthening following the addition of some atoms (Al, Ti). Wang et al. observed serration behaviour under high strain rate deformation in CoCrFeNiMn, due to grain collapse and thermal softening. Micro-voids form on the boundaries of grains deformed at high strain rate and their coalescence induces cracks (Wang et al. 2016b).

Mohaty and co-workers reported a linear increase of Vicker's hardness with sintering temperature for AlCoCrFeNi, which cannot be ascribed to the relative density of the sintered pellet (Mohanty et al. 2017). In fact, the σ-phase and the ordered Al-Ni-rich $L1_2$ phase have coherent or semi-coherent interfaces, making slip transfer feasible. Moreover, the formation of twins leads to sudden hardness increases for samples sintered at high temperatures, with twin boundaries acting as a barrier for dislocation motion (Mohanty et al. 2017).

2.3.2 Functional Properties

With respect to mechanical properties, functional properties of SPS'ed. HEAs have seldom been investigated. The few studies reported in the literature focus on electrochemical and magnetic properties and are briefly discussed below.

The corrosion of bulk AlBFeNiSi in seawater has been probed by potentiodynamic polarization measurements and immersion tests (Wang et al. 2016c). A HEA sample sintered at 520 °C shows pitting resistance due to the formation of an oxide film on the surface. When the passive film dissolves upon prolonged immersion in seawater, a new one is formed. On the other hand, a sample sintered at 1080 °C shows better anodic corrosion resistance in seawater. The passive film forms gradually on the surface and no fluctuation occurs. Similar results are reported by Ge et al. for the AlCuNiTiZr (Ge et al. 2017). Higher sintering temperature correspond to enhanced corrosion resistance. However, fine corrosion products (Cu chlorides) have been found to redeposit on the alloy surface.

Table 2 Overview of compression properties and Vicker's hardness of the investigated HEAs

High-Entropy Alloy	Yield strength, σ_y (GPa)	Compressive strength, σ_{max} (GPa)	Compression ratio, ε_y	Vicker's Hardness	Reference
AlCoCrCuFe	–	–	–	770(10)	Praveen et al. (2012)
AlCoCrFeNi	–	1.91	–	625	Ji et al. (2014)
	1.63	3.23	29.1%	518	Zhang et al. (2016)[a]
$Al_{0.6}CoCrFeNi$	1.87	2.15	10.6%	594	Fu et al. (2015a)
$Al_{0.75}CoCrFeNi$	1.94	2.22	7.6%	577	Chen et al. (2015)
$Al_{0.7}Co_{0.3}CrFeNi$	2.03	2.63	8.1%	624(26)	Chen et al. (2013b)
$Al_{0.75}CoCrFeNiTi_{0.25}$	1.93	2.38	6.8%	609	Chen et al. (2015)
$Al_{0.3}CoCu_{0.7}FeNi$	1.80	1.94	10.6%	454	Fu et al. (2016)
AlCoFeNiTi	–	2.99	–	704	Fu et al. (2015b)
$Al_{0.5}CoFeNiTi_{0.5}$	–	2.85	–	733	Fu et al. (2014b)
$Al_{0.5}CoCrFeNiTi_{0.5}$	–	2.73	–	640	Fu et al. (2014b)
$Al_{0.6}CoCrFeNiTi_{0.4}$	2.08	2.52	11.5%	573	Fu et al. (2013a)
	2.73	3.17	10.1%	712(12)	Chen et al. (2013a, b)
AlCoCrFeNiMn	–	2.14	–	–	Wang et al. (2014)
CoCrCuFeNi	–	–	–	400(1)	Praveen et al. (2012)
	0.87	1.86	32.2%	321(3)	Wang et al. (2016c)
$CoCr_{0.2}CuFeNi_{1.8}$	0.75	1.92	32.4	322(5)	Wang et al. (2016c)
$CoCr_{0.5}CuFeNi_{1.5}$	1.01	1.96	29.5%	376(5)	Wang et al. (2016c)
$CoCr_{0.8}CuFeNi_{1.2}$	1.15	1.94	26.9%	374(3)	Wang et al. (2016c)
CoCrFeNiMn	–	1.99	–	646	Ji et al. (2015a)
$Co_{0.5}CrFeNiTi_{0.5}$	2.65	2.69	10.0%	846	Fu et al. (2013a)

[a]Results for the alloy produced via SPS of elemental powder mixture.

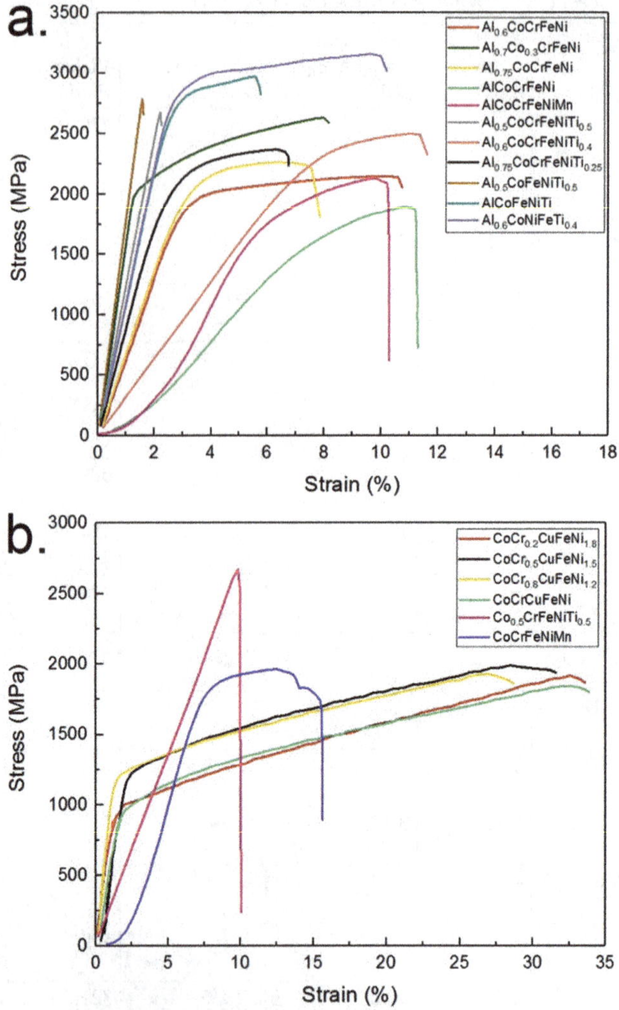

Fig. 4 Room temperature compression stress-strain curves of MA-SPS HEAs containing Al (**a**) or without Al (**b**). (Data from Chen et al. 2013a, b; Chen et al. 2015; Fu et al. 2015a, b; Fu et al. 2014a, b; Fu et al. 2013a, b; Ji et al. 2014; Ji et al. 2015a, b; Wang et al. 2016a; Wang et al. 2014)

The formation of surface oxides (Fe_2O_3, Fe_3O_4, Cr_2O_3, Al_2O_3 and TiO_2) has been reported as improving oxidation and corrosion resistance of the AlCoCrFeTi HEA (Raphael et al. 2017). The slowly growing passive film blocks the outward diffusion of other alloy elements and the inward diffusion of gaseous impurities.

Ji and co-workers reported the room temperature magnetic hysteresis curve for SPS'ed. CoCrFeNiMn. The mechanically alloyed powder shows higher saturated magnetization and lower remanence ratio in comparison with cast alloys of similar composition. The superparamagnetic behaviour displayed by the powder, which can be ascribed to it being made of magnetic nanoparticles of diameter ~10 nm, disappears after sintering due the corresponding grain coarsening and phase evolution (Ji et al. 2015a).

3 High-Entropy Alloys Composites

Metal matrix composites (MMC) are a class of materials often displaying properties unattainable with conventional alloys. Their features often entail tuneable thermal and mechanical properties, thus MMCs are frequently implemented for the thermal management of electronic components and for applications in which high hardness, yield and tensile strength are pivotal. MMCs rely on the addition of a reinforcing material to a metal matrix. For the additive to be effective, a careful balance of wettability and chemical affinity is essential, and the new phase should be homogeneously distributed. These requirements are difficult to satisfy with complex multicomponent alloy matrices. The development of HEA composites (HEACs) is thus still in its early stages, and medium-entropy alloys consisting of four components are more often exploited.

Unlike the pure sintering of metal powders, the preparation of HEACs might be hampered by low densification. The incomplete densification of large elemental clusters during SPS has been reported as the result of entrapped gases and/or localized increases in electrical resistivity. Furthermore, the limited electrical conductivity of certain additives or their low homogeneity can hinder the formation of bonds between reinforcing material and matrix. These results have been reported for the addition of hexagonal and cubic boron nitride, silicon carbide and functionalized carbon nano-tubes to the $Al_2CoCrFeNi$ HEA (Riva et al. 2018).

On the other hand, the addition of oxides to a HEA matrix has been increasingly successful (see Table 3).

For example, the addition of Al_2O_3 to $Al_{0.4}CoCrFeNi_{1.2}Ti_{0.3}$ resulted in a smooth distribution of oxide nanoparticles in the matrix and the formation of strong interfacial bonding on grain boundaries (Yang et al. 2016). As a result of the pinning effect of the additive, HEACs display fine microstructures with uniform grains as shown in Fig. 5, Hadraba et al. highlighted the absence of a preferably oriented microstructure (Hadraba et al. 2017).

Tan et al. overcome the densification issues by using gas-atomised HEA as additive, thus obtaining a homogeneous distribution of spherical particles in an amorphous matrix (Tan et al. 2016). The investigation of the interdiffusion layer surrounding the HEA particles suggested that the localized high temperatures involved in the SPS process could lead to the rapid, non-homogeneous crystallization of the amorphous powder, as well as enhance the diffusion between amorphous powder and HEA particles (Tan et al. 2016). The following sintering model, mediated from the one established to investigate the temperature distribution and neck formation of WC-Co combined particles during SPS, was thus proposed:

Table 3 Summary of sintering conditions, indexed phases, compression properties and Vicker's hardness of the investigated multi-principal component composite systems

Matrix	Additive	Sintering conditions	Phases	σ_y (MPa)	σ_{max} (MPa)	ε_p, %	HV	Reference
AlCoCrFe	0.8 wt.% Y_2O_3:Ti (2:1)	1000 °C 50 MPa 5 min	*bcc* σ	–	–	–	1070 (20)	Praveen et al. (2013a)
$Al_{0.5}Co_{0.3}CrFeNi$	1.2 wt.% C	1000 °C 30 MPa 8 min	*fcc* *bcc* $Cr_{23}C_6$	–	2131	3.0	617(25)	Fang et al. (2014)
$Al_{65}Cu_{16.5}Ti_{18.5}$	50 vol.% $Al_{0.6}CoCrFeNi$	550 °C 400 MPa –	*bcc* *fcc* AlCu Al3Ti $AlCu_2Ti$	–	3120(80)	–	–	Tan et al. (2016)
$Al_{0.4}CoCrFeNi_{1.2}Ti_{0.3}$	12 wt.% Al_2O_3	1000 °C 30 MPa 10 min	*fcc* *bcc* Al_2O_3	–	2250(10)	20.0 (0.5)	743(12)	Yang et al. (2016)
$Al_2CoCrFeNi$	2 wt.% nanodiamonds	850 °C 50 MPa 10 min	B2 Diamond	–	–	–	337(30)	Riva et al. (2018)
$CoCr_{0.25}CuFeNi_{1.5}V_{0.25}$	0.26 wt.% carbides	1000 °C 50 MPa 6 min	*fcc* (V, $Cr)_7C_3$	1045	–	>35.0	335(2)	Wang et al. (2017)
$CoCuFeNi_{1.5}V_{0.5}$	0.29 wt.% carbides	1000 °C 50 MPa 6 min	*fcc* V_2C	1130	–	>35.0	339(2)	Wang et al. (2017)
$CoCrFe_5Mn_3$	0.2 wt.% graphite	900 °C 40 MPa 5 min	*fcc*	–	–	–	–	Liu et al. (2017b)

CoCrFeNi	5 wt.% Ni-coated graphite (3:1) 8 wt.% Ni-coated MoS_2 (3:1)	1150 °C 30 MPa 20 min	*fcc* Graphite Ni MoS_2	610(13)	920(16)	7.5(1.2)	221(14)	Zhang et al. (2017a)
CoCrFeNi	10 wt.% Ag 10 wt.% BaF_2/CaF_2 eutectic	1100 °C 30 MPa 20 min	*fcc* Ag BaF_2 CaF_2	468	787	15.8	151	Zhang et al. (2017b)
CoCrFeNiMn	0.3 wt.% (Ti, Y oxides)	1150 °C 50 MPa 5 min	*fcc*	–	–	–	–	Hadraba et al. (2017)
TiC	0.3 vol.% $MoNb_{0.2}Ta_{0.2}V_{0.4}W$	1500 °C 30 MPa 30 min	*bcc* TiC	–	3060	–	–	Liu et al. (2017a)

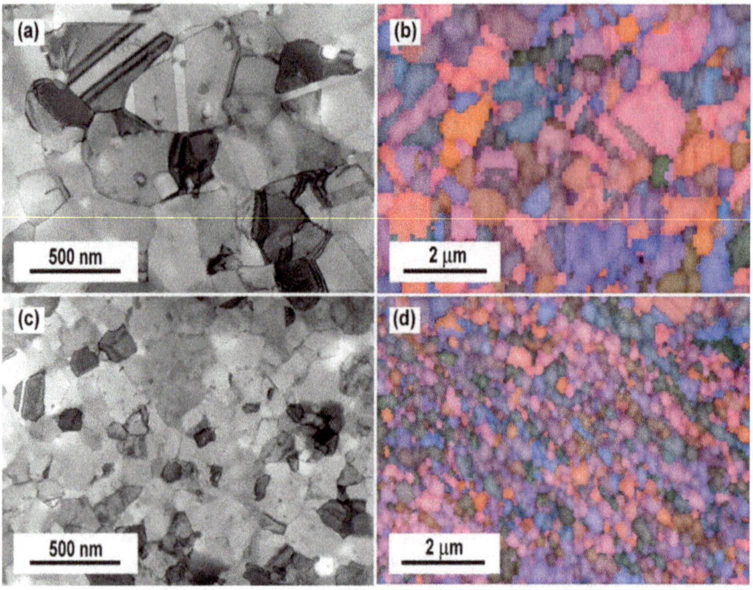

Fig. 5 Grain structure of SPS'ed CoCrFeNiMn (**a, b**) and oxide-containing CoCrFeNiMn (**c, d**) by STEM (*left*) and EBSD (*right*). (Reproduced with permission from Hadraba et al. 2017)

$$\frac{\Delta T_a^{\text{lateral}}}{\Delta T_{HEA}} = \frac{C_{HEA}\nu_{HEA}\left(r_{HEA}^2 - h^2\right)^2\left(\log_e \frac{2r_{HEA}-0.02}{0.02}\right)}{4\pi^2 r_{HEA}^2 C_a \nu_a \left(r^2 - r_{HEA}^2\right)^4} \tag{1}$$

Where $\Delta T_a^{\text{lateral}}$ is the temperature increase from the lateral part of the amorphous matrix; ΔT_{HEA} is the temperature increase of the inner HEA particle; C_{HEA}, C_a and ν_{HEA}, ν_a are the specific heat capacity and electrical resistivity of the HEA and of the amorphous powder respectively; r_{HEA} is the radius of the HEA particle and r is the radius of the HEA particle plus the interdiffusion layer (Tan et al. 2016).

Finally, Liu and coworkers developed a novel reactive sintering method aiming to form the composite in situ, by reaction between the starting materials. In the process, carbide powders (WC, Mo_2C, VC, NbC and TaC) reacted with Ti to form TiC and a refractory HEA phase. All reactions occurring during the sintering have similar negative ΔG, thus the activation energy is controlled by carbon diffusion rate (Liu et al. 2017a).

More often, however, carbides represent the chosen additive to the HEA and are prepared in situ either during mechanical alloying or during sintering. MA creates several favourable conditions for carbide formation: local temperature rise following high-energy milling; creation of high diffusivity path due to lattice deformation; formation of highly reactive sites on particle surfaces. High localized temperatures, on the other hand, might aid the formation of carbides during SPS (Wang et al. 2017).

3.1 Mechanical Properties

The reported HEACs have been investigated as promising candidates for a variety of applications. The diverse set of tested properties, matrices and additives makes it difficult to identify trends.

Nevertheless, as in the case of SPS'ed. HEAs, HEACs have mostly been characterised for their mechanical properties and Vicker's hardness. Composites exhibit higher hardness in comparison with the original alloys, as expected by the strengthening agent addition. However, the hardness increase is seldom matched by good plasticity and elasticity (see Table 3). Riva and co-workers reported an 8% decrease in Young's modulus following the addition of the strengthening phase (Riva et al. 2018).

The search for new self-lubricating materials is the basis for the interest in fracture toughness, friction and wear investigations. Two works reported the presence of dimples on the fractured surface, indicating intense plastic deformation occurring in the HEA prior to fracture (Zhang et al. 2017a). Meanwhile, the additive interrupted the continuity of the matrix and acted as a potential source for crack propagation. Investigation of the wear mechanism by the same authors revealed a combination of abrasive adhesive wear. Friction-induced oxidation in the worn surface was also reported, with metal oxides proving crucial in enhancing tribological properties at high temperatures (Zhang et al. 2017a).

Conversely, the limitation of current state-of-the-art fusion structural materials led to the development of oxides containing HEACs and the investigation of their tensile strength and creep resistance. The higher strength of the composite with respect to the original HEA is connected with the finer microstructure of the first, resulting in grain boundaries strengthening and on the pinning of dislocation movements on the additive's particles (Wang et al. 2017; Hadraba et al. 2017). The yield strength enhancement can be related to a mismatch in the coefficients of thermal expansions and in the elastic moduli of additive and matrix (Wang et al. 2017; Liu et al. 2017b).

Finally, the search for materials suitable for the thermal management of electronic components led to the only investigation of thermoelectric properties of a HEAC. The $Al_2CoCrFeNi$ HEA containing 2 wt.% nanodiamonds shows high electrical resistance due to the combination of lattice distortion typical of multi-principal component alloys and the loss of mean free path following the introduction of the additive. Unlike the original HEA, however, the electrical resistivity of the composite decreases with increasing temperature (Riva et al. 2018).

4 Emerging Trends

Recent works have shown the appeal of HEAs as a new kind of sintering additive for ceramics. They could lower the sintering temperature and inhibit the grain growth of ceramic. Moreover, HEAs possess very good wettability with ceramics.

Table 4 Synthetic conditions and Vicker's hardness of TiB_2-HEA ceramics

HEA Additive	Sintering conditions	Phases	Hardness, HV	Flexural strength (MPa)	Reference
10 wt.% $CoCrFeNiMn_{0.5}Ti_{0.5}$	1500 °C 30 MPa 5 min	TiB_2 *bcc* TiO Ti_9O_{17}	2175	427.69	Zhao et al. (2017)
5 wt.% AlCoCrFeNiTi	1600 °C 30 MPa 5 min	TiB_2 *bcc* Al_2O_3 Ti(O, N,C)	2402	800.02	Ji et al. (2015b)

Based on this TiB_2-HEA ceramics were prepared by mechanical alloying followed by spark plasma sintering (Zhao et al. 2017; Ji et al. 2015b).

As reported by Zhao et al., when the HEA is added to the TiB_2 ceramic powder, the densification process is sensibly improved. At the high temperatures required for ceramic sintering, the HEA is liquid; which allows for wetting and dissolving of the primary phase. Furthermore, increasing the proportion of HEA inhibits the grain growth of TiB_2 (Zhao et al. 2017).

The synthetic conditions and mechanical properties of the specimens investigated by Zhao and Ji are reported in Table 4.

The flexural strength improvement displayed in both samples following the HEA addition is attributed to three concurrent process (Ji et al. 2015b):

The filling of the material's porosity by the liquid alloy, enhanced by the good wetting ability and bonding between the two phases.

The absence of metal borides, which form for most metallic sintering aids and embrittle the ceramic.

Ductile phase toughening following the choice of a *bcc* structured HEA.

A new avenue of research was recently opened by the creation of High-Entropy Metal diborides, possessing virtually one solid-solution diboride phase of the hexagonal AlB_2 structure. The reported specimens include $(Hf_{0.2}Zr_{0.2}Ta_{0.2}Nb_{0.2}Ti_{0.2})B_2$, $(Hf_{0.2}Zr_{0.2}Ta_{0.2}Mo_{0.2}Ti_{0.2})B_2$, $(Hf_{0.2}Zr_{0.2}Mo_{0.2}Nb_{0.2}Ti_{0.2})B_2$, $(Hf_{0.2}Mo_{0.2}Ta_{0.2}Nb_{0.2}Ti_{0.2})B_2$, $(Mo_{0.2}Zr_{0.2}Ta_{0.2}Nb_{0.2}Ti_{0.2})B_2$, and $(Hf_{0.2}Zr_{0.2}Ta_{0.2}Cr_{0.2}Ti_{0.2})B_2$.

These materials represent a new class of high-entropy systems, as well a new type of ultra-high temperature ceramics (Gild et al. 2016).

5 Conclusion

Although research is still at a relatively early stage, it appears that powder metallurgy processing routes for producing HEA-based materials has sensible advantages over the casting route. The increased homogenization often associated with sintered alloys tends to provide materials with superior mechanical properties compared to cast ones. However, sintering of individual metals to form HEAs seems to be very challenging. HEAs should be premelted or mechanically alloyed before sintering, to be able to obtain homogeneous samples – especially if metals with different melting points and diffusion rates are used. Only when these preliminary steps are taken, are dense samples obtained.

High Entropy Alloy Composites with ceramic additives seems to be promising for further development, especially in systems where the HEA melts during processing, effectively wets the ceramic additive surface, can flow and thereby reduce porosity.

Finally, apart from mechanical properties, other functional properties of HEAs such as electrochemical and magnetic properties are still unexplored and should be investigated to obtain materials with promising properties.

Acknowledgements This work was partially supported by the Welsh Government and Higher Education Funding Council for Wales through the Sêr Cymru National Research Network in Advanced Engineering and Materials and by the Materials Advanced Characterization Centre (MACH1) at Swansea University. The authors also gratefully acknowledge the helpful comments and suggestions of the reviewers, which have improved the presentation.

References

Brif Y, Thomas M, Todd I (2015) The use of high-entropy alloys in additive manufacturing. Scr Mater 99:93–96

Cantor B, Chang ITH, Knight P, Vincent AJB (2004) Microstructural development in equiatomic multicomponent alloys. Mater Sci Eng A 375-377:213–218

Chen W, Fu Z, Fang S, Wang Y, Xiao H, Zhu D (2013a) Processing, microstructure and properties of Al0.6CoNiFeTi0.4 high entropy alloy with nanoscale twins. Mater Sci Eng A 565:439–444

Chen W, Fu Z, Fang S, Xiao H, Zhu D (2013b) Alloying behaviour, microstructure and mechanical properties in a FeNiCrCo0.3Al0.7 high entropy alloy. Mater Des 51:854–860

Chen Z, Chen W, Wu B, Cao X, Liu L, Fu Z (2015) Effects of Co and Ti on microstructure and mechanical behaviour of Al0.75FeNiCrCo high entropy alloy prepared by mechanical alloying and spark plasma sintering. Mater Sci Eng A 648:217–224

Chen H, Kauffmann A, Gorr B, Schliephake D, Seemüller C, Wagner JN, Christ HJ, Heilmaier M (2016) Microstructure and mechanical properties at elevated temperatures of a new Al-containing refractory high-entropy alloy Nb-Mo-Cr-Ti-Al. J Alloys Compd 661:206–215

Eissman N, Klöden B, Weissgärber T, Kieback B (2017) High-entropy alloy CoCrFeMnNi produced by powder metallurgy. Powder Metall 60(3):184–197

Fang S, Chen W, Fu Z (2014) Microstructure and mechanical properties of twinned Al0.5CrFeNiCo0.3C0.2 high entropy alloy processed by mechanical alloying and spark plasma sintering. Mater Des 54:973–979

Fu Z, Chen W, Fang S, Zhang D, Xiao H, Zhu D (2013a) Alloying behavior and deformation twinning in a CoNiFeCrAl0.6Ti0.4 high entropy alloy processed by spark plasma sintering. J Alloys Compd 553:316–323

Fu Z, Chen W, Xiao H, Zhou L, Zhu D, Yang S (2013b) Fabrication and properties of nanocrystalline Co0.5FeNiCrTi0.5 high entropy alloy by MA–SPS technique. Mater Des 44:535–539

Fu Z, Chen W, Chen Z, Wen H, Lavernia EJ (2014a) Influence of Ti addition and sintering method on microstructure and mechanical behavior of a medium-entropy Al0.6CoNiFe alloy. Mater Sci Eng A 619:137–145

Fu Z, Chen W, Fang S, Li X (2014b) Effect of Cr addition on the alloying behaviour, microstructure and mechanical properties of twinned CoFeNiAl0.5Ti0.5 alloy. Mater Sci Eng A 597:204–211

Fu Z, Chen W, Wen H, Chen Z, Lavernia EJ (2015a) Effects of Co and sintering method on microstructure and mechanical behavior of a high-entropy Al0.6NiFeCrCo alloy prepared by powder metallurgy. J Alloys Compd 646:175–182

Fu Z, Chen W, Wen H, Morgan S, Chen F, Zheng B, Zhou Y, Zhang L, Lavernia EJ (2015b) Microstructure and mechanical behavior of a novel Co20Ni20Fe20Al20Ti20 alloy fabricated by mechanical alloying and spark plasma sintering. Mater Sci Eng A 644:10–16

Fu Z, Chen W, Wen H, Zhang D, Chen Z, Zheng B, Zhou Y, Lavernia EJ (2016) Microstructure and strengthening mechanism in an *fcc* structured single-phase nanocrystalline Co25Ni25Fe25Al7.5Cu17.5 high-entropy alloy. Acta Mater 107:59–71

Ganji RS, Karthik PS, Sankara Rao KB, Rajulapati KV (2017) Strengthening mechanisms in equiatomic ultrafine grained AlCoCrCuFeNi high-entropy alloy studied by micro- and nanoindentation methods. Acta Mater 121:58–68

Ge W, Wang Y, Shang C, Zhang Z, Wang Y (2017) Microstructures and properties of equiatomic CuZr and CuZrAlTiNi bulk alloys fabricated by mechanical alloying and spark plasma sintering. J Mater Sci 52:5726–5737

German RM, Suri P, Park SJ (2009) Review: liquid phase sintering. J Mater Sci 44:1–39

Gild J, Zhang Y, Harrington T, Jiang S, Hu T, Quinn MC, Mellor WM, Zhou N, Vecchio K, Luo J (2016) High-entropy metal diborides: a new class of high-entropy materials and a new type of ultrahigh temperature ceramics. Sci Rep 6:37946

Gludovatz B, Hohenwarter A, Thurston KVS, Bei H, Wu Z, George EP, Ritchie RO (2016) Exceptional damage-tolerance of a medium-entropy alloy CrCoNi at cryogenic temperatures. Nat Commun 7:10602

Hadraba H, Chlup Z, Dlouhy A, Dobes F, Roupcova P, Vilemova M, Matejicek J (2017) Oxide dispersion strengthened CoCrFeNiMn high-entropy alloy. Mater Sci Eng A 689:252–256

Hemphill MA, Yuan T, Wang GY, Yeh JW, Tsai CW, Chuang A, Liaw PK (2012) Fatigue behaviour of Al0.5CoCrCuFeNi high entropy alloys. Acta Mater 60(16):5723–5734

Ji W, Fu Z, Wang W, Wang H, Zhang J, Wang Y, Zhang F (2014) Mechanical alloying synthesis and spark plasma sintering consolidation of CoCrFeNiAl high-entropy alloy. J Alloys Compd 589:61–66

Ji W, Wang W, Wang H, Zhang J, Wang Y, Zhang F, Fu Z (2015a) Alloying behavior and novel properties of CoCrFeNiMn high-entropy alloy fabricated by mechanical alloying and spark plasma sintering. Intermetallics 56:24–27

Ji W, Zhang J, Wang W, Wang H, Zhang F, Wang Y, Fu Z (2015b) Fabrication and properties of TiB2-based cermets by spark plasma sintering with CoCrFeNiTiAl high-entropy alloy as sintering aid. J Eur Ceram Soc 35:879–886

Kingery WD (1959) Densification during sintering in the presence of a liquid phase. I. Theory. J Appl Phys 30:301–306

Laplanche G, Kostka A, Horst OM, Eggeler G, George EP (2016) Microstructure evolution and critical stress for twinning in the CrMnFeCoNi high-entropy alloy. Acta Mater 118:152–163

Liu B, Wang J, Chen J, Fang Q, Liu Y (2017a) Ultra-high strength TiC/refractory high-entropy-alloy composite prepared by powder metallurgy. JOM 69(4):651–656

Liu X, Zhang L, Xu Y (2017b) Microstructure and mechanical properties of graphene reinforced Fe50Mn30Co10Cr10 high-entropy alloy composites synthesized by MA and SPS. Appl Phys A Mater Sci Process 123:567

Maulik O, Kumar D, Kumar S, Fabijanic DM, Kumar V (2016) Structural evolution of spark plasma sintered AlFeCuCrMgx (x = 0, 0.5, 1, 1.7) high entropy alloys. Intermetallics 77:46–56

Miracle DB, Senkov ON (2017) A critical review of high entropy alloys and related concepts. Acta Mater 122:448–511

Mohanty S, Gurao NP, Biswas K (2014) Sinter ageing of equiatomic Al20Co20Cu20Zn20Ni20 high entropy alloy via mechanical alloying. Mater Sci Eng A 617:211–218

Mohanty S, Samal S, Tazuddin A, Tiwary CS, Gurao NP, Biswas K (2015) Effect of processing route on phase stability in equiatomic multicomponent Ti20Fe20Ni20Co20Cu20 high entropy alloy. Mater Sci Tech 31(10):1214–1222

Mohanty S, Maity TN, Mukhopadhyay S, Sarkar S, Gurao NP, Bhowmick S, Biswas K (2017) Powder metallurgical processing of equiatomic AlCoCrFeNi high entropy alloy: microstructure and mechanical properties. Mater Sci Eng A 679:299–313

Moravcik I, Cizek J, Gavendova P, Sheikh S, Guo S, Dlouhy I (2016) Effect of heat treatment on microstructure and mechanical properties of spark plasma sintered AlCoCrFeNiTi0.5 high entropy alloy. Mater Lett 174:53–56

Moravcik I, Cizek J, Zapletal J, Kovacova Z, Vesely J, Minarik P, Kitzmantel M, Neubauer E, Dlouhy I (2017) Microstructure and mechanical properties of Ni1,5Co1,5CrFeTi0,5 high entropy alloy fabricated by mechanical alloying and spark plasma sintering. Mater Des 119:141–150

Mridha S, Samal S, Khan PY, Biswas K, Bajargan G (2013) Processing and consolidation of nanocrystalline Cu-Zn-Ti-Fe-Cr high-entropy alloys via mechanical alloying. Metall Mater Trans A 44A:4532–4541

Muddle BC (1984) Interphase boundary precipitation in liquid phase sintered W-Ni-Fe and W-Ni-Cu alloys. Metall Trans A 15:1089–1098

Poulia A, Georgatis E, Lekatou A, Karantzalis AE (2016) Microstructure and wear behavior of a refractory high entropy alloy. Int J Refract Met Hard Mater 57:50–63

Praveen S, Murty BS, Kottada RS (2012) Alloying behavior in multi-component AlCoCrCuFe and NiCoCrCuFe high entropy alloys. Mater Sci Eng A 534:83–89

Praveen S, Anupam A, Sirasani T, Murty BS, Kottada RS (2013a) Characterization of oxide dispersed AlCoCrFe high entropy alloy synthesized by mechanical alloying and spark plasma sintering. Trans Indian Inst Metals 66(4):369–373

Praveen S, Murty BS, Kottada RV (2013b) Phase evolution and densification behavior of nano-crystalline multicomponent high entropy alloys during spark plasma sintering. JOM 65 (12):1797–1804

Raphael A, Kumaran S, Kumar KV, Varghese L (2017) Oxidation and corrosion resistance of AlCoCrFeTi High Entropy Alloy. Mat Today Proc 4(2A):195–202

Riva S, Tudball A, Mehraban S, Lavery NP, Brown SGR, Yusenko KV (2018) A novel High-Entropy Alloy-based composite material. J Alloys Compd 730:544–551

Senkov ON, Scott JM, Senkova SV, Miracle DB, Woodward CF (2011) Micro-structure and room temperature properties of a high-entropy TaNbHfZrTi alloy. J Alloys Compd 509 (20):6043–6048

Tan Z, Wang L, Xue Y, Zhang P, Cao T, Cheng X (2016) High-entropy alloy particle reinforced Al-based amorphous alloy composite with ultrahigh strength prepared by spark plasma sintering. Mater Des 109:219–226

Tsai KY, Tsai MH, Yeh JW (2013) Sluggish diffusion in Co–Cr–Fe–Mn–Ni high- entropy alloys. Acta Mater 61(13):4887–4897

Tsau CH, Lee PY (2016) Microstructures of Al7.5Cr22.5Fe35Mn20Ni15 high-entropy alloy and its polarization behaviours in sulfuric acid, nitric acid and hydrochloric acid solutions. Entropy 18:288

Wang C, Ji W, Fu Z (2014) Mechanical alloying and spark plasma sintering of CoCrFeNiMnAl high-entropy alloy. Adv Powder Technol 25(4):1334–1338

Wang B, Fu A, Huang X, Liu B, Liu Y, Li Z, Zan X (2016b) Mechanical properties and microstructure of the CoCrFeMnNi high entropy alloy under high strain rate compression. JMEPEG 25:2985–2992

Wang HL, Gao TX, Niu JZ, Shi PJ, Xu J, Wang Y (2016c) Microstructure, thermal properties, and corrosion behaviour of FeSiBAlNi alloy fabricated by mechanical alloying and spark plasma sintering. Int J Min Metal Mater 23(1):77–82

Wang P, Cai H, Cheng X (2016a) Effect of Ni/Cr ratio on phase, microstructure and mechanical properties of NixCoCuFeCr2−x (x = 1.0, 1.2, 1.5, 1.8 mol) high entropy alloys. J Alloys Compd 662:20–31

Wang P, Cai H, Zhou S, Xu L (2017) Processing, microstructure and properties of Ni1.5CoCuFeCr0.5 xVx high entropy alloys with carbon introduced from process control agent. J Alloys Compd 695:462–475

Waseem OA, Ryu HJ (2017) Powder metallurgy processing of a WxTaTiVCr High-Entropy Alloy and its derivative alloys for fusion material applications. Sci Rep 7:1926

Yang S, Yan X, Yang K, Fu Z (2016) Effect of the addition of nano-Al2O3 on the microstructure and mechanical properties of twinned Al0.4FeCrCoNi1.2Ti0.3 alloys. Vacuum 131:69–72

Yeh JW, Chen SK, Lin SJ, Gan JY, Chin TS, Shun TT, Tsau CH, Chang SY (2004) Nanostructured high-entropy alloys with multiple principal elements: novel alloy design concepts and outcomes. Adv Eng Mater 6(5):299–303

Yu DY, Zhang Y (2016) The ultrahigh charpy impact toughness of forged AlxCoCrFeNi high entropy alloys at room and cryogenic temperatures. Intermetall 70:24–28

Yusenko KV, Riva S, Carvalho PA, Yusenko MV, Arnaboldi S, Sukhikh AS, Hanfland M, Gromilov SA (2017) First hexagonal close packed high-entropy alloy with outstanding stability under extreme conditions and electrocatalytic activity for methanol oxidation. Scr Mater 138:22–27

Yusenko KV, Riva S, Crichton WA, Spektor K, Bykova E, Pakhomova A, Tudball A, Kupenko I, Rohrbach A, Klemme S, Mazzali F, Margadonna S, Lavery NP, Brown SGR (2018) High-pressure high-temperature tailoring of High Entropy Alloys for extreme environments. J Alloys Compd 738:491–500

Zhang KB, Fu ZY, Zhang JY, Wang WM, Lee SW, Niihara K (2010) Characterization of nanocrystalline CoCrFeNiTiAl high-entropy solid solution processed by mechanical alloying. J Alloys Compd 495(1):3–38

Zhang Y, Yang X, Liaw PK (2012) Alloy design and properties optimization of high- entropy alloys. JOM 64(7):830–838

Zhang A, Han J, Meng J, Su B, Li P (2016) Rapid preparation of AlCoCrFeNi high entropy alloy by spark plasma sintering from elemental powder mixture. Mater Lett 181:82–85

Zhang A, Han J, Su B, Li P, Meng J (2017a) Microstructure, mechanical properties and tribological performance of CoCrFeNi high entropy alloy matrix self-lubricating composite. Mater Des 114:253–263

Zhang A, Han J, Su B, Meng J (2017b) A novel CoCrFeNi high entropy alloy matrix self-lubricating composite. J Alloys Compd 725:700–710

Zhao K, Niu B, Zhang F, Zhang J (2017) Microstructure and mechanical properties of spark plasma sintered TiB2 ceramics combined with a high-entropy alloy sintering aid. Adv Appl Ceram 116 (1):19–23

Zheng ZY, Li XC, Zhang C, Li JC (2015) Microstructure and corrosion behaviour of the FeCoNiCuSnx high entropy alloys. Mater Sci Tech 31(10):1148–1152

Spark Plasma Sintering of High Entropy Alloys

Surekha Yadav, Krishanu Biswas, and Arvind Kumar

1 Introduction

High Entropy Alloys (HEAs), being true multi-component multi-principal element alloys are considered new 'Avatar' in the field of metallurgy and materials engineering. The very first publication on HEAs came in 2004, when two independent groups of researchers led by Yeh and Cantor reported their work on multimetallic CoCrFeMnNi alloy (Cantor et al. 2004; Yeh et al. 2004b). Conventional alloys are based on mainly one or two elements or one compound. Though some of these conventional alloys exhibit exceptional properties and contributed to the civilization but since their compositions are based on only corners of the phase diagram, they limit in their degree of freedom to choose composition and develop superior properties, special microstructure and applications. On the other hand; HEAs allow discovering the central unexplored region of the hyper dimensional complex space in the phase diagram (Cantor et al. 2004; Yeh et al. 2004b) and thus, open up new vistas for design of novel alloys with exceptional properties.

HEAs can be defined in two ways based on either entropy of mixing or composition. According to Yeh et al., the HEAs are defined as multi-component solid solution alloys, which contain five or more elements in the concentration range of 5–35 atomic % (Yeh et al. 2004b). Some minor elements can also be added in the HEAs, though their concentration should be less than 5 atomic %. Alternatively, based on entropy, HEAs are defined as alloys having configurational entropy of mixing greater than 1.5R (R = Universal gas constant). Since conventional alloys are based on one or two major elements, their properties are dictated by the host element.

S. Yadav · K. Biswas (✉)
Department of Materials Science and Engineering
Indian Institute of Technology Kanpur, Kanpur, India
e-mail: kbiswas@iitk.ac.in

A. Kumar
Department of Mechanical Engineering, Indian Institute of Technology Kanpur, Kanpur, India

© Springer Nature Switzerland AG 2019

P. Cavaliere (ed.), *Spark Plasma Sintering of Materials*,
https://doi.org/10.1007/978-3-030-05327-7_19

However, in case of HEAs, the properties are dominated by number and type of elements added. HEAs being multi-principal multi-component alloys exhibit fascinating physical and mechanical properties, including high strength at room as well as elevated temperature, high hardness, resistance to wear and corrosion and better oxidation resistance etc. (Murty et al. 2014). These properties make high entropy alloys an interesting field of study both from scientific and technological view point.

HEAs in the bulk form can be synthesized by different fabrication routes including arc melting, induction melting and mechanical alloying followed by sintering. The solid state processing has advantage over other fabrication routes as alloys prepared by arc and induction melting contain many structural defects such as porosity, inhomogeneity, segregation etc. Compositional inhomogeneity and segregation are serious issues in conventionally cast product. Solid state processing involves formation of alloys using mechanical alloying (MA) followed by consolidation of the milled powder. The MA powder can be consolidated using various sintering techniques like hot isostatic pressing (HIP), hot forging, extrusion, conventional sintering and advanced sintering technique including spark plasma sintering (SPS) (Suryanarayana 2001). High heating rate, limited grain growth, controlled environment, short sintering time and ability to synthesize different types of materials make the SPS the most widely used processing route for HEAs (Munir et al. 2006). Desired microstructure and high sintered density can be attained by proper choice of SPS parameters, ultimately leading to the desired properties. HEAs synthesized by SPS route exhibit high hardness, better wear resistance properties and homogenous nanocrystalline microstructure compared to HEAs synthesized by other routes. Most importantly, SPS allows retaining nanocrystalline structure in the MA powder due to high heating rate, shorter sintering time and application of high electric current. It also allows engineering various microstructures, single phase and multiphase.

In this present chapter, a brief summary will be presented of all the research work done till now in the field of HEAs using SPS as a fabrication route, challenges faced during sintering and plausible solutions. The chapter is aimed to provide the avid reader a comprehensive idea about processing of bulk components via MA followed by SPS route. First the basic concepts about HEAs and their derivatives are discussed in brief. We shall discuss the effect of SPS parameters on densification behavior, phase formation and different physical and mechanical properties in HEAs. SPS processing of HEAs will be discussed in three parts: SPS of 3d transition metal HEAs, SPS of refractory HEAs and SPS of HEA composites. This will be followed by discussion on potential applications of sintered HEAs and future perspective in terms of technological advancement.

2 High Entropy Alloys

High entropy alloys are a new class of materials. As mentioned earlier HEAs are defined as solid solution alloys containing five or more elements in the range of 5–35 atomic % (Yeh et al. 2004b). An alternative definition based on entropy of mixing was provided by Cantor (Cantor et al. 2004), according to which HEAs are alloys having configurational entropy of mixing greater than 1.5 R. HEAs can further be divided into equiatomic (all elements having equal atom%) and non-equiatomic depending upon the composition of the alloying elements. If an arbitrary number of 13 elements are selected for equiatomic composition, it leads to 7099 number of possible combinations of systems (Murty et al. 2014). Varying the composition in non-equiatomic manner further leads to tremendous number of possible alloy system. Thus, number of alloy systems that can be studied in the non-equiatomic HEA is huge and it opens up possibilities of design and development of novel alloys.

2.1 Emergence of High Entropy Alloy

Alloy development has played a crucial role in development of human civilization, dating back to Bronze Age to present time. Most of the metallic materials we use today are in the form of alloys and therefore alloys play major role in development of human civilization. Over the years, alloys design strategy has evolved from simple single element composition to complex multi-element compositions (Yeh et al. 2004b). Earliest example of multi-component alloy system is found in antiquity in which Cu, Au, Ag, Fe, Pb, Sn, Hg and Zn; eight different elements were found (Murty et al. 2014). Though alloy design and development existed from the start of civilization, intensive research during last 100 years has led to much progress in the development of new alloys and alloy composites having exceptional properties. However, all these conventional alloys were based on the one or two principal elements or one compound component and were centered on the corners of phase diagram that limits the compositional development, microstructural and special properties. The quest to develop new alloys with special microstructure and properties led to the exploration of unexplored central region of phase diagram and discovery of High Entropy Alloys (HEA) (Miracle and Senkov 2017). Earliest example of HEA is found in a book "Recherchessur les Propriétés des Alliages Métallique," written by German scientist Karl Franz Achard (Achard 1788). In this book; he has compiled the 900 different alloy compositions (binary, ternary, quinary, sexinary and septenary alloys) of 11 metals including Fe, Cu, Pb, Sn, Bi, Sn, Ag, Co, Pt and As. However, this work has been ignored and came to light in 1963 by Prof. Cyril Stanley Smith (Murty et al. 2014). The recent emergence and development of HEAs can be attributed to two independent group of researcher Prof. J.W. Yeh and Prof. B. Cantor. In 1981, Prof Cantor with his undergraduate students initiated the work on different multi-element compositions. First paper was published in 2004 in which they proposed a new single phase FCC equiatomic CoCrFeMnNi HEA, also known as Cantor alloy (Cantor et al. 2004). It has also been concluded that in HEA; maximum number of phases is always less than the phases predicted by Gibbs phase rule. In 1995, Prof Yeh has independently explored the field of multi-component alloys (MCAs) based on concept that high mixing entropy, which might play an important role in reducing the number of phases in the multi-component

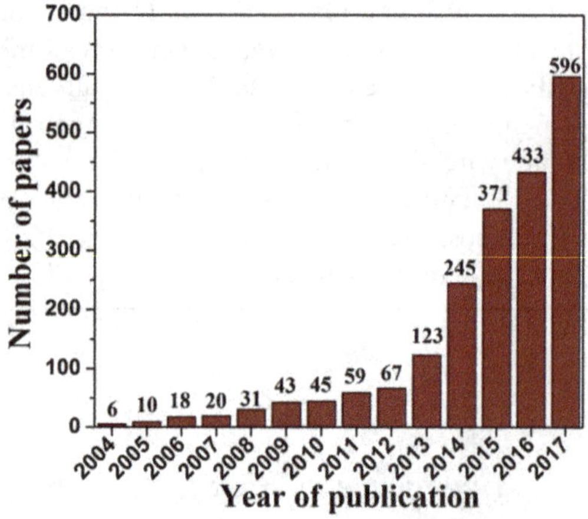

Fig. 1 Number of papers published in recent decade. (Gao et al. 2016)

alloys. Paper titled "Nanostructured high entropy alloys with multi-principal elements – novel alloy design concepts and outcomes" was the first paper to elucidate the high entropy concept correlating experimental results and theory (Yeh et al. 2004b). In the same year 2004, Yeh et al. have also published the other papers on various properties of HEA (Chen et al. 2004; Hsu et al. 2004; Huang et al. 2004; Yeh et al. 2004a). In 2003, in an independent work; Prof. S. Ranganathan reviewed the bulk metallic glasses, gum metals and HEAs (Ranganathan 2003). The progress made in HEAs in terms of number of papers published is shown in Fig. 1 (Gao et al. 2016). It is evident HEAs and HEA-based materials have been expanding very rapidly and this field has become one of the most "happening" areas in materials science and engineering.

2.2 *Design Strategies*

The quest for new alloy compositions for property enhancement is never ending in the field of materials science and engineering. Traditionally, phase diagrams have been utilized to design and develop novel alloys. Only binary and ternary phase diagrams are available in the literature, which make it difficult to design the HEAs composition based on phase diagrams. In addition, not the all combination of elements results in solid solution phase, some even leading to amorphous phase. To overcome these challenges and develop effectively design alloys different empirical rules have been proposed by group of researchers for HEAs. Design strategy for HEAs can be divided into two groups: (i) Parametric approach and (ii) Computational approach.

2.2.1 Parametric Approach

Parametric approach considers the thermodynamic and geometric effect to predict the phases in HEAs. Phase formation in an alloy by mixing of different elements is governed by thermodynamic parameter, Gibbs free energy (G), given by (Prince 1966)

$$\Delta G_{\mathrm{mix}} = \Delta H_{\mathrm{mix}} - T\Delta S_{\mathrm{mix}} \tag{1}$$

where ΔG_{mix} is Gibbs free energy of mixing, ΔH_{mix} is enthalpy of mixing, ΔS_{mix} is entropy of mixing and T is temperature at which elements are mixed. Gibbs free energy is a function of enthalpy and entropy and hence, the competition between ΔH_{mix} and $T\Delta S_{\mathrm{mix}}$ determines the phase formation in an alloy. Phases with the lowest Gibbs free energy are considered as stable one and are most likely to form among all the possible phases; ΔH_{mix} and ΔS_{mix} are provided by:

$$\Delta H_{\mathrm{mix}} = \sum_{i=1,\, i\neq j}^{n} \left(4\Delta H_{\mathrm{mix}}^{AB} C_i C_j\right) \tag{2}$$

$$\Delta S_{\mathrm{mix}} = -R\sum_{i=1}^{n} \left(C_i \ln C_i\right) \tag{3}$$

where C_i, C_j are atomic concentration of ith and j^{th} element, H_{mix}^{AB} is mixing enthalpy of binary equiatomic alloys calculated using Miedema's model (Miedema 1973; Takeuchi and Inoue 2000) and R is universal gas constant.

On the other hand, Hume-Rothery rules are considered as essential parameters required to be followed for selection of elements during alloy design (Calvo-Dahlborg and Brown 2017). These rules for the formation of binary solid solution are stated as follows:

(a) The difference between the radii of solute and solvent atoms should not be greater than 15%.
(b) Solute and solvent atoms should have similar or same crystal structure.
(c) Valence electron concentration (VEC) should be same or similar for both the elements.
(d) To avoid the compound formation, electronegativity should also be similar for both the solute and solvent atoms.

These rules in combination of thermodynamic parameters formed the basis of empirical rules for phase formation rules in HEAs. Additionally, Zhang et al. (2008) have proposed a geometric parameter, the atomic size difference for multi-component alloys, given by:

$$\delta = \sqrt{\sum_{i=1}^{n} c_i \left(1 - \frac{r_i}{\sum_{j=1}^{n} c_j r_j}\right)^2} \quad (4)$$

where r_i and r_j is the atomic radius of the ith and j^{th} component, n is the number of elements.

Based on experimental observation, they proposed that simple solid solution forms when $\delta \leq 6.4\%$, $-20 \leq \Delta H_{mix} \leq 5$ kJ/mol and $12 \leq \Delta S_{mix} \leq 17.5$ J/K.mol (Zhang et al. 2008).

Zhang et al. (Yang and Zhang 2012; Zhang et al. 2012) have also proposed a new thermodynamic parameter Ω, which combines the effect of ΔH_{mix} and ΔS_{mix} as:

$$\Omega = \left|\frac{T_m \, \Delta S_{mix}}{\Delta H_{mix}}\right| \quad (5)$$

where T_m is hypothetical melting temperature calculated by rule of mixture.

$$T_m = \sum_{i=1}^{n} X_i (T_m)_i \quad (6)$$

Based on values of Ω and δ, Zhang et al. (2012) proposed that solid solution forms when $\Omega \geq 1.1$ and $\delta \leq 6.6\%$. Guo et al. (Sheng 2013) discretely identified the values of δ and ΔH_{mix} for the formation of solid solutions ($\delta \leq 6.6\%$, $-11.6 \leq \Delta H_{mix} \leq 3.2$ kJ/mol), amorphous phase ($\delta > 6.4\%$, $\Delta H_{mix} < -12.2$ kJ/mol) and intermetallics.

Effect of valence electron concentration on phase formation in HEAs was first observed by Guo at el. (Sheng and Liu 2011). Guo et al. (Sheng and Liu 2011) have studied a series of $Al_xCrCuFeNi_2$ alloy to study the effect of electron concentration on phase formation. They observed the FCC phases stabilize at higher VEC value (≥8), BCC phase at lower VEC value (<6.87) and FCC and BCC both stabilizes at intermediate value ($6.87 \leq$ VEC < 8). These values of VEC are for reference which can vary for different alloy system.

Otto et al. (2013) found that configurational entropy maximization alone does not result in single phase solid solution formation and hence the repulsive or attractive interactions and atomic size difference between elements are also need to be considered. Accordingly, additional parameters have been proposed dealing with these interactions along with atomic size mismatch. Takeuchi et al. (2013) proposed new parameter mismatch entropy, $\frac{\Delta s_\sigma}{k_\beta}$, considering the misfit entropy term Δs_σ to describe the entropy of mixing of atoms of different sizes. It is given by following equation:

$$\frac{\Delta s_\sigma}{k_\beta} = \left[\frac{3}{2}(\zeta - 1)y_1 + \frac{3}{2}(\zeta - 1)^2 y_2 - \left\{\frac{1}{2}(\zeta - 1)(\zeta - 3) + ln\zeta\right\}(1 - y_3)\right] \quad (7)$$

where, $\zeta = \frac{1}{1-\rho}$, ρ is the packing fraction (0.64 for dense random packing), k_β is the Boltzmann constant, y_1, y_2 and y_3 are dimensionless parameter whose sum is equal to one.

Here,

$$y_1 = \frac{1}{\sigma^3}\sum\nolimits_{j>i=1}^{n}\left(d_i + d_j\right)\left(d_i - d_j\right)^2 c_i c_j \tag{7a}$$

$$y_2 = \frac{\sigma^2}{\left(\sigma^3\right)^2}\sum\nolimits_{j>i=1}^{n} d_i d_j \left(d_i - d_j\right)^2 c_i c_j \tag{7b}$$

$$y_3 = \frac{\left(\sigma^2\right)^3}{\left(\sigma^3\right)^2} \tag{7c}$$

$$\sigma^k = \sum\nolimits_{i=1}^{n} c_i d_i^k \tag{7d}$$

Here, d_i and c_i are the atomic diameter and mole fraction of the ith component respectively.

Therefore, a set of parameters can be used to predict multi-component alloy compositions to design HEAs. However, these parameters are based on the empirical analysis of available data and hence do not provide sufficiently high success rate. Thus, parametric approach leads to large number of possible combinations, from which it is a mundane task to obtain HEA forming systems and compositions.

2.2.2 Computational Approach

Due to large number of possible combination available for HEAs, it is not feasible to study all the possible combination experimentally. In this case, computational methods like Density Functional Theory (DFT), Molecular dynamics (MD) and Monte Carlo (MC) simulations can be used for better identification of combination for solid solution phase formation in HEAs (Ye et al. 2016). However, significant time and cost required limits their use for high throughput alloy screening. An alternative, semi-empirical mathematical method, known as CALculation of PHAse Diagrams (CALPHAD) has been widely used to predict the possible phases in an alloy system and other thermodynamic properties. CALPHAD method uses the Gibbs free energy of a phase to derive all other thermodynamic properties. In case of HEAs, CALPHAD combines and extrapolates the binary and ternary thermodynamic descriptions to derive thermodynamic properties of multi-component system (Murty et al. 2014). Tazuddin et al. (Gurao and Biswas 2016) have successfully utilized the CALPHAD for successful prediction of number of phases in 1287 equiatomic HEAs systems. It was predicted using CALPHAD that out of 1287 number of systems; 10 will result in single phase. Experimental investigation reveals that 6 out of 10 were forming single phase HEA. Thus, a tremendous amount of time and cost can be saved by proper application of CALPHAD and things can be planned in a better way.

2.3 Four Core Effects

It has been observed that the microstructure and properties of HEAs are mainly affected by interaction among different elements present in the alloy system. In 2006, four core effects were proposed by Yeh et al. (Jien-Wei 2006), each for thermodynamics (High entropy effect), kinetics (Sluggish diffusion effect), structure (Lattice distortion effect) and properties (Cocktail effect). These four effects and their influence on structure and properties are shown in Fig. 2 (Gao et al. 2016).

2.3.1 High Entropy Effect

High entropy effect is the first important core effect because of its role in the enhancement of possibility of solid solution phase formation having simple microstructure by minimizing the Gibbs free energy of formation (Yeh 2013). According to maximum entropy production principal (MEPP), high entropy of mixing tend to stabilize the solid solution phases than intermetallic phases (Zhang et al. 2014). Table 1 shows the values of ΔG_{mix}, $\Delta H_{mix,}$and ΔS_{mix} for compounds, solid solution phases and elemental phases without considering the strain energy effect to enthalpy. As can be seen from Table 1, the solid solution phases have medium negative ΔH_{mix} and high ΔS_{mix}, while compound phases have high negative ΔH_{mix} and small ΔS_{mix}. This makes the solid solution phases favorable for formation of equilibrium phases at high temperature in comparison to compounds. Smaller difference in value of ΔH_{mix} for unlike pair of atoms favors the solid solution phases. As reported in the literature; alloy system containing elements with small difference in enthalpy of mixing shows single solid solution phase even at high temperature (Otto et al. 2013; Senkov et al. 2011a; Tsai et al. 2013). On contrary, the alloys with higher difference in enthalpy of mixing for unlike pairs have more than two phases (Zhang et al. 2008). Effect of atomic size difference on the phase formation was also studied by researchers and it was found that smaller size difference, smaller enthalpy of mixing

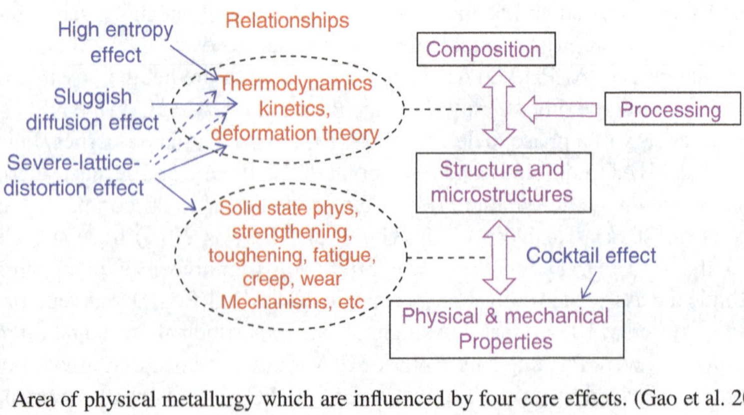

Fig. 2 Area of physical metallurgy which are influenced by four core effects. (Gao et al. 2016)

Table 1 Comparison of ΔH_{mix}, ΔS_{mix} and ΔG_{mix} for elemental, intermediate, compound and random solid solution phases

Comparative states	Elemental phases	Compounds	Intermediate phases	Random solid solutions
ΔH_{mix}	~0	Large −ve	Less large −ve	Medium −ve
ΔS_{mix}	~0	~0	Medium	$\Delta S_{mix} = -R\sum_{i=1}^{n}(C_i \ln C_i)$
ΔG_{mix}	~0	Large −ve	Larger −ve	Larger −ve

and higher entropy of mixing favors the formation of disordered solid solution phases (Sheng and Liu 2011; Yang and Zhang 2012; Zhang et al. 2008).

2.3.2 Severe Lattice Distortion Effect

Difference in atomic sizes of neighbouring atoms, different crystal structure and bonding energy of surrounding atoms in HEAs lead to lattice stress and strain; causing higher lattice distortion compared to that observed in the conventional alloys in which solvent atoms are surrounded by same kind of neighboring atoms (Jien-Wei 2006). This severe lattice distortion contributes to superior hardness, decrease in X-ray diffraction peak intensity due to diffuse scattering on distorted atomic planes (Jien-Wei 2006; Yeh et al. 2004a, 2007). Increased electron and phonon scattering in distorted lattice leads to decrease in electrical and thermal conductivity (Jien-Wei 2006; Kao et al. 2011), and increase in hardness can be contributed to large solution hardening in distorted lattice (Jien-Wei 2006; Senkov et al. 2010). In one of interesting studies, Wang et al. has used Monte Carlo simulation in combination with principles of maximum entropy to study the effect of entropy force on maximization of entropy of mixing for solid solution phase (Wang 2013). Equiatomic BCC and FCC system having four to eight elements were studied. Results are shown in Table 2. It was observed that in case of quaternary and quinary equiatomic BCC alloys, the majority of the same elements are found in the second nearest cell. While majority of the same atoms are found in first nearest neighboring cell in case of quaternary and quinary FCC. Because larger number of unlike pairs between center atoms and dissimilar atoms in the first cell would provide larger distortion and hence this suggest that BCC solid solution would have larger solution hardening effect than FCC solid solution by the same set of elements with same proportions (Wang 2013).

2.3.3 Sluggish Diffusion Effect

The most widely discussed effect in the literature is sluggish diffusion in HEAs. It is one of the salient features among the four core effects of HEAs which helps in improving properties by obtaining supersaturated state, fine precipitates with slower grain growth, increased recrystallization temperature, reducing particle coarsening

Table 2 Distribution of the shortest distances between the same-element atoms on the nearest neighbour sites in the created FCC and BCC MaxEnt model. (Wang 2013)

Phases	Cell type	Distance distribution in nearest neighbor sites (%)					
		1	2	3	4	5	6
Quaternary	BCC	8.5	83	6.9	1.6	0	0
	FCC	74.3	23.8	1.9	0	0	0
Quinary	BCC	0	30.6	3.4	0.5	0	0
	FCC	47.3	7.5	0.2	0	0	0
Senary	BCC	0	41.4	52.4	6.0	0.2	0
	FCC	17.2	64.9	17.7	0.2	0	0
Septenary	BCC	0	19.2	65.3	15.0	0.3	0.2
	FCC	3.9	50.4	44.4	1.1	0.2	0
Octonary	BCC	0	2.5	70.9	24.3	2.1	0.2
	FCC	0.2	27.1	69.8	2.7	0.2	0

Distance between successive nearest neighbor site for BCC and FCC are $\sqrt{3}a_0/2$, a_0, $\sqrt{2}a_0$,…,$\sqrt{2}a_0/2$, a_0, $\sqrt{3}/2a_0$ respectively

rate and increasing creep resistance (Hsu et al. 2004, 2011; Liu et al. 2013; Senkov et al. 2011b; Tsai et al. 2008, 2009). Phase formation and transformation in HEAs require the cooperative diffusion of different elements to attain the phase separation in HEAs. As mentioned earlier HEA is a random solid solution in which solvent atoms and solute atoms cannot be differentiated and matrix is regarded as whole solute matrix. In the solute matrix, the vacancies are surrounded by different kind of elements which compete with the vacancy during diffusion. Since each neighboring atom is different than the other, it leads to fluctuation in the lattice potential energy (LPE) sites and thus, needs higher activation energy for diffusion to complete. These large numbers of LPE sites found in the HEAs causes the hindrance in diffusion of atoms leading to sluggish diffusion (Tsai et al. 2013).

First direct study on the sluggish diffusion in HEAs was performed by Tsai et al. (2013) in the equiatomic single phase FCC CoCrFeMnNi HEA system. In order to study the diffusion effect, four quasi-binary diffusion couples were made in such a way that only two elements were differed in the two end members of each diffusion couple, shown in Table 3 (Tsai et al. 2013). These diffusion couples were than fixed in Mo tube, sealed in a vacuum quartz tube. Subsequently, diffusion experiments were carried out at four different temperatures 1173 K, 1223 K, 1273 K and 1323 K and diffusion coefficient and activation energy were calculated for each conditions. It was found that diffusion rates were in the following orders: Mn > Cr > Fe > Co > Ni.

Diffusion coefficients of each element at T/T_m in different matrices are shown in Fig. 3 (Tsai et al. 2013). The diffusion coefficient was the smallest in HEA matrix for all the elements. Q/T_m (Melting point normalized activation energy) was highest in the case of HEA. In addition, it was also observed that Q/T_m value for same element is dependent on the number of elements in the matrix.

Table 3 The concentrations of end members of three diffusion couples (Tsai et al. 2013)

Couple	Alloy	Composition (at %)				
		Co	Cr	Fe	Mn	Ni
Cr-Mn	1	22	29	22	5	22
	2	22	17	22	17	22
Fe-Co	3	33	23	11	11	22
	4	11	23	33	11	22
Fe-Ni	5	23	24	30	11	12
	6	23	24	12	11	30

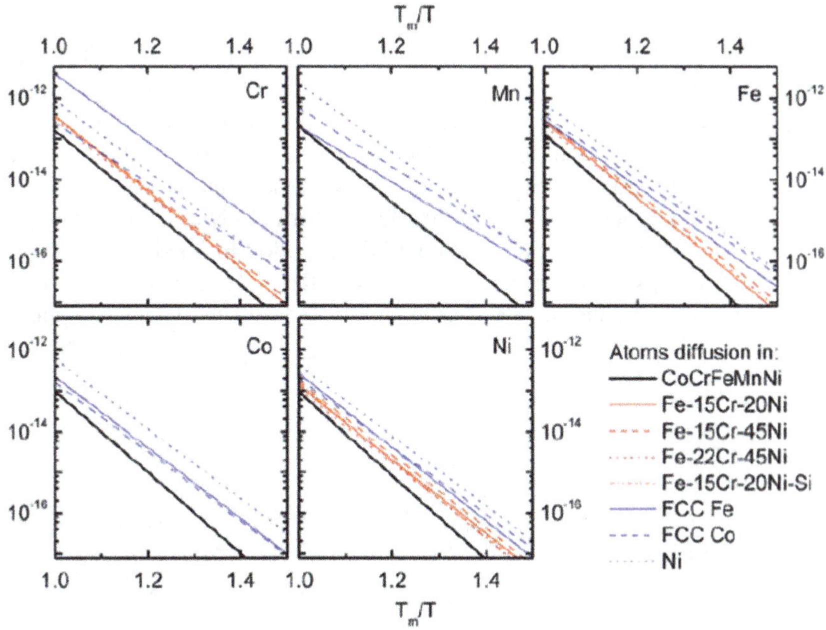

Fig. 3 Dependence of diffusion coefficient on temperature for Cr, Mn, Fe, Co and Ni in different matrices. (Tsai et al. 2013)

In an interesting study Vaidya et al. (2016, 2018) has studied the Ni tracer diffusion and bulk tracer diffusion in CoCrFeNi and CoCrFeMnNi HEAs in two different investigations. It was found that Ni diffusion in HEA was two order of magnitude slower in compared to Ni diffusion in Pure Ni. It was noted that sluggish diffusion in HEAs simply cannot be attributed to presence of increasing number of elements rather also depends on type of elements. Diffusivity variation in CoCrFeNi expressed as Ni < Co < Fe < Cr at higher temperatures and Ni < Co < Fe < Cr at lower temperature.

2.3.4 Cocktail Effect

The term "Multimetallic cocktails" was first proposed by Ranganathan (2003). In HEAs, the cocktail effect is used to emphasize the non-linear effect of different elements on property enhancement. The properties of HEA are considered as a combination of basic elemental properties along with effect of mutual interaction among all the elements. Thus,the cocktail effect can be considered as ranging from atomic scale to microscale multiphase cocktail effect. Refractory quaternary MoNbTaW and quinary MoNbTaVW HEAs developed by Air Force Research Laboratory of USA have melting point higher than Ni and Co based superalloys. The both the alloys exhibit higher softening resistance than superalloys and yield strength of 400 MPa at 1873 K as shown in Fig. 4 (Senkov et al. 2011b). These properties are the results of non-linear synergistic effect, where properties are synergistic mixture of different elements leading to unpredictable properties. In another study Zhang et al. have studied the FeCoNi $(AlSi)_{0-0.8}$ alloy to find the alloy composition having optimum combination of mechanical, magnetic and electric properties (Zhang et al. 2013). They found that $FeCoNi(AlSi)_{0.2}$ alloy composition shows the best properties with 1.15 T saturation magnetization, 1400 A/m coercivity, 69.5 μΩm electrical resistivity, 342 MPa yield strength and 50% strain without fracture, making them a suitable candidate for many potential applications (Zhang et al. 2013).

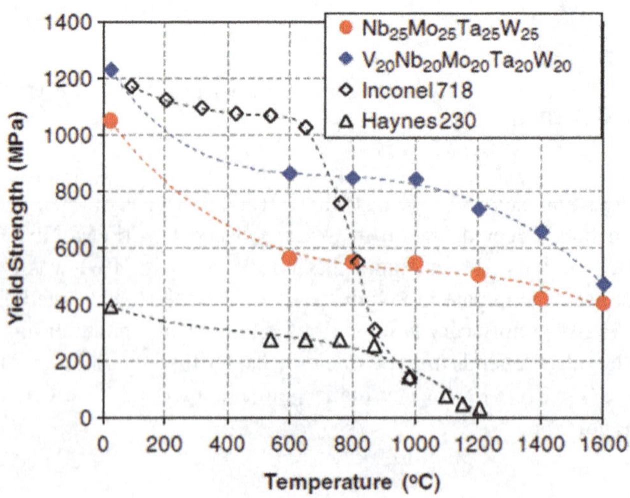

Fig. 4 Yield strength as a function of temperature. (Senkov et al. 2011b)

3 Spark Plasma Sintering of HEAs

The structural applications of HEAs demand these materials to be synthesized in the bulk form with required microstructure and suitable properties. A variety of routes such as liquid state route (includes arc melting, induction melting, laser melting and electric resistance melting) and solid state route (includes mechanical alloying and sintering) can be used for the synthesis of bulk HEAs. However, sintering has the advantage over melting route as alloys having different melting temperature elements, with difference in density can be synthesized using SPS route without any considerable segregation, porosity or inhomogeneity (Stachowiak et al. 2016). Spark plasma sintering uses simultaneous application of high current (800–1500 A) and pressure (30–100 MPa) to achieve the desired dense microstructure with limited grain growth in a controlled environment with proper choice of sintering parameters.

In this section first, we will discuss in brief about the solid state processing route, their advantage and disadvantage over other processing routes, followed by literature review of HEAs synthesized by solid state route and their properties. Most of the HEA forming systems reported have been synthesized using casting technique; with only very few HEAs (approximately 10% of total reported HEAs) have been synthesized using powder metallurgy route (Koch 2007). However, the field is gradually expanding and MA followed by SPS is considered as the prime processing route to obtain various HEAs. The route consisting MA followed by SPS has been found to the most effective route in obtaining nanostructured bulk HEAs. MA has widely been utilized to obtain alloys phases, crystalline, quasicrystalline and amorphous etc. (Murty and Ranganathan 1998). MA route acts as alloying and mechanical activation step prior to SPS. It helps in achieving optimum contribution of density and microstructure in the subsequent steps.

3.1 Mechanical Alloying of HEAs

Mechanical alloying (MA) is a non-equilibrium high energy solid state processing route that uses repeated fracture and cold welding of the powder particles to achieve the alloying (Murty and Ranganathan 1998; Suryanarayana 2001). Preparation via consolidation of nanocrystalline HEA requires mechanical activation. Mechanical activation was best achieved by high energy ball milling. MA is a versatile technique that has following advantages (Murty and Ranganathan 1998; Suryanarayana 2001):

(i) A variety of materials such as nanocrystalline metals and alloys, extended solid solutions, quasicrystalline materials, ceramics, brittle intermetallics compounds and amorphous materials can be synthesized using MA(Murty and Ranganathan 1998).
(ii) Different materials having widely varying melting temperature and vapor pressure can also be milled together to achieve the homogenous microstructure.

(iii) Materials synthesized using MA does not have segregation, a common problem with casting and melting route.
(iv) Nanocrystalline microstructure can be routinely obtained by MA. If this nanocrystalline microstructure can be retained after consolidation, this can lead to improvement in properties of the material.

MA also has some disadvantages which include possible contamination of milled powder from the milling media or environment. However, proper choice of milling media, inert or vacuum environment and proper choice of operating parameters can reduce the amount of contamination to a negligible level (Suryanarayana 2001). MA can effectively be used to obtain FCC or BCC solid solution by altering energy transfer during ball milling. In an interesting study, Salemi et al. (2016) have investigated the effect of energy transfer during ball milling on solid solution formation in single phase FCC equiatomic CuNiCoZnAl HEA. MA with two different conditions was performed. At first they have used balls with similar radii and rotation speed of 300 RPM while in second case balls of different radii with milling speed of 350 RPM was used as milling parameters. Energy analysis shows that balls with different radii and milling speed of 350 RPM results in 1.7 times higher energy transfer during milling. It was concluded that HEA formation takes place under high energy transfer conditions. XRD, SEM and TEM investigations along with thermodynamic modeling confirms the presence of FCC solid solution phase.

3.2 SPS of 3D Transition Metal HEAs

Mechanically alloyed powder needs to be subsequently consolidated to obtain HEAs in the bulk form in order to study the properties of material (Munir et al. 1998, 2001, 2006). SPS, a class of pulsed electric current sintering (PECS) technique has widely been utilized to obtain the bulk form of MA powder. SPS has several advantage over conventional sintering techniques like Hot Isostatic Pressing (HIP), Hot Pressing (HP) or atmospheric furnace such as ease of operation and cost, high reproducibility, high heating rate, limited grain growth, higher densification and lower sintering temperature compared to conventional sintering. Also variety of materials like composites, ceramics, metallic alloys, polymer etc. can be synthesized using SPS making it a versatile sintering technique (Tokita 1999).

3.2.1 Single-phase HEAs

High entropy effect plays a major role in formation of single phase in HEAs. Most of the single phase HEAs reported have simple crystal structure such as BCC, FCC or HCP depending upon the constitute elements (Gao et al. 2016). Depending upon the type of crystal structure, these single phase HEAs exhibit different properties such as high temperature superior mechanical properties (required in aerospace industry) or high strength and ductility (useful for low temperature applications). These properties can further be improved by fine tuning of scale of microstructure, defects etc. via suitable processing route. Thus, it will be interesting to discuss the single phase HEAs and their properties reported in the literature. A detailed literature survey indicates only few reports are available for single phase HEA via MASPS route.

Liu et al. have (2016) studied the room temperature tensile properties of CoCrFeMnNi single FCC phase HEA synthesized using MA and SPS. The powder was first prepared by gas atomization which was further processed using MA. Powder processed using MA has finer microstructure compared to gas-atomized powder. TEM study of sintered HEA depicts that HEA sintered using MA powder has nanocrystalline microstructure as compared to gas atomized sintered HEA. Microhardeness of milled powder as a function of milling time was measured and it was found that hardness increases with milling time. Comparative study of room temperature tensile properties of sintered HEA and as cast HEA shows that sintered HEAs have higher tensile strength (1000 MPa) and reasonable ductility (6%) as compared to the cast HEA. The property enhancement was attributed to nanocrystalline microstructure, which could be achieved by MA followed by SPS route.

Wang et al. (2017a) have studied the effect of Mo addition on the flow behavior and microstructure during high temperature deformation (973 K to 1373 K temperature and $10^{-3} - 1\ s^{-1}$ strain rate) of FCC $CrFeCoNiMo_{0.2}$HEA prepared by powder metallurgy route. Mo addition is expected to provide high temperature strength and stability of the Cantor alloy. The XRD pattern of gas atomized $CrFeCoNiMo_{0.2}$ HEA powder consolidated at 1273 K and 30 MPa pressure using SPS shows the single FCC phase solid solution. Dynamic recovery is the dominant softening mechanism for sample compressed at 973 K, while for sample compressed above 973 K dynamic recrystallization was dominant mechanism. It was found that 1373 K and $1\ s^{-1}$ strain rate is the optimum parameters to achieve completely recrystallized, uniform and fine microstructure with high amount of annealing twin grain boundary. Also, Mo addition leads to enhancement in activation energy and lattice distortion. However, excess Mo addition leads to formation of Mo rich σ phase precipitates in HEA matrix, helping in grain refinement during dynamic recrystallization. MA followed by SPS route can effectively be utilized to achieve the right combination of strength and ductility in this non-equiatomic single phase HEA.

In an interesting study, Fu et al. (2016) have reported the mechanical behavior and strengthening mechanism of novel non-equiatomic nanocrystalline FCC $Co_{25}Ni_{25}Fe_{25}Al_{7.5}Cu_{17.5}$ HEA prepared by MA and SPS. Nanocrystalline FCC phase (average grain diameter~ 24 nm) remains intact after spark plasma sintering with average grain diameter of 95 nm. The SPSed $Co_{25}Ni_{25}Fe_{25}Al_{7.5}Cu_{17.5}$ HEA depicts the 834.9% and 251.9% increase in yield strength (1795 MPa) and hardness (454 HV) respectively, compared to as cast $Co_{25}Ni_{25}Fe_{25}Al_{7.5}Cu_{17.5}$ HEA. Ultra high strength of $Co_{25}Ni_{25}Fe_{25}Al_{7.5}Cu_{17.5}$ HEA is attributed to dislocation strengthening as well as grain boundary strengthening.

3.2.2 Multi-phase HEAs

Most of the HEAs reported in the literature are multiphase. Multiphase HEAs inherit the properties of all the phases present in the microstructure. For example, the dual phase HEAs have soft phase to enhance ductility and hard phase for strength improvement and hence, simultaneous improvement of strength and ductility can be achieved. Properties of these multiphase HEAs mainly depends on the microstructure. Depending upon the particular application, by proper choice of alloying elements, the desired microstructure with required phase can be synthesized. Structural application of these alloys require the proper combination of properties such as yield strength, hardness, elastic modulus, ultimate strength, fatigue, creep etc. The properties are sensitive to microstructure and strongly affected by the grain size, presence of second phase (intermetallics, dispersoids, precipitates etc.), shape, size and distribution of second phase, dislocations, twinning, annealing time and temperature. Scientific understanding of structure-property correlation is much needed for future application of HEAs. Depending upon the number of phases present, the multiphase HEAs can be divided into two broad categories: (i) multiphase HEAs with two phases and (ii) multiphase HEAs with more than two phases.

Multiphase HEAs with Two Phases

MASPS is a solid state process and hence the two phase microstructure can be engineered by a variety of mechanisms. The MA powder consisting of multi-component system is predominantly single FCC or BCC phase, which are considered as metastable supersaturated solid solutions. Hence, these solid solutions can undergo a variety of phase transformations during sintering, leading to formation of two phase microstructure. Eutectoid and Precipitation are two most common phase transformations, observed in the common alloys. Both types of transformations involve the rearrangement of atoms to form the new phase. Depending upon whether the transformed product is a precipitate, sigma/Laves phase or eutectoid or spinodal decomposition, multiphase HEAs can further be divided into following subcategory:

HEAs with Precipitate Phases

The strength and hardness of a metallic alloy can be enhanced by formation of uniformly dispersed fine grained second phase particles in the matrix phase (Mondet et al. 2017). It is a general tendency of metastable phase to transform into more stable lower energy phase when an external stimulus is provided. One such transformation can result in formation of precipitates from the metastable phases. Depending upon the location of precipitation, whether along the grain boundary or within grain, precipitate can impart useful properties to the alloy system. Properties are also influenced by the type of second phase precipitate (ordered/disordered phases, sigma phase, Laves phase). One conventional method to achieve precipitation hardening is by subsequent heat treatment and ageing. However, in an interesting study Mondet et al. (2017) first time successfully explored the use of SPS as a tool for controlled in-situ precipitation along with consolidation in magnesium alloys to achieve the enhanced mechanical properties. This study confirms the potential of SPS technique not just as mere consolidation technique but also as an effective means to achieve the combination of sintering and heat treatment with controlled heating and cooling rate. Thus, one step process is an important advantage over conventional casting route for which this in-situ combination is hardly possible (Mondet et al. 2017). Thus, by proper tuning of heating rate and temperature, spark plasma sintering can be used to achieve the desired precipitation. This opens up the possibility for multi-component system where controlled microstructure can be attained by multistage sintering with controlled heating/cooling rate. Some examples from HEAs literature are reported here:

Mohanty et al. (2017a) had studied the effect of ageing on microstructural evolution and hardness of equiatomic $Al_{20}Co_{20}Cu_{20}Ni_{20}Zn_{20}$ HEA. $Al_{20}Co_{20}Cu_{20}Ni_{20}Zn_{20}$ HEA was prepared by mechanical alloying and then sintered at 1223 K using spark plasma sintering. X-ray diffraction of sintered samples reveals the presence of β (FCC) solid solution phase and α ($L1_2$) phase. Solutionizing treatment of HEA at 1433 K for 96 hr. leads to formation of single phase β (FCC) solid solution. Ageing treatment of the solutionized samples was performed at 673, 773, 873, 1073 and 1273 K temperature for 48 hr. XRD analysis of the aged samples reveal the presence of α ($L1_2$) and γ (FCC) phase. Three- fold increase in hardness was reported for aged samples compared to solutionized samples. The sample aged at 773 K shows the highest hardness of 6.24 GPa. Increase in the hardness was attributed to the combination of solid solution strengthening; grain boundary strengthening, precipitation hardening and Orowan dislocation bypass mechanism.

In another interesting study, Ganji et al. (2017) have studied the strengthening mechanism in AlCoCrCuFeNi HEA synthesized via MA and SPS, where sintering led to precipitation of ordered BCC phase in FCC matrix.. MA powder consists of FCC solid solution phase with average grain size of 10 nm. Sintered microstructure consists of FCC and B2 phase with bimodal grain size distribution with average grain size of 112 ± 46 nm and 1550 ± 500 nm, respectively. Dislocations and twins were also observed by TEM analysis of sintered HEA. Hardness of 6.5 GPa and 8.13 GPa was reported from microindentation and nanoindentaion, respectively. Detailed analysis of strengthening mechanism suggests strain hardening and grain boundary strengthening as dominant strengthening mechanism with some contribution from twin boundary strengthening, frictional stress and solid solution strengthening.

Similar study was also done by Pohan et al. (2017) in which effect of Al addition on microstructural evolution and mechanical properties of equiatomic CoCrFeMnNi. It is to be noted that Al addition leads to formation of ordered B2 precipitates in FCC matrix, which is confirmed by TEM study. Presence of B2 precipitates leads to high hardness due to precipitation hardening and grain boundary hardening.

Fu et al. (2015) have done the comparative study of sintering route on $Al_{0.6}NiFeCrCo$and $Al_{0.6}NiFeCr$ HEA fabricated by mechanical alloying and SPS or HP (hot pressing). MA $Al_{0.6}NiFeCrCo$ and $Al_{0.6}NiFeCr$ HEA consist of minor FCC (15 vol %) and major BCC (85 vol %) solid solution phases, however BCC phase fraction was higher in case of $Al_{0.6}NiFeCr$ HEA. After SPS and HP treatment major FCC (69 vol %) and minor BCC (31 vol %) phase fraction was observed for both $Al_{0.6}NiFeCrCo$ and $Al_{0.6}NiFeCr$ HEAs. SPS'ed. samples show better densification, hardness and strength than HP'ed. samples. Thus, sintering leads to phase evolution of metastable supersaturated solid solution (major FCC and minor BCC phase) to stable solid solution (major BCC and minor FCC) phase.

Zhang et al. (2010) have studied the phase transformation in equiatomic CoCrFeNiTiAl HEA prepared by MA and SPS. After mechanical alloying metastable supersaturated BCC and FCC solid solution phases were observed in MA powder with average crystallite size of 40 nm. The annealing at 873 K for 1 hr. leads metastable phases transformed to two BCC phases minor FCC phase precipitates are observed when MA powder further annealed at higher temperature (973–1273 K). High relative density of 98% and Vickers hardness of 432 HV is observed for samples sintered at 1073. Ordered BCC and FCC phases were observed in case of sintered samples.

Spinodal Decomposition in HEAs

Spinodal decomposition of metastable solid solution phase formed via MA can lead to formation of nanoscaled product phase. Mohanty et al. (2015) have reported the comparative study on the effect of processing route (casting and SPS) in equiatomic TiFeNiCoCu HEA. Intermetallic Ti_2Ni Laves phase along with FCC and BCC solid solution was observed in case of arc melted samples. While spinodal decomposition during SPS leads to formation of dual BCC and FCC phase structure. Thus, microstructure can be tuned to obtain desired phases by choice of processing conditions.

Type of transformation also depends upon the alloy system chosen. Not all the transformation occurs in all the alloy system. In some alloy system other than solid-liquid transformation no other phase transformation was observed. Some examples of HEAs are listed below in which phase transformation took place during sintering; however this transformation cannot be categorized under particular type.

In another study from present authors group, Mirdha et al. (2013) have investigated the microstructural evolution of $Cu_xZn_yTi_{20}Fe_{20}Cr_{20}$ (x/y = 1/0, 3/1, 1/1, where $x + y = 40$) as a function of sintering temperature (873 K, 973 K, 1073 K, and 1173 K). X-ray diffraction and TEM investigation reveals the presence of duplex microstructure with FCC and BCC phase. It is to be noted that highest densification (95%) and hardness (6GPa) was observed for $Cu_{30}Zn_{10}Ti_{20}Fe_{20}Cr_{20}$ sintered at 1173 K.

Wu et al. (2016) have studied the effect of Ti addition on microstructure and mechanical behavior of $Fe_{30}Ni_{30}Co_{30}Mn_{10}$ HEA prepared by MA and SPS. MA alloyed powder of with Ti and without Ti HEA shows the single FCC phase in XRD pattern. After SPS, the microstructure of $Fe_{27}Ni_{27}Co_{26}Mn_{10}Ti_{10}$ HEA shows of

FCC1 and FCC2 phase with similar lattice parameters. Ti addition improves the yield strength and hardness of HEA, though strain to failure ratio decreases with Ti addition. This change in the mechanical behavior with Ti addition is attributed to enhanced grain boundary strengthening and solid solution strengthening. $Fe_{27}Ni_{27}Co_{26}Mn_{10}Ti_{10}$ HEA microstructure contains large number of nanoscale twins which helps in grain boundary strengthening.

Multiphase HEAs with More than Two Phases

MASPS route can lead to formation of multiphase microstructure consisting of more than two phases due to phase transformation of supersaturated solutions leading to formation of different microstructures. In the following, we shall discuss salient results on investigations on the multiphase HEAs.

HEAs with Precipitates

Fu et al. (2013) have studied the alloying behavior, microstructural evolution and mechanical properties of non-equiatomic $CoNiFeCrAl_{0.6}Ti_{0.4}$ HEA prepared by MA and SPS. MA leads to formation of stable FCC and metastable BCC phases. Detailed TEM investigation shows that after sintering at 1273 KBCC phase, FCC1 (Fe-Cr rich) and FCC 2 phases are observed. During sintering BCC phase partially transforms to FCC 2 (Al-Ni-Ti rich) phase and new BCC phase. Sintered alloy shows the high hardness and compressive strength of 573 HV and 2.52 GPa respectively. Detailed study on the fractured surface indicates the plastic mode of fracture and intergranular fracture as main fracture mechanism. In another study, Fu et al. (2014) have studied the phase evolution, mechanical properties and alloying behavior of $CoFeNiAl_{0.5}Ti_{0.5}$ and $CrCoFeNiAl_{0.5}Ti_{0.5}$HEAs prepared by MASPS route. MA results in formation of metastable FCC solid solution in $CoFeNiAl_{0.5}Ti_{0.5}$ alloy and supersaturated BCC and FCC solid solution phases in $CrCoFeNiAl_{0.5}Ti_{0.5}$ HEA. TEM/EDS analysis of sintered samples suggests that sintered $CrCoFeNiAl_{0.5}Ti_{0.5}$ HEA is made-up of FCC1 (Fe-Cr rich) and FCC2 (Al-Ni-Ti rich) phase of same lattice parameters while $CoFeNiAl_{0.5}Ti_{0.5}$ HEA consists of FCC (major phase), BCC (minor phase) and an unknown phase (shown in Fig. 5). Room temperature compression study shows the compressive strength of 2850 MPa and 2726 MPa for $CoFeNiAl_{0.5}Ti_{0.5}$ and $CrCoFeNiAl_{0.5}Ti_{0.5}$HEA respectively. Cr addition also leads to slight decrease in hardness value from 733 HV to 640 HV. This decreases in hardness and compressive strength is attributed to weakening effect of solid solution strengthening due to Cr addition.

Praveen et al. (2012) have studied the phase formation and mechanical properties of quinary equiatomic AlCoCrCuFe and NiCoCrCuFe prepared by MA and SPS. After 15 h. of milling BCC and FCC phases were observed. After sintering at 1173 KBCC and FCC phases evolved to B2 phase, FCC phase (Cu-rich) and σ phase in case of AlCoCrCuFe HEA and FCC1, FCC2 and Cr rich σ phase in case of NiCoCrCuFe HEA. In another study Praveen et al. (Sathiyamoorthi et al. 2017) investigated the thermal stability of CoCrFeNi HEA prepared via ball milling and

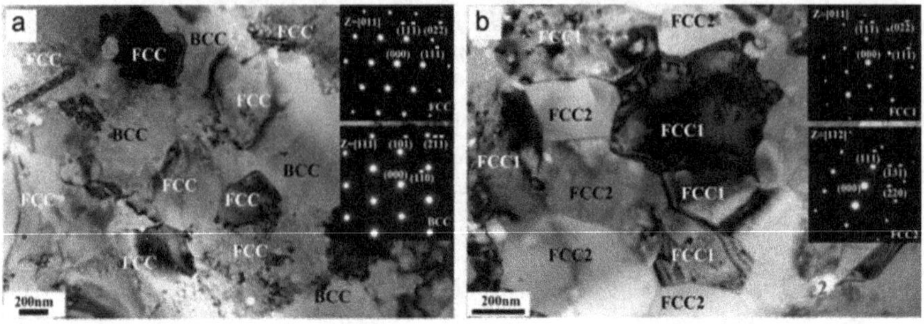

Fig. 5 TEM image and corresponding SAED patterns of bulk HEAs after SPS. (**a**) Bright field image of $CoFeNiAl_{0.5}Ti_{0.5}$, the corresponding SAED pattern of FCC and BCC phase along [011] and [111] zone axis shown in the insets, (**b**) bright field image of $CrCoFeNiAl_{0.5}Ti_{0.5}$, the corresponding SAED pattern of FCC1 and FCC2 phase along [011] and [112] zone axis shown in the insets. (Fu et al. 2013)

SPS at 1173 K having FCC phase with Cr_7C_3carbide formation and oxide inclusions as observed in the XRD and TEM analysis. Sintered HEA were annealed at a range of temperature (0.56 T_m to 0.68 T_m) for 600 hr. However, no significant change in hardness and grain size was observed after prolonged time of exposure suggesting excellent thermal stability of the alloy. Thermal stability was attributed to grain boundary strengthening.

Chen et al. (2013) synthesized the $FeNiCrCo_{0.3}Al_{0.7}$ HEA by MA followed by SPS at 1273 K. A supersaturated BCC solid solution phase was formed during MA. TEM study reveals transformation of BCC phase into two BCC phase of same lattice parameter of (0.301 nm) BCC_A2 (Fe-Cr rich), BCC2_B2 (Ni-Al rich) and a new FCC phase (0.372 nm) during sintering. $FeNiCrCo_{0.3}Al_{0.7}$HEA exhibit high yield strength (2033 ± 41 MPa), compressive strength (2635 ± 55 MPa) and Vickers hardness (624 ± 26Hv). Inter-crystalline fracture and quasi-cleavage fracture were the dominant fracture mechanism.

Wang et al. (2016) have prepared the non equiatomic $Ni_xCoCuFeCr_{2\text{-}x}$ ($x = 1.0$, 1.2, 1.5, 1.8 mol) HEAs by proper combination of mechanical alloying and SPS. XRD analysis of MA powder reveals that FCC phase fraction increases and BCC phase fraction decreases with increasing Ni/Cr ratio. Sintering of milling powder facilitates the transformation of FCC and BCC phase to FCC1 (Ni (Co, Fe)) phase, FCC2 (Cu-rich) and σ phase (Cr (Fe) phase). Hardness and compressive strength of alloys first increases with increasing Ni/Cr ratio and then decreases, reverse trend is observed for plasticity. Among all the compositions $Ni_{1.5}CoCuFeCr_{0.5}$ shows the optimum strength and plasticity. In an another study Wang et al. (2017b) $Ni_{1.5}CoFeCu_{1-x}Al_xV_{0.5}$ ($x = 0.1$, 0.2, 0.3 mol) HEAs were fabricated using MA followed by consolidation using SPS at 1273 K and 50 MPa pressure. XRD analysis reveals $Ni_{1.5}CoFeCu_{1-x}Al_xV_{0.5}$ MA powder consists of FCC and BCC phase, as the Al content increases FCC phase fraction reduces and BCC phase fraction increases in milled HEA powder. However, during SPS phase evolution takes place, a new FCC phase forms along with some carbide phases and BCC phase disappears. With increasing Al concentration yield strength of HEAs increases which is attributed to grain boundary strengthening.

In an interesting study Moravcik et al. (2017) investigated the tensile behavior of nonequiatomic

$Ni_{1.5}$ $Co_{1.5}CrFeTi_{0.5}$ single phase FCC HEA fabricated via MASPS route. MA powder consists of FCC1, FCC2 and BCC phase with lattice parameters of 0.352, 0.358 and 0.287 nm, respectively. SPS of MA HEA powder was done in multiple steps over a temperature range (1273 K–1473 K) with varying heating rate (100 K min^{-1} to 20 K min^{-1}) and holding time. Microstructure of the sintered samples contains fine grain FCC solid solution phase with twins and oxide inclusions (as shown in Fig. 6). $Ni_{1.5}$ $Co_{1.5}CrFeTi_{0.5}$ HEA shows excellent tensile properties compared to cast HEAs with tensile strength of 1384 MPa, 4.01% elongation to fracture and elastic modulus of 216 GPa (shown in Fig. 7).

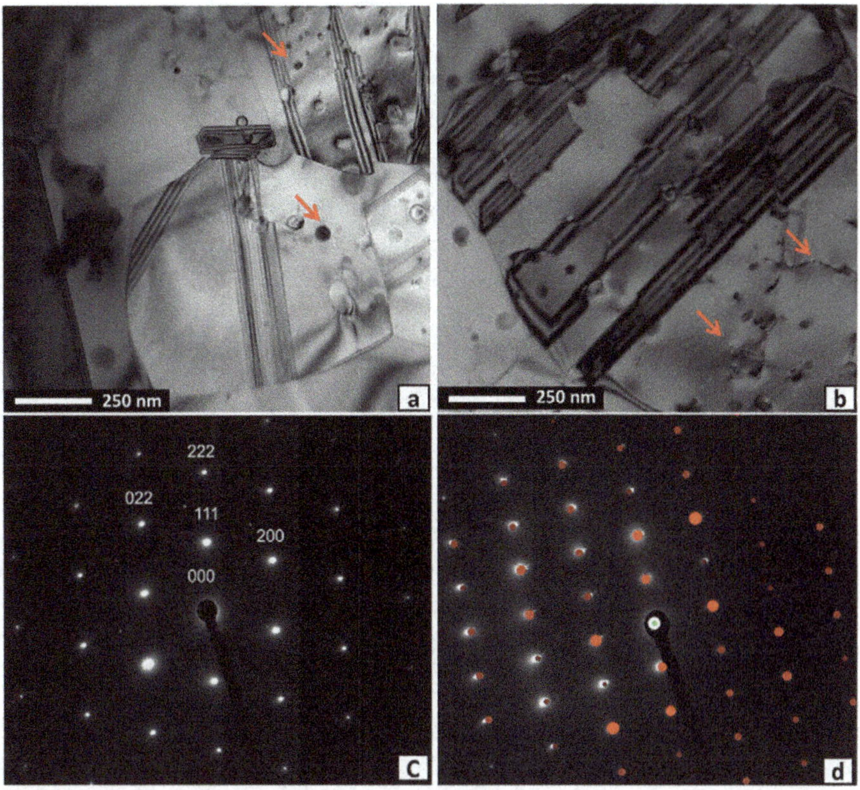

Fig. 6 TEM micrographs of (**a**) FCC grain structure (**b**) nano-twinning within a single FCC grain. (**c**) SAED pattern of FCC grain in 110 orientation. (**d**) SAED pattern of an oxide particle matching the calculated reflections for aluminum oxide. Arrows in (**a**) indicate oxide inclusions, arrows in (**b**) indicate dislocations pinned at oxide inclusions. (Wang et al. 2016)

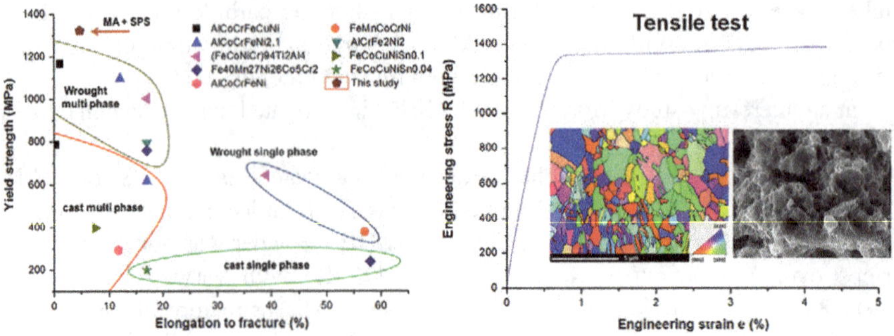

Fig. 7 (**a**) Engineering yield strength values and elongations to fracture of the SPS compacted alloy presented in this study and different HEA and MPEA alloys with comparable densities produced by traditional casting processes and (**b**) Engineering stress – strain curve from tensile test. EBSD orientation maps with pole figures showing one phase FCC bulk material with no significant preferred orientation of its grains shown in inset. (Wang et al. 2016)

Fang et al. (2014) have studied the mechanical properties of $Al_{0.5}CrFeNiCo_{0.3}C_{0.2}$ HEA synthesized via powder metallurgy route. Microstructure of MA powder consists of supersaturated BCC and FCC solid solution phases. Sintered $Al_{0.5}CrFeNiCo_{0.3}C_{0.2}$ HEA consists of Ni-Al rich BCC phase, Al-rich ordered BCC phase and Fe-Ni rich FCC phase with minute amount of $Cr_{23}C_6$ carbide phase. Nanoscaled twins were also observed in sintered FCC phase. $Al_{0.5}CrFeNiCo_{0.3}C_{0.2}$ HEA exhibited high compressive strength (2131 MPa) and hardness (617 HV).

HEAs with Eutectoid Transformation

Eutectoid transformation is a commonly observed phenomenon in steel resulting in pearlite formation. Layered pearlite structure provides the steel much needed strength and ductility. A similar phenomenon was also observed in some high entropy alloys. Mohanty et al. (2017b) have studied the microstructural evolution and mechanical properties of equiatomic AlCoCrFeNi high entropy alloy prepared via MASPS route. Mechanically alloyed powder reveals the presence of FCC and BCC solid solution phases. Sintering at different temperatures (973–1273 K) leads to phase transformation leading to $L1_2$, α' and σ phase formation from eutectoid transformation of BCC solid solution phases, though FCC phase remains intact after sintering. Increase in sintering temperature leads to improvement in hardness and highest hardness of 8 GPa was observed in the case of samples sintered at 1273 K. Lower fracture toughness of HEA was attributed to presence of brittle σ phase.

3.3 SPS of Refractory HEAs

Refractory HEAs have attracted the attention of researchers due to their exceptional high temperature thermal stability, superior mechanical properties at high temperature, high oxidation and corrosion resistance and improved wear resistance and thus, making them a potential candidate for high temperature applications. The refractory HEAs mainly consists of at least 4 out of 9 refractory elements: Cr, Hf, Mo, Nb, Ta, V, Ti, W and Zr. Preliminary results on refractory high entropy alloys shows their potential to replace superalloys (Waseem and Ryu 2016, 2017). In the present section, some case studies from the literature have been reported to better understand the refractory alloys. The preparation of bulk refractory HEA via SPS processes challenges, as the refractory metals have high melting temperature, low diffusivity. Hence, good densification requires utilization of high temperature, slower heating rate and multi-stage sintering schedule.

Waseem et al. (2018) have studied the effect of Ti addition on microstructure and mechanical properties of refractory $Ti_xWTaVCr$ ($x = 0, 4$ and 7 atomic %) fabricated by MA and followed by consolidation at 1773 K temperature and 50 MPa pressure using SPS. Single phase BCC solid solution was formed at 7 atomic % Ti content. Ti_xWTaVC ($x = 7$ atomic %) HEA exhibit the better room temperature and high temperature compressive yield strength compared to pure W and other refractory HEAs owing to solid solution strengthening and presence of Cr and V. The increased Ti addition leads to improvement in oxidation resistance at 1273 K.

Kang et al. (2018) have investigated the microstructural evolution and mechanical properties of equiatomic single phase BCC WNbMoTaV HEA fabricated via mechanical alloying and SPS in the temperature range of 1773–1973 K. Small fraction of Ta_2VO_6 inclusions were observed in sintered samples which were probably inherited during oxidation process. Sample sintered at 1773 K exhibit the high compressive strength and plastic strain of 2612 MPa and 8.8%, respectively were attributed to solid solution strengthening, Orowan strengthening, interstitial solid solution strengthening and grain boundary strengthening.

In one such study, Kumar et al. (2017) have studied the property-structure correlation using conventional strengthening mechanism in AlCuCrFeMnW refractory high entropy alloy. SEM microstructure reveals the presence of ordered B2 phase, sigma phase and a BCC phase in the alloy. This alloy exhibits the high hardness of 891 HV. The high hardness is attributed to grain size strengthening, solid solution strengthening and precipitation hardening. Thermal stability studies were performed at 1273 K. Phase transformation was observed at three temperatures of 1191.17, 1198.23 and 1209 K at different heating rates of 10, 20 and 30 K/min. Respectively.

From above examples it can be concluded that refractory HEAs exhibit superior mechanical properties, high temperature stability and oxidation resistance at high temperatures, making them a suitable candidate for high temperature application. These exceptional properties of refractory HEAs are due to presence of ordered and intermetallic phases.

3.4 SPS of HEAs Based Composites

Metal matrix composites are class of materials which exhibit combined properties of both the matrix and reinforced element. A wide variety of elements such as carbides, oxides, borides, graphite and nonmetallic fibers can be chosen as reinforcement depending upon the properties desired. Selection of these second phase reinforcement materials depends upon the physical (thermal expansion coefficient, diffusivity, interfacial energy between matrix and reinforcement, thermal conductivity etc.) and chemical (solubility) properties of these alloys. Thus, depending upon the reinforcement, these composites can be used for thermal, electrical and mechanical properties. Some examples from literature have been discussed in this section to shed the light on microstructure and properties of HEA-based composites. There exists large number of ways of mixing two or more phases to obtain composites. The choice of second phase dictates the processing route, microstructure and properties.

Tan et al. (2016) have prepared the $Al_{65}Cu_{16.5}Ti_{18.5}$ amorphous alloy by mechanical alloying and then this amorphous powder mixture was mechanically mixed with $Al_{0.6}CoCrFeNi$ HEA prepared by gas atomization. Amorphous alloys exhibit exceptional strength, hardness whereas HEAs are ductile and tough. Thus, HEA-amorphous composite will provide exceptional combination of properties. However, amorphous phase is metastable and sintering schedule needs to be developed properly to retain it in the bulk form. HEA-amorphous alloy composite mixture was sintered at low temperature (823 K) using spark plasma sintering. Detailed TEM investigation of the sintered composite reveals that the microstructure consist of amorphous/nanocrystalline matrix, inter-diffusion layer (ID) made up of grains of size 200–400 nm and dual phase (BCC + FCC) HEA particles (As shown in Fig. 8).

The addition of second phase oxides/borides/nitrides particles has widely been adopted for HEA-based composites. Yang et al. (2017) have reinforced the Al_2O_3 nanoparticles into Al $_{0.4}FeCrCo_{1.5}NiTi_{0.3}$ HEA matrix via MA and SPS. Detailed TEM investigation reveals the homogenous dispersion of Al_2O_3 particles into FCC HEA matrix. HEA/Al_2O_3 composite exhibited the high hardness (654 HV), fracture strength (2.14 GPa) and yield stress (2.05 GPa). In another study, Yang et al. (2016) have studied the effect of 12 wt% Al_2O_3 dispersion in $Al_{0.4}FeCrCoNi_{1.2}Ti_{0.3}$prepared via powder metallurgy route. FCC, BCC and Al_2O_3 phases were observed in sintered microstructure. High hardness (743 HV), fracture strength (2250 MPa) and plastic strain (20%) were observed in sintered composites.

Ji et al. (2015) have used mechanical alloying and SPS route to synthesize the CoCrFeNiTiAl/TiB_2 HEA composite. HEA-TiB_2 composite sintered at 1873 K shows high relative density (99.12%), hardness (2356 HV_5) and flexural strength (800 MPa). High temperature microscopy reveals the contact angle of 16.73° at 1693 K showing excellent wettability between TiB_2 and HEA.

In an interesting study, Hadraba et al. (2017) studied the effect of oxide dispersion on microstructural evolution and mechanical properties of single phase

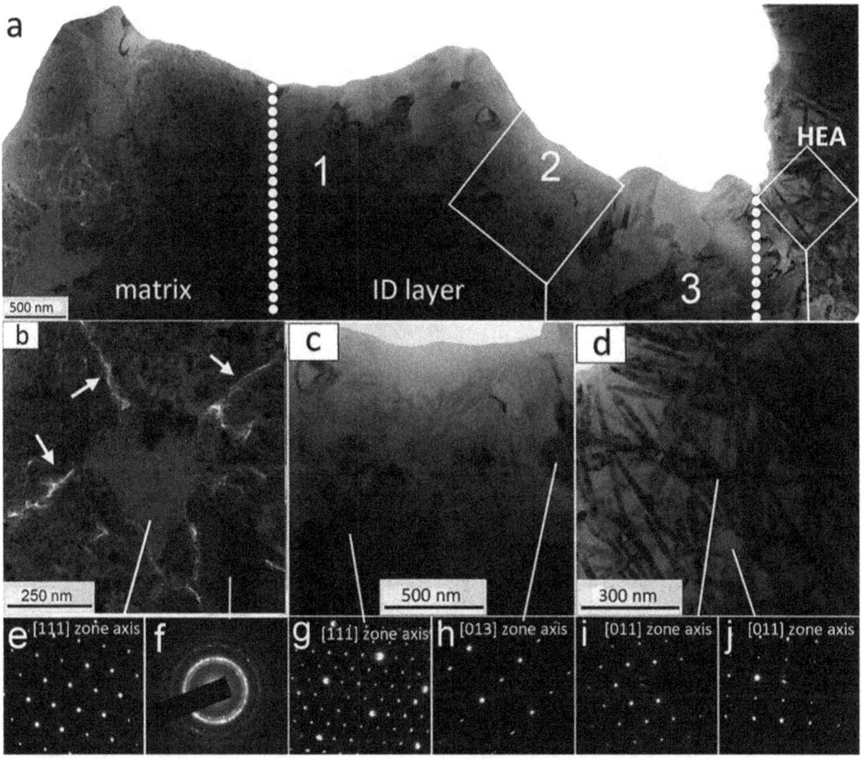

Fig. 8 Bright filed TEM image corresponding to (**a**) the composite; (**b**) the matrix; (**c**) the interdiffusion (ID) layer and (**d**) HEA particle respectively; SAED patterns corresponding to the matrix (**e**, **f**), the ID layer (**g**, **h**), and the HEA particle (**i**, **j**) respectively. (Tan et al. 2016)

CoCrFeMnNi alloy prepared using MA and SPS. Oxide dispersion strengthened (ODS) HEA was prepared by addition of Y, O_2 and Ti during milling. The milled powder was sintered using SPS at 1423 and 50 MPa pressure. Microstructural and phase composition study of sintered ODS HEA was done using XRD, STEM and TEM. XRD of SPSed ODS HEA shows the peaks of FCC only. However, STEM-SAD patterns and TEM EDS analysis shows ODS HEA matrix contain Ti rich Y_2O_3 precipitates. Oxide dispersion does not change the elastic modulus however tensile properties of HEA significantly changes. Yield strength and ultimate tensile strength of ODS increases from 30% at room temperature to 70% at 1073 K. This 30% increase in tensile properties at room temperature is attributed to finer microstructure, grain boundary strengthening and pinning of dislocation movements on oxides. Grain growth inhibition due to nano-metric oxide, suppressed twinning deformation and limited plastic deformation helps to improve the properties with increasing temperature.

In a recent publication, Riva et al. (2018) prepared the Al_2CoCrFeNiHEA reinforced with different reinforcing elements such as nano-diamonds, carbides

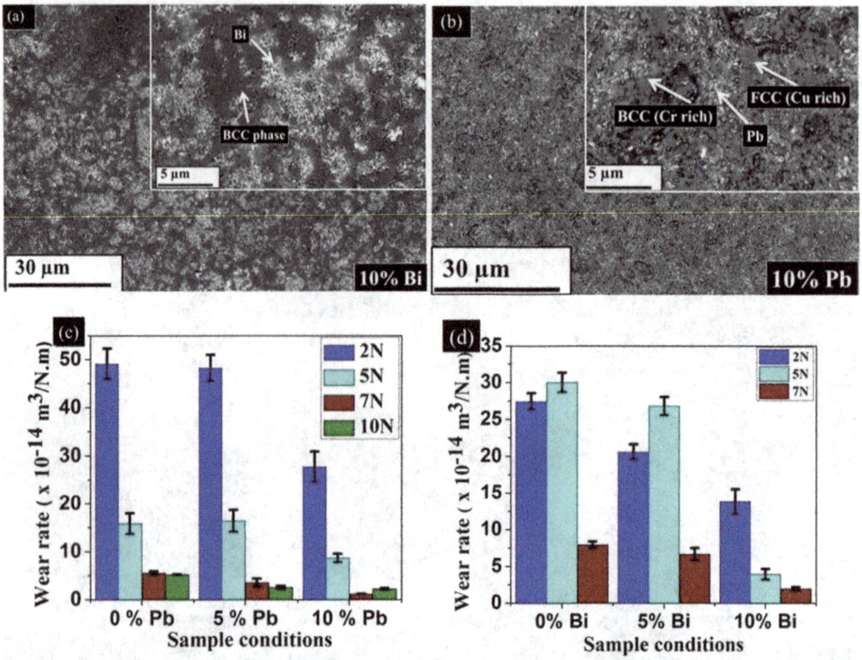

Fig. 9 BSE SEM micrograph exhibiting uniform dispersion of (**a**) $(CuCrFeTiZn)_{90}Pb_{10}$; (**b**) $(AlCrFeMnV)_{90}Bi_{10}$; wear rate as a function of load for (**c**) varying Pb concentration and (**d**) varying Bi concentration. (Yadav et al. 2017)

(SiC), oxides (SC_2O_3), nitrides (h-BN, c-BN and CN) and investigated their effect on thermal, electrical and mechanical properties. It was found that 2 wt% nano-diamonds dispersoid composite exhibit the better densification and properties compared to other reinforcing agent composites. However, properties obtained by nano-diamond composite are not enough for thermoelectric application. Diamond addition also leads to brittleness in composite making it difficult to handle. Although by proper choice of sintering parameters and composition of reinforcing element, microstructure can be tuned to achieve the better properties.

In one such interesting work from authors group, Yadav et al. (2017) have studied the effect of soft dispersoid; Pb and Bi on the tribological properties of dual phase CuCrFeTiZn and single phase AlCrFeMnV HEAs, respectively prepared via powder metallurgy route. Soft dispersoids such as Bi and Pb are considered boundary lubricant and plays an important role in achieving better wear resistance properties. With this objective MASPS with proper processing parameters was used successfully to engineer the microstructure consisting of dispersoid in the harder matrix consisting of HEAs phases. Detailed microstructural and tribological investigation confirms the homogenous dispersion of soft dispersoid phase and improved wear resistance behavior of dispersoid-containing alloys (shown in Fig. 9).

4 Challenges of SPS Processing and Possible Solutions

Some common problems encountered during processing of HEAs and HEA-based composites are listed below:

(i) For spark plasma sintering powders are prepared using mechanical alloying. However, mechanical alloying might lead to contamination in a form of foreign particle incorporated from milling media or oxidation from external environment. These contaminants can be carried over to the final product via SPS and hence can hamper the desired microstructure and properties. Unnecessary temperature rise during milling can influence the alloying of elements leading to unnecessary phase formation. All these challenges need to be overcome by proper choice of milling media, milling environment, milling speed etc. to prevent the contamination from milling media, vial and balls are should be of same material. Dry protective inert atmosphere can prevent the oxidation of alloying powder. Thus, the proper choice of milling parameters can help to achieve the desired alloying.

(ii) Mechanically alloyed powders needs to be consolidated using spark plasma sintering for further characterization. Since, MA is a non-equilibrium process; metastable phases are formed during MA, which can then transform to other phases during sintering. Thus, it is very important to carefully choose the sintering parameters (heating rate, holding time, pressure, and sintering temperature) to retain the alloyed phases during sintering. To avoid the contamination sintering is performed in protective argon atmosphere. Cooling and heating rate might lead to phase transformation. This can be prevented by providing predetermined heating/cooling rate during sintering.

5 Potential Applications of Sintered HEAs

MASPS route has been utilized to prepare novel HEA and HEA-based composites, for different potential applications. In the following, we shall discuss a few examples to highlight the efficacy of SPS route for synthesis of components.

5.1 Irradiation Resistance HEAs

Structural materials used in nuclear reactors needs high thermal stability, high temperature strength, resistance to oxidation and corrosion at high temperature, self-healing ability to withstand the damage caused by continuous ion irradiation (Xia et al. 2015). Conventional materials; austenitic/ferritic steels, zirconium alloys, ceramics and composites used in nuclear reactors are limited in their capacity to prevent the damage in nuclear reactors (Xia et al. 2015). Thus, development of alternative new materials exhibiting the irradiation resistance is the need of the hour. Some preliminary research studies performed on HEAs confirm the potential of HEAs as alternative material for nuclear reactors. However, all the studies reported in the literature are based on cast HEAs and thin film HEAs (Kumar et al. 2016; Nagase et al. 2015; Xia et al. 2015; Yang et al. 2018). There is no reports on the irradiation behavior of SPS based HEAs.

5.2 HEAs as Bearing Materials

Self-lubricating metal matrix composites play an important role as machine parts that operate under harsh environment such as high temperature, under heavy load and oxidative media. These composites are generally made of a harder matrix phase and a second solid lubricant phase. Properties of matrix and solid lubricating phase are crucial for their application as wear resistant materials. Harder matrix phase imparts the strength and oxidation resistance, while softer phase acts as a stress bearing. Thus, proper choice of matrix and lubricant phase is essential to achieve the desired property. Recent studies on high entropy alloys show that HEAs can be promising candidate for matrix phase due to their superior mechanical properties. Homogenous distribution of lubricating phase and better bonding of matrix and lubricant can be achieved by proper choice of fabrication route. Some examples from the literature have been discussed in this section of HEA-dispersoid composites exhibiting excellent wear resistance properties.

Zhang et al. (2017b) have fabricated the novel self-lubricating CoCrFeNi/Ag/BaF_2/CaF_2 HEA composite via SPS at 1373 K from the gas atomized powder of CoCrFeNi, Ag and BaF_2/CaF_2. Microstructure of sintered composite consists of FCC phase, Ag, BaF_2 and CaF_2phase. HEA composite exhibit the good mechanical properties such as hardness (151 HV), fracture toughness (18.3 MPa $m^{1/2}$), yield strength (468 MPa) and compressive strength (787 MPa) come from the CoCrFeNi HEA matrix. Tribological properties of HEA composite was studied over a temperature range RT to 1073 K. Excellent tribological properties were attributed to effect of solid lubricant Ag and BaF_2/CaF_2 (shown in Fig. 10).

In another study Zhang et al. (2017a) studied the tribological properties of CoCrFeNi/Ni-coated graphite/Ni-coated MoS_2 composite prepared via SPS. Sintered microstructure mainly consists of FCC HEA matrix, Ni, graphite and MoS_2 phase. Combination of HEA matrix and solid lubricant result in good mechanical and tribological properties. HEA composite exhibit the high compressive strength (921 MPa), hardness (271 HV), yield strength (610 MPa) and fracture toughness (14.3 MPa $m^{1/2}$).

Raphel et al. (2017) have reported the oxidation and corrosion resistance of AlCoCrFeTi HEA prepared via MA and SPS. Sintered HEA consist of BCC phase with minor amount of FCC phase. High hardness value of 1074 HV was observed for sintered samples. AlCoCrFeTi HEA also shows the excellent corrosion and oxidation resistance.

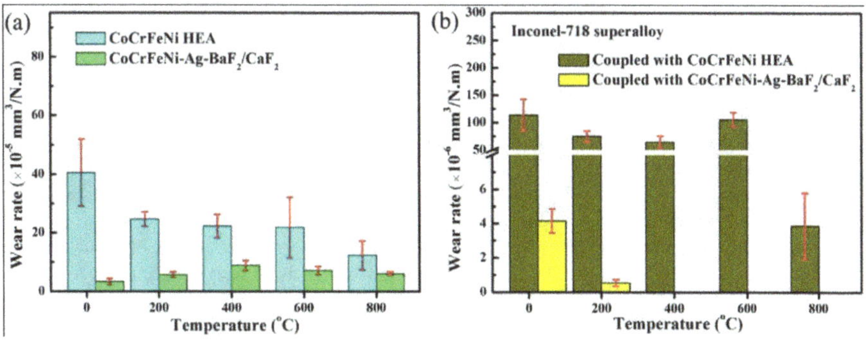

Fig. 10 The variation of wear rates of as a function of temperature for (**a**) the CoCrFeNi HEA and CoCrFeNi-Ag-BaF_2/CaF_2 self-lubricating composite and (**b**) the coupled Inconel-718 superalloy balls. (Zhang et al. 2017b)

6 Summary and Future Prospective

The present chapter is specifically aimed at discussing the design and development of HEAs with enhanced properties prepared via spark plasma sintering route. The chapter starts with a brief introduction about high entropy alloys, phase formation rules followed by core effect which are responsible for many physical properties of HEAs. A brief literature review of single phase, multiphase metallic and refractory HEAs has been provided in order to elucidate the supremacy of SPS route over other conventional processing techniques. In summary, following can be listed:

(i) Mechanical alloying (MA) followed by spark plasma sintering (SPS) is the best route to obtain bulk nanocrystalline HEA and HEA-based composites.
(ii) MA of multicomponent HEA leads to formation of single/multi-phase metastable solid solutions. During SPS processing, these solid solutions can undergo phase transformation leading to formation of variety of microstructure.
(iii) Addition of second phase, oxides, nitrides, carbides, borides, diamond in the HEA matrix leads to property enhancement. However, sintering of HEA with the ceramic particles requires utilization of multistage sintering, higher sintering temperature to obtain the desired microstructure.
(iv) SPS of refractory HEAs is challenging and needs usages of higher temperature, lower heating rate and multistage sintering schedule. In a nutshell, SPS can be used effectively to engineer variety of microstructure in HEAs and thus opens up new possibilities of design and development of novel materials.

Acknowledgement The authors would like to thank DST, CARE IIT K for funding to setup SPS facility at IIT Kanpur. The authors would like to thank DST, ISRO and BRNS for funding the research activities in HEAs.

References

Achard FC (1788) Recherches sur les propriétés des alliages métalliques. Imprimé chez George Jacques Decker & Fils, Imprimeurs Prives Du Rol.

Calvo-Dahlborg M, Brown S (2017) Hume-Rothery for HEA classification and self-organizing map for phases and properties prediction. J Alloys Compd 724:353–364

Cantor B, Chang I, Knight P, Vincent A (2004) Microstructural development in equiatomic multicomponent alloys. Mater Sci Eng A 375:213–218

Chen T-K, Shun T-T, Yeh J-W, Wong M-S (2004) Nanostructured nitride films of multi-element high-entropy alloys by reactive DC sputtering. Surf Coat Technol 188:193–200

Chen W, Fu Z, Fang S, Xiao H, Zhu D (2013) Alloying behavior, microstructure and mechanical properties in a FeNiCrCo0. 3Al0. 7 high entropy alloy. Mater Des 51:854–860

Fang S, Chen W, Fu Z (2014) Microstructure and mechanical properties of twinned Al 0.5 CrFeNiCo 0.3 C 0.2 high entropy alloy processed by mechanical alloying and spark plasma sintering. Mater Des (1980–2015) 54:973–979

Fu Z, Chen W, Fang S, Li X (2014) Effect of Cr addition on the alloying behavior, microstructure and mechanical properties of twinned CoFeNiAl0. 5Ti0. 5 alloy. Mater Sci Eng A 597:204–211

Fu Z, Chen W, Fang S, Zhang D, Xiao H, Zhu D (2013) Alloying behavior and deformation twinning in a CoNiFeCrAl0. 6Ti0. 4 high entropy alloy processed by spark plasma sintering. J Alloys Compd 553:316–323

Fu Z, Chen W, Wen H, Chen Z, Lavernia EJ (2015) Effects of Co and sintering method on microstructure and mechanical behavior of a high-entropy Al 0.6 NiFeCrCo alloy prepared by powder metallurgy. J Alloys Compd 646:175–182

Fu Z et al (2016) Microstructure and strengthening mechanisms in an FCC structured single-phase nanocrystalline Co25Ni25Fe25Al7. 5Cu17. 5 high-entropy alloy. Acta Mater 107:59–71

Ganji RS, Karthik PS, Rao KBS, Rajulapati KV (2017) Strengthening mechanisms in equiatomic ultrafine grained AlCoCrCuFeNi high-entropy alloy studied by micro-and nanoindentation methods. Acta Mater 125:58–68

Gao MC, Yeh J-W, Liaw PK, Zhang Y (2016) High-entropy alloys. Springer, AG Switzerland

Gurao N, Biswas K (2016) In the quest of single phase multi-component multiprincipal high entropy alloys. J Alloys Compd 697:434–442

Hadraba H, Chlup Z, Dlouhy A, Dobes F, Roupcova P, Vilemova M, Matejicek J (2017) Oxide dispersion strengthened CoCrFeNiMn high-entropy alloy. Mater Sci Eng A 689:252–256

Hsu C-Y, Juan C-C, Wang W-R, Sheu T-S, Yeh J-W, Chen S-K (2011) On the superior hot hardness and softening resistance of AlCoCrxFeMo0. 5Ni high-entropy alloys. Mater Sci Eng A 528:3581–3588

Hsu C-Y, Yeh J-W, Chen S-K, Shun T-T (2004) Wear resistance and high-temperature compression strength of Fcc CuCoNiCrAl 0.5 Fe alloy with boron addition. Metall Mater Trans A 35:1465–1469

Huang PK, Yeh JW, Shun TT, Chen SK (2004) Multi-principal-element alloys with improved oxidation and wear resistance for thermal spray coating. Adv Eng Mater 6:74–78

Ji W, Zhang J, Wang W, Wang H, Zhang F, Wang Y, Fu Z (2015) Fabrication and properties of TiB2-based cermets by spark plasma sintering with CoCrFeNiTiAl high-entropy alloy as sintering aid. J Eur Ceram Soc 35:879–886

Jien-Wei Y (2006) Recent progress in high entropy alloys. Ann Chim Sci Mater 31:633–648

Kang B, Lee J, Ryu HJ, Hong SH (2018) Ultra-high strength WNbMoTaV high-entropy alloys with fine grain structure fabricated by powder metallurgical process. Mater Sci Eng A 712:616–624

Kao Y-F, Chen S-K, Chen T-J, Chu P-C, Yeh J-W, Lin S-J (2011) Electrical, magnetic, and Hall properties of AlxCoCrFeNi high-entropy alloys. J Alloys Compd 509:1607–1614

Koch CC (2007) Structural nanocrystalline materials: an overview. J Mater Sci 42:1403–1414

Kumar D, Maulik O, Kumar S, Prasad Y, Kumar V (2017) Phase and thermal study of equiatomic AlCuCrFeMnW high entropy alloy processed via spark plasma sintering. Mater Chem Phys. https://doi.org/10.1016/j.matchemphys.2017.08.049

Kumar NAPK, Li C, Leonard KJ, Bei H, Zinkle SJ (2016) Microstructural stability and mechanical behavior of FeNiMnCr high entropy alloy under ion irradiation. Acta Mater 113:230–244. https://doi.org/10.1016/j.actamat.2016.05.007

Liu W, Wu Y, He J, Nieh T, Lu Z (2013) Grain growth and the Hall–Petch relationship in a high-entropy FeCrNiCoMn alloy. Scr Mater 68:526–529

Liu Y, Wang J, Fang Q, Liu B, Wu Y, Chen S (2016) Preparation of superfine-grained high entropy alloy by spark plasma sintering gas atomized powder. Intermetallics 68:16–22

Miedema A (1973) The electronegativity parameter for transition metals: heat of formation and charge transfer in alloys. J Less Common Met 32:117–136

Miracle D, Senkov O (2017) A critical review of high entropy alloys and related concepts. Acta Mater 122:448–511

Mohanty S, Gurao N, Padaikathan P, Biswas K (2017a) Ageing behaviour of equiatomic consolidated Al 20 Co 20 Cu 20 Ni 20 Zn 20 high entropy alloy. Mater Charact 129:127–134

Mohanty S, Maity T, Mukhopadhyay S, Sarkar S, Gurao N, Bhowmick S, Biswas K (2017b) Powder metallurgical processing of equiatomic AlCoCrFeNi high entropy alloy: microstructure and mechanical properties. Mater Sci Eng A 679:299–313

Mohanty S, Samal S, Tazuddin A, Tiwary C, Gurao N, Biswas K (2015) Effect of processing route on phase stability in equiatomic multicomponent Ti20Fe20Ni20Co20Cu20 high entropy alloy. Mater Sci Technol 31:1214–1222

Mondet M, Barraud E, Lemonnier S, Allain N, Grosdidier T (2017) Optimisation of the mechanical properties of a Spark Plasma Sintered (SPS) magnesium alloy through a post-sintering in-situ precipitation treatment. J Alloys Compd 698:259–266

Moravcik I et al (2017) Microstructure and mechanical properties of Ni1, 5Co1, 5CrFeTi0, 5 high entropy alloy fabricated by mechanical alloying and spark plasma sintering. Mater Des 119:141–150

Mridha S, Samal S, Khan PY, Biswas K (2013) Processing and consolidation of nanocrystalline Cu-Zn-Ti-Fe-Cr high-entropy alloys via mechanical alloying. Metall Mater Trans A 44:4532–4541

Munir Z, Anselmi-Tamburini U, Ohyanagi M (2006) The effect of electric field and pressure on the synthesis and consolidation of materials: a review of the spark plasma sintering method. J Mater Sci 41:763–777

Munir ZA, Charlot F, Bernard F, Gaffet E (2001) One-step synthesis and consolidation of nanophase materials. US Patent 6,200,515

Munir ZAR, Shon I-J, Yamazaki K (1998) Simultaneous synthesis and densification by field-activated combustion. US Patent 5,794,113

Murty B, Ranganathan S (1998) Novel materials synthesis by mechanical alloying/milling. Int Mater Rev 43:101–141

Murty BS, Yeh J-W, Ranganathan S (2014) High-entropy alloys. Butterworth-Heinemann, AG Switzerland

Nagase T, Rack PD, Noh JH, Egami T (2015) In-situ TEM observation of structural changes in nano-crystalline CoCrCuFeNi multicomponent high-entropy alloy (HEA) under fast electron irradiation by high voltage electron microscopy (HVEM). Intermetallics 59:32–42

Otto F, Dlouhý A, Somsen C, Bei H, Eggeler G, George EP (2013) The influences of temperature and microstructure on the tensile properties of a CoCrFeMnNi high-entropy alloy. Acta Mater 61:5743–5755

Pohan RM et al (2017) Microstructures and mechanical properties of mechanically alloyed and spark plasma sintered Al0.3CoCrFeMnNi high entropy alloy. Mater Chem Phys. https://doi.org/10.1016/j.matchemphys.2017.09.013

Praveen S, Murty B, Kottada RS (2012) Alloying behavior in multi-component AlCoCrCuFe and NiCoCrCuFe high entropy alloys. Mater Sci Eng A 534:83–89

Prince A (1966) Alloy phase equilibria. Elsevier, AG Switzerland

Ranganathan S (2003) Alloyed pleasures: multimetallic cocktails. Curr Sci 85:1404–1406

Raphel A, Kumaran S, Kumar KV, Varghese L (2017) Oxidation and Corrosion resistance of AlCoCrFeTi High Entropy Alloy. Mater Today 4:195–202

Riva S, Tudball A, Mehraban S, Lavery NP, Brown SG, Yusenko KV (2018) A novel High-Entropy Alloy-based composite material. J Alloys Compd 730:544–551

Salemi F, Abbasi M, Karimzadeh F (2016) Synthesis and thermodynamic analysis of nanostructured CuNiCoZnAl high entropy alloy produced by mechanical alloying. J Alloys Compd 685:278–286

Sathiyamoorthi P, Basu J, Kashyap S, Pradeep K, Kottada RS (2017) Thermal stability and grain boundary strengthening in ultrafine-grained CoCrFeNi high entropy alloy composite. Mater Des 134:426–433

Senkov O, Scott J, Senkova S, Miracle D, Woodward C (2011a) Microstructure and room temperature properties of a high-entropy TaNbHfZrTi alloy. J Alloys Compd 509:6043–6048

Senkov O, Wilks G, Miracle D, Chuang C, Liaw P (2010) Refractory high-entropy alloys. Intermetallics 18:1758–1765

Senkov O, Wilks G, Scott J, Miracle D (2011b) Mechanical properties of Nb25Mo25Ta25W25 and V20Nb20Mo20Ta20W20 refractory high entropy alloys. Intermetallics 19:698–706

Sheng G (2013) Phase selection rules for complex multi-component alloys with equiatomic or close-to-equiatomic compositions. Nature 35:85–96

Sheng G, Liu CT (2011) Phase stability in high entropy alloys: formation of solid-solution phase or amorphous phase. Prog Nat Sci 21:433–446

Stachowiak G, Basu B, Kalin M (2016) Tribology of ceramics and composites: a materials science perspective. In: Springer, AG Switzerland

Suryanarayana C (2001) Mechanical alloying and milling. Prog Mater Sci 46:1–184

Takeuchi A, Amiya K, Wada T, Yubuta K, Zhang W, Makino A (2013) Entropies in alloy design for high-entropy and bulk glassy alloys. Entropy 15:3810–3821

Takeuchi A, Inoue A (2000) Calculations of mixing enthalpy and mismatch entropy for ternary amorphous alloys. Mater Trans JIM 41:1372–1378

Tan Z, Wang L, Xue Y, Zhang P, Cao T, Cheng X (2016) High-entropy alloy particle reinforced Al-based amorphous alloy composite with ultrahigh strength prepared by spark plasma sintering. Mater Des 109:219–226

Tokita M (1999) Mechanism of spark plasma sintering. In: Proceeding of NEDO International Symposium on Functionally Graded Materials, Japan, pp 1–13

Tsai C-W, Chen Y-L, Tsai M-H, Yeh J-W, Shun T-T, Chen S-K (2009) Deformation and annealing behaviors of high-entropy alloy Al0. 5CoCrCuFeNi. J Alloys Compd 486:427–435

Tsai K-Y, Tsai M-H, Yeh J-W (2013) Sluggish diffusion in Co–Cr–Fe–Mn–Ni high-entropy alloys. Acta Mater 61:4887–4897

Tsai M-H, Yeh J-W, Gan J-Y (2008) Diffusion barrier properties of AlMoNbSiTaTiVZr high-entropy alloy layer between copper and silicon. Thin Solid Films 516:5527–5530

Vaidya M, Pradeep KG, Murty BS, Wilde G, Divinski SV (2018) Bulk tracer diffusion in CoCrFeNi and CoCrFeMnNi high entropy alloys. Acta Mater 146:211–224

Vaidya M, Trubel S, Murty BS, Wilde G, Divinski SV (2016) Ni tracer diffusion in CoCrFeNi and CoCrFeMnNi high entropy alloys. J Alloys Compd 688:994–1001

Wang J, Liu Y, Liu B, Wang Y, Cao Y, Li T, Zhou R (2017a) Flow behavior and microstructures of powder metallurgical CrFeCoNiMo0. 2 high entropy alloy during high temperature deformation. Mater Sci Eng A 689:233–242

Wang P, Cai H, Cheng X (2016) Effect of Ni/Cr ratio on phase, microstructure and mechanical properties of NixCoCuFeCr2− x (x= 1.0, 1.2, 1.5, 1.8 mol) high entropy alloys. J Alloys Compd 662:20–31

Wang P, Cheng X, Cai H, Xue Y, Zhang Y (2017b) Influence of increasing Al concentration on phase, microstructure and mechanical behaviors of Ni1. 5CoFeCu1− xAlxV0. 5 high entropy alloys. Mater Sci Eng A 708:523–536

Wang S (2013) Atomic structure modeling of multi-principal-element alloys by the principle of maximum entropy. Entropy 15:5536–5548

Waseem OA, Lee J, Lee HM, Ryu HJ (2018) The effect of Ti on the sintering and mechanical properties of refractory high-entropy alloy TixWTaVCr fabricated via spark plasma sintering for

fusion plasma-facing materials. Mater Chem Phys 210:87–94. https://doi.org/10.1016/j.matchemphys.2017.06.054
Waseem OA, Ryu HJ (2016) Tungsten-based composites for nuclear fusion applications. In: Nuclear material performance. InTech. https://doi.org/10.5772/62434
Waseem OA, Ryu HJ (2017) Powder metallurgy processing of a WxTaTiVCr high-entropy alloy and its derivative alloys for fusion material applications. Sci Rep 7. https://doi.org/10.1038/s41598-017-02168-3
Wu B, Chen W, Jiang Z, Chen Z, Fu Z (2016) Influence of Ti addition on microstructure and mechanical behavior of a FCC-based Fe30Ni30Co30Mn10 alloy. Mater Sci Eng A 676:492–500
Xia S-Q, Zhen W, Yang T-F, Zhang Y (2015) Irradiation behavior in high entropy alloys. J Iron Steel Res Int 22:879–884
Yadav S, Kumar A, Biswas K (2017) Wear behaviour of high entropy alloys containing soft dispersoids (Pb, Bi). Mater Chem Phys. https://doi.org/10.1016/j.matchemphys.2017.06.020
Yang S, Yan X, Yang K, Fu Z (2016) Effect of the addition of nano-Al2O3 on the microstructure and mechanical properties of twinned Al0. 4FeCrCoNi1. 2Ti0. 3 alloys. Vacuum 131:69–72
Yang S, Zhang Y, Yan X, Zhou H, Pi J, Zhu D (2017) Deformation twins and interface characteristics of nano-Al2O3 reinforced Al0. 4FeCrCo1. 5NiTi0. 3 high entropy alloy composites. Mater Chem Phys. https://doi.org/10.1016/j.matchemphys.2017.11.037
Yang T et al (2018) Effects of temperature on the irradiation responses of Al0.1CoCrFeNi high entropy alloy. Scr Mater 144:31–35. https://doi.org/10.1016/j.scriptamat.2017.09.025
Yang X, Zhang Y (2012) Prediction of high-entropy stabilized solid-solution in multi-component alloys. Mater Chem Phys 132:233–238
Ye Y, Wang Q, Lu J, Liu C, Yang Y (2016) High-entropy alloy: challenges and prospects. Mater Today 19:349–362
Yeh J-W (2013) Alloy design strategies and future trends in high-entropy alloys. JOM 65:1759–1771
Yeh J-W, Chang S-Y, Hong Y-D, Chen S-K, Lin S-J (2007) Anomalous decrease in X-ray diffraction intensities of Cu–Ni–Al–Co–Cr–Fe–Si alloy systems with multi-principal elements. Mater Chem Phys 103:41–46
Yeh J-W et al (2004a) Formation of simple crystal structures in Cu-Co-Ni-Cr-Al-Fe-Ti-V alloys with multiprincipal metallic elements. Metall Mater Trans A 35:2533–2536
Yeh JW et al (2004b) Nanostructured high-entropy alloys with multiple principal elements: novel alloy design concepts and outcomes. Adv Eng Mater 6:299–303
Zhang A, Han J, Su B, Li P, Meng J (2017a) Microstructure, mechanical properties and tribological performance of CoCrFeNi high entropy alloy matrix self-lubricating composite. Mater Des 114:253–263
Zhang A, Han J, Su B, Meng J (2017b) A novel CoCrFeNi high entropy alloy matrix self-lubricating composite. J Alloys Compd 725:700–710
Zhang K, Fu Z, Zhang J, Wang W, Lee S, Niihara K (2010) Characterization of nanocrystalline CoCrFeNiTiAl high-entropy solid solution processed by mechanical alloying. J Alloys Compd 495:33–38
Zhang Y, Yang X, Liaw P (2012) Alloy design and properties optimization of high-entropy alloys. JOM 64:830–838
Zhang Y, Zhou YJ, Lin JP, Chen GL, Liaw PK (2008) Solid-solution phase formation rules for multi-component alloys. Adv Eng Mater 10:534–538
Zhang Y, Zuo T, Cheng Y, Liaw PK (2013) High-entropy alloys with high saturation magnetization, electrical resistivity, and malleability. Sci Rep 3:1455
Zhang Y, Zuo TT, Tang Z, Gao MC, Dahmen KA, Liaw PK, Lu ZP (2014) Microstructures and properties of high-entropy alloys. Prog Mater Sci 61:1–93

Part VII
SPS of Ni-Based Alloys

Spark Plasma Sintering as a Route for Producing In-Demand Microstructures: Application to the Tensile-Ductility Enhancement of Polycrystalline Nickel

P. Langlois, D. Tingaud, and G. Dirras

1 Introduction

Spark plasma sintering (SPS) constitutes one of the main bottom-up techniques that powder metallurgy offers to produce in-demand microstructures of various types; its main advantages, compared to conventional consolidation methods, reputedly lie with both its much shorter duration and its lower temperature of processing, which limit grain growth during consolidation. This versatile technique is of prime interest for addressing challenging issues such as improving both the strength and the ductility of bulk materials basically designed for Hall-Petch strengthening (Fedorov et al. 2002; Wang et al. 2002a). Different strategies have yet been proposed to offset their lower tensile ductility (elongation to failure) at ambient temperatures (Wang and Ma 2004; Zhao et al. 2008); one of them deals with the incorporation of coarser grains, which are prone to store dislocations, in a fine-grained matrix. Since this incorporation can be achieved in different manners, this chapter, although devoted to nickel only, is divided in accordance with the sintering of increasingly complex structures; Sect. 2, which is already in line with the current trend for improving mono-modal structures, provides the benchmark for dealing with multimodal structures, as described in Sect. 3.

Irrespective of the much more numerous applications for nickel-based alloys that will anyway benefit from the outcomes on nickel in terms of understanding the deformation mechanisms involved, potential large-scale applications in domains pertaining, for instance, to microelectromechanical systems or to nuclear steam generators are expected for nickel itself. On the broader front, the improvement of nickel processing is obviously in line with the concept of "functionally graded materials" (Miyamoto et al. 1997) which generally refer to composite materials formed of two or more constituents whose local compositions or microstructures

P. Langlois (✉) · D. Tingaud · G. Dirras
LSPM, Université Paris 13, Sorbonne Paris Cité, CNRS, Villetaneuse, France
e-mail: patrick.langlois@lspm.cnrs.fr; guy.dirras@lspm.cnrs.fr

© Springer Nature Switzerland AG 2019
P. Cavaliere (ed.), *Spark Plasma Sintering of Materials*,
https://doi.org/10.1007/978-3-030-05327-7_20

vary continuously in accordance with their intended use; that tailoring concept has evolved into the "harmonic structure" design of metallic materials having a three-dimensional network consisting of a periodic array of soft coarse grains connected to a continuous high-strength ultrafine-grained skeleton through decreasing-size entities (grains or substructures) whose function is to make the interface mechanically compatible (Sekiguchi et al. 2010).

The present compilation, which is mainly based on results accumulated in our research group on that topic over the years, under a strategy similar to the one developed in close collaboration with Prof. K. Ameyama's group at Ritsumeikan University, Japan, on the harmonic structure concept applied to titanium alloys (Dirras et al. 2017b; Liu et al. 2018), is put in perspective with the data of the literature.

2 Mono-modal Structures

Information relating to general considerations for processing nickel by SPS is hereby provided and followed, after a necessary comparison with hot isostatic pressing (HIP) which is another major route for processing powders, by specific considerations in regard to the grain size of the initial powder.

2.1 General Considerations for Nickel

Since the main characteristic of SPS compared to conventional sintering techniques consists in Joule heating, considerations start off with the influence of the current flow and continue with those of the other processing parameters.

2.1.1 Influence of the Current Flow

The SPS technique relies on the control of the electrical power delivered to a system in which the temperature of the sample has to be monitored through measurements as thoroughly accurate as possible since high heating rates, especially when combined with short dwell times, can cause temperature gradients that would lead to some structural and mechanical inhomogeneity of the sintered material; decisive progress in this regard can be achieved by taking into account the thermal and electrical properties of each constitutive part for modelling the entire system (sample, punches, and die). Such an optimal approach has been investigated (Vanmeensel et al. 2007) and experimentally validated for ZrO_2-TiN powder mixtures that advantageously transform from electrically insulating powder compacts into electrically conductive composites once percolation has occurred; one peculiar point of interest relates to the significant temperature measurement difference (220 °C) observed between a central pyrometer and an external one, respectively, at the bottom of a blind borehole in the upper punch and on the outer die wall surface, once densification has occurred or when a fully dense compact is processed. For nickel, which is an electrically conductive material, temperature is therefore often underestimated, even when monitored by a thermocouple set in the die wall as close as possible to the sample; surrounding the die with porous graphite felt insulation proves to reduce the radiation heat losses and to decrease the radial temperature gradient inside the sample.

The influence of the electrical characteristics of the mould on the system resistance has been experimentally investigated (Minier et al. 2010) by processing a nickel powder (Alfa Aesar, 99.9%, 4.3 ± 1.5 μm) under the same parameters (temperature, heating rate, dwell time, and pressure) in different moulds: (i) with die and punches in graphite, (ii) again with graphite die and punches but the latter being separated from the powder by alumina spacers, and (iii) with alumina die and graphite punches, the last two layouts distinctively preventing the current from flowing either axially (ii) or radially (iii) around the sample. The die was surrounded by a carbon felt, to reduce the radiation heat losses from the outer wall, and graphite foils were inserted between the powder and the punches. Temperature was monitored by a K-type thermocouple set in the die wall 2 mm far from the sample and half height of the die. The electrical power which has to be delivered includes amounts covering both the energy released by Joule effect from the electrically conducting elements of the system, the thermal energy transferred by conduction between the different system parts, and the one transferred by radiation; the entire system (punches, die, and sample) can thus be assimilated, based on electrical characteristics (Table 1), to an association of resistances either in parallel or in series.

During the heating rise, the electrical power in the setups (i) and (iii) were similar and higher than that in the setup (ii). During the dwell time, the electrical power delivered in the setup (iii) was higher than the one in the setup (ii) but lower than the one in the setup (i). The highest density and grain size were obtained when sintering in the alumina die, (iii), whereas the lowest ones were obtained when sintering in the graphite die with alumina spacers, (ii), which brought conclusive evidence about the larger densification by Joule effect rather than by thermal conduction, as well as about the correlation between the as-processed microstructures and the electrical and thermal characteristics of the system (Table 1).

The separate influence on the system resistance of the anisotropic electrical characteristics of graphite foils (Table 1), which had been neglected, has also been

Table 1 Electrical and thermal characteristics of the system components

	Graphite die/punches	Alumina die/spacers/ powder	Graphite foils		Nickel powder
			Along [100]	Along [001]	
Electrical resistivity, Ω.cm	1.6×10^{-3}	1×10^{14}	1×10^{-3}	5×10^{-2}	6.9×10^{-6}
Thermal conductivity, $W.m^{-1}.K^{-1}$	8×10^{1}	$2.6–3.5 \times 10^{1}$	1.6×10^{2}	4×10^{0}	9.09×10^{1}

Adapted from Minier et al. (2010, 2012)

Table 2 Influence of graphite-foil setup and sintering conditions on relative density, d, and grain size, D

	200 °C/min, 30 MPa		200 °C/min, 70 MPa	
Setup	d, %	D, μm	d, %	D, μm
a	93.8 ± 0.1	11 ± 1	97.7 ± 0.1	10 ± 1
b	93.8 ± 0.1	9 ± 1	98.1 ± 0.1	12 ± 1
c	94.0 ± 0.1	15 ± 1	97.8 ± 0.1	13 ± 1

Adapted from Minier et al. (2010)
(a) Graphite foils between sample and mould
(b) Graphite foils between sample and punches only
(c) No graphite foil

investigated in a graphite mould whether 0.2 mm-thick graphite foils had been inserted between the sample and the punches or between the sample and the die, with their [100] direction perpendicular and parallel to the mould axis, respectively; the bulk microstructure of the as-processed samples has, however, shown no significant dependence on the graphite foil setup, irrespective of pressure, even though resulting relative densities did depend on pressure (Table 2).

The influence of the geometrical characteristics of the mould on the system resistance has then been experimentally investigated (Minier et al. 2012) by comparatively processing the previous powder and an alumina one, for the heating volume to be limited to the punches and the sample, and to the punches and the die, for conductive nickel and insulating alumina, respectively. The outer diameter of two 10 mm inner diameter dies has been increased from 20 to 27.8 mm, the corresponding dies being designated as (10/20) and as (10/27.8), respectively, while the 30 mm outer diameter of a 15 mm inner diameter die, designated as (15/30), has been defined so that the die- to sample-volume ratios be constant in comparison with the (10/20) die on the one hand and the die volume be constant in comparison with the (10/27.8) die on the other hand.

In contrast with the delivered electrical powers which proved similar with the (10/20) and (10/27.8) dies only, the delivered electrical power densities, referring to the heating volume, proved similar with all the dies. The highest relative density and the coarsest grains have been observed for the nickel sample sintered in the die with the smallest outer diameter, (10/20), while a finer microstructure has been obtained with the (10/27.8) die due to the higher thermal transfer from the sample to the die, in relation with the larger die volume to heat. This observation is in sharp contrast with the results obtained for the alumina samples where increased die dimensions proved to favour grain growth (Table 3). In order to obtain similar microstructures when elaborating larger nickel samples, it seems therefore necessary to keep the die volume constant, as opposed to keeping the die- to sample-volume ratio constant when dealing with insulating samples.

2.1.2 Influence of Processing Temperature and Pressure

The influence of SPS processing parameters, namely, temperature and pressure, on the microstructure and mechanical properties of nickel samples has been

Table 3 Influence of die dimensions on relative density, d, and grain size, D

Die	Heating rate, °C. min^{-1}	Nickel samples	Alumina samples		
		d, %	D, μm	d, %	D, μm
10/20	50	98.9 ± 0.1	37 ± 5	99.7 ± 0.2	2.0 ± 0.2
	200	98.4 ± 0.1	14 ± 2	99.5 ± 0.2	2.0 ± 0.2
10/ 27.8	50	98.1 ± 0.1	13 ± 2	99.8 ± 0.1	4.8 ± 0.6
	200	97.9 ± 0.1	9 ± 3	99.6 ± 0.1	2.8 ± 0.3
15/30	50	97.9 ± 0.1	17 ± 2	99.8 ± 0.1	2.2 ± 0.3
	200	98.1 ± 0.1	12 ± 2	99.3 ± 0.1	2.2 ± 0.2

Adapted from Minier et al. (2012)

Table 4 Influence of processing parameters, p, T, on relative density, d, and grain size, D

p, T	700 °C		850 °C		1000 °C	
	d, %	D, μm	d, %	D, μm	d, %	D, μm
50 MPa	96.8	5.0	97.8	27.8	98.1	47.8
65 MPa	97.4	5.4	98.2	25.8	98.3	42.5
80 MPa	97.7	5.7	99.0	26.7	99.2	44.8

Adapted from Borkar and Banerjee (2014)

Table 5 Influence of processing parameters, p, T, on yield strength, σ_y, and elongation to failure, ε_f

p, T	700 °C		850 °C		1000 °C	
	σ_y, MPa	ε_f, %	σ_y, MPa	ε_f, %	σ_y, MPa	ε_f, %
50 MPa	278	43	154	59	134	67
65 MPa	265	49	158	56	130	65
80 MPa	253	50	166	59	150	58

Adapted from Borkar and Banerjee (2014)

investigated in a systematic manner (Borkar and Banerjee 2014). Nine bulk polycrystalline samples have been fabricated from a high-purity microstructured powder (Alfa Aesar, 99.99%, 1–5 μm), by using combinations of three different pressures and temperature. Table 4 illustrates the strong effect that temperature had on the grain size, while relative densities increased with increasing temperature for a given pressure and with increasing pressure at a given temperature. Taking a closer look at Table 4, it can, however, be noted that the increase in relative densities reduced from ~1.3% to ~0.2% between the samples processed at 700 and 850 °C and at 850 and 1000 °C, respectively, although the temperature increase remained the same. Also, for any given temperature, increasing pressure from 50 to 80 MPa appeared to have little influence, ~ 1%, on the increase of relative density.

The tensile yield strength of the SPS-processed samples decreased from 278 to 130 MPa with increasing temperature, whereas elongation to failure increased from 43% to 67% (Table 5). Reduction in tensile yield strength was mainly associated with the significant grain growth observed for higher temperatures. During tensile deformation, dislocation pile-up took place at the grain boundaries between deforming and nondeforming grains, leading to stress concentrations eventually activating slip in the neighbouring nondeforming grains. For larger grain sizes, a larger stress concentration therefore led to a lower tensile yield strength.

With respect to annealing twin boundaries (TBs) which typically belong to the Σ3 category (60°/<111>) in face-centred cubic (fcc) metals and alloys (Bhattacharjee et al. 2007), the results indicated that, irrespective of pressure, their density increased with temperature. This increase is in agreement with the fact that nickel possesses a relatively low stacking-fault energy, compared to other metallic materials such as aluminium, and therefore has an increased susceptibility to form annealing twins. Furthermore, during grain growth, the formation of coherent Σ3 boundaries, which require much less energy as compared to any random high-angle boundaries, is quite advantageous from the perspective of total energy of the system. All samples indeed exhibited substantial ductility due to the presence of a large fraction of Σ3 boundaries whose density increased with temperature; the results also clearly indicated that there was no substantial influence of the processing pressure on tensile yield strength nor on ductility, at any given temperature.

Before leaving this subsection, it is pointed out that, as many of the authors referred in this chapter do, a specific procedure shall appropriately be followed when processing under high-operating pressure, due to the mechanical strength of the graphite die that increases with temperature; for instance, an initial pressure of 50 MPa may be applied at room temperature, and the pressure only increased to the higher level when the temperature reaches 350–600 °C (Dutel et al. 2012).

2.2 Comparison with HIP

In our research group, ultrafine-grained samples have been consolidated by SPS and hot isostatic pressing (Gubicza et al. 2009), starting from the same Ni nanopowder (Argonide, 100 nm). Before HIP processing under 140 MPa at 700 °C for 150 min, the powder initially contained in a high-vacuum glass capsule had continuously been handled in order to prevent oxidation, whereas the glass capsule had simply been broken in air, and the powder rapidly poured into the graphite mould prior to SPS processing under 150 MPa at 500 °C for 1 min. Energy-filtered transmission electron microscopy (EFTEM) imaging of the initial powder has in this regard revealed a significant amount of oxygen in the surface layer (approximately 2 nm thick) of the initial particles (Fig. 1), presumably contained in the NiO phase that was identified by X-ray diffraction (XRD); the intensity ratio for the initial powder was found to be $0.5 \pm 0.1\%$, which did not change during HIP processing, whereas the SPS procedure resulted in a material having a significantly higher oxide content (intensity ratio of $0.9 \pm 0.1\%$).

High-temperature consolidation resulted in both cases, HIP and SPS, in grain coarsening (Table 6); the more moderated grain growth of the SPS-processed sample may, however, be clearly attributed to the lower temperature as well as to the shorter duration of that procedure. The higher degree of oxidation of the particle surfaces

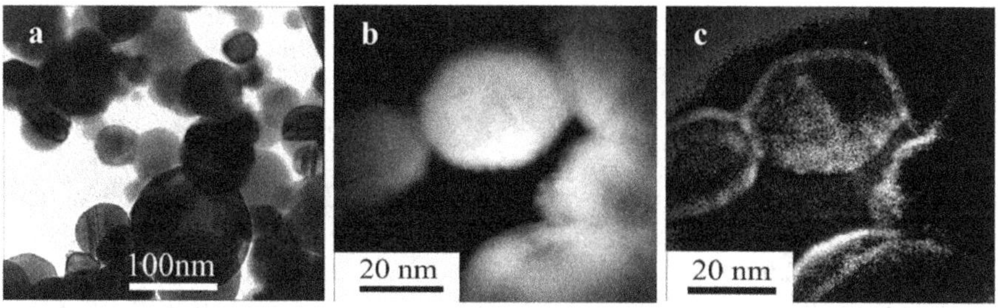

Fig. 1 Bright-field TEM image of the initial powder (**a**) and EFTEM images showing the element maps for nickel (**b**) and oxygen (**c**), respectively. (Gubicza et al. 2009)

Table 6 Main characteristics obtained for the samples processed by HIP and SPS

Sample	Grain size, nm	Dislocation density, 10^{14} m^{-2}	Twin probability, %	Yield strength, MPa
HIP	403	5.6 ± 0.5 (16 ± 2)	0.32 ± 0.03 (0.02 ± 0.02)	542 ± 14
SPS	294	5.1 ± 0.5 (21 ± 3)	0.29 ± 0.03 (0.18 ± 0.02)	682 ± 15

Adapted from Gubicza et al. (2009)

may also have had a retarding effect on grain coarsening during SPS consolidation. The relative densities were 95.5 ± 0.1 and $95.4 \pm 0.1\%$ for the HIP- and SPS-processed samples, respectively, which means that the remaining porosity was nearly the same for both samples.

The SPS-processed sample had a much higher yield strength than that of the HIP-processed sample (Table 6); this may have resulted from the higher oxide dispersoid content and/or from the smaller grain size. However, the HIP-processed sample exhibited a strain-hardening behaviour up to failure, whereas the SPS-processed sample exhibited an early saturation ($\varepsilon \approx 0.04$) of its flow stress, followed by a strain softening due to cracking as a most likely consequence of high flow-stress level and weak boundary strength in presence of oxides. During compression, the dislocation density, already significant after consolidation, increased for both samples, whereas the twin probability, which had already decreased from 2–3 to ~0.3% during consolidation, again decreased (Table 6); this is in agreement with the fact that the plastic deformation is basically controlled by dislocations. The majority of the twin boundaries in the as-processed samples were most probably growth twins, and the prevailing reduction of the twin density, often referred to as "detwinning" and mainly occurring in large grains, a result of the interaction between gliding dislocations and pre-existing twin boundaries; the drastic detwinning observed for the HIP-processed sample was therefore in agreement with its grain-size distribution. The larger strain to failure in the HIP-processed sample (0.35 vs 0.22) may also have contributed to the higher extent of detwinning.

Another opportunity to compare HIP and SPS processing has been provided, as a subsequent step in regard to the elaboration of nickel ferromagnetic nanoparticles (~100 nm) by a modified polyol process, on the competing assumptions that the former would favour the full densification of a nanostructured material and the latter the consolidation of the powder with a limited grain growth (Bousnina et al. 2010). HIP and SPS processing have been performed under 120 MPa at 650 °C for 60 min and under 53 MPa at 600 °C for 5 min, respectively. Although the nickel powder had initially been heated under hydrogen to clear it of organic impurities and to thermally reduce all traces of nickel oxides, the HIP- and SPS-processed samples still contained ~8 and ~6% of nickel phosphide (Ni_3P), respectively. The grain sizes increased up to 0.62 and 0.20 μm after HIP and SPS processing, respectively. No residual porosity could be observed despite the values of relative density measured after HIP and SPS processing, 87.9 and 86.0 MPa, respectively; it was then suggested that those values originated with the Ni_3P impurity. Also, compressive testing resulted in a weak hardening for the HIP-processed sample and in a higher yield strength but a brittle behaviour for the SPS-processed sample, with clear evidence of interparticle de-cohesion.

2.3 *Specific Considerations in Regard to Grain Size*

Although the content of the previous subsection validated the assumption of a more limited grain growth through SPS, which correlated with a higher yield strength, the presence of the Ni_3P impurity that resulted, in the latter case, from the initial powder processing had a detrimental effect which prevented any further comparison with HIP. For that reason, and since grain size alone neither defines the microstructure nor characterises the mechanical behaviour of metallic polycrystals, this subsection starts with additional information on the influence of initial powder processing.

2.3.1 Influence of Initial Powder Processing

Nickel samples have been processed by SPS both directly from an elemental Ni powder (Alfa Aesar, 99.9% purity, APS 3–7 μm) and after mechanical activation of that powder by a short-duration treatment in a high-energy planetary ball mill (Naimi et al. 2013). The mechanically activated powder did not retain its spherical morphology and formed large agglomerates as a result of successive fracture/welding processes; although the median grain size increased from 8 to 200 μm, the crystallite size decreased from 250 to 52 nm, compared to the initial powder. Ball milling also introduced a larger particle size distribution and a significant defect density. Since dilatometric measurements showed that the sintering of the elemental powder occurred in two steps, additional batches of both powders have been sintered with two different subsequent heating rates. In order to improve the densification, the applied pressure has also been increased to 200 MPa; due to the poor flexural strength of the graphite, specific tungsten carbide-cobalt tools (die and punches) have been implemented, and SPS processing conditions have been optimised in relation to their distinct electrical and thermal conductivities. A final batch of

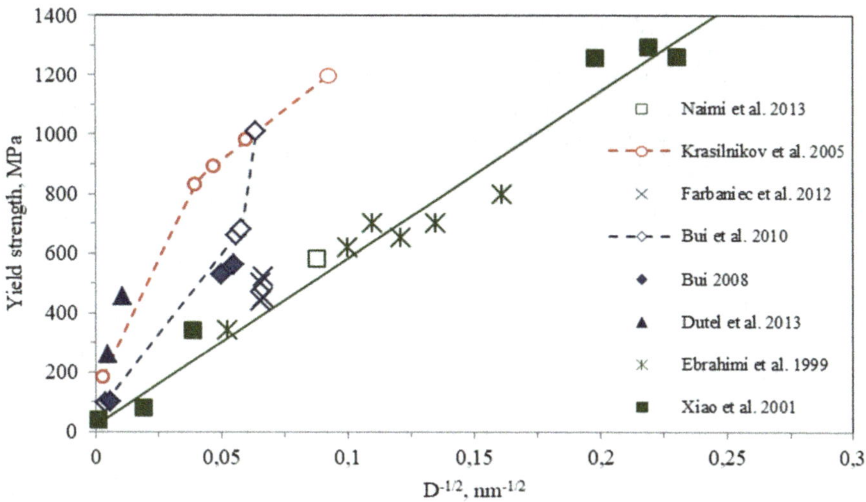

Fig. 2 Hall-Petch plot for nickel processed by different routes

mechanically activated powder which had resulted in a higher yield strength has consequently been sintered under 200 MPa at 550 °C for 10 min with a heating rate of 200 $°C.min^{-1}$ up to 450 °C and of 50 $°C.min^{-1}$ afterwards; it resulted in an ultrafine-grained sample with an average crystallite size of 130 nm and a density of 98 ± 0.5% that exhibited a higher yield strength (~590 MPa) than the previous sample, together with an improved ductility of 30%.

The yield strength of 590 MPa is approximately consistent with the Hall-Petch relationship (Fig. 2), in the same way as data reported for electrodeposited nickel (Ebrahimi et al. 1999; Xiao et al. 2001); the interesting feature lies in the ductility of 30% that was superior to the level of plastic strain sustained by most nanograined nickel-based alloys. In the framework of this chapter, the study presented hereby incidentally allows for the insertion of Fig. 2 that will be reappraised later on.

A more striking evidence of the influence of the initial powder processing can emerge from the direct observation of the distinct sintering behaviours of powders with the same grain size. Two commercially available nanograined nickel powders with a similar mean grain size of ~50 nm have therefore been considered: one powder (Argonide) produced by electro-explosion of wire (EEW) and the next one (Tekna) by inductively coupled plasma torch (ICPT). In the EEW process, a very short high-voltage electrical pulse converts a metallic wire into a plasma which is confined by the field created by this pulse; when the field collapses, the explosive release of the high-pressure metal vapour leads to the rapid adiabatic cooling of the metal aerosol (Tepper 1998). In the ICPT process, micrometre-sized particles are coaxially injected into the centre of the discharge, where they vaporise, and are entrained towards the downstream end of the torch where a quench gas flow is radially injected, which allows for nanoparticles to nucleate and coalesce up to their final dimension (Dolbec et al. 2008).

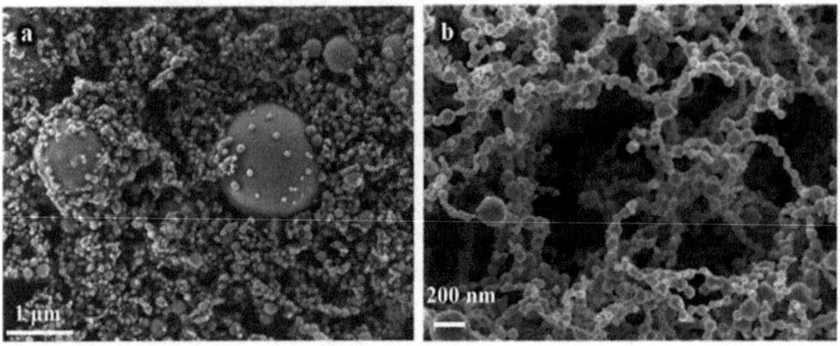

Fig. 3 SEM images of EEW- (**a**) and ICPT-processed (**b**) powders

In order to give full insight on the influence of the initial powder processing, the microstructural state of the powders prior to SPS processing has been thoroughly characterised through TEM and scanning electron microscopy (SEM) observations (Fig. 3). Spherical shapes clearly predominated. It also appeared that the ICPT-processed powder had started to sinter, with frequent necking between particles leading to some "pearl-necklace" kind of arrangement, in direct relation with a too long residence time of the nanoparticles in the downstream end of the torch. The grain-size distributions extracted from the TEM micrographs showed log-normal distributions centred on 48 and 55 nm for the EEW- and ICPT-processed powders, respectively, whereas the ICPT-processed powder only contained nanometre-sized or ultrafine particles, the EEW-processed one presented a multimodal distribution with a significant amount (≈12% area) of grains whose size ranged between 0.25 and 5 μm.

Inductively coupled plasma optical emission spectrometry analyses showed the presence of less than 0.3 and 0.03at% of impurities in EEW- and ICPT-processed Ni, respectively; the samples had to be dissolved in nitric and fluorhydric acids, prior to spectrometric analysis, and some black residual precipitate, assumed to be carbon, however remained in the case of ICPT-processed Ni. XRD analyses performed on both powders only revealed the characteristic diffraction peaks of the fcc structure of nickel. At room temperature in air, XRD revealed a fast oxidisation of EEW-processed Ni, also underscored by EFTEM, while no NiO peaks appeared on the diffraction pattern of ICPT-processed Ni, even after days in air; the presence of carbon within this powder might have prevented or at least reduced the oxidisation rate of the nickel particles.

Prior to sintering, the nanograined powders had been surrounded by a classical papiex sheet (carbon) in order as usual to limit reactions between the powder and the graphite die. SPS processing experiments have been performed under a maximum applied pressure of 100 MPa with a heating rate of ~100 °C.min^{-1} up to a temperature ranging between 450 °C and 850 °C and a dwell time of 1, 15, or 30 min. Figure 4 presents the relative density evolutions of the sintered samples. EEW-processed powders exhibited a classical behaviour with an increase of density

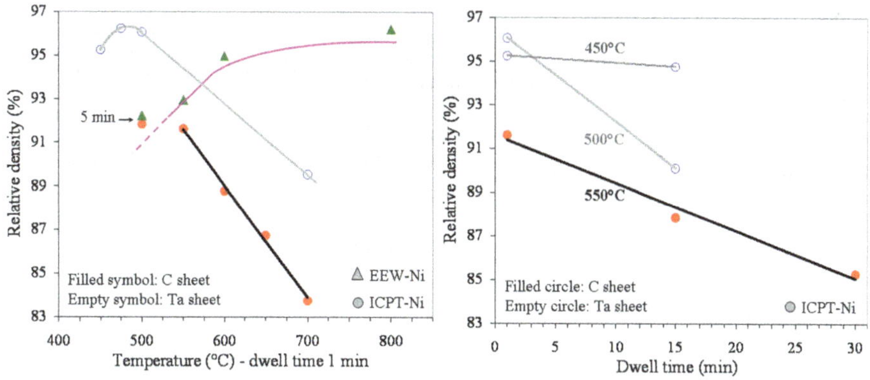

Fig. 4 Evolution of the relative density of the EEW- and ICPT-processed powders as a function of temperature (**a**) and dwell time (**b**)

as well as of grain size with temperature and dwell time. On the opposite, the relative density of ICPT-processed powders decreased linearly with temperature and dwell time beyond 500 °C. This negative densification rate has been observed up to a temperature around 725 °C where the sample melted well below the melting temperature of nickel (1453 °C).

Sintering of nanograined powders does generally occur at much lower temperatures than with their coarse-grained counterparts, and it has indeed been reported that nickel nanograins can melt at lower temperature than coarse-grained polycrystalline nickel; the decrease, even if subject to reappraisal (Wang et al. 2002b), is however much lower than the present one, even when considering an experimental uncertainty, not in excess of ~100 °C, due to the temperature measurement in the graphite die. More importantly, if this melting were a general trend for nanograined nickel powders or if it were due to the experimental setup, the EEW-processed powder would also have melted. Similarly with self-propagating high-temperature syntheses where, after ignition, the energy released by the exothermic reaction itself heats the sample instead of being simply lost by thermal conduction, it has then been suspected that an exotherm might have led to the melting of the sample, in possible relation with carbon contamination. Additional sintering experiments have therefore been performed on ICPT-processed Ni under the same experimental conditions, except that the papiex layer had been replaced by a tantalum sheet in order to prevent carbon contamination. It appears (Fig. 4) that much higher relative densities have been obtained, but a negative densification rate was still observed, and melting again occurred at a low temperature. It has eventually been assumed that, under the unexpected circumstance of porosity increasing with temperature and owing to the necklace-like arrangement of electrically conductive nanoparticles within the ICPT-processed powder, an effect similar to the electro-explosion of nanowires may have occurred. As a partial conclusion, it can be stated that the EEW-processed powders, although inherently having a presumably high density of crystal defects, exhibited a multimodal structure that may have contributed to their more suitable behaviour. It

Table 7 Microstructural characteristics of the as-processed samples

				Grain boundaries, %	
	Sintering temperature, °C	Relative density, %	Grain size, μm	LAGBs	Σ3
A	700	97.4	5.4	6	20
B	850	99.0	17.1	5	41
C	1000	99.2	45.0	3	52

Adapted from Dutel et al. (2013)

shall also be noted that preparing the powder in the mould with a tantalum sheet instead of a papiex layer proved beneficial in this case.

2.3.2 Influence of Grain Size

A large range of grain sizes, namely, 5–45 μm, has been obtained through SPS processing of a high-purity (99.9 wt%) Ni powder (Sigma-Aldrich) consisting of 1–5 μm multi-crystalline entities (Dutel et al. 2013). SPS processing has been performed under 100 MPa during 1 min at three dwell temperatures of 700, 850, and 1000 °C, the corresponding samples being hereafter referred to as *A*, *B*, and *C*, respectively. A thorough microstructural study has been carried out by means of electron backscattering diffraction (EBSD); the striking feature was the presence of high fractions of Σ3 boundaries (Table 7), many of them being twin boundaries as seen from inverse pole figure (IPF) images, which were in compliance with stacking-fault energy considerations. Conversely, the amount of low-angle grain boundaries (LAGBs) which have misorientations between 5° and 15° was less than 6% for the three cases investigated.

Both the average grain size (area fraction calculation) and the relative density increased with the processing temperature (Table 7), which means that grain growth was indeed correlated with diffusion mechanisms. However, the increase of the relative density between samples *A* and *B* was ~1.6%, while the one between samples *B* and *C* was only ~0.2% for the same temperature increase, which means that some saturation state had been reached in terms of attainable relative density.

Sample *B* exhibited a quasilinear hardening after the onset of yielding, during quasi-static compression at room temperature, whereas sample *C* exhibited a parabolic hardening behaviour. Yield strengths were ~435 and ~200 MPa for samples *B* and *C*, respectively, which confirmed that their level was influenced by grain size. The deformation micro-mechanisms occurring during compression have been investigated at different fixed strain amounts through the microstructural evolution of sample *B*. Both grain boundary migration, grain formation, and TB disruption have been observed at a strain of ~5% but were no longer observed later on. At a strain of ~8%, TBs had almost completely vanished, and a new microstructure was formed instead, along with a change of grain orientation probably due to lattice rotation during plastic deformation. At a strain of ~50%, the presence of the Σ3 boundaries was very remote. This occurred probably as a consequence of intense dislocation

Table 8 Evolution of grain boundaries with compressive strain

Strain, %	0	5	8	50
LAGBs, %	5	9	28	52
Σ3, %	41	30	8	2
TB/Σ3, %	51	48	24	0

Adapted from Dutel et al. (2013)

Table 9 Influence of SPS processing parameters, T, p, and t, on relative density, d, and mean grain size, D

Initial powder, nm		SPS parameters			d, %	D, nm
		T, °C	p, MPa	t, min		
SPS-A	50	500	150	1	94.47	250
SPS-B	100	500	150	1	95.53	294
SPS-C	100	500	100	4	94.00	306
SPS-D	100	500	100	2	96.75	318
SPS-CG	5000	550	150	1.5	98.00	50,000

Adapted from Bui et al. (2010)

activity and grain rotations that induced a marked <011> fibre texture. Quantitatively (Table 8), after a strain of 5%, the number fraction of Σ3 boundaries had decreased. After 8%, the density of Σ3 boundaries had dropped even more and got close to zero after 50%. The evolution of the TB fraction followed that of Σ3. These observations depart from the observed stability of TBs in electrodeposited Ni containing a controlled fraction of nanotwins. Most probably, this behaviour was correlated to the processing route or to the subsequent deformation route and the amount of deformation introduced.

2.3.3 Deviations from the Hall-Petch Relationship

A relatively large range of grain sizes has been obtained through SPS processing of two high-purity (~99.99%) Ni nanopowders (Argonide) whose spherical particles had nominal sizes of ~50 and ~100 nm (Bui et al. 2010). EFTEM investigations showed the presence of an oxygen-containing phase, forming an external 2 nm-thick layer, which was identified by XRD as a NiO structure. Table 9 summarises the processing parameters used for producing the samples designated as *SPS-A*, *SPS-B*, *SPS-C*, *SPS-D*, and *SPS-CG*, the latter one having been consolidated from a coarse-grained powder (5 μm) for comparison purpose.

The relative densities are also listed in Table 9; sample *SPS-CG* exhibited the highest one, whereas sample *SPS-A* exhibited the smallest one, which could be attributed to its higher NiO content. The volume fraction of the NiO phase has thus been estimated through the intensity ratio of the NiO and Ni XRD peaks. Since smaller particles have a higher specific surface area, and since NiO has a lower density (6.67 g.cm^{-3}) compared to Ni (8.902 g.cm^{-3}), the results were coherent with the measured relative densities.

Table 10 Mechanical characteristics of the HIP- and SPS-processed materials

	HIP-A	HIP-B	HIP-CG	SPS-A	SPS-B	SPS-C	SPS-D	SPS-CG
Yield strength, MPa	559	530	96	1022	682	660	659	95
Compressive strength, MPa	784	789	566	1178	873	878	880	566
Strain at maximum stress	0.22	0.34	0.46	0.01	0.035	0.28	0.35	0.46
Strain at failure	0.22	0.34	>0.5	0.04	0.21	0.33	0.36	>0.5

Reproduced from Bui et al. (2010)

It was also found that the mean grain sizes for the consolidated samples were about 3–5 times larger than initially (Table 9), indicating that the high-temperature consolidation did result in grain coarsening. The microstructure of the as-processed samples was heterogeneous and consisted of a matrix of ultrafine grains (UFG) and embedded coarse grains (CG) themselves sometimes subdivided into smaller grain blocks. Additionally, the microstructure exhibited a random crystallographic texture, and the fraction of high-angle grain boundaries (HAGBs) which have misorientations higher than 15° was expectedly high (about 85%).

The bulk SPS-processed samples have been tested at room temperature under quasi-static compression until failure; their resulting mechanical characteristics are listed in Table 10. It was observed that:

- Sample *SPS-A* had the highest yield strength (1022 MPa), probably due to its highest oxide content and smallest grain size. Its ductility was however very low.
- The yield strength and the ductility of sample *SPS-B* were about 1.5 times smaller and 5 times higher, respectively, than those of sample *SPS-A*. These differences have been correlated to its higher fraction of coarse grains and its lower NiO content.
- As expected from their microstructural characteristics, the mechanical behaviour of samples *SPS-C* and *SPS-D* was identical. Their strain to failure was very satisfying, ~35%, due to their higher fraction of coarse grains compared to samples *SPS-A* and *SPS-B*. In contrast, they emphasise the transition in the strain-hardening behaviour, from work hardening to softening, that occurred for the SPS-processed materials having a mean grain size smaller than ~300 nm, namely, *SPS-A* and *SPS-B*.

In order to evaluate the influence of the processing route on the mechanical properties of these samples, they were compared with those of HIP-processed samples consolidated from the same 50 nm, 100 nm, and coarse-grained Ni powders and designated as *HIP-A*, *HIP-B*, and *HIP-CG*, respectively (Table 10). Samples *HIP-A* and *HIP-B* yielded average grain sizes of ~333 and ~403 nm, respectively, and their oxide contents were the same as those of the initial powders. Both HIP-processed samples exhibited work hardening up to a large plastic strain value, in contrast to the samples *SPS-A* and *SPS-B*, for which the deposition of oxide dispersions within the boundaries or the triple junctions might have weakened them and resulted in softening due to intergranular failure.

The contribution of grain refinement to the strength of materials can generally be expressed by the Hall-Petch relationship (Hall 1951; Petch 1953); Fig. 2 however shows a strong deviation (Krasilnikov et al. 2005) that has been attributed to an additional strengthening effect produced by the presence of substructures, such as geometrically necessary boundaries and trapped lattice dislocations, formed during the severe plastic deformation (SPD) processing route that had been followed. For the samples processed by HIP (Bui 2008) and SPS (Bui et al. 2010), such a deviation has also been observed and was even more pronounced for the sample *SPS-A* processed from the powder having the smallest particle size. This effect is most probably due to the higher oxide content at lower grain sizes, as already stressed. Assuming a simple linear addition of the strength contributions from the grain size and the oxide particles, the oxide strengthening can be estimated by subtracting the value obtained from the Hall-Petch plot for oxide-free samples from the measured yield strength. It can be concluded that this contribution led to a deviation from the Hall-Petch relationship, which solely reflects the grain-size effect.

2.4 Possible Combinations

This subsection discusses a technique and an approach, routinely used by our research group that could be more generally used in combination with SPS, namely, cold isostatic pressing and severe plastic deformation.

2.4.1 Combination with CIP

Cold isostatic pressing (CIP) has already been explored for comparison purposes as an optional step for processing the initial powders prior to SPS (Dutel et al. 2017). In the framework of this section, which deals with mono-modal structures, only the part of the study relating to the nanograined powder and to the coarse-grained one is reported; the former one (Tekna) was made of 35-nm-sized spherical particles, and the latter one (Sigma-Aldrich) of 5–13 μm agglomerates containing ~1 μm particles. No preferential orientation was observed by IPF imaging. The CIP processing has been performed with a pressure of 1.1 GPa for the coarse-grained powder and a pressure of 1.9 GPa for the nanograined one (Langlois et al. 2010). SPS processing was performed under a controlled argon atmosphere, with an uniaxial pressure of 100 MPa, a dwell time of 1 min, a heating rate of ~90 K.min^{-1}, and a temperature of 500 °C for the nanograined powder and of 850 °C for the coarse-grained powder. Texture-free samples have been obtained.

For the nanograined powder, relative densities were measured at 95.8% after direct SPS processing, at 87% after CIP processing at 1.9 GPa, and at 93.9% after both CIP and SPS processing. Grain sizes increased from 250 nm after direct SPS to

Table 11 Influence of pre-compaction by CIP on SPS-processed microstructures

	Relative density, %		Grain size, μm		Grain boundaries, %			
Powder type	CIP	SPS			LAGBs		Σ3	
Nanograined		95.8	0.25		2.0		20.0	
	(87.0)	93.9	1.75		3.0		30.0	
Coarse-grained		99.0		12.0		15.6		26.6
	(91.0)	98.5		41.0		3.4		63.5
Bimodal		99.4	0.58	3.5	2.5	1.4	17.0	40.0
	(87.0)	96.3	0.52	2.8	3.0	1.4	23.0	52.0

Adapted from Dutel et al. (2017)

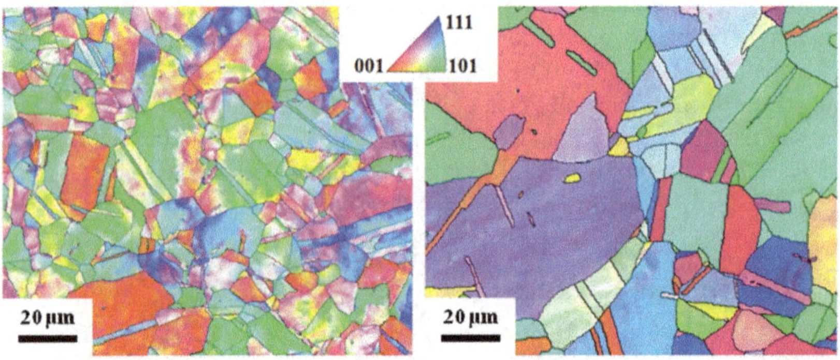

Fig. 5 EBSD IPF images of samples consolidated from coarse-grained powders (**a**) after direct SPS; (**b**) after CIP and SPS. (Dutel et al. 2017)

1.75 μm after both CIP and SPS. For comparison purposes, SPS processing parameters had remained constant, which means that this grain coarsening could be offset in a fully optimised process since CIP would allow sintering temperature to be reduced. Σ3 fraction numbers increased from 20.0% after direct SPS to 30.0% after both CIP and SPS, and LAGB fraction numbers increased from 2.0% to 3.0%. These data are summarised in Table 11.

For the coarse-grained powder, relative densities were measured at 99% after direct SPS processing, at 91% after CIP processing at 1.1 GPa, and at 98.5% after both CIP and SPS processing. Grain sizes increased from 12 μm after direct SPS to 41 μm after both CIP and SPS (Fig. 5). Σ3 fraction numbers increased from 26.6% after direct SPS to 63.5% after both CIP and SPS, whereas LAGB fraction numbers decreased from 15.6% to 3.4%. These data are also summarised in Table 11.

Table 12 summarises the room-temperature compression data at the strain rate of $2\ 10^{-3}\ s^{-1}$ for the different samples. For the samples sintered from nanograined powders, a yield strength of 900 MPa was measured after CIP and SPS, followed by a sharp strain hardening with a measured true stress to failure of 1.57 GPa for a plastic strain of about 4%, whereas the directly SPS-processed sample exhibited a typical brittle behaviour without measurable plastic deformation before failure. For the samples sintered from coarse-grained powders, yield strengths of 400 MPa for

Table 12 Influence of pre-compaction by CIP on mechanical properties after SPS

Powder type	Process	Yield stress, MPa	Compressive strength, MPa	Strain to failure, %
Nanograined	SPS	(1050)	(1050)	(brittle)
	CIP + SPS	900	1570	4
Coarse-grained	SPS	400	(620)	(> 50)
	CIP + SPS	200	(380)	(> 50)
Bimodal	SPS	530	730	17
	CIP + SPS	500	680	34

Reproduced from Dutel et al. (2017)

the directly SPS-processed sample and 200 MPa after CIP and SPS were measured. Both samples depicted a similar strain-hardening behaviour, up to 50% true plastic strain, at which point the capacity of the testing machine was reached and the test was interrupted.

It can be concluded that pre-compaction by CIP, combined to SPS, could serve as a means for Σ3 boundary engineering; present data, as well as data mentioned in the above subsections, indeed show that such grain boundaries allow for a good compromise between compressive strength and strain to failure.

2.4.2 Combination with SPD

Although an optimal level of synergy between SPS and severe plastic deformation has not yet been met on Ni, their combination is afoot. On the one hand, there is an already rich literature on Ni by SPD (Krasilnikov et al. 2005; Zhao et al. 2008). On the other hand, a lot of studies have already been performed where SPD and SPS are combined, not to mention the instances where the genuine qualities of SPS-processed materials make them selected for further SPD processing when defect formation shall be limited for some reason, such as in the case of providing Cu precursors perfectly fit for cold working with a minimised degree of grain refinement in order to produce stronger yet lighter conducting wires (Arnaud et al. 2016). It can therefore be predicted that SPS processing of previously SPD-processed Ni samples will become increasingly common; Ni powders have, for instance, as mentioned in Sect. 2.3, already been mechanically activated by ball milling prior to SPS processing (Naimi et al. 2013).

3 Multimodal Structures

A dedicated subsection focuses first on the case of bimodal structures typically made from powder blends with controlled volume fractions of high-purity nanometre- and micrometre-sized particles. A separate subsection devoted to generic multimodal structures supplements this case.

3.1 Bimodal Structures

As mentioned in the introduction, bimodal grain structures have been most notably investigated for addressing the issue of improving both the strength and the ductility of materials basically designed for Hall-Petch strengthening.

3.1.1 Inherited Bimodal Structure

Bulk polycrystalline nickel samples may directly inherit their bimodal structure through consolidation from heterogeneous powders consisting of a mixture of nanometre- and micrometre-sized particles; a high-purity (N99.8 wt%) Ni powder (Argonide) fabricated by electro-explosion of wire (Fig. 3a) has been processed by SPS under 100 MPa at 600 °C for 15 min (Tingaud et al. 2015), which resulted in samples composed of~45 vol.% ultrafine-grained (with an average grain size of ~550 nm) and ~55 vol.% coarse-grained (D > 1 μm) components, with similar misorientation-angle distributions in the UFG and CG regions. High fractions of Σ3 boundaries that include coherent twin boundaries have been observed in both regions.

Residual porosity was mostly found in the UFG volumes, in relation with the increase of the oxygen content in the areas containing nanograins (~10–20 nm), due to a higher amount of NiO phase originating from the native NiO layer on the surface of the nanoparticles. This oxide phase, with an XRD intensity ratio of 3.0 ± 0.5% compared to Ni, hindered the consolidation of the nanoparticles, thus leaving a large local porosity (up to 10%) and an incomplete sintering.

The macroscopic mechanical behaviour of the SPS-processed samples was studied through uniaxial compression at room temperature at two different strain rates, namely, 2×10^{-4} and 2×10^{-2} s^{-1}, in order to better understand the contributions of the UFG and CG components to the mechanical performance. The higher strain rate resulted in a significant strain hardening, whereas the lower one resulted in a marked softening beyond ~10% strain, which led the compressive strength to decrease from 930 to ~600 MPa, with the corresponding strain even decreasing from ~70% to ~10%. These observations are in good agreement with literature; both the increase of the strain rate and the processing of a bimodal microstructure have been suggested as possible strategies to postpone the onset of deformation instability, thus improving the ductility of UFG materials (Wang and Ma 2004).

EBSD investigations carried out on the transverse face (parallel to the compression direction) of the samples revealed differences; for the lower strain rate, the grains were only slightly elongated, whereas they were heavily flattened with a high aspect ratio for the higher strain rate. EBSD data also showed dislocation patterning within coarse grains and revealed the increase of the fraction of low-angle grain boundaries with compression, thus suggesting dislocation activity. In many areas, nanograins were forming thin layers following the contours of the ultrafine grains, whose orientation was not always uniform, suggesting a high density of dislocations presumably arranged into LAGBs. A large number of the smaller grains retained their spherical shape, suggesting that grains rolled, rotated, and/or glided collectively along their boundaries in order to accommodate plastic strain. It has also been suspected that the pores acted as "free volumes" which were facilitating the deformation. The rheological behaviour of the nanometre-sized component was then thought to have had a significant impact on the mechanical response, especially on the work-hardening behaviour. Further characterisations showed that the dislocation density increased by about one order of magnitude, while the twin boundary probability decreased during compression, which indicated that the deformation in the consolidated samples had mainly occurred by the motion and multiplication of dislocations. The decrease of the twin boundary probability due to the interaction between gliding dislocations and TBs, resulted in a detwinning process, as already described in Sect. 2.2 (Gubicza et al. 2009). The crystallite-size distributions of the initial sample and of its deformed counterparts also showed a significant reduction in the mean crystallite size by a factor of about four that occurred during compression at both strain rates.

In situ TEM investigation of the microstructure during deformation confirmed that various mechanisms were involved:

- In the UFG volume, dislocation-based mechanisms of plasticity prevailed; dislocation emission from both grain boundaries and intragranular sources was observed. At higher strain rate, the annihilated dislocation density per unit strain decreased, which yielded higher dislocation density, contributing to the larger strain hardening.
- Besides dislocation activity, rolling and sliding of nanograins were observed during deformation, which have a considerable contribution to the observed high strain-rate sensitivity of the bimodal microstructure.
- At high stress levels, the required level for deformation twinning was locally achieved, especially at crack tips, whereas growth twins which already existed within the as-consolidated microstructure were subject to a detwining process in the course of straining, with an overall decrease of the TB probability as a result.
- Cracks nucleated and easily propagated in the volumes where the porosity was large; crack paths clearly followed the regions with mainly nanograins (Fig. 6). These volumes indeed had smaller resistance against crack opening due to the weaker particle bonding and to the higher porosity caused by their incomplete sintering.

This last point emphasises that, although dislocations had a significant contribution to plastic deformation in the consolidated bimodal microstructures, sintering considerations alone may have a strong and sometimes decisive impact on the mechanical properties.

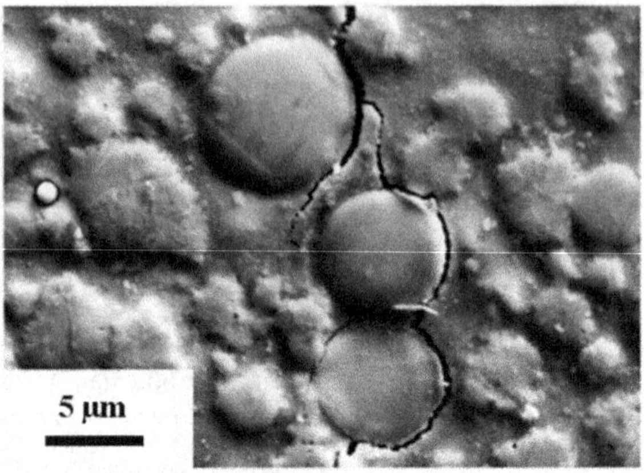

Fig. 6 Postmortem observation of crack propagation. (Dutel 2013)

Table 13 Relative density of the bimodally structured samples. (Dutel 2013)

Dwell temperature	500 °C						850 °C
Blend[a], wt%	(100:0)	90:10	70:30	40:60	10:90	(0:100)	(0:100)
Relative density, %	(96.0)	97.2	97.9	99.4	91.5	(81.2)	(99.0)

[a]Percentage of nanometre- and micrometre-sized components, respectively

3.1.2 Controlled Volume Fractions and Combination with CIP

As stated in Sect. 2.4, CIP has been explored for comparison purposes as an optional step for processing the initial powders prior to SPS. A previous study had led to the choice of the mixture that produced the higher relative density (Table 13), namely, 40% nanograined powder and 60% coarse-grained powder, to be processed under the same parameters as for the initial powders (Dutel et al. 2017).

Relative densities were measured at 99.4% after direct SPS and at 96.3 after both CIP and SPS, with a relative density of the green sample of 87% after CIP at 1.1 GPa (Table 11). After consolidation, no significant variation of the volume fractions was observed, with a calculated value of 44% nanograins and 56% coarse grains for the sample directly processed. Average grains sizes were measured separately for coarse grains and nanograins; after direct SPS, the grain sizes were measured at 3.5 µm and 580 nm, respectively, and after both CIP and SPS, at 2.8 µm and 520 nm, respectively. Σ3 grain boundaries, including TBs, are depicted in red on the EBSD micrographs obtained with a step size of 50 nm, shown in Fig. 7; after direct SPS, they represented 40% of the grain boundaries in the coarse grains and 17% in the nanograins. After CIP and SPS, these proportions were 52 and 23%, respectively. LAGBs are represented in green in Fig. 7; after SPS, they represented 1.4% of the boundaries in the coarse grains and 2.5% in the nanograins. After CIP and SPS, these proportions were 1.4 and 3%, respectively. Large isolated nanograined domains were observed for the directly SPS-processed sample (Fig. 7a), contrariwise to the case of the pre-compacted sample (Fig. 7b). In the latter case, a more or less

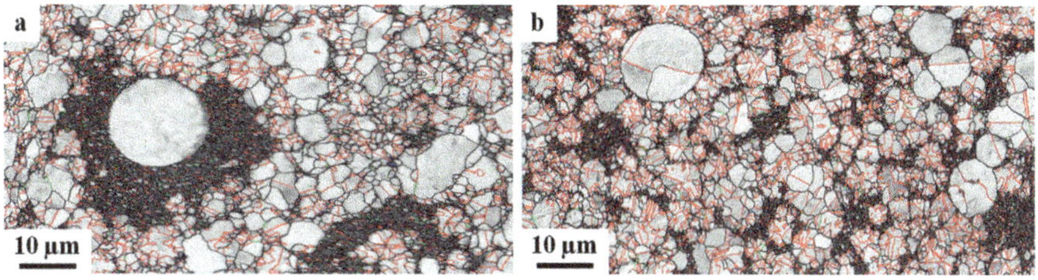

Fig. 7 EBSD grain boundary map images of bimodally structured samples (**a**) after direct SPS and (**b**) after CIP and SPS, respectively. (Dutel et al. 2017)

connected nanograined shell enclosing coarser grains was observed; CIP allowed the final structure to be more homogeneous, which is worth noticing in the framework of this chapter.

Data from the compression tests are summarised in Table 12; the yield strength was within the 500–530 MPa range. The directly SPS-processed sample presented a sharper strain hardening, especially in the beginning of the plastic deformation, and reached a compressive strength of 730 MPa and a strain to failure of 17%. After both CIP and SPS, these values were 680 MPa and 34%, respectively.

3.1.3 Comparison with HIP

Near fully dense polycrystalline nickel with a random crystallographic texture has been consolidated by hot isostatic pressing of blends of nanopowder and conventional microcrystalline powders with different volume fractions (Dirras et al. 2010). This study has had the twofold advantage to provide, on the one hand, an interesting reference for comparison purposes between HIP- and SPS-processed bimodal structures and, on the other hand, useful considerations on the very phenomena at stake in the bimodal structures as regards elaboration and deformation, respectively.

Bulk samples were processed under specific conditions (Table 14) from blends of a high-purity (~99.9%) Ni nanopowder (Argonide) with an average size of ~100 nm and of a conventional coarse-grained powder (Goodfellow) agglomerated into ~50 μm-sized clusters of ~5 μm-sized particles. The relative amount of coarse-grained (CG) powder was ~33 and ~17 vol.% for samples designated as *A* and *B*, respectively. The relative densities after HIP processing were 96, 98, 97, and 94% for the UFG and CG specimens and samples *A* and *B*, respectively. In both samples *A* and *B*, the majority of the boundaries, 92 and 95%, respectively, were high-angle grain boundaries. The fractions of the CG component were 61 and 39% for samples *A* and *B*, respectively; the CG clusters were homogeneously distributed within the UFG matrix, and their average size remained the same as in the initial CG powder. At the same time, the mean grain size inside the CG clusters increased with decreasing UFG fraction; their values were 13 and 7 μm for samples *A* and *B*, respectively, which indicated that the UFG component acted against the coarse grain

Table 14 HIP processing parameters for the different powder compositions

	T, °C	p, MPa	t, min
UFG	700	140	150
CG	700	160	150
Sample *A*	700	156	155
Sample *B*	640	145	60

Reproduced from Dirras et al. (2010)

coalescence, in a sort of "barrier effect" that would have to be reassessed if the UFG component were oxide-free.

After HIP processing, both the dislocation density and the twin probability in the UFG volumes of samples *A* and *B* were much lower than in the UFG sample; this was explained by a "shielding effect" of CG grains on the UFG fraction during consolidation, meaning that at the contact places between the nanopowder and coarse-grained powder particles, the deformation of the softer particles was larger than in the nanopowder particles when they fill the empty spaces in the composite material during consolidation. Moreover, the nanopowder particles in the pockets between CG clusters were most probably protected from external stresses. As a consequence of the moderate deformation of the nanopowder particles, the dislocation density and the twin probability were smaller in the UFG grains after HIP processing. Despite the lower CG content of sample *B*, which should have resulted in a weaker "shielding effect" on the UFG component, the dislocation density and the twin probability of the latter were not higher than the values for sample *A*; this was assumedly attributed to the shorter sintering time in the case of sample *B*.

The mechanical properties of the HIP-processed samples have been studied by nanoindentation and quasi-static compressive testing at room temperature, at a strain rate of $1.6 \times 10^{-4}\ s^{-1}$. The resulting yield strengths were 510, 80, 200, and 270 MPa for the UFG and CG specimens and samples *A* and *B*, respectively; actually, the yield-strength values for composite-like samples *A* and *B* were smaller than the values linearly interpolated from those of the fully CG and UFG specimens. This was assumedly attributed to the fact that the microstructure in either the CG or UFG components had been strongly affected by the volume fraction of the other component ("barrier and shielding effects"). In the same manner, the nanohardness values for sample *A* and *B* were smaller than those linearly extrapolated; this was assumedly explained by the decrease of the grain size in the CG component with increasing UFG volume fraction ("barrier effect"). As a result of the interplay between the coarse-grained and ultrafine-grained components during sintering, the mechanical behaviour of the composite materials could not be obtained by a linear interpolation between the characteristics of their fully ultrafine-grained and coarse-grained counterparts.

3.2 Other Multimodal Structures

Multimodal structures can, as for the bimodal ones, result from the sintering of Ni powder blends of controlled volume fractions of particles with a large range of grain size. A multimodal grain structure may, however, directly result from the elaboration process of the initial powder.

3.2.1 Inherited Multimodal Structure

As previously observed, powders fabricated by electro-explosion of wires often result in a large range of grain sizes; in the case of the high-purity (99.98%) nanograined Ni powder (Argonide) that has been chosen for this study, the initial particle sizes were encompassing nearly three orders of magnitude (10 nm–5 μm), and SPS processing temperature has been varied in order to maximise microstructure control (Dutel et al. 2012). Processing at three dwell temperatures of 550, 600, and 800 °C under a pressure of 100 MPa during 1 min with a heating rate of 100 °C. min^{-1} indeed led to multimodal microstructures.

Combined SEM and TEM investigations showed that the powder initially contained ultrafine-grained ($0.1 < D < 1$ μm) and coarse-grained ($1 < D < 5$ μm) particles whose surface area fractions were evaluated to be ~11 and 7%, respectively. The remainder of the powder was composed of spherical nanoparticles with an average area size of ~70 nm. Spherical multi-crystalline aggregates (MCAs) had a variable size ($5 < D < 120$ μm) and crystallographic orientation; the more the aggregate was subdivided, the more random the orientation was. Their median size was ~10 μm, and they were occupying a surface area fraction of ~10% (Table 15). The rapid grain growth that occurred during SPS processing thus led to a final grain size two orders of magnitude larger than that of the initial powder particles; because the growth and shrinkage of each grain are governed by its curvature, large grains developed at the expense of small grains to reduce the free energy of the material. The initial nanoparticles were prone to be consumed in this growth process. In addition, the excess energy that is stored in nanograined powders produced by EEW facilitates abnormal grain growth which is temperature-driven (Fu and Shearwwod 2004); in this case, it was observed at 500 °C (0.34 T_m). Although the overall size of the MCAs did not evolve with increasing sintering temperatures in the

Table 15 Relative density, d, component fraction and average grain size, D, of SPS-processed multimodal Ni samples

	T, °C	d, %	MCA		CG		UFG	
			% (area)	D, μm	% (area)	D, μm	% (area)	D, nm
A	550	95.5	≈10	5–120	17	2.0	73	390
B	600	97.4			20	1.8	70	450
C	800	97.7			25	1.7	65	525

Reproduced from Dutel et al. (2012)

range 550–800 °C, the average size of their constituent grains increased from 5 to 8 μm, in the same manner as already seen for CG clusters after HIP processing of bimodal blends in Sect. 3.1 (Dirras et al. 2010).

Besides MCAs, the microstructure was comprising dispersed single-crystalline entities and small agglomerates that were micrometre-sized grains (1 < D < 5 μm) and UFG grains. The microstructural characteristics of these coarse-grained and UFG components evolved linearly with the sintering temperature. The average area size of the UFGs increased from 390 to 525 nm, while their surface area fraction decreased from 73% to 65% at 550 and 800 °C, respectively (Table 15). Conversely, the surface area fraction of the CGs increased from 17% to 25%, while their average area grain size remained constant at ~1.8 μm. It can therefore be concluded that adjusting the processing temperature did allow the control of the final microstructure.

Compression tests at a strain rate of $2 \times 10^{-4}\ s^{-1}$ and at room temperature have been carried out on the samples with the higher density, i.e. processed at 600 °C (sample *B*) and 800 °C (sample *C*). Sample *B* possessed a higher strength than sample *C*, while the latter had a higher strain at failure. In contrast to the behaviour of a mono-modally structured coarse-grained counterpart, samples *B* and *C* exhibited strain softening after the initial rapid but brief strain hardening that occurred after the application of a small strain; this softening behaviour was more pronounced for sample *B* than for sample *C*. Because the surface area fraction (10%) and size (average number diameter of 10 μm) of the MCAs remained constant, the main difference between samples *B* and *C* has been attributed to the behaviour of the UFG particles; their average grain sizes were ~450 and 525 nm and their fractions ~70 and 63% in samples *B* and *C*, respectively (Table 15).

The main observations on the evolution of misorientation across boundaries in the MCA, CG, and UFG components of the microstructure of sample *C*, in its initial state and after compression up to failure, were as follows (Table 16): (i) in the initial state, a large fraction of Σ3 boundaries was present in the CG and UFG components compared to the MCA one; (ii) upon straining, the number fraction of LAGBs increased more in the MCA and CG components than in the UFG one, suggesting that plastic deformation occurred mainly via grain rotation in the UFG component due to the smaller grain size; and (iii) upon straining, a sharp decrease (almost down to zero) of the Σ3 boundaries was observed in all the components, as a result of their interaction with the intense intragranular dislocation activity as the main plastic-deformation mechanism in the MCA and CG components due to their coarser grain size. In addition, since grain rotations induced the deformation incompatibilities at

Table 16 Evolution of the grain boundary misorientations

	Initial state		Postmortem	
	LAGBs, %	Σ3, %	LAGBs, %	Σ3, %
UFGs	2.1	34.1	23.6	5.3
CGs	6.9	44.3	75.8	1.2
MCAs	6.1	8.4	76.3	1.4

Adapted from Dutel et al. (2012)

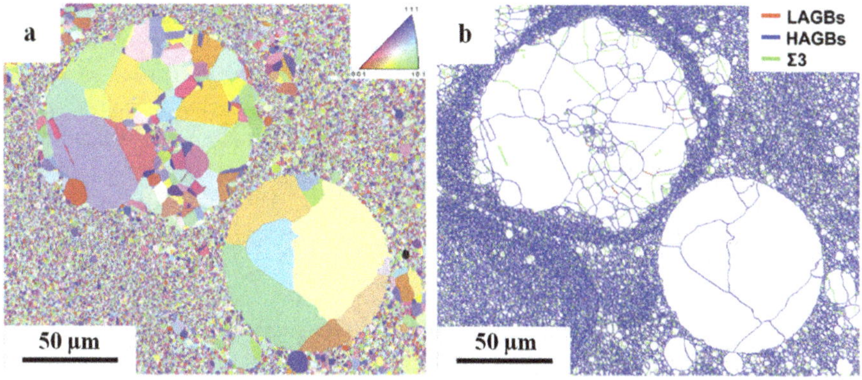

Fig. 8 EBSD IPF map (**a**) and EBSD grain boundary map (**b**) of a SPS-processed multimodally structured sample. (Farbaniec et al. 2014)

grain boundaries which caused them to crack, the level of grain boundary cracking was depending on the amount of the UFG component, which explains why the early strain softening behaviour proved more pronounced for sample *B* than for sample *C*.

3.2.2 Controlled Volume Fractions

Bulk multimodally structured samples have been processed from a blend of three powders: two heterogeneous (fabricated by EEW) high-purity (> 99.7%) nanograined powders (Argonide) having a mean particle size of ~50 and ~100 nm, respectively, mixed with a coarse-grained powder (Alfa Aesar) having particle sizes in the range of 100–150 μm (Farbaniec et al. 2014). The blend was made of 200, 300, and 130 g of these three powders, respectively. SPS processing was performed at a sintering temperature of 750 °C with a heating rate of 50 $°C.min^{-1}$ and a dwell duration of 15 min, under an uniaxial pressure of 70 MPa.

Figure 8 shows the microstructure of the SPS-processed samples that were comprising coarse spherical multi-crystalline entities, the "core", embedded in the "shell", a fine-grained matrix or a thick rim distinguishable from the matrix. In comparison with studies where only one type of powder had been used (Bui et al. 2010; Dutel 2013), the larger and highly subdivided CG entities were coming from the coarse-grained powder, while the smaller and less subdivided ones were coming from the two heterogeneous nanograined powders. Based on SEM analysis, the size of the CG core was about 135 ± 50 μm, but it was subdivided into fine-grained (FG) multi-crystals (average grain size of ~11.4 μm). Only the grain sizes of the FG matrix were following a log-normal distribution, with an average grain size of ~1.4 μm, which is about one order of magnitude lower than that of the CG entities, whereas the thick shell which was often surrounding the larger CG entities had its own grain size distribution with an average grain size of 5.3 μm. Actually, such a shell ever featured around CG entities, neither during the previous sintering experiments of powder under the same SPS processing parameters (Dutel 2013) nor during the HIP processing of a blend of heterogeneous nanograined powder and commercial purity CG Ni powder (Dirras et al. 2010); it has therefore been suspected that the current flow might have locally been impacted by some difference in the powder resistivity as a consequence of the more complex multimodal structure.

Tensile testing at a strain rate of 2×10^{-3} s^{-1} has been performed on both samples due to their slightly different but high enough relative densities of 97.8 and 98.4%, respectively. After the onset of yielding, two regimes of hardening have been observed; the first one occurred up to about 0.7% plastic deformation where the stress sharply increased from 470 MPa at yielding to about 550 MPa, whereas the second one corresponded, up to the failure point, to a limited stress increase of less than 20 MPa within a moderate plastic deformation range of 0.7–2.8%. Compared to a conventional Ni sample with an average grain size of ~25 μm, there was an obvious increase of the yield strength, as a consequence of the high fraction of the FG component (Zhang et al. 2014). In addition, contrariwise to the transition from work hardening to softening, reported in Sect. 2.3, that occurred for the SPS-processed Ni samples having a grain size smaller than ~300 nm (Bui et al. 2010), an early strain softening has not been observed, most probably because the intergranular microcrack growth in the FG component and their subsequent propagation were delayed (or blunted) by the CG entities (Dirras et al. 2014).

Dislocation emission and annihilation at grain boundaries and twinning at crack tip were the main deformation mechanisms that took place within the fine-grained component and contributed to the increased yield strength. Final failure was accounted for a crack-network buildup, in agreement with the mechanism-based plastic model used to describe the strength and ductility of bimodally structured metals (Zhu et al. 2012). In the framework of the harmonic structure design, it has been suggested that the early stage of deformation is dominated by the plastic deformation of the stronger FG shell region (Vajpai et al. 2016); the harmonic structure would deform in the same manner as the homogeneous FG microstructure and then by a typical deformation mode for CG materials in the later stages with a large uniform deformation. In this process, the CG component should undergo plastic deformation to adjust the shape change of the FG shell in order to maintain the integrity of the material; in the present case, however, the non-homogeneous grain-size distribution generated strong deformation incompatibilities which resulted in cracking and cavitation in the FG component. Debonding at interfaces between the matrix and the coarse entities was indeed viewed as a consequence of large deformation incompatibilities due to a large difference in the grain sizes, since no crack nor debonding was observed between the fine-grained matrix and the rim whose average grain sizes were of the same order of magnitude. Control of grain-size gradient between the core and the shell therefore appears to be of great importance for tensile-ductility optimisation. Circular-like cracks also occurred in the coarse-grained entities near their fractured surface, probably due to the complex stress state.

3.2.3 Harmonic Structures

In regard to the "harmonic structure" design, it may be pointed out that the state of the art is not yet as developed for nickel as for some other materials such as titanium alloys (Dirras et al. 2017a, b) for which a three-dimensional multi-scale polycrystalline plasticity model coupled with damage has already been developed in agreement with experimental data (Liu et al. 2018). It must therefore be realised that there is scope for an even greater progress; similar to titanium alloys or copper (Sawangrat et al. 2014), polycrystalline nickel exhibiting harmonic structures could indeed be produced by tailoring the parameters of severe plastic deformation, through either mechanical milling or jet milling, and those of subsequent SPS.

4 Conclusions

The versatility of the SPS technique has been demonstrated; in many instances, it provided a unique opportunity to consolidate powders that could not have been implemented through conventional techniques. Such instances occurred with the replication of either the fine-grained or the extremely heterogeneous structure of the initial powder that required a limited grain growth.

In regard to the specific expectation treated in this chapter, and although most mechanical behaviours, due to relative-density considerations, have so far been determined through quasi-static compressive rather than tensile testing, the overall goal remains the enhancement of tensile ductility of polycrystalline nickel. The comparison with HIP, which might have sound disappointing at first glance since SPS- and HIP-processed samples often showed distinctively higher yield strengths and higher strains, respectively, enabled to capitalise on it to better understand the deformation mechanisms involved. For instance, the reciprocal roles of the UFG and CG components in a "barrier effect" and a "shielding effect" have been emphasised, as well as the role of the Σ3 grain boundaries to allow for a good compromise between compressive strength and strain to failure, which, in turn, highlighted the otherwise possibly neglected benefit of such a step as pre-compaction by CIP as a means for Σ3 grain boundary engineering.

More decisively, SPS has proved to be highly suitable, in terms of process control, for implementing the increasingly complex structures that have been sought for enhancing tensile ductility, in terms of material design, from the bimodal to the multimodal ones.

Acknowledgements Amongst the too numerous collaborators for them all to be named here, the authors willingly extract the names of Drs Q.H. Bui, L. Farbaniec, and G.D. Dutel. They are also grateful for the support provided by the French research funding agency ANR through grant ANR-09-BLAN-0010.

References

Arnaud C, Lecouturier F, Mesguich D, Ferreira N, Chevallier G, Estournès C, Weibel A, Peigney A, Laurent C (2016) High strength–high conductivity nanostructured copper wires prepared by spark plasma sintering and room-temperature severe plastic deformation. Mater Sci Eng A 64:209–213. https://doi.org/10.1016/j.msea.2015.09.122

Bhattacharjee PP, Sinha SK, Upadhyaya A (2007) Effect of sintering temperature on grain boundary character distribution in pure nickel. Scr Mater 56:13–16. https://doi.org/10.1016/j.scriptamat.2006.09.003

Borkar T, Banerjee R (2014) Influence of spark plasma sintering (SPS) processing parameters on microstructure and mechanical properties of nickel. Mater Sci Eng A 618:176–181. https://doi.org/10.1016/j.msea.2014.08.070

Bousnina MA, Dakhlaoui Omrani A, Schoenstein F, Madec P, Haddadi H, Smiri LS, Jouini N (2010) Spark plasma sintering and hot isostatic pressing of nickel nanopowders elaborated by a modified polyol process and their microstructure, magnetic and mechanical characterization. J Alloys Compd 504S:S323–S327. https://doi.org/10.1016/j.jallcom.2010.02.142

Bui QH (2008) Polycristaux à grains ultrafins élaborés par métallurgie des poudres: microstructures, propriétés mécaniques et modélisation micromécanique. Thesis (PhD), Université Paris 13

Bui QH, Dirras G, Ramtani S, Gubicza J (2010) On the strengthening behavior of ultrafine-grained nickel processed from nanopowders. Mater Sci Eng A 527:3227–3235. https://doi.org/10.1016/j.msea.2010.02.003

Dirras G, Gubicza J, Ramtani S, Bui QH, Szilágyi T (2010) Microstructure and mechanical characteristics of bulk polycrystalline Ni consolidated from blends of powders with different particle sizes. Mater Sci Eng A 527:1206–1214. https://doi.org/10.1016/j.msea.2009.09.050

Dirras G, Tingaud T, Csiszár G, Gubicza J, Couque H, Mompiou F (2014) Characterization of bulk bimodal polycrystalline nickel deformed by direct impact loadings. Mater Sci Eng A 601:48–57. https://doi.org/10.1016/j.msea.2014.02.043

Dirras G, Tingaud T, Ueda D, Hocini A, Ameyama K (2017a) Dynamic Hall-Petch versus grain-size gradient effects on the mechanical behavior under simple shear loading of β-titanium Ti-25Nb-25Zr alloys. Mater Lett 206:214–216. https://doi.org/10.1016/j.matlet.2017.07.027

Dirras G, Ueda D, Hocini A, Tingaud T, Ameyama K (2017b) Cyclic shear behavior of conventional and harmonic structure-designed Ti-25Nb-25Zr β-titanium alloy: Back-stress hardening and twinning inhibition. Scr Mater 138:44–47. https://doi.org/10.1016/j.scriptamat.2017.05.033

Dolbec R, Bolduc M, Fan X, Guo J, Jurewicz J, Labrot T, Xue S, Boulos M (2008) Nanopowders synthesis at industrial-scale production using the inductively coupled plasma technology. In: Technical proceedings of NSTI-Nanotech 2008 conference, Boston, 1–5 June 2008

Dutel GD (2013) Comportement mécanique et mécanismes de déformation et d'endommagement de polycristaux de nickel mono- et bi-modaux élaborés par SPS. Thesis (PhD), Université Paris 13

Dutel GD, Tingaud D, Langlois P, Dirras G (2012) Nickel with multimodal grain size distribution achieved by SPS: microstructure and mechanical properties. J Mater Sci 47:7926–7931. https://doi.org/10.1007/s10853-012-6670-1

Dutel GD, Langlois P, Tingaud D, Dirras G (2013) Room-temperature deformation micromechanisms of polycrystalline nickel processed by spark plasma sintering. Mater Charact 79:76–83. https://doi.org/10.1016/j.matchar.2013.02.013

Dutel GD, Langlois P, Tingaud D, Vrel D, Dirras G (2017) Data on the influence of cold isostatic pre-compaction on mechanical properties of polycrystalline nickel sintered using spark plasma sintering. Data Brief 11:61–67. https://doi.org/10.1016/j.dib.2017.01.009

Ebrahimi F, Bourne GR, Kelly MS, Matthews TE (1999) Mechanical properties of nanocrystalline nickel produced by electrodeposition. Nanostruct Mater 11:343–350. https://doi.org/10.1016/S0965-9773(99)00050-1

Farbaniec L, Dirras G, Krawczynska A, Mompiou F, Couque H, Naimi F, Bernard F, Tingaud D (2014) Powder metallurgy processing and deformation characteristics of bulk multimodal nickel. Mater Charact 94:126–137. https://doi.org/10.1016/j.matchar.2014.05.008

Fedorov AA, Gutkin MY, Ovid'ko IA (2002) Triple junction diffusion and plastic flow in fine-grained materials. Scr Mater 47:51–55. https://doi.org/10.1016/S1359-6462(02)00096-9

Fu Y, Shearwwod C (2004) Characterization of nanocrystalline TiNi powder. Scr Mater 50:319–323. https://doi.org/10.1016/j.scriptamat.2003.10.018

Gubicza J, Bui HQ, Fellah F, Dirras GF (2009) Microstructure and mechanical behavior of ultrafine-grained Ni processed by different metallurgy methods. J Mater Res 24:217–226. https://doi.org/10.1557/JMR.2009.0010

Hall EO (1951) The deformation and ageing of mild steel: III discussion of results. Proc Phys Soc Lond B 64:747–753

Krasilnikov N, Lojkowski W, Pakiela Z, Valiev R (2005) Tensile strength and ductility of ultrafine grained nickel processed by severe plastic deformation. Mater Sci Eng A 397:330–337. https://doi.org/10.1016/j.msea.2005.03.001

Langlois P, Fagnon N, Dirras G (2010) Microstructure engineering from metallic powder blends for enhanced mechanical properties. J Phys Conf Ser 240:012016. https://doi.org/10.1088/1742-6596/240/1/012016

Liu J, Li J, Dirras G, Ameyama K, Cazes F, Ota M (2018) A three-dimensional multi-scale polycrystalline plasticity model coupled with damage for pure Ti with harmonic structure design. Int J Plast 100:192–207. https://doi.org/10.1016/j.ijplas.2017.10.006

Minier L, Le Gallet S, Grin J, Bernard F (2010) Influence of the current flow on the SPS sintering of a Ni powder. J Alloys Compd 508:412–418. https://doi.org/10.1016/j.jallcom.2010.08.077

Minier L, Le Gallet S, Grin J, Bernard F (2012) A comparative study of nickel and alumina using spark plasma sintering (SPS). Mater Chem Phys 134:243–253. https://doi.org/10.1016/j.matchemphys.2012.02.059

Miyamoto Y, Niino M, Koizumi M (1997) FGM research programs in Japan – from structural to functional uses. In: Shiota I, Miyamoto Y (eds) Functionally graded materials 1996. Elsevier Amsterdam

Naimi F, Minier L, Le Gallet S, Couque H, Bernard F (2013) Dense nanostructured nickel produced by SPS from mechanically activated powders: enhancement of mechanical properties. J Nanomater:674843. https://doi.org/10.1155/2013/674843

Petch NJ (1953) The cleavage strength of polycrystals. J Iron Steel Res Int 175:25–28

Sawangrat C, Kato S, Orlov D, Ameyama K (2014) Harmonic-structured copper: performance and proof of fabrication concept based on severe plastic deformation of powders. J Mater Sci 49:6579–6585. https://doi.org/10.1007/s10853-014-8258-4

Sekiguchi T, Ono K, Fujiwara H, Ameyama K (2010) New microstructure Design for Commercially Pure Titanium with outstanding mechanical properties by mechanical milling and hot roll sintering. Mater Trans 51:39–45. https://doi.org/10.2320/matertrans.MB200913

Tepper F (1998) Electro-explosion of wire produces nanosize metals. Met Powder Rep 53:31–33. https://doi.org/10.1016/S0026-0657(98)80193-8

Tingaud D, Jenei P, Krawczynska A, Mompiou F, Gubicza J, Dirras G (2015) Investigation of deformation micro-mechanisms in nickel consolidated from a bimodal powder by spark plasma sintering. Mater Charact 99:118–127. https://doi.org/10.1016/j.matchar.2014.11.025

Vajpai SK, Sawangrat C, Yamaguchi O, Ciuca OP, Ameyama K (2016) Effect of bimodal harmonic structure design on the deformation behaviour and mechanical properties of co-Cr-Mo alloy. Mater Sci Eng C 58:1008–1015. https://doi.org/10.1016/j.msec.2015.09.055

Vanmeensel K, Laptev A, Van der Biest O, Vleugels J (2007) Field assisted sintering of electro-conductive ZrO_2-based composites. J Eur Ceram Soc 27:979–985. https://doi.org/10.1016/j.jeurceramsoc.2006.04.142

Wang YM, Ma E (2004) Three strategies to achieve uniform tensile deformation in a nanostructured metal. Acta Mater 52:1699–1709. https://doi.org/10.1016/j.actamat.2003.12.022

Wang YM, Chen M, Zhou F, Ma E (2002a) High tensile ductility in a nanostructured metal. Nature 419:912–915. https://doi.org/10.1038/nature01133

Wang L, Tan Z, Meng S, Liang D, Liu B (2002b) Low temperature heat capacity and thermal stability of nanocrystalline nickel. Thermochim Acta\ 386:23–26. https://doi.org/10.1016/S0040-6031(01)00724-9

Xiao C, Mirshams RA, Whang SH, Yin WM (2001) Tensile behavior and fracture in nickel and carbon doped nanocrystalline nickel. Mater Sci Eng A 301:35–43. https://doi.org/10.1016/S0921-5093(00)01392-7

Zhang Z, Vajpai SK, Orlov D, Ameyama K (2014) Improvement of mechanical properties in SUS304L steel through the control of bimodal microstructure characteristics. Mater Sci Eng A 598:106–113. https://doi.org/10.1016/j.msea.2014.01.023

Zhao Y, Topping T, Bingert JF, Thornton JJ, Dangelewicz AM, Li Y, Zhu Y, Zhou Y, Lavernia EJ (2008) High tensile ductility and strength in bulk nanostructured nickel. Adv Mater 20:3028–3033. https://doi.org/10.1002/adma.200800214

Zhu L, Shi S, Lu K, Lu J (2012) A statistical model for predicting the mechanical properties of nanostructured metals with bimodal grain size distribution. Acta Mater 60:5762–5772. https://doi.org/10.1016/j.actamat.2012.06.059

Nickel-Tungsten Composite-Like Microstructures Processed by Spark Plasma Sintering for Structural Applications

D. Tingaud, T. Sadat, and G. Dirras

1 Introduction

Obtaining, via microstructure optimization, materials with high strength and ductility, in order to broaden their potential application fields, is still challenging. Indeed, combining both, mechanical strength and ductility, has been a longtime part of the optimization route for structural metallic materials. In such an approach, the elaboration of a composite-like material via the incorporation of a coarse-grained phase within a fine-grained matrix is a part of the optimization scheme. It is indeed considered that micrometer-sized grains will impart the necessary storage of dislocations during straining. In the context of this quest for better properties, we have extended the approach to the inclusion of chemical elements having different crystallographic structures. Indeed, powder metallurgy route allows innovative microstructures to be fabricated by alloying Ni, a face-centered cubic (fcc) element, with W, a body-centered cubic (bcc) one. While the former is known to be soft and ductile, the latter is hard and brittle. Spark plasma sintering (SPS) was specifically chosen among other powder metallurgy techniques because it is sufficiently fast to avoid formation of intermetallic phases, contrariwise to classical sintering methods such as hot isostatic pressing (HIP) or hot pressing (HP) that usually lead to a complex and brittle intermetallic interface, as predicted by the Ni-W phase diagram (Massalski 1990).

The structure of this chapter is as follows: in the first part, the SPS behaviors of nanometer (average grain size of about 50 nm)-, ultrafine (ufg, 0.8–1 μm)-, and micrometer (few tens of μm)-sized powders of commercially pure W (Sect. 3.1), as well as that of a W nanopowder elaborated by self-propagating high-temperature synthesis (SHS) method, will be presented along with the corresponding microstructure characteristics and selected mechanical property features (Sect. 3.2). Following

D. Tingaud (✉) · T. Sadat · G. Dirras
Université Paris 13, Sorbonne Paris Cité, LSPM-CNRS, Villetaneuse, France
e-mail: david.tingaud@univ-paris13.fr; dirras@univ-paris13.fr

© Springer Nature Switzerland AG 2019

P. Cavaliere (ed.), *Spark Plasma Sintering of Materials*,
https://doi.org/10.1007/978-3-030-05327-7_21

the same methodology, the second part of the chapter (Sect. 3.3) is devoted to analyzing the effect of alloying W with a controlled fraction of fine-grained Ni powder. The SPS behavior of Ni-96W alloy will be analyzed and its mechanical properties discussed. In the last part of the chapter (Sect. 4), we will describe and discuss the SPS behaviors and the microstructures and the mechanical properties of Ni-W alloys (Ni-10W to Ni-65W).

2 Experimental Procedures

2.1 *Sintering Parameters*

SPS experiments have been carried out using a Dr. Sinter 515S Syntex (Fuji Electronic Industrial Co., Japan) belonging to the "Plateforme CNRS de Frittage Ile de France" (ICMPE, Thiais, France). Graphite dies with an internal diameter of 8.4 mm and 20.4 mm were used for the W samples and the Ni-W alloys, respectively. These molds can be used up to 2000 °C and 100 MPa (graphite 2333). The die wall was lined with a laminated graphite foil of 200 μm thick in order to restrict reactions between the powder and the die but also to facilitate the release of the sintered sample from the mold. Disks of the same foil were also placed between the punches and the powder for the same reasons. To limit heat loss by radiation, the mold is also covered by graphite felt. A primary vacuum of about 20 Pa was systematically made in the chamber before placing it in a slight Ar under pressure to avoid powder emission outside the chamber in case of die breakage. For the sintering over 1000 °C, the temperature was controlled by a pyrometer focused on a hole placed at mid height in the outer wall of the die. So, due to the emissivity of graphite, the temperature is only monitored by the PID parameters below 570 °C. For the sintering below 1000 °C, a K thermocouple placed in the same hole was used instead of the pyrometer to control the temperature. Common parameters for every SPS were as follows: an on-off 3.3 ms pulse sequence of 12:2 was used; a pressure of about 50 MPa was applied at room temperature and then increased up to 100 MPa at 500–700 °C and maintained throughout the sintering; the heating rate was always fixed at 100 °C/min and then reduced to 20 °C/min 2 min before the sintering temperature to avoid overheating; and after the soaking time, there was no cooling control.

2.2 *Material Characterizations*

After sintering, the graphite foil was removed, and the upper and lower faces were mechanically ground using SiC papers (600, 1000, 2400, and 4000 grit). Prior to X-ray diffraction (XRD) and electron backscattering diffraction (EBSD) investigations, all the W samples (except shs-W, Sect. 3.2; see Dine et al. 2018 for details) were electropolished using and 2% NaOH solution (8 V, 30 s), while the Ni-W alloys were polished using an OPS solution from Struers. XRD experiments were performed on INEL diffractometers, equipped with a curved position sensitive detector that employs the Bragg-Brentano geometry using monochromatic Cu $K\alpha_1$ radiation (1.540598 Å). Rietveld method refinement on XRD pattern was done using MAUD software. Scanning electron microscopy (SEM) and EBSD investigations were carried out on a ZEISS SUPRA 40VP electron microscope equipped with a field effect gun. EBSD was performed at 20 kV. The step size was adapted to the size of the observed entities, and the number and size of EBSD scans were adjusted in order to get enough statistical data. OIM software version 5 (TexSEM Laboratories) software was used to analyze the data. A grain is defined by at least three neighboring pixels with a misorientation below 15°. Boundaries with misorientation angles between 2° and 15° are regarded as low-angle grain boundaries (LAGBs), while boundaries with misorientation above 15° are considered as high-angle grain boundaries (HAGBs). To complement the SEM analysis of the starting nanopowders, transmission electron microscopy (TEM) investigations were carried out by means of a JEOL 2010 electron microscope operating at 200 kV. The apparent mass density was measured at least three times in *ortho-xylene* using the method based on the Archimedes' principle, enabling the calculation of the relative density, measured density divided by theoretical density (19.30 for W and 8.59 for Ni) which is calculated based on a law of mixture.

Hardness was measured using a Duramin 20 Vickers apparatus with a 19.6 N force applied for 10 s for the W samples, while hardness measurements of the Ni-W alloys were done on an HMV microhardness tester at a load of 0.98 N for 10 s. In all cases, every measurement was repeated at least ten times to get a good statistical evaluation of the sample hardness. Compression and tensile tests were also performed using a Deben™ MICROTEST machine (Deben UK Ltd.) with a load cell with a maximum capacity of 5 kN. The compression tests were done at room temperature in the same direction as that of the load applied during SPS on $2 \times 2 \times 3\ mm^3$ samples, while tensile tests were done in a radial direction on dog bone-shaped samples of 18 mm in length (4 mm gage length). The stress-strain plots were corrected by subtracting the machine stiffness.

2.3 Powder Characteristics

The micrometric commercial W powder (provided by Alfa Aesar, thereafter named mW) has a purity of 99% (metal basis) and is made of cuboid particles of about 150 microns (Fig. 1a) which are subdivided into several micrometric grains. The ufg powder, provided by Sigma-Aldrich (thereafter named uW), has a purity greater than 99.9%. It is constituted of polyhedron single-crystalline particles with a sharp grain size distribution that ranged between 0.6 and 1.0 μm (Fig. 1b). As for the micrometric W powder, uW only exhibits the classical α-W structure (A1, bcc). The ufg W powder presented here was used to fabricate the Ni-W alloys presented in

Fig. 1 SEM and TEM observations of the micro- (**a**), ultrafine (**b**) and nano- (**c**) grained size W powders. The grain size distribution (**d**) stands for W nanopowder

Sects. 3.3 and 4. Figure 1c also displays a typical TEM image of the W nanopowder provided by Sky Spring (thereafter named nW). This nanopowder, whose purity is 99.9%, has been elaborated by electro-explosion of wires (EEW). A grain size distribution, based on more than 400 grains, using TEM image processing, shows that the average grain size in number is about 50 nm (150 nm in area). About 10% of the grains are in the ufg regime (100 nm < Φ <1 μm). SEM observations also reveal that some spherical particles with a diameter of several microns are also embedded within this nanopowder. Besides, XRD combined with Rietveld method points out that nW powder contains about 30% of β-W (A15, Cr_3Si prototype, see Fig. 2). This metastable structure is generally observed after the reduction of W oxides or in thin films (Singla et al. 2013). As reported (Yih and Wang 1979), a differential thermal analysis (DTA) performed on nW indicates that the A15 structure is unstable above 700 °C and turns into α-W. In order to avoid additional processing steps which can lead to contamination, this allotropic transformation has been done in situ during SPS. A W nanopowder synthesized by SHS was also sintered by SPS. Its elaboration process and its characteristics will be presented in Sect. 3.2.

A fine nickel powder (99.9% purity) provided by Sigma-Aldrich (St. Louis, MO, USA) was used to produce the Ni-W alloys. This powder is made of agglomerates of an average size of about 5 μm containing grains of about 1 μm as it can be clearly seen on EBSD performed on these agglomerates dispersed in conductive carbon resin (Fig. 3). XRD only reveals the classical fcc Ni pattern.

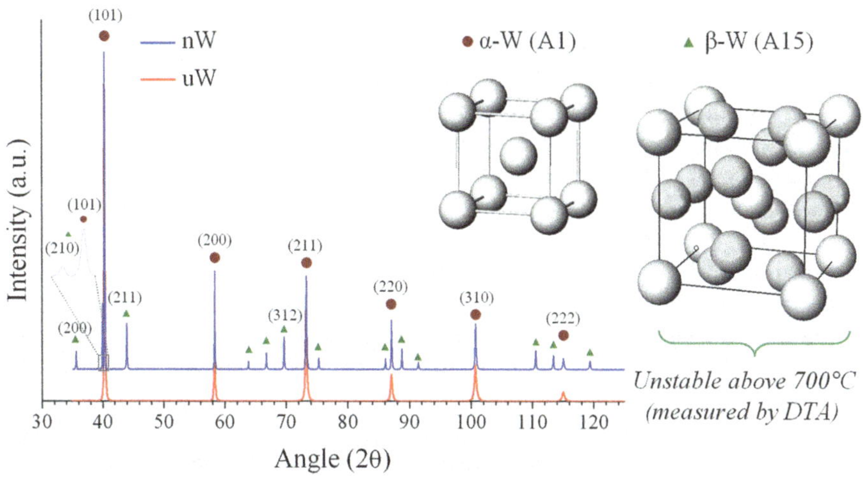

Fig. 2 XRD of nW and uW powders and reference crystalline structures. nW powder is a mixture of α and β (≈30%) phases

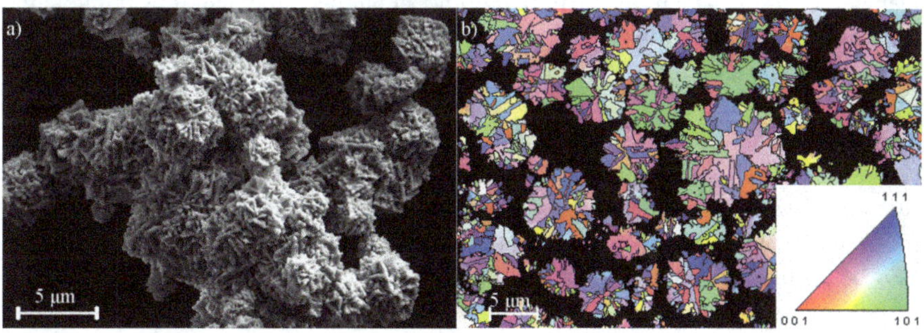

Fig. 3 (**a**) SEM image and (**b**) IPF map from EBSD of the fine Ni powder

3 Bulk W Samples Elaborated by SPS

Due to its density, hardness, low thermal expansion, and excellent mechanical properties at high temperatures (Lassner and Schubert 1998), W is appropriate for many industrial applications such as lightning and vacuum tube filaments, heating elements, plasma-facing materials in Tokamak, and aeronautic and military uses like rocket thermal protection shield or dense inert metal explosives. Owing to its very high melting point (≈3695 K), W parts are usually processed by powder metallurgy. Even when using very high sintering temperatures and/or long soaking time, a full dense W sample is rarely obtained. Starting with fine grain size W powders, a density of about 90% is thus reported after natural sintering at 1800 °C for 50 h and at 2500 °C for 1 h in hydrogen atmosphere (Bewlay and Briant 1995; Kim et al. 2006). Pressure-assisted sintering methods are known to promote densification. HP and HIP were thus also used to sinter W. A relative mass density of 92.7% was reached with HIP at 1277 °C for 2 h under 195 MPa (Monge et al. 2009), while a value of about 94% was obtained after HP at 2300 °C for 30 min under 250 MPa (Karpinos et al. 1970). Nevertheless, all these techniques involving a long processing time lead to coarse grain microstructures while keeping fine grains promote ductility. In fact, W exhibits a ductile-to-brittle transition temperature (DBTT) at 200–400 °C depending on the microstructural parameters and chemical composition. Since W single crystal exhibits ductile properties (Allen et al. 1961), W brittleness at low temperature is assumed to be due to the weakness of grain boundaries which mainly depend on segregation of light elements (C, N, O). With larger grains, the grain boundary length is lower, and thus the impurity concentrations increase which play in favor of brittleness. Dense W parts with a fine-grained microstructure are thus needed to tend toward improved mechanical properties and sufficient machinability. SPS is an advanced sintering method similar to HP except for the heating source which is a low-voltage direct current passing through the die in order to heat by Joule effect. This specific heating mode involves fast heating rates (several hundred degrees per minute) and many specific effects like plasma, local overheating at the grain contacts, electromigration, impurity removal, etc. (Groza and Zavaliangos 2000; Deng et al. 2018) which lower sintering time and temperature and thus promote the densification of fine-grain microstructures.

In this section, starting with three commercially available pure W powders (nW, uW, and mW), we will present their behaviors under SPS, the resulting microstructures and the mechanical characteristics (hardness, compression tests) for selected compositions. These results will be subsequently compared to that of bulk W samples obtained from the sintering of a W nanopowder produced by SHS (Sect. 3.2) and to that of a Ni-96 W alloy to emphasize the effect of a Ni alloying on the uW SPS behavior and properties (Sect. 3.3).

3.1 SPS of Commercial W Powders

3.1.1 Sintering Conditions

All the sintering conditions are gathered in Table 1. Typical pressure and temperature cycle at 1800 °C for 1 min as well as the derivative of the piston displacement with time which can be associated with the densification rate are proposed on Fig. 4 for the three commercial powders. After densification due to grain rearrangement induced by the compaction up to 100 MPa, the sintering starts at ≈800, 900 and 1375 °C, respectively, for the uW, nW, and mW powders. Besides, for mW, the high densification rate occurs around 1800 °C, and it remains important until the end of the dwell time, meaning that enhanced densification could be obtained with a higher sintering temperature and/or soaking time. Conversely, for the finer powders, the higher densification rate is obtained at about 1050 °C and is very low above 1400 °C. That clearly evidences the effect of grain size on the sintering behavior. Besides, the

Table 1 Sintering conditions, relative density, average grain size, LAGB fraction, and Vickers hardness of pure W starting with micrometer-, ultrafine-, or nanometer-grained size powders (Sect. 3.1) or nanopowder processed by SHS (Sect. 3.2)

Powder	$T_{sintering}$ (°C)	Time (min)	$d_{rel.}$ (%)	Grain size (μm)		Growth factor	LAGB (%)	Hv2 (GPa)
				Number	Area			
mW	1800	1	90.3	31	110	–	≈ 50	2.8 ± 0.4
uW	1400	1	90.1	0.5	1.0	1	<5	4.0 ± 0.2
		10	91.1	1.2	1.5	2		4.9 ± 0.2
		30	94.1	2.8	4.5	5		4.7 ± 0.3
		60	95.5	3.7	5.4	6.5		4.3 ± 0.2
	1800	1	94.1	4.0	6.6	7.5		3.5 ± 0.2
nW	1400	1	93.8	2.3	5.5	≈40	<10	4.0 ± 0.3
	1800	1	92.3	3.4	8.2	≈60		3.7 ± 0.2
shs-W	2000	5	≈99	0.4	6.9	–	≈ 50	4.3 ± 0.2

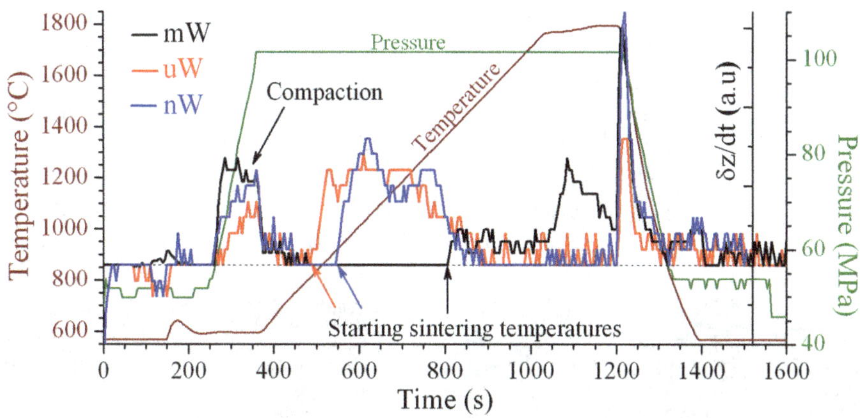

Fig. 4 Typical SPS cycle at 1800 °C for 1 min and densification rates of the W powders

allotropic transformation from β-W to α-W of the nW that occurs above 700 °C does not appear on the densification rate because this transformation only leads to a volume expansion lower than 1%. Based on these first results, the other sintering cycles were done at 1400 °C with dwelling times ranging from 1 to 60 min.

After sintering, the remaining graphite foils surrounding the sample were removed by grinding and polishing with SiC papers. If the thickness of the removed layer is too low, different tungsten carbides can be detected by XRD. The layer thickness containing these tungsten carbides increases with sintering temperature and dwell time. At least, a 300 μm layer was removed on both sample sides. Finally, XRD was used to confirm that the remaining parts of all the W samples are only made of pure α-W. This C contamination is very difficult to avoid since it occurs even when there is no direct contact between the W powder and graphite foil or die. Even using boron nitride, known to block carbon diffusion, instead of the classical graphite foil to cover the inner walls of the graphite die (Lee et al. 2016), only reduces the thickness of the WC surface layer from 253 to 191 μm. Besides, for identical sintering parameters, the grain growth was reported to be 2–3 times higher with no significant effect on density when BN is employed. No explanation was given, but BN, as an insulating material, should probably strongly modify the current lines and current density during SPS. For similar reasons, we also previously observed strong effects on microstructures when replacing C foils by tantalum in an attempt to reduce C contamination during the SPS of Ni nanopowders (Dutel 2013).

3.1.2 Microstructural Characterizations

The mW sample exhibits the lower density after 1 min at 1800 °C which is in agreement with the densification rate that is still significant at the end of the dwell time. SEM images thus clearly evidence very large porosity (up to several tenths of microns) between large spherical particles, a classical feature of incomplete sintering. EBSD analysis reveals a homogeneous microstructure with large micrometric grains, about 31 μm (110 microns in area) with random grain orientation and high LAGB fraction (≈50%) located around the large pores.

Regarding the nW, a higher mass density is obtained at 1400 °C-1 min (see Table 1). Increasing the sintering temperature does not reduce the porosity. Besides, heterogeneous microstructures are revealed by EBSD (Fig. 5a). Large spherical

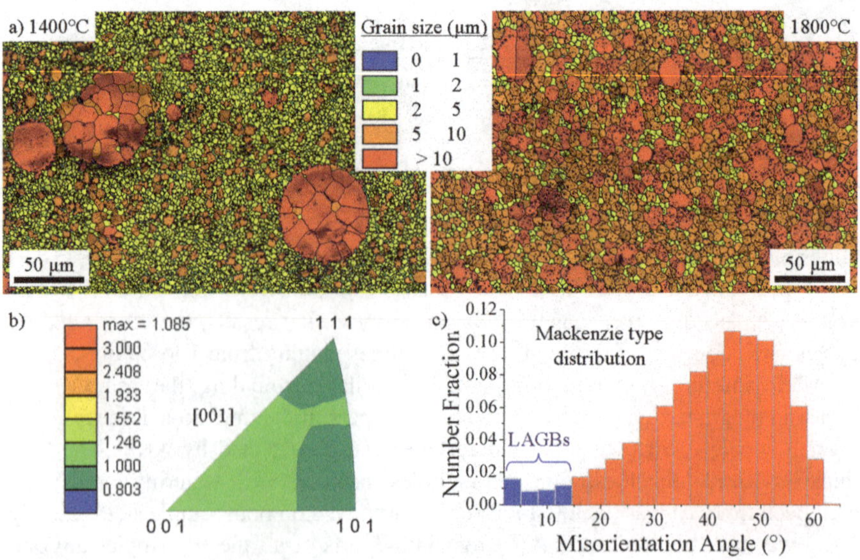

Fig. 5 (**a**) Grain size maps with HAGBs and image quality (IQ) for nW sintered at 1400 °C and 1800 °C, (**b**) typical triangular representation of IPF, and (**c**) grain boundary misorientation distribution for nW sintered at 1400 °C

agglomerates of micrometric grains with a diameter up to several tenths of micrometers are embedded in a fine-grained matrix (average grain size of about 2 μm). These agglomerates represent about 15% of the sample area, roughly the equivalent of the micrometric and ultrafine grains present in the starting nW. With the sintering temperature, these agglomerates tend to grow but not as fast as the fine grains of the matrix. Thus, with a sintering temperature of 1800 °C, the agglomerates are difficult to discern from the matrix, and the microstructure appears to be more homogeneous. As previously reported (Chaim et al. 2009), very fast grain growth can occur during SPS. After a sintering at 1400 °C and 1800 °C for 1 min, the grains are, respectively, about 40 and 60 times larger than the initial grain size of the nanopowder, and abnormal grain growth is also observed. The grain size distribution of the nW powder spreads over almost three orders of magnitude (10 nm – several microns). Since the growth and shrinkage of each grain are dictated by its curvature, the smaller grains tend to disappear to reduce the material's free energy. Thus, the micrometric and ultrafine grains tend to grow faster by consuming the nanoparticles that surround them. Besides, due to this very large grain size distribution, a current line concentration can also occur and act in favor of abnormal grain growth. These abnormal grain growth and heterogeneous microstructure seem to be a general feature of EEW nanopowders sintered by SPS since this behavior has already been reported for other metals like Ni and Zn (Dutel et al. 2012; Dirras et al. 2013). A typical misorientation angle distribution is shown on Fig. 5c. It displays a classical Mackenzie-type distribution, a perfectly random misorientation distribution. In addition, the LAGB fraction is always low, below 10%. So there are a few sub-boundaries and a low dislocation concentration within the grains defined by EBSD. A triangular representation of the inverse pole figure (IPF), displayed on Fig. 5b, also underscores a perfect random grain orientation. Whatever the starting powder, this random grain texture will be observed for all the W sintered samples, meaning that no development of preferential crystallographic orientation was induced by the applied pressure during SPS.

Starting with a narrow ufg size distribution, the uW leads to homogeneous microstructures whatever the sintering conditions, stressing the key factor of this parameter compared to the nW behavior. Figure 6 shows the average grain size according to relative density for different investigations of ufg W powders processed by SPS (for the sake of clarity, only some optimized results, with a density above 90%, were represented). Our results, in the lower part of Fig. 6, exhibit an interesting combination of grain size and density, especially for the sintering performed at 1400 °C. At this temperature with a dwell time of 1 min, the growth factor is close to 1, and an ufg microstructure is conserved with a mass density of more than 90%. With soaking time at 1400 °C, the mass density and the grain size slightly increase. Even after 60 min at 1400 °C, a fine microstructure still remains along with a relative mass density of more than 95%. Besides, sintering the uW powder at 1800 °C for only 1 min leads to a similar mass density as that obtained at 1400 °C for 30 min but with an average grain size 50% higher. The other investigations with a sintering temperature above 1400 °C always lead to a worse combination (above the dotted blue line on Fig. 6) of density and grain size whatever the other sintering parameters.

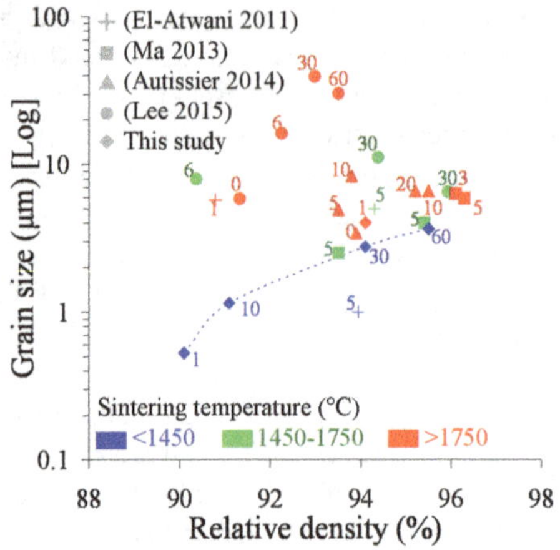

Fig. 6 Average grain size according to relative density. Numbers stand for dwell time in min (blue dot line joins the uW sintered at 1400 °C for different dwell times)

These results are quite in agreement with a recent investigation on the densification and grain growth kinetics of an ufg W processed by SPS under 40 MPa (Deng et al. 2017). The sintering process is reported to occur in two following steps: densification without or limited grain growth between 900 and 1450 °C and a grain growth without much increase in mass density for temperature above 1450 °C. It is interesting to note that this transition temperature corresponds to the secondary recrystallization temperature of W. The densification is otherwise reported to be mainly controlled by grain boundary diffusion and dislocation motion, while the grain growth stage is controlled by surface diffusion. On Fig. 6, the highest noticeable densities, about 96%, are obtained while using a two-step SPS cycle in order to avoid the pore-boundary separation (Ma et al. 2013). Besides, a better combination of grain size/density compared to our results (El-Atwani et al. 2011, point below the blue dot line on Fig. 6) is obtained when the ufg W powder is sintered at a temperature below 1450 °C but at a pressure of 200 MPa (instead of 100 MPa). The applied pressure is thus another key factor, and the improvement of graphite tools, since dies can now be used up to 200 MPa or higher with SiC punches, will allow to tend toward full dense W. However, a full pure dense W (≈99%) with ufg structure (0.2–0.4 µm) has already been processed (Yao et al. 2011) with a complex combination of two elaboration techniques: SPS at a temperature between 1000 at 1400 °C of a W nanopowder (50 nm) followed by resistance sintering under ultrahigh pressure treatment (9 GPa).

3.1.3 Mechanical Properties

Mechanical tests in uniaxial compression were performed at room temperature and at a strain rate of the order of 10^{-3} s^{-1} (Fig. 7). The three samples exhibit high yield strength and some ductile behavior which might arise from the right combination of grain size/density but also to the cleaning effect of SPS that reduces the oxygen

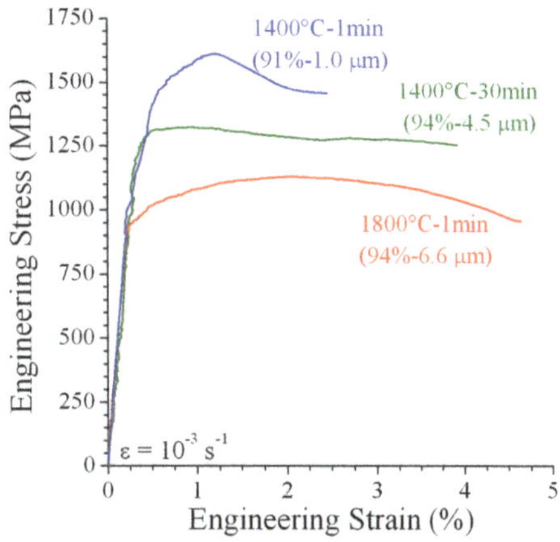

Fig. 7 Compression tests at room temperature of uW elaborated under different SPS conditions

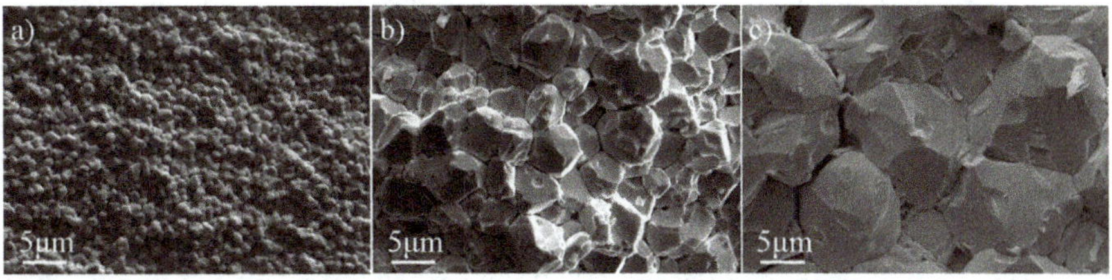

Fig. 8 Fracture surfaces of uW sintered at 1400 °C for 1 min (**a**), 30 min (**b**) and at 1800 °C for 1 min (**c**)

content and thus strengthens the grain boundaries. For the two materials consolidated at 1400 °C, the elastic limit values are 1470 MPa and 1240 MPa, while hardening corresponds to 140 and 80 MPa for a dwell time of 1 and 30 min, respectively. In addition, for these two materials, the extent of the plastic deformation is 1.5% and 3.5%, respectively. For the material sintered at 1800 °C for 1 min dwell time, the yield strength is 915 MPa. 215 MPa, 2.6%, and 4% are obtained, for work-hardening, the deformation interval to achieve this hardening, and extent of the plastic deformation, respectively. Due to the average grain size (area) about 7.5 times higher than for the material sintered at 1400 °C for 1 min, the yield strength decreases of about 38%, but, on the contrary, a work-hardening increase of more than 54% is found, within a deformation interval of 2.6%, probably stemming from a greater structural defect storage capacity during the deformation, which also results in a plastic domain twice as high, of the order of 4%.

For the three samples, the main cracks propagate parallel to the compression direction, and then perpendicular secondary cracks also develop starting from these main cracks. SEM fracture surface observations (Fig. 8) clearly evidence intergranular fractures between polyhedron grains. Pores are clearly visible on the surface of these grains with a diameter depending on the sintering parameters: one micrometer or less for the uW-1400 °C-30 min sample and several microns for uW-1800 °C-1 min.

3.2 SPS of a W Nanopowder Elaborated by SHS

During SHS, the reaction of a powder mixture is initiated by a local supply of thermal energy. The energy released by the reaction is then used to heat the surrounding layers of powder mixture, so that the reaction propagates as a combustion front with a high velocity (up to 25 cm/s). The temperature reached during this process is very high (up to 5000 °C), close to the adiabatic reaction temperature. SHS is mainly used for extracting metals from ore or for welding, but it can also be used for producing nanopowders with controlled grain size. This processing method is also suitable for producing large nanopowder quantities (tens of grams per synthesis) (Dine et al. 2017). W nanopowder was elaborated by the reduction of tungsten trioxide by magnesium (Mg) in the presence of NaCl, a reaction moderator:

$$WO_3 + (3 + x)\,Mg + y\,NaCl \rightarrow W + 3\,MgO + x\,Mg + y\,NaCl$$

Mg is in excess in order to promote the complete tungsten oxide reduction, while NaCl reduces the overall temperature (1800 °C) and is also known to act in favor of the production of nanoparticles. The resulting mixture is lixiviated in HCl, dried and filtered in order to only keep the W powder. Details of the overall process can be found elsewhere (Dine et al. 2015, 2017).

The shs-W powder has a platelet shape with an average grain size of 50–60 nm. XRD (Fig. 9a) reveals that this powder mainly comprises the α-W phase, but traces of tungsten oxides remain. These oxides might be removed by a thermal treatment

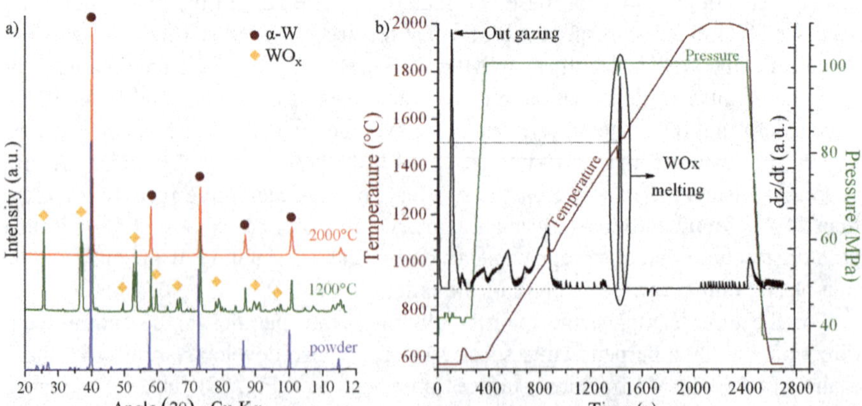

Fig. 9 (**a**) XRD of shs-W powder and after SPS at 1200 °C and 2000 °C; (**b**) SPS cycle at 2000 °C and densification rate of shs-W. WO_x melting at 1500 °C induces fast shrinkage

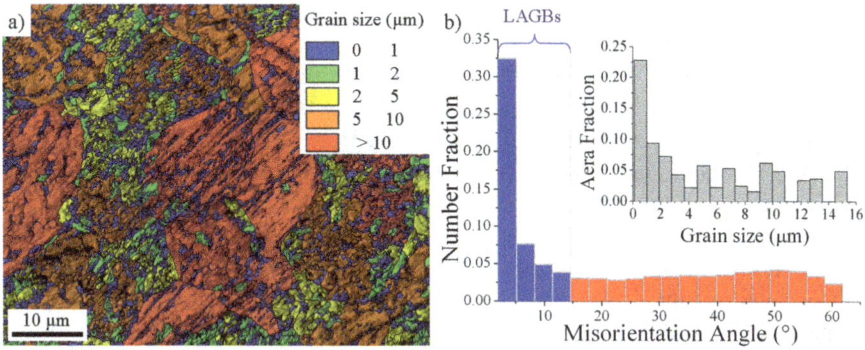

Fig. 10 (**a**) Grain size map with HABGs and IQ; (**b**) Grain boundary misorientation distribution and grain size distribution from EBSD of shs-W sintered at 2000 °C

under hydrogen atmosphere but in order to restrict the processing steps, this nanopowder was directly sintered by SPS using similar experimental parameters as those described in Sect. 2, except for the heating rate which is 50 °C/min instead of 100 °C/min (Dine et al. 2018). An SPS cycle and the derivative of the pistons' displacement are depicted on Fig. 9b. Compared to Fig. 4, a strong peak appears below 600 °C which might be associated with the release of residual species due to incomplete drying. Besides, the sintering starts just after the cold compaction at 600 °C where the nW, which has a similar grain size compared to the shs-W, only starts to release porosity at about 900 °C, showing a higher sintering reactivity. However, the main difference is a very intense densification peak at about 1500 °C, beyond which there is nearly no more piston displacement. To explain this phenomenon, an additional sintering was conducted before this densification peak at 1200 °C. As revealed by XRD (Fig. 10a), the obtained sample contains a large fraction of tungsten oxides which appears due to in situ recrystallization of amorphous tungsten oxides. These amorphous oxides have formed during the nanopowder elaboration process due to very fast cooling during SHS. This sample is also very porous (≈30%) because the oxides restrict the diffusion processes and thus the shrinkage. Since the melting temperatures of WO_3 and WO_2 are 1473 °C and 1700 °C, the formation of a liquid phase that induces very fast densification probably explains the intense densification peak at 1500 °C. In addition, at 2000 °C, no more oxides are detected by XRD (Fig. 9a), meaning that liquid oxides decompose in contact with carbon. This sample exhibits a mass density about 99%. However, due to the large amount of liquid phase, a large part of the sample is expelled from the mold, and only a thin sample is finally obtained. Its microstructure (Fig. 10) consists of micrometric and ultrafine grains leading to a multimodal grain size distribution with average grain sizes of about 0.35 μm and close to 7 μm, in number and in area, respectively. A very high LAGB fraction (close to 50%) subdivides these grains defined by EBSD in smaller crystallographic homogeneous domains. Therefore, nanoscale crystallite size of about 70 nm is measured by Rietveld method refinement on XRD pattern. Thus, near full dense W with an ufg microstructure can be obtained starting with nanopowder produced by SHS and sintered by SPS, using the melting of remaining tungsten oxides, the cleaning effect, and fast shrinkage of SPS. According to the densification rate (Fig. 9b), this near full density can probably be reached at a lower sintering temperature, down to 1600 °C which may still reduce the grain/crystallite size. However, a better control of oxide content in the SHS nanopowder is needed to process large W sample parts through the adjustments of the SHS parameters or by adding a thermal treatment in reducing atmosphere prior to sintering. Another way to reduce the oxide fraction is to elaborate nanopowder of W alloys with corrosion protective elements. In fact, a recent investigation on W alloys combining SHS and SPS processes already gives some promising results, the amount of oxides being sufficiently restricted by the alloying elements to get dense W alloy without matter flow outside the graphite die (D. Vrel, private communication).

3.3 *Ni-Activated SPS of W*

Instead of reducing the grain size, another classical metallurgical way to improve the mechanical performance of W is via alloying. Rhenium is an element known to reduce the DBTT (Lassner and Schubert 1998; Mutoh et al. 1995) that is partially explained by oxygen removal from grain boundaries and dislocations. Among the other rare earth elements used as alloying elements, Yttria allows to fabricate fully dense W resulting in a fine-grained microstructure (Zhao et al. 2015). However, since rare earth elements are very expensive to be considered for the production of large W alloy parts, many other refractory alloying elements have been investigated (Cr, Mo, Ta, V, etc.) to keep a high density. Low contents of Cr, Ta, or V ($\leq$10 wt%) tend to decrease the W grain growth and lead to dense samples (>96%) with a fine-grained microstructure and improved mechanical properties (Arshad et al. 2015; Zhou et al. 2014; Wang et al. 2017). The influence of addition of dispersed ceramics such as Y_2O_3, ZrN, or TiC has also been investigated (Autissier et al. 2014; Ding et al. 2015; Lee et al. 2012; Liu et al. 2016). In all cases, these particles limit the grain growth and strengthen the grain boundaries, inducing higher ductility and machinability.

Ni has been also used in a small amount to activate W sintering. In general, Ni salts are added to W powders in liquid media in order to cover the grains with a thin Ni layer prior to drying and sintering (Hayden and Brophy 1963; Li et al. 2017). It was shown that the addition of 1 at% of Ni increases the density from about 70% to 89% during a natural sintering at 1400 °C for 2 h (Gupta et al. 2007). However, few attempts were made starting from a Ni/W powder mixture (Genç et al. 2010). In this section, the effect of Ni alloying and its subsequent effects onto the SPS behavior and on the post-processed microstructural characteristics and corresponding mechanical properties in compression of W will be presented and discussed.

Table 2 Sintering conditions, relative density (d_{th} = 19.092), average grain size, LAGB fraction, and Vickers hardness of Ni-96W samples

$T_{sintering}$ (°C)	Dwell time (min)	$d_{relative}$ (%)	Grain size (μm)		Growth factor	LAGB (%)	Hv2 (GPa)
			Number	Area			
1400	1	97.5	3.2	6.1	6.0	<10	4.0 ± 0.3
	30	98.3	3.5	6.5	6.5		4.0 ± 0.2
1800	1	99.1	18.6	30.9	39		3.8 ± 0.1

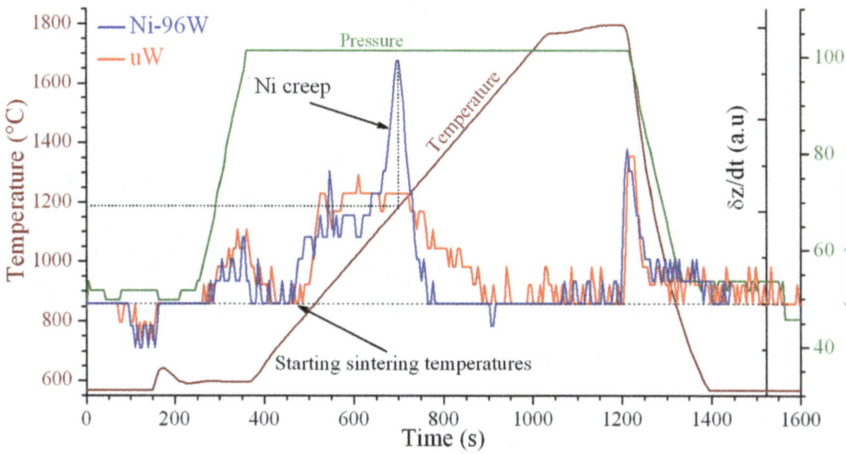

Fig. 11 SPS cycle at 1800 °C-1 min and densification rate of Ni-96W compared to uW

3.3.1 Sintering Conditions

Only 4.2% of weight (2% in volume) of fine Ni powder was mixed with the ufg W powder (uW) by using a Turbula mixer (15 h, 96 cycles/min) in order to tend toward a dense W alloy with high density. The blend (named thereafter Ni-96W, 96 wt% of uW) was sintered by SPS using the same conditions as those used for the uW. The choice of this mixing method as well as SPS as the sintering process will be discussed in Sect. 4. All the sintering conditions are gathered in Table 2. The densification rate is similar for all the sintering. The case of the sintering performed at 1800 °C-1 min is displayed on Fig. 11. Compared to the behavior of pure uW powder, the sintering starts at a slightly lower temperature (≈780 against 800 °C) after compaction, but the main difference is observed at about 1200 °C where an intense densification peak is observed. This temperature is well below the melting temperature of Ni (1456 °C), and this fast densification is probably due to Ni plastic flow. After this peak, even if Ni or Ni(W) solid solution has to melt according to the Ni-W phase diagram (Massalski 1990) at a higher temperature, nearly no densification occurs, meaning that the optimized sintering temperature of this powder blend is about 1300 °C.

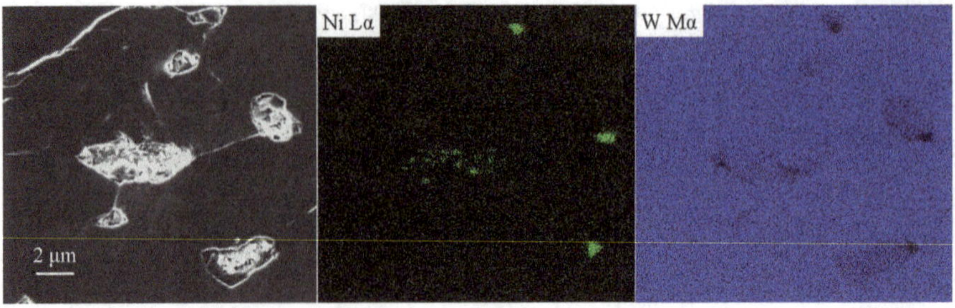

Fig. 12 SEM/EDS analyses of Ni-96W sample sintered at 1800 °C

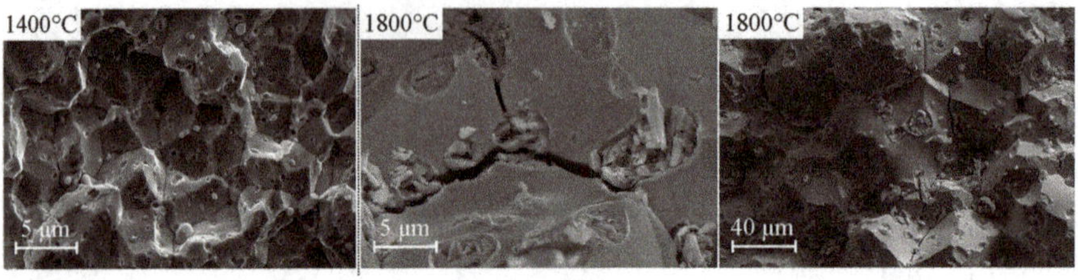

Fig. 13 Fracture surfaces of Ni-96W samples sintered at 1400 and 1800 °C

3.3.2 Microstructural Characterizations

In all the sintered samples, the α-W structure is the major phase, and no trace of fcc Ni structure is detected by XRD. In addition, for the samples sintered at 1400 °C, some peaks with very low intensities appear and might be due to traces of complex (Ni,W) oxides. These oxides are no longer visible after a sintering at 1800 °C, and, as observed for tungsten oxide during the SPS of shs-W (Sect. 3.2), these oxides have probably been removed due to their melting and the cleaning effect of SPS. However, for all the samples, bright areas, identified as Ni-rich zones by EDS (Fig. 12), are clearly visible on fracture surfaces for all the Ni-96W samples (Fig. 13). These zones might be Ni(W) since this solid solution exists under a wide range of composition as it can be seen on the Ni-W phase diagram (Massalski 1990). In addition, the plastic strain of the Ni-rich phase associated with a fast cooling during SPS can lead to an amorphous or poorly crystallized phase that could explain why this fcc solid solution is not detected by XRD. Complementary investigations are underway to fully determine the chemical/structural nature of these bright areas.

In all cases, for density measurements, a rule of mixtures of the starting phases was used to define the theoretical density (19.092) which gives some uncertainty on the computed relative density. However, even if the theoretical density of pure W (19.30) was taken into account, the same conclusions on the effect of Ni addition will be drawn. As it can be seen in Table 2 (to compare with Table 1), the addition of Ni clearly promotes the shrinkage, and a near full dense sample is obtained after a sintering at 1800 °C-1 min. In addition, homogeneous microstructures with normal

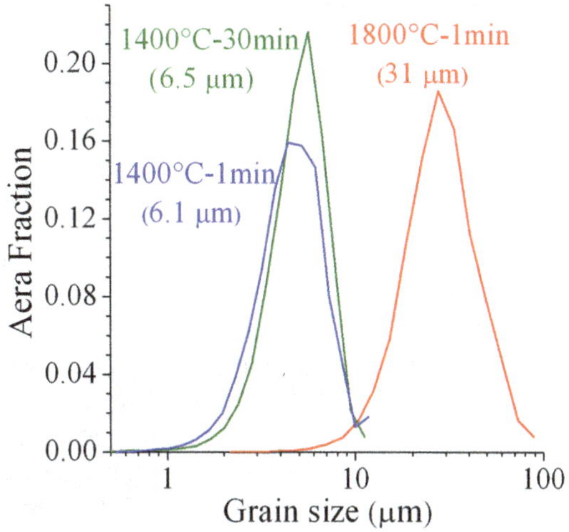

Fig. 14 Grain size distributions in area for Ni-96W samples from EBSD analyses

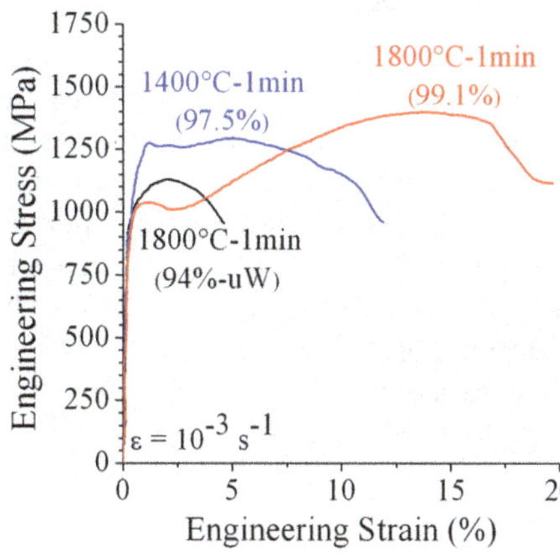

Fig. 15 Compression tests at room temperature of Ni-96W samples. uW (Sect. 3.1) is added for the sake of comparison

grain size distributions are preserved but shifted toward higher values (Fig. 14). The W grain size is thus larger in presence of Ni than that observed for pure uW sintered in the same conditions. As for pure uW, a longer dwell time leads to a slight increase in density and grain size, while increasing the sintering temperature induces a moderate shrinkage but a very fast grain growth. These different effects of Ni on sintering behavior of W might be explained by plastic flow of the Ni-rich phase which leads to grain rearrangement under pressure and fills the residual porosity observed in pure uW (as illustrated by comparison of the fracture surfaces presented on Figs. 8 and 13). Besides, the grain boundary diffusion in W is known to be faster in the presence of Ni (Kaysser et al. 1987) leading to a lower sintering temperature and a higher grain growth rate. As for the pure uW, EBSD analyses show that for all the Ni-96W samples, there is no preferential orientation and a typical Mackenzie misorientation distribution arises with a low LABG fractions (<10%).

3.3.3 Mechanical Properties

Figure 15 shows the mechanical behavior in terms of engineering stress-strain plots of the SPS processed samples. The influence of the sintering conditions is clearly evidenced and can be summarized as follows.

The sample sintered at 1400 °C has higher yield strength of the order of 1000 MPa, against 800 MPa for the sintered sample at 1800 °C. A yield point of 1275 MPa is followed by a work-hardening of about 275 MPa after a deformation increase of 0.8% for the lowest temperature. Afterward, a stress decrease is observed (15 MPa) followed immediately by a slight hardening resuming of the order of 35 MPa. Plastic deformation amount is about 6%. Comparatively, in the case of the material sintered at 1800 °C, considerable changes are observed. The initial hardening is of the same order of magnitude, about 240 MPa, as well as the deformation range (0.8%). As before, a yield point is observed. The ensuing softening is of the order of 30 MPa. However, a high work-hardening then follows, which brings about a hardening of nearly 390 MPa, over a deformation range of nearly 6%. Plastic deformation amounts to about 18%.

As for uW (Sect. 3.1), the main cracks propagate in a direction parallel to the compression load and intergranular fractures of polyhedron grains, on which ones Ni-rich zones are clearly visible, remain (Fig. 13). But many slip bands and local cracks between W grains mostly due to strain incompatibilities are observed by SEM for the Ni-96W samples and evidence intense plastic strain of W grains. In order to insist on the strong effect of Ni addition on ductile behavior, the compression curve of uW sintering at 1800 °C for 1 min is added on Fig. 15. Since W brittleness is due to grain boundary weakness, Ni-rich phase may remove oxygen, a weakening element, which would explain the (Ni,W) oxide formation detected by XRD at 1400 °C and the probable strengthening of the grain boundary, thus allowing the strain to be transferred to the W grains.

4 Ni-W Alloys Processed by SPS

As predicted by the seminal Hall-Petch law, reducing the grain size leads to higher mechanical properties (Hall 1951; Petch 1953). However, this improvement is generally made at the expense of ductile properties. Several alternatives have been proposed in order to take advantage of this effect including nanotwinning (Lu et al. 2009) or the elaboration of bi- or multimodal grain size microstructure. In this latter case, the small grains are supposed to improve the mechanical resistance, while the larger grains preserve ductility. The usual way to elaborate such microstructures is via thermomechanical treatment. To get a better control of the respective fraction of each component of the microstructure (grain size fractions), powder metallurgy routes can be used, starting with multimodal powder (Dutel et al. 2012; Tingaud et al. 2015) or a blend of nano- and micro-grained size powders (Dirras et al. 2014).

A key parameter to preserve important ductility is the homogeneous repartition of the fine and large grains throughout the entire sample in order to delay the plastic strain localization. For instance, cold isostatic pressing (CIP) prior to SPS of a Ni powder blend has been shown to promote this homogeneity (Dutel et al. 2017).

In order to go one step further, in our research group, the concept of bimodal structure has been, for the first time, extended to a multiphase material processed by combination of powders based on soft micrometer-sized polycrystalline Ni and hard ufg W.

4.1 Sintering Conditions

In order to elaborate the Ni-W alloys, the Ni and uW powders described in Sect. 2 have been blended in a Turbula mixer (15 h, 96 cycles/min) in various proportions, from 0 to 65% in weight (Ni-xW). Generally, powder blending was done using ball milling process in order to insure homogeneous mixture but also to reduce the grain size and to store energy within the particles by plastic strain. Due to a reduced volume to surface ratio and a high defect concentration, the powder reactivity and the diffusion processes increase, resulting in faster sintering at lower temperatures. However, it also leads to contamination and promotes the diffusion of impurities. Thus, by carefully tracking carbide formation during SPS of Ni-W powder blends, it was possible to show that ball milling prior to SPS promotes the formation of WC and Ni_2W_4C, not only on the surface but throughout the sample due to fast C diffusion in the strained Ni phase (Dudina et al. 2016). On the contrary, by using a horizontal mixer, these carbides only appear at the sample surfaces as observed for the uW (Sect. 3.1). Besides, increasing the stored energy in the powder blend will promote the formation of brittle Ni-W intermetallic phases that must be avoided to preserve ductility. For instance, Ni-40W mixture of fine-grained (≈1–2 μm) powders has been reported being thermally stable up to 1400 °C, while the same blend after ball milling turns into an intermetallic mixture beyond an annealing temperature of 730 °C (Maweja et al. 2012). Soft blending process seems thus a wise choice as well as SPS compared to the other sintering techniques since SPS can be fast enough to prevent the formation of these undesirable intermetallics. In the W-Cr system, it has been shown that after ball milling, the SPS is fast enough to preserve the W-Cr solid solution where HIP leads to the formation of Cr-rich phase (Vilémová et al. 2018). Besides, for the sake of comparison, a Ni-50W mixture was sintered by HP in similar conditions as those used for SPS except for the heating rate (15 °C/min instead of 100°/min). This sample exhibits a complex intermetallic mixture and a tensile brittle behavior where no intermetallic and ductile behavior are obtained for the sample blend processed by SPS (see next sections), stressing that the intermetallic formation is due to kinetic effects.

The Rietveld refinement method on XRD patterns was used to make sure that the crystallite size and the strain are similar to that of the powder blends before and after the mixing. Different sintering cycles were used, but the optimized sintering

Table 3 SPS of Ni and Ni-W alloys performed at 1000 °C for 1 min, except for Ni-65W sintered at 1000 °C for 30 min: relative density, average grain size in area, atomic fraction of W in Ni(W) solid solution, area fraction of pure W dispersed in Ni(W) matrix, Σ3 boundary fraction in the fcc phase, and Vickers hardness

Sample	$d_{relative}$ (%)	Grain size (μm)		W (at.%) in Ni(W)	α-W (area %)	Σ3 (%)	Hv0.1 (GPa)
		Ni(W)	W				
Ni	99.2	61.0	–	–	–	52.0	–
Ni-10W	99.8	8.4	–	2.0	–	52.4	1.9 ± 0.1
Ni-30W	99.5	5.4	0.85	12.2	2	48.0	2.7 ± 0.1
Ni-50W	98.9	3.5	0.83	16.5	17	30.0	4.1 ± 0.3
Ni-65W	96.7	3.9	0.80	18.6	20	35.8	4.9 ± 0.7

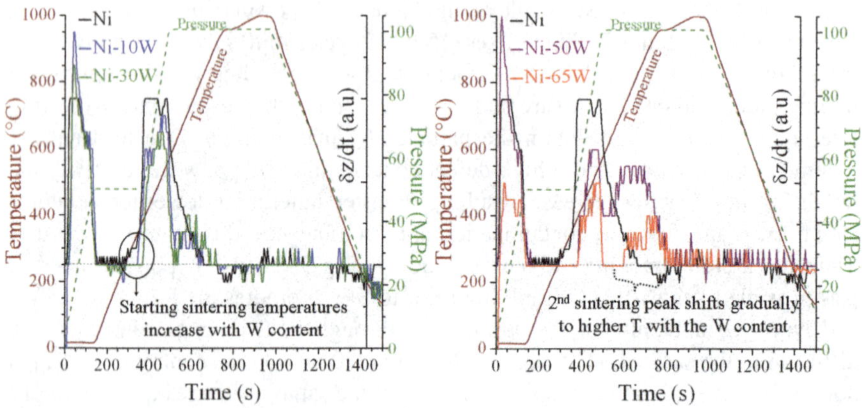

Fig. 16 SPS cycle at 1000 °C and densification rates for the Ni-xW alloys ($x = 0$–65 wt%)

conditions were obtained at 1000 °C for 1 min except for pure nickel and for the mixture with the higher W content (Table 3). For pure Ni, an intense densification peak occurs at around 500 °C which also corresponds to the moment when pressure increases up to 100 MPa (Fig. 16). Beyond this peak, above 800–850 °C, no more shrinkage occurred, and the pure Ni was in fact already fully dense at 850 °C-1 min (Dutel et al. 2013). When the fraction of W increases in the initial powder blend, the sintering temperature gradually increases, and a second densification peak appears at a higher temperature, meaning that the blends need more and more energy to shrink. The maximum of this second peak as well as its surface gradually increase with the W content, so that the mixture with the lower W content (Ni-10W) behaves almost like pure Ni, while for the higher W content (Ni-65W), a second peak maximum is only obtained at 1000 °C, and therefore the full densification is not achieved for this sample (≈88.3%). Instead of increasing the sintering temperature that can promote grain growth, a dwell time of 30 min at 1000 °C was considered for this sample in an attempt to reduce the residual porosity.

4.2 Microstructural Characterizations

As for pure W sample, carbides (WC and mainly Ni_6W_6C) were formed during SPS of Ni-W alloys but only at the sample surface and can be totally removed by a sufficient grinding/polishing. Synchrotron experiments performed on tensile specimens of the different Ni/W alloys, machined in the center of the SPS disks, thus reveal no more trace of these carbides (Sadat 2015). In fact, the carbide formation is very difficult to completely avoid (Marvel et al. 2015) even if metal foils are used to separate the powder blend from the graphite die. It has been shown during SPS of transparent ceramics which makes it easy to observe the contamination (Morita et al. 2015) that carbon can vaporize under high current density. To reduce this contamination, current density can be lowered by reducing the heating rate. However, by doing so, SPS would tend to be similar to HP and fast sintering, and many other interesting properties specific to SPS would be lost. In addition, the Ni-W system is very sensitive to carbon contamination. The Ni-W-C phase diagram shows an equilibrium state between the fcc Ni, cc W, and Ni_6W_6C phase at 1300 °C, and the low solubility of C in Ni and W involves that carbides are prone to precipitate at their interfaces even with a very low C content (Cury et al. 2009).

XRD patterns of all the Ni-W alloys are shown on Fig. 17. As expected, SPS is fast enough to completely avoid the formation of brittle Ni/W intermetallic phases but does not totally prevent W diffusion in the Ni phase. Indeed, all the diffraction peaks of Ni gradually moved to lower diffraction angle according to the starting W content, meaning an increase of the lattice parameter of the fcc phase. This evolution is due to the formation of Ni(W) solid solution which progressively enriches in W. Assuming that the peak shifts are only due to W substitution in the Ni fcc lattice,

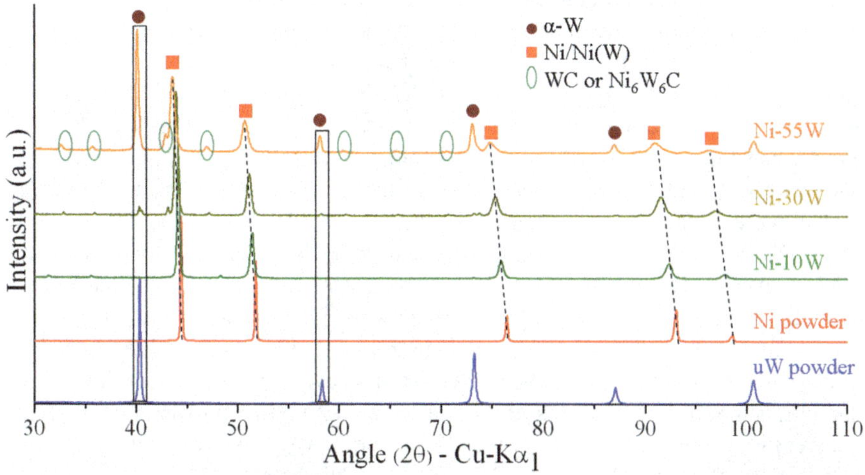

Fig. 17 XRD of Ni-xW alloys ($x = 10$–55 wt%) and starting powders. Dotted lines point out the shift of the diffraction peaks of fcc phase

the average composition of the Ni(W) solid solution can be estimated using Vegard's law (Table 3). A linear correlation was found between the amount of W that diffuses into the Ni structure and the starting fraction of W in the powder blend, in line with a previous study (Genç et al. 2012a). The composition of the Ni(W) solid solutions of Ni-50W and Ni-65W samples is close to the maximum solubility of W in fcc Ni (≈40 wt%) foreseen by the phase diagram, meaning that the Ni(W) solid solution is close to saturation. Conversely, pure α-W phase remains since the diffraction peaks of W are always found at the same diffraction angles. The fraction of the α-W phase is estimated to be about 2, 17, and 20% in area for the Ni-30W, Ni-50W, and Ni-65W samples, respectively. For, the Ni-10W sample, W completely disappears, and this sample is thus only made of pure Ni(W) solid solution. To summarize these results, confirmed by EDS analyses, all the Ni-W samples consist of a fcc Ni (W) solid solution and a fraction of pure α-W. The W content of the Ni(W) solid solution and the fraction of the remaining W phase increase with the fraction of W powder in the starting powder mixture.

The relative density of the Ni-W samples was estimated assuming a rule of mixtures of pure Ni and W phases. Taking into account the Ni(W) formation will only slightly modify the results since the theoretical density variation is estimated to be less than 1%. All the Ni-W alloys are close to full density (>99%) except for the richer W sample (Ni-65W) for which a longer dwell time helps but do not totally suppress the porosity. Besides, the measured density evolves linearly with the initial W amount (Sadat et al. 2016a). Compared to previous studies (Genç et al. 2012a, b) on the sintering of Ni-W blends, full densification is achieved here for a large range of Ni-W composition because of the good dispersion of the ufg W particles within the blend which allows the Ni to embed these particles by plastic flow during SPS. Besides, the sintering of Ni-W blends cannot be done at high temperature due to Ni creep which occurs at about 1200 °C under 100 MPa for our blend (see Fig. 11) which is a low sintering temperature for W. Thus, the chosen W powder must also start to sinter at low temperature as the nano- or ultrafine-grained powders used in Sect. 3.1 (Fig. 4) in order to restrict porosity between the W particles. Some attempts (not shown here) were done to sinter in similar SPS conditions a blend of the same Ni powder but with the nW powder. Nanoparticles remain very difficult to disperse, even when the blends were done in ethanol under ultrasounds, and these Ni-W samples always exhibit a low density due to important porosities close to W clusters.

No preferential grain orientation is revealed by EBSD in the Ni-W alloys (Fig. 18). However, with a more statistical analysis of grain orientation using XRD, a partial <110> fiber texture was found in Ni and Ni-30W samples with a volume fraction of 11.9 and 13.69%, respectively (Sadat et al. 2016b). For the W phase, random grain boundary misorientation distributions are obtained as for the pure W samples with a low LAGB fraction. The average W grain size for all the samples is the same and close to the starting grain size of the powder, about 0.8 μm (in area) since the grain growth of the uW powder is very low even with a sintering temperature of 1400 °C (see Table 1). For pure Ni sample, the sample is already dense (99.0%) after a sintering at 850 °C for 1 min and exhibits an average grain size of about 17.1 μm. Increasing the temperature from 850 to 1000 °C leads to nearly no

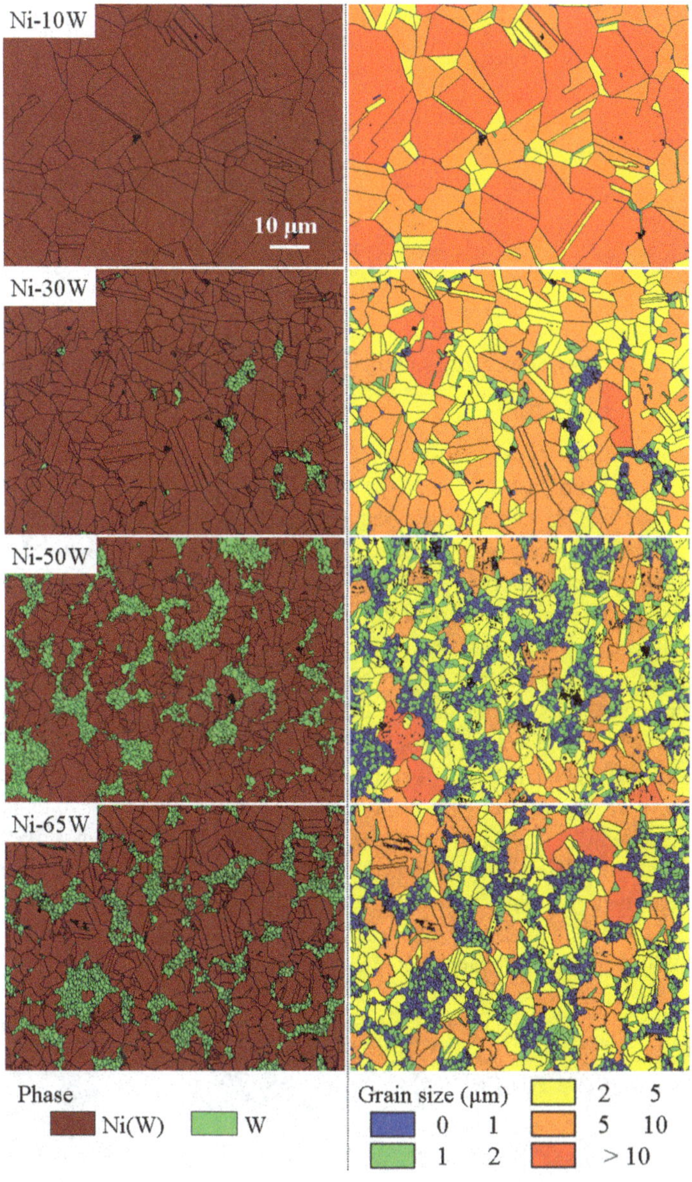

Fig. 18 Phase maps (left) and grain size maps (right) of Ni-xW alloys from EBSD (the same scale stands for all the maps)

more densification (increase of 0.2%) and very fast grain growth, leading to very large grains of Ni of about 61 μm. In the Ni-W samples, the presence of ultrafine W particles surrounding the agglomerates of Ni in the starting blend strongly reduces the Ni grain growth. In fact, the Ni grain size tends to be close to the starting

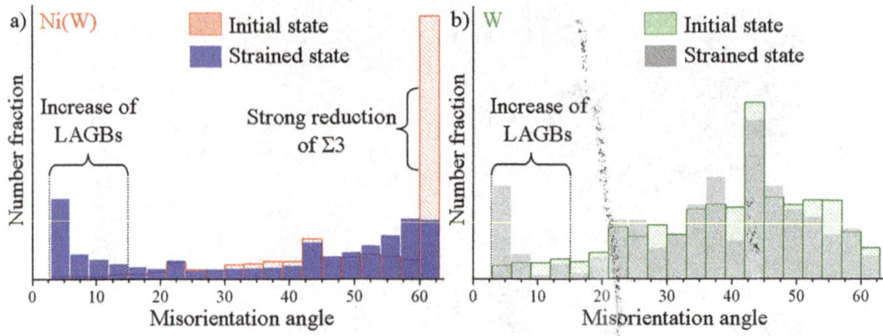

Fig. 19 Grain boundary misorientation distributions of Ni(W) (a) and W (b) phases of Ni-30W alloy in the as-processed state and after tensile test

agglomerate size (≈4–5 μm), meaning that the grain growth is restricted within these agglomerates. In the case of the Ni-10W sample, the size of the Ni(W) phase is slightly higher (≈8 μm) because as soon as the W is fully dissolved in the Ni phase, the Ni(W) grain is no longer blocked. However, the grain growth of Ni(W) is probably lower than that observed in Ni due to the fact that Ni diffusion is slowed down by the W atoms in substitution positions. A typical grain boundary misorientation distribution obtained for the Ni-rich phase in the as-processed state is displayed on Fig. 19a which evidences a low LAGB fraction (<9%) and a very high Σ3 boundary fraction that essentially includes twins. The stacking fault energy is low for Ni which induces the formation of thermal twins. It was reported (Tiearney and Grant 1982) that the stacking fault energy decreases in Ni(W) solid solution with the amount of W with a yielding point at about 12 at% which is close to the measured value for the Ni-30W alloy. Above this composition, the Σ3 fraction thus decreases. The formation of Σ3 is also promoted by higher sintering temperature or longer dwell time, which explains, respectively, why more twins are observed for pure Ni sintered at 1000 °C instead of 850 °C (52 and 41%, respectively) and why the Ni-65W exhibits more Σ3 than Ni-50W.

4.3 Mechanical Properties

The Vickers hardness (Hv) of the Ni-W alloys is displayed in Table 3. The comparison of the Ni-W alloy hardnesses is done with the pure Ni sintered at 850 °C in order to reduce the effect of grain size on Hv evolution. Hv gradually increases with the starting amount of W from 1.01 GPa for pure Ni sintering at 850 °C to 4.86 GPa for the Ni-65W alloy which is consistent with previous work (Genç et al. 2012a). In addition, the change in hardness induced by the amount of W in the Ni(W) solid solution was previously reported for nanocrystalline (NC) and coarse grain (CG) size (Rupert et al. 2011; Pelloux and Grant 1960). The strengthening of NC is reported to evolve linearly with the amount of W, while the CG Ni(W) fit with the power law of

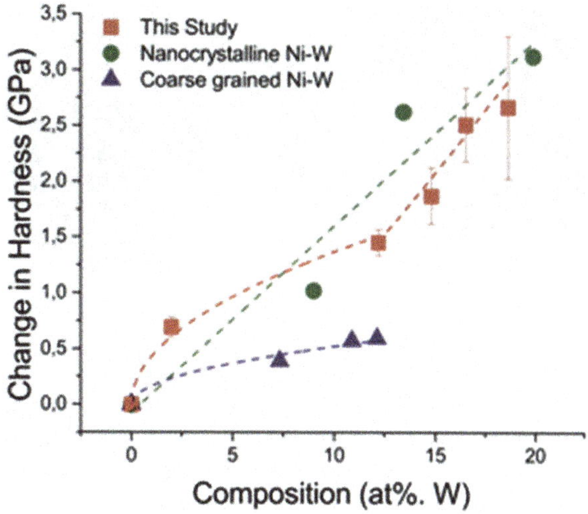

Fig. 20 Evolution of the change in Hv as a function of W content for Ni(W). (Additional data taken from Rupert et al. 2011; Pelloux and Grant 1960. Adapted from Sadat et al. 2016b)

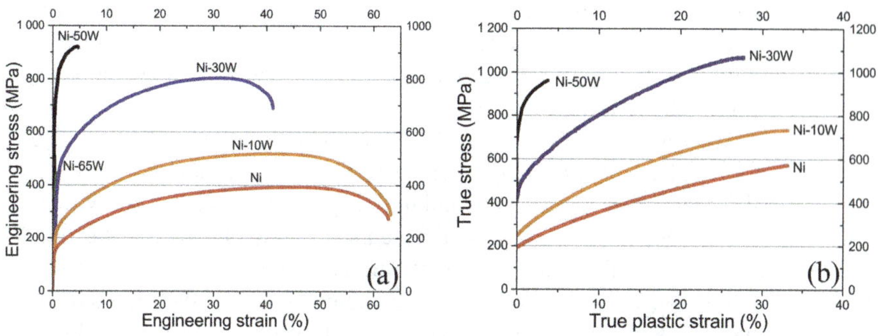

Fig. 21 Engineering stress-strain curves (left) and true stress-strain curves (right) of tensile tests at room temperature of Ni-W alloys. (Reproduced from Sadat et al. 2016b)

Fleischer (Fig. 20). The order of magnitude of the strengthening of our fine-grain Ni (W) solid solution is close to those obtained for the NC Ni(W). Besides, it seems to follow a Fleischer-like relationship up to 12.2 at% (corresponding to the Ni-30W alloy) above this point, a linear behavior is established. However, for alloys, the Hv evolution is not only due to W enrichment of the Ni(W) solid solution but more likely to a combined effect with the grain size reduction.

Tensile tests were performed on these Ni-W alloys at room temperature and at a strain rate of 5.6 10^{-4} s^{-1} (Fig. 21). Except for the Ni-65W sample that exhibits a brittle behavior, maybe due to a too high level of residual porosity, increasing the initial W amount leads to an enhanced mechanical strength. The Ni-10W exhibits a higher yield strength and hardening without loss of ductility compared to pure Ni because of the formation of a pure Ni(W) solid solution during sintering. The substitution of Ni by large W atoms in the fcc lattice slows down the dislocation motion and thus increases the mechanical properties. In addition to this solid solution

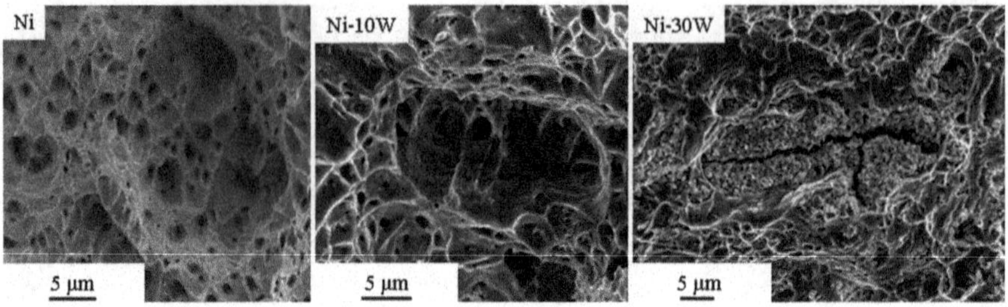

Fig. 22 Fracture surfaces obtained by SEM of Ni and Ni-W alloys after tensile tests

effect, the mechanical resistance of the Ni-30W and Ni-50W still increase due to the presence of hard ultrafine pure W grains dispersed within the Ni(W) matrix. The Ni-50W exhibits the higher engineering stress (≈940 MPa) with an elongation to failure about 5%, but the best combination of strength (≈805 MPa) and plastic strain (≈41% at failure) is clearly obtained for the Ni-30W alloy. Similar conclusions rise from the true stress versus true strain curve. The Ni-30W alloy possesses the largest strength and uniform strain (1070 MPa, 28%, respectively) as well an important strain hardening rate. Similar work-hardening rate was observed during compression tests (Sadat et al. 2016a).

Regarding the fracture surfaces (Fig. 22), profound dimples are clearly visible in the Ni or Ni(W) phase. These dimples are formed by growth and coalescence of porosity which then induce necking and failure. They demonstrate that Ni(W) acts as pure Ni phase and that intense plastic strain occurs within the fcc phase. Besides, W phase behaves like a brittle material with intergranular cracks that propagate within the W ufg cluster. However, these cracks are confined inside the W clusters because they are blunted by the surrounding Ni(W) solid solution. This behavior was previously reported in bimodal microstructures where the propagation of cracks initiated in the more brittle fine-grain component was blocked by the large grains (Billard et al. 2006; Dutel et al. 2012; Dirras et al. 2014). Because of the stress concentration at the cracks tips, large plastic strain occurs in the Ni(W) matrix and contributes to the strain hardening observed for the Ni-W alloys.

Postmortem SEM/EBSD investigations were made on the Ni-30W sample which exhibits the best mechanical properties. Surface sample observations reveal crack formation in W clusters, as observed in the fracture surfaces, and elongation of Ni (W) grains parallel to the tensile direction. Besides, the evolution of grain boundary misorientation distributions (Fig. 19) shows an important increase in the LAGB fraction within the W phase, meaning that some plastic strain occurs in the hard phase. Thus, the Ni(W)/W interface is strong enough to allow the strain to be transferred from one phase to the other. For the Ni(W) phase, a strong increase in LAGBs underscores an intense plastic strain. In addition, the Σ3s fall down to a very low value. A large fraction of these grain boundaries probably turns into lower misorientation angle owing to grain rotation during deformation. This behavior was reported previously in pure Ni elaborated by SPS (Dutel et al. 2012, 2013). The interaction between grain rotation mechanism and the disruption of semi-coherent twin boundaries with gliding dislocations may contribute to the overall hardening.

To sum up, the Ni(W) phase behaves like pure Ni and is not only able to heavily deform but benefits from a higher mechanical strength in relation with w solid solution effect. Besides, a strong interface between Ni(W)/W allows to transfer the strain from one component to the other and a composite-like effect due to the presence of W ufg dispersed in the Ni(W) matrix contributes to strongly increase the mechanical resistance, leading to high overall mechanical properties.

5 Conclusions

Spark plasma sintering, due to its specific properties such as fast sintering, cleaning effects, etc., allows to fabricate innovative Ni-W alloys with enhanced mechanical properties.

- Regarding the pure W material, SPS allows to elaborate a fine-grained microstructure (several μm) having relative density above 94% which leads to high mechanical strength over 1 GPa, along with non-negligible ductility in compression, 2–4% depending on the composition. The results show that a powder having average grain sizes within ultrafine or nanocrystalline domain is required to ensure a low sintering temperature and restrict the grain growth that mainly occurs above 1400 °C. A tight grain size distribution is also necessary to obtain a homogeneous microstructure. Besides, it is shown that a fraction of tungsten oxides in the starting W powder can lead to a very fast shrinkage due to the melting of the oxides at about 1500 °C. These oxides are then totally removed by cleaning SPS effect, leading to a promising fully dense (≈99%) pure W material.
- Investigations on the effect of small addition of Ni (4 wt%, Ni-96W) in the starting ufg W powder lead to dense W samples (97–99%) with homogeneous microstructures. The presence of Ni-rich phase at the W grain boundaries strongly increases the ductile properties, and an elongation to failure up to 18% has been obtained which is 4–5 times the value obtained for pure W elaborated in the same conditions. Further analyses are however required to fully comprehend the mechanical behavior of this alloy.
- For the Ni/W alloys, compared to other sintering techniques, SPS is sufficiently fast to avoid the formation of brittle intermetallic phases but does not completely prevent the W diffusion in the Ni fcc phase. Thus, the dense (>99%) Ni-W alloys obtained in a large range of composition (10–55 wt% of W) are made of pure ufg W grains embedded in Ni(W) solid solution. The fraction of α-W phase and the W ration in the Ni(W) increase with the starting ratio of W in the powder mixture. Through the combined actions of solid solutions and composite-like effect, high mechanical strength (1070 MPa) and uniform strain (≈28%) have been both attained for the optimized Ni-30W alloy.

- Considering SPS experiments conducted here, the remaining difficulty to get rid of is the formation of carbides. Investigations carried out here confirm that a low energy milling process is to be favored to restrict the C contamination. However, although it is beyond the scope of this work, it should also be noticed that the formation of fine, well-dispersed complex nickel/tungsten carbides such as Ni_6W_6C might be used to enhance mechanical properties with moderate loss of ductility by adjusting the processing parameters.

Acknowledgments T. Sadat is grateful to École Doctorale Galilée and SREI of University Paris 13 (Grant n°: SMV/JRM/749-2014) for their financial support.

The authors are grateful to Mr. B. Villeroy for the technical support for SPS, to Dr. Housaer François for the sintering of Ni-W alloys by HP (UMET, CNRS UMR 8207 – Université de Lille), to Ms. Sarah Dine for the elaboration of shs-W, and also to Dr. P. Langlois for fruitful discussion and manuscript improvement.

References

Allen BC, Maykuth DJ, Jaffee RI (1961) The recrystallization and ductile-brittle transition behaviour of tungsten-effect of impurities on polycrystals prepared from single crystals. J Inst Met 90:120–128

Arshad K, Guoa W, Wang J et al (2015) Influence of vanadium precursor powder size on microstructures and properties of W–V alloy. Int J Refract Met Hard Mater 50:59–64

Autissier E, Richou M, Minier L et al (2014) Spark plasma sintering of pure and doped tungsten as plasma facing material. Phys Scr 159:014034

Bewlay BP, Briant CL (1995) The formation and the role of potassium bubbles in NSdoped tungsten. In: Lassner LB, Schubert WD, Lux B (eds) The chemistry of non-sag tungsten. Pergamon, Oxford, pp 137–159

Billard S, Fondère JP, Bacroix B, Dirras G (2006) Macroscopic and microscopic aspects of the deformation and fracture mechanisms of ultrafine-grained aluminum processed by hot isostatic pressing. Acta Mater 54:411–421

Chaim R, Shlayer A, Estournes C (2009) Densification of nanocrystalline Y2O3 ceramic powder by spark plasma sintering. J Eur Ceram Soc 29:91–98

Cury R, Joubert JM, Tusseau-Nenez S, Leroy E, Allavena-Valette A (2009) On the existence and the crystal structure of Ni_4W, NiW and NiW_2 compounds. Intermetallics 17:174–178

Deng S, Yuan T, Li R et al (2017) Spark plasma sintering of pure tungsten powder: densification kinetics and grain growth. Powder Technol 310:264–271

Deng S, Li R, Yuan T et al (2018) Direct current-enhanced densification kinetics during spark plasma sintering of tungsten powder. Scr Mater 143:25–29

Dine S, Aïd S, Ouaras K et al (2015) Synthesis of tungsten nanopowders: comparison of milling, SHS, MASHS and milling-induced chemical processes. Adv Powder Technol 26:1300–1305

Dine S, Kentheswaran V, Vrel D, Couzinié JP, Dirras G (2017) Synthesis of nanometric MoNbW alloy using self-propagating high-temperature synthesis. Adv Powder Technol 28:1739–1744

Dine S, Bernard E, Herlin Boime N et al (2018) SHS synthesis and SPS densification of Nanometric tungsten. Adv Eng Mater 20:1701138

Ding XY, Luo LM, Chen HY et al (2015) Fabrication of W–1 wt.% TiC composites by spark plasma sintering. Fusion Eng Des 92:29–34

Dirras G, Gubicza J, Couque H et al (2013) Mechanical behaviour and underlying deformation mechanisms in coarse- and ultrafine-grained Zn over a wide range of strain rates. Mater Sci Eng A 564:273–283

Dirras G, Tingaud D, Csiszár G, Gubicza J, Couque H, Mompiou F (2014) Characterization of bulk bimodal polycrystalline nickel deformed by direct impact loadings. Mater Sci Eng 601:48–57

Dudina DV, Bokhonov BB, Ukhina AV et al (2016) Reactivity of materials towards carbon of graphite foil during Spark Plasma Sintering: a case study using Ni–W powders. Mater Lett 168:62–67

Dutel GD (2013) Comportement Mécanique et Mécanismes de Déformation et d'Endommagement de polycristaux de Nickel mono- et bi-modaux élaborés par SPS. Dissertation, Université Paris 13 (France)

Dutel GD, Tingaud D, Langlois P et al (2012) Nickel with multimodal grain size distribution achieved by SPS: microstructure and mechanical properties. J Mater Sci 47:7926–7931

Dutel GD, Langlois P, Tingaud D, Dirras G (2013) Room-temperature deformation micro-mechanisms of polycrystalline nickel processed by spark plasma sintering. Mater Charact 79:76–83

Dutel GD, Langlois P, Tingaud D, Vrel D, Dirras G (2017) Data on the influence of cold isostatic pre-compaction on mechanical properties of polycrystalline nickel sintered using Spark Plasma Sintering. Data Brief 11:61–67

El-Atwani O, Quach DV, Efe M, Cantwell PR et al (2011) Multimodal grain size distribution and high hardness in fine grained tungsten fabricated by spark plasma sintering. Mater Sci Eng A 528:5670–5677

Genç A, Coşkun S, Öveçoğlu ML (2010) Fabrication and properties of mechanically alloyed and Ni activated sintered W matrix composites reinforced with Y2O3 and TiB2 particles. Mater Charact 61:740–748

Genç A, Ayas E, Öveçoglu ML, Turan S (2012a) Fabrication of *in situ* Ni(W)–WC nano composites via mechanical alloying and spark plasma sintering. J Alloys Comp 542:97–104

Genç A, Öveçoglu ML, Baydogan M, Turan S (2012b) Fabrication and characterization of Ni–W solid solution alloys via mechanical alloying and pressureless sintering. Mater Des 42:495–509

Groza JR, Zavaliangos A (2000) Sintering activation by external electrical field. Mater Sci Eng A 287:171–177

Gupta VK, Yoon DH, Meyer HM III, Luo J (2007) Thin intergranular films and solid-state activated sintering in nickel-doped tungsten. Acta Mater 55:3131–3142

Hall EO (1951) The deformation and ageing of mild steel: III Discussion of results. Proc Phys Soc Lond B 64:747–753

Hayden HW, Brophy JH (1963) The activated sintering of tungsten with group VIII elements. J Electrochem Soc 110:805–810

Karpinos DM, Kravchenko AA, Pilipovskii YL, Tkachenko VG, Shamatov YM (1970) Hot pressing of tungsten and its pseudoalloys part I. Sov Powder Metall Met Ceram 9:287–291

Kaysser WA, Hofmann-Amtenbrink M, Pezew G (1987) Sintering'85. Plenum Press, New York

Kim Y, Hong MH, Lee S et al (2006) The effect of yttrium oxide on the sintering behavior and hardness of tungsten. Met Mater Int 12:245–248

Lassner E, Schubert WD (1998) Tungsten: properties, chemistry, technology of the element, alloys and chemical compounds. Springer Science+Business Media, New York

Lee D, Umer MA, Shin Y et al (2012) The effect of sintering conditions and ZrN volume fraction on the mechanical properties of spark plasma sintered W/ZrN composites. Mater Sci Eng A 552:481–485

Lee G, McKittrick J, Ivanov E et al (2016) Densification mechanism and mechanical properties of tungsten powder consolidated by spark plasma sintering. Int J Refract Met Hard Mater 61:22–29

Li B, Sun Z, Hou G, Ding F, Hu P, Yuan F (2017) The effects of alumina reinforcement and nickel activated sintering on nanosized tungsten matrix. J Alloys Compd 692:420–426

Liu R, Xie ZM, Fang QF et al (2016) Nanostructured yttria dispersion-strengthened tungsten synthesized by sol-gel method. J Alloys Compd 657:73–80

Lu L, Chen X, Huang X, Lu K (2009) Revealing the maximum strength in nanotwinned copper. Science 323:607–610

Ma J, Zhang J, Liu W, Shen Z (2013) Suppressing pore-boundary separation during spark plasma sintering of tungsten. J Nucl Mater 438:199–203

Marvel JM, Cantwell PR, Harmer MP (2015) The critical influence of carbon on the thermal stability of nanocrystalline Ni-W alloys. Scr Mater 96:45–48

Massalski TB (1990) Binary alloy phase diagrams, 2nd edn. ASM, Metals Park

Maweja K, Phasha MJ, Choenyane LJ (2012) Thermal stability and magnetic saturation of annealed nickel–tungsten and tungsten milled powders. Int J Refract Met Hard Mater 30:78–84

Monge MA, Auger MA, Leguey T, Ortega Y, Bolzoni L, Gordo E, Pareja R (2009) Characterization of novel W alloys produced by HIP. J Nucl Mater 386:613–617

Morita K, Kim BN, Yoshida H (2015) Spectroscopic study of the discoloration of transparent MgAl2O4 spinel fabricated by spark-plasma-sintering (SPS) processing. Acta Mater 84:9–19

Mutoh Y, Ichikawa K, Nagata K et al (1995) Effect of rhenium addition on fracture toughness of tungsten at elevated temperatures. J Mater Sci 30:770–775

Pelloux RMN, Grant NJ (1960) No title. Trans Am Inst Min Metall Eng: 218–232

Petch NJ (1953) The cleavage strength of polycrystals. J Iron Steel Res Int 175:25–28

Rupert TJ, Trenkle JC, Schuh CA (2011) Enhanced solid solution effects on the strength of nanocrystalline alloys. Acta Mater 59:1619–1631

Sadat T (2015) Alliages Ni-W : de la mise en oeuvre par frittage flash aux micro-mécanismes de déformation et d'endommagement. Dissertation, Université Paris 13 (France)

Sadat T, Dirras G, Tingaud D et al (2016a) Bulk Ni–W alloys with a composite-like microstructure processed by spark plasma sintering: microstructure and mechanical properties. Mater Des 89:1181–1190

Sadat T, Hocini A, Lilensten L et al (2016b) Data on the impact of increasing the W amount on the mass density and compressive properties of Ni–W alloys processed by spark plasma sintering. Data Brief 7:1405–1408

Singla G, Singh K, Pandey OP (2013) Structural and thermal properties of in-situ reduced WO_3 to W powder. Powder Technol 237:9–13

Tiearney TC, Grant NJ (1982) Measurement of structural parameters important in creep of Ni-Mo and Ni-W solid solutions. Metall Trans A 13:1827–1836

Tingaud D, Jenei P, Krawczynska A, Mompiou F, Gubicza J, Dirras G (2015) Investigation of deformation micro-mechanisms in nickel consolidated from a bimodal powder by spark plasma sintering. Mater Charact 99:118–127

Vilémová M, Illková K, Lukáč F, Matějíček J, Klečka J, Leitner J (2018) Microstructure and phase stability of W-Cr alloy prepared by spark plasma sintering. Fus Eng Des 127:173–178

Wang Z, Yuan Y, Arshad K et al (2017) Effects of tantalum concentration on the microstructures and mechanical properties of tungsten-tantalum alloys. Fus Eng Des 125:496–502

Yao M, Zhangjian Z, Jun T, Ming L (2011) Fabrication of ultra-fine grain tungsten by combining spark plasma sintering with resistance sintering under ultra high pressure. Rare Metal Mater Eng 40:4–8

Yih SWH, Wang CT (1979) Tungsten, sources, metallurgy, properties and application. Plenum Press, New York

Zhao M, Zhou Z, Ding Q et al (2015) Effect of rare earth elements on the consolidation behavior and microstructure of tungsten alloys. Int J Refract Met Hard Mater 48:19–23

Zhou Y, Sun QX, Xie ZM et al (2014) The microstructure and microhardness of W–5wt% Cr alloy fabricated by spark plasma sintering. J Alloys Compd 585:771–775

Spark Plasma Sintering of NiTi Shape Memory Alloy

V. Senthilkumar and C. Velmurugan

1 Introduction of Shape Memory Alloys and Its Unique Properties

1.1 Introduction

In this section, the basic functioning of NiTi-based SMA has been briefly explained. The working principle on which it is working and transformation kinetics from austenite to martensite phase have been detailed; the unique properties in shape memory mechanism have been presented. The applications of NiTi and its classification based on the compositions have been focused.

1.2 Shape Memory Alloy

Shape memory alloys are smart engineering materials which exhibit the unique property of solid to solid phase transformations such as austenite to martensite and vice versa with external stimuli. The SMAs are subjected to deformation due to externally applied load, which can retain their parent structure after the removal of the load. The non-diffusive transformation is occurred between the austenite and martensite with the reference of transformation behaviour of crystalline structure (Elahinia 2016; Velmurugan et al. 2017). NiTi SMAs possessed two different phases known as low- and high-temperature phases with different crystal structures and properties. The low-temperature martensite and high-temperature austenite phases

V. Senthilkumar (✉) · C. Velmurugan
National Institute of Technology, Tiruchirappalli, India
e-mail: vskumar@nitt.edu

© Springer Nature Switzerland AG 2019

P. Cavaliere (ed.), *Spark Plasma Sintering of Materials*,
https://doi.org/10.1007/978-3-030-05327-7_22

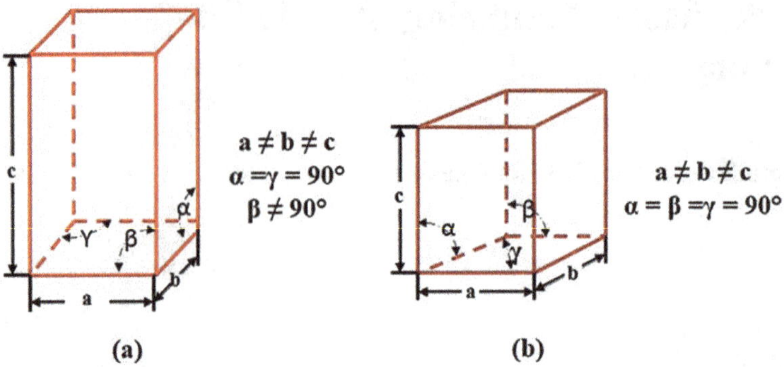

Fig. 1 (**a**) Monoclinic and (**b**) orthorhombic crystal structure

are used as terminals to execute the unique properties such as shape memory effect (SME) and superelasticity (SE). The austenite phase consists of orthorhombic crystalline and the martensite phase with monoclinic crystalline structure as shown in Fig. 1. Upon cooling and in the absence of an applied load, phase transformation that happened from austenite to martensite is called as forward transformation. The regained phase transformation of austenite from martensite during heating is defined as a reverse transformation of SMAs (Kumar and Lagoudas 2008).

The forward transformation is induced by a cooling of austenite structure or an increment of stress which causes the multivariant or twinned structure of martensite. The reverse transformation is prompted by a heating or diminishing of stress which induced single variant or detwinned structure of austenite. The phase transformation of NiTi is occurred via four temperature limits: martensite start (M_s), martensite finish (M_f), austenite start (A_s) and austenite finish (A_f). The twinned martensite structure is achieved, when the phase transformation is occurred from M_s to M_f. The path transformation from this M_s to M_f is called as forward transformation. Moreover, the phase transformation from A_s to A_f is defined as the reverse transformation as shown in Fig. 2.

The unique properties such as superelasticity and shape memory effect enable NiTi as an important class of material in medical, aerospace and industrial fields. Superelasticity is offered on SMA when it is in completely austenite phase (A_f).

1.3 Shape Memory Effect and Superelasticity

SMAs can be able to memorize the parent phase after the deformation due to heating over the transformation range known as shape memory effect. Originally, SMAs are existing in two different phases with three crystal structures and six phase transformations as presented in Fig. 3 (Sun and Huang 2009).

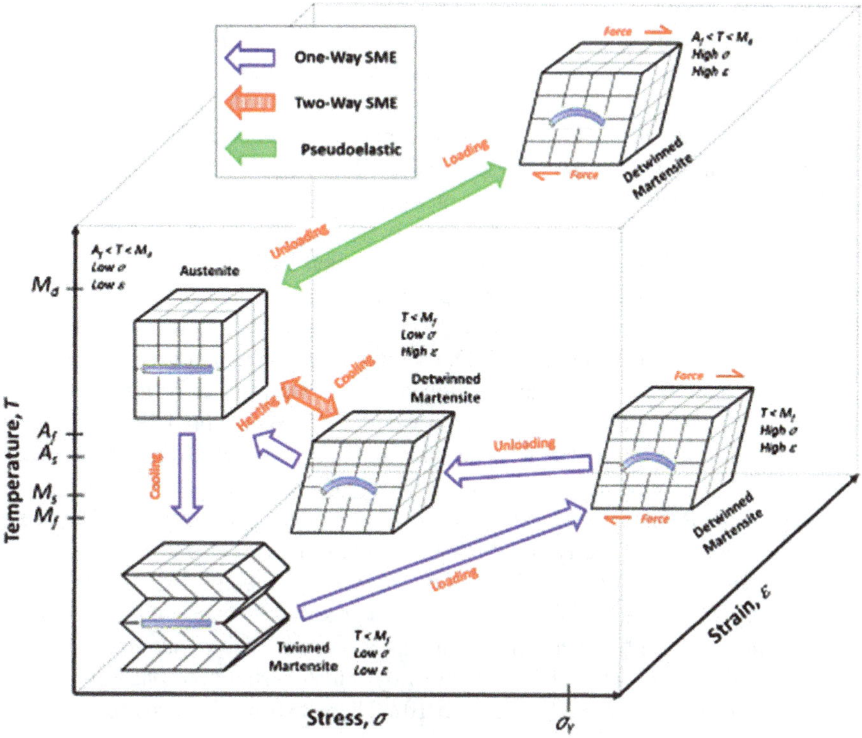

Fig. 2 SMA phases and crystal structures (Jani et al. 2014)

Fig. 3 Three phases with six transformations: (A) austenite; (TM) twinned martensite; (DM) detwinned martensite; the subscripts "l" and "k" are used for the orientation variants of martensite

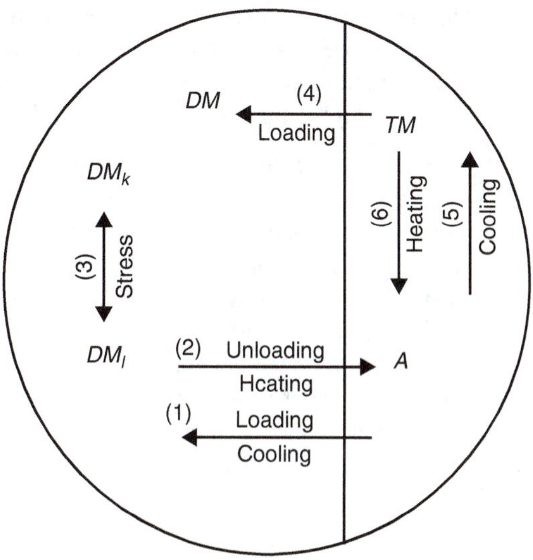

1. Austenite transforms into detwinned martensite upon loading (A→DM).
2. Detwinned martensite undergoes reverse transformation into austenite after unloading at a high temperature or after heating without load or with a very low load (DM→A).
3. Detwinned martensite (variant k) transforms into another detwinned martensite (variant l) after application of load (DM k→ DM l).
4. Twinned martensite transforms into detwinned martensite upon loading (DM →TM).
5. Austenite transforms into twinned martensite upon cooling (A→TM).
6. Twinned martensite transforms into austenite upon heating (TM →A).
7. The parent phase of austenite is transformed into twinned martensite when the SMA is subjected to external load. The detwinned austenite can be achieved again when the external applied load is removed. These path transformations have been attained by the property of shape memory effect (SME). The predetermined shape of SMA has been encrypted at A_f which is attained after the deformation between M_f and A_s by removing the stress acting on it. Thus the parent shape of the austenite phase is regained by shape memory effect (Fremond and Miyazaki 2014). The SME property of SMA is executed in two ways such as one-way shape memory effect (OWSME) and two-way shape memory effect (TWSME). The intrinsic property of OWSME is executed to recover high-temperature austenite phase only where the reorientation of martensite phase is restricted. The mechanism of TWSME is achieved by remembering both phases which is not an intrinsic property (Ma et al. 2010). These two mechanisms of SMA are presented in Fig. 2.
8. The transformation of martensite can be induced by the mechanical loading and unloading from the austenite phase of SMA. The ability to recover the large strains of ~ 8% in stress-strain hysteresis during the mechanical loading and unloading under isothermal conditions is referred as superelasticity (SE) or pseudoelasticity characteristic of SMAs. This effect can be observed from the temperature higher than A_f (Rao et al. 2015).

The transformation from austenite to martensite and the reverse transformation from martensite to austenite do not take place at the same temperature. A schematic representation of hysteresis loop of transformation temperatures is shown in Fig. 4a. The complete transformation cycle can be executed by the martensite start and finish (M_s, M_f) and austenite start and finish (A_s, A_f). The martensite phase can be induced easily when the stress induced between the temperatures of austenite finish and a maximum temperature of M_d. Austenite is thermally stable phase in this temperature (A_f) under no-load condition. Therefore, the stress-induced martensite can retain into austenite when the load is removed. The induced stress-strain curves under different temperatures are shown in Fig. 4b (Stoeckel 1995).

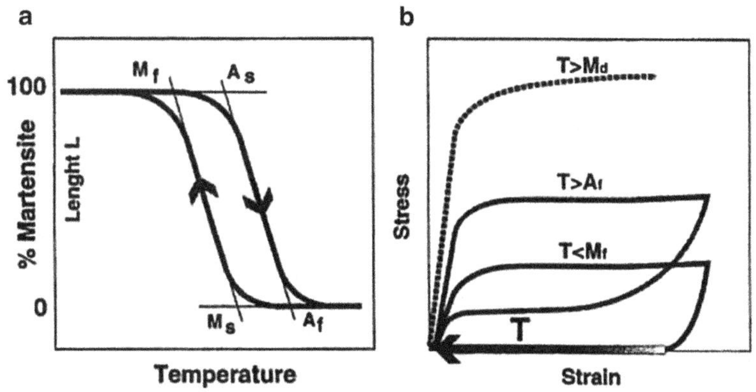

Fig. 4 (**a**) Schematic view of hysteresis loop, (**b**) stress-strain curves at different temperatures (Rao et al. 2015)

1.4 Classification of SMAs

SMAs are grouped under different categories based on the compositional elements presented in the alloy system. The characteristics of SMAs are found in different ranges such as metals, ceramics, polymers, composites and hybrid systems. The commercial applications were satisfied using mainly four systems of SMA such as NiTi, Cu-Al-Ni, NiTiCu/ Nb and Fe-Mn-Si (Duerig and Pelton 2002).

NiTi SMAs are established in the applications where the shape memory effect is required for commercial purposes due to their high percentage shape recovery amidst other engineering properties. Superelasticity is the ability to deform reversibly to a very high strain of ~ 6–10% and recover the predetermined shape upon unloading. These characteristics were found in some other materials. The unique shape memory effect was found initially from equiatomic NiTi alloy system (Buehler et al. 1963). Similar effect was found in some other alloys such as Au-Cd and In-Ti which had not much attention for its commercial usage. In contrast, NiTi alloy system became more popular among researchers due to their tremendous properties such as SME, SE, corrosion resistant, low elastic anisotropy and fatigue strength. Based on the composition presented in the alloy system which exhibited the shape memory, characteristics are classified in the following:

- *Cu-Al-based SMAs:*
  - Cu-Al
  - Cu-Al-Ni
  - Cu-Al-Mn
  - Cu-Al-Be
- *Cu-Zn-based SMAs:*
  - Cu-Zn-Al

  - Cu-Zn-Ni
  - Cu-Zn-Sn
  - Cu-Sn
- *Fe-based SMAs:*
  - Fe-Mn
  - Fe-Mn-Si
  - Fe-Mn-Si-Cr
  - Fe-Ni-Co-Ti
  - Fe-Pt
  - Fe-Mn-Al
  - Fe-Ni-C
- *Other SMA systems:*
  - Ni-Al
  - Ti–Nb
  - Ti–Mo
  - Ti–V

The beneficial effects by the addition of the Cu-based alloys are high damping coefficient, wide transformation temperature range, small hysteresis and large superelasticity effect. All of these properties have widened the potentials of Cu based in shape memory applications where the NiTi alloy is currently explored. The deformation in the lower temperature martensite phase can be restricted by the Cu element which tends to make the materials with brittleness induced by coarse grains, elastic anisotropy and high degree of order (Tadaki 1998).

The third mostly applied system of SMAs is Fe-based alloys after NiTi- and Cu-based alloys. The Fe-based SMAs have more economical advantage than NiTi SMA system due to relatively cheap cost of constituents and ease of fabrication. Therefore, the conventional facilities have been used for steelmaking. Fe-based SMAs are popularly called as shape memory steel (SMS) as listed above. The shape memory capacity exhibited by the Fe- based SMAs is very less than the NiTi- and Cu- based SMAs. Moreover, a large transformation hysteresis could be achieved by the Fe-based SMAs which limits their area of application (Wen et al. 2014).

1.5 Applications of SMAs

NiTi SMAs are used in the extrinsic two-way effect which offers the solutions for existing problems in a range of engineering fields. Moreover, the SMAs have been used in a diverse range of commercial robotic systems, especially as micro-actuators or artificial muscles. Many applications can be implemented by NiTi SMAs in the fields of biomedical due to their unique property of biocompatibility (Predki et al.

2008). The Ni oxidation can be prevented by NiTi SMAs because of the strong bond that existed between Ti and Ni. Thus the contamination of clinical equipment by the Ni leakage has been restricted using NiTi alloy. The NiTi SMA is implemented in medical applications such as drug delivery systems, stent delivery systems, guide wires, implant devices, self-expanding stents and atrial occlusion devices due to the properties of substantial moduli differences between martensite and austenite, MRI compatibility, corrosion resistance and kink resistance (Rao et al. 2015).

The main challenges in the robotic field are to increase the performance and miniaturization of the hardware platform and to increase the intelligence of the integrated system. The clamping difficulties, control issues, miniature electrical connection, low electrical resistance, small strain output and very low efficiency are some of the technical issues which can be resolved by selecting the suitable feedback sensor by NiTi SMAs. The concept of biocompatibility has been focused on the field of biomedical due to known toxic, allergenic and carcinogenic properties of nickel. Moreover, the NiTi SMAs are used in the field of aeronautical specifically in the components such as winglet, landing gear, vortex generator, fuselage, aileron, rudder and fin (Jani et al. 2014). Therefore, the applications of NiTi are extending in all the engineering fields.

2 Phase Transformation Behaviour of NiTi SMAs

2.1 Introduction

The unique properties of SMAs have been decided by the martensite transformation temperature. This transformation temperature is used to determine the stage at which such effects had occurred. The practical applications have been set by these transformation temperatures. Therefore, the control of phase transformation is essential in the design of SMAs to alter the SME and SE to suit required applications. In this section, the influence of composition and process temperature on phase transformation has been discussed.

2.2 Effect of Composition (% of Ni, Ti) on Phase Transformation Behaviour of NiTi SMAs

The atomic percentage of elemental compositions such as Ni and Ti and the addition of elements with NiTi system influenced the transformation behaviour of NiTi SMAs. The unique properties of NiTi SMAs can be determined using the martensite transformation. The applications of such SMAs are being decided by the transformation temperature at which the phases have been changed. Therefore, the control of transformation temperatures is often more stringent (Seelecke and Muller 2004).

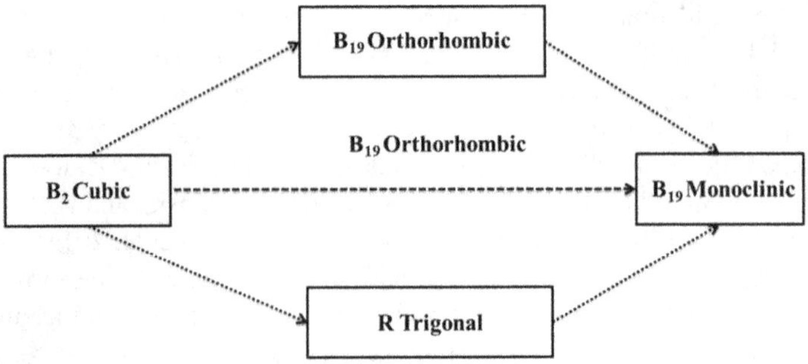

Fig. 5 Different path transformation during martensite transformation

During the phase transformation, the path is important which can alter the transformation temperature. Basically, there are three types of path that can be used to attain the martensite phase in SMAs as shown in Fig. 5.

In the first case, the parent phase of austenite (B_2) has been transferred to martensite ($B_{19'}$) directly without any intermediate stage. This transformation is called as one-stage transformation which can be executed by quenching the NiTi alloy from high-temperature phase. The second case of phase transformation occurs through "R" phase during martensite transformation. Upon cooling from the austenite phase, the resistivity was found to be increased and caused a small sharp peak temperature which is termed as "R" phase transformation. When the NiTi alloy is subjected to cold working followed by an annealing process at optimum temperature (673 K), the high density of dislocations has been produced. The thermomechanical treatment at proper temperature induces the precipitates on NiTi alloy. In this state, the martensite transformation induced the Ni_4Ti_3 precipitate which tends to suppress the $B_{19'}$ phase. The third type of path transformation is achieved through orthorhombic structure (B_{19}). The austenite phase is transferred to martensite phase (monoclinic structure) through the intermediate phase of orthorhombic structure (B_2-B_{19}-B_{19}'). This transformation can be achieved in NiTi by the replacement of Cu in Ni position. Moreover, there is a difficulty to achieve the orthorhombic structured B_{19} phase. When the replacement of Cu content in Ni position exceeds 10%, the B_{19} doesn't capture in path transformation. Therefore, the optimum level (0 > Cu <10) of Cu replacement accelerated the two-stage transformation in binary NiTi alloy (Gou et al. 2015; Wagner et al. 2010; Mitchell and Jerina 2007).

The replacement of ternary element in NiTi alloy composition influences the martensitic transformation. Moreover, the replacement of NiTi chemical compositions by the ternary element is based on electronic configuration. The element which is planned to replace in NiTi binary alloy should have an equal electron configuration. The configuration in the outer shells of Ti can be expressed in terms of nd_2 which is similar for zirconium (Zr) and hafnium (Hf). Moreover, the configurations of Co and Fe are approximately similar to Ni configuration. Hence, the replacement of elemental composition in Ni and Ti atomic position has been declared by the electron configuration. Based on electron configuration, some of the elements were

Table 1 Occupancy of different alloying elements in NiTi site

X	$Ti_{0.5-c}Ni_{0.5}X_c$	$Ti_{0.5-c/2}Ni_{0.5-c/2}X_c$	$Ti_{0.5}Ni_{0.5-c}X_c$
Sc	0.01 ± 0.08	−0.08 + 0.11	0.04 + 0.09
V	0.13 ± 0.10	0.54 ± 0.06	0.78 ± 0.06
Cr	0.51 ± 0.12	0.50 ± 0.11	0.89 ± 0.05
Mn	0.24 ± 0.07	0.55 ± 0.11	0.66 ± 0.09
Fe	0.91 ± 0.08	0.89 ± 0.07	–
Co	1.01 ± 0.06	1.00 ± 0.07	–
Cu	0.34 ± 0.23	0.52 ± 0.18	1.11 ± 0.13
Pd	0.92 ± 0.06	0.94 ± 0.06	0.90 ± 0.08
Au	0.14 ± 0.07	0.55 ± 0.07	0.64 ± 0.06

experimentally tested to replace Ni and Ti positions. The elements Fe, Pd, Co and Sc were found to replace Ni; whereas Cr, Mn and Cu were found to replace both Ni and Ti positions. The percentage of the additional ternary element can be calculated using relations 1, 2 and 3 (Nakata et al. 1998).

$$Ti_{0.5}Ni_{0.5-C}X_C, \quad (1)$$

$$Ti_{0.5-C}Ni_{0.5}X_C, \quad (2)$$

$$Ti_{0.5-C/2}Ni_{0.5-C/2}X_C, \quad (3)$$

Equation (1) is used to replace the ternary element in Ni position. Similarly, Eqs. (2) and (3) can be expressed to add the elements in Ti side and both Ni and Ti sides, respectively. The ternary elements were listed in Table 1 grouped under the above-said equations.

Similarly, the martensite phase transformation is intensely influenced by the composition of Ni and Ti atomic percentage. The shape memory characteristics were found from the nearly equiatomic NiTi alloy. The transformation temperature can be modified up to 100 k when the composition changed by 1%. Therefore, the efficient way to adjust the transformation temperature is to change the composition in NiTi binary elements rather than the replacement of it with ternary element.

Figure 6 shows that the superfluous Ni on Ni-rich side subjected to solubility could not be observed from the Ti-rich side. The solubility limit of NiTi on Ti-rich side was seen to be a straight vertical line. The NiTi intermetallic can be preserved on Ni-rich side when it was subjected to quenching from higher temperature to room temperature as shown in Fig. 6. An elastic constant (c') existed prior to martensite transformation during reduction in temperature. The critical value of this elastic constant was observed at a point where the transformation occurred (Manosa et al. 1994). The critical point was not affected by the composition of NiTi alloy. But the elastic constant involved is sensitive for composition and temperature during martensite transformation. Thus, the martensite transformation depends on the composition of Ni in NiTi alloy. The experimental results were plotted in Fig. 7 which show that the increase of Ni content causes a decrease in transformation temperature. Moreover, the changes in transformation temperature were observed till the 51.5

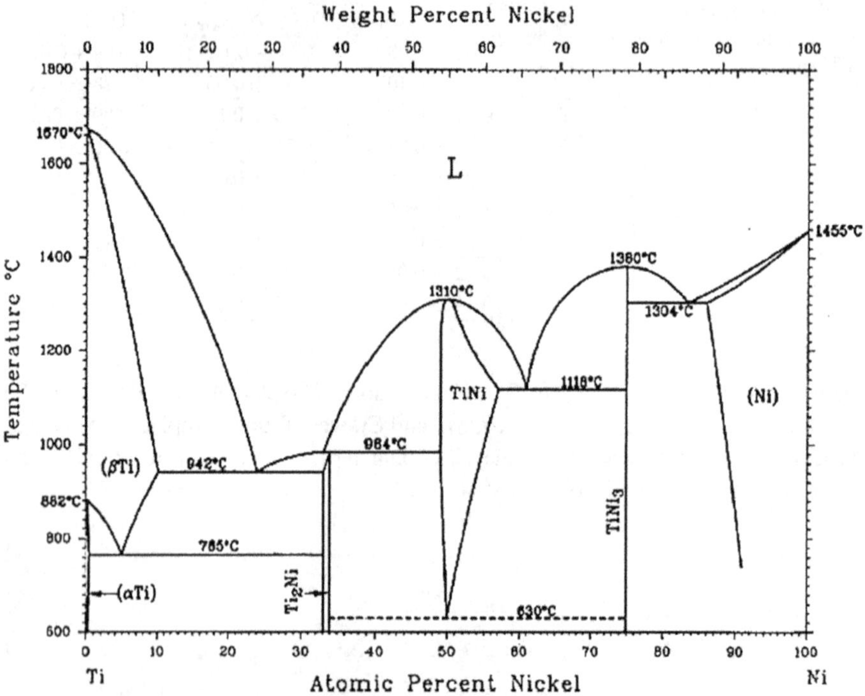

Fig. 6 Phase diagram of NiTi alloy (Bram et al. 2002)

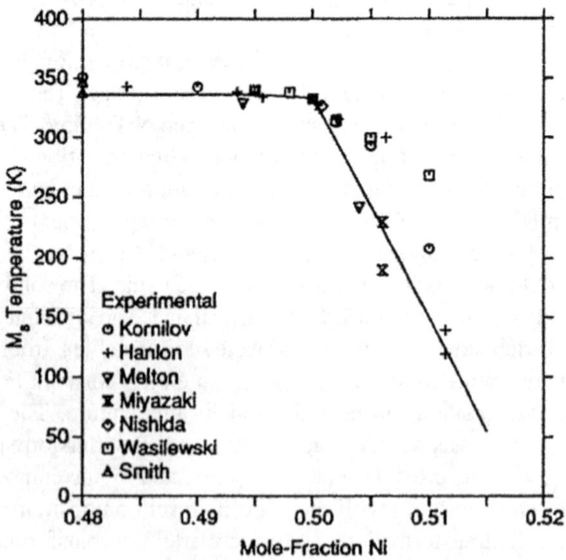

Fig. 7 Influence of Ni content on martensite transformation temperature (Tang 1997)

at. % of Ni. Further, increasing Ni caused the same transformation temperature; no change was observed. Hence, the optimum atomic % of Ni has been determined at 51.5% for altering the transformation temperature of NiTi alloy (Frenzel et al. 2010).

2.3 *Effect of Process Temperature on Phase Transformation Behaviour of NiTi SMAs*

Rather than the changing of composition in NiTi alloy, the changing of processing temperature also influences the martensite transformation temperature. The control of transformation temperature in NiTi alloy is difficult once the alloy has been synthesized. The processes of aging, annealing and solution treatment were used to control the transformation temperature of NiTi alloy irrespective of the number of cycles.

The secondary phase elements of precipitates have played an important role in transformation behaviour of NiTi SMAs. The precipitates such as Ti_4Ni_3, $TiNi_3$ and Ti_2Ni_3 were formed at different processing temperatures. The phase of Ti_4Ni_3 occurs at lower aging temperature and shorter aging time, $TiNi_3$ phase appears at a higher aging temperature and longer time, and Ti_2Ni_3 phase exists at intermediate temperature and time (Otsuka and Ren 2005). The processing temperature and time influence the size and numbers of precipitates formed in NiTi alloy. The "R" phase causes the multistage (two or three stages) transformation in NiTi alloy that influences the occurrence of precipitates. The size of the Ni_4Ti_3 precipitates increases with an increase of aging temperature and time. This precipitate has been coherent with B_2 austenite matrix that has found to occur with a tiny size (10 nm). With the increase of temperature, the sizes of precipitate increases and lose its coherency with parent phase (Jiang et al. 2013).

As the temperature increases, the size of precipitates increases with a lose in coherency in the parent structure. The martensite start temperature was increased with the annealing temperature which caused local stress. Due to this local stress on NiTi, the precipitates were found in coherence or semi coherence with the matrix which produces a tendency for the formation of martensitic variants. The increase of annealing temperature causes the augmentation of M_s and detraction of M_f as shown in Fig. 8. Moreover, the endothermic and exothermic heat can be enhanced with an increase of annealing temperature (Tadayyon et al. 2016).

From the experiment results obtained by Karimzadeh et al. (2016), the transformation temperature has been increased with the occurrence of R on NiTi after aging at 300–400 °C. When the temperature increased to 700 °C, the austenite $B_{19'}$ phase had occurred. Moreover, similar results could be obtained on NiTi when it was treated by solution annealing. Therefore, the amendment of transformation temperature in NiTi alloy has been achieved through aging, annealing and solution treatment at various temperatures.

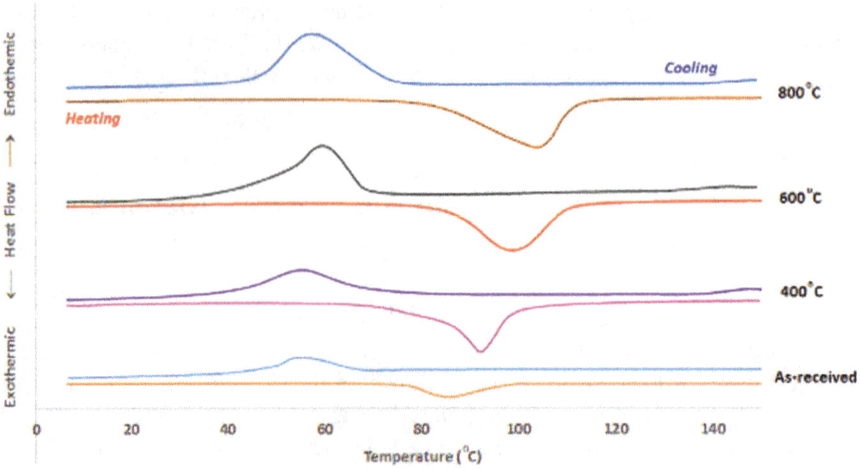

Fig. 8 DSC curves of samples annealed at different temperatures for 45 min (Tadayyon et al. 2016)

3 Manufacturing Methods of SMAs

3.1 Introduction

In this section, the various methods of NiTi fabrication process are discussed. In recent days, the applications of NiTi alloy has been extended in all fields of engineering such as aerospace, medical, automobile, actuators, dampers and grippers. The main advantages and challenges of various manufacturing processes are discussed.

3.2 Manufacturing of NiTi Components

The manufacturing processes are significantly affecting the properties of NiTi alloy. Basically, manufacturing methods are carried out at different temperatures which influence the transformation behaviour of NiTi. Therefore, based on the factors such as raw material and temperature used, the manufacturing methods can be classified as follows (Fig. 9).

The NiTi alloys have been fabricated using a conventional technique of casting and melting which is carried out under vacuum arc remelting (VAR), vacuum induction melting (VIM) and electron beam welding (EBM). These methods are accomplished with high temperature which enhances the secondary phases and impurities such as $Ni_2Ti_4O_x$ and TiC. The functional properties such as SE and SME in NiTi are degraded due to these impurities. Further, in the casting process, the cast NiTi raw material needs machining processes to obtain desired shapes.

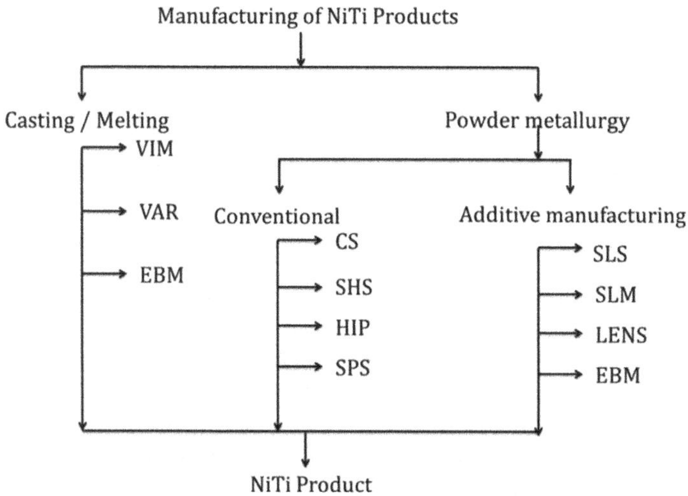

Fig. 9 Illustration of NiTi manufacturing techniques

Machining of NiTi SMAs is another challenging one which affects its unique properties. Moreover, the tools used in machining process consume much wear and require some special features such as coating and textures (Elahinia et al. 2012). The purity of raw materials highly influences the homogeneity and uniformity of the casted NiTi products. The NiTi alloy is subjected to melts and dissolves with carbon and formed TiC during solidification in graphite crucible. The presence of TiC enhances the concentration of Ni content in NiTi alloy which affected the phase transformation temperature. Moreover, the content of carbon particles in NiTi depends on the melting temperature. When it is exceeds the melting temperature of 1723 K, the use of graphite crucible is not possible. Fortunately, the melting point of the stoichiometric Ni–Ti alloy is 1510 K so that the melting procedure can be carried out at relatively low temperatures. The carbon content in the ingot prepared under a pertinent operation lies between 200 and 500 ppm. Such a small amount of carbon does not affect the functional properties of shape memory alloy (Patoor et al. 2006).

Another conventional technique for fabricating the NiTi products is powder metallurgy (PM) which can be used to produce near-net-shape devices. The PM techniques are broadly grouped into two ways such as conventional processes and additive manufacturing processes. Moreover the convention process of PM can be classified into conventional sintering (CS), self-propagating high-temperature synthesis (SHS), hot isostatic pressing (HIP), spark plasma sintering (SPS) and metal injection moulding (MIM). Similarly, the additive manufacturing process is further grouped under selective laser sintering (SLS), selective laser melting (SLM), laser-engineered net shape (LENS) and electron beam welding (EBM). All the PM methods can be processed using the raw materials with required chemical composition and structure of the target product. The prior step of powder preparation is required for PM process. The problems associated in casting and melting like segregation and extensive grain growth can be avoided in PM route. The secondary operations such as machining and forging are not required during the finishing of final products. Moreover, PM can be used to control the chemical composition which influences the phase transformation behaviour of NiTi. One of the major disadvantages of this method is the high impurity pick-up that is caused by the large surface area of the powder particles. Additionally, these methods are limited in the complexity of the resulting parts and in directing the size and shape of porosity, when desired (Elahinia et al. 2016).

The attention on fabricating of porous NiTi alloy has been increased due to its potential applications in medical field such as tissue implants and bone replacement. The porous structures exhibit the peculiar mechanical properties, low relative density and damping capacity. In addition to that of porous structure, fabrication of NiTi with nanostructure has substantial attention in their fabrication. An advance in nanotechnology has led to NiTi SMA being used in dental and orthopaedic applications because of its high success rate. This kind of nanocrystalline NiTi is implemented in implantation due to higher atomic concentration at the interfacial structures. Consequently, many recent investigations have focused on the fabrication of nanocrystalline NiTi using different fabrication methods. The nanocrystalline grains are induced in NiTi alloy by the severe plastic deformation processes such as local canning compression, cold drawing, cold rolling, and high-pressure torsion (Arunkumar et al. 2018; Chu et al. 2005).

Another one technology of PM is additive manufacturing (AM) which gained significant attention for processing the NiTi because they have prevented many of the challenges associated in conventional methods. These processes depend on the CAD data and entails adding material in consecutive layers made of powders and melted. The ratio of Ni and Ti elements in preparing NiTi determines the unique properties of the final product in AM method. The powders can be created either by pre-alloying or by elementally blending the Ni and Ti particles. The second important requirement in AM is the processing parameters which are to be added to an optimum limit for making highly densified NiTi products (Shishkovsky et al. 2012; Elahinia et al. 2016).

4 Spark Plasma Sintering of NiTi

4.1 Introduction

One of the conventional PM routes of NiTi manufacturing method is spark plasma sintering technique. In this section, the mechanism of SPS and SPS of NiTi SMAs is discussed. The fundamental working principle of SPS including the parameters used in SPS and their control has been discussed briefly. The selection of SPS parameters for sintering the NiTi alloy to obtain the better functional behaviour is discussed.

4.2 Working Principle of SPS Process

A rapid synthesis technique of SPS is fabricating the products by sintering the powder particles with high density condition. High-density products are synthesized with the utilization of electric current which tends to heat and densify the powder elements together with the pressure and joule heating of either the powders or die and punch or both. Spark plasma sintering (SPS) is also known as field-assisted sintering technique (FAST), pressure-assisted densification (CAPAD), pulsed current hot pressing (PCHP), electric current-assisted sintering (ECAS), current-activated, plasma-activated sintering (PAS), pulsed electric current sintering (PECS) and several other names (Tokita 2013; Orru et al. 2009).

The manufacturing of NiTi products using conventional methods is normally associated with serious contamination, grain growth and complex post-fabrication and thermomechanical processing operations for further characteristics. Hence, a novel, cost-effective and efficient process is highly in demand for NiTi manufacturing. Powder metallurgy is a near-net-shape manufacturing method that is being used for advanced functional material production. At the same time, conventional powder metallurgy has been done in two steps such as compaction followed by sintering process which may create contamination on products. To avoid such problems in conventional sintering, SPS is used to fabricate the samples with high densification. The compaction and sintering processes can be done simultaneously in SPS with the application of pressure and temperature. The powder particles after mixing and ball milling operations (if needed) are kept in graphite crucible which was made in the shape of punch and die as shown in Fig. 10. The field section is associated with electric current and passed through powder particles presented in graphite crucible.

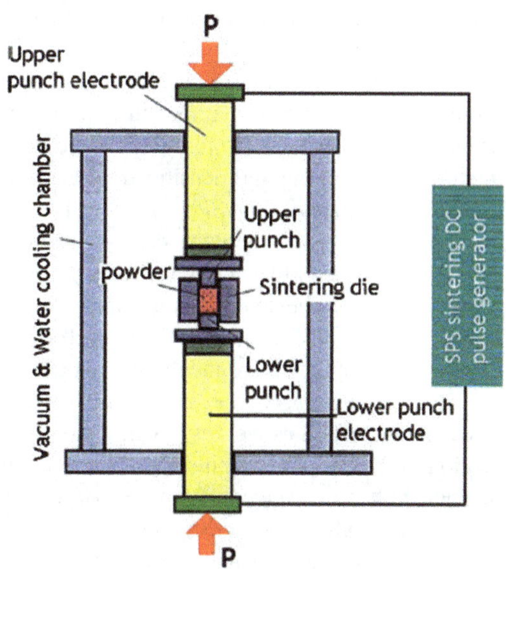

Fig. 10 Schematic of SPS process setup

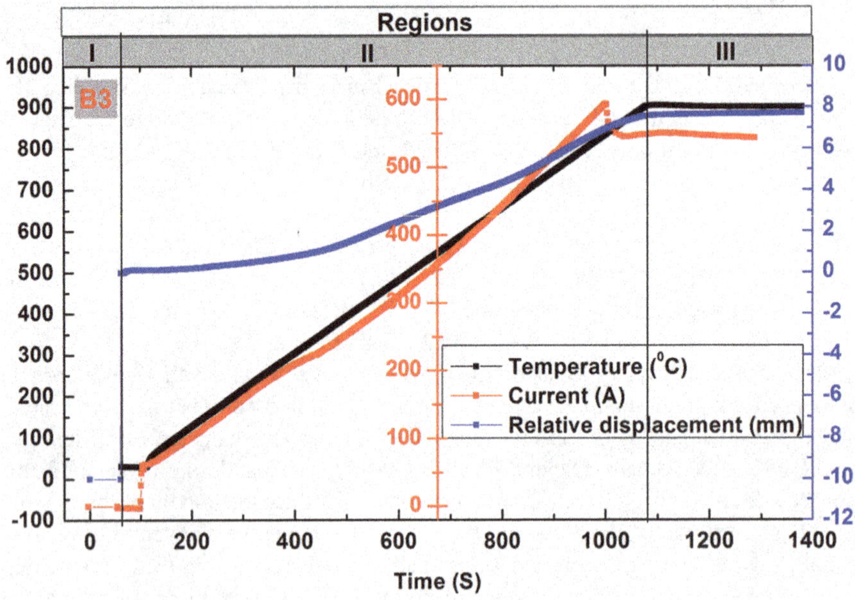

Fig. 11 Densification curve for NiTi sintered at 900 °C using SPS

The electrical contact between the electrodes, specimen and the die has been generated in SPS with the application of pressure which creates compacts. The compaction of powder particles is executed at much higher heating rates. Therefore, the SPS method has advantage over its conventional sintering methods and investigated widely in the past few years (Laptev 2007). The parameters such as pressure, temperature and current field are used to control the SPS process on powder materials.

The sintering temperature can be monitored continuously using digital pyrometer. Moreover, the measuring temperature is not the direct temperature of compacted powders. There is a chance for temperature difference between the die and powder mixture. The alloying and densification kinetics can be monitored continuously by the data collected during SPS process. The relation between the time and SPS parameters such as temperature, pressure and current can be plotted using that data.

Figure 11 shows that the densification curve for NiTi alloy which exhibits the relation between the temperature, current and relative punch displacement with respect to time. These parameters could be recorded during SPS process and can be used to draw the above-mentioned temporal plots. The sintering process can be executed in three stages: mechanical movement and rearrangement of the particles, Joule heating effect as well as plastic deformation of the particles and rapid densification of the body. The first step occurred at lower temperature with lower time in seconds. In this first stage, the powder particles could be rearranged and pores could be removed slightly due to external pressure. The punch has been displaced slightly at this stage because of smaller temperature and current (Bonifacio et al. 2012).

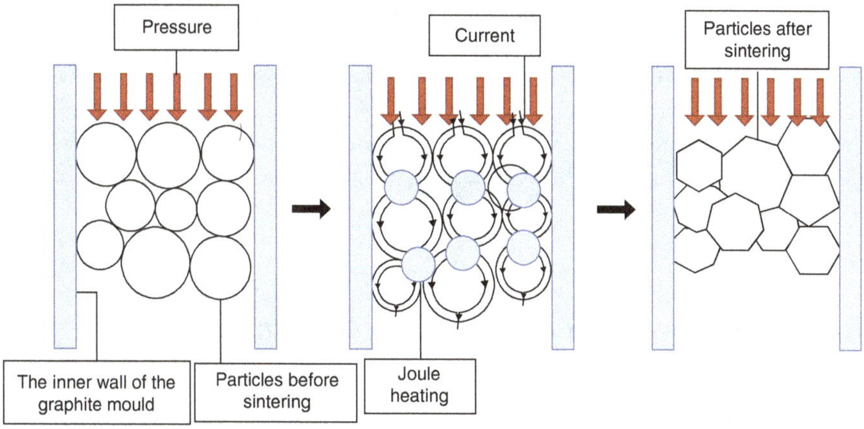

Fig. 12 Sintering of the powder particles during SPS process (Cheng et al. 2017)

The second stage of SPS is executed or nearly 500 s where the punch displacement augmented rapidly as shown in Fig. 11. In this stage, current and temperature enhanced quickly and plasma generated at a gap between the powder particles. The high-pulse DC discharge caused the cleaning of particle surface and activated. The current and temperature of particles increased rapidly due to strong spark discharge effect, and the necking that occurred between particles is facilitated by the softening of the particle surface.

The current passed directly through the neck of the particles when the two particles connected as shown in Fig. 12. The Joule heating is a main heating source to make particles connect together replacing the spark discharge effect. Large amount of Joule heating produced high temperature to make the particle surface melt and evaporate, resulting in growth of the sintering neck in the contact point, and the densification of powder can be enhanced. The densification can further be enhanced through the plastic deformation due to the application of external pressure. The final stage of densification using SPS occurred during soaking and mass transportation under the same pressure of second stage. Thus the final densified products have been sintered after the SPS process (GurtSantanach et al. 2011; Cheng et al. 2017).

5 Effect of SPS Parameters on Functional Properties of NiTi SMA

5.1 Introduction

The SPS parameters such as sintering temperature, pressure and current are considered with respect to time for plotting the temporal plots during SPS process. These graphs can be used to identify the flow of current and pressure which desired the density of the sintered products. Therefore, the study of such parameters is essential for sintering process with respect to sintered product characteristics such as mechanical, phase transformation behaviour and microstructural characteristics. The influences of SPS parameters on mechanical properties such as density, hardness and phase transformation with microstructural characteristics are discussed in this section.

5.2 *Influences of SPS Parameters on Mechanical Characteristics of NiTi SMA*

5.2.1 Influences of SPS Parameters on Density of NiTi SMA

Over the past decades, shape memory alloys have fascinated great attention in various applications extending from aerospace and naval to surgical instruments, medical implants and fixtures. Among SMAs, NiTi alloy has been used most comprehensively due to large flow stress and shape memory effect (SME) strain (Shao et al. 2017). Moreover, the attention on porous NiTi is increasing in all possible application of medical implant devices and as high-energy absorption structural material. Such kind of NiTi has been developed using the methods such as conventional sintering, self-propagating high-temperature synthesis and hot isostatic press. The stress-strain curve in these manufacturing techniques exhibited brittleness in nature (Li et al. 2000, 2002). Hence, it is necessary to develop a better processing method which can provide the porous structure of NiTi with higher ductility. Therefore, the spark plasma sintering (SPS) method was introduced for obtaining NiTi with ductility behaviour.

The pores in sintered NiTi influence the density behaviour and its applications where it can be used. In addition to that, the fabrication of NiTi SMAs with functional properties is difficult in conventional sintering. The SPS process can sinter the NiTi alloy with desired functional properties and high density. The NiTi powders with desired ratio are allowed for blending followed by milling operation to obtain the raw material for SPS process. The spherical granulated shapes of NiTi elemental powders were converted into slenderer flattened shapes after the milling at certain duration as shown in Fig. 13. The flattened shapes could be used to avoid the pores and enhance the density of sintered products (Arunkumar et al. 2018).

The milled powders are allowed to a pressure assisted pulsed current sintering process (SPS) by creating momentary discharge in the gaps between the powder particles. Thus, electrical discharge activated the surface of the powder particles which lead to form a soft and neck formation. With the assistance of pressure acting on the powder surface, it starts to diffuse with each other. Normally, the sintered NiTi using SPS method consists of very low porosity compared with the products from conventional sintering. Besides, the porosity is formed in spark plasma sintering of NiTi due to improper diffusion of particular particles and the presence

Fig. 13 NiTi powders (**a**) mixed (**d**) milled condition (Arunkumar et al. 2018)

of secondary phases (Ni4Ti3, Ni3Ti and Ni2Ti3) (Lee et al. 2014; Zhang et al. 2015).

In SPS, the mechanical property is heavily influenced by porosity which is affected by the sintering temperature. Earlier investigation reported that the NiTi alloy has been sintered at various temperatures (700–9000 °C). The porosity was obtained as 50% for the specimen sintered at 7000 °C. Higher relative density (95%) has been achieved on the samples sintered at 8000 °C and more than that.

The pores and unsintered structure of particles can be observed from specimen sintered at lower temperature as shown in Fig. 14a. The ductile morphology of NiTi can be observed in the samples when the sintering temperature was increased from 700 °C as shown in Fig. 14d. The enhancement of temperature led to clustering of the particles which in turn promote abnormal grain growth. Thus the rapid densified products can be sintered with an increase of sintering temperature (Fu et al. 2004).

The addition of ternary alloy of NH_4HCO_3 was added with NiTi alloy to investigate the density behaviour of NiTi. The sintering and pore-forming process during SPS were recorded incessantly and shown in Fig. 15. In stage 1, the vacuum was increased significantly during the sintering time from 0 to 200 s and further maintained in a steady state for the entire process. The chance of occurrence of pores was observed in stage 2, when the time was increased to 700 s.

The target temperature of 1000 °C was attained in this stage. The pressure and displacement were found to be varying, very low in stage 3 which caused the highly densified NiTi. Thus the density of sintered NiTi can be increased and porosity decreased with an increase of temperature as shown in Fig. 16. The porous structure was increased with an increase of the content of NH_4HCO_3 with NiTi alloy. The pores presented in NiTi without any additional element was observed to be less, whereas it can be increased with sintering temperature combined with NH_4HCO_3 as shown in Fig. 17. The initial size of pores for base NiTi was 21 μm which has increased to 415 μm with an increase of NH_4HCO_3 up to 20%. The density of NiTi is affected by the addition of NH_4HCO_3 with enhancement of sintering temperature. This kind of porous structure of NiTi with 30–50% can be used in tissue growth which caused a suitable characteristic in bone implant.

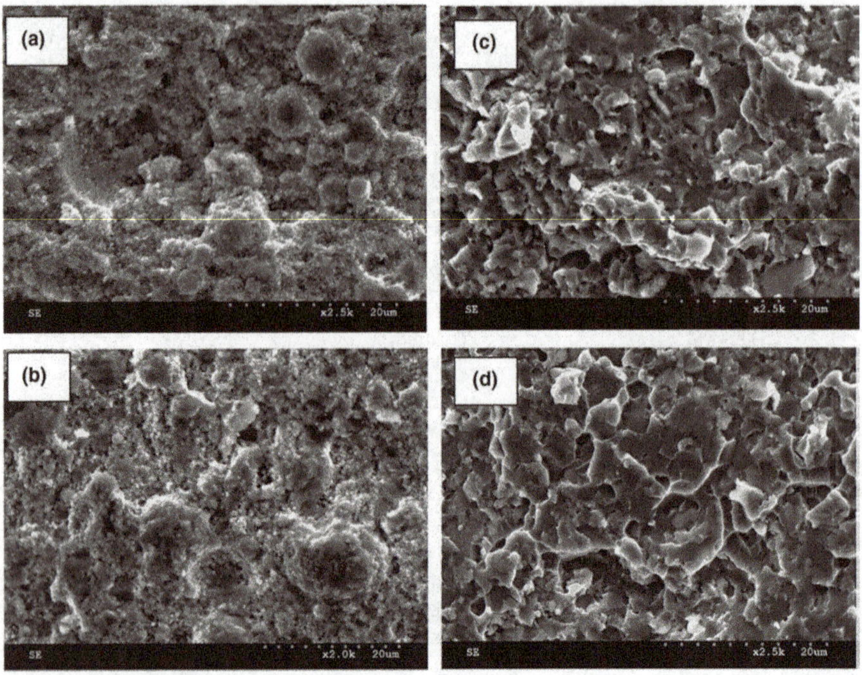

Fig. 14 Scanning electron micrographs of fracture morphology of NiTi sintered at temperatures of (**a**) 700, (**b**) 800, (**c**) 900 and (**d**) 1000 °C (Shearwood et al. 2005)

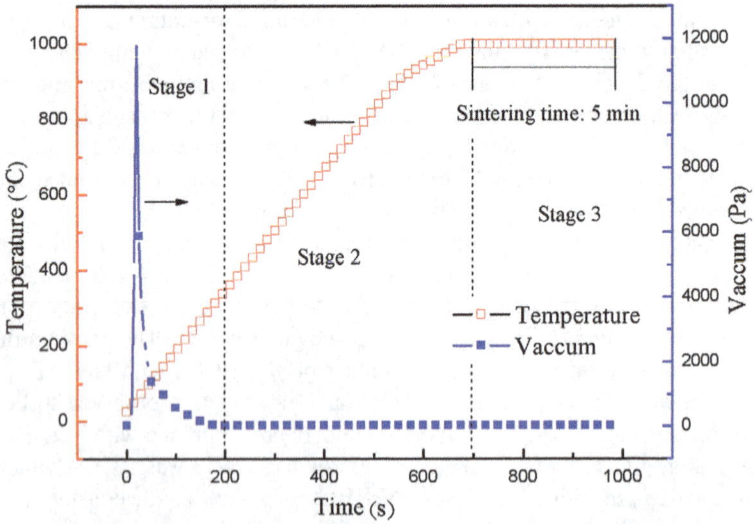

Fig. 15 Temperature-vacuum-sintering time profile during SPS

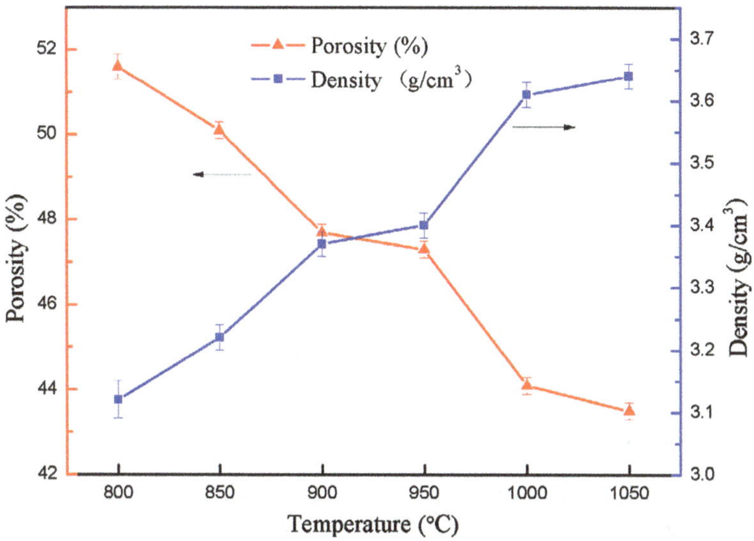

Fig. 16 Porosities and densities of the porous NiTi alloys prepared by SPS with different sintering temperatures (Zhang et al. 2015)

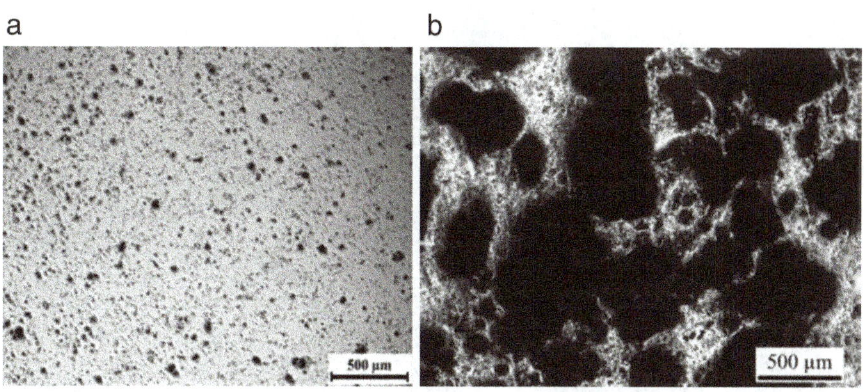

Fig. 17 Micrographs of the porous NiTi synthesized using SPS with various contents of NH4HCO3 (**a**) 0 (**b**) 20 wt. % (Zhang et al. 2015)

5.2.2 Influences of SPS Parameters on Hardness (Vickers and Nanoindentation) of NiTi SMA

The fundamental mechanical property of NiTi SMAs is hardness which has been extensively investigated using nanoindentation technique. The SMA-based micro-electro-mechanical systems (MEMS), micro-grippers and micro-actuators require the characterization of material with small-scale systems instead of investigating their performance on bulk materials scale. The indentation technique is used to extract small-scale mechanical properties of metals, whereas the indenter has been

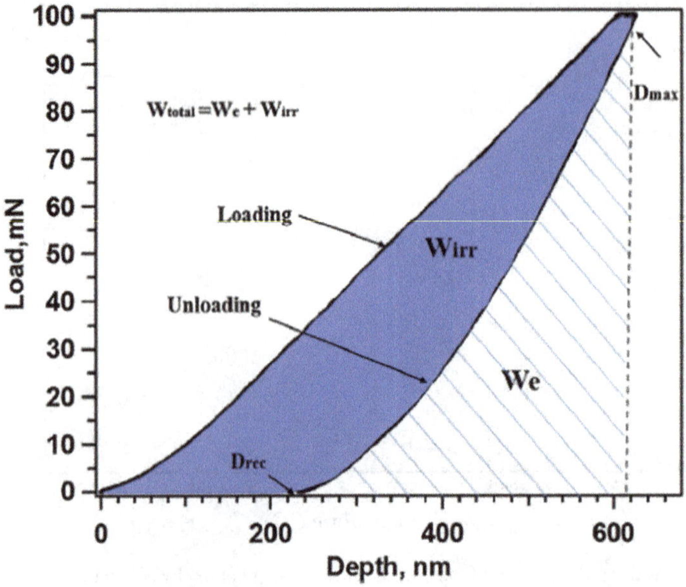

Fig. 18 Indentation loading and unloading curve with important indentation parameters

selected with known material properties (elastic modulus and hardness) (Fischer-Cripps 2011a). The spherical and pyramidal indenters are used to measure the hardness of materials in indentation technique. Spherical indenters are used widely because they can yield a smooth conversion from elastic to elastic-plastic contact and also create well-suited and repeatable indents on a single location. In recent years, spherical indentation technique has been deliberated as a standard method of measuring the hardness of functional materials (Li et al. 2015).

Normally, NiTi have large depth recoverable ratio due to its superelasticity property. The depth recoverable ratio had attained for NiTi about 45% for Berkovich (~200 mN) and 95% for spherical (~2.75 N) indenters. Hence, higher superelasticity can be identified for NiTi when spherical indenter was used in nanoindentation. The narrow indents in austenitic NiTi can recover fully upon unloading the indentation load, while martensite of NiTi can recover partially upon unloading and fully retains after heating to austenite phase. This effect can be observed in NiTi-based SMAs and known as "self-healing" effect (Ni et al. 2002).

Figure 18 shows the loading and unloading curve plotted using the data obtained from nanoindentation technique. This loading and unloading curve can be plotted in load versus depth parameters. The recoverable work energy can be calculated by integrating the unloading curve. The ratio of the recoverable energy to the total energy defined as ηw can be expressed as

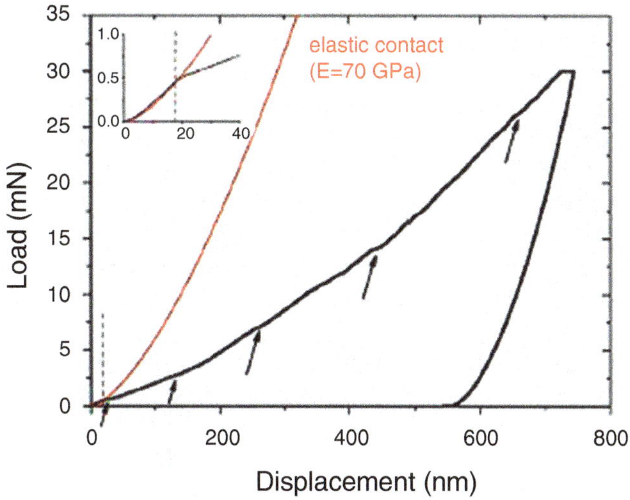

Fig. 19 Load displacement curves of nanoindentation experiment in the selected <211> grain

$$\eta_w = \frac{W_e}{W_t} = \frac{\int_{h_r}^{h_{\max}} F dh}{\int_0^{h_{\max}} F dh}$$

The nanoindentation technique can be investigated on a particular grain which exhibits the bulk metal characteristics of NiTi. The grain of <211> has been investigated using nanoindentation, and its loading and unloading curve is shown in Fig. 19.

The pop-ins or discontinuities was observed during loading on a particular grain and marked using arrows. These can be avoided during unloading due to large remnant deformation and poor recovery. A load displacement relationship can be defined using the Hertzian relationship as follows.

$$P = \frac{4}{3} E_r R^{1/2} h^{3/2}$$

where P is the indentation load, E_r is the indentation modulus, R is the indenter radius, and h is the indentation depth (Johnson 1985; Fischer-Cripps 2011b). The indentation modulus (E_r) can be calculated using the following relation.

$$\frac{1}{E_r} = \frac{(1 - \nu^2)}{E} + \frac{(1 - \nu_i^2)}{E_i}$$

where ν is the Poisson ratio of NiTi and ν_i and E_i are the Poisson ratio and Young's modulus, respectively.

The SPS parameters such as sintering temperature and pressure increase the relative density of sintered NiTi which causes the enhancement of hardness. A uniform hardness profile can be obtained for NiTi sintered using SPS which provides the uniform temperature profile during the sintering process. This hardness profile also can be improved with the addition of ternary alloy with NiTi SMA. The substitution of Cu in NiTi system has increased the hardness. Similarly, the increase of sintering temperature causes the formation of precipitates in NiTi alloy. The formation of secondary phases enhanced the hardness of NiTi alloy. The precipitate that formed during SPS process such as Ni4Ti3 strengthens the NiTi matrix by creating strain fields which improve shape memory recoverability (Cristea et al. 2015).

The NiTi powder particles can be rearranged with effective packing and rupturing the existing large agglomerates due to the enhancement of pressure in SPS process. Thus the better homogeneity obtained by sintering process destructs the agglomerates into finer sizes. Further, increment of pressure yields to achieve more contact points between the powder particles which reduce the diffusion distance and enhance the hardness of the compacted specimens. Increase in sintering temperature gradually accelerates mass transfer through lattice diffusion, grain boundary diffusion and viscous flow of liquid phase. The viscous flow of liquid phase is reduced with an increase of sintering temperature which enhanced the mass transfer. The pressure also is applied at this higher temperature which leads to particle rearrangement with highly densified products (Groza 2007). Therefore, higher hardness can be achieved by increasing the SPS parameters.

Moreover, the single-step sintering temperature facilitates to enhance the hardness property of NiTi SMAs. When the NiTi elements allowed to second step heating in certain temperature, the hardness of sintered NiTi could be affected. The second step heating in SPS process slackened the arrangement of powder particles which are packed in the first stage of single step heating. The hardness value was measured for the NiTi samples (A1, B1 = NiTi single-step heating, A2, B2 = sintered NiTi + second step heating, S1, S2 = NiTiCu) which were produced using SPS process. Addition of ternary element such as Cu enriched the hardness as shown in Fig. 20. Moreover, the sintered NiTi exhibits lower hardness when they allowed to second step heating process (Cristea et al. 2015).

The hardness property of NiTi SMA has been increased with an increase in sintering temperature up to certain limit; thereafter, the value of hardness is strained to maintain a constant for SPS-sintered NiTi as shown in Table 2. As the sintering temperature increased from 1050 to 1150 °C, there is no much difference in the hardness of NiTi alloy. The homogenized hardness can be attained at a certain temperature, whereas the homogeneous phase of NiTi could be achieved.

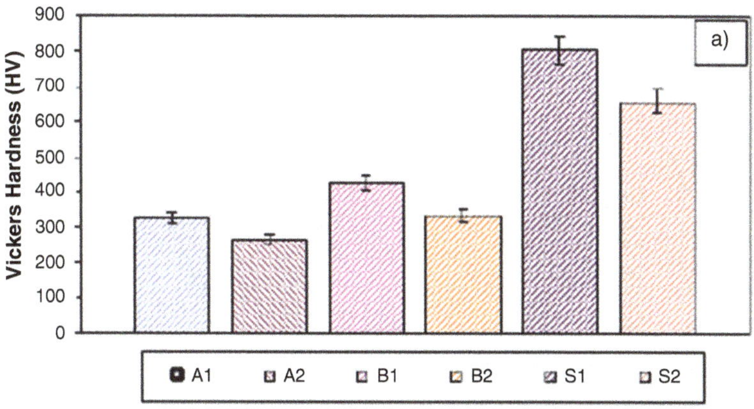

Fig. 20 Hardness values of spark plasma-sintered NiTi and NiTiCu (Cristea et al. 2015)

Table 2 Vickers hardness of the SPS-processed and SPS-homogenized NiTi alloys

Materials	1050 °C/4 min (SPS)	1050 °C/4 min (SPS)-1150 °C/2 h
HV_{10} (GPa)	5.3 ± 0.1	5.5 ± 0.1

5.2.3 Influences of SPS Parameters on Mechanical Stress-Strain Behaviour of NiTi SMA

The high damping characteristic of NiTi SMA can play an important role in diverse fields such as naval, civil, aerospace and automotive engineering. This prospective role stalks from certain properties related to fatigue resistance, corrosion, stiffness and damping. Besides, these properties heavily depend on loading conditions such as temperature strain amplitude, frequency, time (loading history) and thermal treatment received (De Santis et al. 2001). The investigation on all of these mechanical properties can execute when the fundamental stress-strain behaviour is well defined.

NiTi shape memory alloys have been extensively used in medical, aerospace, automotive and energy absorption applications as compared with other types of SMAs like Cu-based and Fe-based SMAs. Understanding of the mechanical properties and deformation mechanisms are precarious for NiTi SMAs. The stress-strain relation of NiTi can be found through uniaxial tension and compression test. The stress-strain can be plotted for NiTi SMAs at a strain rate and environmental temperature range. Also, it can be plotted under (1) the dynamic stress-strain curves of NiTi SMAs exhibited a work-hardening behaviour; (2) the specimen temperature changed during dynamic deformation, which was monitored using a small thermocouple; and (3) the higher the deformation temperature (Qiu et al. 2017).

An earlier investigation on compressive strain was achieved for NiTi (49.5 at. % Ni and 50.5 at. % Ti) between 4.8% and 9.6% under high strain rate loading (1200 s^{-1}) as shown in Fig. 21. The phase transformation was limited at this strain

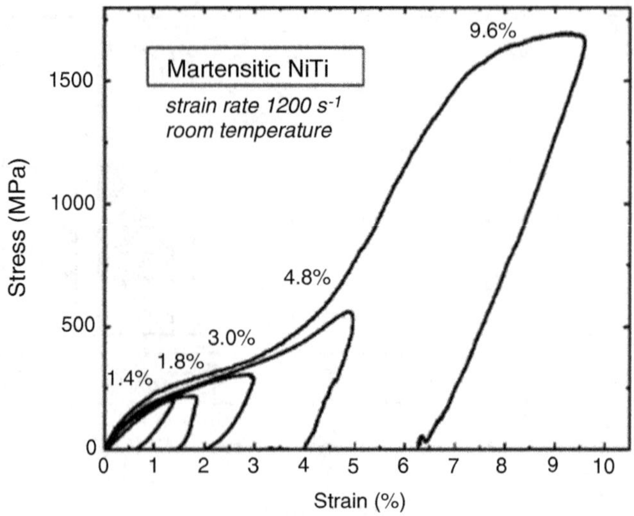

Fig. 21 Series of compressive stress-strain curves of martensitic NiTi SMA (Qiu et al. 2015)

level for NiTi, thereby shape memory effect could not be obtained. The phase transformation vanished gradually as the specimen was compressed to higher-level strains and fully recoverable after the sample was annealed at 873 K (600 °C) for 30 min. However, the deformation of more than 7% of strain leads to permanent plastic deformation which ultimately lowers the phase transformation temperatures (Qiu et al. 2015).

Based on this result, the mechanical stress-strain characteristics depend on the strain rate and process temperature. The density or porous structure of sintered NiTi is influenced by the sintering temperature. On the other hand, the stress-strain behaviour of NiTi can be influenced by that porosity obtained during sintering process. When the NiTi was sintered at low temperature (750 °C) and pressure (5 MPa) in SPS, a porosity level of 25% is achieved. However, a fully densified NiTi product can be produced when the sintering temperature and pressure increased to 850 °C and 50 MPa, respectively. The reason for good superelasticity behaviour of SPS-processed NiTi is continuous connectivity of powder particles. The connectivity cannot be established among NiTi powders in the case of the 25% porosity. In addition to that, the large NiTi powder particles are clustered, and some converted into unwanted brittle intermetallics. This would have been produced by excess local high temperature during the SPS process. Thus the strength of NiTi alloy could be enhanced with an increase of SPS parameters such as sintering temperature and pressure through porosity as shown in Fig. 22 (Zhao et al. 2005).

As discussed above, one of the major applications of NiTi is biomedical field with products such as bone and tissue implant. Therefore, the compressive strength and elastic modulus of NiTi need to be achieved nearly to the value of human bone. The compressive strength of NiTi can be evaluated using cyclic loading and unloading at

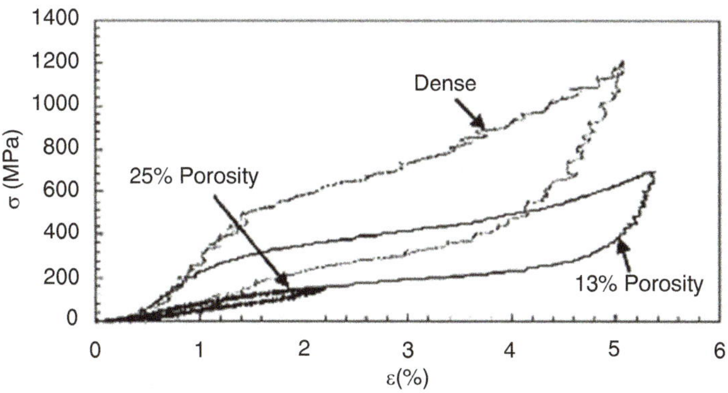

Fig. 22 Strength of sintered NiTi alloy

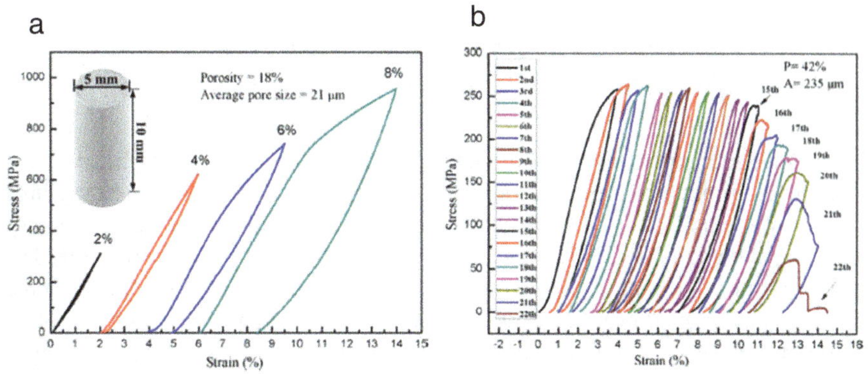

Fig. 23 (**a**) Stress-strain curves of loading-unloading compressive tests of the porous NiTi with different pre-strain of 2%, 4%, 6% and 8%. (**b**) The cycling stress-strain curves of NiTi at pre-strain of 4% under 22 loading-unloading (Zhang et al. 2015)

different strain levels. The superelastic recovery ratio can be deduced with an increase of pre-strain level of NiTi.

The stress-strain curve of porous NiTi is plotted under loading and unloading of compressive test under different pre-strain levels as shown in Fig. 23a. The tremendous superelasticity can be observed at the strain level of 2%. The superelastic recovery ratio of the porous NiTi alloys can be diminished by increasing the porous structure and pore size. The superelastic recovery strain of the porous NiTi alloy could be developed up to 90% by training, but with further increasing of training cycles (22 cycles), the stress-strain curves obtained an obvious ladder-shaped stress plateau, which indicated the collapse of the pores (Fig. 24a). Additionally, increasing the porosity and pore size of NiTi decreased the elastic modulus and compressive strength of the porous NiTi alloys (Zhang et al. 2015). Hence, the selection of strain and the sintering temperature and pressure which determine the level of porosity are the important factors for calculating the mechanical stress-strain relationship.

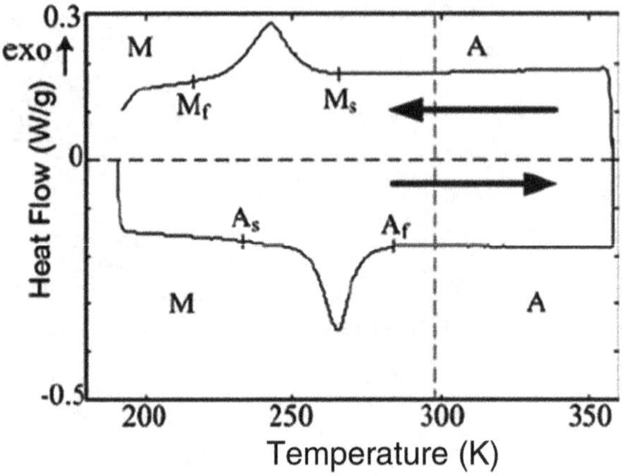

Fig. 24 Schematic of DSC curve for NiTi alloy

5.3 *Influences of SPS Parameters on Phase Transformation and Microstructural Characteristics of NiTi SMA*

The unique properties of NiTi alloy such as SME and SE can be determined by the phase transformation temperatures of NiTi SMAs. The transformation temperatures such as M_s, M_f, A_s and A_f can be measured using differential scanning calorimetry (DSC) as shown in Fig. 24. The path transformation between these temperatures has been executed the martensite and austenite transformation of NiTi alloy. Three different phases have been observed throughout the martensitic transformation of NiTi such as B_2 (cubic), R (rhombohedral) and $B_{19'}$ (monoclinic). The high-temperature austenitic B_2 phase undergoes a deformed structure transformation to intermediate R phase and transforms to the low-temperature martensite $B_{19'}$ phase. The occurrence of intermediate phases depends on the composition and processing temperature of NiTi alloy which are explained in the previous section. Generally, the phase transformation in NiTi alloy is influenced by SPS process parameters such as temperature and pressure through the formation of secondary phases. The transformation temperature influenced by SPS parameters such as temperature and pressure decomposes the atomic structure. The transformation temperature can be amended by the atomic distance shrinkage with the help of pressure (Yamamoto et al. 2007).

The DSC analysis of nanostructured NiTi powders is shown in Fig. 25. The one broad peak was observed in the heating curve which leads the martensite (monoclinic) to austenite (cubic) transformation, and another one broad peak found on the cooling curve relates to the transformations between the martensite and austenite (Shearwood et al. 2005).

The DSC analysis of sintered NiTi at different temperatures is shown in Fig. 26. The stronger phase transformation can be observed from the samples sintered at 700 °C and 800 °C compared to that of powder DSC. The differences between the starting and finishing temperature of transformation of sintered NiTi have considerably confined with an increase of sintering temperature. Moreover, the austenite

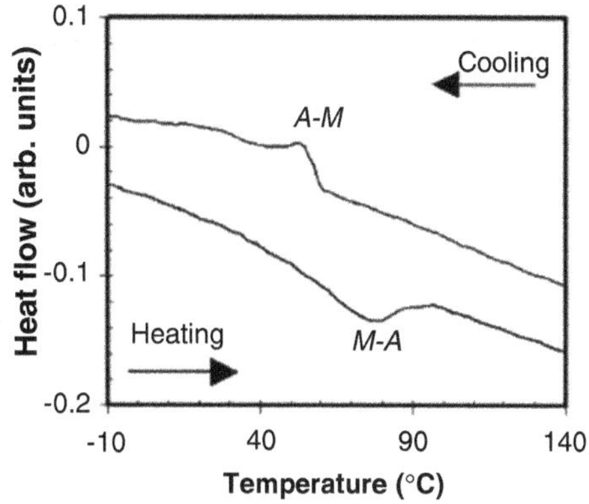

Fig. 25 DSC results for TiNi nano-powder (Shearwood et al. 2005)

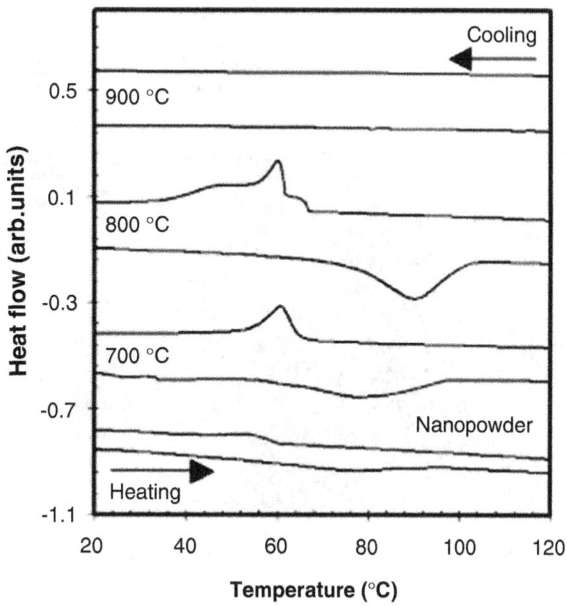

Fig. 26 DSC curves for NiTi sintered at different temperatures

transformation temperature increases with an increase in the sintering temperature. There is no major peak on DSC curve observed in the sample sintered at 900 °C indicating that there is no oxidation at this higher temperature (Shearwood et al. 2005).

The secondary phases such as Ni_3Ti, $NiTi_2$ and Ni_4Ti_3 affect the phase transformation behaviour of NiTi alloy. Particularly, precipitation of Ni_4Ti_3 can be used to determine the martensitic transformation path and temperature in NiTi shape memory alloys. A strong bonding between Ni_4Ti_3 and NiTi during martensite transformation enhances the microstructural characterization as shown in Fig. 27. The transformation path of B_2–$B_{19'}$ and B_2–R–$B_{19'}$ in NiTi alloys consists of Ni_4Ti_3

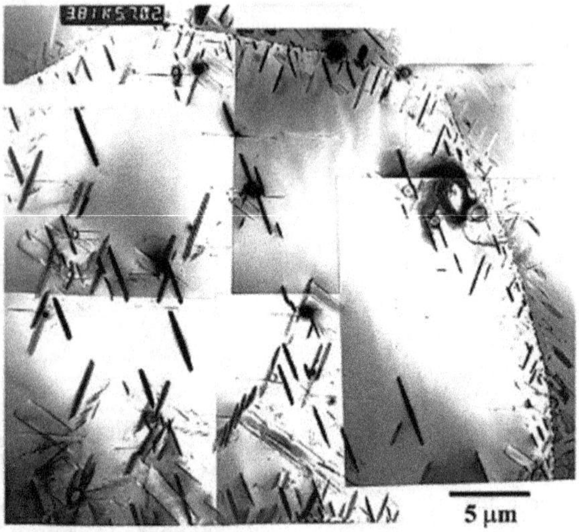

Fig. 27 TEM micrograph showing the presence of Ni4Ti3 precipitates in NiTi alloy (Khalil-Allafi et al. 2002)

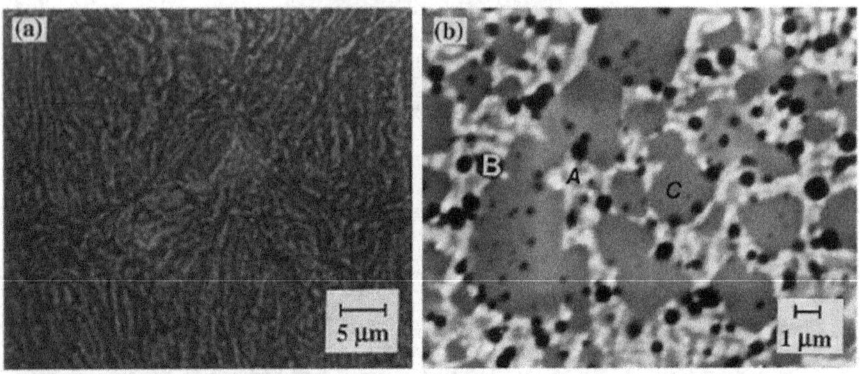

Fig. 28 SEM micrographs of NiTi sintered at (**a**) 900 °C and (**b**) 1100 °C (Ye et al. 1998)

precipitate which can be identified in the B_2 austenite matrix (Otsuka and Ren 2005; Zhou et al. 2010).

The milled NiTi powders were sintered using SPS at 900 and 1100 °C. The microstructures of these two are different and shown in Fig. 28. The structure of the sample sintered at 900 °C shows the lamella-like structure, whereas the 1100 °C sintered sample shows the parent phase with precipitates such as (A) Ni_3Ti, (B) $NiTi_2$ and (C) NiTi. Moreover the presence of precipitates has been ensued in XRD profile of sintered NiTi as shown in Fig. 29 (Ye et al. 1998). In this case, the major phase identified from the sample sintered at 900 °C consists of Ni_3Ti and $NiTi_2$. Similarly, the SPS-sintered sample at 1100 °C contains major phase of Ni_3Ti only. The hardness of this specimen could be reduced from 1076 to 635 Hv which revealed the effect of precipitates.

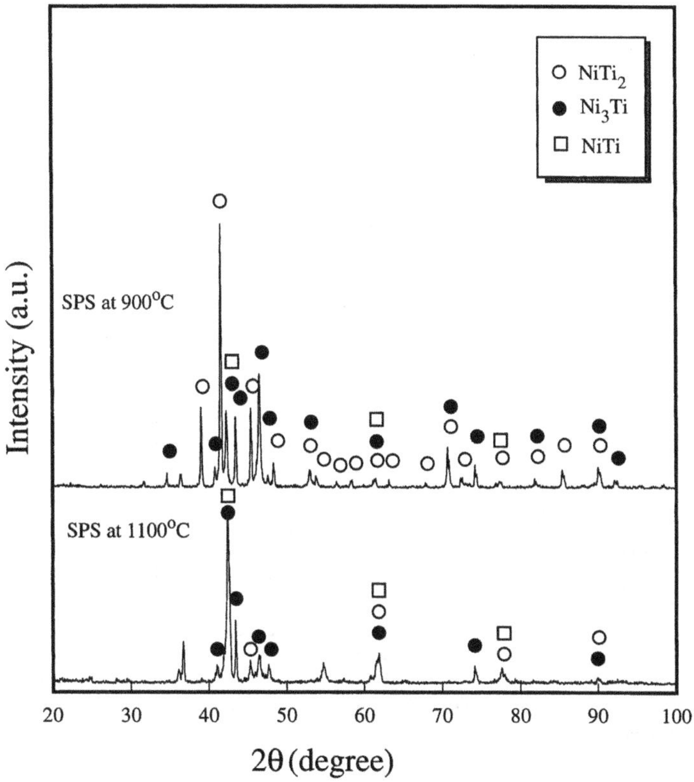

Fig. 29 X-ray diffraction patterns of SPS-sintered NiTi (Ye et al. 1998)

The addition of ternary alloy with NiTi has influenced the phase transformation behaviour. The addition of carbon nanotubes (CNT) with NiTi was investigated at different volume percentage. The DSC curves of Fig. 30 show that the addition of CNT could affect the phase transformation temperature. The phase transformation temperatures can be observed to be decreased due to the reaction CNT with Ti during cooling and heating, corresponding to B_2-$B_{19'}$ and $B_{19'}$-B_2, respectively (Chen et al. 2011; Lee et al. 2014).

6 Summary

In the preceding sections, various aspects of NiTi SMAs have been discussed. The basic characteristics of NiTi SMA and their tremendous properties have been reported. The effects of changing the parent phase composition and addition of external element were deliberated in detail on the phase transformation behaviour of NiTi alloy. The specific applications of NiTi have been mentioned in each section

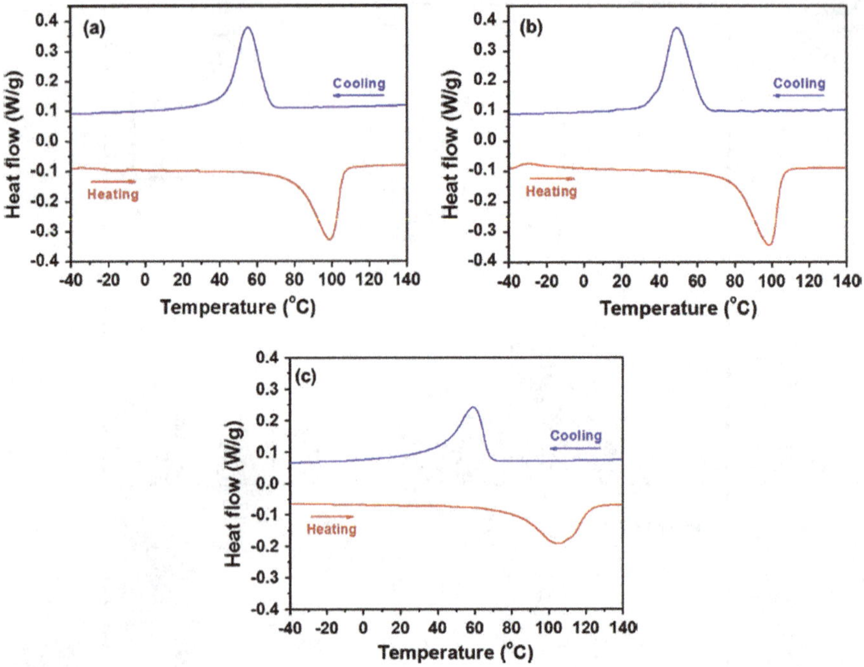

Fig. 30 DSC curves of the (**a**) NiTi, (**b**) 0.5 vol % CNT/NiTi and (**c**) 1.0 vol % TiC/NiTi composites (Lee et al. 2014)

under different properties where it needs to be improved from existing. Moreover, the potential manufacturing methods that produce NiTi with its unique properties have been discussed. The origin of NiTi and its classification based on some other SMAs is detailed. The rapid manufacturing method of SPS and its working mechanism with process parameters have stated especially for NiTi alloy.

The rapid manufacturing technique such as SPS can synthesize the NiTi alloy with higher densification. The parameters that affect mechanical and phase transformation behaviour are sintering temperature and pressure. In Sect. 5, the effects of SPS parameters on mechanical parameters such as density and hardness were investigated. Moreover, the phase transformation behaviour of NiTi and the modification of transformation temperatures through SPS parameters have also been explained. The application of NiTi can be decided by the austenite finish temperature which could be modified by aging temperature, Ni composition and addition of ternary alloy. In addition to these existing methods, SPS sintering temperature and pressure also can be modified in the transformation phase. The evolution of mechanical and microstructural characteristics has been done with the effects of precipitates as such Ni_3Ti, $NiTi_2$ and Ni_3Ti_4. Based on this study, SPS process is found to be most suitable for NiTi fabrication with required functional properties. The amendment of phase transformation has been achieved successfully by SPS parameters.

The alteration of precipitates that can provide favourable properties in NiTi system can be made through SPS process.

7 Future Direction

The maximum available existing works for NiTi fabrication are conventional melting processes and powder synthesis processes. These processes have reported about the porosity and density percentage by varying Ni composition. Limited studies are carried out to enhance the microstructural and mechanical characteristics of NiTi alloy. Very few investigations were made on the alteration of phase transformation temperatures of NiTi SMAs. The studies are required for NiTi-based SMAs using SPS with the variations of parameters other than temperature and pressure such as punch displacement, holding time and electrical voltage. With very few research works being carried out to evaluate the evolution of microstructure as a function of SPS parameters, a comprehensive analysis focusing on the interaction among SPS input parameters, microstructure and mechanical properties in the case of NiTi SMA will provide necessary inputs to the industry. Most of the SPS studies were done for cylindrical-shaped NiTi samples which restricted the testing of them in tensile loading. In order to investigate the superelasticity property, investigation is necessary to enhance the size of the sample which can adopt in uniaxial tensile mode. Currently, superelasticity has been calculated using compression and rarely micro-tensile analysis. Moreover, it is necessary to investigate the ternary alloys such as Hf, Pd, Si, Cu and Zn which can modify the phase transformation temperatures of NiTi binary alloy. The SPS-processed samples were used only for analysis purposes, which needs to be improved into the making of products. Therefore, a special attention is required on traditional and nontraditional machining processes which can be used to make the products from NiTi SMAs fabricated through SPS process.

References

Arunkumar S, Kumaravel P, Velmurugan C, Senthilkumar V (2018) Microstructures and mechanical properties of nanocrystalline NiTi intermetallics formed by mechanosynthesis. Int J Miner Metall Mater 25(1):80–87

Bonifacio CS, Rufner JF, Holland TB, Van Benthem K (2012) In situ transmission electron microscopy study of dielectric breakdown of surface oxides during electric field-assisted sintering of nickel nanoparticles. Appl Phys Lett 101:4583–4586

Bram M, Ahmad-Khanloua A, Heckmannb A, Fuchsa B, Buchkremera HP, Stovera D et al (2002) Powder metallurgical fabrication processes for NiTi shape memory alloy parts. Mater Sci Eng A 337:254–263

Buehler WJ, Gilfrich JV, Wiley RC (1963) Effect of low-temperature phase changes on the mechanical properties of alloys near composition TiNi. J. Appl. Phys., 34(5):1475–1477

Chen F, Tong YX, Lu XL, Wang X, Tian B, Li L et al (2011) Effect of graphite addition on martensitic transformation and damping behavior of NiTi shape memory alloy. Mater Lett 65 (7):1073–1075

Cheng Y, Cui Z, Cheng L, Gong D, Wang W (2017) Effect of particle size on densification of pure magnesium during spark plasma sintering. Adv Powder Technol 28(4):1129–1135

Chu CL, Chung PH, Lin WSD (2005) Fabrication and properties of porous NiTi shape memory alloys for heavy load-bearing medical applications. J Mater Process Technol 169(1):103–107

Cristea CD, Lungu M, Balagurov AM, Marinescu V, Culicov O, Sbarcea G, Cirstea V (2015) Shape memory NiTi and NiTiCu alloys obtained by spark plasma sintering process. In: Advanced engineering forum, vol. 13. Trans Tech Publications, pp 83–90

De Santis S, Trochu F, Ostiguy G (2001) Stress-strain hysteresis and damping in MnCu and NiTi alloys. Metall Mater Trans A 32(10):2489–2498

Duerig TW, Pelton AR (2002) An overview of superelastic stent design. In: Materials science forum, vol 394. Trans Tech Publications, Aedermannsdorf, pp 1–8

Elahinia M (2016) Shape memory alloy actuators: design, fabrication and experimental evaluation, 1st edn. John Wiley and Sons ltd, Chichester

Elahinia MH, Hashemi M, Tabesh M, Bhaduri SB (2012) Manufacturing and processing of NiTi implants: a review. Prog Mater Sci 57(5):911–946

Elahinia M, Moghaddam NS, Andani MT, Amerinatanzi A, Bimber BA, Hamilton RF (2016) Fabrication of NiTi through additive manufacturing: A review. Prog Mater Sci 83:630–663

Fischer-Cripps AC (2011a) Factors affecting nanoindentation test data. In: Nanoindentation. Springer, New York, pp 77–104

Fischer-Cripps AC (2011b) Contact mechanics. In: Nanoindentation. Springer, New York, p 2

Fremond M, Miyazaki S (2014) Shape memory alloys, vol 351. Springer, New York

Frenzel J, George EP, Dlouhy A, Somsen C, Wagner MX, Eggeler G (2010) Influence of Ni on martensitic phase transformations in NiTi shape memory alloys. Acta Mater 58(9):3444–3458

Fu Y, Moochhala S, Shearwood C (2004) Spark plasma sintering of TiNi nanopowders. In: BioMEMS and nanotechnology, vol. 5275. International Society for Optics and Photonics, pp 9–18

Gou L, Liu Y, Ng TY (2015) Effect of Cu Content on Atomic Positions of $Ti_{50}Ni_{50\text{-}x}Cu_x$ Shape Memory Alloys Based on Density Functional Theory Calculations. Metals 5(4):2222–2235

Groza JR (2007) Nanocrystalline powder consolidation methods. In: Koch CC (ed) Nanostructured materials, 2nd edn. William Andrew Publishing, Norwich, pp 173–233

GurtSantanach J, Weibel A, Estourne's C, Yang Q, Laurent C, Peigney A (2011) Spark plasma sintering of alumina: study of parameters, formal sintering analysis and hypotheses on the mechanism(s) involved in densification and grain growth. Acta Mater 59:1400–1408

Jani JM, Leary M, Subic A, Gibson MA (2014) A review of shape memory alloy research, applications and opportunities. Mater Des 56:1078–1113

Jiang SY, Zhao YN, Zhang YQ, Li HU, Liang YL (2013) Effect of solution treatment and aging on microstructural evolution and mechanical behavior of NiTi shape memory alloy. Trans Nonferrous Met Soc China 23(12):3658–3667

Johnson KL (1974) Contact mechanics 1985. Cambridge University Press, Cambridge, pp 57–63

Karimzadeh M, Aboutalebi MR, Salehi MT, Abbasi SM, Morakabati M (2016) Adjustment of aging temperature for reaching superelasticity in highly Ni-rich Ti-51.5 Ni NiTi shape memory alloy. Mater Manuf Process 31(8):1014–1021

Khalil-Allafi J, Dlouhy A, Eggeler G (2002) Ni_4Ti_3-precipitation during aging of NiTi shape memory alloys and its influence on martensitic phase transformations. Acta Mater 50 (17):4255–4274

Kumar PK, Lagoudas DC (2008) Introduction to shape memory alloys. In: Shape memory alloys. Springer, New York, pp 1–51

Laptev AV (2007) Theory and technology of sintering, thermal and chemicothermal treatment. Powder Metall Met Ceram 46(7–8):317–324

Lee J, Hwang J, Lee D, Ryu HJ, Hong SH (2014) Enhanced mechanical properties of spark plasma sintered NiTi composites reinforced with carbon nanotubes. J Alloys Compd 617:505–510

Li BY, Rong LJ, Li YY (2000) Stress–strain behavior of porous Ni–Ti shape memory intermetallics synthesized from powder sintering. Intermetallics 8(5–6):643–646

Li YH, Rong LJ, Li YY (2002) Compressive property of porous NiTi alloy synthesized by combustion synthesis. J Alloys Compd 345(1–2):271–274

Li P, Karaca HE, Cheng YT (2015) Spherical indentation of NiTi-based shape memory alloys. J Alloys Compd 651:724–730

Ma J, Karaman I, Noebe RD (2010) High temperature shape memory alloys. Int Mater Rev 55 (5):257–315

Manosa L, Jurado M, Planes A, Zarestky J, Lograsso T, Stassis C (1994) Elastic constants of bcc. Cu-Al-Ni alloys. Phys Rev B 48:9969

Mitchell MR, Jerina KL (2007) Fatigue and fracture of medical metallic materials and devices (No. 1481). ASTM International, p 18

Nakata Y, Tadaki T, Shimizu K (1998) In: Inoue K, Mukherjee K, Otsuka K, Chen H (eds) Displacive phase transformations and their applications in materials engineering. TMS, Warrendale, p 187

Ni W, Cheng YT, Grummon DS (2002) Recovery of microindents in a nickel–titanium shape-memory alloy: a self-healing effect. Appl Phys Lett 80(18):3310–3312

Orru R, Licheri R, Locci AM, Cincotti A, Cao G (2009) Consolidation/synthesis of materials by electric current activated/assisted sintering. Mater Sci Eng R 63(4–6):127–287

Otsuka K, Ren X (2005) Physical metallurgy of Ti–Ni-based shape memory alloys. Prog Mater Sci 50(5):511–678

Patoor E, Lagoudas DC, Entchev PB, Brinson LC, Gao X (2006) Shape memory alloys, Part I: General properties and modeling of single crystals. Mech Mater 38(5–6):391–429

Predki W, Knopik A, Bauer B (2008) Engineering applications of NiTi shape memory alloys. Mater Sci Eng A 481:598–601

Qiu Y, Young ML, Nie X (2015) High strain rate compression of martensitic NiTi shape memory alloys. Shape Memo Superelast 1(3):310–318

Qiu Y, Young ML, Nie X (2017) High Strain Rate Compression of Martensitic NiTi Shape Memory Alloy at Different Temperatures. Metal Mater Trans A 48(2):601–608

Rao A, Srinivasa AR, Reddy JN (2015) Design of shape memory alloy (SMA) actuators, vol 3. Springer, Heidelberg

Resnina N, Belyaev S, Shelyakov A, Ubyivovk E (2017) Violation of the sequence of martensite crystals formation on cooling and their shrinking on heating during B2↔B19 martensitic transformation in $Ti_{40.7}Hf_{9.5}Ni_{44.8}Cu_5$ shape-memory alloy. Phase Transit 90(3):289–298

Seelecke S, Muller I (2004) Shape memory alloy actuators in smart structures: Modeling and simulation. Appl Mech Rev 57(1):23–46

Shao Y, Guo F, Ren Y, Zhang J, Yang H, Jiang D et al (2017) NiTi-enabled composite design for exceptional performances. Shape Memo Superelast 3(1):67–81

Shearwood C, Fu YQ, Yu L, Khor KA (2005) Spark plasma sintering of TiNi nano-powder. Scr Mater 52(6):455–460

Shishkovsky I, Yadroitsev I, Smurov I (2012) Direct selective laser melting of nitinol powder. Phys Procedia 39:447–454

Stoeckel D (1995) The shape memory effect-phenomenon, alloys and applications. In: Proceedings: Shape Memory Alloys for Power Systems EPRI (1), Fremont, pp 1–13

Sun L, Huang WM (2009) Nature of the multistage transformation in shape memory alloys upon heating. Met Sci Heat Treat 51(11):573–578

Tadaki T (1998) Cu-based shape memory alloys. In: Otsuka K, Wayman CM (eds) Shape memory materials. Cambridge University Press, Cambridge, pp 97–116

Tadayyon G, Mazinani M, Guo Y, Zebarjad SM, Tofail SA, Biggs MJ (2016) The effect of annealing on the mechanical properties and microstructural evolution of Ti-rich NiTi shape memory alloy. Mater Sci Eng A 662:564–577

Tang W (1997) Thermodynamic study of the low-temperature phase B19' and the martensitic transformation in near-equiatomic Ti-Ni shape memory alloys. Metall Trans 28A:537

Tokita M (2013) Handbook of advanced ceramics: Chapter 11.2. 3. Spark Plasma Sintering (SPS) method, systems, and applications. Elsevier Inc. Chapters

Velmurugan C, Senthilkumar V, Dinesh S, Arulkirubakaran D (2017) Machining of NiTi-shape memory alloys-A review. Mach Sci Technol. https://doi.org/10.1080/10910344.2017.1365894

Wagner MX, Dey SR, Gugel H, Frenzel J, Somsen C, Eggeler G (2010) Effect of low-temperature precipitation on the transformation characteristics of Ni-rich NiTi shape memory alloys during thermal cycling. Intermetallics 18(6):1172–1179

Wen YH, Peng HB, Raabe D, Gutierrez-urrutia I, Chen J, Du YY (2014) Large recovery strain in Fe–Mn–Si-based shape memory steels obtained by engineering annealing twin boundaries. Nat Commun 5:4964

Yamamoto T, Sakuma T, Uchida K, Sutou Y, Yamauchi K (2007) Effect of heat aging on thermal and mechanical properties of Ti-Ni-Nb shape memory alloy. Mater Trans 48(3):439–444

Ye LL, Liu ZG, Raviprasad K, Quan MX, Umemoto M, Hu ZQ (1998) Consolidation of MA amorphous NiTi powders by spark plasma sintering. Mater Sci Eng A 241(1–2):290–293

Zhang L, Zhang YQ, Jiang YH, Zhou R (2015) Superelastic behaviors of biomedical porous NiTi alloy with high porosity and large pore size prepared by spark plasma sintering. J Alloys Compd 644:513–522

Zhao Y, Taya M, Kang Y, Kawasaki A (2005) Compression behavior of porous NiTi shape memory alloy. Acta Mater 53(2):337–343

Zhou N, Shen C, Wagner MX, Eggeler G, Mills MJ, Wang Y (2010) Effect of Ni_4Ti_3 precipitation on martensitic transformation in Ti–Ni. Acta Mater 58(20):6685–6694

Part VIII
SPS of Ti-Based Alloys

Spark Plasma Sintering of Titanium-Based Materials

O. O. Ayodele, M. B. Shongwe, B. A. Obadele, and P. A. Olubambi

1 Introduction

In the past years, there has been an increase in consumption of steel, aluminum (Al), and magnesium (Mg) as engineering materials for industrial applications because of their good properties even though they fail to meet up with some specific service condition at high operating temperatures (Henriques et al. 2005). Thus, the search for materials that would replace Al and Mg and surpass their functional requirements has been a matter of interest. The recent development in titanium (Ti) as the fourth most plentiful structural material after aluminum, iron, and magnesium (Lutjering and Williams 2003), has led to the drive in industrial applications (Malinov et al. 2004). Ti have been extensively utilized in aerospace, mechanical, chemical, biomedical, and automobile industries due to its excellent resistance to corrosion, high strength to weight ratio, biocompatibility, toughness, and high melting point properties.

As a result of these notable properties, Ti has been highly demanded, but it displays poor wear resistance and low hardness (Revankar et al. 2017) when compared with other materials. However, these properties have been further improved through alloying with other metals like Al, V, and Sn (Moraes et al. 2014). Titanium-based materials can be processed into near net shape of multiplex profile, although the process or fabrication routes tend to pose a lot of challenges in the titanium industry (Dehghan-Manshadi et al. 2017). Aside from the complex routes in titanium-based productions, it has been observed that the fabrication

O. O. Ayodele (✉) · B. A. Obadele · P. A. Olubambi
Center for Nanoengineering and Tribocorrosion, School of Mining, Metallurgy and Chemical Engineering, University of Johannesburg, Johannesburg, South Africa

M. B. Shongwe
Institute for Nanoengineering Research, Department of Chemical, Metallurgy and Materials Engineering, Tshwane University of Technology, Pretoria, South Africa
e-mail: shongwemb@tut.ca.za

© Springer Nature Switzerland AG 2019

P. Cavaliere (ed.), *Spark Plasma Sintering of Materials*,
https://doi.org/10.1007/978-3-030-05327-7_23

inefficiencies in the production of these components account for their high cost. They are very sensitive to oxygen and nitrogen (Moiseyev 2006). They tend to oxidize which makes the geometry process of titanium and its alloys demanding, and this contributes to the excessive cost of the final components (Lütjering and Williams 2007). Manufacturing intricate components of titanium-based alloys through machining sets a difficult task and results in a huge loss of alloys as waste. In order to minimize this waste, near net shape methods of fabricating these alloys have been utilized via powder metallurgy techniques (Qian 2010).

Powder metallurgy (PM) process has been used to fabricate titanium-based materials as a result of its low cost, good surface finish, increase in material yield, and variation of composition (Tang et al. 2013). It offers great advantage over conventional manufacturing process such as casting due to ease of near net shape fabrication. Nanostructure powders can be consolidated into bulk material through PM with the aid of low sintering temperature and fast solidification (Chen et al. 2009). Some innovative sintering techniques such as hot pressing (HP), microwave sintering, and spark plasma sintering (SPS) have been reported for fabrication of engineering materials (Obadele et al. 2017). Spark plasma sintering utilizes pulsed on-off direct current that causes a quick consolidation of powders, and it has been observed to be an efficient method of fabricating nanostructured materials (Weston et al. 2015). Spark plasma sintering (SPS) is a newly adaptable method to manufacture a number of materials including titanium-based alloys (Feng et al. 2008). Research has proven spark plasma sintering as a better alternative as compared to conventional sintering methods (Lee et al. 2003) because it eliminates grain growth. The aim of this book chapter is to synthesize titanium-based materials with the aid of SPS.

2 Spark Plasma Sintering Techniques

Spark plasma sintering is well known as field-assisted sintering techniques (FAST) or pulsed electric current sintering (PECS) which is used in fabricating titanium-based materials at a lower sintering temperature (Wei et al. 2012). In this technique titanium powders are charged in a graphite die, and a uniaxial force is exerted during sintering. The heating is accomplished by spark discharges in voids between the particles. As a result of this discharge, the particle surface is refined and energized, and a joule heating is generated between the particles by the electric current (Munir et al. 2006). The electric current application facilitates a fast heating rate ensuing in a very short sintering cycle; densification is achieved in a short sintering time for both conductive and nonconductive powder, with this method grain growth is minimized (Shongwe et al. 2017).

SPS is utilized to manufacture nanocrystalline materials by suppressing the grain size due to its fast heating rate (Matizamhuka 2016), and it is an improvement on the hot pressing technique (Kessel and Hennicke 2007; Suárez et al. 2013). Research shows that in hot pressing technique, heat is being dissipated, and the material gets indirect heat which tends to affect its density, but SPS is able to attain 100% theoretical density during sintering without the application of binders (Khalil 2012; Aalund 2008; Rajeswari et al. 2010; Kessel et al. 2009). SPS are characterized with homogeneous sintering, fabrication of several materials, rapid heating rate, high-energy efficiency, minimized grain growth, and better activation of powder surface.

2.1 History and Working Principle of Spark Plasma Sintering

Studies have shown that SPS was patented in 1906 (Bloxam 1906) for the fabrication of metallic powders. After a few years, additional work was conducted to improve the SPS (Inoue 1966), but as technology advances, there was a great development on the sintering ability by Lockheed Missile and Space Co., having worked on Inoue's research. At this time, near net shape components were manufactured, and Sumitomo Coal mining Co. Ltd. (Japanese company) began the commercial fabrication of spark plasma machines (Boesel and Goetzel 1971; Grasso et al. 2009). After a while, FCT Systeme GmBH based in Germany and Thermal Technology LLC based in the USA began manufacturing close equipment based on pulsed DC current (Guillon et al. 2014). Spark plasma sintering comprises of a graphite die where the powders are placed, upper and lower punches, a water-cooled reaction chamber, a vacuum or gas atmosphere (environment), a pulsed DC generator, and pyrometer which measures the temperature with a system monitor (Khalil 2012; Aalund 2008; Sergueeva et al. 2009; Kessel and Hennicke 2007). Figure 1 represents the fundamental arrangement of SPS showing the operational parts of the machine.

The SPS machine generates high electric current due to the application of a low voltage. Temperature of about 2400 °C can be obtainable with the use of graphite tools and specified cooling rates ranging from 150 °C/min to 400 °C/min under auxiliary gas cooling. SPS operates with the utilization of loads of 50–250 KN under vacuum which increases the rate of densification process (Guillon et al. 2014). It has been reported to possess a rapid heating rate (1000 °C/min) which facilitates the fabrication of materials within a shorter cycle (Suarez et al. 2010). The mechanism of the SPS is very unique due to the passage of electric current between the powder particles. The particle surfaces are heated, and plasma is generated. The particle surface is refined and energized as a result of the temperature, this leads to the vaporization of the impurities on the powder particles. The particles later melt and joined together forming a neck. In the second stage, joule heat is produced by the electrical current which moves through the particles through the neck. Joule heat raises the diffusion of the particles in the neck, facilitating their growth. The action of even temperature distribution causes the growth in the material to diminish. Finally, plastic deformation occurs as a result of pressure when the heated material has become soft. However, the compact is obtained due to densification. The illustration below indicates the on-off direct current path and how current flows through the powder particles in the SPS techniques using conductive materials, die and punches.

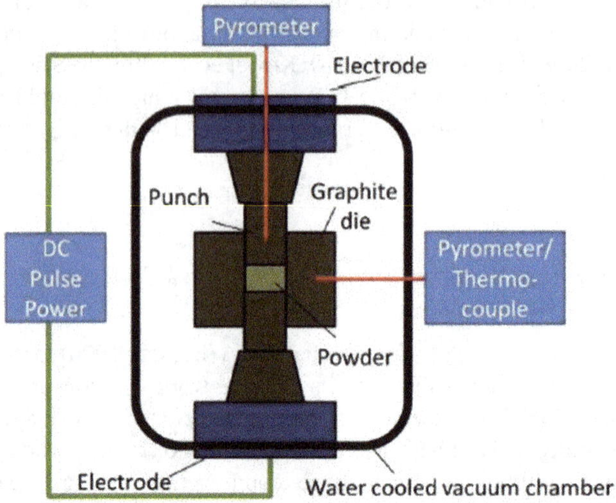

Fig. 1 Schematic of spark plasma sintering (Guillon et al. 2014)

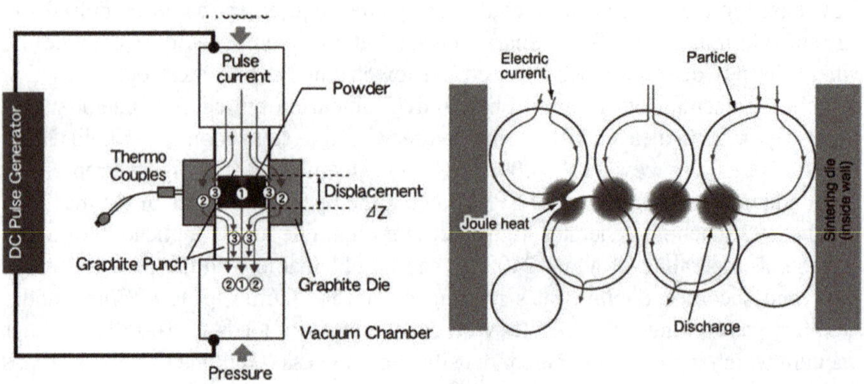

Fig. 2 On-off current path and flow between powder particles (Chen et al. 2005)

Figure 2 shows the path at which current passes between the particles and the possible effect of the joule heat on the particles.

Consequently, the SPS have some features that are of advantage as compared to other conventional or traditional sintering techniques. SPS has a rapid sintering process with homogeneous sintering, the particle surfaces are purified, binders are not required, operation is easy, several materials can be fabricated, and it minimized grain growth. Nevertheless, the temperature profiles observed during simulation indicates those small thermal gradients are produced at fast heating rate in SPS as compared to other sintering techniques (Suárez et al. 2013). The temperature profiles in Fig. 3 show the differences in the sintering operations of hot pressing and SPS. There is a little change in temperature (Δ T), at the edge (TA), and core of the SPS

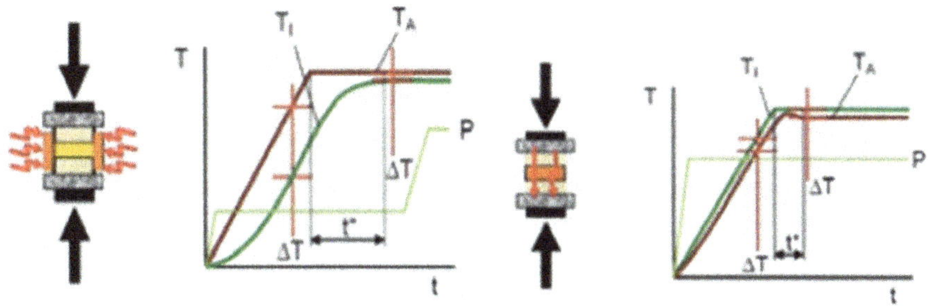

Fig. 3 Temperature profiles in hot pressing (left) and SPS (right) (Suárez et al. 2013)

material as compared to the hot-pressed material (Suárez et al. 2013). Big thermal gradients in the hot pressing techniques could lead to internal stresses and grain growth which affect the mechanical properties of the material, while the SPS eliminates grain growth due to little thermal gradient (Matizamhuka 2016). This method has become a dire need in the fabrication of titanium and titanium-based materials.

3 Titanium and Titanium-Based Materials

Titanium was discovered by William Gregor in 1791. It exists as mineral deposits in the form of rutile (TiO2) and ilmenite (FeTiO3). Titanium is being extracted from its ores by Kroll process; it belongs to the family of transition metals with the melting point 1668 °C and density of 4.50 g/cm^3 (Reis de Vasconcellos et al. 2012). Ti has the atomic weight of 47.90, elastic modulus of 115GPa, and the tensile strength between 210 and 1380 MPa. Titanium exhibits hexagonal close packed structure (HCP) at room temperature called α phase or titanium while at elevated temperature (883 °C); it exists as body-centered cubic called β phase or titanium. These crystal structures account for various development of titanium-based materials or properties as a result of alloying with other materials (Veiga et al. 2012). The alloying metals for titanium can be classified into neutral, α- stabilizers, and β-stabilizers (Peters et al. 2003). The neutral phase has a little effect on the β-transus temperature, meanwhile the β-transus temperature is lowered by the β-stabilizers, and it is increased by α-stabilizers. Aluminum (Al), oxygen, and nitrogen fall under α-stabilizers elements, while iron, vanadium, molybdenum, and chromium are β-stabilizers elements. However, the β- stabilizers are subdivided into β-isomorphous and β-eutectiod elements (Veiga et al. 2012) (Figs. 4 and 5).

Ti alloys can be improved through heat treatment in order to optimize the fracture, toughness, and fatigue strength properties. Ti is further grouped into main categories which include alpha alloys, beta alloys, and alpha-beta alloys.

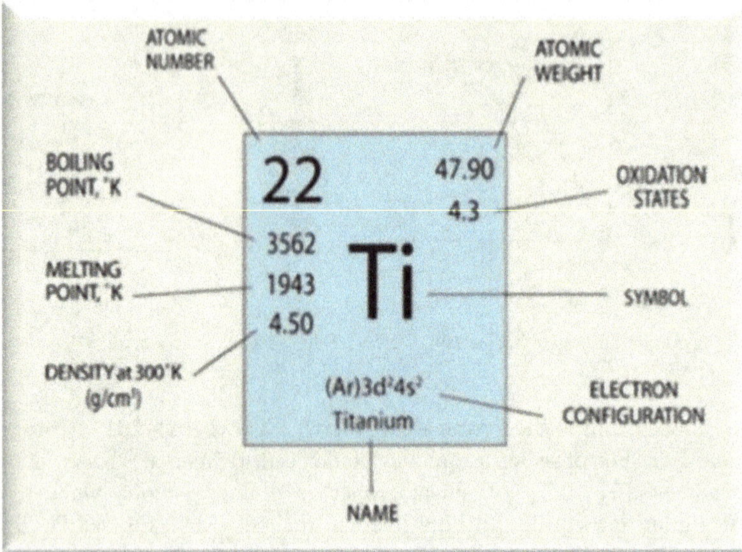

Fig. 4 Physical and chemical properties of titanium (www.dierk-raabe.com)

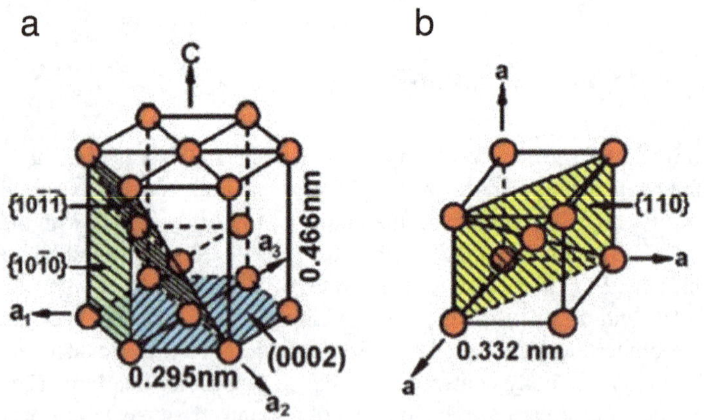

Fig. 5 Structure of (**a**) HCP α and (**b**) BCC β phase (Lütjering and Williams 2007)

- Alpha alloys are non-heat treatable and weldable and possess good ductility, low strength, toughness, and cryogenic features. An example of elements includes aluminum (Al), oxygen (O_2), and nitrogen (N_2).
- The beta alloys are easily heat treatable, weldable, and with high strength; this includes iron (Fe), vanadium (V), chromium (Cr), and manganese (Mn).

Alpha-beta alloys are heat treatable with medium strength (Boyer and Briggs 2005). They possess excellent combination of strength and ductility; microstructural changes can be attained which provides strengthening of the material. Other heat

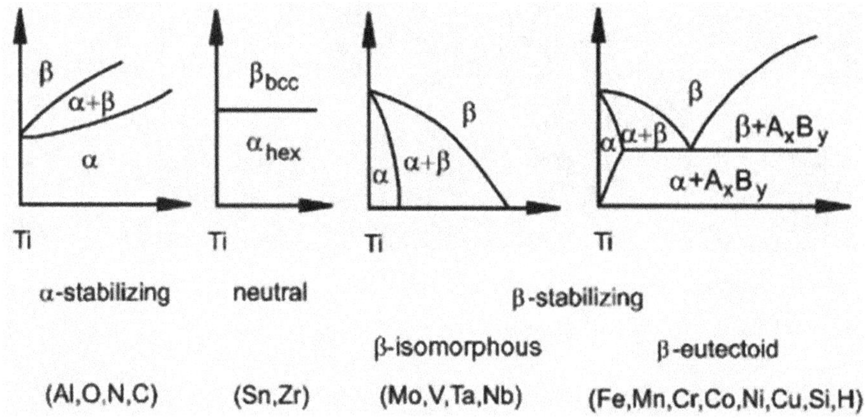

Fig. 6 Effect of alloying elements on phase diagram of titanium alloys (Leyens and Peters 2003)

treatment methods are utilized in this group. α and β alloys are commonly utilized due to their attractive mechanical display. The class of elements or materials under this group includes Ti-6Al-4V. Alloying elements has played a role in formation of titanium-based materials or composites. Examples of titanium-based materials include Ti-6Al-4V, TiAl, Ti-10V-2Fe-3Al, Ti-6Al-2Sn-4Zr-3Mo, etc. The goal of this book chapter is to highlight the spark plasma sintering of titanium-based alloys (Fig. 6).

3.1 Spark Plasma Sintering of Graphitized MWCNT Reinforced Ti-6Al-4V

Ti-6Al-4V alloys are widely used in titanium alpha-beta group due to their beautiful thermomechanical features, which have made them relevant in biomedical components like dental implants and surgical instruments (Veiga et al. 2012). They also contribute to 45% of titanium fabrication (Henriques et al. 2005), although this alloy is limited in use due to their poor tribological properties such as poor wear resistance (Adegbenjo et al. 2017), and in order to improve on this properties, the alloys are reinforced with other materials. However powder metallurgy techniques have been used to fabricate these alloys because of their low cost, among these techniques is spark plasma sintering (SPS) which fabricates at lower temperature due to its fast heating rate. The synthesis of Ti-6Al-4V was investigated using the spark plasma sintering by reinforcing with carbon nanotubes (CNT). Different weight percentage of CNT (1, 2, 3 wt. %) were dispersed in Ti-6Al-4V via high-energy ball milling, and it was homogeneously dispersed in order to avoid damages to the CNT, also avoiding interfacial reaction between the matrix and the CNT. The admixed powders were sintered between the temperature ranges of 850–1000 °C with pressure of

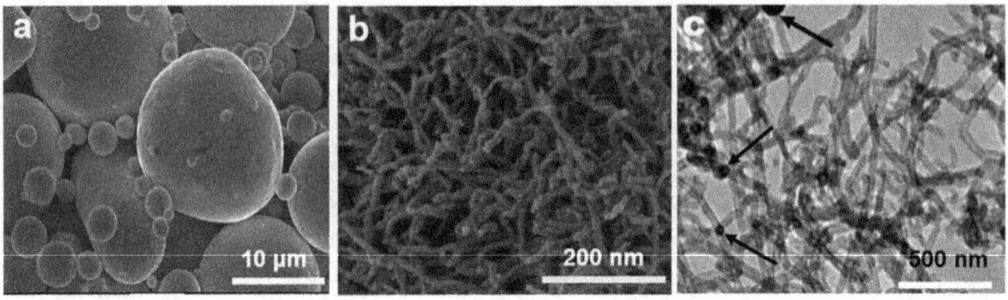

Fig. 7 SEM evolutions of starting powders: (**a**) Ti6Al4V, (**b**) MWCNTs, and (**c**) TEM image of MWCNTs (Adegbenjo et al. 2017)

50 MPa and heating rate at 100 °C/min. The microstructural and phase evolutions of the sintered powder were observed through scanning electron microscopy (SEM), x-ray diffractions (XRD), and transmission electron microscope (TEM).

Figure 7 connotes the evolution of the starting powders of Ti6Al4V and CNTs prior to mixing and sintering. The CNTs are highly entangled as seen from the TEM analysis. It has been observed from literature that during milling of MWCNTs +Ti64 via milling balls, MWCNTs could be structurally deficient due to the energy absorbed, and this later results in sp3 C-C phases (Munir et al. 2015). One of the significance of using SPS is to fabricate at lower temperatures and ability to retain the powder's composition after sintering. However, the XRD results observed after sintering of the composites are shown in Fig. 8.

Figure 8 further expatiates various sintering temperatures and weight fractions of MWCNTs of the composites. This connotes that sintered Ti64 (unreinforced) has α and β-Ti phases with matrix being α phase, and there was no indication of peak broadening as the temperature increases. Also, the XRD pattern which showed reinforced Ti64 with various weight fractions indicate the presence of TiC phases, α and β phases which were dependent on the sintering temperatures and MWCNT weight fractions. The microstructures of the sintered composites are shown in Fig. 9a–h. Figure 9 indicated the microstructural morphologies of the sintered composites at various sintering temperature with different weight fractions of MWCNTs. Unreinforced Ti-6Al-4V (a-b) sintered at 850 and 1000 °C has been observed with a near lamellar and fully lamellar structure, widmanstatten (basket weave) α + β structure network of Ti-6Al-4 V.

However, more pores were seen in the composites (a) sintered at 850 °C as compared to composites (b) sintered at 1000 °C with significant widmanstatten structure at β phase and lesser pores. Reinforced Ti-6Al-4V/MWCNTs with different weight fractions show equiaxed microstructures with α + β phases. It has similar pores as compared to composites (a-b). It was observed that the pores increase with weight fraction at lower sintering temperature due to inadequate dispersion of the MWCNTs in the metal matrix which resulted in retained CNT as seen in the composites from Fig. 9e–g. Notwithstanding, the composite sintered at 1000 °C with 1 wt % of MWCNTs characterized a fully dense, fine equiaxed α + β structure with no indication of pores and retained CNT which connotes homogeneous

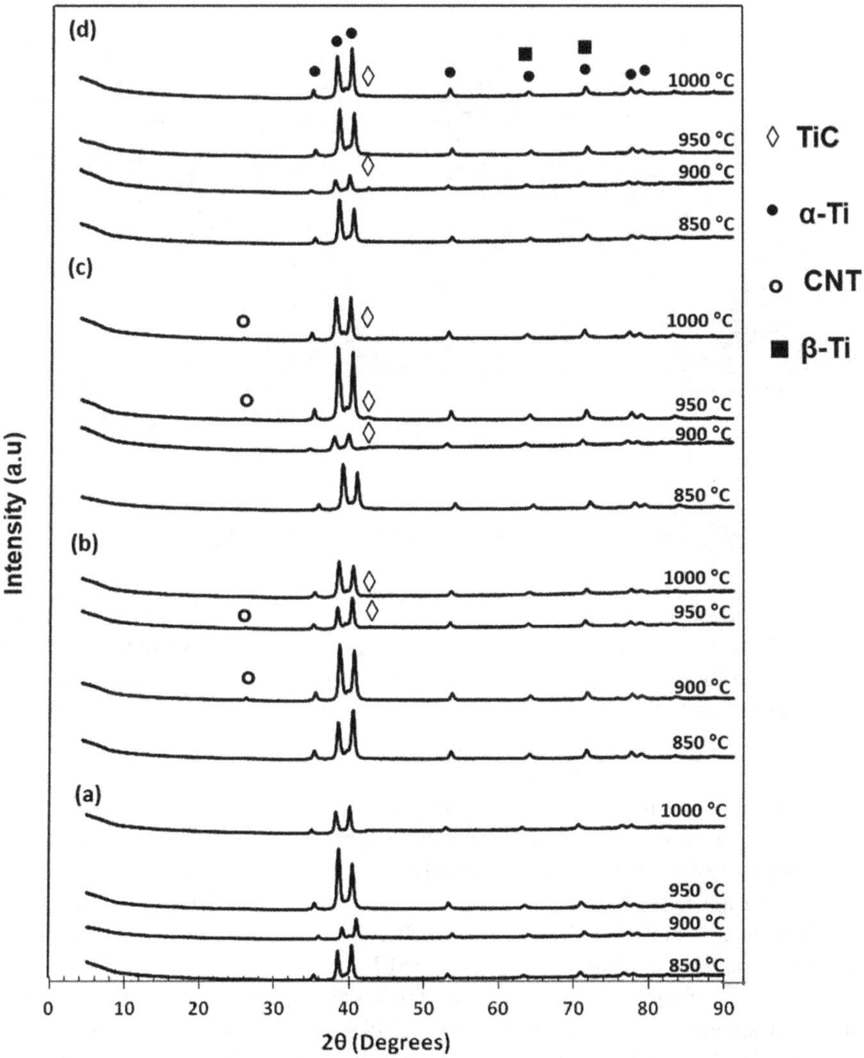

Fig. 8 XRD of Ti64 and $MWCNT_{Gr}$/Ti64 sintered at different temperatures (**a**) Ti64, (**b**) Ti64–1% MWCNTGr, (**c**) Ti64–2%$MWCNT_{Gr}$, and (**d**) Ti64–3%$MWCNT_{Gr}$ (Adegbenjo et al. 2017)

dispersion of CNTs in the metal matrix. A composite with 1 wt% of MWCNTs shows a better relative density as compared to other weight fractions. The hardness of the composites increases with the increase in the weight fractions of the MWCNTs, due to reduction in the pores and improved densification as the sintering temperature increases. The presence of TiC interfacial product can also improve the hardness of the composites (Adegbenjo et al. 2017).

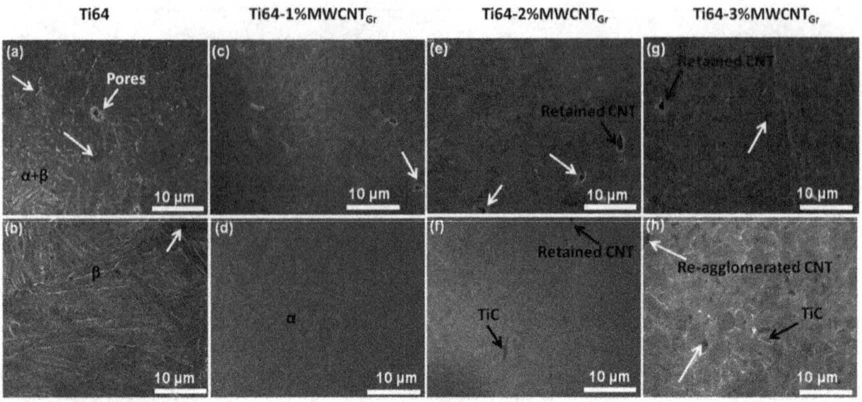

Fig. 9 Micrographs of Ti-6Al-4V and Ti-6Al-4V/MWCNT$_{Gr}$ composites at 850 °C and 1000 °C (Adegbenjo et al. 2017)

3.2 *Spark Plasma Sintering of Ti-22Al-25Nb Alloy*

SPS was utilized in accessing the effect of ball milling time on the microstructure and mechanical properties of sintered and ball milled Ti-22Al-25nb alloy in order to fabricate a composite with high strength and ductility. The powder was milled from composition (at %) of 53 Ti, 22 Al, and 25 Nb using high-energy ball milling (HEBM) at the speed of 300 rpm under a closed atmosphere for 10, 20, 40, and 60 h, respectively, with interval of 20 min/h in order to avoid undue increase in temperature between the powder and the balls. The sintering of the powder was carried out at various sintering temperatures (900, 950, and 1000 °C) with the sintering time of 10 min under the pressure of 50 MPa. The density of the sintered sample was measured by Archimedes method in ethanol; also the microstructure of the high-energy ball milled powder and SPSed sample was analyzed with field emission scanning electron microscope (FESEM).

Figure 10 shows the micrograph of the starting powder (raw) with a large spherical smooth shape, and a dendritic cellular network was observed. Consequently Fig. 11a denotes high-energy ball milled powder at 10 h with spherical shape, but the network has become invisible due to collision. After 20 h of milling as seen in Fig. 11b, the shape has been transformed to flake-like shape due to plastic deformation that occurs. A coarse flake shape was broken down to finer particles, and it becomes brittle at the 40 h milling time as observed in Fig. 11c, and finally at 60 h, agglomeration occurred and the particle size increases (Sim et al. 2018). The mechanical properties of the sintered samples were examined, and it was observed that increase in HEBM time increases both tensile strength (TS) and yield strength. However, the tensile strength increases from 888 MPa to 1105 MPa, while the yield strength increases from 827 MPa to 1092 MPa as the milling time moved from 0 h to 20 h. There was a significant effect of the HEBM time (0–60 h) as it progresses; Vickers hardness of the sintered samples increases from 311 HV to 383 HV. This is

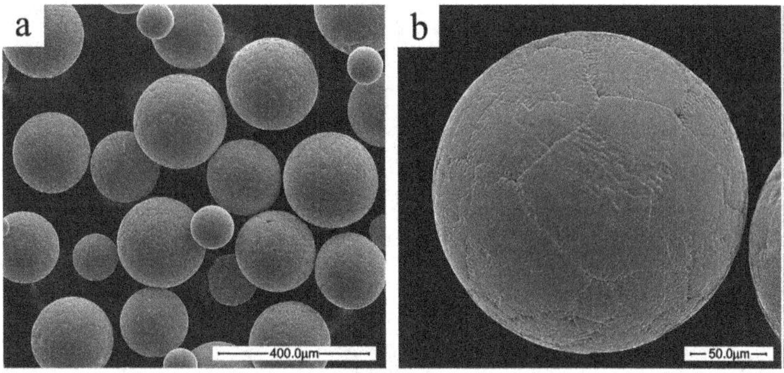

Fig. 10 Microstructure of the starting powder: (**a**) and (**b**) low and high magnification (Sim et al. 2018)

Fig. 11 SEM images of the Ti-22Al-25Nb alloy powder after HEBM at (**a**) 10 h, (**b**) 20 h, (**c**) 40 h, and (**d**) 60 h (Sim et al. 2018)

as a result of the finer grain sizes and higher density of the residual dislocation. Finally it was observed that 20 h HEBM has the best properties (Sim et al. 2018).

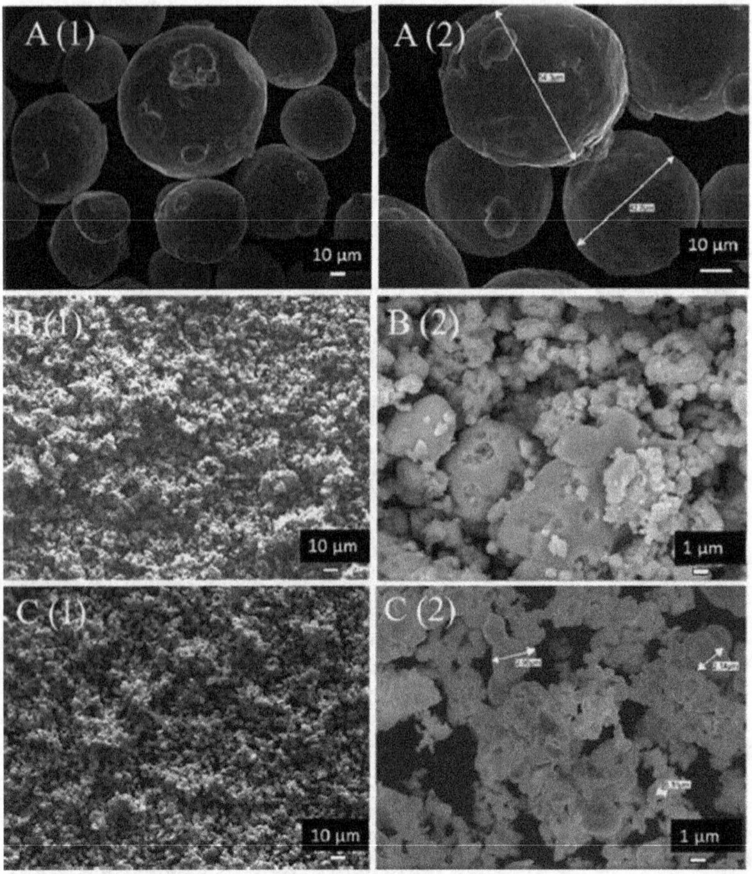

Fig. 12 SEM images of the starting powders: Ti (A1 and A2), Zr (B1 and B2), and Ta (C1 and C2) (Bayode et al. 2017)

3.3 Spark Plasma Sintering of Ti-Zr-Ta Powders

Ti-Zr-Ta alloy finds its application in biomedical and aerospace where high-temperature shape memory alloys are utilized due to its high hardness and corrosion resistance. In this work, the densification and microstructural characteristics of Ti-Zr-Ta powders were studied through the consolidation of SPS. Ti-Zr-Ta powders were mixed in a dry state in a tubular mixer at a speed of 49 rpm with aluminum balls in order to prevent the powders from clustering during mixing. Figure 12 shows the image of the starting powders, Ti powder particles are spherical with $\leq$90 μ, Zr powders possess irregular shapes, while Ta powders have spheroidal shape. The admixed powders were sintered with the following conditions: sintering temperature of 1100 °C, heating rate of 100 °C/min, applied pressure of 50 MPa, and dwelling time between 5 and 10 min. In this work, it was observed that sintering operations

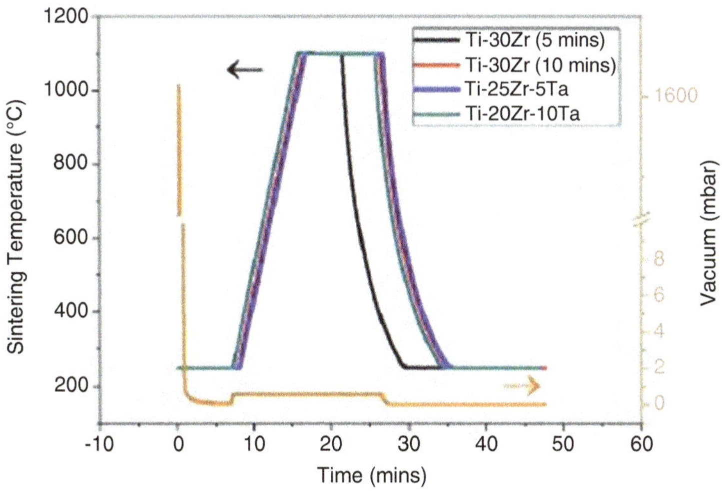

Fig. 13 SPS profile of the both binary and ternary alloys (Bayode et al. 2017)

were carried out on the binary alloys (Ti-30Zr) with the holding time of 5 min and 10 min, also on ternary alloys (Ti-25Zr-5Ta and Ti-20Zr-10Ta) for 10 min as shown in Fig. 13.

The densification behavior of both binary and ternary alloys has been observed in Fig. 14a, b; there was an occurrence of a small protrusion on the shrinkage rate curve (Fig. 14a) between 0 and 3 min, at a temperature of 250 °C, while on Fig. 14b, there was no evidence of shrinkage**.** As the time increases, there was a noticeable level of displacement at 250 °C. The curve reaches its utmost before a sharp fall in shrinkage displacement was noticed. It was observed that the designated curve (Ti-20Zr-Ta) demonstrated the topmost degree of shrinkage as compared to other alloys. As the temperature increases from 250 °C to 580 °C, there was no evidence of shrinkage within the powder under compression (Fig. 14b), but there were occurrences of shrinkage when the temperature was between 580 °C and 950 °C due to the activity of lattice diffusion (Panigrahi et al. 2005). The binary alloys exhibit close shrinkage rates as compared to the ternary alloys that had the addition of Ta which marginally shifted the shrinkage rate for Ti-25Zr-Ta and Ti-20Zr-Ta alloys. Between 580 °C and 600 °C, maximum shrinkage rate was observed in both Ti-25Zr-5Ta and Ti-20Zr-10Ta alloys, but Ti-20Zr-10Ta achieved the maximum shrinkage displacement of about 2.75 mm due to the increase in Ta content (Bayode et al. 2017). The microstructures of the sintered alloys were observed in Fig. 15a–d, and there were little or no evidence of porosity in the specimen.

From Fig. 15, it was observed on the image that some portions are rich in Ti and Zr phases. The dark gray areas are rich in Ti, while the bright areas are rich in Zr. Also, the microstructures of the sintered specimen at dwelling time of 10 min resulted in a uniform composition with a defined martensite plates (Fig. 15a, b). The presence of increased Ta results into undissolved phases as seen in Fig. 15c, d due to

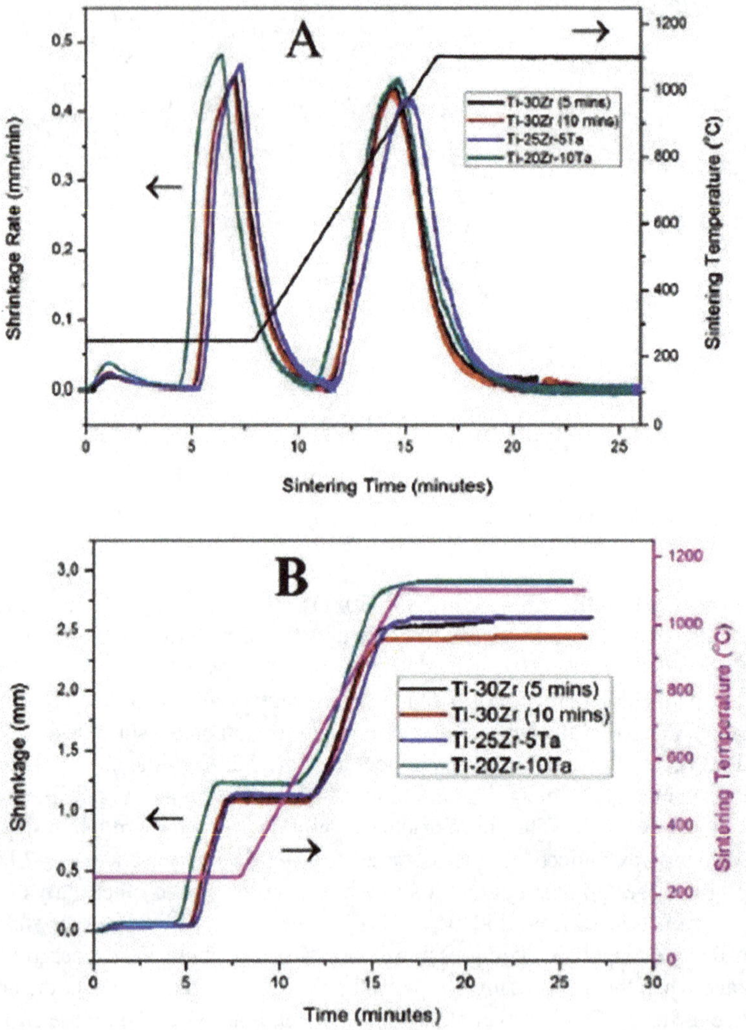

Fig. 14 (**a**) Shrinkage rate vs temperature and time. (**b**) Shrinkage vs temperature and time (Bayode et al. 2017)

agglomeration. The formation of martensite with twinned features was observed in all the sintered samples. The hardness value and relative density of the alloys were analyzed; full densification was also noticed due to the addition of Ta and reduction in Zr content. Ti-30Zr alloy attained the highest microhardness value of 516 HV, and Ti-25Zr-5Ta achieved the lowest value with 362.72 (Bayode et al. 2017).

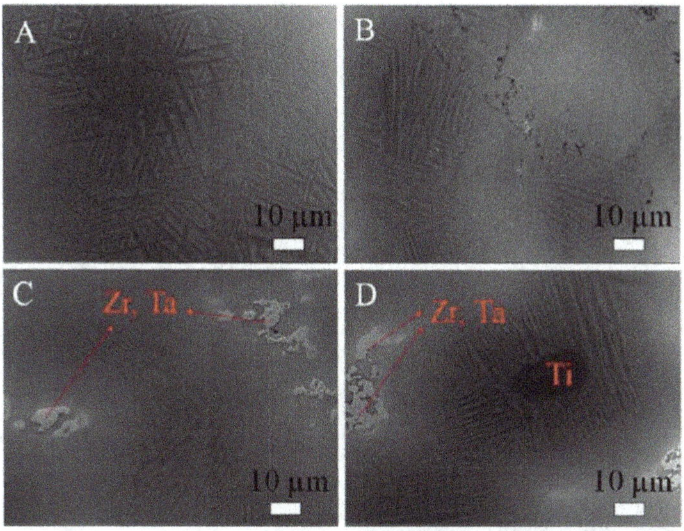

Fig. 15 SEM images of sintered: (**a**) Ti-30Zr at 5 min dwelling time (**b**) Ti-30Zr at 10 min dwelling time (**c**) Ti-25Zr-5Ta, and (**d**) Ti-20Zr-10Ta (Bayode et al. 2017)

3.4 Spark Plasma Sintering of Ti2AlNb-Based Alloy

In this study the effect of mechanical alloying on the microstructures and mechanical properties of Ti2AlNb alloy was investigated using SPS. Elemental powders of Ti, Al, and Nb with high purity were mixed for 10 h at 200 rpm speed before the admixed powders were high-energy ball milled for 5 h, 10 h, 20 h, and 40 h using 300 rpm and 400 rpm speed at 15 min/h interval in order to avoid cold welding. The admixed powder was prepared by spark plasma sintering with the following sintering parameters: sintering temperature of 1150 °C and applied pressure of 35 MPa. Figure 16 represents the starting powders; Ti and Nb particles are polyhedron in shape, while the Al powder particles have a spherical shape. The XRD of the admixed powders after 10 h of milling at 200 rpm shows there is no homogeneous mixture between the powder particles. The formation of (Ti, Al) and (Nb, Al) was observed when the milling speed was increased to 400 rpm which indicated the diffusion of Al into the lattice of Ti and Nb. Also, a fine crystalline bcc solid solution was formed with increased milling time of 40 h at 400 rpm.

Furthermore, Fig. 17a–d shows the backscattered SEM image of the sintered compact at various mechanical alloying times through SPS at 1150 °C. The microstructure of the sintered Ti2AlNb from MA at 10 h and 20 h is not even due to improper dispersion, and Nb phase was retained. However, increase in MA time at 40 h resulted in the formation of B2-matrix (light gray area), O-phase (dark gray), and $\alpha 2$-phase (black equiaxed area); also the microstructure becomes more homogeneous, and the grains are finer. This established the formation of a bcc solid solution without the evidence of pores, with the relative density of 99.81% close to

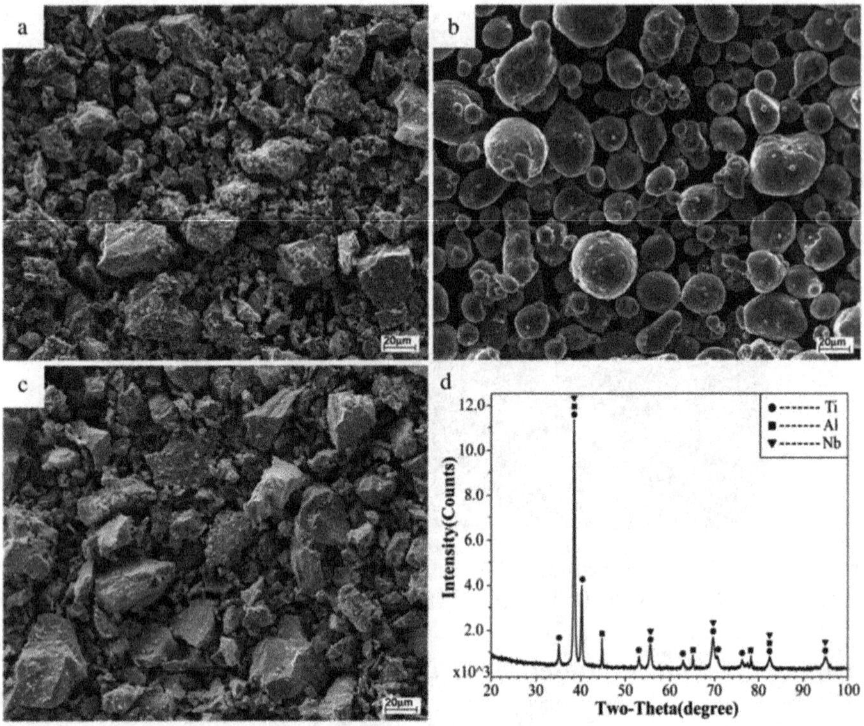

Fig. 16 SEM images of the starting powder: (**a**) Ti, (**b**) Al, (**c**) Nb, and (**d**) XRD pattern of the powder mixture blended for 10 h at 200 rpm (Sim et al. 2017)

the theoretical density. The synthesis of SPS and MA has resulted in fabrication of ultrafine compact with the yield strength (1621 MPa) and ultimate strength (2542 MPa) (Sim et al. 2017).

3.5 Spark Plasma Sintering of Ti-Ni-ZrO_2

Ti-Ni-ZrO_2 composites were fabricated through spark plasma sintering, and the effect of zirconia on the properties was examined by Obadele et al. 2017. In this work, commercially Ti powder, ZrO_2 powder, and Ni powders were used as starting materials. The images of the starting powders are represented in Fig. 18a–c; Ti and Ni are spherical, while ZrO2 are with doughnut-like shape. The composites were mixed into two stoichiometry ratios (Ti-45Ni-5 ZrO_2 and Ti-40Ni-10 ZrO_2) in a tubular mixer at a speed of 49 rpm for 8 h in dry environment. The admixed composites were consolidated by SPS at the sintering temperature range from 850 °C to 950 °C, dwelling time of 10 min, heating rate at 100 °C/min, and pressure of 50 MPa. After sintering, relevant characterization studies were carried out on the

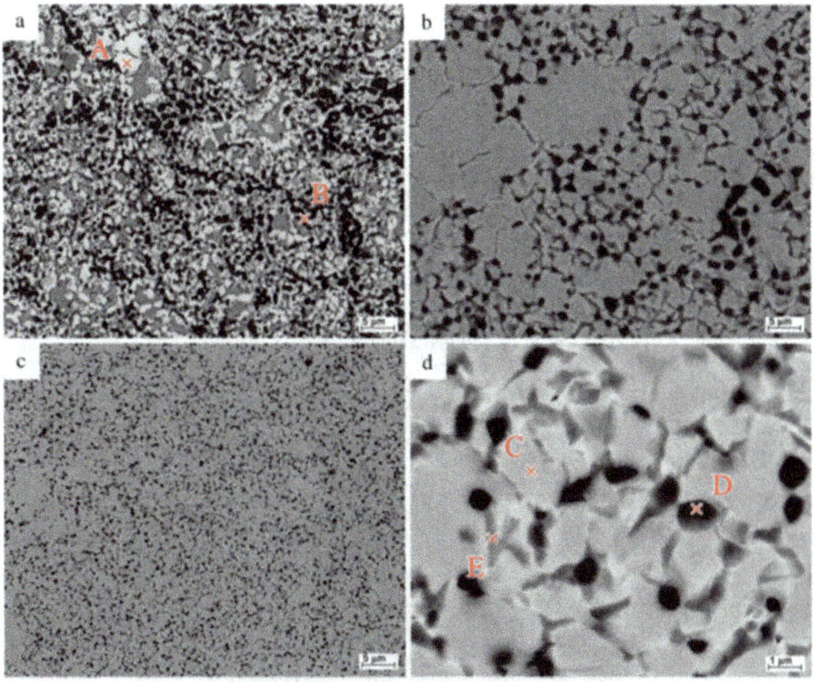

Fig. 17 BSE images of the compacts fabricated from the MAed powder at the following time: (**a**) 10 h, (**b**) 20 h, (**c**) 40 h, and (**d**) the high magnification of (**c**) (Sim et al. 2017)

fabricated composites. The relative density of the composites with various volume ratios and cp-Ti, sintered between 850 °C and 950 °C, was investigated. The relative density of the cp-Ti attained 98.48% densification at a sintered temperature of 850 °C, while Ti-45Ni-5ZrO_2 indicates a decrease in relative density at 850 °C due to the presence of pores. Also, it was noticed that relative density increases with increase in temperature for both percentages of ZrO_2 content due to refinement of the particles as the temperature increases. However, relative density increases with the temperature and increase in ZrO_2 content from 5% to 10% as shown in Fig. 19.

The hardness values of the sintered composites at different sintering temperatures were obtained through the means of microhardness tester. The sintered composite at the sintering temperature of 850 °C with 5% volume fraction of ZrO_2 was low in hardness value as compared to sintered composite at higher sintering temperature with more volume fraction of ZrO_2 because of their low relative sintered density. However, the sintered composites (Ti-Ni- ZrO2) between the temperatures of 850 °C and 950 °C attained a higher hardness value as compared to cp-Ti with hardness value of 480 HV as shown in Fig. 20.

Furthermore, Fig. 21a–g illustrates the microstructural evolutions and metallurgical interactions of the sintered composites with different volume ratios and cp-Ti. Fig. 21a connotes the image of sintered cp-Ti with α phase, thereby showing the

Fig. 18 SEM images of (**a**) cp-Ti, (**b**) Ni, and (**c**) ZrO_2 (Obadele et al. 2017)

presence of round pores. As seen in the micrograph below, Fig. 21b–g connotes the composites with different volume ratios, and they are well represented with the presence of Ni which facilitates the creation of TiNi due to its affinity for Ti. Ni particles are homogeneously arranged within the Ti matrix with spherical globules, and the white spot in the micrograph is rich in particles of undiffused ZrO_2 (Obadele et al. 2017).

3.6 *Spark Plasma Sintering of TiB Reinforced Ti6Al4V Composite*

In this research, in situ reinforced titanium-matrix composite with TiB whiskers was consolidated through press-and-sintering and spark plasma sintering techniques; their effects on the composites were investigated as well as the changes in size,

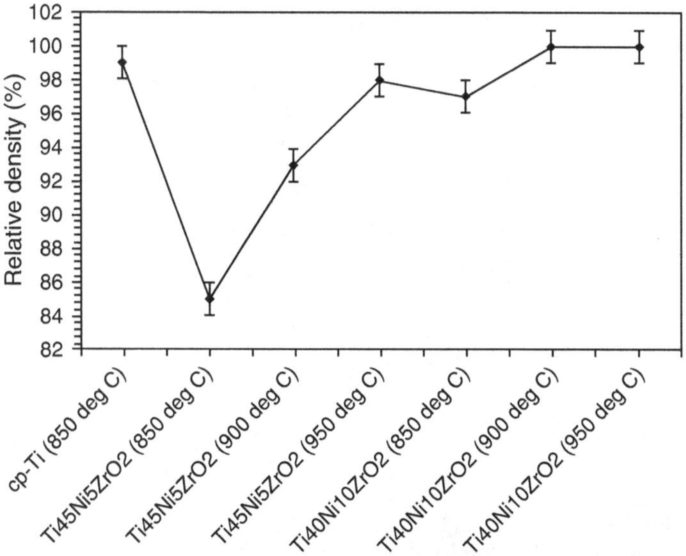

Fig. 19 Sintered relative density of Ti-Ni- ZrO_2 and cp-Ti (Obadele et al. 2017)

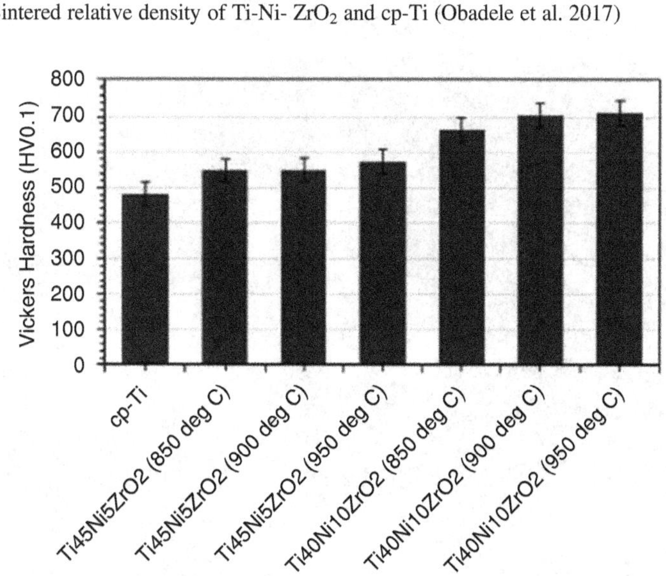

Fig. 20 Average hardness value of the sintered composites (Obadele et al. 2017)

aspect ratio, and distribution of the in situ whiskers in different composites. Ti64 and B4C powders were used as the starting materials; both have a particle size of 63 μm and 4–6 μm, respectively. The powders were mixed and milled in a jar mill with ceramic balls for 2 h, admixed powders was cold pressed (press-and-sintering) at the pressure of 420 MPa, and it was sintered in the furnace at 1250 °C for 150 min using

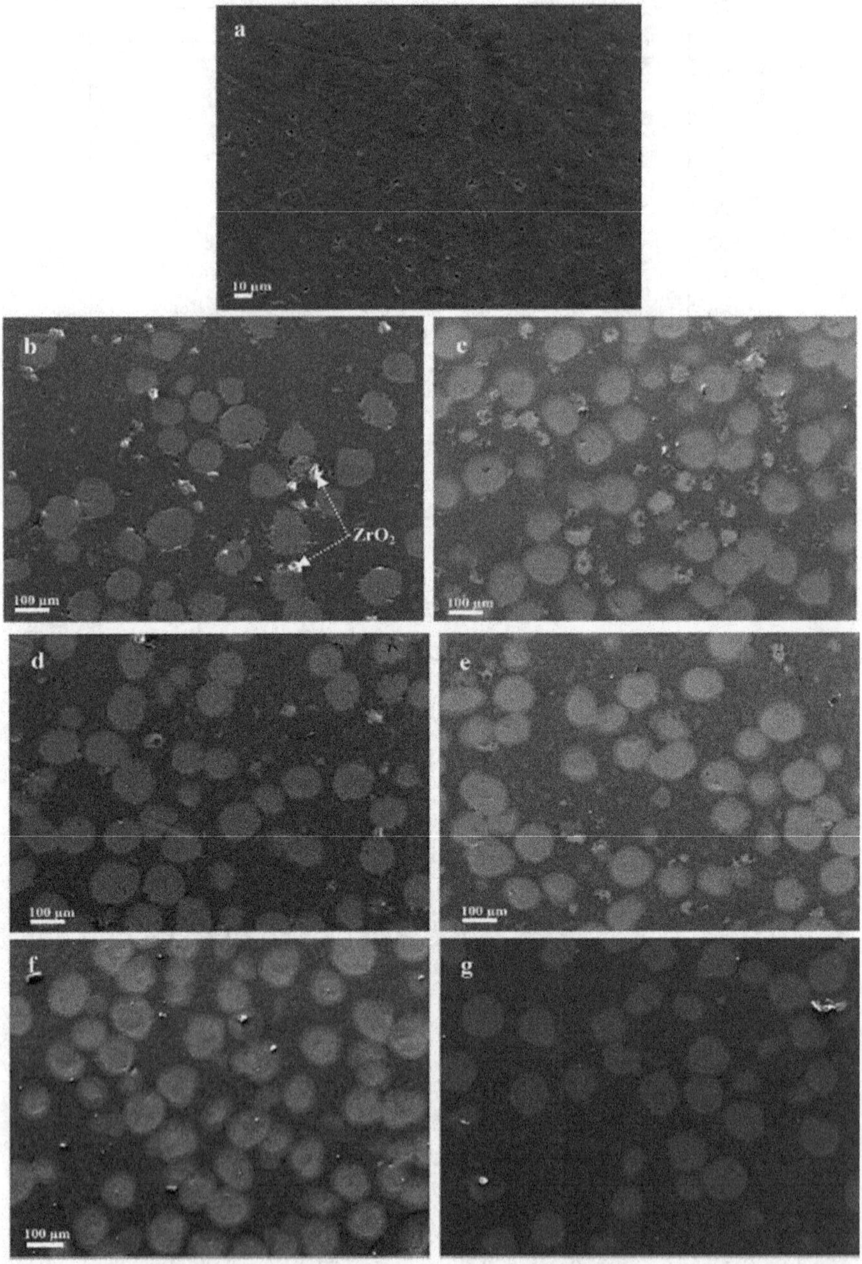

Fig. 21 SEM images of sintered composites (**a**) cp-Ti, (**b**, **c**, **d**) Ti-45Ni-5ZrO_2, and (**e**, **f**, **g**) Ti-40Ni-10ZrO_2 (Obadele et al. 2017)

Table 1 Parameters of the composite

Sample	TiB vol. %	Processing type	Sintering temperature (°C)	Holding time (min)
PM 1	1	CV	1250	150
PM 2	2	CV	1250	150
L-SPS2	2	SPS	900	10
H-SPS0.5	0.5	SPS	1100	10
H-SPS 1	1	SPS	1100	10
H-SPS 2	2	SPS	1100	10

Stoichiometric prediction; *CV* cold pressing + vacuum sintering (Tabrizi et al. 2015)

a vacuum of 10–2 Pa. Consequently, the admixed powder was consolidated by SPS at the sintering temperature at 900 and 1100 °C, pressure of 20 MPa, heating rate at 100 and 25 °C/min, and dwelling time of 10 min. The composites are designated in Table 1 showing the volume fraction of TiB, sintering temperature, and holding time used for the experimentation.

After experimentation procedure, the microstructural evolutions of the composites were examined using the scanning electron microscope (SEM) and optical microscope (OM). Figure 21a, b represents the OM image of the cold-pressed and vacuum-sintered composites with different volume fractions of TiB at the same sintering temperature and holding time as observed in Table 1. It was observed that the in situ TiB_W had a gray needle-like shape. Titanium matrix comprises of bright phase (α) and brown phase (β) as seen in Fig. 22b, while the dark regions with round shape are the remained porosities observed after fabrication. The porosities observed in Fig. 22 would have an impact on the mechanical properties of the sample. Also, the SEM images of PM 1 and PM2 were examined at higher magnification in Fig. 23a–c.

From Fig. 23c, some clusters of TiB_W were observed in the porosities of PM1when examined with higher magnification SEM. This is due to the nucleation of several TiB whiskers near the previous B4C particles. Furthermore, the EDS investigations on sample PM1 and PM2 indicate that both single and clustered whiskers consist of TiB. Figure 23b connotes that some TiB_W cuts perpendicular to the longitudinal axis at point A and parallel to the longitudinal axis at point B. The whiskers are 40–70 μm in length and 2–5 μm in thickness with the aspect ratio of (8–20) for press-and-sintered samples due to the long sintering cycles. From Fig. 24a, b, which represents the OM images of sample L-SPS2 with a well-dispersed equal particle size of 6–10 μm, the particles are the remainder of B4C additives particles that are deficient of activation energy. Also, in situ TiB whiskers are noticed at higher magnification in Fig. 24b, while Fig. 25 is characterized with uneven particle size. Additionally, Fig. 26a, b connotes the OM images of L-SPS2 and HSPS0.5 after etching; it was observed that the widmanstatten plates of α phase are changed into β phase. The SEM images of sample L-SPS2 fabricated at 900 °C show a little number of porosities in Fig. 27, but the porosities are minimized and less visible at the sintering temperature of 1100 °C in Fig. 28 for sample H-SPS2. The size of TiB_W fabricated by SPS at 900 °C was studied to have the maximum length of 15 μm, while the in situ whiskers sintered at 1100 °C were measured to be

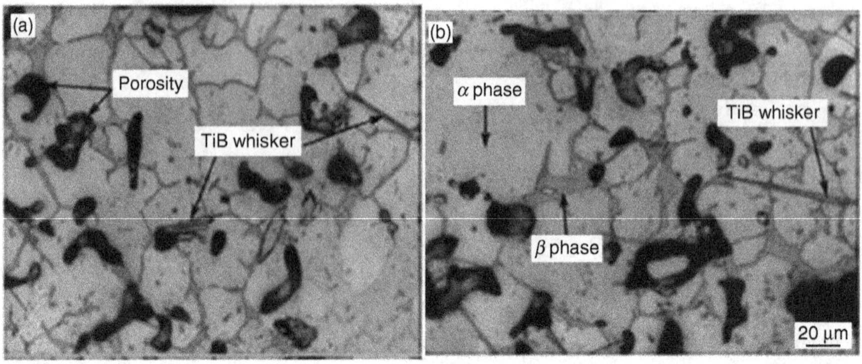

Fig. 22 Optical microstructures of press-and-sintered samples: (**a**) PM 1 and (**b**) PM 2 (Tabrizi et al. 2015)

10–20 μm in length. Thus, the aspect ratio of the whiskers in sample L-SPS2 had a ratio of 7:1, while H-SPS varies from 10:1 to 25:1 as compared to press-and-sintered samples PM1 and PM2, the aspect ratio of TiB_W in H-SPS samples is 20% larger. SPS has contributed in minimizing the clustering of TiB whiskers due to the increase in temperature from 900 °C to 1100 °C. This led to an increase in aspect ratio of TiB_W with increase in the length of whiskers from 7 μm to 10–25 μm. However, the press-and-sintered samples are characterized with defects because of the remained porosities and clustering of TiB_W noticed (Tabrizi et al. 2015).

3.7 *Effects of Process Parameters on Spark Plasma Sintering*

Consolidation of titanium-based materials through SPS can be influenced through the process parameters in order to achieve full densification of materials. Process or sintering parameters such as sintering temperature, pressure, holding time, and heating rate contribute to ultrafine microstructures, better phase evolutions of sintered samples, which further improved the thermal, physical, and mechanical properties of the materials.

1. Heating rate: SPS generates a rapid heating rate that is high as 1000 °C/min which helps in attaining full densification of the material. The effect of the fast heating rate has enhanced in minimizing metallurgical interactions such as grain growth. According to Shen et al. (2002), it was investigated that the heating rate of alumina had slight change in the density to a value of 350 °C/min and resulted in a negative effect in the density of the material at higher heating rate of 600 °C/min, i.e., the density decreased with increased heating rate. Likewise, the grain size of the material also decreased with increase in heating rate. Lui et al. (2013) investigated that a fast heating rate is necessary for the densification of materials through SPS. Also, the fabrication of cp-Ti powder via SPS by Yang and Qian

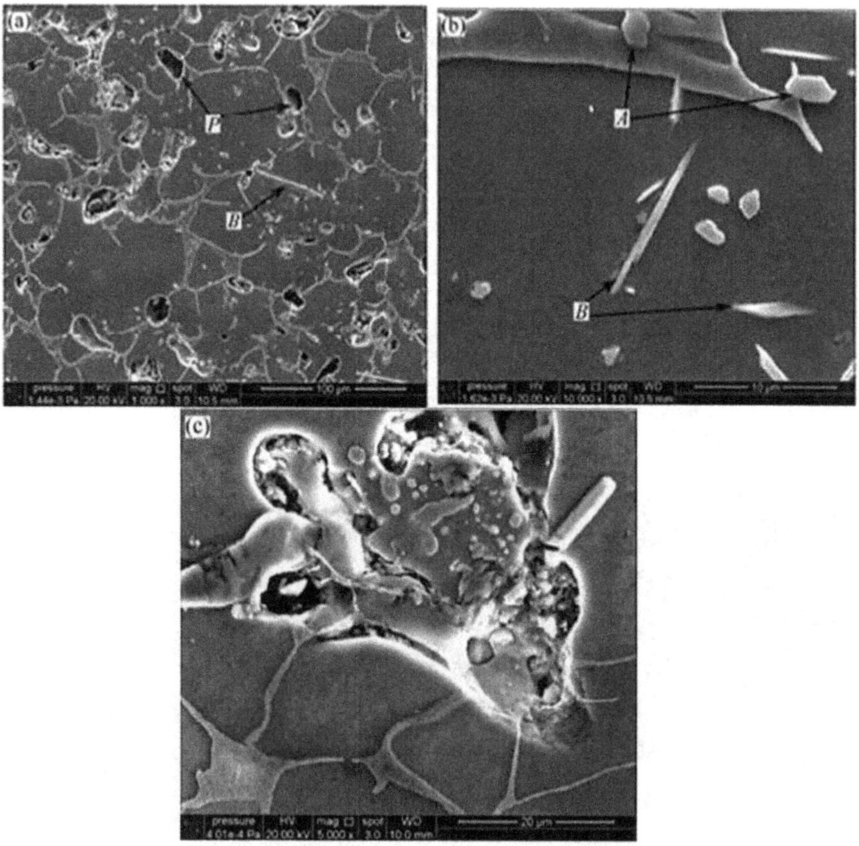

Fig. 23 SEM images: (**a**) two-phase ($\alpha + \beta$) matrix of PM2, in situ TiB_W at point B, porosities at point P; (**b**) point A (transverse) and point B (longitudinal) regions of TiB_W in PM2; (**c**) TiB_W cluster in PM1 (Tabrizi et al. 2015)

et al. (2015) observed that the heating rate affected the densification process without much effect on the final sintered density.

2. Sintering pressure: Sintering pressure of a sample increases from low to medium and high values as the relative density of the samples increases which can lead to increase in grain growth; also the increase in pressure can lead to increase in the microhardness value of a sample. The utilization of high pressure can lead to densification during heating; furthermore it favors the development of necks which subdues the electric discharge. This tends to influence the joule effect through diffusion and followed by grain growth (Garbiec et al. 2016).
3. Sintering temperature: The sintering temperature used in fabricating materials during SPS can greatly affect the outcome of the material. It has been observed that increase in sintering temperature will increase the density of the sintered sample. Garbiec et al. (2016) reported that Ti-6Al-4V attained a near full

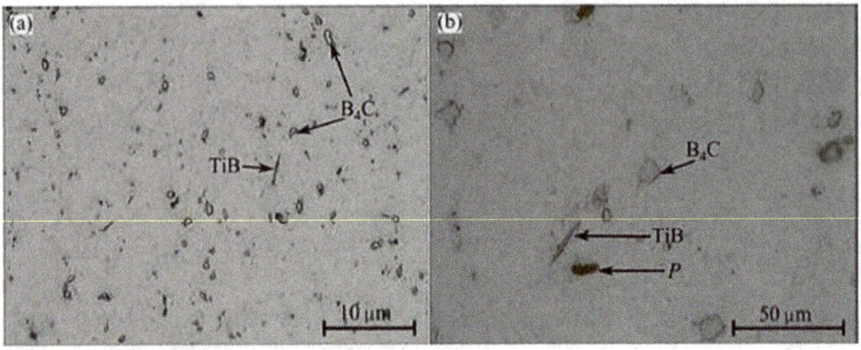

Fig. 24 As-polished OM images of L-SPS2 sintered at 900 °C: (**a**) unreacted B_4C particles dispersed in matrix; (**b**) in situ TiB_W produced during sintering, remained B_4C particles and porosities (point P) (Tabrizi et al. 2015)

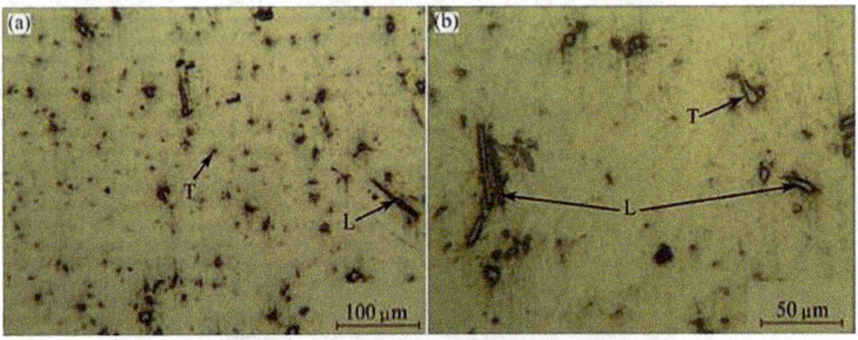

Fig. 25 As-polished OM images of H-SPS1 sintered at 1100 °C with longitudinal (L) and transverse (T) regions of TiB whiskers (Tabrizi et al. 2015)

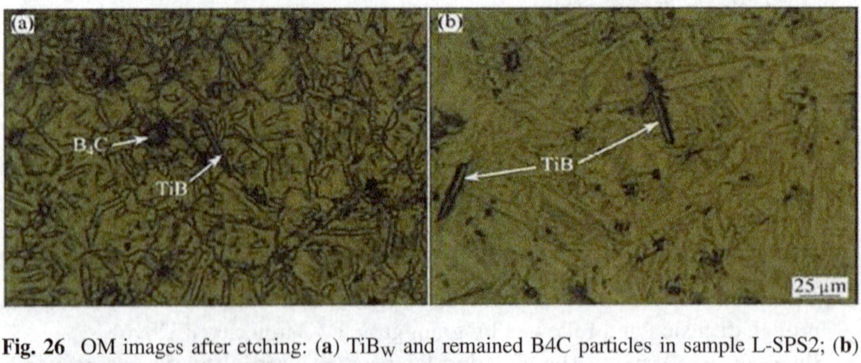

Fig. 26 OM images after etching: (**a**) TiB_W and remained B4C particles in sample L-SPS2; (**b**) TiB_W in H-SPS0.5 sample (Tabrizi et al. 2015)

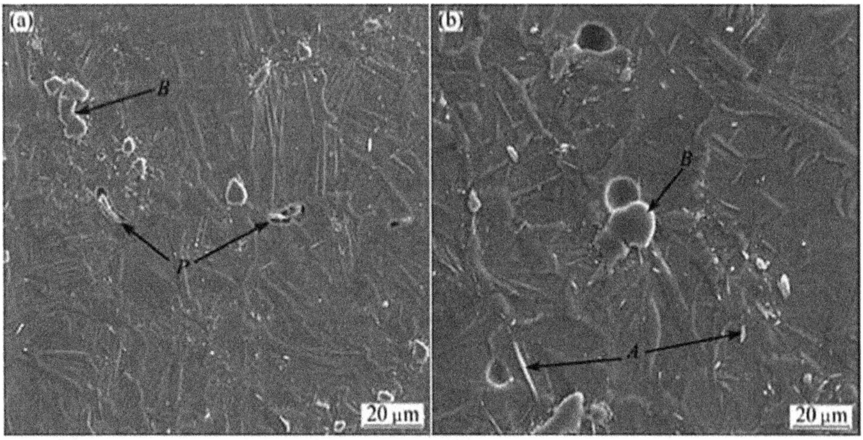

Fig. 27 SEM images of sample L-SPS2 sintered at 900 °C with in situ TiB_W (point A), remained B4C particles (point B), and porosities (point P) (Tabrizi et al. 2015)

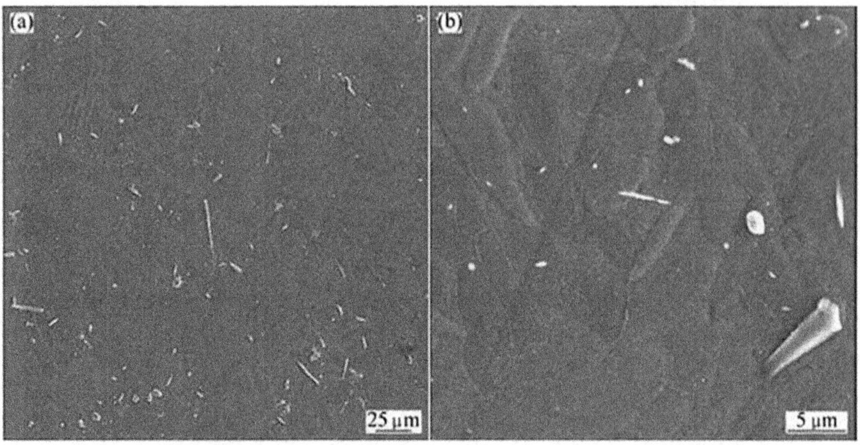

Fig. 28 SEM images of sample H-SPS2 sintered at 1100 °C with in situ TiB_W in matrix free of remained B4C porosities (Tabrizi et al. 2015)

densification at 1000 °C. Also, Teber et al. (2012) investigated that an increase in density results in decrease in the level of porosity with increase temperature. According to Falodun et al. (2018), the relative density of sintered Ti-6Al-4V increases from 97.29% to 99.54% as the sintering temperature increases from 800 °C to 1000 °C as shown in Fig. 29.

4. Holding time: The holding time of a sintered sample can influence the densification of the sample which variably affects the mechanical properties of the samples. According to Yamanoglu et al. (2018), it was investigated that the holding time affects the relative density of sintered Ti5Al2.5Fe alloys. The holding time increases as the relative density of the sample increases.

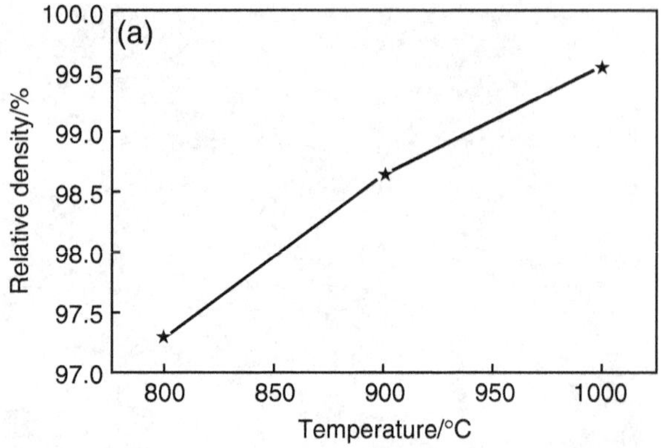

Fig. 29 Relative density of sintered compacts of Ti-6Al-4V (Falodun et al. 2018)

4 Conclusions

Spark plasma sintering has proven to be a novel sintering technique in effective fabrication of titanium-based alloys due to its ability to control metallurgical variations in microstructures and phases, grain growth that is associated with other sintering techniques. SPS offers a very fast heating rate and short sintering cycles in consolidating titanium-based alloys. Also, the utilization of SPS of titanium-based materials has led to development of materials that are highly desired in the industries especially in aerospace and chemical applications due to the fabrication of materials that are ultralight and toughened, which are readily used in high-temperature environment. However, SPS has been reported in literature as a better consolidation technique as compared to conventional techniques.

Acknowledgments The author would like to appreciate the National Research Foundation (South Africa), Centre for Nanoengineering and Tribocorrosion, University of Johannesburg, South Africa, and Institute for Nanoengineering Research, Tshwane University of Technology, South Africa.

References

Aalund R (2008) Spark plasma sintering. Ceramic Industry Magazine, May

Adegbenjo AO, Olubambi PA, Potgieter JH, Shongwe MB, Ramakokovhu M (2017) Spark plasma sintering of graphitized multi-walled carbon nanotube reinforced Ti6Al4V. Mater Des 128:119–129

Bayode BL, Lethabane ML, Olubambi PA, Sigalas I, Shongwe MB, Ramakokovhu MM (2017) Densification and micro-structural characteristics of spark plasma sintered Ti-Zr-Ta powders. Powder Technol 321:471–478

Bloxam AG (1906) Improved manufacture of electric incandescence lamp filaments from tungsten or molybdenum or an alloy thereof. British patent 27,002

Boesel RW, Goetzel CG (1971) Powder Metallurgy 3:38

Boyer RR, Briggs RD (2005) The use of β titanium alloys in the aerospace industry. J Mater Eng Perform 14(6):681–685

Chen W, Anselmi-Tamburini U, Garay JE, Groza JR, Munir ZA (2005) Fundamental investigations on the spark plasma sintering/synthesis process: I. Effect of dc pulsing on reactivity. Mater Sci Eng A 39:132–138

Chen YY, Yu HB, Zhang DL, Chai LH (2009) Effect of spark plasma sintering temperature on microstructure and mechanical properties of an ultrafine grained TiAl intermetallic alloy. Mater Sci Eng A 525:166–173

Dehghan-Manshadi A, Bermingham MJ, Dargusch MS, StJohn DH, Qianb M (2017) Metal injection moulding of titanium and titanium alloys: challenges and recent development. Powder Technol 319:289–301

Falodun OE, Obadele BA, Oke SR, Ige OO, Olubambi PA, Lethabane ML, Bhero SW (2018) Influence of spark plasma sintering on microstructure and wear behaviour of Ti-6Al-4V reinforced with nanosized TiN. Trans Nonferrous Metals Soc China 28(1):47–54

Feng P, He Y, Xiao Y, Xiong W (2008) Effect of VC addition on sinterability and microstructure of ultrafine Ti (C, N)-based cermets in spark plasma sintering. J Alloys Compd 460(1–2):453–459

Garbiec D, Siwak P, Mroz A (2016) Effect of compaction pressure and heating rate on microstructure and mechanical properties of spark plasma sintered Ti6Al4V alloy. Arch Civil Mech Eng 16(4):702–707

Grasso S, Sakk Y, Maizza G (2009) Electric current activated/assisted sintering (ECAS): a review of patents 1906-2008. Sci Technol Adv Mater 10:1–24

Guillon O, Gonzalez-Julian J, Dargatz B, Kessel T, Schierning G, Räthel J, Herrmann M (2014) Field-assisted sintering technology/spark plasma sintering: mechanisms, materials, and technology developments. Adv Eng Mater 16:830–848

Henriques ARV, Campos P, Cairo AC, Bressiani CJ (2005) Production of titanium alloys for advanced aerospace systems by powder metallurgy. Mat Res 8(4). São Carlos. https://doi.org/10.1590/S1516-14392005000400015

http://www.dierk-raabe.com/titanium-alloys

Inoue K (1966) Electric discharge sintering. US patent 3,241,956

Kessel HU, Hennicke J (2007) Aspects concerning the super-fast sintering of powder metallic and ceramic materials. High-Perform Ceram 56(3):164–166

Kessel HU, Hennicke J, Kirchner R, Kessel T (2009) Rapid sintering of novel materials by FAST/SPS – further development to the point of an industrial production process with high cost efficiency. FCT Systeme GmbH, Rauenstein, Germany. http://fct-systeme.de/download/20100225123420/FCT-Sintered-Materials.pdf

Khalil KA (2012) Advanced sintering of Nano-ceramic materials. In: Ceramic materials – Progress in modern ceramics. InTech, Rijeka

Lee YI, Lee JH, Hong SH, Kim DY (2003) Preparation of nanostructured TiO2 ceramics by spark plasma sintering. Mater Res Bull 38(6):925–930

Leyens C, Peters M (2003) Titanium and titanium alloys. Fundamentals and applications. WILEY-VCH Verlag GmbH & Co. KGaA, Weinheim. ISBN: 3-527-30534-3

Lui L, Hou Z, Zhang B, Ye F, Zhang Z, Zhou Y (2013) A new heating route of spark plasma sintering and its effect on alumina ceramic densification. Mater Sci Eng A 559:462–466

Lutjering G, Williams JC (2003) Titanium. Springer, Berlin

Lütjering G, Williams JC (2007) Titanium. Springer Science & Business Media, Berlin

Malinov S, Zhecheva A, Sha W (2004) Relation between the microstructure and properties of commercial titanium alloys and the parameters of gas nitriding. Met Sci Heat Treat 46 (7–8):286–229

Matizamhuka WR (2016) Spark plasma sintering (SPS)-an advanced sintering technique for structural nanocomposite materials. J South Afr Inst Min Metall 116(12):1171–1180

Moiseyev VN (2006) Titanium alloys: Russian aircraft and aerospace applications. CRC press, Boca Raton

Moraes PEL, Contieri JR, Lopes SN, Robin A, Caram R (2014) Effects of Sn addition on the microstructure, mechanical properties and corrosion behavior of Ti–Nb–Sn alloys. Mater Charact 96:273–281

Munir ZA, Anselmi-Tamburini U, Ohyanagi M (2006) The effect of electric field and pressure on the synthesis and consolidation of materials: a review of the spark plasma sintering method. J Mater Sci 41(3):763–777

Munir KS, Kingshott P, Wen C (2015) Carbon nanotube reinforced titanium metal matrix composites prepared by powder metallurgy—a review. Solid State Mater Sci 40(1):38–55. https://doi.org/10.1080/10408436.2014.929521

Obadele BA, Ige OO, Olubambi PA (2017) Fabrication and characterization of titanium-nickel-zirconia matrix composites prepared by spark plasma sintering. J Alloys Compd 710:825–830

Panigrahi BB, Godkhindi MM, Das K, Mukunda PG, Ramakrishnan P (2005) Sintering kinetics of micrometric titanium powder, Mater Sci Eng A 396:255–262

Peters M, Hemptenmacher J, Kumpfert J, Leyens C (2003) Titanium and titanium alloys. Leyens C, Peters M, ed. Weinheim, Wiley-VCH, p 1

Qian M (2010) Cold compaction and sintering of titanium and its alloys for near-net-shape or preform fabrication. Int J Powder Metall 46(5):29–44

Rajeswari K, Hareesh US, Sbairi R, Chakravarty D, Johnson R (2010) Comparative evaluation of spark plasma (SPS), microwave (MWS), two stage sintering (TSS) and conventional sintering (CRH) on the densification and micro structural evolution of fully stabilized zirconia ceramics. Sci Sinter 42:259–267

Reis de Vasconcellos LM, Carvalho YR et al (2012) Porous titanium by powder metallurgy for biomedical application: characterization, cell citotoxity and in vivo tests of osseointegration. InTech, Rijeka, pp 47–73

Revankar DG, Shetty R, Shrikanth RS, Gaitonde NV (2017) Wear resistance enhancement of titanium alloy (Ti–6Al–4V) by ball burnishing process. J Mater Res Technol 6(1):13–32

Sergueeva AV, Hulbert DM, Mara NA, Mukharjee AK (2009) Mechanical properties of nanocomposite materials. In: Wilde G (ed) Nanostructured materials. Elsevier, New York. Ch. 3, pp 127–171

Shongwe MB, Ramakokovhu MM, Lethabane ML, Olubambi PA (2017) Comparison of spark plasma sintering and hybrid spark plasma sintering of Ni-Fe alloys. Proceedings of the World Congress on Engineering. Vol II WCE

Shen Z, Johnsson M, Zhao Z, Nygren M (2002). Spark Plasma Sintering of Alumina. J Am Ceram Soc 85(8):1921–1927

Sim KH, Wang G, Son RC, Choe SL (2017) Influence of mechanical alloying on the microstructure and mechanical properties of powder metallurgy Ti2AlNb-based alloy. Powder Technol 317:133–141. London

Sim KH, Wang G, Kim TJ, Ju KS (2018) Fabrication of a high strength and ductility Ti–22Al–25Nb alloy from high energy ball-milled powder by spark plasma sintering. J Alloys Compd 741:1112–1120

Suárez M, Fernández A, Menéndez JL, Torrecillas R, Kessel HU, Hennicke J, Kessel T (2013) Challenges and opportunities for spark plasma sintering: a key technology for a new generation of materials. In: Sintering applications. Intech, Rijeka

Tabrizi SG, Sajjadi SA, Babakhani A, Lu W (2015) Influence of spark plasma sintering and subsequent hot rolling on microstructure and flexural behavior of in-situ TiB and TiC reinforced Ti6Al4V composite. Mater Sci Eng A 624:271–278

Tang CY, Wong CT, Zhang LN et al (2013) In situ formation of Ti alloy/TiC porous composites by rapid microwave sintering of Ti6Al4V/MWCNTs powder. J Alloys Compd 557:67–72

Teber A, Schoenstein F, Têtard F, Abdellaoui M, Jouini N (2012) Effect of SPS process sintering on the microstructure and mechanical properties of nanocrystalline TiC for tools application. Int J Refract Met Hard Mater 30(1):64–70

Veiga C, Davim JP, Loureiro AJR (2012) Properties and applications of titanium alloys: a brief review. Rev Adv Mater Sci 32(2):133–148

Wei S, Zhanga Z et al (2012) Effect of Ti content and sintering temperature on the microstructures and mechanical properties of TiB reinforced titanium composites synthesized by SPS process. Mater Sci Eng A 560:249–255

Weston SN, Derguti F, Tudball A, Jackson M (2015) Spark plasma sintering of commercial and development titanium alloy powders. J Mater Sci 50:4860–4878. https://doi.org/10.1007/s10853-015-9029-6

www.substech.com/dokuwiki/doku

Yamanoglu R, Daoud I, Olevsky EA (2018) Spark plasma sintering versus hot pressing- densification, bending strength, microstructure, and tribological properties of Ti5Al2.5Fe alloys. Powder Metall 61(2):178–186

Yang YF, Qian M (2015) Spark plasma sintering and hot pressing of titanium and titanium alloys. In: Titanium powder metallurgy. Butterworth-heinemann, Boston, pp 219–235

The Use of SPS for High-Rate Diffusion Welding of High-Strength Ultrafine-Grained α-Titanium Alloy Ti-5Al-2V

Aleksey Nokhrin, Maksim Boldin, Aleksandr Piskunov, Nataliya Kozlova, Mikhail Chegurov, Vladimir Kopylov, Nataliya Tabachkova, Vladimir Chuvil'deev, and Petr Tryaev

1 Introduction

Today ultrafine-grained (UFG) metals and alloys obtained through equal channel angular pressing (ECAP) (Segal et al. 2010) generate particular interest among researchers. This is because UFG materials are found to have a unique combination of high strength and ductility (Valiev and Langdon 2006; Wang et al. 2002; Chuvil'deev et al. 2008), corrosion resistance (Kim et al. 2011; Balyanov et al. 2004; Kim and Kim 2014; Chuvil'deev et al. 2012, 2017b) and radiation resistance (Andrievskii 2011), high-rate superplasticity (Valiev and Langdon 2006; Chuvil'deev et al. 2004, 2010; Perevezentsev et al. 2002), etc.

One of the key challenges that currently prevent industrial introduction of UFG materials is the welding issue: traditional technologies such as argon arc welding or electron beam welding accompanied by metal meltdown fail to preserve the UFG structure with high physicochemical properties and performance characteristics in a weld seam. Solid phase technologies, such as classical diffusion welding or superplastic forming, require high temperatures and prolonged hold time which result in

A. Nokhrin (✉) · M. Boldin · A. Piskunov · N. Kozlova · M. Chegurov · V. Chuvil'deev
Lobachevsky State University of Nizhny Novgorod, Nizhny Novgorod, Russia
e-mail: nokhrin@nifti.unn.ru

V. Kopylov
Lobachevsky State University of Nizhny Novgorod, Nizhny Novgorod, Russia

Physics and Technology Institute, National Academy Science of Belarus, Minsk, Belarus

N. Tabachkova
National University of Science and Technology "MISiS", Moscow, Russia

P. Tryaev
Afrikantov Experimental Design Bureau for Mechanical Engineering JSC (Afrikantov OKBM JSC), Nizhny Novgorod, Russia

© Springer Nature Switzerland AG 2019

P. Cavaliere (ed.), *Spark Plasma Sintering of Materials*,
https://doi.org/10.1007/978-3-030-05327-7_24

recrystallization processes and subsequent loss of any unique properties found with UFG materials.

The technology of spark plasma sintering (SPS) offers great opportunities to obtain high-strength weld seams. SPS is based on the idea of high-rate heating (up to 2500 °C/min) of materials (powders, bulk materials, composites) in vacuum or inert atmosphere by means of passing millisecond high-power pulses (up to 5000 A) of direct pulse current through samples and simultaneous pressure application (Tokita 2013; Munir et al. 2006; Munir and Quach 2011). Such a combination enables high-density structures in metals, alloys, and ceramics (Orlova et al. 2014; Chuvil'deev et al. 2015a, b, 2017a; Munir et al. 2006; Munir and Quach 2011). Advantages of SPS technology include an opportunity to reduce optimal sintering temperature for UFG materials (Orlova et al. 2014; Chuvil'deev et al. 2013, 2015a, b). High heating rates and an opportunity to ensure significant diffusion acceleration at reduced heating temperatures are essential to limit the process of grain growth and preservation of the original UFG structure in materials.

It shall be noted that ensuring a high-density weld seam with strength properties similar to those of the core material is a scientific challenge: a high-density structure presupposes prolonged hold time of the material at elevated temperatures, while a UFG structure may be preserved only if grain growth is limited by means of reducing the heating temperature. This problem may be fixed if the conditions for high-rate heating of UFG materials are extended to include additional mechanisms of diffusion mass transfer related to nonequilibrium grain boundaries.

The goal of the paper is to study the SPS potential for diffusion welding of high-strength corrosion-resistant UFG titanium alloys used in nuclear engineering.

2 Materials and Methods

Samples of UFG α-titanium alloy Ti-4.73Al-1.88V (Russian industrial alloy PT3V) served as the materials of research. The UFG structure in materials under study was formed through equal channel angular pressing (ECAP). The number of ECAP cycles ranged from 1 to 4, while ECAP temperature varied from 450 °C to 475 °C. The rate of deformation during ECAP was 0.2–0.4 mm/s.

Diffusion welding was performed using a Dr. Sinter model SPS-625. Samples sized 7 × 7 × 3.5 mm obtained through spark cutting from UFG metal pieces sized 14 × 14 × 100 mm were welded. The level of surface roughness varied as a result of polishing with a diamond paste of various dispersion level (20–28 μm, 10–14 μm, 3–5 μm).

Diffusion welding was performed at the heating rate varying from 10 °C/min to 350 °C/min, at the welding temperature ranging from 600 °C to 1140 °C and under the pressure applied within the interval from 50 to 100 MPa. Diffusion welding time (at the temperature selected) varied from 0 to 90 min. The value and rate of shrinkage were controlled with a dilatometer being a part of a Dr. Sinter model SPS-625. The temperature was measured with a pyrometer Chino IR-AH focused on the side surface of samples.

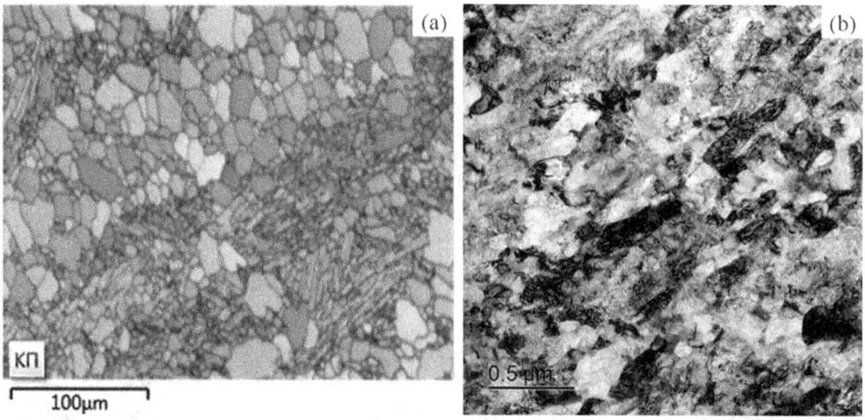

Fig. 1 Titanium alloy structure in the original state (**a**) and after ECAP (**b**)

Structure studies were carried out using Jeol JSM-6490 scanning electron microscope (SEM) and Jeol JEM-2100 transmission electron microscope (TEM). Pore size in weld seams was determined with Leica DM IRM metallographic microscope and Jeol JSM-6490 SEM.

Microhardness was measured using Duramin Struers-5 microhardness tester with a 2 kg load. Electrochemical studies of samples were carried out in aqueous solution 10%HNO_3 + 0.2%HF using R-8 potentiostat-galvanostat (Russia) at room temperature. Analysis of Tafel areas of anodic potentiodynamic dependences "potential–current density" in semilogarithmic coordinates helped to determine corrosion current density (i_{cor}) which is proportionate to material corrosion rate in the given environment.

3 Experimental Results

In the initial state, the alloy is characterized by heterogeneous size distribution of grains. The average grain size varies from 10–20 μm to 50–100 μm (see Fig. 1a). The average grain size after N = 4 ECAP cycles is 0.2–0.5 μm (see Fig. 1b).

Energy-dispersive (EDS) analysis reveals two types of grain boundaries (GB) in the structure of a coarse-grained alloy. Type one includes grain boundaries that we have termed "clean." These grain boundaries form an absolute majority. Average local concentration of aluminum in such GB is 3.8 ± 0.9 wt.%, while average vanadium concentration is 1.9 ± 0.2 wt.%. These figures differ little from concentrations of these elements in the material volume: average aluminum concentration in GBs is 3.6 ± 0.9 wt.%, while vanadium concentration approximates ~1.4 ± 0.2 wt.%.

The second GB type includes grain boundaries the local concentration of vanadium in which may reach 10 wt.% (Fig. 2a). There are also single boundaries the local concentration of vanadium in which reaches 16–18 wt.%, while aluminum

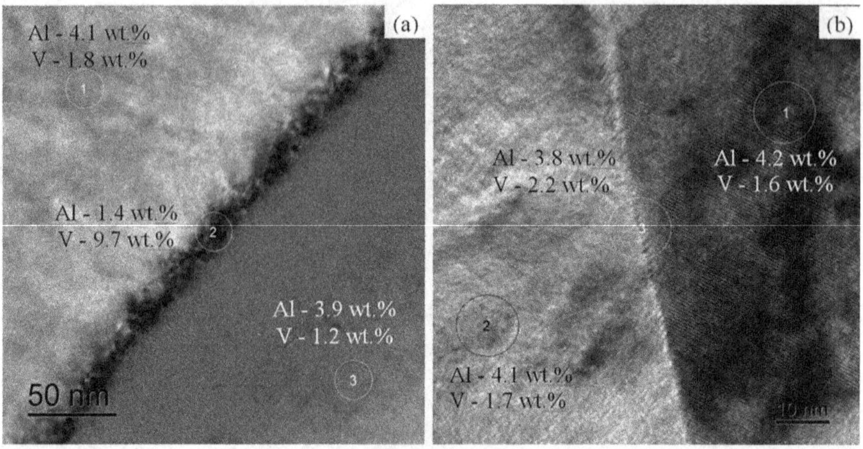

Fig. 2 Results of EDS analysis of grain boundaries in a coarse-grained alloy (**a**) and UFG titanium alloy (**b**)

concentration is under 1 wt.%. The average value of aluminum concentration in the crystal lattice near such boundaries approximates ~4 wt.%, while vanadium concentration is ~1.6 wt.%.

It shall be emphasized that vanadium-rich GB are found in the areas of the titanium alloy structure where the average grain size substantially exceeds the average size of a fine-grained matrix (Fig. 1a).

Grain boundaries in UFG alloys are clean with no excess vanadium concentrations detected (Fig. 2b). Average vanadium concentration on UFG alloy grain boundaries is 1.9 ± 0.3 wt.%, which is close to vanadium concentration in a crystal lattice 1.5 ± 0.3 wt.%.

Analysis of dependences of the shrinkage value (L) and the shrinkage rate (S) on the heating time prove that the welding process in UFG samples is more intensive: the shrinkage value and shrinkage rate in UFG titanium alloys appear to be much higher than similar characteristics in coarse-grained samples. The shrinkage value in coarse-grained alloys may be smaller than the thermal expansion contribution of the plant and remain unregistered by a dilatometer.

When summarized, the findings prove that under identical temperature and heating rate welding conditions, the average size and volume ratio of pores in UFG samples appear to be much smaller than in coarse-grained materials (see Fig. 3).

The studies conducted show that the rate of weldability (shrinkage rate) in UFG alloys tends to increase if the heating rate grows. With coarse-grained alloys, we observe an opposite effect: an increase in the heating rate leads to a decrease in the shrinkage rate (intensity of the welding process) (see Fig. 4). We reckon that this effect arises from the following: high-rate heating (caused by overall reduction in the process time) brings down the intensity of grain boundary migration and enables a more fine-grained structure (as compared to slower heating). This in turn decreases the typical

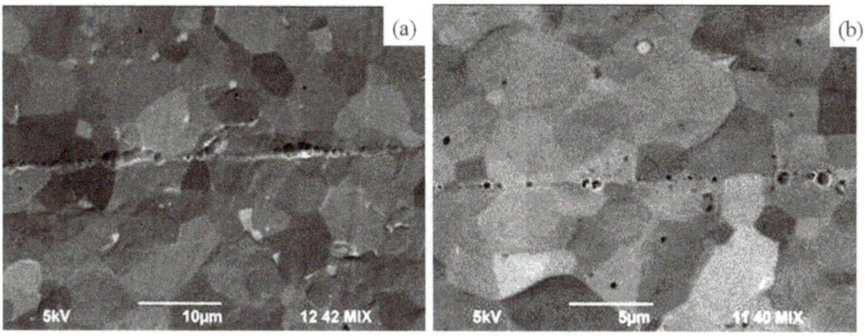

Fig. 3 Weld seam produced from coarse-grained (**a**) and UFG titanium alloys (**b**) obtained under identical temperature and heating rate welding conditions. Welding temperature 1030 °C (above recrystallization temperature). Heating rate V = 50 °C/min, P = 50 MPa

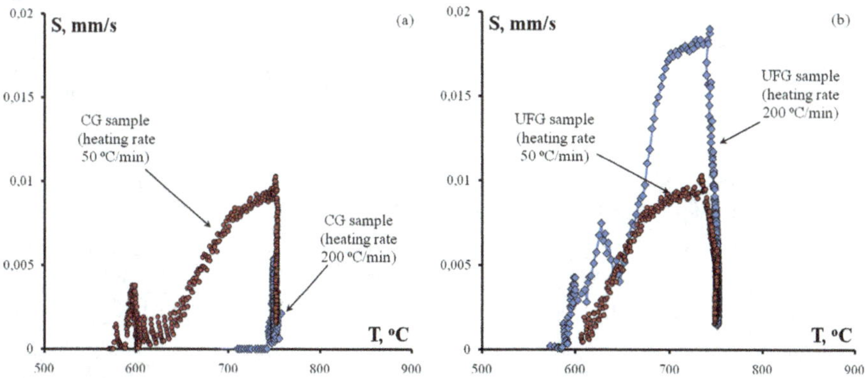

Fig. 4 Comparison of temperature dependences on shrinkage rates for coarse-grained (CG) alloy samples (**a**) and UFG titanium alloy samples (**b**) with initial roughness level of 3–5 μm welded at heating rates of 50 °C/min and 200 °C/min under 80 MPa

scale of diffusion mass transfer *x*, the value of which in case of grain boundary diffusion is proportionate to grain size.

The research conducted into mechanical properties proves that the hardness of weld seams obtained though SPS (Fig. 5a) at temperatures below the temperature at the start of recrystallization process corresponds to the hardness of the core material measured at a distance of no less than the thickness of three walls of a welded item from the seam edge (Fig. 5b).

For the purposes of comparison, Fig. 5c, d contains photos showing weld seams in UFG alloy samples obtained through argon arc welding (Fig. 5c) and electron beam welding (Fig. 5d).

When comparing anodic potentiodynamic dependences for coarse-grained and UFG alloy samples after diffusion welding, it should be noted that corrosion current density of weld seams from UFG titanium alloy within the temperature range of

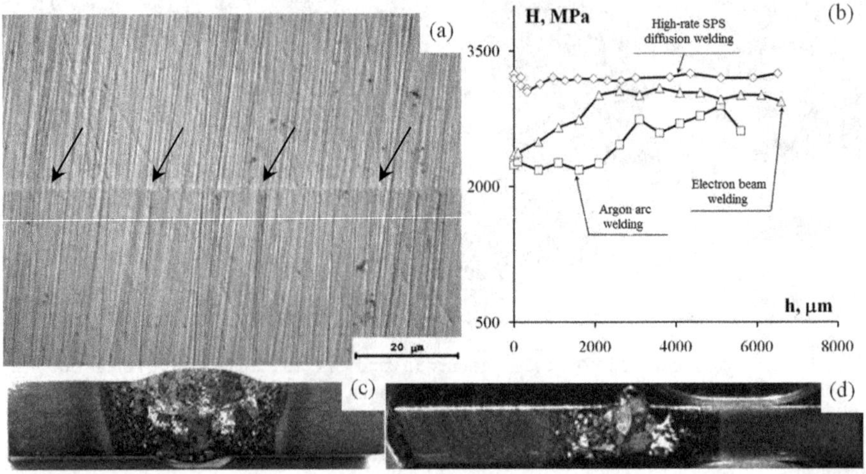

Fig. 5 Microsection of a weld seam obtained though SPS (**a**), argon arc welding (**c**) and electron beam welding (**d**) of UFG alloy samples, microhardness distribution (**b**) in a cross section of UFG alloy weld seams obtained through a variety of welding technologies (*h* distance from the seam center)

Table 1 Results of electrochemical studies of weld seams produced from coarse-grained and UFG titanium alloy (heating rate 10 °C/min, pressure 50 MPa)

State	SPS welding temperature (°C)	Maximal current density (mA/cm^2)
Initial state (coarse-grained)	600	1.15
	700	1.70
	800	0.25
UFG state (ECAP, N = 4)	600	0.70
	700	1.00
	800	0.12

600–800 °C appears to be less than current density of coarse-grained alloy weld seams ensured through similar welding modes.

With an increase in diffusion welding temperature, the difference in corrosion resistance between weld seams from coarse-grained and UFG titanium alloys grows: after diffusion welding at the temperature of 600 °C, the ratio of corrosion current density in weld seams produced from coarse-grained and UFG alloys was ~1.6; however, when the welding temperature rose to 700 °C and 800 °C, this ratio increased to 1.7 and ~2.1, respectively (see Table 1).

In our opinion, this effect is associated with recrystallization processes which underlie formation of a more homogeneous equiaxed structure in UFG α-titanium alloys, as well as with reduced local concentration of corrosive additives (V, Al) on UFG α-titanium alloy grain boundaries (see Fig. 2).

Analysis of anodic potentiodynamic dependences for UFG alloy weld seams obtained at different heating rates shows that irrespective of the original roughness, corrosion current density in samples obtained at a high heating rate (200 °C/min) appears to be ~1.5 times less than in samples obtained at the heating rate of 50°C. Thus, corrosion current density i_{cor} for UFG alloy weld seams (with initial roughness of 3–5 μm) obtained at the heating rate of 50 °C/min is 0.98 mA/cm^2, and at the heating rate of 200 °C/min, it equals 0.5 mA/cm^2. Current density for coarse-grained alloy weld seams obtained under similar conditions appears to be higher: for weld seams obtained at the heating rate of 50 °C/min, the i_{cor} ~2.5 mA/cm^2, while for those obtained at the heating rate of 200 °C/min, the i_{cor} ~2.8 mA/cm^2 (see Table 1).

Thus, the corrosion resistance of weld seams in UFG metal samples appears to be higher than the corrosion resistance of weld seams in coarse-grained materials.

4 Conclusions

1. SPS is proven to be an effective technology for high-rate diffusion welding of UFG titanium alloys. Optimization of temperature and rate conditions for SPS allows to produce UFG titanium alloy weld seams the hardness of which corresponds to that of the core material, whereas corrosion resistance is superior to that of the coarse-grained alloy weld seams.
2. Porosity of weld seams from UFG titanium alloys is found to be less than the porosity of weld seams from coarse-grained titanium alloys obtained under similar conditions. An increase in the heating rate leads to higher intensity of diffusion welding with UFG titanium alloy samples; however, with coarse-grained alloys, higher heating rates result in reduced intensity of diffusion welding.

This effect is associated with the formation of UFG structure in a weld seam during high-rate diffusion welding, which helps to reduce the typical scale of diffusion mass transfer and typical time of diffusion "exit" of pores from the weld seam due to the contribution of grain boundary diffusion, the intensity of which significantly exceeds the diffusion intensity in a crystal lattice limiting the rate of diffusion welding of coarse-grained materials.

Acknowledgments The authors are grateful for the support to the Russian Science Foundation (grant No. 16-13-00066).

References

Andrievskii RA (2011) Radiation stability of nanomaterials. Nanotechnol Russ 6:357–369. https://doi.org/10.1134/S1995078011030037

Balyanov A, Kutnyakova J, Amirkhanova N, Stolyarov V, Valiev R, Liao X, Zhao Y, Jiang Y, Xu H, Lowe T, Zhu Y (2004) Corrosion resistance of ultra fine-grained Ti. Scr Mater 51:225–229. https://doi.org/10.1016/j.scriptamat.2004.04.011

Chuvil'deev VN, Nieh TG, Gryaznov MY, Sysoev AN, Kopylov VI (2004) Low-temperature superplasticity and internal friction in microcrystalline Mg alloys processed by ECAP. Scr Mater 50:861–865. https://doi.org/10.1016/j.scriptamat.2003.12.003

Chuvil'deev VN, Kopylov VI, Gryaznov MY, Sysoev AN, Ovsyannikov BB, Flyagin AA (2008) Doubling of the strength and plasticity of a commercial aluminum-based alloy (AMg6) processed by equal channel angular pressing. Dokl Phys 53:584–587. https://doi.org/10.1134/S1028335808110086

Chuvil'deev VN, Shavleva AV, Nokhrin AV, Pirozhnikova OE, Gryaznov MY, Lopatin YG, Sysoev AN, Melekhin NV, Sakharov NV, Kopylov VI, Myshlayev MM (2010) Influence of the grain size and structural state of grain boundaries on the parameter of low-temperature and high-rate superplasticity of nanocrystalline and microcrystalline alloys. Phys Solid State 52:1098–1106. https://doi.org/10.1134/S1063783410050422

Chuvil'deev VN, Kopylov VI, Bakmet'ev AM, Sandler NG, Nokhrin AV, Tryaev PV, Lopatin YG, Kozlova NA, Piskunov AV, Melekhin NV (2012) Effect of simultaneous enhancement in strength and corrosion resistance of microcrystalline titanium alloys. Dokl Phys 57:10–13. https://doi.org/10.1134/S1028335812010119

Chuvil'deev VN, Nokhrin AV, Baranov GV, Moskvicheva AV, Boldin MS, Kotkov DN, Sakharov NV, Blagoveshensky YV, Shotin SV, Melekhin NV, Belov VV (2013) Study of the structure and mechanical properties of nano- and ultradispersed mechanically activated heavy tungsten alloys. Nanotechnol Russ 8:108–121. https://doi.org/10.1134/S1995078013010047

Chuvil'deev VN, Boldin MS, Dyatlova YG, Rumyantsev VI, Ordan'yan SS (2015a) A comparative study of the hot pressing and spark plasma sintering of Al_2O_3-ZrO_2-Ti(C,N) powders. Inorg Mater 51:1047–1053. https://doi.org/10.1134/S0020168515090034

Chuvil'deev VN, Panov DV, Boldin MS, Nokhrin AV, Blagoveshensky YV, Sakharov NV, Shotin SV, Kotkov DN (2015b) Structure and properties of advanced materials obtained by spark plasma sintering. Acta Astronaut 109:172–176. https://doi.org/10.1016/j.actaastro.2014.11.008

Chuvil'deev VN, Kopylov VI, Nokhrin AV, Tryaev PV, Kozlova NA, Tabachkova NY, Lopatin YG, Ershova AV, Mikhaylov AS, Gryaznov MY, Chegurov MK (2017a) Study of mechanical properties and corrosive resistance of ultrafine-grained α-titanium alloy Ti-5Al-2V. J Alloys Compd 2017:354–367. https://doi.org/10.1016/j.jallcom.2017.06.220

Chuvil'deev VN, Blagoveshchenskiy YV, Nokhrin AV, Boldin MS, Sakharov NV, Isaeva NV, Shotin SV, Belkin OA, Popov AA, Smirnova ES, Lantsev EA (2017b) Spark plasma sintering of tungsten carbide nanopowders obtained through DC arc plasma synthesis. J Alloys Compd 708:547–561. https://doi.org/10.1016/j.jallcom.2017.03.035

Kim H, Kim W (2014) Annealing effects on the corrosion resistance on ultrafine-grained pure titanium. Corros Sci 89:331–337. https://doi.org/10.1016/j.corsci.2014.08.017

Kim H, Yoo S, Ahn J, Kim D, Kim W (2011) Ultrafine grained titanium sheets with high strength and high corrosion resistance. Mater Sci Eng A 528:8479–8485. https://doi.org/10.1016/j.msea.2011.07.074

Munir Z, Quach D (2011) Electric current activation of sintering: a review of the pulsed electric current sintering process. J Am Ceram Soc 94:1–19. https://doi.org/10.1111/j.1551-2916.2010.04210.x

Munir Z, Anselmi-Tamburini U, Ohyanagi M (2006) The effect of electric field and pressure on the synthesis and consolidation materials: a review of the spark plasma sintering method. J Mater Sci 41:763–777. https://doi.org/10.1007/s10853-006-6555-2

Orlova AI, Volgutov VY, Mikhailov DA, Bykov DM, Skuratov VA, Chuvil'deev VN, Nokhrin AV, Boldin MS, Sakharov NV (2014) Phosphate $Ca_{1/4}Sr_{1/4}Zr_2(PO_4)_3$ of the $NaZr_2(PO_4)_3$ structural type: synthesis of a dense ceramic material and its radiation testing. J Nucl Mater 446:232–239. https://doi.org/10.1016/j.jnucmat.2013.11.025

Perevezentsev VN, Chuvil'deev VN, Kopylov VI, Sysoev AN, Langdon TG (2002) Developing high strain rate superplasticity in Al-Mg-Sc-Zr alloys using equal-channel angular pressing. Ann Chim Sci Mater 27:99–109

Segal VM, Beyerlein IJ, Tome CN, Chuvil'deev VN, Kopylov VI (2010) Fundamentals and engineering of severe plastic deformation. Nova Science Publishers, New York

Tokita M (2013) Spark plasma sintering (SPS) method, systems, and applications (Chapter 11.2.3). In: Somiya S (ed) Handbook of advanced ceramics, 2nd edn. Academic Press, Oxford

Valiev R, Langdon T (2006) Principles of equal-channel angular pressing as a processing tool for grain refinement. Prog Mater Sci 51:881–981. https://doi.org/10.1016/j.pmatsci.2006.02.003

Wang Y, Chen M, Zhou F, Ma E (2002) High tensile ductility in nanostructured metal. Nature 419:912–915. https://doi.org/10.1038/nature01133

Near-Net Shaping of Titanium-Aluminum Jet Engine Turbine Blades by SPS

Thomas Voisin, Jean-Philippe Monchoux, and Alain Couret

1 Introduction

SPS has shown an undeniable ability to produce a variety of high-quality materials (Orrù et al. 2009). For metals, this technique offers three main advantages: (i) a rapid heating that can stabilize microstructures in a short time and thus limit grain growth; (ii) a limited temperature and pressure gradients in samples upon sintering, resulting in homogeneous, dense materials; and (iii) a processing in solid state, preventing materials from being melted and thus avoiding contamination, segregation, and all problems associated with metals in liquid state. This presents a great potential for metals that conventional processing techniques fail to produce cost-effectively.

For decades, intermetallic titanium aluminides (TiAl), composed of around 48% titanium, 48% aluminum, and several alloying elements, have been subject to intensive development for high temperature structural applications. Initially developed at the Air Force Wright-Patterson Base, OH, in the early 1980s for military jet engines (Kothari et al. 2012), they are now designed and implemented to replace the so-called nickel-based superalloys currently used for low-pressure turbine blades (Pollock 2016). In fact, TiAl alloys offer high strength and high oxidation and corrosion resistance at high temperature (Kim 1989) with a density twice lower than superalloys (3.9 g/cm^3). These properties also make this material a good choice in the automobile industry for injection valves or turbocharger wheels (Shouren et al. 2008). However, after more than 30 years, even though tremendous scientific

T. Voisin (✉)
Lawrence Livermore National Laboratory, Livermore, CA, USA

CEMES/CNRS, Toulouse, France
e-mail: Voisin2@llnl.gov

J.-P. Monchoux · A. Couret
CEMES/CNRS, Toulouse, France

© Springer Nature Switzerland AG 2019

P. Cavaliere (ed.), *Spark Plasma Sintering of Materials*,
https://doi.org/10.1007/978-3-030-05327-7_25

improvements made their industrial production effective, TiAl still suffers from important limitations: a low plastic elongation in tension at room temperature (typically around 1%), and a difficult and thus expensive manufacturing (Clemens and Mayer 2016).

Still, engineers at General Electric (GE), USA, have adapted the conventional gravity casting and near-net shape casting toward the production of TiAl turbine blades that successfully equip two stages of the GEnx™ engine low-pressure turbine in the Boeing's 787 Dreamliner flying since 2012 (Bewlay et al. 2016). The alloy used is the well-studied GE 48-2-2 (Ti-48Al-2Cr-2Nb) (Huang 1988). In Europe, Leistritz Turbinentechnik GmbH, Germany, supplies MTU Aero Engines with turbine blades forged from the TNM-V4 alloy (Ti-45Al-5Nb-0.2B-0.2C) (Janschek 2015). They are used in the PW1134G engines, in service in the A320neo jets since late 2014. Since 2016, SNECMA-SAFRAN produces GE 48-2-2 blades for the LEAP® engines (product of SAFRAN Aircraft Engines and GE Aviation, under the 50/50 joint company CFM) in service in the Airbus A320neo, Boeing 737 MAX, and COMAC C919 (Sallot et al. 2017).

However, these successful commercial implementations do not mean the intrinsic limitations of TiAl were overcome. Long and complex heat treatments and machining are necessary to obtain the final product. As-cast and as-forged materials are not suitable due to undesirable microstructure and texture or residual porosity. This presents a major obstacle to implementing TiAl in blade stages of the turbine closer to the combustion chamber where the temperature is higher than 800 °C.

In this context, and in an effort to propose a viable alternative that would prevent the long thermomechanical treatments and strongly reduce the final machining, we have adapted the Spark Plasma Sintering (SPS) process to the production of TiAl near-net shape turbine blades to be used at higher temperatures.

In a first part, we will describe what are the microstructures and mechanical properties achievable by SPS with the well-studied TiAl alloys 48-2-2. Then, we will present the development of the new IRIS TiAl alloy, adapted to the SPS process, which exhibits outstanding mechanical properties. In a second part, we will depict the work done toward the control of both temperature field and densification mechanisms. Finally, a third part will concern the production of near-net shape turbine blades by SPS in one cycle, without any post-process thermal treatments.

2 Experimental

2.1 SPS Apparatus

Three different SPS machines were used, each one offering different current capacities. The first one is a SYNTEX 2080 (Syntex Inc., Kanagawa, Japan), present at the PNF2 (Plateforme Nationale de Frittage Flash, Toulouse, France). It allows sintering samples up to 50 mm in diameter with a maximum current output of 8 kA and a maximum force of 200 kN. This machine was used for most of the material design and microstructure/mechanical properties' optimization. The second machine, a

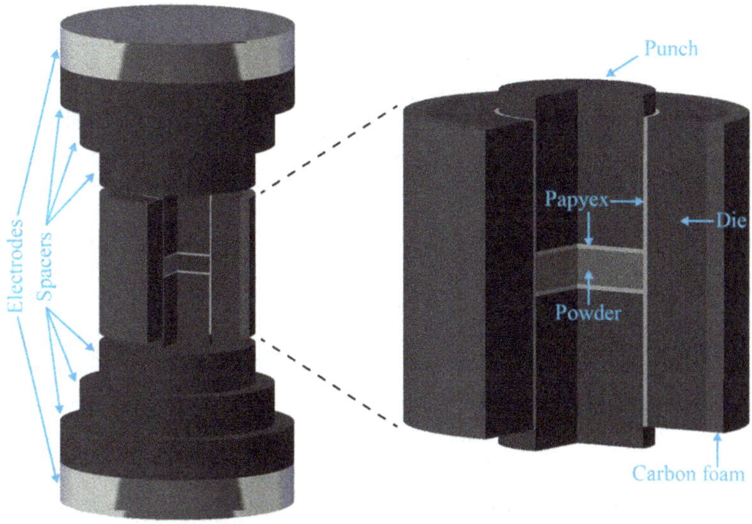

Fig. 1 Set-up used for sintering a disk in the SPS

SYNTEX 9.40, from the same manufacturer, supplies a 20 kA current and a force of 3000 kN. With this machine, located at the Mecachrome company, Vibraye, France, we densified the full-scale turbine blades as it allows sintering samples up to 300 mm in diameter. The last machine is a FCT HPD-125 (FCT System GmbH, Frankenblick, Germany). It provides a higher current of maximum 24 kA but a lower force up to 1250 kN. It allows sintering samples as large as 200 mm. It was also used to make near-net shape blades.

2.2 *Set-Up Configuration and SPS Cycle*

Temperature field control, alloy development, and material property optimization were conducted using a conventional graphite set-up as shown in Fig. 1. The powder is poured into a cylindrical graphite die with the inner surface covered of a 0.5 mm thick graphite foil (Papyex) and closed by two graphite punches. The outer surface of the die is covered by carbon foam to limit the heat loss by convection. This set-up is then positioned into the vacuum chamber, surrounded by heat deflectors to protect the inside of the chamber. The electrodes are water cooled. The three machines are equipped with a pyrometer located outside the chamber and aiming at the center of the die, on the outer surface.

Otherwise described, the SPS cycle consisted of applying gradually a pressure of 100 MPa in the first 2 min while the temperature was raised at rate of 100 °C/min until the dwell temperature was reached (between 1000 °C and 1400 °C depending on the alloy and the targeted microstructure). To avoid a temperature overshoot, the last 3 min before reaching the dwell temperature, the heating rate was dropped to

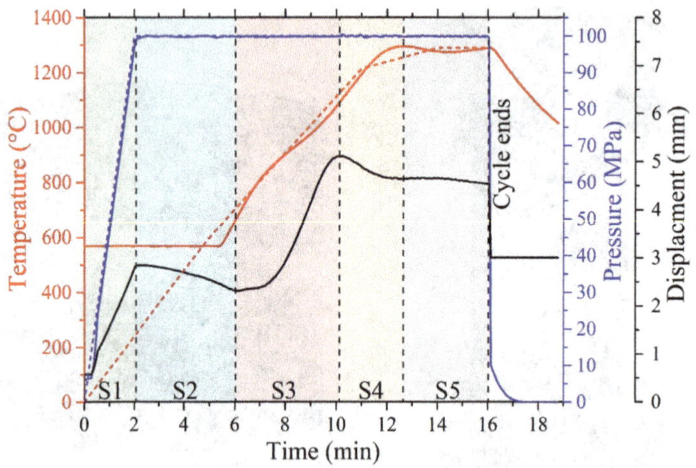

Fig. 2 Temperature, pressure, and displacement as a function of time during a conventional SPS cycle. S1 to S5 correspond to the five stages of densification described in the text. For temperature and pressure, the dashed lines are the parameters' set-up and the full lines are the actual measurements upon sintering

25 °C/min (Fig. 2). The type of pyrometer used did not allow reading temperatures below around 600 °C, explaining the absence of measurement the first 5 min. The displacement was also recorded during the cycle. A positive displacement means the electrodes get closer to each other; in other words, the sample is being compressed. When sintering a conventional disk, the displacement (black curve) follows five stages (Fig. 2). First, the displacement increases linearly with the applied pressure. Second, the displacement decreases. This means that the set-up (powder and graphite tools) are expending as a function of the temperature which causes to move the electrodes away from each other. Third, sintering occurs and the powder is being compacted to full density. Fourth, the sample is fully dense, but the temperature keeps rising, and so the set-up keeps expanding with the temperature increase. Fifth, when the dwell temperature is reached, the displacement stabilizes, with minor fluctuations due to slight variations in temperature. Last, the SPS cycle is ended, meaning both the power and the pressure are brought back to zero. Note the cooling rate in this example is similar to the heating rate, but this will strongly depend on the size of the set-up and machine.

2.3 *TiAl Powder*

Several different alloys were used in this study; all powder was made by gas atomization by the Crucible Industries, NY, USA. For each alloy, the powder particles presented a highly non-equilibrium dendritic microstructure, due to the rapid solidification encountered during the atomization process. This microstructure was characterized in our previous work (Voisin 2014). The powder particle size ranges from 50 to 200 μm.

2.4 *Material Characterization*

Microstructures were analyzed by Scanning Electron Microscopy (SEM) using a Jeol 6490 (Tungsten filament) and by Transmission Electron Microscopy (TEM) using a Jeol 2010 (LaB_6 filament) working at 200 keV. For SEM observations, samples were mechanically polished with SiC papers with particle sizes ranging from 200 to 10 μm and diamond suspensions with sizes from 6 to 1 μm. Samples for TEM were mechanically polished down to 25 μm with SiC papers and then electropolished with an A3-type solution (25% 2-butoxyethanol in methanol). Mechanical properties were characterized by tensile testing at 10^{-4}/s with a Zwick/Roell Z030 and creep testing with an ADAMEL tester under constant load and temperature.

3 Titanium Aluminides by SPS

The TiAl phase diagram was first established by (Murray 1987). Its central part (between 20% and 70% of aluminum), which shows the best industrial potential, was then more accurately refined by (McCullough et al. 1989). It exhibits a great complexity, allowing solidifying through the bcc β phase or the hcp α phase, depending on the aluminum content and the nature of the alloying elements. Upon cooling, TiAl alloys overcome several phase transformations that can be controlled, or avoided, by adjusting the chemistry and thermal treatments (Appel and Wagner 1998). This presents a good opportunity to tune the microstructure toward a given application.

3.1 *Phase Diagram and Microstructure of TiAl*

The 48-2-2 alloy ($TiAl_{48}$-Cr_2-Nb_2) is the most documented TiAl alloy as a result of the extensive work led by GE for aeronautic applications (Bewlay et al. 2016). It is satisfactorily described by the binary phase diagram. Thus, it is the most suited material to start with for identifying how the SPS can affect the microstructure of TiAl alloys. The three main microstructures obtained via conventional manufacturing routes has been perfectly reproduced by SPS (Voisin et al. 2013) and are reported in Fig. 3a. Their nature depends on the dwell temperature during the SPS cycle, as given by the phase diagram (Fig. 3b). When samples are heated in the $\alpha + \gamma$ field (orange region), the resulting microstructure at room temperature is composed of

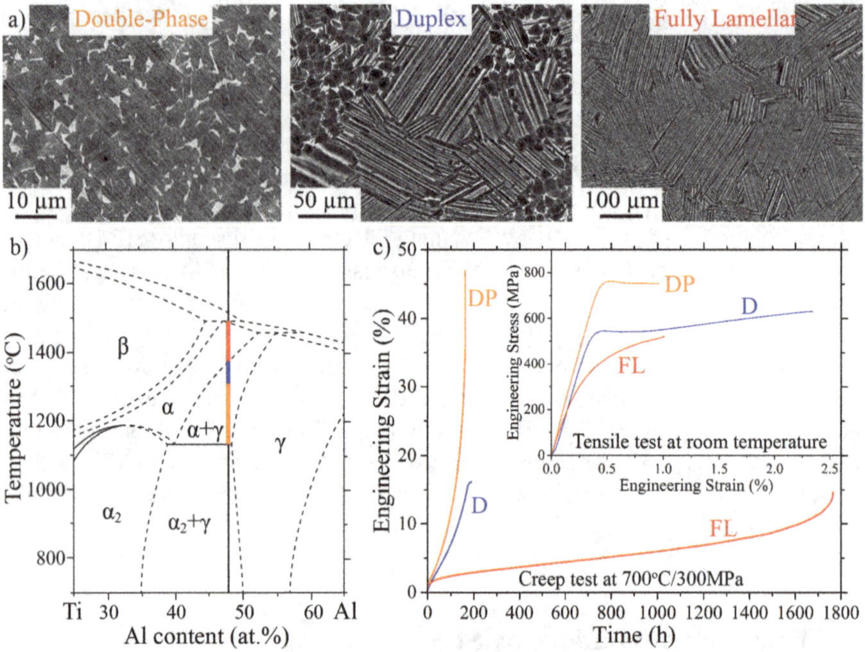

Fig. 3 TiAl 48-2-2 by SPS. (**a**) Three main microstructures presenting an industrial interest. (**b**) Phase diagram of the TiAl system. The three highlighted regions give the microstructures described in (**a**). (**c**) Corresponding mechanical properties at room and high temperature. DP, D, and FL stand for double-phase, duplex, and fully lamellar, respectively

two phases, α_2 and γ (Fig. 3a), one hcp, the other one fcc. If samples are heated in the single-phase α field at higher temperature (red region), after cooling, the microstructure will be the so-called fully lamellar structure, where each grain is an alternation of thin α_2 and γ lamellae. For samples heated just below the α-transus (blue region), the microstructure will be duplex, which consists of a combination of equiaxed γ grains and small lamellar grains. Each microstructure exhibits specific mechanical properties (Fig. 3c). The double-phase is very resistant at room temperature but suffers from a poor ductility and a lack of strength at high temperature. On the contrary, the fully lamellar structure has a low strength at room temperature but offers high strength at high temperature. Note that the ductility is very limited for this type of structure. Finally, the duplex is in the middle range concerning strength at room temperature. However, it is not resistant at high temperature. Its main advantage is a reasonable plastic strain at room temperature. As can be observed here, one of the main challenges with TiAl alloys is to design an alloy that would combine high strength at high temperature and decent ductility at room temperature.

3.2 TiAl Alloy Design for SPS

At room temperature, depending on the aluminum content, cooling rate, and alloying elements, the three main phases achievable in TiAl alloys are the α_2, β, and γ (classified here from harder to softer). The γ phase is thus the one deforming the most TiAl alloys at room temperature. This fcc structure deforms largely by dislocation slip; however, due to a long range order (alternation of aluminum and titanium atoms in the stacking sequence along the {100} direction) that limits the available slip systems, twinning is the second deformation system, allowing to match the Von Mises criteria (Groves and Kelly 1963). The bcc β phase is very ductile at high temperature, being thus an advantage for processes such as forging but undesirable in the final product as it is detrimental to high temperature mechanical properties. Also, it is usually considered brittle at room temperature (Appel et al. 2011) even though it was shown recently that β grains can deform due to dislocation pileup at their grain boundaries (Voisin et al. 2012). The α_2 phase, hcp, provides the strength at both room and high temperature. Deformation of α_2 grains is rarely observed in normal testing conditions. Unfortunately, this is the main cause of the limited ductility in TiAl alloys as it leads to strong strain incompatibility between phases (Voisin 2014).

We have shown that the addition of heavy elements greatly strengthens SPS TiAl alloys at high temperature when in solution in the γ phase (Jabbar et al. 2011, 2014). Likewise, we have demonstrated that the addition of boron in small quantity reduces the grain size by pinning grain boundaries during the SPS cycle (Luo et al. 2013). Based on these studies, we have developed the IRIS alloy (Ti-48Al-2W-0.08B) (Couret et al. 2013). This material, successfully sintered by SPS to full density, has undergone an in-depth microstructure optimization resulting in the near-lamellar microstructure depicted Fig. 4.

As shown in Fig. 4a, the microstructure mainly contains lamellar grains. When looking closer (inset), one can see that these lamellar grains are surrounded by γ grains containing β precipitates (bright particles). The TEM pictures in Fig. 4b tell more about these borders. It appears that γ grains are actually the continuity of at least one of the adjacent lamellar grain, where basically α_2 is not present anymore. This is a very important feature of this microstructure as γ grains are present all around lamellar grains as a thin layer, allowing for more deformation to occur and thus increasing the ductility. At the meantime, the thickness of these borders is limited, avoiding large clusters of γ grains as observed in the duplex structure (Fig. 3a) that are detrimental to the high temperature mechanical properties. Also, the γ lamellae thickness is small enough to prevent dislocation pileup but big enough to permit dislocations to move (Voisin et al. 2016), and the volume fraction and thickness of α_2 lamellae are low, allowing more homogeneous deformation of lamellar grains.

This near-lamellar microstructure exhibits outstanding mechanical properties at both room and high temperature. At room temperature (Fig. 4c), a great balance between strength and ductility is obtained, comparable to the duplex microstructure

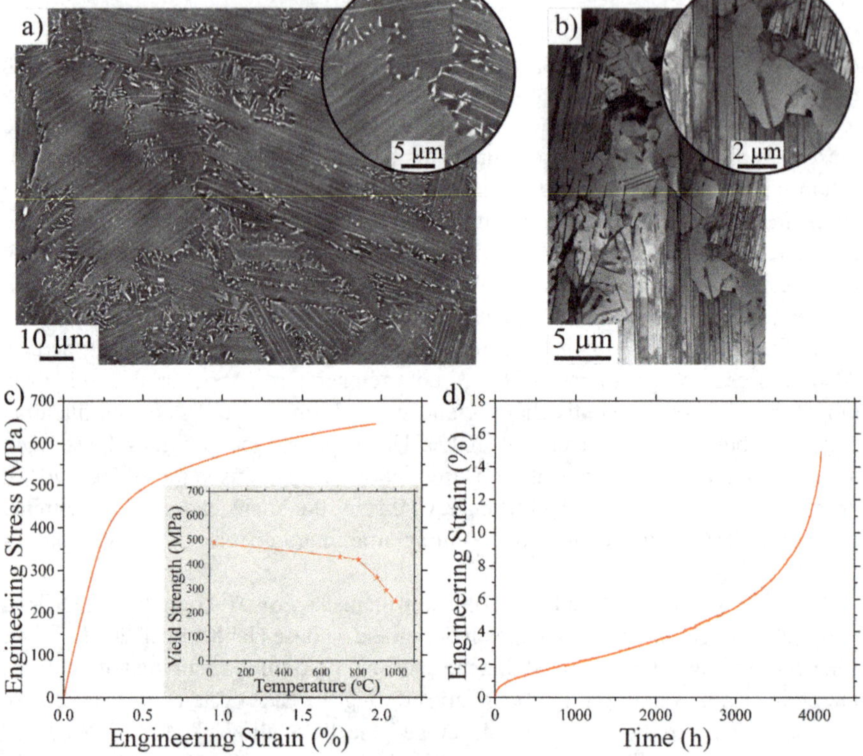

Fig. 4 IRIS alloy by SPS. (**a**) and (**b**) show the microstructure as seen by SEM and TEM, respectively. The insets are zoomed in from the same region. (**c**) Tensile curve at room temperature. The inset shows the change in yield strength at 0.2% as a function of the tensile testing temperature. (**d**) Strain/time curve from a creep test conducted at 700 °C/300 MPa

of the 48-2-2 (Fig. 3c). At the same time, the IRIS alloy maintains a very low strain rate of 3.5×10^{-9}/s under creep at 700 °C/300 MPa (three times slower than the fully lamellar 48-2-2), with a lifetime greater than 4000 h (more than two times longer than the fully lamellar 48-2-2). The stability of the strength as a function of temperature is also very impressive as only when samples are tested above 800 °C the yield strength starts dropping slowly, remaining higher than 250 MPa at 1000 °C (Fig. 4c – inset).

The combination of a high γ-content inside and around lamellar grains, promoted by a high aluminum content, with a strong γ phase reinforced by the addition of tungsten (Voisin et al. 2017; Couret et al. 2017), and an optimized processing cycle has allowed the as-SPS near-lamellar IRIS alloy being among the most performing TiAl alloys at high temperatures (Clemens and Mayer 2016) while still offering a large ductility at room temperature.

4 Control of the Densification: Temperature and Porosity

Outstanding mechanical properties were obtained using the SPS process due to the specific microstructure of the IRIS alloy. However, this was possible mainly because both temperature gradient through the sample and final density were controlled so that the as-sintered material reaches full density as well as a homogeneous microstructure. This becomes even more important when complex shapes are to be given to the as-SPS component to ensure reproducible properties within one part as well as between several parts sintered in the same conditions. In this section, we first present a study of the densification as a function of the location in the sample during the SPS cycle. Then, we depict the work done toward measuring and predicting the temperature field in samples of different level of complexity.

4.1 Sample Density Mapping as a Function of Temperature

4.1.1 Method

A great method to map out pores in a sample would have been the 3D X-ray micro-computed tomography (micro-CT) technique; however, it is not easy to access, and more importantly, the "limited" resolution of around 1 μm per pixel, even when using high energy X-ray, may have cause missing some of the smaller pores (Voisin et al. to be published). Also, the energy of the X-ray limits the size of samples that can be studied. We have chosen here to map the center cross section of cylindrical samples in a SEM, with the assumption that pore distribution and location are the same in any vertical section that would cross the middle of the cylinder. Figure 5 depicts the method. The SPS cycle is interrupted after a chosen amount of

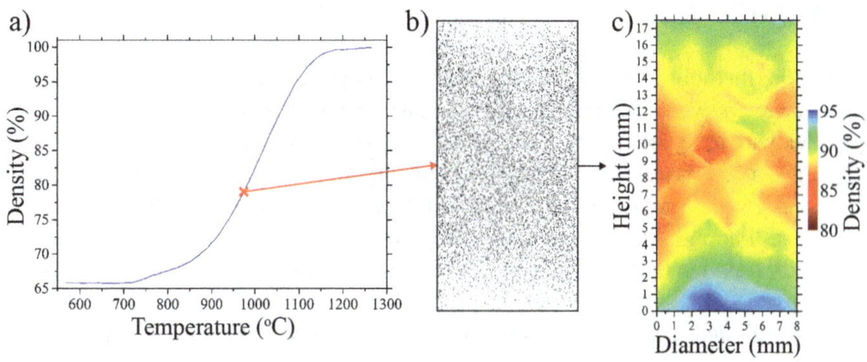

Fig. 5 Mapping the material density in an as-SPS cylinder. (**a**) Density of a sample as a function of temperature during a SPS cycle. The red cross illustrates an interrupted test. (**b**) Middle cross section of a sample for which the SPS cycle was interrupted, analyzed by SEM, and made binary. Black color comes from the pores. (**c**) Density map corresponding to the sample shown in (**b**)

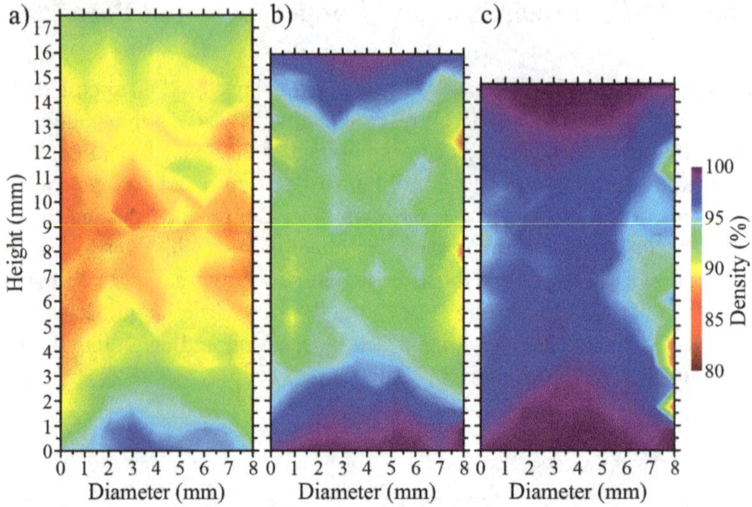

Fig. 6 Density maps in high, narrow samples (dimensions after full densification are diameter 8 mm and height 13 mm). Maps were made in samples with an average overall density of (**a**) 90%, (**b**) 94.4%, and (**c**) 97.2%. Their SPS cycles were interrupted at temperatures of 1000 °C, 1020 °C, and 1070 °C, respectively

compaction (Fig. 5a). The sample is then recovered and vertically sectioned at the middle. The cut surface is mechanically polished, entirely mapped out in the SEM, and binarized (Fig. 5b). The resulting picture is then gridded, and in each square of the grid, the fraction of pores is quantified. This allows us to build a map of the sample density as shown in Fig. 5c.

4.1.2 Density Mapping of Cylinders of Various Sizes

Using the method described in the previous section, cylinders of different heights and diameters were sintered to various densities and mapped out. Note that TiAl powder was used for direct quantitative analysis. Three sets of dimensions were chosen to study the effect of the sample geometry on the densification behavior: first, high and narrow samples (Fig. 6); second, intermediary height and width (Fig. 7a, b); and third, flat and wide (Fig. 7c, d). They are intended to reproduce variations of geometry found in turbine blades, with the two extremes being the bulky root and the thin curved sail (Appel et al. 2011).

The high and narrow samples exhibit three different regions upon densification (Fig. 6):

- Underneath the punches (top and bottom) is the region that sees its density increasing faster, reaching full density when the average sample density is around 90% (Fig. 6b).

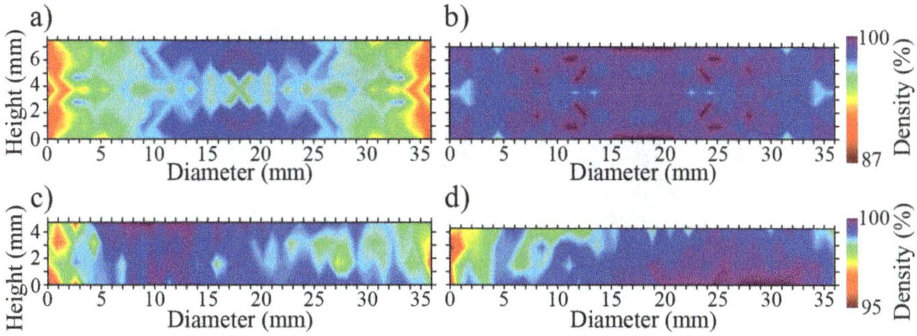

Fig. 7 Density maps of wider samples. (**a**) and (**b**) Dimensions after full densification are diameter 36 mm and height 6 mm. Average sample densities are 95% and 98%, obtained by stopping the SPS cycles at 1000 °C and 1100 °C, respectively. Note that, for these two samples, only a quarter of their surface was mapped due to their big size. The analyzed quarter was then reflected vertically and horizontally to construct the full sample. (**c**) and (**d**) Dimensions after full densification are diameter 36 mm and height 3 mm. Average sample densities are 98% and 99.5%, obtained by stopping the SPS cycles at 1050 °C and 1110 °C, respectively

- Along the die inner surface, where the papyex is lying (Fig. 1), is the region where compaction is the most difficult. This is due to the friction between powder particles and immobile surfaces. There, density can drop locally below 90% when the overall sample density is above 97% (Fig. 6c). We will refer to this mechanism as "friction-effect."
- In the middle of the samples, i.e., anywhere but underneath the punches or along the die vertical inner surface, the powder particles show intermediary density, usually slightly below the average sample density.

Following the same principle, samples with increased diameter (36 mm) and various heights were partially sintered and their density maps were constructed (Fig. 7). These new dimensions did not seem to change the trend observed in the smaller samples. Edges (along the die vertical inner surface) are more difficult to densify than underneath the punches. However, it can be seen that for thinner samples (Fig. 7c, d), the densification appears less homogeneous, with the fully dense region not lying at the middle anymore. For example, the sample in Fig. 7d shows one side that densifies before the other one. This is ascribed to a higher sensitivity to nonparallel punches.

To study how powder particles move and rearrange during densification, we have analyzed a funnel-like sample positioned vertically in the SPS, with the bottleneck facing down as illustrated in Fig. 8. The cross section was taken at the middle (Fig. 8a) of a sample sintered to an average density of 96.5%. The resulting density map shows a higher density underneath the top punch (Fig. 8b) as expected; however, the powder particles in the bottleneck behaved differently. First, due to the limited width of 4 mm, the friction-effect became predominant. Second, the much higher density of the bulk part of the funnel quickly prevented the punches to keep densifying the bottleneck. This shows that, even at high temperature when TiAl

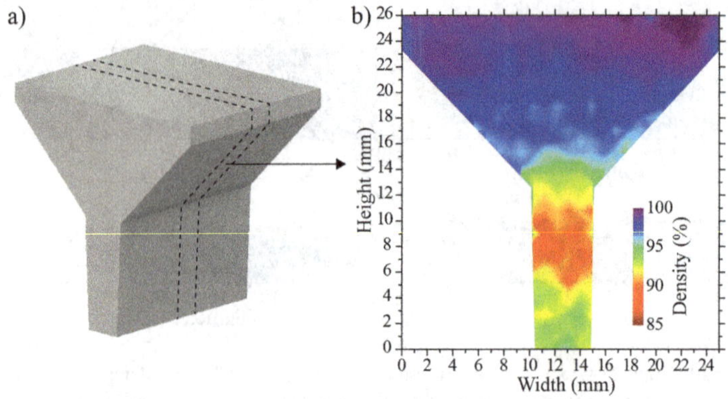

Fig. 8 Density maps of a funnel sample sintered up to an average density of 96.5%. (**a**) Geometry of the as-SPS sample. It was position vertically in the SPS as shown here. Interfaces with the punches are on the top and bottom. (**b**) Density map of the cross section highlighted by the dashed lines in (**a**)

becomes very ductile, there is only a limited transfer of material throughout the sample.

In conclusion, in a complex shape, thin, flat regions will densify first, as they require less displacement of the punches to reach full density. Regions elongated along the compaction direction (vertically) will become even harder to densify when they are narrower. As the sample diameter reduces for a given height, the friction-effect will be more important, requiring higher temperature or pressure to reach full density. Also, thin, flat regions become more sensitive to punches misalignment, which will lead to heterogeneous densification. In a long, thin, and flat sample, regions that first reach full density could stop the punches motion and prevent a full densification as the compacted powder particles do not circulate enough between the punches. These quantitative measurements for TiAl alloys were used for accurate modeling of the compaction behavior in SPS (Mondalek et al. 2011).

4.2 *Temperature Control in Complex Shapes*

Inhomogeneous densification in a sample can be due to geometry effects as shown in the previous section. However, a heterogeneous temperature field will also affect the density as well as the microstructure and resulting mechanical properties. We have developed a model to predict the maximum temperature reached during the SPS cycle in every location of samples using the finite element method (FEM) (Voisin et al. 2013). Results from simulations are validated by experimental observations. We use the sudden change in microstructure from a fine $\alpha + \gamma$ double-phase to a coarse single phase α microstructure (corresponding to duplex and fully lamellar at room temperature, respectively) that occurs in TiAl over a very short range of

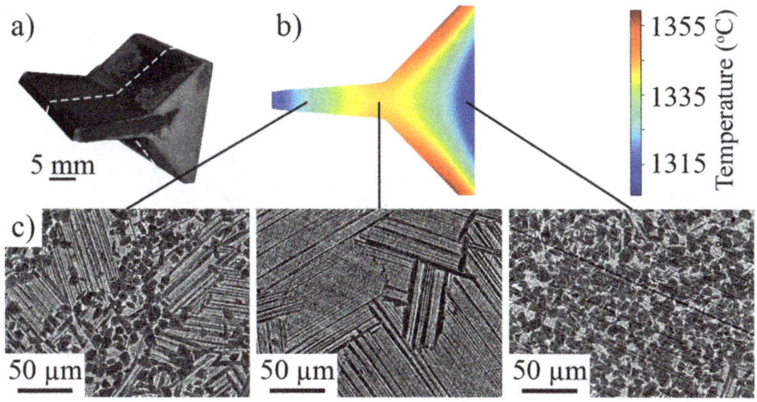

Fig. 9 Temperature field in complex shapes. (**a**) Component as-SPS. (**b**) Cross section highlighted by the white dashed line in (**a**) where temperature was simulated by FEM. (**c**) Microstructures as observed by SEM in the regions designated by the black lines

temperatures (Denquin and Naka 1996) as soon as the α-transus is crossed upon heating (Fig. 3). We measured this temperature to be 1335 ± 5 °C in the GE 48-2-2 (Voisin et al. 2013). By sintering several samples of the same geometry but at different temperatures, it is possible to cross the α-transus in different locations of the sample and thus deduce the temperature field (Voisin et al. 2013). An example of temperature field modeling taken from (Voisin et al. 2013) is given in Fig. 9. The sintered sample presents a thin and a bulk region to account for variations in geometry encountered in turbine blades (Fig. 9a). The temperature gradient is simulated in this sample by FEM (Fig. 9b) and compared to microstructure analysis by SEM (Fig. 9c). As can be seen, modeling and microstructure are in a good agreement. Regions of the sample that have a temperature above 1335 °C exhibit a fully lamellar microstructure, proof that the α-transus was crossed. Regions showing temperatures just below 1335 °C have a duplex structure, while colder regions show a finer double-phase microstructure containing few lamellar grains. While all experimental temperature measurements were conducted with the GE 48-2-2 powder as it presents a faster and more obvious α-transus, the final material used for turbine blades is the IRIS alloy. We have found that its optimized near-lamellar microstructure and corresponding mechanical properties are stable over a temperature range of around 80 °C (Voisin et al. 2016), which is greater than the total temperature gradient of 60 °C predicted in Fig. 9. Thus, complex shapes made with the IRIS alloy exhibit homogenous structures and properties as-SPS.

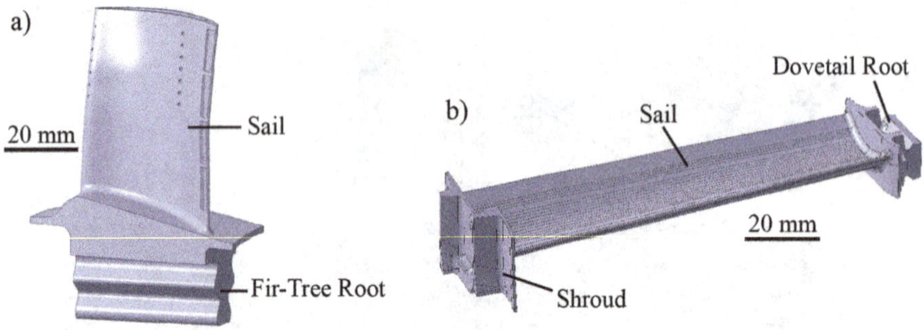

Fig. 10 Jet engine turbine blades of interest in this work. (**a**) High-pressure blade close to the combustion chamber. (**b**) Blade located in the last stage of low-pressure turbine

5 Near-Net Shapes: Turbine Blades

With a better understanding and control of densification mechanisms and temperature gradient in components sintered by SPS, an important effort has been put toward producing near-net shape turbine blades. Two types of blade are of interest in this framework, a high-pressure blade, close to the combustion chamber, and a low-pressure blade, located in the last stage of low-pressure turbine (Fig. 10). They both present several challenges, the main ones being strong changes in thickness between the root and the sail, and fine features that may encounter important friction that would resist the densification, as observed in (Fig. 8).

Note that, in this work, the internal cooling channels usually present in the high-pressure blade (Fig. 10a) that allow for the colder air coming from the high-pressure compressor to enter through the root and exit through the leading and trailing edges are not taken in account. Due to their complexity, they will be the subject of a dedicated forthcoming study.

5.1 Methodological Increase in Sample Geometry Complexity

To sinter a complex shape, the graphite mold is designed accordingly (Fig. 11a). Punches are machined, so they give the final shape to the sample when full densification is reached (blue parts in Fig. 11b). Inserts are used to fill the gaps between punches and die (green parts in Fig. 11a, b). A step-by-step strategy was adopted to study the densification of different features of the blades. Simple cases were first investigated such as in Fig. 11c where the sample is sintered horizontally as shown in (Fig. 11a, b). It presents a bulk and a thin region to recreate the difference in geometry encountered between the root and the sail.

It appeared that an important factor to take in account when designing such molds is the parallelism between horizontal surfaces belonging to the two punches. As highlighted by the study of densification mechanisms in Sect. 4.1.2, misalignments

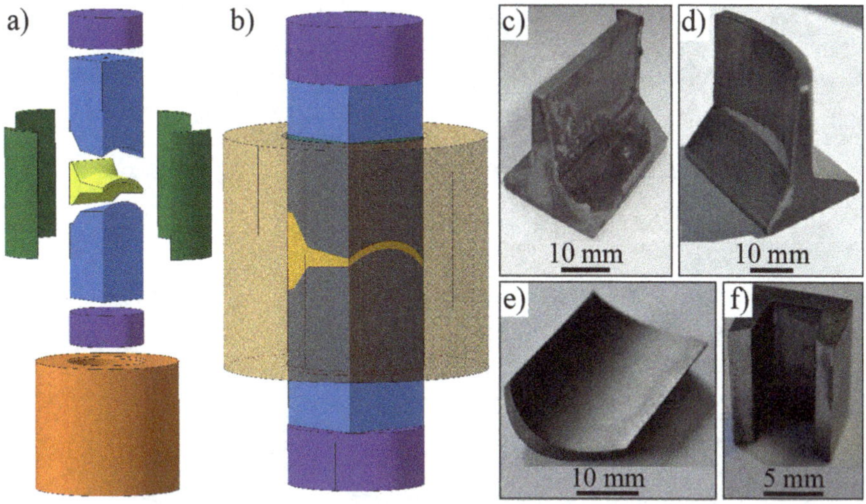

Fig. 11 Sintering of individual features of the blades. (**a**) and (**b**) Graphite set-up used to sinter a complex shape (in yellow). Here, one can see an exploded view and a configuration after full densification, respectively. (**c**) and (**d**) As-SPS sample containing a thin and a bulk region, with and without curvature, respectively. (**e**) As-SPS sample presenting the exact curvature and dimensions of the sail of the blade seen in (Fig. 10b). (**f**) As-SPS sample having fine vertical regions as encountered in the blade root (Fig. 10b)

of punches can cause one part of the sample to reach full density earlier than the rest of the sample and stop the densification by preventing punches to move further (Fig. 7c, d). For the first try to sinter the sample shown in Fig. 11c, the surfaces that would shape the sail (thin region) were designed to be perfectly parallel, however, punches slightly rotated in the set-up, which generated a negative angle as depicted in Fig. 12a. This caused the thin region to densify first and prevented the bulk region to reach full densification. A simple solution is to include a draft angle in the design of punches (Fig. 12b) to compensate a possible misalignment and force the material to move from the thin region toward the bulk. This design modification proved to enable full densification.

Moving forward, a similar sample was sintered with a curved thin region to full density (Fig. 11b). Then, a sample presenting the exact curvature and size of a portion of the low temperature blade sail (Fig. 10b) was fully densified (Fig. 11e) with no particular difficulty. Finally, a sample with fine vertical features as can be found in the root of the low temperature blade (Fig. 10b) was sintered (Fig. 11f). It appeared to require a higher pressure, for a same temperature, than the sail (Fig. 11e). This is due to the important friction-effect highlighted in Figs. 6 and 8. Note that several samples of this geometry but with various thicknesses of the arms were sintered. The one presented here in Fig. 11f has the thinnest arms possible. Thinner arms require unreasonable pressure or temperature.

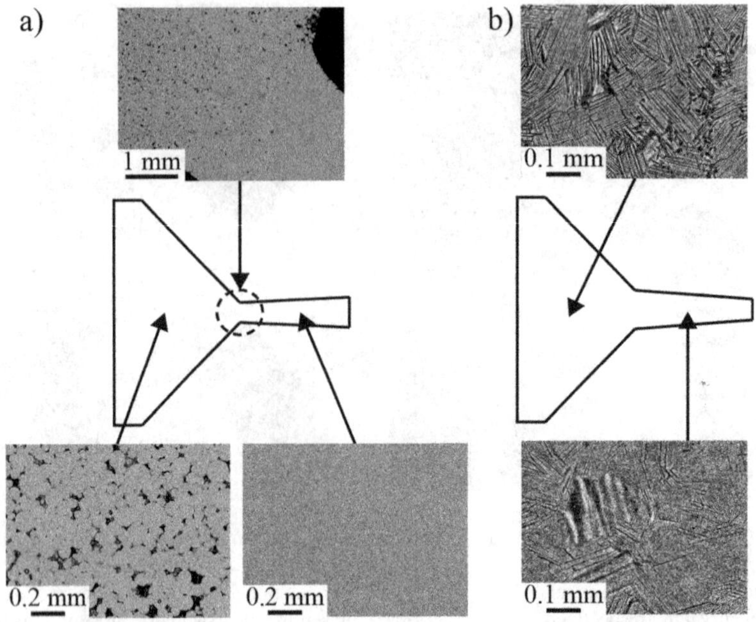

Fig. 12 Effect of the punches misalignment during the SPS cycle. (**a**) A slight rotation of the punches causes the originally designed parallel surfaces to create a negative angle. As seen in the SEM picture, the bulk region is highly porous. (**b**) Same set-up and SPS cycle but punches included a positive angle in the thin region to compensate possible punch rotation. Sample is now fully dense

5.2 Small-Scale Blades

First attempts in producing by SPS a component that would gather all challenging features, i.e., a curved thin sail and a bulk root containing a narrow platform and fine features, were made at a small scale (1/3) (Couret et al. 2011) and are presented Fig. 13. As depicted in Fig. 13a, b, the punches are now separated in several subparts that give the shape to the sample. Separating parts allows them to slightly move against each other, while the pressure is applied at the beginning of the SPS cycle and thus redistribute stresses on the whole set-up to avoid failure due to stress concentrations. Note that the section of the sail progressively increases toward the root to ensure full densification by avoiding premature local densification as shown in Fig. 12. The resulting sample (Fig. 13c) reached full density.

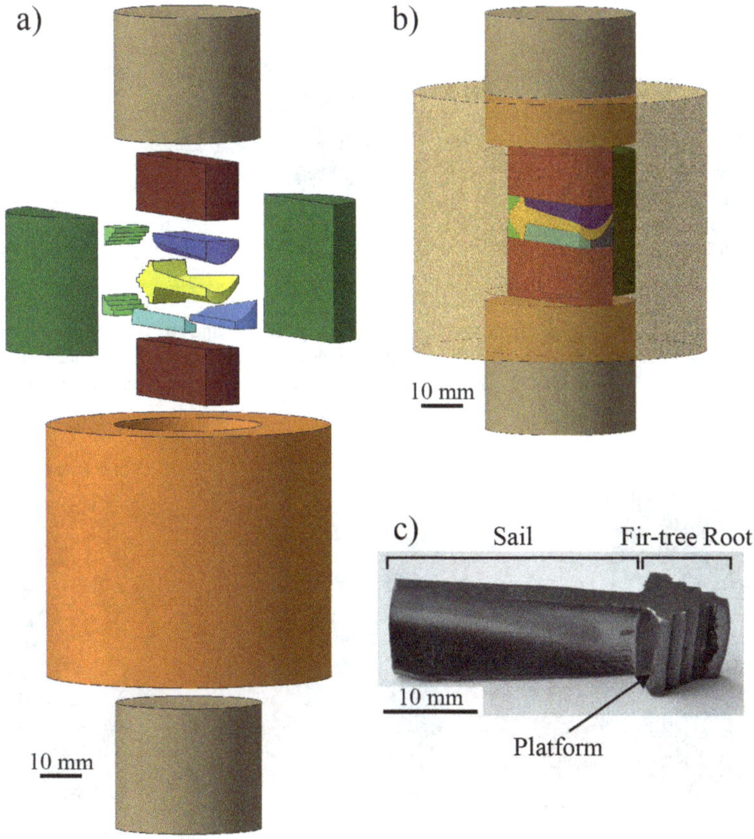

Fig. 13 Down-scaled complex shape presenting a combination of the challenging features of turbine blades. (**a**) and (**b**) 3D views (exploded and after densification, respectively) of the set-up used to densify by SPS such sample. (**c**) Sample as-SPS

5.3 *Full-Scale Near-Net Turbine Blades*

5.3.1 High-Pressure Turbine Blade

Following the same principles, mainly avoiding stress concentrations in the set-up and preventing blocking of the punches motion due to local premature densification, full-scale near-net shape high-pressure turbine blades were successfully sintered by SPS. An example is given in Fig. 14. The set-up is presented in Fig. 14a, b. It is composed of several punches and inserts that directly shape the sample upon densification. The resulting near-net shape blade is shown in Fig. 14c. This fully dense blade is 80 mm high and 50 mm wide. It presents sharp edges and accurately matches the initial design. FEM simulation shows a temperature gradient below 80 °C, which ensure a homogeneous near-lamellar microstructure with the IRIS alloy.

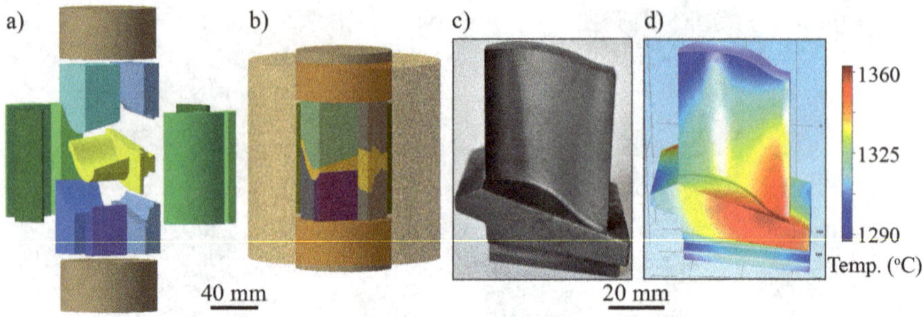

Fig. 14 Full-scale near-net shape high-pressure turbine blade. (**a**) and (**b**) Graphite set-up used to sinter such sample, exploded and after densification views, respectively. Note that in (**a**), the die is not shown. (**c**) As-SPS near-net shape blade. (**d**) FEM model of the temperature gradient

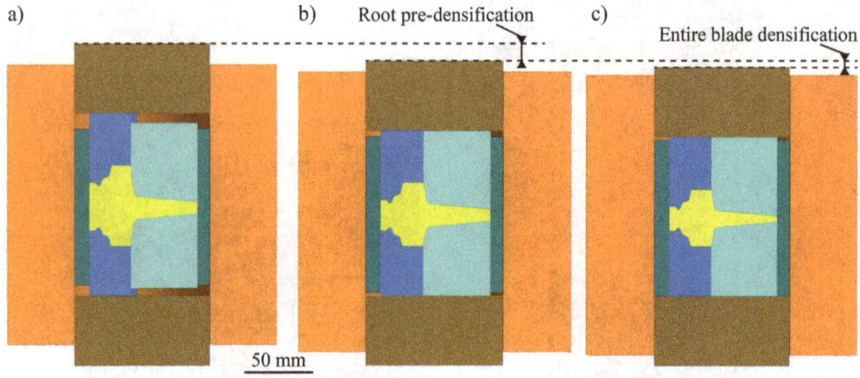

Fig. 15 Mechanism to fully densify the low-pressure turbine near-net shape blade. (**a**) Set-up at the beginning of the SPS cycle, (**b**) after pre-densification of the root, and (**c**) at the end of the cycle when the entire near-net shape blade is dense

Note that full density in the root of the high-pressure blade was obtained using an additional mechanism (Fig. 15) as the difference in size between the root and the sail is quite larger than in the low-pressure blade. The root is pre-densified first (Fig. 15a, b) by allowing only one region of the punches to move (in dark blue). Then the entire blade is compacted at a same pace (Fig. 15c) until full densification.

5.3.2 Low-Pressure Turbine Blade

In addition to a long and thin sail, the low-pressure turbine blade presents a greater challenge: a shrouded tip (Fig. 10b). Upon cooling, the TiAl blade shrinks much more than the graphite due to the difference in thermal expansion coefficient. The geometry of the set-up (Fig. 16a, b) is such that the middle punches (in dark blue in Fig. 16a, b) will prevent the blade from shrinking by opposing a force on both the

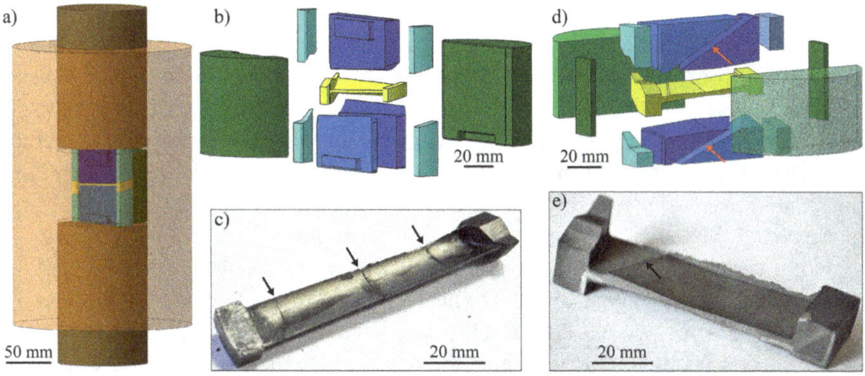

Fig. 16 Full-scale near-net shape low-pressure turbine blade. (**a**) and (**b**) Graphite set-up used to sinter such sample, after densification and exploded views, respectively. (**c**) Resulting as-sintered near-net shape blade. Arrows highlight cracks that occurred during cooling. (**d**) Updated set-up that includes a separation of the middle punches as highlighted by the red arrows. (**e**) Resulting as-sintered sample. The arrow points out an artefact coming from the new punch design (see text for more details)

root and the shroud during cooling. This leads inevitably the blade sail to break due to the too high residual stress (Fig. 16c – black arrows).

To prevent the sail failure to happen, the design of the punches needs to be optimized so that it compensates the thermal shrinking differential. By separating middle punches in two complementary parts as pointed out by the red arrows in Fig. 16d, they can slip against each other upon cooling when the blade shrinks and avoid blade failure. Note that, here, the curvature of the sail was not taken in account to reduce the cost of set-up machining. As can be seen in Fig. 16e, the resulting sample is intact. Only a small edge is present on the sail surface (black arrow) that is a side effect of the new punch geometry. FEM simulations predict a thermal gradient even smaller than in the high-pressure near-net shape blade.

6 Discussion

6.1 High-Quality as-SPS Material

Processing metallic powder by SPS has proven to be remarkable. In the case of TiAl, SPS permits obtaining texture-free microstructures that can be accurately optimized by adjusting the dwell temperature according to the phase diagram. This means that the as-SPS samples do not require any further heat treatment. For TiAl alloys, this presents a huge gain in time (and thus cost) in the production line. Also, it allows better optimizing the microstructure as the temperature window that can be used is quite larger compare to other techniques (Tetsui et al. 2005; Yanqing et al. 2009) as powder particles are not melted such as in casting or maintained at a given high temperature to allow malleability such as in forging. In fact, upon processing, the temperature needs to be high enough to allow for full densification and remain below the melting point. This is obviously a function of material, but for TiAl, it appears to be from around 1150 to 1450 °C. This 300 °C processing window enables three main microstructures that are double-phased, duplex, and lamellar, with quite different properties (Fig. 3). This means this same material can be used for different applications; moreover, each microstructure can be finely tuned (mainly grain size and phase content) by slightly changing the SPS temperature.

In our case, the optimized near-lamellar IRIS alloy allowed to match and greatly outperform the requirements for jet engine low-pressure turbine blades (Voisin et al. 2016, 2017; Appel et al. 2011). This is attributed to the high control of the homogeneous microstructure and also the rapid processing due to the fast heating, which greatly limited the formation of β phase, detrimental to the ductility (Voisin et al. 2012). Note that the temperature and environment in the high-pressure turbine would not allow for the use of TiAl but such blade was used in this work anyway as it presents an interesting geometry for process development purpose. Also, as it will be demonstrated in Sect. 6.3.2, the graphite set-up can be successfully used for other materials that can withstand such harsh conditions.

6.2 Near-Net Shaping by SPS

A metal is as good as its pore content is low. Pores have a major impact on metals' mechanical properties (Haynes 1971, 1977). Thus, porosity has to be greatly limited when it cannot be avoided. The high-pressure and temperature that can be reached during the SPS cycle allows obtaining fully dense samples (Munir et al. 2006). However, samples with complex geometries require understanding what the limits are. Porosity maps, as shown in Figs. 6, 7 and 8, have greatly helped quantifying the effect of the sample shape on the densification behavior. For example, fine features that elongate vertically, such as in the blade shrouded tip, will require higher pressure to overcome the friction coming from the vertical interfaces between powder and set-up. Our studies also highlighted the lack of material lateral motion at high temperature that requires its distribution in a complex mold to be forced by incorporating draft angles to the punches. This mainly prevents premature local densification of some regions of the part, which results in blocking further densification of the entire sample.

Controlling the temperature gradient in the set-up during the SPS cycle allowed us obtaining homogeneous samples. A FEM model was developed using simple cases such as sintering disks of different diameters and thicknesses and validated with experimental observations. It proves to be a necessary tool to predict temperature gradient in near-net shape blades; in particular it helped designing the graphite set-up so that a homogeneous near-lamellar microstructure can be maintained throughout as-SPS components when using the IRIS alloy. For example, a direct result of modeling is the oversized die in the low-pressure blade set-up in Fig. 16. By allowing punches to go deep in the die, the path of the current is maximized, and instead of forcing the current to go directly through the smaller punches, it can spread in the die and allow a more homogeneous heating of the sample, drastically reducing the temperature gradient.

Two types of near-net shape blades were successfully densified by SPS, each presenting its own challenges that could be overcome by controlling densification behavior and temperature gradient but also by separately identifying difficulties generated by each shape feature. For example, we found that the sail is the easiest region of both types of blade to sinter. Due to its limited thickness, it requires lower pressure to reach full density. Also, it requires less displacement from the punches. This leads to the sail being dense before the rest of the blade, which stops the punches motion, resulting in the root and shroud (in the low-pressure blade) to be porous. As mentioned earlier, the addition of draft angles in the punch design greatly improved the densification by forcing the material to move from the center of the sail to the root and shroud in the low-pressure blade.

6.3 *Moving Forward*

6.3.1 Functionally Gradient Metals (FGM)

SPS also offers the possibility to obtain TiAl samples, and more generally metals, with a gradient of microstructure. This can be obtained by designing the graphite set-up, in agreement with FEM simulations, so that large temperature gradients are generated in a same sample (Voisin et al. 2013). For a more controlled and clear interface between microstructures, we are investigating another option that consists in sintering a sample in multiple steps. The set-up is presented in Fig. 17a. In this example, the final sample is a cylinder, and the material used is the GE 4822 TiAl alloy. Half of the sample is sintered at a high temperature that allows a full transformation into lamellar microstructure at room temperature. This dense "half-cylinder" is then placed in another set-up and maintained in position using an insert. Powder is poured in the remaining half and sintered at a lower temperature, which leads to a double-phase microstructure. Note the presence of a draft angle on the moving punch that forces the powder to move toward the dense half cylinder for a sufficient joining. Resulting microstructures are shown in Fig. 17b, c. On the right-hand side is the fully lamellar microstructure, and a clear vertical interface is visible at the middle. Note that the pre-sintered fully dense half-cylinder needs to be cleaned from any contamination using acetone to limit the amount of oxygen trapped on the surface, which can lead to the formation of an undesired layer of α grains along the interface as it is the case in Fig. 17b. Figure 17c shows the result of a sufficient cleaning; the interface is clear of α grains.

Mechanical properties of such samples remain to be tested, and the strength of such interfaces needs to be more investigated, but this presents a unique opportunity

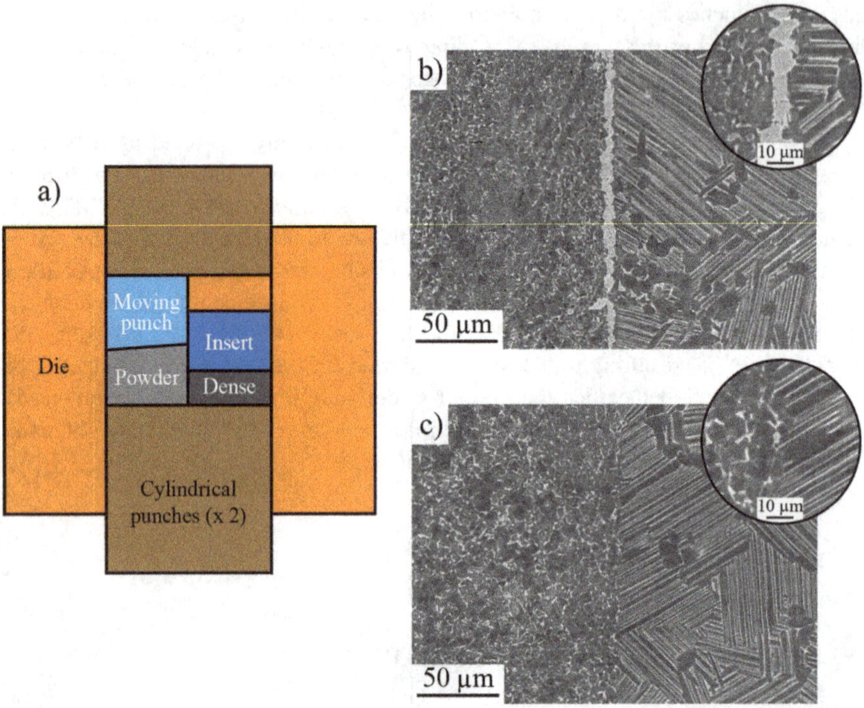

Fig. 17 FGM in TiAl made by SPS. (**a**) Set-up used to sinter a sample with two different microstructures. (**b**) Resulting microstructures on both sides of the interface. Note the presence of a layer of α grains along the interface due to a contaminated surface retaining too much oxygen. (**c**) Similar microstructure where surface contamination was removed before sintering

to fabricate samples with complex microstructures. FGM could be made also by using different metals and alloys in different regions of a same sample.

6.3.2 Niobium Silicide

Successful densification of complex near-net shape blade has been demonstrated with TiAl alloys. However, as mentioned earlier, this alloy is not suitable for the temperature and pressure faced in the high-pressure turbine. Contrariwise, niobium silicide-based alloys (NbSi) are presenting great potential to withstand the harsh environment encountered in the high-pressure turbine (temperature higher than 1200 °C) and replace the existing single crystal nickel alloys (Bewlay et al. 2002). However, conventional processing routes suffer from important limitations in term of microstructure homogeneity and porosity. Powder metallurgy techniques appear more promising (Drawin and Justin 2011), and SPS has proven to produce high-quality NbSi parts (Drawin et al. 2011).

Fig. 18 Front and rear view of a NbSi near-net shape high-pressure turbine blade produced by SPS

The same set-up presented in (Fig. 14) was used to densify NbSi alloys (Nb-Ti-Hf-Cr-Al-Si). As the temperature to full densify metals by SPS is material dependent, in this case, a maximum temperature of 1500 °C was maintained for a few minutes. The resulting fully dense near-net shape high-pressure turbine blade is presented in Fig. 18. Note that, besides the dwell temperature, successfully sintering this new material by SPS did not require any change in the process compared to TiAl. This highlights the great potential of such a technique that allows processing very different materials without particular challenge.

7 Conclusions

To summarize, we have shown in this work that SPS can be successfully used to sinter complex shapes with a high control of the microstructure. This was possible by understanding the powder particles densification behavior during the SPS cycle and modeling the temperature gradient to predict the final microstructure. This enables us to design accordingly the graphite set-up and promote a fully dense material with homogeneous microstructure. Meanwhile, we have shown that an optimized TiAl alloy has been made by SPS, with mechanical properties exceling expectations.

In addition, our preliminary studies have shown that functionally gradient materials can be sintered by SPS, which opens the door to numerous applications. Likewise, we have shown that complex shaping by SPS is not limited to TiAl but can be used for very different materials.

Densification of complex components by SPS, such as jet engine near-net shape turbine blades, is now a reality. Importantly, this allows fabricating materials or shapes that conventional manufacturing techniques may struggle or fail to produce.

Acknowledgments This study has been conducted in the framework of the cooperative project "IRIS-ANR-09-MAPR-0018-06" supported by the French Agence Nationale de la Recherche (ANR), which is acknowledged. A part of this work was performed under the auspices of the US

Department of Energy by Lawrence Livermore National Laboratory under the contract No. DE-AC52-07NA27344. Authors thank the PNF2 for providing SPS facilities (Plateforme Nationale de Frittage Flash/CNRS in Toulouse, France). Frédéric Bernard (ICB), Nikhil Karnatak (Mecachrome), and their teams are greatly acknowledged for providing support on sintering large components. Marc Thomas (ONERA) is thanked for the fruitful discussions and Lise Durand (CEMES) for developing FEM models. Yannick Le Berre (Chastagner-Delaize) and his team are thanked for producing graphite parts.

References

Appel F, Wagner R (1998) Microstructure and deformation of two-phase γ-titanium aluminides. Mater Sci Eng R Rep 22:187–268

Appel F, Paul JDH, Oehring M (2011) Gamma titanium aluminide alloys: science and technology. Wiley-VCH, Weinheim

Bewlay BP, Jackson MR, Gigliotti MFX (2002) Niobium silicide high temperature in situ composites. In: Westbrook JH, Fleischer RL (eds) Intermetallic compounds – principles and practice. John Wiley & Sons, Hoboken

Bewlay BP, Nag S, Suzuki A, Weimer MJ (2016) TiAl alloys in commercial aircraft engines. Mater High Temp 33:549–559

Clemens H, Mayer S (2016) Intermetallic titanium aluminides in aerospace applications – processing, microstructure and properties. Mater High Temp 33:560–570

Couret A, Monchoux JP, Durand L, Jabbar H, Voisin T (2011) Method for manufacturing a part having a complex shape by flash sintering, and device for implementing such a method. US patent. Centre National de la Recherche Scientifique CNRS

Couret A, Monchoux JP, Thomas M, Voisin T (2013) Method for manufacturing a titanium-aluminum alloy part. US patent. Centre National de la Recherche Scientifique CNRS

Couret A, Voisin T, Thomas M, Monchoux JP (2017) Development of a TiAl alloy by spark plasma sintering. JOM 69(12):2576–2582

Denquin A, Naka S (1996) Phase transformation mechanisms involved in two-phase TiAl-based alloys – II. Discontinuous coarsening and massive-type transformation. Acta Mater 44:353–365

Drawin S, Justin JF (2011) Advanced lightweight silicide and nitride based materials for turbo-engine applications. High Temp Mater – J AerospaceLab 3:1–13

Drawin S, Monchoux JP, Raviart J, Couret A (2011) Microstructural properties of Nb-Si based alloys manufactured by powder metallurgy. Adv Mater Res 278:533–538

Groves GW, Kelly A (1963) Independent slip systems in crystals

Haynes R (1971) Effect of porosity content on the tensile strength of porous materials. Powder Metall 14:64–70

Haynes R (1977) A study of the effect of porosity content on the ductility of sintered metals. Powder Metall 20:17–20

Huang SC (1988) Titanium aluminum alloys modified by chromium and niobium and method of preparation. General Electric Co., Schenectady

Jabbar H, Monchoux JP, Thomas M, Couret A (2011) Microstructures and deformation mechanisms of a G4 TiAl alloy produced by spark plasma sintering. Acta Mater 59:7574–7585

Jabbar H, Monchoux JP, Thomas M, Pyczak F, Couret A (2014) Improvement of the creep properties of TiAl alloys densified by spark plasma sintering. Intermetallics 46:1–3

Janschek P (2015) Wrought TiAl Blades. Mater Today Proc 2:S92–S97

Kim YW (1989) Intermetallic alloys based on gamma titanium aluminide. JOM 41:24–30

Kothari K, Radhakrishnan R, Wereley NM (2012) Advances in gamma titanium aluminides and their manufacturing techniques. Prog Aerosp Sci 55:1–16

Luo JS, Voisin T, Monchoux JP, Couret A (2013) Refinement of lamellar microstructures by boron incorporation in GE-TiAl alloys processed by spark plasma sintering. Intermetallics 36:12–20

McCullough C, Valencia JJ, Levi CG, Mehrabian R (1989) Phase equilibria and solidification in Ti-Al alloys. Acta Metall 37:1321–1336

Mondalek P, Silva L, Bellet M (2011) A numerical model for powder densification by SPS technique. Adv Eng Mater 13:587–593

Munir ZA, Anselmi-Tamburini U, Ohyanagi M (2006) The effect of electric field and pressure on the synthesis and consolidation of materials: a review of the spark plasma sintering method. J Mater Sci 41:763–777

Murray JL (1987) Phase diagrams of binary titanium alloys. ASM International, Metals Park

Orrù R, Licheri R, Locci AM, Cincotti A, Cao G (2009) Consolidation/synthesis of materials by electric current activated/assisted sintering. Mater Sci Eng R Rep 63:127–287

Pollock TM (2016) Alloy design for aircraft engines. Nat Mater 15:809

Sallot P, Martin G, Knittel S (2017). Implementation of γ-TiAl alloys for low pressure turbine blades: opportunities and new challenges. In TMS2017

Shouren W, Peiquan G, Liying Y (2008) Centrifugal precision cast TiAl turbocharger wheel using ceramic mold. In: Proceedings of the international conference on the advanced materials processing technology, 2001, vol 204, pp 492–97

Tetsui T, Shindo K, Kaji S, Kobayashi S, Takeyama M (2005) Fabrication of TiAl components by means of hot forging and machining. Intermetallics 13:971–978

Voisin T (2014) Exploration of the SPS route to produce turbine blades for aeronautical applications: development of an efficient TiAl alloy and densification of near-net shapes. PhD Thesis, Paul Sabatier-CEMES/CNRS

Voisin T, Monchoux JP, Clemens H, Couret A (2012) First investigations on a TNM TiAl alloy processed by spark plasma sintering. MRS Proc 1516:17–22

Voisin T, Durand L, Karnatak N, Le Gallet S, Thomas M, Le Berre Y, Castagné JF, Couret A (2013) Temperature control during spark plasma sintering and application to up-scaling and complex shaping. In: Proceedings of the international conference on the advanced materials processing technology, 2001, vol 213, pp 269–78

Voisin T, Monchoux JP, Thomas M, Deshayes C, Couret A (2016) Mechanical properties of the TiAl IRIS alloy. Metall Mater Trans A 47:6097–6108

Voisin T, Monchoux JP, Thomas M, Couret A (2017) High creep resistance of titanium aluminides sintered by SPS. In: Antoun B, Arzoumanidis A, Jerry Qi H, Silberstein M, Amirkhizi A, Furmanski J, Hongbing L (eds) Challenges in mechanics of time dependent materials, volume 2: proceedings of the 2016 annual conference on experimental and applied mechanics. Springer International Publishing, Cham

Voisin T, Calta NP, Khairallah SA, Forien J-B, Balogh L, Cunningham R, Rollett AD, Wang Y Defects-dictated tensile properties of selective laser melted Ti-6Al-4V. Materials & Design 158(2018) 113–126

Yanqing S, Xicong Y, Jingjie G, Hengzhi F (2009) Study on vacuum suction casting for TiAl-based alloys. Rare Metal Mater Eng 38:1505–1508

Spark Plasma Sintering of Ti-PSZ Composites and Functionally Graded Materials for Biocompatible Materials

T. Fujii and K. Tohgo

1 Introduction

Much attention has been paid to titanium (Ti) and its alloys as high mechanical and corrosion-resistant materials for many applications (Boyer 1996; Chunxiang et al. 2011; Banerjee and Williams 2013). In the aerospace industry, they have been widely used, e.g., compressor, turbine, and pressure piping, in an engine as they exhibit high corrosion resistance, high strength, and relatively light weight. Moreover, they are used as a fasteners for carbon fiber-reinforced plastic (CFRP) used for aircraft fuselage because their physical properties are comparable to the CFRP (Kolesnikov et al. 2008). In the medical field, as they exhibit biocompatibility due to their high corrosion resistance, they are used as artificial bones and teeth, stent, and so on (Wang 1996).

To improve the mechanical and physical properties of Ti and its alloys, they are used as a matrix of composites. As for particulate- and whisker-reinforced composites, the use of the reinforcement of TiC (Tjong and Mai 2008), TiB (Li et al. 2016), SiC (Tjong and Chen 1997), Al_2O_3 (Fujii et al. 2018), zirconia (Tohgo et al. 2015), and so on was proposed. As for the fiber-reinforced composites, the use of the reinforcement of B (Akechi and Hara 1979), B/SiC (Zhang et al. 2014), B/B_4C (Park et al. 1984), SiC (Park et al. 1984), and so on was also proposed. As it is difficult to fabricate them by the dissolution technique, the fabrication of Ti-based composites by a powder metallurgy technique has been focused on. The spark plasma sintering (SPS) has been attracting attention as a powder metallurgy technique. SPS has advantages over some conventional technique such as the hot pressing (HP) and hot isostatic pressing (HIP), e.g., its sintering time and temperature are much shorter and lower due to its heating mechanism. SPS originates from a pulsed direct electrical current through a graphite die and powder under uniaxial pressure, and

T. Fujii (✉) · K. Tohgo
Department of Mechanical Engineering, Shizuoka University, Naka-ku, Hamamatsu, Japan
e-mail: fujii.tomoyuki@shizuoka.ac.jp; tohgo.keiichiro@shizuoka.ac.jp

© Springer Nature Switzerland AG 2019
P. Cavaliere (ed.), *Spark Plasma Sintering of Materials*,
https://doi.org/10.1007/978-3-030-05327-7_26

the powder is heated by the Joule heating of powder and graphite die and spark plasma generated among conductive particles (Tokita 1993). It has been reported that the mechanical properties, optical properties, and physical properties of materials fabricated by SPS are improved because grain growth is prevented (Khor et al. 2003; Su et al. 2004).

In this article, we introduce the fabrication of monolithic Ti and composite and functionally graded materials (FGMs) consisting of Ti and partially stabilized zirconia (PSZ) by SPS technique to produce a biocompatible material for implant treatment. Firstly, sintering process of monolithic Ti by SPS and its mechanical properties were discussed. Secondly, Ti-PSZ composites (Tohgo et al. 2015) and FGMs (Fujii et al. 2017) were fabricated, and their mechanical properties, such as bending properties and fracture toughness, were evaluated. Then, densification process of Ti-PSZ composites and the behavior of atom diffusion were examined (Shinohara et al. 2017).

2 Sintering Process of Ti and Its Mechanical Properties

2.1 Sintering Technique for Monolithic Ti

Raw materials used were pure titanium (Ti, Toho Titanium Co. Ltd., average particle diameter of 23 μm). Prior to sintering, the powder was milled for 7 h in vacuum by a vibrational ball milling apparatus. The milled powder became small and smooth by removing rough edges, compared with the as-received powder, as shown in Fig. 1. The powder was placed in a graphite die with the inner diameter of 10.5 mm to fabricate a fully sintered compact with the diameter of 10 mm and the thickness of 5 mm. Graphite sheet with the thickness of 0.25 mm was inserted between the powder, graphite die, and plungers to easily eject a sintered compact after sintering. The sintering was conducted using an SPS apparatus (SPS-211Lx, Fuji Electronic Industrial Co. Ltd.).

(a) As-received particles **(b) Ball-milled particles**

Fig. 1 SEM image of Ti powder. (**a**) As-received particles. (**b**) Ball-milled particles

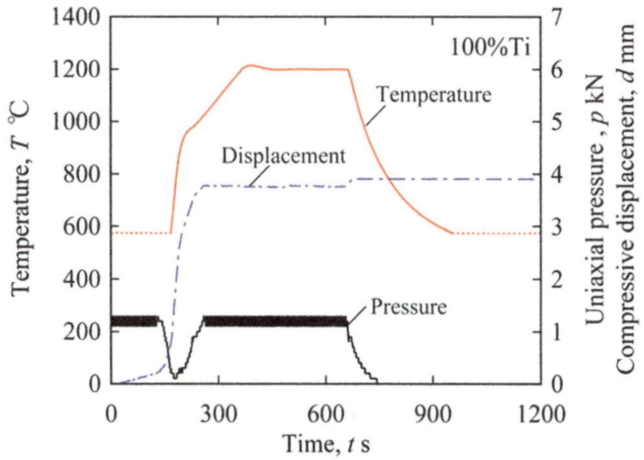

Fig. 2 Change in temperature, pressure, and displacement during sintering

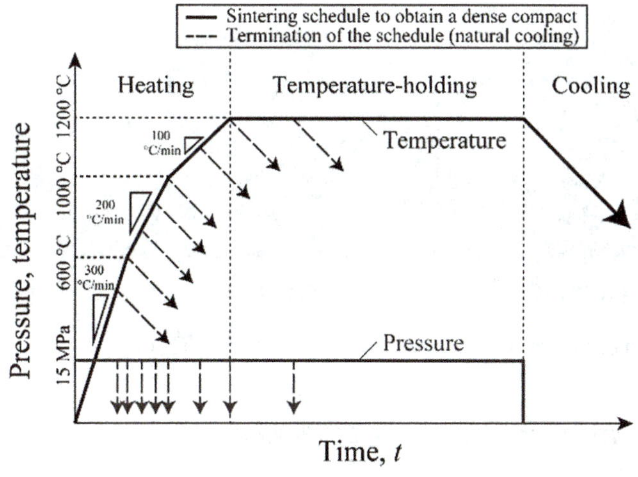

Fig. 3 Sintering schedule

Figure 2 shows the relationship between temperature, uniaxial pressure, and compressive displacement of the plungers during sintering under the condition of 1200 °C and 5-min hold with the heating rate of 300 °C/min up to 600 °C, 200 °C/min from 600 °C to 1000 °C, and 100 °C/min from 1000 °C to 1200 °C. The Ti powder was densified up to 1000 °C in the heating process as shown by the change of displacement. Hence, it was confirmed that a fully sintered Ti compact was fabricated under the condition. To investigate the sintering process of Ti powder, sintering was terminated at given stages in the sintering schedule for the fully sintered compact, and semi-sintered compacts were fabricated, as shown in Fig. 3.

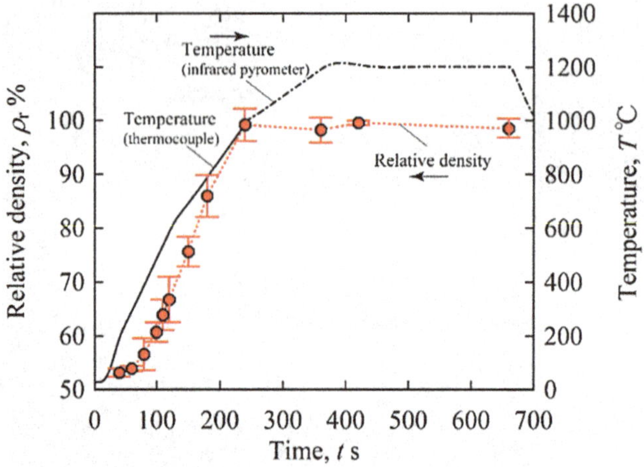

Fig. 4 Relative density as a function of sintering condition

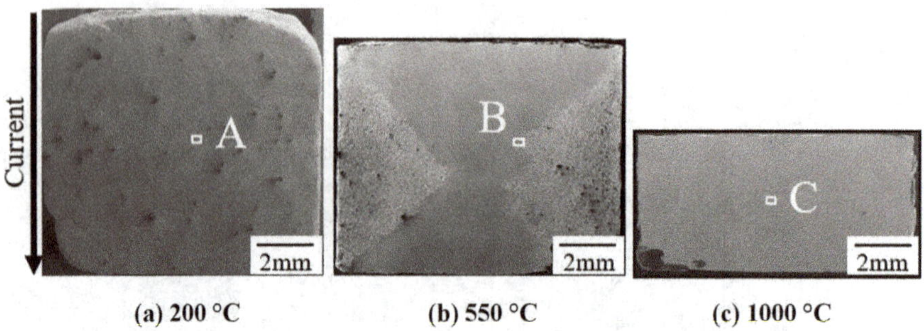

Fig. 5 Sintering process of a Ti compact. (**a**) 200 °C. (**b**) 550 °C. (**c**) 1000 °C

2.2 *Densification Process of Ti Powder*

Figure 4 shows the relative density and temperature as functions of time, and the scatter bars denote the maximum and minimum relative densities of five specimens for each condition. The relative density increases from 300 °C with elevating temperature and becomes almost 100% at 1000 °C in the heating process.

Figures 5 and 6 show the microstructure of compacts in each stage of the sintering. The necking between each Ti particle is observed in the Ti compact sintered at 200 °C. Then, the densification occurs from the vicinity of contact area of Ti powder and the upper and lower plungers, and smooth polished surface areas where the sintering is finished are partially observed at 550 °C. The sintering progresses, and then no large voids are observed, and the sintering is almost completed at 1000 °C.

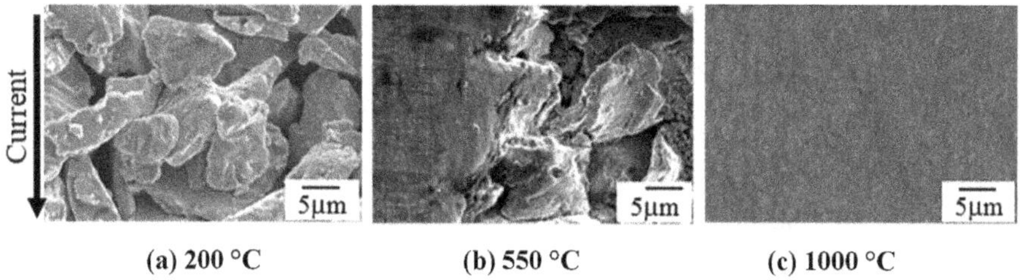

Fig. 6 Microscopic view of the regions denoted by A, B, and C on each sintering process in Fig. 5. (**a**) 200 °C. (**b**) 550 °C. (**c**) 1000 °C

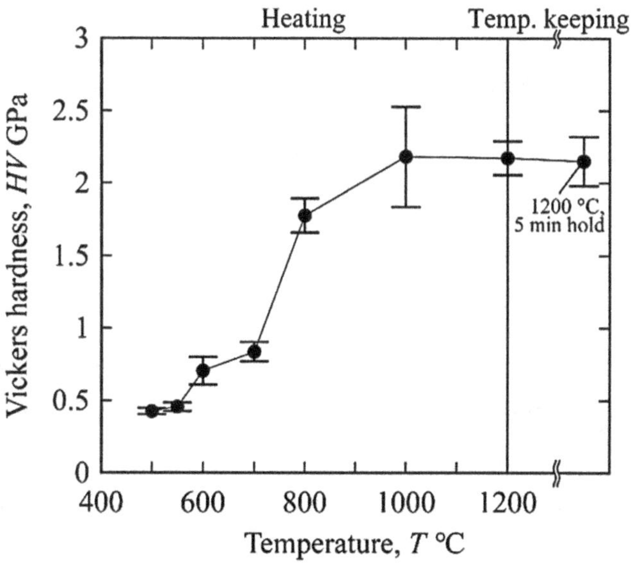

Fig. 7 Vickers hardness as a function of sintering condition

2.3 *Mechanical Properties of Monolithic Ti*

The Vickers hardness tests were conducted at smooth polished surface of each specimen fabricated under the condition of an applied load of 1.96 N and holding time of 30 s. Figure 7 shows the Vickers hardness as a function of sintering condition. The hardness increases with increasing temperature up to 1000 °C, and then the hardness becomes constant at around 2.2 GPa.

Three-point bending tests were conducted on a rectangular smooth specimen fabricated at 1200 °C and 5-min hold with the thickness h of 2 mm, width b of 2 mm, and length of 10 mm at room temperature (R.T.) in air. The bending stress was calculated by the following equation, and the strain was measured by the strain gauge attached at the back face against a loading point.

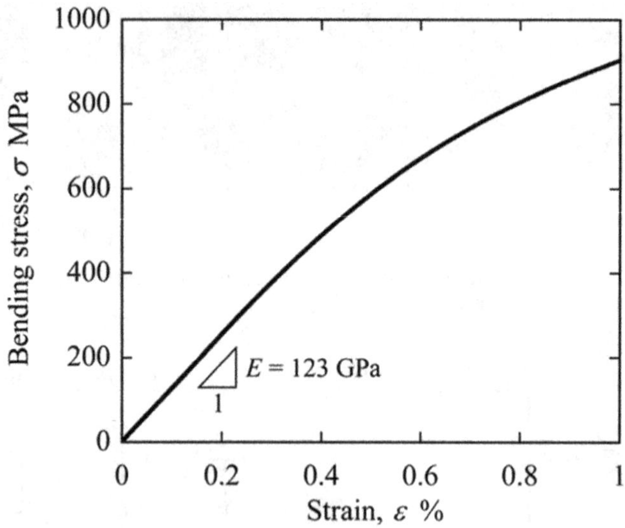

Fig. 8 Bending stress-strain relation

$$\sigma = \frac{3PS}{2bh^2} \tag{1}$$

where P and S are an applied load and span of 8 mm. Figure 8 shows the bending stress-strain relationship. It shows a large plastic deformation, and no fracture occurs up to the tensile strain of 1%. Its ductile behavior and Young's modulus of 123 GPa are comparable to that of monolithic Ti fabricated by the dissolution technique. It is found that the monolithic Ti can be successfully fabricated by SPS technique.

3 Sintering of Ti-PSZ Composites and Their Mechanical Properties

3.1 Sintering of Ti-PSZ Composites (Tohgo et al. 2015)

Raw materials used were Ti shown in Fig. 1 and partially stabilized zirconia (PSZ, KCM Corporation Co. Ltd., average particle diameter of 0.26 μm) shown in Fig. 9. The sintering was conducted under the condition of 15 MPa pressure in vacuum at 1200 °C and 5-min hold, similar to monolithic Ti in Sect. 2. Here, the specimens are referred to with their Ti volume fraction, except for the monolithic PSZ (100%PSZ).

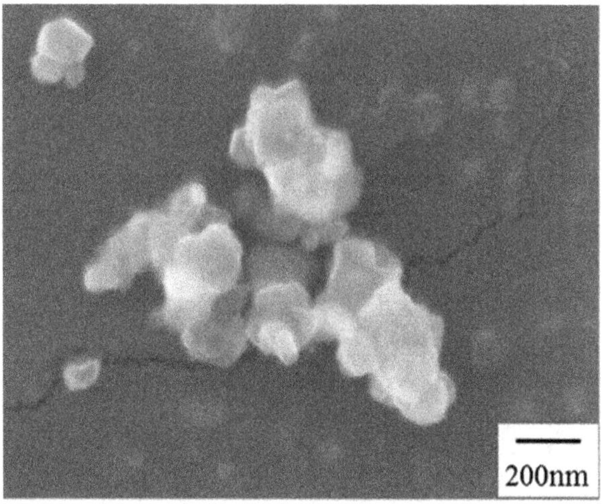

Fig. 9 As-received PSZ particles

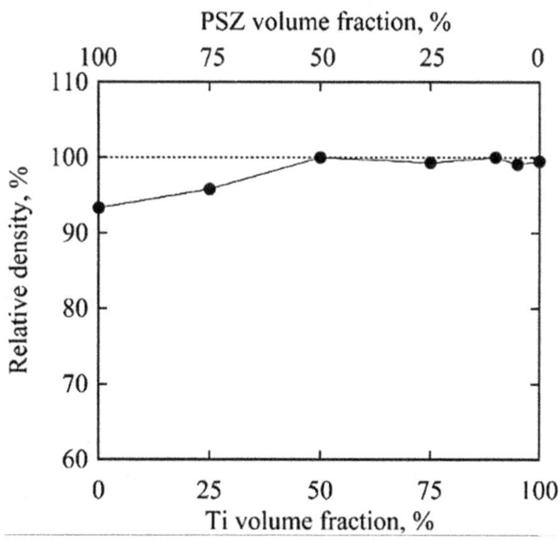

Fig. 10 Relative density as a function of Ti volume fraction

3.2 *Sintering State of Ti-PSZ Composites*

Figures 10 and 11 show the relative density as a function of Ti volume faction of the Ti-PSZ composites and their microstructure, respectively. The relative density of the composites is around 93% for PSZ and increases up to 100% with increasing Ti volume fraction. From Fig. 11, the coarse Ti phase is surrounded with the PSZ matrix in the 25%Ti, while the small PSZ particles are distributed in the region between Ti grains in the 75%Ti. In both composite, no large defects are observed.

Figure 12 shows the XRD patterns of each composite, which were measured using a Cu target at an accelerating voltage of 40 kV and an accelerating current of

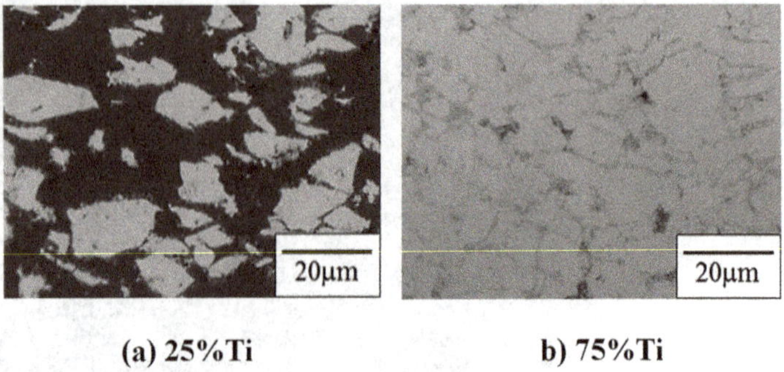

Fig. 11 Microstructure of Ti-PSZ composites. (**a**) 25%Ti. (**b**) 75%Ti

Fig. 12 XRD patterns of Ti-PSZ composites

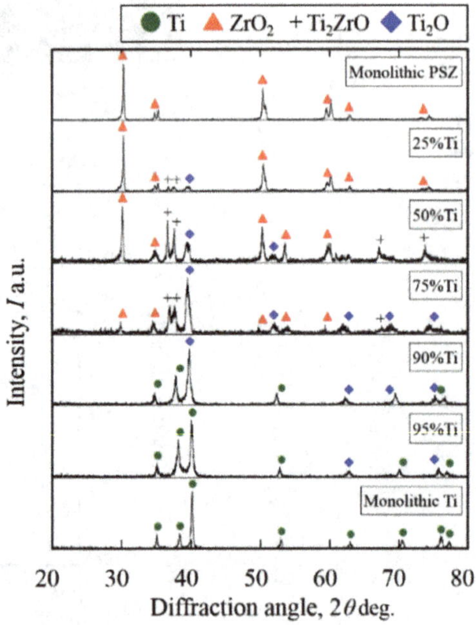

40 mA. The marks denote the peaks for phases which are identified based on the Powder Diffraction File (PDF). As peaks for Ti-PSZ, Ti_2O, and Ti_2ZrO in the XRD pattern are very close each other, the phases are identified based on the closest peak. The XRD patterns for the 100%PSZ and 100%Ti exhibit only the peaks corresponding to PSZ and Ti, respectively. As for the composites with 25%Ti, 50%Ti, and 75%Ti, the peaks for reaction compounds of Ti_2O and Ti_2ZrO are detected in addition to PSZ, but no peaks for pure Ti are detected. The Ti-rich composites of 90%Ti and 95%Ti, the peaks for reaction compounds of Ti_2O are detected in addition to pure Ti, but no peaks for PSZ are detected. Hence, the reaction compounds of Ti_2O and Ti_2ZrO in the composites are produced in SPS process. It should be noted that the pure Ti remained in the sintered composites with

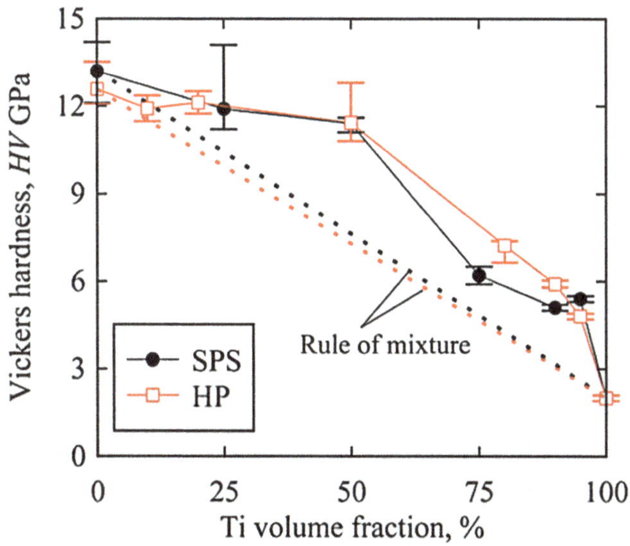

Fig. 13 Vickers hardness as a function of Ti volume fraction

Ti volume fraction more than 90%. The same reaction products were created in the composites fabricated by HP, as in reference (Fujii et al. 2010).

3.3 *Mechanical Properties*

The Vickers hardness tests were conducted under the condition of an applied load of 196 N and holding time of 30 s. Figure 13 shows the Vickers hardness as a function of sintering condition, and the dotted lines denote the prediction by the rule of mixture. The Vickers hardness of the composites fabricated by HP (Fujii et al. 2010) is also plotted. Both composites exhibit the almost same characteristics on the Vickers hardness, and the hardness of the composites is higher than the prediction by the rule of mixture due to creation of the reaction phases.

Cracks occurred at the edges of Vickers indent in the case of the composites from the 100%PSZ to 95%Ti fabricated by SPS and from 100%PSZ to 90%Ti fabricated by HP. Their fracture toughness K_C was measured by an indentation fracture (IF) method based on JIS R 1607 (Japan Industrial Standard) using the following equation:

$$K_C = 0.026\frac{E^{1/2}P^{1/2}a}{C^{3/2}},\ C/a > 2.5 \tag{2}$$

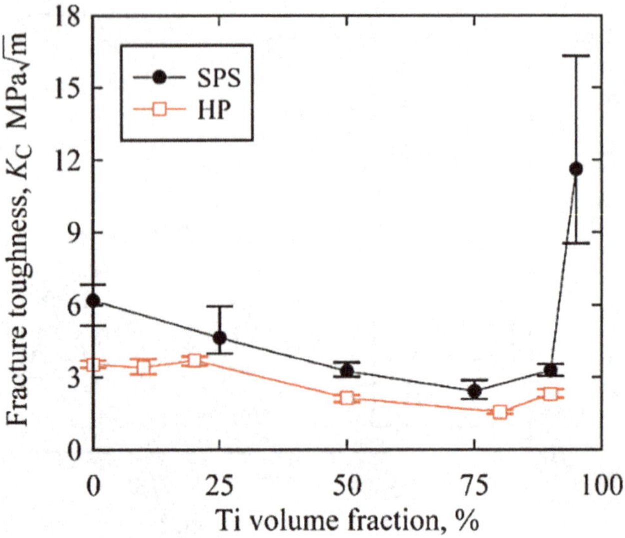

Fig. 14 Fracture toughness as a function of Ti volume fraction

where *E*, *P*, *a*, and *C* are Young's modulus, applied load, average half crack length, and average diagonal indent length, respectively. Figure 14 shows the fracture toughness of the composites fabricated by SPS and HP. Note that as the measured values of *C*/*a* were within 1.08–1.31 in the 95%Ti by SPS, the fracture toughness calculated by Eq. (2) is not valid, but the fracture toughness is plotted as a reference. The fracture toughness of both composites decreases with increasing Ti volume fraction up to the Ti volume fraction of 70%. The fracture toughness of the composites by SPS improves, compared with the composites by HP, irrespective of Ti volume fraction.

Three-point bending tests were conducted, similar to the monolithic Ti explained in Sect. 2.3. Figure 15 shows the bending stress-strain relationship of the composites, and the bending stress-strain relationship of the monolithic Ti is also shown. Brittle fracture occurs in all composites including the 100%PSZ after elastic deformation. Figure 16 shows the fracture surface of three-point bending specimens of 25%Ti and 75%Ti. The river patterns are observed at Ti phase in both composites, and the composites are fractured in a brittle manner, irrespective of Ti volume fraction.

Figure 17 shows the elastic modulus as a function of Ti volume fraction in composites fabricated by SPS, and those by HP were also shown. The dotted lines denote the prediction of the rule of mixture. The elastic modulus of both composites is higher than the rule of mixture due to occurrence of the reaction products. Note that the elastic modulus of the composites including monolithic Ti and SPS fabricated by SPS is a little higher than those by HP, and this might be attributed to the difference in the testing techniques. That is, four-point bending tests were conducted for the composites fabricated by HP, while three-point bending tests were conducted for the composites fabricated by SPS because the specimens by SPS are smaller than

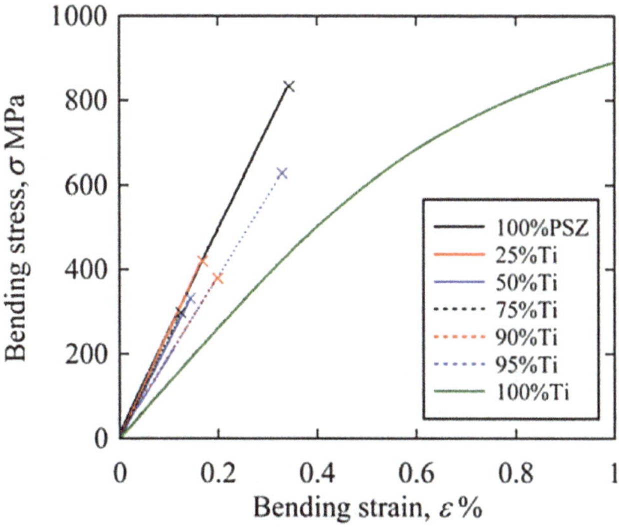

Fig. 15 Bending stress-strain relationship of Ti-PSZ composites

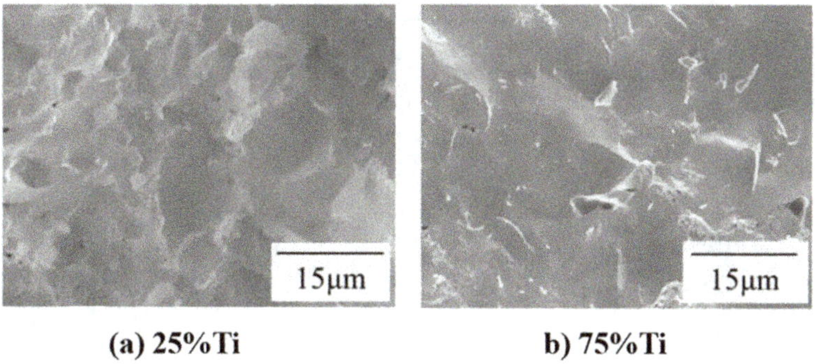

Fig. 16 Fracture surface. (**a**) 25%Ti. (**b**) 75%Ti

those by HP. It is well known that the elastic modulus measured by the four-point bending test is more accurate than that by three-point bending test as the bending moment is constant within an inner span of a four-point bending specimen.

Figure 18 shows the bending strength as a function of Ti volume fraction in composites fabricated by SPS, and that by HP is also shown. As for the both composites, the strength decreases with increasing Ti volume fraction up to the 75%Ti by SPS and the 80%Ti by HP and then increasing Ti volume fraction. Consequently, the composites were fabricated to derive the ductility of Ti phase and the hardness of PSZ phase, but they exhibited brittleness because the Ti phase was oxidized during sintering, irrespective of sintering technique.

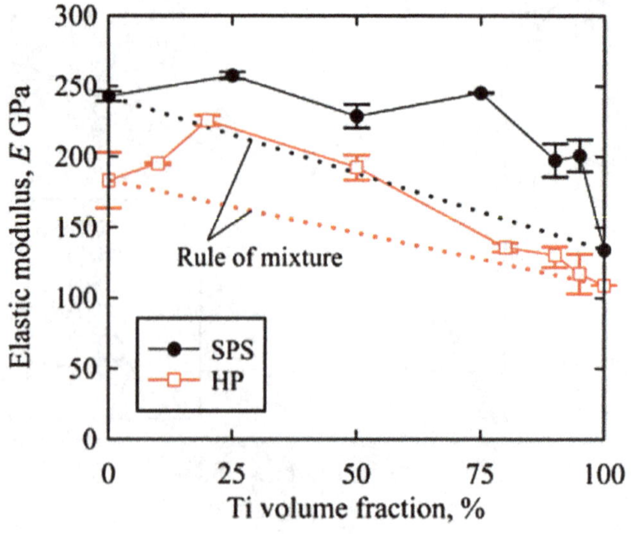

Fig. 17 Elastic modulus as a function of Ti volume fraction

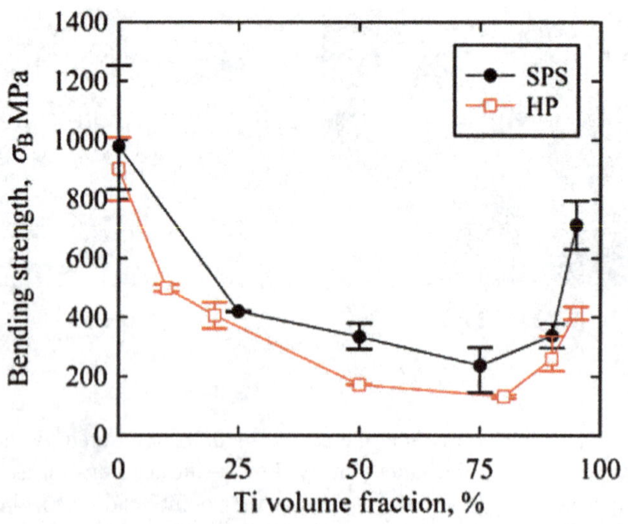

Fig. 18 Bending strength as a function of Ti volume fraction

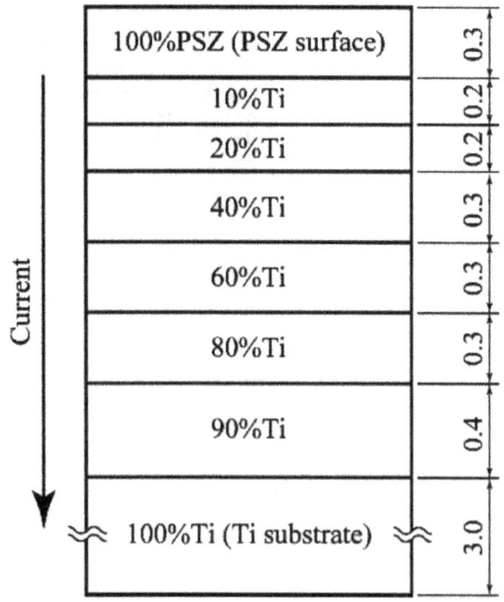

Fig. 19 Dimensions of Ti-PSZ FGM

4 Sintering of a Ti-PSZ FGM and Its Fracture Toughness

4.1 Sintering Technique for Ti-PSZ FGMs (Fujii et al. 2017)

Raw materials used were Ti and PSZ explained in the previous section, and these powders were mixed in various volume ratios covering from 100% PSZ to 100% Ti at 10% or 20% intervals by a vibrational ball milling apparatus. The mixed powders were placed in a graphite die and plungers in ascending order according to PSZ volume fraction from the Ti substrate, as shown in Fig. 19. The sintering was conducted under the condition of 15 MPa pressure in vacuum at 1200 °C and 5-min hold, similar to the monolithic Ti and Ti-PSZ composites. After sintering, a sintered compact was cut with a diamond saw, and its cross section was ground and polished with grit sandpaper and diamond paste.

4.2 Sintering State of FGM

Figure 20 shows microstructure of the FGM. No delamination between each layer is observed, and no large voids are observed in the Ti substrate and PSZ surface. The Ti particles are well dispersed in PSZ matrix in PSZ-rich layer, and the PSZ phase is also dispersed in Ti matrix in Ti-rich layer. However, the traces of drop-off of Ti particles are observed in the vicinity of interface between 80%Ti and 60%Ti layer. This drop-off would occur during surface grinding using sandpaper. All of the specimens exhibited almost the same microstructure, and the Ti-PSZ FGM was successfully fabricated by SPS.

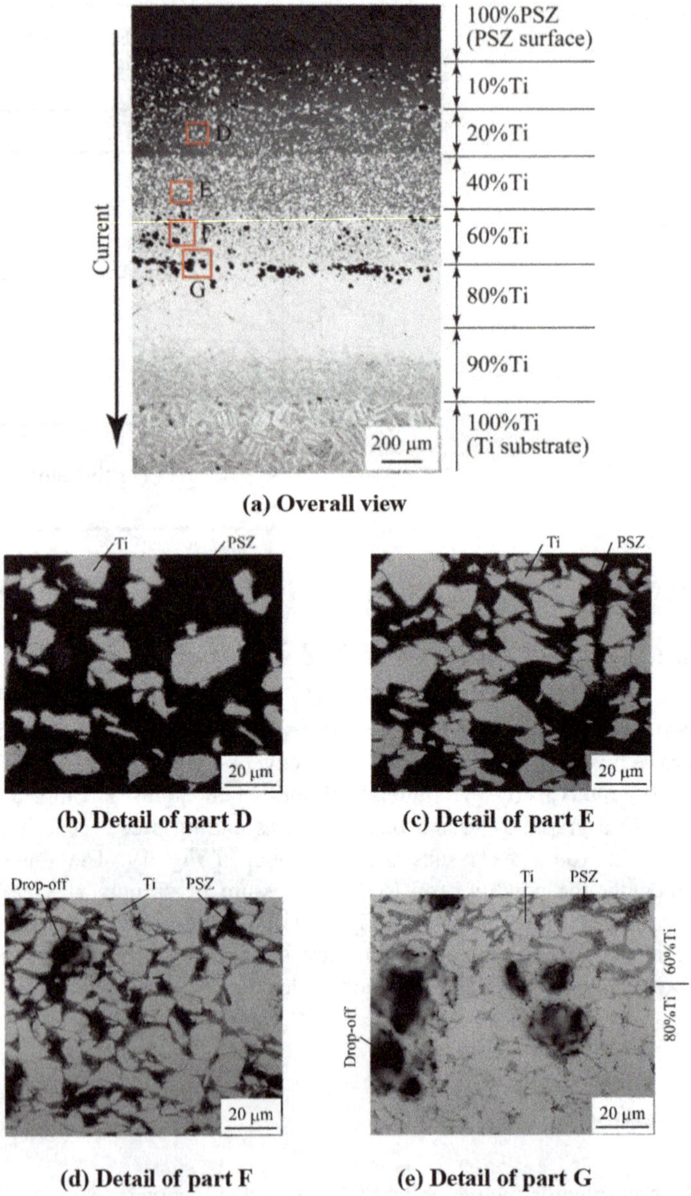

Fig. 20 Microstructure of a PSZ-Ti FGM. (**a**) Overall view. (**b**) Detail of part D. (**c**) Detail of part E. (**d**) Detail of part F. (**e**) Detail of part G

Figure 21 shows the distribution of oxygen concentration on the cross section in the FGM measured by an electron probe microanalyzer (EPMA). In this figure, oxygen concentration of the monolithic PSZ fabricated by SPS and Ti oxide is

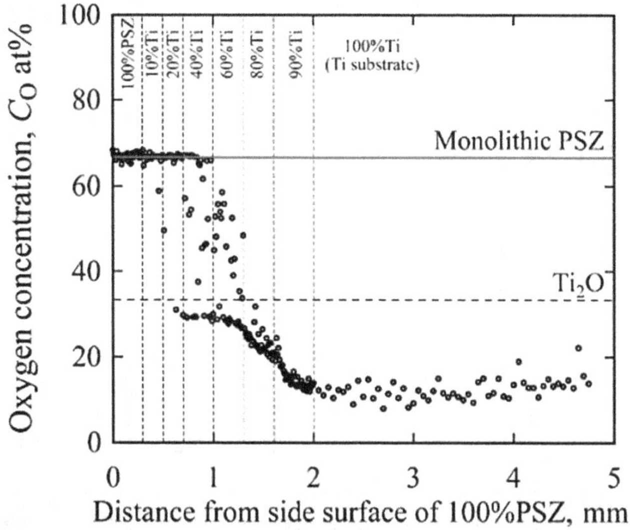

Fig. 21 Oxygen distribution in the FGM

shown by lines. Note that the oxygen concentration in the Ti substrate is about 10%. This would be caused by the vacuum condition during sintering. When Ti is sintered under medium vacuum condition (several Pa), Ti is easy to react with oxygen remaining in a furnace because Ti is chemically active at elevated temperature. Therefore, it was considered that a small amount of oxygen in the furnace affected the oxygen concentration in the Ti substrate. The oxygen in the graded layer ranges from about 66% to 30%, and it implied that Ti phase in the graded layer was completely oxidized, similar to the Ti-PSZ composites.

Figure 22 shows the Vickers hardness distribution on the cross section in the FGM. The Vickers hardness of the monolithic Ti and PSZ fabricated by SPS is also shown by lines. The Vickers hardness is comparable to that in the monolithic PSZ in the PSZ surface and decreases with decreasing PSZ volume fraction in the graded layer. The Vickers hardness in the Ti substrate is high in the region near the graded layer and the same level in the region apart from the graded layer as compared with that of the monolithic Ti. The Vickers hardness would be affected by residual stress created during sintering due to mismatch of thermal expansion of Ti and PSZ and volume change by phase transformation.

4.3 Distribution of Fracture Toughness

Three-point bending tests were carried out on a single-cracked rectangular specimen with thickness h of 3 mm and width b of 4 mm. The cracks on the PSZ surface were introduced by indentation technique as a crack-starter. Note that the distribution of

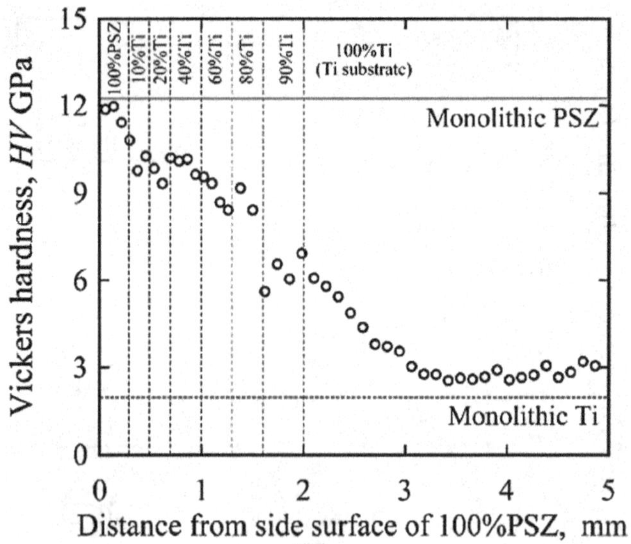

Fig. 22 Vickers hardness distribution in the FGM

fracture toughness in a FGM can be obtained using one specimen because the stress intensity factor during stable crack growth exhibits fracture resistance for crack growth at crack tip region in a specimen (Tohgo et al. 2005, 2008). The tests were conducted at a crosshead speed of 0.05 mm/min at R.T. in air. The relationship between load and load-point displacement was measured, and fracture process on the side surface was recorded with a video microscope. The crack length was measured from the image recorded by the video microscope, the relationship between load and crack length was obtained, and the stress intensity factor was calculated. It is well known that elastic properties in a FGM vary with the change of material constitution and the stress intensity factor is affected by distribution of the elastic properties in a graded layer (Tohgo et al. 2008). For the sake of simplicity, the stress intensity factor was estimated under the assumption that the Ti-PSZ FGM was a homogeneous material. The stress intensity factor is given by the following equation:

$$K = \frac{3}{2}\frac{PS}{bh^2}F\left(\frac{a}{h}\right) \tag{3}$$

where P, S, and a are an applied load, span, and crack length of a specimen, respectively. The function of F is a correction factor for its geometry. In this study, the following equation proposed by Fett (Fett 1991) was used:

Fig. 23 Crack path in the Ti-PSZ FGM

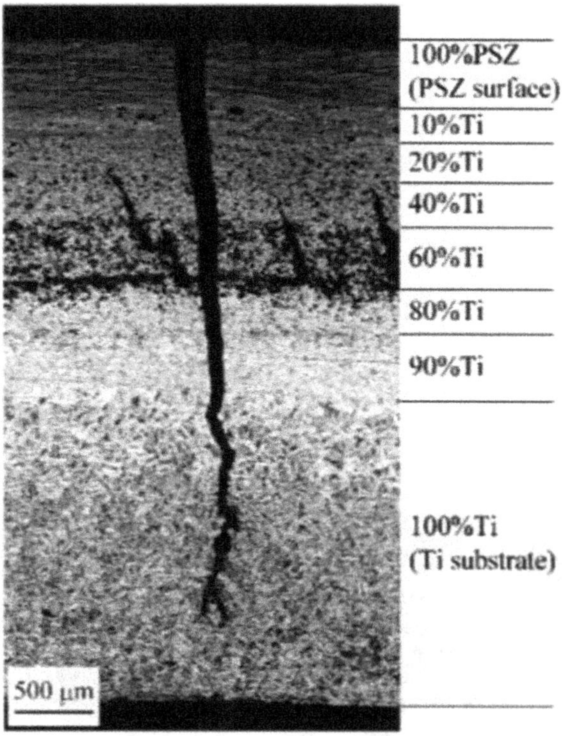

$$F(\alpha) = \frac{1}{(1-\alpha)^{\frac{3}{2}}}\left[A_0\alpha + (1\text{-}\alpha)\left(A_1 + A_2\alpha + A_3\alpha^2 + A_4\alpha^3 + A_5\alpha^4\right)\right] \quad (4)$$

where $A_0 = 0.3738$, $A_1 = 1.0426$, $A_2 = -2.2336$, $A_3 = 3.9625$, $A_4 = -3.3469$, and $A_5 = 1.0121$. After the test, the specimen was forcedly separated to observe its fracture surface.

Figure 23 shows the side surface of the FGM after testing. Note that some scratches which were created by diamond saw are observed in the graded layer and these scratches could not be removed by grinding and polishing. A crack path was not affected by these scratches and straight in the graded layer and zigzag in the Ti substrate. The crack propagated within the Ti and PSZ phase, irrespective of Ti and PSZ volume fraction in the graded layer, and the interfacial debonding between PSZ and Ti phase is not observed. The crack propagated without delamination between each layer, and the interfacial strength of each layer appears to be relatively high. Figure 24 shows the fracture surface of the FGM. Flat fracture surface is observed in the grade layer from PSZ surface to interface between 90%Ti and 100% Ti, while fracture surface in the Ti substrate becomes rough. Quasi-cleavage, river pattern, and ductile fracture surface were intermixed.

Figure 25 shows load and crack length as functions of load-point displacement for the three-point bending test. The solid and dotted lines in the crack length-

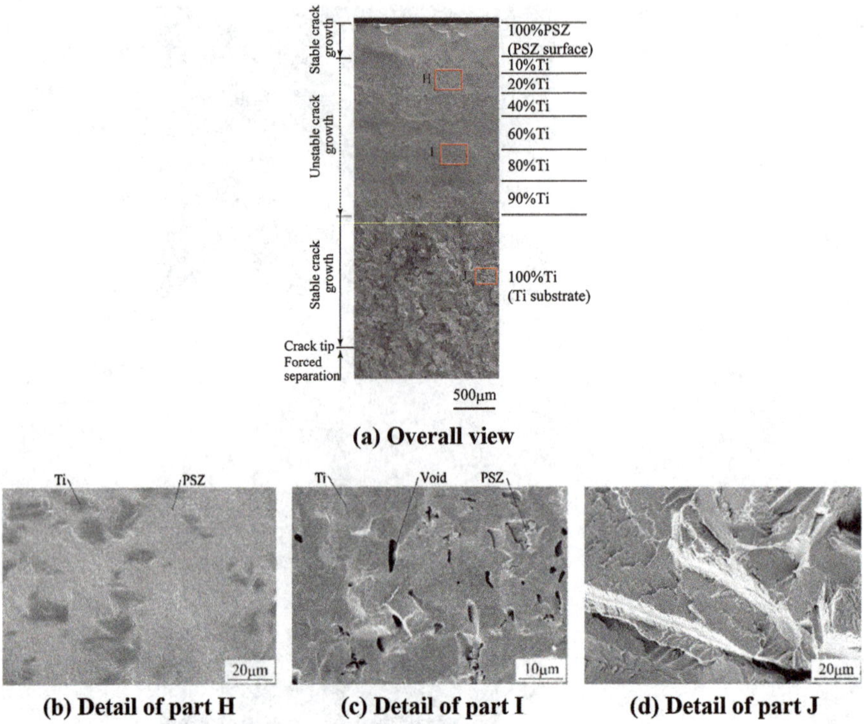

Fig. 24 Fracture surface. (**a**) Overall view. (**b**) Detail of part H. (**c**) Detail of part I. (**d**) Detail of part J

displacement relation denote the stable and unstable crack growth, respectively. At the point K in the load-point displacement relation, the median cracks induced by indentation would coalesce to be a through-thickness crack because the applied load decreased, but any cracks could not be observed on the side surface. At the point L, a crack was firstly observed in the PSZ surface. Then, the crack stably grew up to 10% Ti, and the crack unstably grew up to the interface between the 90%Ti and Ti substrate at the point M. After that, the crack stably grew in the Ti substrate. From the relationship between the load and crack length during stable crack growth, the fracture toughness at the crack tip region was obtained. Figure 26 shows the fracture toughness distribution in the FGM. The fracture toughness of the composites corresponding to each layer which was estimated in Sect. 3 was also plotted. The unstable crack growth occurred from 10%Ti layer to the interface between 90%Ti layer and Ti substrate. This means that the fracture toughness of each layer was low, irrespective of Ti volume fraction. This phenomenon was caused by the existence of Ti oxide created during sintering in each layer. In the Ti substrate, the fracture toughness increased with crack growth due to pure Ti phase. The roughness of

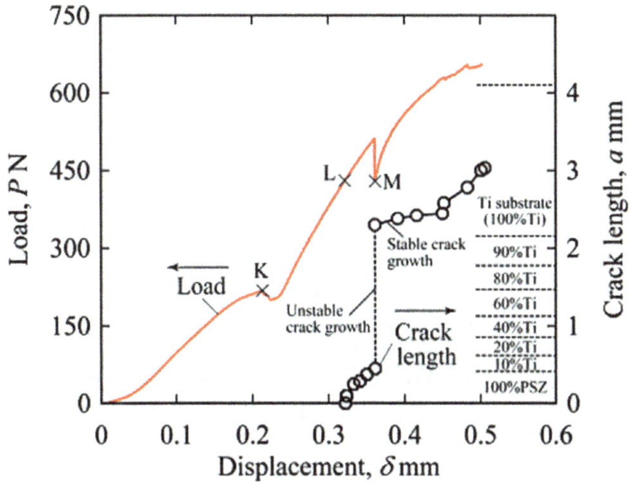

Fig. 25 Change in load and crack length during bending

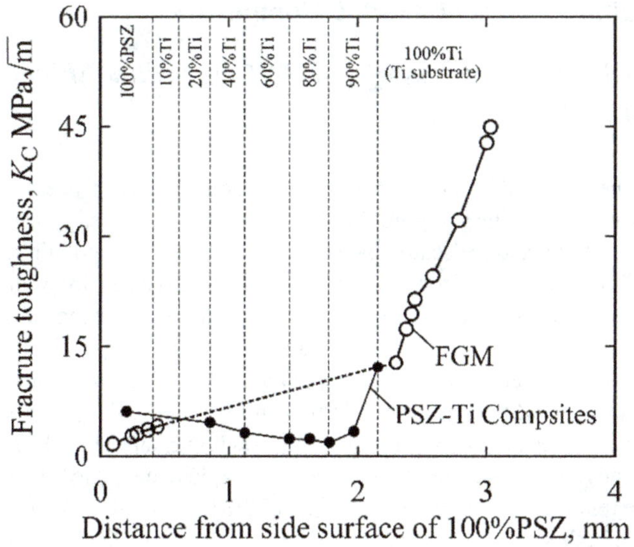

Fig. 26 Fracture toughness distribution

fracture surface in the Ti substrate increases with being apart from the graded layer, and the gradual increase of fracture toughness in the Ti substrate might be caused by the crack path deflection inducing roughness of fracture surface.

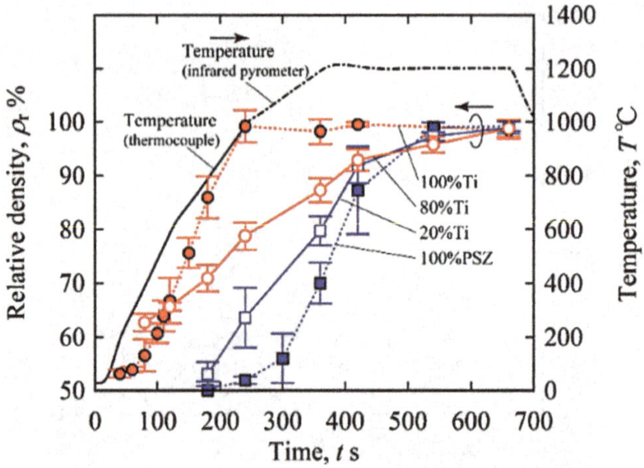

Fig. 27 Relative density and temperature as a function of sintering time

5 Sintering Process of Ti-PSZ Composites

5.1 Densification Process of Ti-PSZ Composites (Shinohara et al. 2017)

Similar to the monolithic Ti in Sect. 2.2, the densification processes of Ti-PSZ composites were examined. Figure 27 shows the relative density and temperature as functions of time, and the scatter bars denote the maximum and minimum relative densities of five specimens for each condition. The relative density of the 100%PSZ also increased from 800 °C with elevating temperature and became over 98% at 1200 °C and 3-min hold. For the 80%Ti and 20%Ti, their densification progressed in an intermediate manner between the 100%Ti and 100%PSZ based on their volume fraction of Ti and PSZ up to 1200 °C. In the temperature-holding process, however, the densification became slower in the composites than in the 100%Ti and 100% PSZ. Especially, the densification process was more delayed in the 80%Ti than in the 20%Ti. The relative density of the 80%Ti and 20%Ti sintered at 1200 °C and 5-min hold became higher than 98%.

Figure 28 shows the microstructure of 100%PSZ in the sintering process. Each PSZ particle was connected by creating necking between particles in the 100%PSZ sintered at t 800 °C, and no large voids are observed in the 100%PSZ sintered at 1200 °C and 3-min hold. Figures 29 and 30 show the microstructure of 20%Ti and 80%Ti in the sintering process, respectively. From Fig. 29, the coarse Ti particle was surrounded with the fine PSZ particles in the 20%Ti sintered at 800 °C, and no large voids are also observed in the 100%PSZ sintered at 1200 °C and 3-min hold. From Fig. 30, some large voids are observed in the 80%Ti sintered at 800 °C, and then the sintering was almost completed at 1200 °C and 3-min hold.

(a) 800 °C (b) 1200 °C, 3 min holding

Fig. 28 Sintering process of 100%PSZ. (**a**) 800 °C. (**b**) 1200 °C, 3-min holding

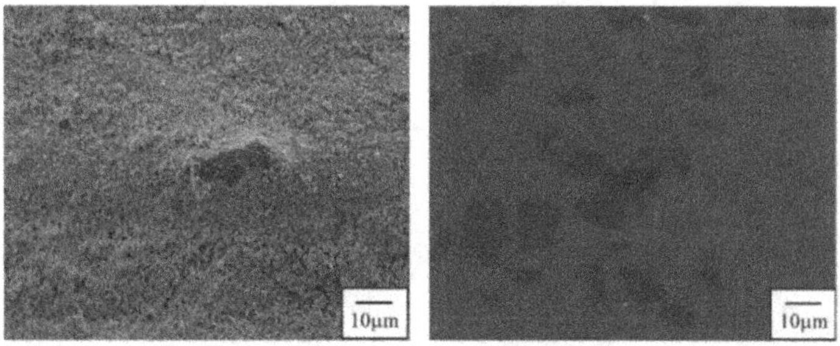

(a) 800 °C (b) 1200 °C, 3 min holding

Fig. 29 Sintering process of 20%Ti. (**a**) 800 °C. (**b**) 1200 °C, 3-min holding

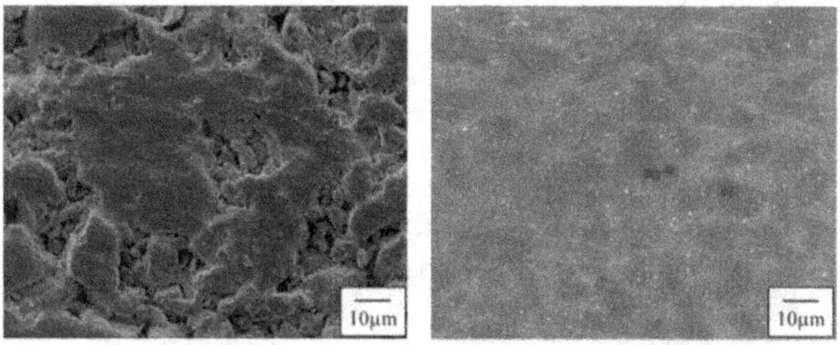

(a) 800 °C (b) 1200 °C, 3 min holding

Fig. 30 Sintering process of 80%Ti. (**a**) 800 °C. (**b**) 1200 °C, 3-min holding

From the relative density and microstructure, the densification process is explained as follows; the early stage of densification process in the composites proceeded in an intermediate manner between the monolithic PSZ and Ti. Then, their densification was delayed, compared with that of the monolithic PSZ and Ti. The sintering of composites was finished at 1200 °C and 5-min hold, while the

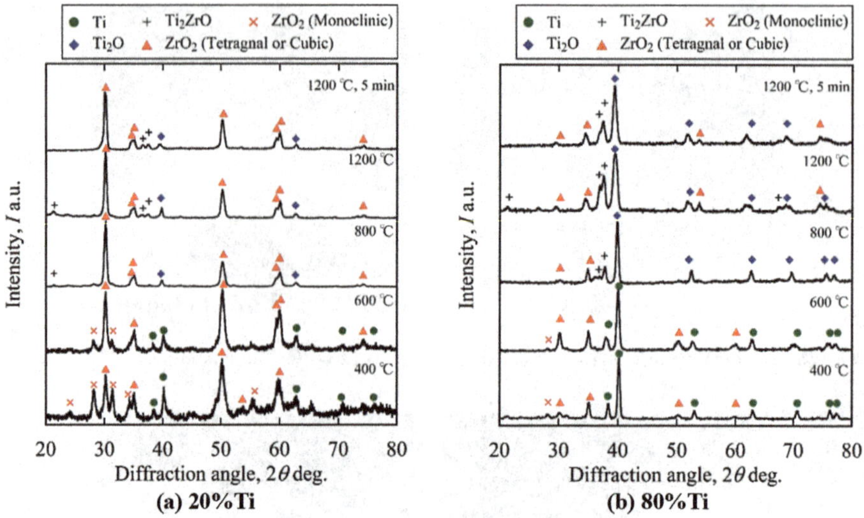

Fig. 31 XRD profile of PSZ-Ti composites. (**a**) 20%Ti. (**b**) 80%Ti

sintering of the monolithic Ti and PSZ was finished at 1000 °C and 1200 °C and 3-min hold, respectively.

The sintering state in the composites was examined from a viewpoint of the creation of reaction compounds. Figure 31 shows XRD profiles for the 20%Ti and 80%Ti. Note that the PSZ phases were identified as tetragonal/cubic ZrO_2 or monoclinic ZrO_2, as it was impossible to distinguish between tetragonal ZrO_2 and cubic ZrO_2 because their peaks are very close to each other. As for the 20%Ti, the monoclinic ZrO_2, tetragonal/cubic ZrO_2, and pure Ti were detected up to 600 °C, and the tetragonal ZrO_2, Ti_2O, and Ti_2ZrO were detected over 800 °C, while the monoclinic ZrO_2 and pure Ti were not detected. For the PSZ phase, the raw PSZ powder consists with tetragonal/cubic and monoclinic ZrO_2, and the same phases of raw PSZ powder were detected because no phase transformation occurred up to 600 °C. Only tetragonal/cubic ZrO_2, however, was detected over 800 °C because phase transformation from monoclinic ZrO_2 to tetragonal ZrO_2 occurred over about 700 °C according to the ZrO_2-Y_2O_3 phase diagram (Scott 1975). For the Ti phase, no chemical reaction or phase transformation occurred up to 600 °C, and pure Ti reacted with Zr and/or O in the PSZ phase over 800 °C. This trend was similar for the 80% Ti; the phases in raw PSZ and Ti powders were detected below 600 °C, while the reaction phases of Ti_2O and Ti_2ZrO and tetragonal ZrO_2 were detected over 800 °C.

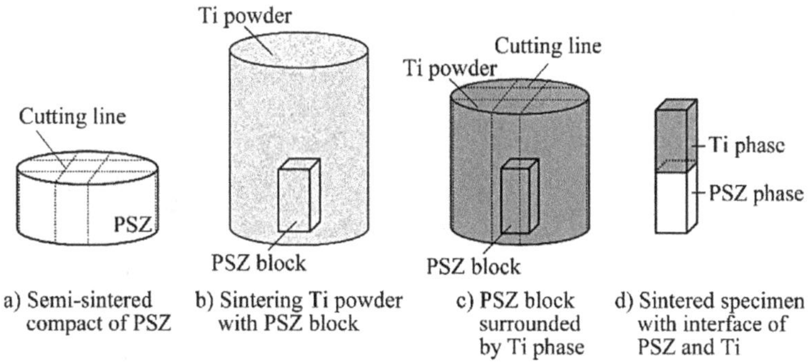

Fig. 32 Fabrication of pillar-shaped specimen with PSZ-Ti interface

5.2 *Diffusion Process and Mechanical Properties of Interface Between PSZ and Ti Phase*

It was found that the Ti phase was completely oxidized due to oxygen diffusion from the PSZ phase to Ti phase during sintering in the previous section. In this section, oxygen diffusion behavior at Ti-PSZ interface is discussed. Figure 32 shows the fabrication procedure of a pillar-shaped specimen with a Ti-PSZ interface. First, a semi-sintered PSZ compact was sintered under the condition of 1200 °C and 5-min hold and was cut into pillar-shaped block. Second, Ti powder was filled around the PSZ block in a graphite die and plungers. Third, the Ti powder with PSZ block was also sintered. To examine the diffusion behavior near the interface, the three sintering conditions were selected between the temperature of 1000 °C and the temperature of 1200 °C and 5-min hold in the sintering schedule. Finally, a pillar-shaped specimen was cut from a sintered Ti phase with PSZ block. All side surfaces of the specimen were ground with sandpaper up to #800. Note that we failed to fabricate the specimen with an interface which was directly made from powders of PSZ and Ti due to residual stress based on the mismatch of their thermal expansion coefficients in our preliminary experiment.

Figure 33 shows the overall view and magnified view near the Ti-PSZ interface. The pillar-shaped specimen with an intact interface could be fabricated. Alteration phases were formed in both PSZ and Ti phases. Atomic concentrations of Ti, O, and Zr were measured by an EPMA to evaluate the phases. Figure 34 shows the atomic concentrations near the interface of the specimen under the condition of 1200 °C and 5-min hold. Oxygen diffused from PSZ to Ti phase, while Ti diffused from Ti to PSZ phase up to 50 μm. According to the XRD analysis explained in Sect. 5.1, the reaction phases such as Ti_2ZrO and Ti_2O would be formed in the diffusion phases in the PSZ phase. The thickness of diffusion phase in Ti phase was defined by the O concentration of 15 at% which was average value of monolithic Ti, while that in PSZ phase was defined by the Ti concentration of 0 at%.

Figure 35 shows the thickness of diffusion phases in each phase. The thickness of diffusion phase in both phases increased with increasing sintering temperature and

Fig. 33 Pillar-shaped specimen. (**a**) Overall view. (**b**) Detail near Ti-PSZ interface

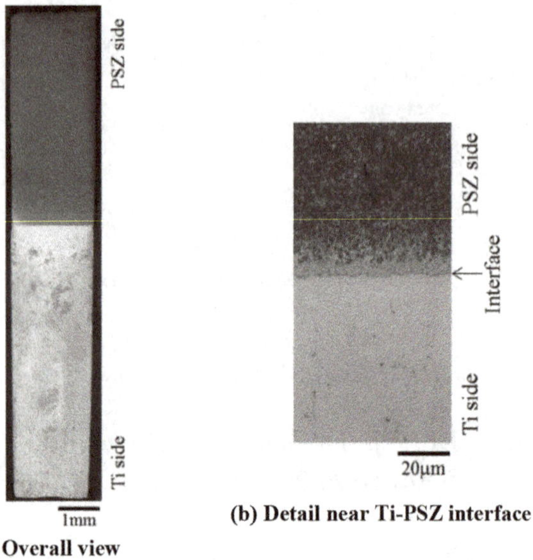

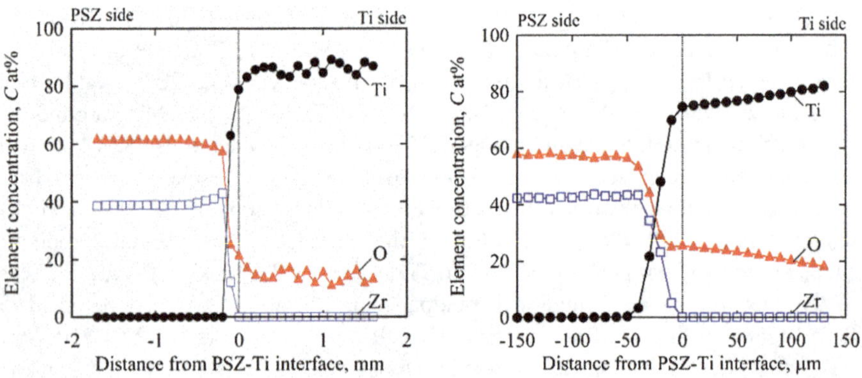

Fig. 34 Atomic concentration, 1200 °C and 5-min hold. (**a**) Entire region. (**b**) Near PSZ-Ti interface

holding time, and this means the sintering progressed for holding time at elevated temperature.

5.3 *Mechanical Properties of Ti-PSZ Interphase*

To evaluate the strength of specimen with Ti-PSZ interface, we developed a compressive-bending testing technique for a tiny pillar-shaped specimen, as shown in Fig. 36. Two metal plates fix a Ti-PSZ specimen, and the compressive load is

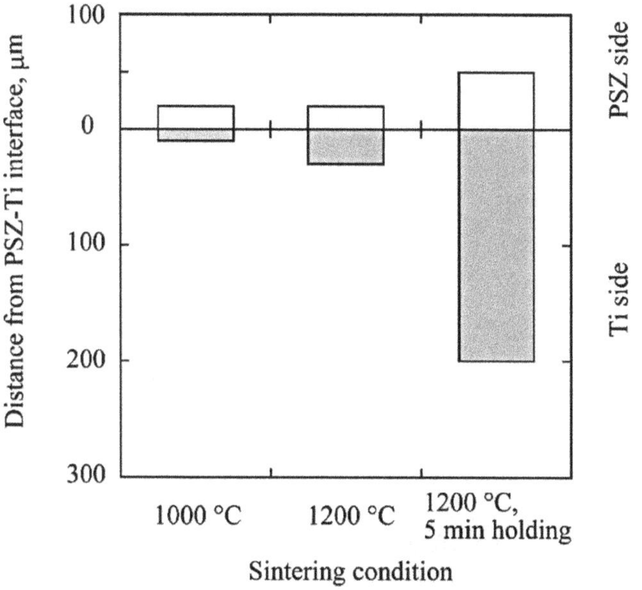

Fig. 35 Thickness of reaction phase in each phase of PSZ and Ti

Fig. 36 Bending testing for a pillar-shaped specimen

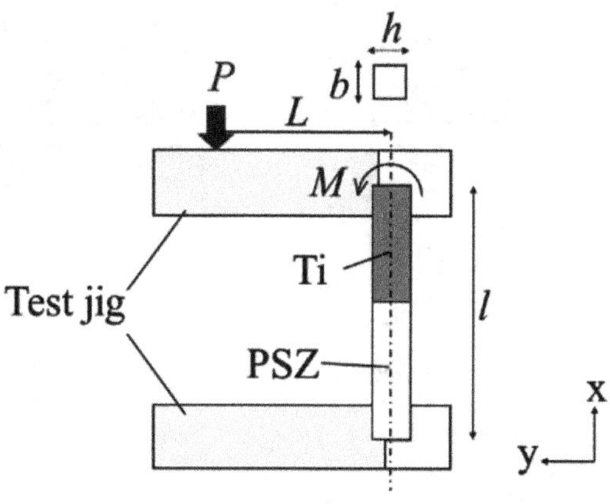

applied to the upper plate. The specimen is subjected to bending moment and compressive load, simultaneously. The bending stress at the tensile side of the specimen is given by:

$$\sigma = \frac{PL}{bh^3/12} \cdot \frac{h}{2} - \frac{P}{bh} \tag{5}$$

where P, L, b, and h are an applied load, distance between specimen and load point, thickness, and height of a specimen, respectively, as shown in Fig. 36. The

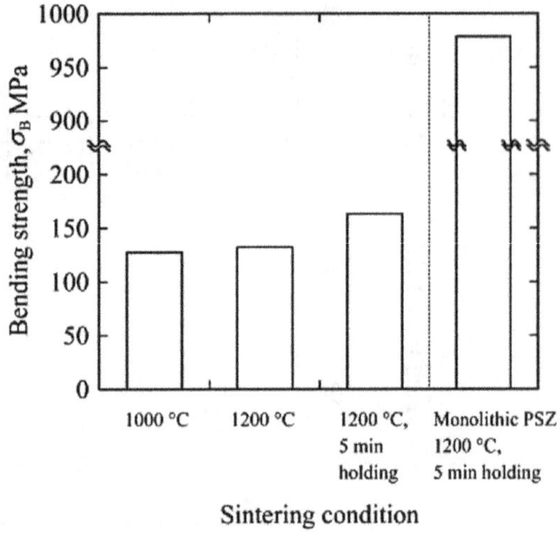

Fig. 37 Bending strength of Ti-PSZ interface

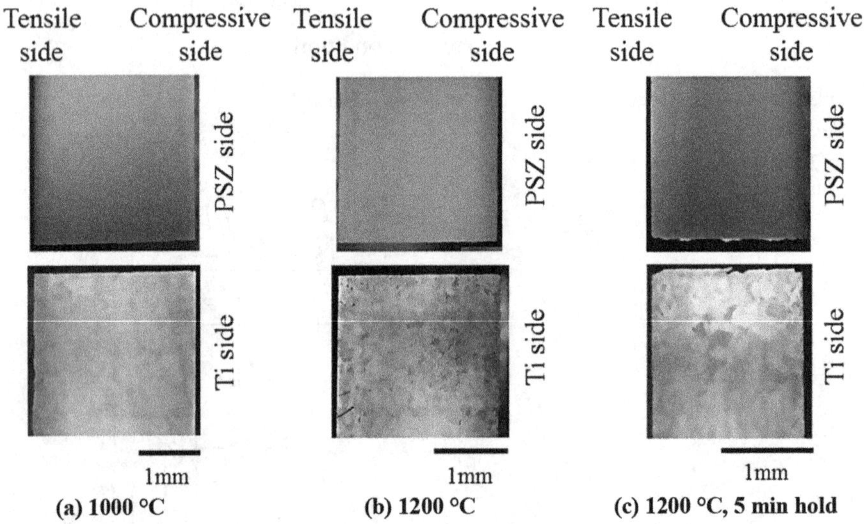

Fig. 38 Fracture morphology for pillar-shaped specimens. (**a**) 1000 °C. (**b**) 1200 °C. (**c**) 1200 °C, 5-min hold

experiments were conducted under the condition of the crosshead speed of 0.1 mm/min, and the distance L was set to be 20 mm. Figure 37 shows the bending strengths as a function of sintering condition. The bending strength increases with sintering progress, but the strength is very low compared with monolithic PSZ shown in Fig. 18.

Figure 38 shows the fracture morphology of the specimens. As for the specimen sintered at 1000 °C and 1200 °C without holding, fracture occurred at the interface between Ti-PSZ phases, as shown in Fig. 38a, b. For the specimen at 1200 °C and

5-min holing shown in Fig. 38c, fracture occurred at the diffusion phase in Ti phase and the interface. The bending strength increases because the interfacial strength increases with sintering progress. The diffusion phase in Ti phase in which the brittle Ti oxide was created becomes thick with sintering progress, the strength of the diffusion phase would be low, compared with the interface, and fracture partially occurred at the diffusion phase, not the interface. Hence, the specimen with Ti-PSZ interface fractured in the interphase including diffusion phase and interface, irrespective of sintering condition.

The reason why the Ti-PSZ composites and FGM exhibit brittleness and low strength is explained based on the diffusion and reaction behavior in Ti phase. Ti powder begins to be densified at 300 °C, while the PSZ powder begins to be densified at 1200 °C. During sintering, Ti phase is reacted into brittle Ti oxide with oxygen from PSZ powder over 800 °C, and the interphase between Ti and PSZ phases including their reaction phase is created. From Fig. 37, the strength of interphase would be lower than that of Ti and PSZ phases in the composites. Hence, as the weak interphase can be fracture origin controlling the strength of brittle composites, the strength of PSZ-Ti composites is low as compared with that of PSZ and Ti. Moreover, when the volume fraction of Ti is comparable to that of PSZ, the strength of composites becomes lower as shown in Fig. 18 because the Ti-PSZ interphase increases due to formation of interpenetrating microstructure in which the Ti and PSZ phases are co-continuous.

6 Conclusions

In this article, the Ti-PSZ composites and FGM were aimed to be fabricated by SPS technique to produce biocompatible materials with high mechanical performance. First, sintering process of monolithic Ti and its mechanical property were discussed. Then, Ti-PSZ composites and FGM were fabricated; their mechanical properties and sintering process were examined. Finally, the sintering process of Ti-PSZ interphase and its mechanical properties were examined to investigate the diffusion behavior between Ti and PSZ phases. The obtained results are summarized as follows:

(1) The sintering of monolithic Ti was completed under the condition of 1000 °C, while the sintering of Ti-PSZ composites was completed under the condition of 5-min hold at 1200 °C.
(2) According to the XRD analysis, reaction phases, such as Ti_2O and Ti_2ZrO, were created over 800 °C in the sintering process in the Ti phase of the composites and FGM, except for the Ti volume fraction more than 90%.
(3) In the composites, the bending strength and fracture toughness decreased up to 75%Ti and then increased from 75%Ti to 100%Ti with increasing Ti content. In the FGM, the fracture toughness in the graded layer was low and increased in the Ti substrate. These results might be caused by hardening and embrittlement due to the reaction phases.

(4) In the Ti-PSZ interphase sintered under the condition of 5-min hold at 1200 °C, oxygen diffused from PSZ to Ti phase in Ti phase over several hundreds of microns, while Ti diffused from Ti to PSZ phase up to several tens of microns. The fracture occurred at the interface and reaction phases in Ti phase near the interface.

Acknowledgments This work is partially supported by the Grant-in-Aid for Scientific Research (B)15H03891 and (B)23360053 from the Ministry of Education, Culture, Sports, Science and Technology of Japan.

References

Akechi K, Hara Z (1979) Production of boron fibre-titanium composite material by resistance sintering process. J Jpn Soc Powder Powder Metall 26:174–179

Banerjee D, Williams JC (2013) Perspectives on titanium science and technology. Acta Mater 61:844–879

Boyer RR (1996) An overview on the use of titanium in the aerospace industry. Mater Sci Eng A 213:103–114

Chunxiang C, BaoMin H, Lichen Z, Shuangjin L (2011) Titanium alloy production technology, market prospects and industry development. Mater Des.32:1684–1691

Fett T (1991) An analysis of the three-point bending bar by use of the weight function method. Eng Fract Mech 40:683–686

Fujii T, Tohgo K, Araki H, Wakazono K, Ishikura M, Shimamura Y (2010) Fabrication and strength evaluation of biocompatible ceramic-metal composite materials. J Solid Mech Mater Eng 4:1699–1710

Fujii T, Tohgo K, Isono H, Shimamura Y (2017) Fabrication of PSZ-Ti functionally graded materials by spark plasma sintering and its fracture toughness. Mater Sci Eng A 682:656–663

Fujii T, Tohgo K, Iwao M, Shimamura Y (2018) Fabrication of alumina-titanium composites by spark plasma sintering and their mechanical properties. J Alloys Compd. https://doi.org/10.1016/j.jallcom.2018.02.142

Khor KA, Cheng KH, Yu LG, Boey F (2003) Thermal conductivity and dielectric constant of spark plasma sintered aluminum nitride. Mater Sci Eng A 347:300–305

Kolesnikov B, Herbeck L, Fink A (2008) CFRP/titanium hybrid material for improving composite bolted joints. Compos Struct 83:368–380

Li S, Kondoh K, Imai H, Chen B, Jia L, Umeda J, Fu Y (2016) Strengthening behavior of in situ-synthesized (TiC/TiB)/Ti composites by powder metallurgy and hot extrusion. Mater Des 95:127–132

Park Y, Narayen D, Schmerling M, Marcus HL (1984) Fatigue crack growth behaviour of Ti-6 Al-4V metal matrix/continuous SiC and B4C/B fibre composites. J Mater Sci 19:2239–2245

Scott HG (1975) Phase relationships in the zirconia-yttria system. J Mater Sci 10:1527–1535

Shinohara T, Fujii T, Tohgo K, Shimamura Y (2017) Densification process in fabrication of PSZ-Ti composites by spark plasma sintering technique. Mater Charact 132:230–238

Su X, Wang P, Chen W, Shen Z, Nygren M, Cai Y, Yan D (2004) Effects of composition and thermal treatment on infrared transmission of Dy-α-sialon. J Eur Ceram Soc 24:2869–2877

Tjong SC, Chen F (1997) Wear behavior of as-cast ZnAl27/SiC particulate metal-matrix composites under lubricated sliding condition. Metall Mater Trans A 28A:1951–1955

Tjong SC, Mai YW (2008) Processing-structure-property aspects of particulate- and whisker-reinforced titanium matrix composites. Compos Sci Technol 68:583–601

Tohgo K, Suzuki T, Araki H (2005) Evaluation of R-curve behavior of ceramic-metal functionally graded materials by stable crack growth. Eng Fract Mech 72:2359–2372

Tohgo K, Iizuka M, Araki H, Shimamura Y (2008) Influence of microstructure on fracture toughness distribution in ceramic-metal functionally graded materials. Eng Fract Mech 75:4529–4541

Tohgo K, Fujii T, Harada M, Isono H, Shimamura Y (2015) Fabrication of PSZ-Ti composites by spark plasma sintering and their mechanical properties. Mater Sci Eng A 621:166–172

Tokita M (1993) Trends in advanced SPS spark plasma sintering systems and technology. J Soc Powder Technol Jpn 30:790–804

Wang K (1996) The use of titanium for medical applications in the USA. Mater Sci Eng A 213:134–137

Zhang W, Yang YQ, Zhao GM, Feng ZQ, Huang B, Luo X, Li MH, Chen YX (2014) Interfacial reaction studies of B_4C-coated and C-coated SiC fiber reinforced Ti-43Al-9V composites. Intermetallics 50:14–19

Index

© Springer Nature Switzerland AG 2019
P. Cavaliere (ed.), *Spark Plasma Sintering of Materials*,
https://doi.org/10.1007/978-3-030-05327-7

E

F

CPSIA information can be obtained
at www.ICGtesting.com
Printed in the USA
BVHW010428050719
552608BV00021B/12/P

9 783030 053260